Robert Schaefer Carlos Cotta
Joanna Kołodziej Günter Rudolph (Eds.)

Parallel Problem Solving from Nature – PPSN XI

11th International Conference
Kraków, Poland, September 11-15, 2010
Proceedings, Part II

Springer

Volume Editors

Robert Schaefer
AGH University of Science and Technology
Faculty of Electrical Engineering, Automatics, Computer Science
and Electronics, Department of Computer Science
Mickiewicza 30, 30-059 Kraków, Poland
E-mail: schaefer@agh.edu.pl

Carlos Cotta
Universidad de Málaga
Departamento de Lenguajes y Ciencias de la Computación
ETSI Informática, Campus Teatinos, 29071 Málaga, Spain
E-mail: ccottap@lcc.uma.es

Joanna Kołodziej
University of Bielsko-Biala
Department of Mathematics and Computer Science
Willowa 2, 43-309 Bielsko-Biala, Poland
E-mail: jkolodziej@ath.bielsko.pl

Günter Rudolph
Technische Universität Dortmund, Fakultät für Informatik
44221 Dortmund, Germany
E-mail: guenter.rudolph@tu-dortmund.de

Library of Congress Control Number: 2010934114

CR Subject Classification (1998): J.3, I.2, F.1, F.2, I.5, G.2

LNCS Sublibrary: SL 1 – Theoretical Computer Science and General Issues

ISSN 0302-9743
ISBN-10 3-642-15870-6 Springer Berlin Heidelberg New York
ISBN-13 978-3-642-15870-4 Springer Berlin Heidelberg New York

springer.com

Printed in Germany

Typesetting: Camera-ready by author, data conversion by Scientific Publishing Services, Chennai, India
Printed on acid-free paper 06/3180

Lecture Notes in Computer Science 6239

Commenced Publication in 1973
Founding and Former Series Editors:
Gerhard Goos, Juris Hartmanis, and Jan van Leeuwen

Preface

We are very pleased to present to you this LNCS volume, the proceedings of the 11th International Conference on Parallel Problem Solving from Nature (PPSN 2010). PPSN is one of the most respected and highly regarded conference series in evolutionary computation, and indeed in natural computation as well. This biennial event was first held in Dortmund in 1990, and then in Brussels (1992), Jerusalem (1994), Berlin (1996), Amsterdam (1998), Paris (2000), Granada (2002), Birmingham (2004), Reykjavik (2006) and again in Dortmund in 2008.

PPSN 2010 received 232 submissions. After an extensive peer review process involving more than 180 reviewers, the program committee chairs went through all the review reports and ranked the papers according to the reviewers' comments. Each paper was evaluated by at least three reviewers. Additional reviewers from the appropriate branches of science were invoked to review into disciplinary papers. The top 128 papers were finally selected for inclusion in the proceedings and presentation at the conference. This represents an acceptance rate of 55%, which guarantees that PPSN will continue to be one of the conferences of choice for bio-inspired computing and metaheuristics researchers all over the world who value the quality over the size of a conference.

The papers included in the proceedings volumes cover a wide range of topics, from evolutionary computation to swarm intelligence, from bio-inspired computing to real-world applications. Machine learning and mathematical games supported by evolutionary algorithms as well as memetic, agent-oriented systems are also represented. They all are the latest and best in natural computation. The proceedings are composed of two volumes divided into nine thematic sections.

In accordance with the PPSN tradition, all papers at PPSN 2010 were presented as posters. There were nine sessions of posters. Each session consisted of around 15 papers. For each session, we covered as wide a range of topics as possible so that participants with different interests could find some relevant papers at every session.

PPSN 2010 featured three distinguished keynote speakers: John Garibaldi, Zbigniew Michalewicz and Darrell Whitley who delivered lectures entitled: Ensemble Fuzzy Reasoning, Some Thoughts on Wine Production, and Elementary Landscapes Made Easy, respectively.

PPSN 2010 also included eight interesting tutorials. These covered the wide area of natural computing science. The first of them "A Rigorous Theoretical Framework for Measuring Generalization of Co-evolutionary Learning" (X. Yao) was devoted to the genetic algorithm theory while the following two "Foundations of Evolutionary Multi-objective Optimization" (F. Neumann, T. Friedrich) and "Hybrid Optimization Approaches" (G. Raidl) introduced important groups of algorithms inspired by nature. The next tutorials, "Natural Computing and

Finance" (T. Brabazon, M. O'Neill), "Heuristic and Meta-heuristic Approaches for Scheduling in Large Scale Distributed Computing Environments" (F. Xhafa) and "Artificial Immune Systems in Optimization and Classification Problems with Engineering and Biomedical Applications" (T. Burczyński, M. Bereta, W. Kuś), focused on important engineering, business and medical applications. Finally, "Learning to Play Games" (S. M. Lucas) and "The Complexity of Elections: New Domain for Heuristic Computations" (P. Faliszewski) concerned games and social problems.

PPSN 2010 also included four workshops. They made an excellent start to the five-day event. The workshops offered an ideal opportunity for participants to explore specific topics in natural computation in an informal setting. They sowed the seeds for the future growth of natural computation. The first of them "Self-tuning, Self-configuring and Self-generating Search Heuristics (Self* 2010)" (G. Ochoa, M. Schoenauer, D. Whitley) focused on developing automated systems to replace the role of a human expert in the design, tuning and generation of search heuristics. The next pair of workshops "Understanding Heuristics: How Do We Get the Best of Both Theory and Empirical Methods?" (E. Ozcan, A. Parkes, J. Rowe) and "Experimental Methods for the Assessment of Computational Systems (WEMACS)" (T. Bartz-Beielstein, M. Chiarandini, L. Paquete, M. Preuss) concerned two complementary theoretical and experimental approaches to the analysis of heuristic and meta-heuristic algorithms. The last one "Workshop on Parallel and Cooperative Search Methods" (D. Ouelhadj, E. Ozcan, M. Toulouse) dealt with cooperative parallel searches improving performance, especially when dealing with large scale combinatorial optimization problems.

The success of any conference depends on its authors, reviewers and organizers. PPSN 2010 was no exception. We are grateful to all the authors who submitted their papers and to all the reviewers for their outstanding work in refereeing the papers on a very tight schedule. We relied heavily on a team of volunteers, especially those in Kraków, to keep the PPSN 2010 wheel turning.

PPSN XI would not have been possible without the support of Microsoft Poland, Intel and HP.

September 2010

Robert Schaefer
Carlos Cotta
Joanna Kołodziej
Günter Rudolph
Juan J. Merelo
Hans-Paul Schwefel

Organization

PPSN XI was organized and hosted by the Intelligent Information Systems Group of the Department of Computer Science, Faculty of Electrical Engineering, Automatics, Computer Science and Electronics, AGH University of Science and Technology, Poland. The conference took place in the AGH Conference and Teaching Center U-2, Kraków.

Steering Committee

David W. Corne	Heriot-Watt University Edinburgh, UK
Kenneth De Jong	George Mason University, USA
Agoston E. Eiben	Vrije Universiteit Amsterdam, Netherlands
Juan Julián Merelo Guervós	Universidad de Granada, Spain
Günter Rudolph	Technische Universität Dortmund, Germany
Thomas P. Runarsson	University of Iceland, Iceland
Marc Schoenauer	Université Paris Sud, France
Xin Yao	University of Birmingham, UK

Conference Committee

General Chair	Robert Schaefer (AGH Kraków, Poland)
Honorary Chair	Hans-Paul Schwefel (Technische Universität Dortmund, Germany)
Program Chairs	Carlos Cotta (University of Málaga, Spain)
	Joanna Kołodziej (ATH Bielsko-Biała, Poland)
	Günter Rudolph (Technische Universität Dortmund, Germany)
Workshop Chair	Aleksander Byrski (AGH Kraków, Poland)
Tutorial Chair	Krzysztof Cetnarowicz (AGH Kraków, Poland)
E-proceedings Chair	Juan Julián Merelo Guervós (Universidad de Granada, Spain)
E-publicity Chair	Bartłomiej Śnieżyński (AGH Kraków, Poland)
Technical Editorial Chair	Jarosław Koźlak (AGH Kraków, Poland)
Financial Manager	Leszek Siwik (AGH Kraków, Poland)
Local Organization	Jacek Dajda (AGH Kraków, Poland)
	Ewa Olejarz (AGH, Kraków, Poland)
	Rafał Dreżewski (AGH Kraków, Poland)
	Piotra Gawor (ATH Bielsko-Biała, Poland)
	Dawid Kotrys (ATH Bielsko-Biała, Poland)
	Anna Prochownik (ATH Bielsko-Biała, Poland)

Workshops

Self-tuning, Self-configuring and Self-generating Search Heuristics (Self* 2010)
Gabriela Ochoa, Marc Schoenauer and Darrell Whitley

Understanding Heuristics: How Do We Get the Best of Both Theory and Empirical Methods?
Ender Özcan, Andrew Parkes and Jonathan Rowe

Experimental Methods for the Assessment of Computational Systems (WEMACS)
Thomas Bartz-Beielstein, Marco Chiarandini, Luis Paquete and Mike Preuss

Workshop on Parallel and Cooperative Search Methods
Djamila Ouelhadj, Ender Özcan and Michel Toulouse

Tutorials

A Rigorous Theoretical Framework for Measuring Generalization of Co-evolutionary Learning
Xin Yao

Foundations of Evolutionary Multi-objective Optimization
Frank Neumann and Tobias Friedrich

Hybrid Optimization Approaches
Günther Raidl

Natural Computing and Finance
Tony Brabazon and Michael O'Neill

Heuristic and Meta-heuristic Approaches for Scheduling in Large Scale Distributed Computing Environments
Fatos Xhafa

Artificial Immune Systems in Optimization and Classification Problems with Engineering and Biomedical Applications
Tadeusz Burczyński, Michał Bereta and Wacaw Kuś

The Complexity of Elections: New Domain for Heuristic Computations
Piotr Faliszewski

Learning to Play Games
Simon M. Lucas

Program Committee

Alex Agapie
Uwe Aickelin
Enrique Alba
Anna I. Esparcia Alcázar
Jhon Edgar Amaya
Jarosław Arabas
Dirk Arnold
Anne Auger
Dariusz Badura
Thomas Bartz-Beielstein
Peter Bentley
Nicola Beume
Marenglen Biba
Mark Bishop
Christian Blum
Yossi Borenstein
Urszula Boryczka
Peter Bosman
Pascal Bouvry
Anthony Brabazon
Juergen Branke
Dimo Brockhoff
Larry Bull
Tadeusz Burczyński
Juan Carlos Burguillo-Rial
Aleksander Byrski
Erick Cantú-Paz
Uday Chakraborty
Ying-ping Chen
Sung-Bae Cho
Siang-Yew Chong
Carlos Coello Coello
Pierre Collet
David W. Corne
Luis Correia
Carlos Cotta
Peter Cowling
Valentin Cristea
Kenneth DeJong
Benjamin Doerr
Marco Dorigo
Rafał Dreżewski
Stefan Droste
Gusz Eiben
Talbi El-Ghazali
Michael Emmerich
Andries Engelbrecht
Thomas English
Antonio J. Fernández
Jonathan Fieldsend
Carlos M. Fonseca
Tobias Friedrich
Marcus Gallagher
Jos Enrique Gallardo
Jonathan M. Garibaldi
Mario Giacobini
Kyriakos Giannakoglou
Jens Gottlieb
Roderich Gross
Juan Julian Merelo Guervos
Steven Gustafson
Hisashi Handa
Nikolaus Hansen
Emma Hart
Philip Hingston
Jeff Horn
Eyke Huellermeier
Christian Igel
Christian Jacob
Wilfried Jakob
Mark Jelasity
Yaochu Jin
Bob John
Bryant A. Julstrom
Iwona Karcz-Dulęba
Andy Keane
Graham Kendall
Joshua Knowles
Joanna Kołodziej
Wolfgang Konen
Jozef Korbicz
Witold Kosinski
Jarosław Koźlak
Oliver Kramer
Krzysztof Krawiec
Halina Kwaśnicka
Pier Luca Lanzi
Pedro Larranaga
Per Kristian Lehre
Jose A. Lozano
Simon Lucas
Evelyne Lutton
Krzysztof Malinowski
Jacek Mańdziuk
Elena Marchiori
Barry McCollum
Jorn Mehnen
Peter Merz
Silja Meyer-Nieberg
Zbigniew Michalewicz
Ralf Mikut
Alberto Moraglio
Boris Naujoks
Ferrante Neri
Frank Neumann
Ewa Niewiadomska-Szynkiewicz
Lars Nolle
Shigeru Obayashi
Andrzej Obuchowicz
Pietro Oliveto
Ben Paechter
Luis Paquete
Mario Pavone
Pawel Pawlewski
Martin Pelikan
Florin Pop
Mike Preuss
Christian Prins
Domenico Quagliarella
Günther Raidl
Robert Reynolds
Philipp Rohlfshagen
Andrea Roli
Jonathan Rowe
Günter Rudolph

Thomas Runarsson
Thomas A. Runkler
Leszek Rutkowski
Conor Ryan
Michael Sampels
Ruhul Sarker
Jayshree Sarma
Robert Schaefer
Robert Scheffermann
Marc Schoenauer
Oliver Schütze
Bernhard Sendhoff
Franciszek Seredyński
Marc Sevaux
Jonathan Shapiro
Ofer Shir
Moshe Sipper
Leszek Siwik
Jerzy Stefanowski
Thomas Stibor
Catalin Stoean
Thomas Stützle
Dirk Sudholt
Ponnuthurai Suganthan
Ryszard Tadeusiewicz
Kay Chen Tan
Alexander Tarakanov
Olivier Teytaud
Lothar Thiele
Dirk Thierens
Jon Timmis
Julian Togelius
Heike Trautmann
Krzysztof Trojanowski
Andrew Tuson
Massimiliano Vasile
Igor Vatolkin
Pedro Isasi Viñuela
R. Paul Wiegand
Slawomir Wierzchoń
Carsten Witt
Janusz Wojtusiak
Fatos Xhafa
Xin Yao
Tina Yu
Ivan Zelinka
Qingfu Zhang

Sponsoring Institutions

Hewlett-Packard Polska
Intel Corporation
Microsoft

Table of Contents – Part II

Multiobjective Optimization, Models and Applications

Applications, Engineering and Economical Models

Multi-agent Systems and Parallel Approaches

Genetic Computing and Games

Table of Contents – Part I

Theory of Evolutionary Computing (I)

Theory of Evolutionary Computing (II)

Machine Learning, Classifier Systems, Image Processing

Memetic Algorithms, Hybridized Techniques, Meta and Hyperheurisics

Multiobjective Optimization, Theoretical Aspects

A Novel Smart Multi-Objective Particle Swarm Optimisation Using Decomposition

Noura Al Moubayed, Andrei Petrovski, and John McCall

Robert Gordon University, Aberdeen
St. Andrew Street
Aberdeen, AB25 1HG, UK
{n.al-moubayed,a.petrovski,j.mccall}@rgu.ac.uk

Abstract. A novel Smart Multi-Objective Particle Swarm Optimisation method - SDMOPSO - is presented in the paper. The method uses the decomposition approach proposed in MOEA/D, whereby a multi-objective problem (MOP) is represented as several scalar aggregation problems. The scalar aggregation problems are viewed as particles in a swarm; each particle assigns weights to every optimisation objective. The problem is solved then as a Multi-Objective Particle Swarm Optimisation (MOPSO), in which every particle uses information from a set of defined neighbours. The paper also introduces a novel smart approach for sharing information between particles, whereby each particle calculates a new position in advance using its neighbourhood information and shares this new information with the swarm. The results of applying SDMOPSO on five standard MOPs show that SDMOPSO is highly competitive comparing with two state-of-the-art algorithms.

1 Introduction

Real-life optimisation problems usually have several conflicting objectives. This leads to an irregular multi-objective space where the optimisation method must be able to find solutions that represent trade-offs between these objectives.

In [1] [2], MOEA/D introduces a new approach to discover Pareto optimal solutions. This is done by decomposing the original MOP into a number of scalar aggregation problems. These scalar problems are then solved using Genetic Algorithms (GA). The advantages of this approach in terms of mathematical soundness, algorithmic structure and computational cost are explained in [2]. MOEA/D has been applied successfully on several real-life MOPs [3].

Particle Swarm Optimisation (PSO) has proved to be very efficient and capable of providing competitive solutions in many application domains [4], [5], [6]. MOPSO methods have also been developed and demonstrated their ability to provide viable solutions [7], [8]. In [6] and [9], authors have observed that although PSO and GA on average yield the same effectiveness (solution quality), PSO is more computationally efficient and uses fewer evaluations than GA - the claim was supported by two statistical tests, which confirmed similar effectiveness of the methods but superior efficiency of PSO over GA. Also, because PSO

R. Schaefer et al. (Eds.): PPSN XI, Part II, LNCS 6239, pp. 1–10, 2010.

requires less subjective tuning in contrast to GA, the former algorithm proved to be much easier to implement [9].

MOPSO/D [10] is a multi-objective optimization method that uses the MOEA/D framework to solve continuous MOPs. MOPSO/D substitutes the genetic algorithm used to implement MOEA/D with PSO. In our opinion, this method does not fully exploit the salient properties of PSO neighborhood relations and uses a genetic operator for avoiding local optima, which confounds the application of the PSO algorithm.

In this paper, we propose a novel MOPSO method using decomposition (SD-MOPSO). This approach takes the advantage of PSO as a simple, fast, efficient and easy to implement method and defines a smart approach for updating the particles whilst using decomposition to enhance the diversity of the solutions.

The paper is organized as follows. Section 2 introduces the methodology of handling multi-objective optimisation using PSO. Section 3 explains the novelty of our approach to PSO implementation. In Section 4 we explain the experimental setup and present the results. Finally, Section 5 summarises the conclusions we have drawn from the presented research work and discusses its implication.

2 Multi-Objective Particle Swarm Optimisation

The difficulty of multi-objective optimisation is that an improvement in one objective often happens at the expense of deteriorating the performance with respect to other objectives. The optimisation challenge therefore is to find the entire set of trade-off solutions that satisfy the conflicting objectives. The objective of optimisation can be represented as a vector F operating on a solution space $\Omega \subset R^n$.

$$F(X) = \{f_1(X), f_2(X), \ldots, f_m(X)\} \tag{1}$$

where $X \in \Omega$, and m is the number of objectives.

The result of the multi-objective optimisation process is a set of trade-off solutions. When minimizing $F(X)$, a domination relationship is defined between these solutions as follows: let $X, Y \in \Omega$, $X \succ Y$ if and only if $f_i(X) \leq f_i(Y)$ for all $i = \{1, 2, ..., m\}$, and there is at least one j for which $f_j(X) < f_j(Y)$. X is a Pareto optimal solution if there is no other solution $Y \in \Omega$ such that $Y \succ X$.

Pareto optimality of a solution guarantees that any enhancement of one objective would results in worsening of at least one other objective. The image of the Pareto optimal set in the objective space - $F(X^*)$ - is called the Pareto Front (PF)[11].

PSO can be used to find the Pareto optimal solutions or to approximate the PF. Each particle in the swarm represents a potential solution and exchanges positional information with the global leader(s) or best local neighbour(s), as well as consults its own personal memory; then, this information is used to move the particle around the search space [12]. Every particle is characterised by its position and velocity. The position is the point in the solution space, whereas the velocity is a vector representing the positional change. Each particle uses the position of the selected leader and its personal movement trajectory to update the velocity and position values.

3 The SDMOPSO Approach

Similar to MOEA/D [2], SDMOPSO decomposes the MOP into scalar aggregation problems. Decomposition transforms the MOP into a set of distinct scalar aggregation problems. Every particle solves the corresponding problem by applying priorities to each objective according to its weighting vector (λ). This assists the optimisation process to find potential solutions that are evenly distributed along the PF and to mitigate against premature convergence. By associating every particle with a distinct scalar aggregation problem, the exploration activity of each particle will be focused on a specific region in the objective space and aimed at reducing the distance to the reference point.

SDMOPSO introduces a new approach for exchanging information between neighbouring particles without a need for extra evaluations. The motivation behind this is that the same solution in the objective space can have different aggregated values depending on its λ vector; thus, a solution (i.e. a position in the decision space) is assigned to the particle that uses it to give the best aggregation value. If the new calculated position does not enhance the aggregated fitness of one particle, then the particle shares the new position with its neighbours as this could enhance their aggregated fitness. In other words, useless information for one particle can be effectively utilised by other particles, depending on their λ vector. This can lead to performing fewer objective function evaluations and results in wider dissemination of the discovered information, facilitating thereby simultaneous optimisation of the scalar problems. Taking into account the topological structure of the PSO population, sharing the information with neighbours will help relaying the discoveries of one particle to the entire swarm.

Many scalar approaches have been proposed to aggregate the objectives of a MOP. Among these, the weighted Tchebycheff method is widely used [2], which is based on a non-linear metric measuring the distance to a reference point in the objective space [13]:

$$minimize \quad g(x|\lambda, z^*) = max_{1 \leq i \leq m}\{\lambda_i|f_i(x) - z_i^*|\} \tag{2}$$

where $x \in R^n$ is a decision vector, $z^* = (z_1^*, \ldots, z_m^*)^T$ such that $z_i^* = min\{f_i(x)\}$ for each $i = 1, \ldots, m$.

The reference point z^* is determined by SDMOPSO as the vector of best values for each objective found so far by the optimisation process. Each particle then will be evaluated according to Eq. 2 using λ associated with it.

SDMOPSO uses the weighted Tchebycheff approach to decompose the optimisation objective defined by Eq. 1 into N scalar optimisation problems, where N is the swarm's size. By changing the weights and using the reference point defined above, any Pareto optimal solution can be generated.

In addition, SDMOPSO uses the concept of a crowding archive to store the set of swarm leaders. The size of the crowding archive is fixed using ϵ-dominance [14]. At the end of each iteration the crowding archive is updated with the new non-dominated particles in the current population, and the corresponding crowding values are adjusted in accordance with the number of new updates.

This approach limits the size of the crowding archive and determines which particles to be deleted when the maximum size is exceeded. This is done in such a way that the diversity of Pareto optimal solutions is maintained [14].

Algorithm 1. SDMOPSO

Initialize the swarm with N particles and N λ vectors
for $i = 1$ to N **do**
    assign the particle i to its closest λ vector
    initialize $pbest_i$
end for
Initialize velocities $V = \{v_1, \ldots, v_N\}$, archive, neighbourhood and z^*
Crowding(archive)
for $i = 1$ to $MaxIteration$ **do**
    for $j = 1$ to N **do**
        define particle j future Velocity, $v_j(t+1)$
        define particle j future Position, $x_j(t+1)$
        calculate scalar aggregate function for j
        update the current population with the new particle j
        update $pbest_j$, archive, and z^*
    end for
end for
Return the final result in the crowding archive

SDMOPSO's first phase starts by initializing the population and initializing N vectors: $\lambda = \{\lambda_1, \lambda_2, \ldots, \lambda_m\}$, where m is the number of objectives and N is the swarm size. λ vectors are uniformly distributed in $[0,1]^m$ subject to $\sum_{i=1}^{m} \lambda_i = 1$. Every particle is assigned a unique λ vector. This λ vector is selected so that it gives the best aggregated fitness value for the initialized particle. For example, in the case of minimization problems the particle will be assigned to the λ vector that minimizes the aggregated fitness, taking into account that each λ is unique and will be assigned to only one particle of the swarm. The particles' memories *pbest* are then initialized, and the initial velocity of each particle is set to zero. The crowding archive is set to a fixed size, which equals to the swarm size, and then is initialized using the non-dominated particles in the swarm. The reference point z^* is the vector in the objective space with the best objective values found so far. The neighbourhood will be initialized by defining the neighbourhood size $\mathcal{N}$. The neighbourhood of a particle is defined by the $\mathcal{N}$ particles that have associated λ vectors with the closest Euclidean distance to its own λ vector.

The second phase of the optimisation process is repeated for a pre-defined number of iterations. During every iteration each particle defines a local view in the objective space. The particle determines the next move by finding the new velocity and new position using Eq. 3 and Eq. 4. The new velocity is calculated using the *pbest* values of a randomly selected neighbour particle and that of the current particle. The particle will then offer this local information (i.e. the decision vector and the corresponding objective vector) to its neighbours (including itself) so that every particle of the neighbourhood uses the new position and the evaluated objectives to calculate a new aggregated fitness value. If the new position enhances the particle's scaled fitness value, then it is adopted as the new position of the particle. Only up to two particles are allowed to update (as suggested in [2]) their information and take advantages of this local information in order to avoid duplication of particles in the swarm. Evaluating the new information using different λ will not involve additional objective function evaluations as it only reads the stored values.

$$V_i(t) = w * V_i(t-1) + C_1 * r_1 * (x_{pbest_i} - x_i(t)) + C_2 * r_2 * (x_{nbest_i} - x_i(t)) \quad (3)$$

$$x_i(t) = x_i(t-1) + V_i(t) \quad (4)$$

where $pbest_i$ is the personal best performance of $particle_i$, $nbest_i$ is a random neighbouring particle from the set of $\mathcal{N}$ neighbours of $particle_i$, $r_1, r_2 \in [0.1, 1]$ are random values, $w \in [0.1, 0.5]$ is the inertia weight, and $C_1, C_2 \in [1.2, 2.0]$ are the learning factors that take uniformly distributed random values in their defined ranges.

After the particle updates its position and velocity, it has to update its *pbest* as well. The *pbest* value will be replaced with the new position only if the new position dominates *pbest*, or if both are mutually non-dominating. The crowding archive is then updated with new non-dominated particles, if found, subject to the crowding restriction. Finally, the reference point will be updated if needed. The final result of the optimisation will be the content of the crowding archive when the run of SDMOPSO is complete. The pseudo-code of the SDMOPSO Algorithm is listed in Algorithm 1.

SDMOPSO introduces the following improvements to the basic MOPSO:

- SDMOPSO enhances the approximation of the PF for a MOP by decomposing the original MOP into scalar aggregation problems and facilitating simultaneous optimisation of these scalar problems.
- SDMOPSO associates every particle with a λ vector according to the best scalar aggregated fitness value. This will enhance the initial population and, together with the way the information is exchanged between the particles in the swarm, can eventually lead to saving processing time.
- SDMOPSO uses the crowding archive to retain the diversity of the swarm leaders, and hence the distribution of the final solutions. This is done by using crowding-based selection method to choose the solutions to be deleted or replaced when the archive is full.
- SDMOPSO takes advantage of MOPSO's simplicity, but improves the original MOPSO by a better use of particles' local information. Each particle always pre-processes its moves and exchanges the discovered information with the entire swarm in order to facilitate simultaneous optimisation of all scalar problems. This could also help mitigating premature convergence to local optima.

4 Experiments and Results

The SDMOPSO method is tested on several standard problems defined in the test suite [15] - for space limit only 9 problems were chosen (Schaffer, Fonseca, Kursawe, Viennet2, Viennet3, ZDT1-4, and ZDT6 . They cover diverse MOPs with convex, concave, connected and disconnected PFs. The method is then compared to MOEA/D [2] and OMOPSO[14].

In the present work, jMetal Framework [16] is used to implement MOEA/D and OMOPSO because it is a general framework that implements the state-of-art multi-objective algorithms. Each algorithm is run 30 times for each test problem. For the bi-objective problems each algorithm uses 300 iteration per run, and 150 individuals per generation. For the three-objective problems the corresponding values of 600 iterations and 300 individuals were used. All compared algorithms adopt real encoding and perform the same number of objective evaluations. For the sake of a fair comparison, the number of the non-dominated solutions found by each algorithm is limited to a fixed threshold (100 for bi-objective problems and 1000 for the three-objective problems). MOEA/D uses differential evolution crossover (DE) with (probability = 1.0) and (differential weight = 0.5), polynomial mutation with (probability = 1/number of decision variables), the mutation distribution index is equal to 20, and the neighbourhood size is set to 30. OMOPSO uses turbulence probability of 0.5. C_1, C_2 were set to a random value in the range [1.5, 2.0]. r_1, r_2 are set to a random value in [0, 1], and w - to a random value in [0.1, 0.5]. SDMOPSO uses the parameters explained in the previous section and neighbourhood size $\mathcal{N}$ =30. Both OMOPSO and SDMOPSO use ϵ=0.0075, the crowding archive of size 150 for bi-objective problems and 300 for three-objective problems. The PF produced by each algorithm is the union of PFs after 30 runs ($PF_{approximated}$).

To validate our approach, two indicators are used for estimating the convergence and diversity of the solutions. The first performance indicator is a generational distance (GD) [17], [18]. GD calculates the average Euclidean distance in the objective space between $PF_{approximated}$ produced by each algorithm and the actual PF (PF_{true}). The distance is calculated for each point of PF_{true} and the nearest point of $PF_{approximated}$. To apply this measure, all the objective values are scaled to be in the range [0,1].

$$I_{GD}(A,R) = \frac{(\sum_{u \in A}(min_{v \in R} \parallel F(u) - F(v) \parallel^2))^{1/2}}{|R|} \quad (5)$$

The second indicator is the R-metrics [17], [18]. R-metrics is a hybrid indicator that simultaneously measures the convergence and diversity of the found solutions. R-metrics uses a set of utility functions u, which can be any scalar functions. In this paper, we use a weighted Tchebycheff function with a sufficiently large number of evenly distributed normalized weighting vectors λ. R-metrics compare two reference sets in our experiments: A is PF_{true} related to

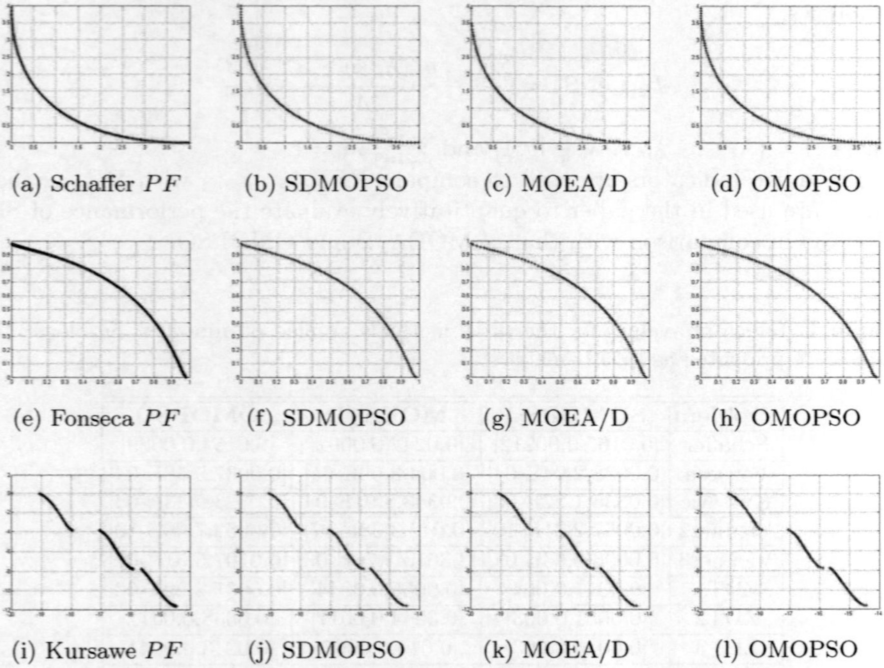

(a) Schaffer PF (b) SDMOPSO (c) MOEA/D (d) OMOPSO

(e) Fonseca PF (f) SDMOPSO (g) MOEA/D (h) OMOPSO

(i) Kursawe PF (j) SDMOPSO (k) MOEA/D (l) OMOPSO

Fig. 1. (a, e, i) are the PF_{true} and the rest are the approximated ones

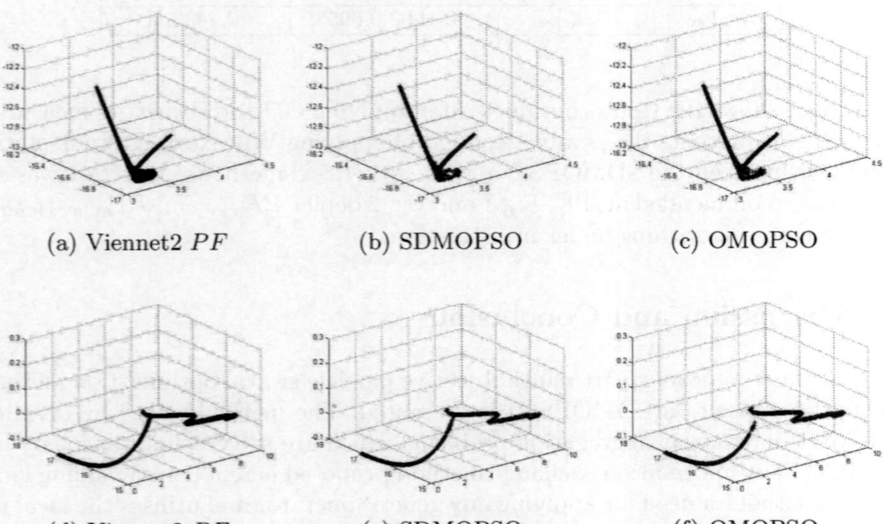

(a) Viennet2 PF (b) SDMOPSO (c) OMOPSO

(d) Viennet3 PF (e) SDMOPSO (f) OMOPSO

Fig. 2. (a, d) are PF_{true} and the rest are the approximated ones the problem under test, and B is $PF_{approximated}$. Then the indicator is defined as follows:

$$I_{R_2}(A,B) = \frac{\sum_{\lambda \in \Lambda} u^*(\lambda, A) - u^*(\lambda, B)}{|\Lambda|} \quad (6)$$

where $\Lambda = \{\lambda_1, \ldots, \lambda_m\}$, $\lambda_i \in [0,1]$ and $\sum_{i=1}^{m} \lambda_i = 1$.

These two indicators are used to compare $PF_{approximated}$ with PF_{true}; their values are used in the paper to quantitatively evaluate the performance of SDMOPSO in comparison with that of MOEA/D and OMOPSO.

Table 1. Indicators values for the three methods applied on nine test problems: the values are presented as [GD,R-metrics]

Problem	SDMOPSO	MOEA/D	OMOPSO
Schaffer	[0.0165,0.00212]	[0.0242,0.0002]	[0.0164,0.0029]
Fonseca	[0.0038,2.94E-04]	[0.004,3.00E-05]	[0.0037,5.48E-04]
Kursawe	[0.0335,1.22E-03]	[0.0343,8.91E-04]	[0.0323,9.43E-04]
Viennet2	[0.0067,2.31E-10]	[0.049,3.38E-07]	[0.0062,7.06E-10]
Viennet3	[0.0096,8.65E-07]	[3.3616,5.76E-03]	[0.0107,5.51E-07]
ZDT1	[0.0044,0.004]	[0.0055,0.0044]	[0.0037,2.29E-03]
ZDT2	[0.0051,0.003]	[0.0044,0.0017]	[0.0038,0.0012]
ZDT3	[0.0043,0.003]	[0.014,0.0067]	[0.0043,0.0041]
ZDT4	[1.4319,0.3068]	[0.7714,0.2338]	[1.4329,0.3744]
ZDT6	[0.003,0.0013]	[0.0029,0.0012]	[0.0031,0.0011]
Average	[0.1519,0.0322]	[0.4271,0.0255]	[0.1517,0.0387]
Std	[0.4498,0.0965]	[1.0581,0.0732]	[0.4503,0.1179]
p-value	[-,-]	[0.1602,0.6953]	[0.4453,1]

Table 1 shows the results obtained after applying GD and R-metrics measures, the last row presents the p-value resulted of applying Wilcoxon sign rank statistical test between the SDMOPSO and the other two methods. This test was selected as recommended in [19] . Fig.1 and Fig.2 depict PF_{true} and $PF_{approximated}$ for the three algorithms under investigation.

5 Discussion and Conclusions

In this paper, a novel smart multi-objective particle swarm optimisation method using decomposition(SDMOPSO) is presented. The method works by dividing the MOP into scalar aggregation problems which are solved simultaneously using PSO. The information exchange method proposed herein helps avoiding local optima without a need for applying any genetic operator and utilises the local information more effectively by facilitating simultaneous optimisation of all scalar problems. In order to maintain the diversity of the final solutions, SDMOPSO uses a crowding archive.

The previous use of PSO within the MOEA/D framework presented in [10] has several limitations. Every particle updates its position using its personal best and global best information without considering the neighborhood best, which can be an important asset for maintaining the diversity of the solutions and avoiding local optima. MOPSO/D uses a mutation operator, which can contribute to the complexity of the method and thus reduce the advantage of PSO over GA. MOPSO/D does not incorporate the non-dominance concept within the optimization process, which could potentially lead to premature convergence.

The results presented in this paper show that OMOPSO, MOEA/D and SD-MOPSO perform similarly on problems with two objectives (Schaffer, Fonseca, Kursawe, ZDT1-4 and ZDT6). When looking at three-objective problems, both SDMOPSO and OMOPSO outperform MOEA/D. However the statistical test over all datasets shows insignificant difference in the indicator values among all the methods. The advantage of the MOPSO-based methods in 3D MOPs could be explained by that EA-based techniques offer advantages in problems where some structure exist in the decision space - the reproduction The advantage of the MOPSO-based methods in 3D MOPs could be explained by that EA-based techniques offer advantages in problems where some structure exist in the decision space - the reproduction operators can exploit this structure very effectively. When, however, such a structure does not exist or is confounded by the interplay of several competing objectives, MOP heuristics aimed at uniform exploration of the solution space can perform better [20]. The results of our experiments support this hypothesis.

For Viennet3, SDMOPSO seems to have better diversity than OMOPSO as OMOPSO does not cover the PF fully (Fig.2e and Fig.2f). For other test problems, SDMOPSO and OMOPSO perform similarly. The major algorithmic difference between SDMOPSO and OMOPSO is that OMOPSO uses mutation[14], whereas SDMOPSO does not apply any genetic operator. Mutation is usually regarded as turbulence that is beyond a particle's own control [7]. The usage of mutation operator by PSO is generally justified because of a very high convergence speed of this method. Such convergence speed could be a disadvantage in solving MOPs, because it may lead to a false PF [21] due to falling into local optima. SDMOPSO, on the other hand, handles this issue by making every particle in the swarm to pre-process its moves and to share this information with its neighbours. This results in a better exploitation of the local information, which alleviates the effect of premature convergence to local optima.

References

1. Zhang, Q., Li, H.: Moea/d: A multi-objective evolutionary algorithm based on decomposition. IEEE Trans. on Evolutionary Computation 11(6), 712–731 (2007)
2. Zhang, Q., Liu, W., Li, H.: The performance of a new version of moea/d on cec09 unconstrained mop test instances. In: CEC 2009: Proceedings of the Eleventh Conference on Congress on Evolutionary Computation, Norway, pp. 203–208. IEEE, Los Alamitos (2009)
3. Awwad Shiekh Hasan, B., Gan, J.Q., Zhang, Q.: Multi-objective evolutionary methods for channel selection in brain-computer interfaces: some preliminary experimental results. In: WCCI, Barcelona, Spain, pp. 3339–3344. IEEE, Los Alamitos (2010)

4. Wang, Z., Durst, G.L., Eberhart, R.C., Boyd, D.B., Ben Miled, Z.: Particle swarm optimization and neural network application for qsar. In: Parallel and Distributed Processing Symposium, International, vol. 10, p. 194 (2004)
5. Jaishia, B., Ren, W.: Finite element model updating based on eigenvalue and strain. Mechanical Systems and Signal Processing 21(5), 2295–2317 (2007)
6. Sudha, B., Petrovski, A., McCall, J.: Optimising Cancer Chemotherapy Using Particle Swarm Optimisation and Genetic Algorithms. In: Yao, X., Burke, E.K., Lozano, J.A., Smith, J., Merelo-Guervós, J.J., Bullinaria, J.A., Rowe, J.E., Tiňo, P., Kabán, A., Schwefel, H.-P. (eds.) PPSN 2004. LNCS, vol. 3242, pp. 633–641. Springer, Heidelberg (2004)
7. Reyes-Sierra, M., Coello, C.A.C.: Multi-objective particle swarm optimizers: A survey of the state-of-the-art. International Journal of Computational Intelligence Research 2(3), 287–308 (2006)
8. Baltar, A.M., Fontane, D.G.: A generalized multiobjective particle swarm optimization solver for spreadsheet models: application to water quality. In: The Twenty Sixth Annual American Geophysical Union Hydrology Days (2006)
9. Hassan, R., Cohanim, B., de Weck, O., Venter, G.: A comparison of particle swarm optimization and the genetic algorithm. In: Structural Dynamics and Materials, Texas, USA (2005)
10. Peng, W., Zhang, Q.: A decomposition-based multi-objective particle swarm optimization algorithm for continuous optimization problems. In: IEEE International Conference on Granular Computing, Hangzhou (2008)
11. Miettinen, K.: Nonlinear Multiobjective Optimization. International Series in Operations Research & Management Science, vol. 12. Springer, Heidelberg (1998)
12. Kennedy, J., Eberhart, R.C., Shi, Y.: Swarm intelligence. Morgan Kaufmann, San Francisco (2001)
13. Jaszkiewicz, A.: On the performance of multiple-objective genetic local search on the 0/1 knapsack problem - a comparative experiment. IEEE Trans. on Evolutionary Computation 6(4), 402–412 (2002)
14. Reyes-Sierra, M., Coello, C.A.C.: Improving PSO-based multi-objective optimization using crowding, mutation and ϵ-dominance. In: Coello Coello, C.A., Hernández Aguirre, A., Zitzler, E. (eds.) EMO 2005. LNCS, vol. 3410, pp. 505–519. Springer, Heidelberg (2005)
15. Lamont, G.B., Veldhuizen, D.A.V.: Evolutionary Algorithms for Solving Multi-Objective Problems. Kluwer Academic Publishers, Norwell (2002)
16. Durillo, J.J., Nebro, A.J., Luna, F., Dorronsoro, B., Alba, E.: jMetal: A Java Framework for Developing Multi-Objective Optimization Metaheuristics. Technical Report ITI-2006-10, Departamento de Lenguajes y Ciencias de la Computación, University of Málaga, E.T.S.I. Informática, Campus de Teatinos (2006)
17. El-Ghazali, T.: Metaheuristics: from design to implementation. John Wiley & Sons, Chichester (2009)
18. Knowles, J., Corne, D.: On metrics for comparing nondominated sets. In: IEEE International Conference on E-Commerce Technology, vol. 1, pp. 711–716 (2002)
19. Desmar, J.: Statistical comparisons of classifiers over multiple data sets. Journal of Machine Learning Research 7, 1–30 (2006)
20. Fleischer, M.: The Measure of Pareto Optima Applications to Multi-objective Metaheuristics. In: Fonseca, C.M., Fleming, P.J., Zitzler, E., Deb, K., Thiele, L. (eds.) EMO 2003. LNCS, vol. 2632, pp. 519–533. Springer, Heidelberg (2003)
21. Coello, C.A.C., Pulido, G.T., Lechuga, M.S.: Handling multiple objectives with particle swarm optimization. IEEE Trans. Evolutionary Computation 8(3), 256–279 (2004)

A Hybrid Scalarization and Adaptive ϵ-Ranking Strategy for Many-Objective Optimization

Hernán Aguirre[1,2] and Kiyoshi Tanaka[2]

[1] International Young Researcher Empowerment Center
[2] Faculty of Engineering, Shinshu University
4-17-1 Wakasato, Nagano, 380-8553 Japan
{ahernan,ktanaka}@shinshu-u.ac.jp

Abstract. This work proposes a hybrid strategy in a two-stage search process for many-objective optimization. The first stage of the search is directed by a scalarization function and the second one by Pareto selection enhanced with Adaptive ϵ-Ranking. The scalarization strategy drives the population towards central regions of objective space, aiming to find solutions with good convergence properties to seed the second stage of the search. Adaptive ϵ-Ranking balances the search effort towards the different regions of objective space to find solutions with good convergence, spread, and distribution properties. We test the proposed hybrid strategy on MNK-Landscapes showing that performance can improve significantly on problems with more than 6 objectives.

1 Introduction

Recently, there is a growing interest on applying multi-objective evolutionary algorithms (MOEAs) to solve *many*-objective optimization problems with four or more objective functions. In general, conventional MOEAs scale up poorly with the number of objectives and new evolutionary algorithms are being proposed [1]. Research has focused mainly on the effectiveness of selection, dimensionality reduction, incorporation of user preferences, and space partitioning.

This work focuses on the effectiveness of selection on many-objective optimization. Some methods have been proposed to improve Pareto selection for many-objective optimization by incorporating indicator functions or extensions of Pareto dominance [2,3,4]. Most of these methods induce a different ranking based on information of how close solutions are to dominate other non-dominated solutions and have been proved effective to improve convergence at the expense of spread or vice-versa. To rank solutions, these methods compare each solution with all other solutions, bringing the computational order to $\mathcal{O}(M|P|^2)$, where M is the number of objectives and $|P|$ the population size. Other methods are based on scalarization functions that map the multi-objective problem to a single-objective one [5]. Since a scalarization function defines a search direction around a single point in objective space, to try to uniformly search the objective space and find good approximations of the Pareto front, very many scalarization functions must be specified. The computational order of ranking

R. Schaefer et al. (Eds.): PPSN XI, Part II, LNCS 6239, pp. 11–20, 2010.

with one scalarization functions is $\mathcal{O}(M|P|)$. However, usually the number of scalarization functions for many-objective optimization is of the same order of the population size making the overall computational order similar to $\mathcal{O}(M|P|^2)$.

In this work, we propose a hybrid strategy in a two-stage search process. The first stage of the search is directed by a scalarization function and the second one by Pareto selection enhanced with Adaptive ϵ-Ranking [6]. The scalarization function provides a computationally fast unifying search direction to drive the population towards central regions of objective space, aiming to find a subset of solutions with good convergence properties. On the other hand, Adaptive ϵ-Ranking uses the local information of the distribution of solutions to balance the search effort towards the different regions of objective space, increasing the discriminatory power of Pareto selection while introducing simultaneously a density estimator that scales-up well on high dimensional spaces, to find solutions with good convergence, spread, and distribution properties.

We study the effects of the scalarization and Adaptive ϵ-Ranking applied independently. Then, we study the effects of the proposed hybrid strategy, showing that it can significantly outperform its individual components, especially on problems with more than 6 objectives. Also, since the hybrid strategy uses just one scalarization function during the first stage, it becomes considerably faster, which is an important issue for scalability on high-dimensional spaces. As benchmark instances we use MNK-Landscapes [7] with $4 \leq M \leq 10$ objectives, $N = 100$ bits, and $0 \leq K \leq 50$ epistatic interactions per bit.

2 Proposed Hybrid Strategy

2.1 Concept

Multi-objective optimizers seek to find trade-off solutions with good properties of convergence to the Pareto front, well spread and uniformly distributed along the front. These three properties are especially difficult to achieve in many-objective problems and most searching strategies compromise one in favor of the other. In addition, larger population sizes are likely to be required in order to create an appropriate approximation of the Pareto front in high dimensional spaces. Both, larger populations and high dimensionality, impose a serious challenge to the computational scalability of current algorithms.

Seeking to find approximations of the Pareto front fulfilling the three properties of convergence, spread, and distribution, rather than expecting a sole strategy to work efficiently for all tasks, it seems reasonable to distribute the search into different strategies that complement each other. The proposed hybrid strategy follows this approach, using one strategy from the domain of scalarization that focus on convergence-only and the other one from the domain of Pareto dominance and its extensions (Adaptive ϵ-Ranking) that in addition to convergence also pays attention to diversity. The strategies are deployed following a two-stage scenario, assigning one strategy to each stage, where the first strategy works to seed the second one, as illustrated in Fig. 1. The expectation for the

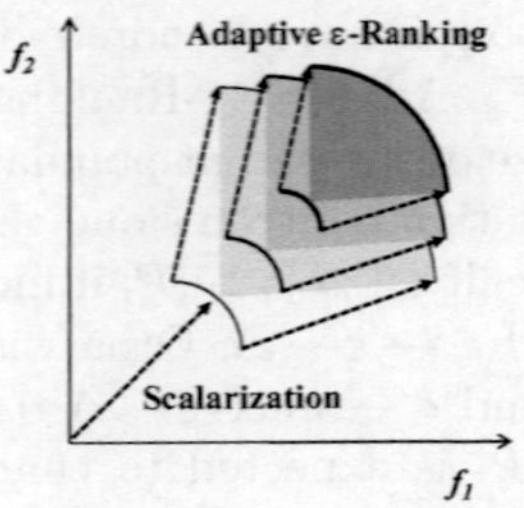

Fig. 1. Hybrid Strategy

hybrid strategy is that better results could be achieved by diversifying the population after some degree of convergence has been achieved than by emphasizing convergence and diversity since the beginning of the search, where the population is random. Also, by simplifying the scalarization strategy to one scalarizing function, it is expected that the hybrid method could speed up the search.

2.2 Scalarization Strategy

The role of the scalarization strategy is to provide a computationally-fast unifying search direction to drive the population towards central regions of objective space, so that solutions with good convergence could be found to seed the second stage of the search. In this work we use $g = \frac{1}{M}\sum_{i=1}^{M} f_i$ as scalarizing function, where f_i denotes the i-th objective value and M the number of objectives.

2.3 Adaptive ϵ-Ranking Strategy $A\epsilon R^E$

Pareto ranking classifies the entire population in one or more sets of equally ranked solutions $\mathcal{F}_i$ ($i = 1, \cdots, N_F$), each set associated to rank i. On many-objective problems the number of Pareto non-dominated solutions increase substantially with the dimensionality of the objective space and $|\mathcal{F}_1|$ usually becomes larger than the size of the parent population $|P|$ from early generations [7].

ϵ-Ranking re-classifies the sets $\mathcal{F}_i$ ($i = 1, \cdots, N_F$) into sets $\mathcal{F}_j^\epsilon$ ($j = 1, \cdots, N_F^\epsilon$), $N_F^\epsilon \geq N_F$, using a randomized sampling procedure that favors a good distribution of solutions based on dominance regions wider than conventional Pareto dominance (ϵ-dominance). The sampling heuristic favors an effective search using the following criteria. (i) Extreme solutions are always part of the sample. (ii) Each (not extreme) sampled solution is the sole sampled representative of its area of influence, which is determined by ϵ-dominance. (iii) Sampling of (not extreme) solutions follows a random schedule. These criteria aim to balance the search effort towards the different regions of objective space, increasing the discriminatory power of Pareto selection while simultaneously introducing a density estimator that scales-up well on high dimensional spaces, to find solutions with good convergence and diversity (spread and distribution) properties.

The number of rank-1 solutions $|\mathcal{F}_1^\epsilon|$ after reclassification depends on the value set to ϵ (≥ 0). Larger values of ϵ imply that sampled solutions ϵ-dominate larger areas, increasing the likelihood of having more ϵ-dominated solutions excluded from the sample that form $\mathcal{F}_1^\epsilon$. Adaptive ϵ-Ranking adapts ϵ at each generation so that $|\mathcal{F}_1^\epsilon|$ is close to the size of the parent population $|\mathcal{P}|$. The adaptation rule takes advantage of the correlation between ϵ and the number of ϵ-nondominated solutions in the sample. Basically, if $|\mathcal{F}_1^\epsilon| > |\mathcal{P}|$ it increases the step of adaptation $\Delta \leftarrow \min(\Delta \times 2, \Delta_{max})$ and $\epsilon \leftarrow \epsilon + \Delta$. Otherwise, if $|\mathcal{F}_1^\epsilon| < |\mathcal{P}|$ it decreases $\Delta \leftarrow \max(\Delta \times 0.5, \Delta_{min})$ and $\epsilon \leftarrow \max(\epsilon - \Delta, 0.0)$. The appropriate value of ϵ that approaches $|\mathcal{F}_1^\epsilon|$ to $|\mathcal{P}|$ is expected to change as the evolution process proceeds, it is problem dependent, and affected by the stochastic nature of the search that alters the instantaneous distributions of solutions in objective space. Adaptation of ϵ and its step of adaptation Δ is important to properly follow the dynamics of the evolutionary process on a given problem.

3 Multiobjective MNK-Landscapes

A multiobjective MNK-Landscape [7] is defined as a vector function mapping binary strings into real numbers $\boldsymbol{f}(\cdot) = (f_1(\cdot), f_2(\cdot), \cdots, f_M(\cdot)) : \mathcal{B}^N \rightarrow \Re^M$, where M is the number of objectives, $f_i(\cdot)$ is the i-th objective function, $\mathcal{B} = \{0, 1\}$, and N is the bit string length. $\boldsymbol{K} = \{K_1, \cdots, K_M\}$ is a set of integers where K_i $(i = 1, 2, \cdots, M)$ is the number of bits in the string that epistatically interact with each bit in the i-th landscape. Each $f_i(\cdot)$ can be expressed as an average of N functions as follows

$$f_i(\boldsymbol{x}) = \frac{1}{N} \sum_{j=1}^{N} f_{i,j}(x_j, z_1^{(i,j)}, z_2^{(i,j)}, \cdots, z_{K_i}^{(i,j)}) \tag{1}$$

where $f_{i,j} : \mathcal{B}^{K_i+1} \rightarrow \Re$ gives the fitness contribution of bit x_j to $f_i(\cdot)$, and $z_1^{(i,j)}, z_2^{(i,j)}, \cdots, z_{K_i}^{(i,j)}$ are the K_i bits interacting with bit x_j in the string $\boldsymbol{x}$. The fitness contribution $f_{i,j}$ of bit x_j is a number between [0.0, 1.0] drawn from a uniform distribution. Thus, each $f_i(\cdot)$ is a non-linear function of $\boldsymbol{x}$ expressed by a Kauffman's NK-Landscape model of epistatic interactions. In addition, it is also possible to arrange the epistatic pattern between bit x_j and the K_i other interacting bits. That is, the distribution $D_i = \{random, nearest\ neighbor\}$ of K_i bits among N. Thus, M, N, $\boldsymbol{K} = \{K_1, K_2, \cdots, K_M\}$, and $\boldsymbol{D} = \{D_1, D_2, \cdots, D_M\}$, completely specify a multiobjective MNK-Landscape.

4 Method of Analysis

In this work, we use the hypervolume $\mathcal{H}$ and the set coverage $\mathcal{C}$ [8] to evaluate the performance of the algorithms, complementing our analysis with the maximum $\max(f_i)$ and minimum $\min(f_i)$ fitness values found in each objective. The measure $\mathcal{C}$ provides information on convergence. $\mathcal{C}(\mathcal{A}, \mathcal{B})$ gives the fraction of solutions in set $\mathcal{B}$ that are dominated at least by one solution in set $\mathcal{A}$.

$\mathcal{H}$ is a measure of convergence and diversity, calculated as the volume of the M-dimensional region in objective space enclosed by the set of non-dominated solutions and a dominated reference point. If the reference point is very close to the Pareto front, non-dominated solutions around the center region of the Pareto front are relatively emphasized in the hypervolume calculation. On the other hand, if the reference point is far from the Pareto front, non-dominated solutions along the extreme regions of the Pareto front are emphasized in the hypervolume calculation. The hypervolume has become widely used to analyze the performance of multi-objective optimizers. However, results on the hypervolume are usually reported using a single reference point, which provides only a partial vision of the results obtained. In many-objective problems, particularly, it is difficult to grasp the convergence and diversity properties of the solutions obtained and reporting results using one reference point could often lead to overstated and sometimes erroneous conclusions about the overall performance of the algorithms. Analysis of hypervolume varying the reference point provides more insights on the distribution of the obtained solutions and helps clarify the relative contribution to the hypervolume of solutions that converge to the central regions of the space and those that contribute to diversity (spread). To enrich our analysis, we compute the hypervolume using different reference points. The reference point $\boldsymbol{r}_{d_R} = (r_1, r_2, \cdots, r_M)$ is calculated by

$$r_i = (1.0 - d_R) \times \min(f_i), i = 1, 2, \cdots, M, \tag{2}$$

where $\min(f_i)$ is the minimum value of the i-th objective function observed in the joined sets of Pareto optimal solutions found by the algorithms we compare, and d_R is a parameter that determines the distance of the reference point to the minimum values found for each objective function $(\min(f_1), \min(f_2), \cdots, \min(f_M))$. In this work, we use $d_R = \{0.01, 0.1, 0.3, 0.5, 0.7, 1.0\}$ to set the reference point $\boldsymbol{r}_{d_R} = (r_1, r_2, \cdots, r_M)$. Note that we maximize all objective functions and the allowed range for all f_i is $[0.0, 1.0]$. Hence, $d_R = 0.01$ means that the reference point is $\boldsymbol{r}_{d_R} = 0.99 \times (\min(f_1), \min(f_2), \cdots, \min(f_M))$ and thus very close to the Pareto front, whereas $d_R = 1.0$ means that $\boldsymbol{r}_{d_R} = (0.0, 0.0, \cdots, 0.0)$ and far from the Pareto front. To calculate $\mathcal{H}$, we use Fonseca et al. [9] algorithm.

5 Experimental Results and Discussion

5.1 Preparation

The performance of the algorithms is verified on MNK-Landscapes with $4 \leq M \leq 10$ objectives, $N = 100$ bits, number of epistatic interactions $K = \{0, 1, 3, 5, 7, 10, 15, 25, 35, 50\}$ $(K_1, \cdots, K_M = K)$, and *random* epistatic patterns among bits in all objectives $(D_1, \cdots, D_M = \textit{random})$. Results presented below show the average performance of the algorithms on 50 different problems randomly generated for each combination of M, N and K. In the plots, error bars show 95% confidence intervals on the mean.

In this work, we implement the proposed hybrid strategy using NSGA-II [10] as a host algorithm, modifying it accordingly to include the scalarization and $A\epsilon R^E$ strategies. During the first stage, selection is based solely on the scalarization function, whereas in the second stage $A\epsilon R^E$ is applied after Pareto dominance. All algorithms used in our study are set with parent and offspring populations of size

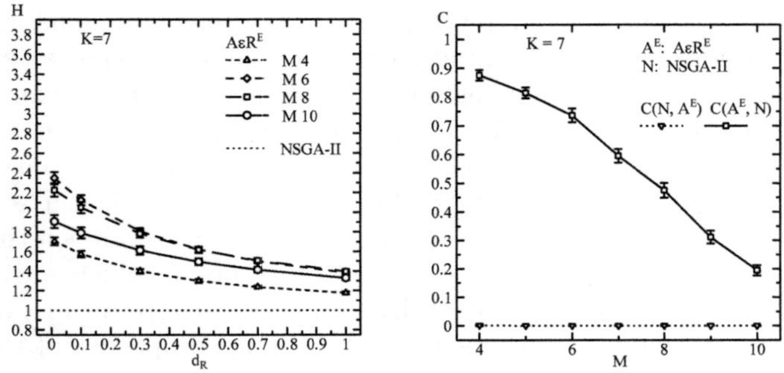

(a) Normalized $\mathcal{H}$ (b) Set coverage $\mathcal{C}$

Fig. 2. Adaptive ϵ-Ranking, $K = 7$, $4 \leq M \leq 10$

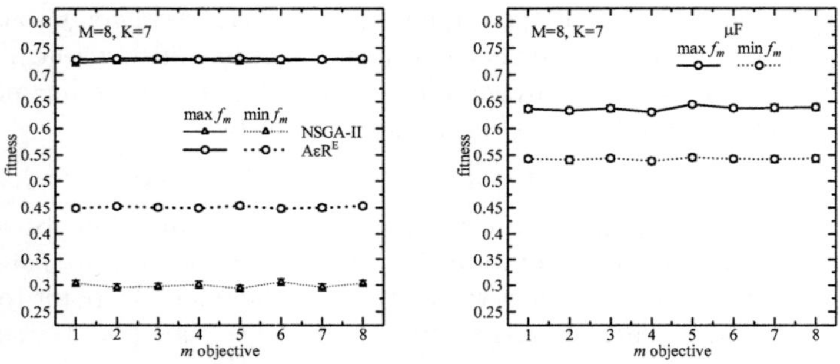

(a) Adaptive ϵ-Ranking (b) Scalarization strategy

Fig. 3. $\max(f_m)$ and $\min(f_m)$, $K = 7$, $4 \leq M \leq 10$

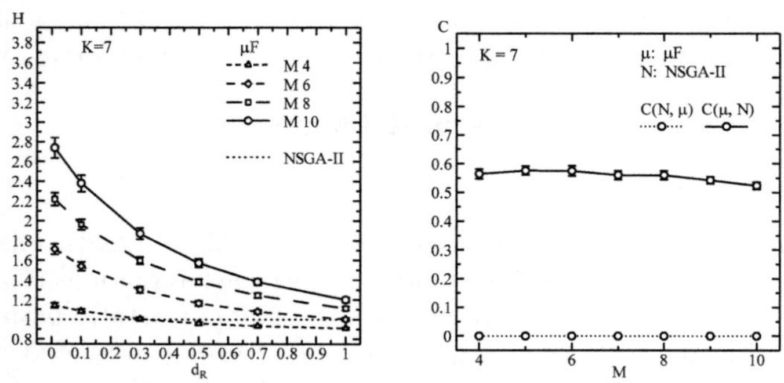

(a) Normalized $\mathcal{H}$ (b) Set coverage $\mathcal{C}$

Fig. 4. Scalarization strategy, $K = 7$ and $4 \leq M \leq 10$

$|\mathcal{P}| = |\mathcal{Q}| = 100$, two point crossover for recombination with rate $p_c = 0.6$, and bit flipping mutation with rate $p_m = 1/N$ per bit. The number of evaluations is set to 3×10^5 ($T = 3000$ generations). In $A\epsilon R^E$, initially $\epsilon = 0.0$, the initial value of the step of adaptation is $\Delta_0 = 0.005$ (0.5%) and its maximum and minimum values are set to $\Delta_{max} = 0.05$ (5%) and $\Delta_{min} = 0.0001$ (0.01%).

5.2 Effects of Individual Components

Adaptive ϵ-Ranking Strategy. In this section we discuss the performance of Adaptive ϵ-Ranking Strategy ($A\epsilon R^E$) using NSGA-II as a reference for comparison. Fig. 2(a) shows the normalized hypervolume $\mathcal{H}$ between $A\epsilon R^E$ and NSGA-II varying the reference point for $K = 7$ and $M = \{4, 6, 8, 10\}$ landscapes. From this figure it can be seen that $A\epsilon R^E$ attains better $\mathcal{H}$ for all values of M regardless of the reference point. Also, note the increasing slopes of the $\mathcal{H}$ curves as the reference point gets closer to the Pareto front, i.e. varying d_R from 1.0 to 0.01. These results suggest that solutions by $A\epsilon R^E$ are better than solutions by NSGA-II particularly in the central regions of the objective space. Notice that the slope of the $\mathcal{H}$ curve becomes steeper by increasing the number of objectives from $M = 4$ to $M = 6$, but it gradually recedes for $M = 8$ and $M = 10$ compared to $M = 6$. This is an indication that the convergence abilities of $A\epsilon R^E$ reduce for $M > 6$, especially to the central regions of objective space.

Fig. 2(b) shows results using the $\mathcal{C}$ coverage measure. Note that $\mathcal{C}(A^E, N)$, the fraction of NSGA-II's solutions dominated by $A\epsilon R^E$'s solutions, is almost 0.9 for $M = 4$ and reduces progressively with M until it approaches 0.2 for $M = 10$. On the contrary, $\mathcal{C}(N, A^E)$ is zero for all M, which means that no solution by $A\epsilon R^E$ is dominated by NSGA-II's solutions. These results confirm the superiority of $A\epsilon R^E$ over NSGA-II and corroborate the decreasing convergence power of $A\epsilon R^E$ for large values of M.

Fig. 3(a) shows the maximum and minimum fitness, $\max(f_m)$ and $\min(f_m)$, of solutions in the Pareto front found by NSGA-II and $A\epsilon R^E$ for $K = 7$ and $M = 8$ landscapes. From this figure, it can be seen that NSGA-II and $A\epsilon R^E$ achieve similar $\max(f_m)$. However, $\min(f_m)$ is lower by NSGA-II. Similar values of $\max(f_m)$ suggest that spread by the algorithms is comparable, but the lower values of $\min(f_m)$ suggest that solutions by NSGA-II seem to be trapped in lower local optima. Another interesting property of $A\epsilon R^E$ is that solutions in the Pareto front are ϵ-nondominated, which gives a good distribution of solutions.

Scalarization Strategy. In this section we analyze the scalarization strategy (μF). Fig. 4(a) shows the normalized $\mathcal{H}$ between μF and NSGA-II varying the reference point on $K = 7$ and $M = \{4, 6, 8, 10\}$ landscapes. Comparing $\mathcal{H}$ by looking at Fig. 4(a) and Fig. 2(a), it can be seen that on $M = \{4, 6\}$ landscapes μF is significantly worse than $A\epsilon R^E$ for any value of d_R. On $M = 8$ landscapes, μF is still worse than $A\epsilon R^E$ for $d_R \geq 0.1$, but similar to $A\epsilon R^E$ for $d_R = 0.01$. However, on $M = 10$ landscapes, μF is better than $A\epsilon R^E$ for $d_R < 0.5$. These results suggest that μF gets better compared to $A\epsilon R^E$ in terms of convergence to central regions when the number of objectives is above eight.

Fig. 4(b) shows $\mathcal{C}$ between μF and NSGA-II. Note that for any number of objectives, the values of $\mathcal{C}(\mu F, N)$ are similar and above 0.5, meaning that more than half of the solutions by NSGA-II are dominated by solutions of μF.

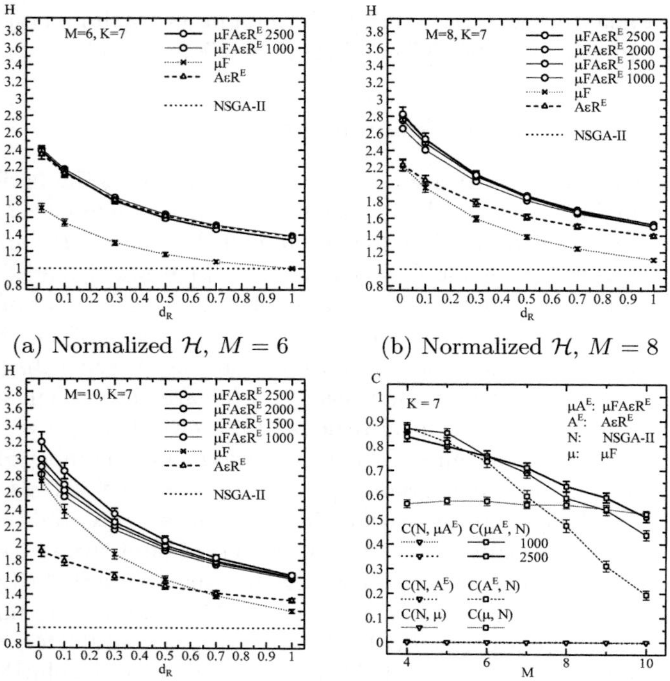

(a) Normalized $\mathcal{H}$, $M = 6$ (b) Normalized $\mathcal{H}$, $M = 8$

(c) Normalized $\mathcal{H}$, $M = 10$ (d) Set coverage $\mathcal{C}$

Fig. 5. Proposed Hybrid Strategy, $K = 7$ and $4 \leq M \leq 10$

Comparing with Fig. 2(b), note that $\mathcal{C}(\mu, N) < \mathcal{C}(A\epsilon R^E, N)$ for $M \leq 7$, but $\mathcal{C}(\mu, N) > \mathcal{C}(A\epsilon R^E, N)$ for $M \geq 8$. These results are in accordance with the observations made for $\mathcal{H}$, and confirm the better convergence properties of μF on landscapes with more than eight objectives. However, note that μF converges to a narrow area, as shown in Fig. 3(b) that plots the $\max(f_m)$ and $\min(f_m)$ of the non-dominated set found by μF.

Overall, these results shows that the scalarization strategy μF converges well, albeit to a narrow region. The similar values of $\mathcal{C}(\mu F, N)$ for all M is an interesting property of μF. It shows that this strategy in terms of convergence can scale up to a large number of objectives, suggesting that it could be useful as part of the hybrid strategy.

5.3 Effects of the Hybrid Strategy

In this section we analyze the hybrid strategy that combines in a two-stage process scalarization and Adaptive ϵ-Ranking ($\mu FA\epsilon R^E$). $\mu FA\epsilon R^E$ first starts with μF and then at generation t_S it switches to $A\epsilon R^E$. Fig. 5 (a)-(c) show $\mathcal{H}$ by $\mu FA\epsilon R^E$ on $M = \{6, 8, 10\}$ landscapes, respectively, varying $t_S = \{1000, 1500, 2000, 2500\}$ and keeping the total number of the generations fixed to $T = 3000$.

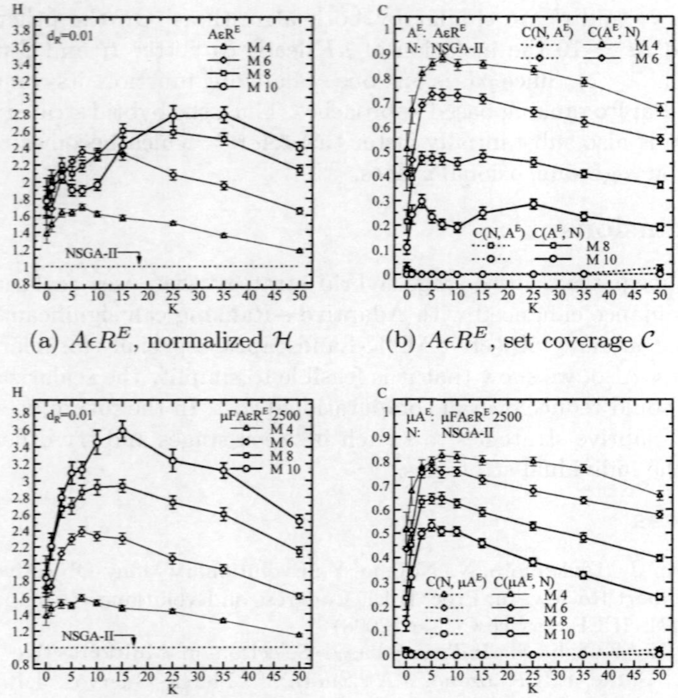

(a) $A\epsilon R^E$, normalized $\mathcal{H}$ (b) $A\epsilon R^E$, set coverage $\mathcal{C}$

(c) $\mu F A\epsilon R^E$, normalized $\mathcal{H}$ (d) $\mu F A\epsilon R^E$, set coverage $\mathcal{C}$

Fig. 6. Adaptive ϵ-Ranking $A\epsilon R^E$ and Proposed Hybrid Strategy $\mu F A\epsilon R^E$, $0 \leq K \leq 50$ and $4 \leq M \leq 10$

The same figures also include results by μF and $A\epsilon R^E$ for comparison. Note that on $M = 6$ the inclusion of μF does not improve $\mathcal{H}$ (actually, on $M = 4$ for which results are not shown, $\mathcal{H}$ reduces by including μF, with larger reductions observed for late switching times t_S). However, switching from μF to $A\epsilon R^E$ during the run can improve $\mathcal{H}$ substantially ($d_R = 0.01$) on $M = 8$ and $M = 10$, with a late switching time ($t_S = 2500$) working better than an early one.

Fig. 5 (d) shows $\mathcal{C}$ values between $\mu F A\epsilon R^E$ and NSGA-II for $t_S = \{1000, 2500\}$. We also include results by μF and $A\epsilon R^E$ for comparison. Results on $\mathcal{C}$ confirm our observations on $\mathcal{H}$ and give a clearer picture of the effects of including μF. Note that $\mu F A\epsilon R^E$ shows relatively better convergence than $A\epsilon R^E$ for $M \geq 7$ and the importance of late switching times t_S for $M \geq 8$. Also, it can be seen that convergence is better than μF for $M \leq 9$ if $t_S = 2500$. For $M = 10$ similar convergence to μF is observed. However, $\mu F A\epsilon R^E$ $t_S = 2500$ shows significantly better $\mathcal{H}$ than μF on $M = 10$ as shown in Fig. 5 (c).

Fig. 6 show results by $A\epsilon R^E$ and $\mu F A\epsilon R^E$ ($t_S = 2500$), varying K from 0 to 50 to observe the scalability of the algorithms on problems of increased epistasis. From these figures, note that similar to $K = 7$, on $M = 4$ performance deteriorates slightly by including μF, especially in terms of convergence, whereas on $M = 6$ $\mathcal{H}$ and $\mathcal{C}$ are similar by both algorithms. On the other hand, on $M = 8$ and $M = 10$ the inclusion of μF leads to better $\mathcal{H}$ and $\mathcal{C}$ on a broad range of K ($K \geq 1$). Since μF is just one scalarazing function, its computation is faster than Pareto ranking based approaches. Thus, the hybrid strategy $\mu F A\epsilon R^E$ ($t_S = 2500$) is also substantially faster than $A\epsilon R^E$, which becomes relevant for scalability on high dimensional spaces.

6 Conclusions

We have shown that a two-stage hybrid strategy that uses scalarization and Pareto dominance enhanced with Adaptive ϵ-Ranking can significantly improve performance on many-objective MNK-Landscapes, especially for a large number of objectives. Also, we show that it is feasible to simplify the scalarization strategy, which could reduce overall computational cost. In the future, we would like to explore adaptive strategies to switch between stages and try other scenarios to deploy the individual strategies.

References

1. Ishibuchi, H., Tsukamoto, N., Nojima, Y.: Evolutionary Many-Objective Optimization: A Short Review. In: Proc. IEEE Congress on Evolutionary Computation, pp. 2424–2431. IEEE Service Center (2008)
2. Zitzler, E., Kunzli, S.: Indicator-based Selection in Multiobjective Search. In: Yao, X., Burke, E.K., Lozano, J.A., Smith, J., Merelo-Guervós, J.J., Bullinaria, J.A., Rowe, J.E., Tiňo, P., Kabán, A., Schwefel, H.-P. (eds.) PPSN 2004. LNCS, vol. 3242, pp. 832–842. Springer, Heidelberg (2004)
3. Emmerich, M., Beume, N., Naujoks, B.: An EMO Algorithm Using the Hypervolume Measure as Selection Criterion. In: Coello Coello, C.A., Hernández Aguirre, A., Zitzler, E. (eds.) EMO 2005. LNCS, vol. 3410, pp. 62–76. Springer, Heidelberg (2005)
4. Koppen, M., Yoshida, K.: Substitute Distance Assignments in NSGA-II for Handling Many-Objective Optimization Problems. In: Obayashi, S., Deb, K., Poloni, C., Hiroyasu, T., Murata, T. (eds.) EMO 2007. LNCS, vol. 4403, pp. 727–741. Springer, Heidelberg (2007)
5. Hughes, E.J.: Evolutionary Many-Objective Optimization: Many Once or One Many? In: Proc. 2005 IEEE Congress on Evolutionary Computation. vol. 1, pp. 222–227. IEEE Service Center (September 2005)
6. Aguirre, H., Tanaka, K.: Adaptive ϵ-Ranking on Many-Objective Problems. Evolutionary Intelligence 2, 183–206 (2009)
7. Aguirre, H., Tanaka, K.: Working Principles, Behavior, and Performance of MOEAs on MNK-Landscapes. European Journal of Operational Research 181(3), 1670–1690 (2007)
8. Zitzler, E.: Evolutionary Algorithms for Multiobjective Optimization: Methods and Applications, PhD thesis, Swiss Federal Institute of Technology, Zurich (1999)
9. Fonseca, C., Paquete, L., López-Ibáñez, M.: An Improved Dimension-sweep Algorithm for the Hypervolume Indicator. In: Proc. 2006 IEEE Congress on Evolutionary Computation, pp. 1157–1163. IEEE Service Center (2006)
10. Deb, K., Agrawal, S., Pratap, A., Meyarivan, T.: A Fast Elitist Non-Dominated Sorting Genetic Algorithm for Multi-Objective Optimization: NSGA-II, KanGAL report 200001 (2000)

pMODE-LD+SS: An Effective and Efficient Parallel Differential Evolution Algorithm for Multi-Objective Optimization

Alfredo Arias Montaño[1,⋆], Carlos A. Coello Coello[1,⋆⋆], and Efrén Mezura-Montes[2,⋆⋆⋆]

[1] CINVESTAV-IPN, Av. IPN 2508, San Pedro Zacatenco, Gustavo A. Madero, México D.F. 07360, Mexico
aarias@computacion.cs.cinvestav.mx, ccoello@cs.cinvestav.mx
[2] LANIA A.C. Rébsamen 80, Centro, Xalapa, Veracruz, 91000, Mexico
emezura@lania.mx

Abstract. This paper introduces a novel Parallel Multi-Objective Evolutionary Algorithm (pMOEA) which is based on the island model. The serial algorithm on which this approach is based uses the differential evolution operators as its search engine, and includes two mechanisms for improving its convergence properties (through local dominance and environmental selection based on scalar functions). Two different parallel approaches are presented. The first aims at improving effectiveness (i.e., for better approximating the Pareto front) while the second aims to provide a better efficiency (i.e., by reducing the execution time through the use of small population sizes in each sub-population). To assess the performance of the proposed algorithms, we adopt a set of standard test functions and performance measures taken from the specialized literature. Results are compared with respect to its serial counterpart and with respect to three algorithms representative of the state-of-the-art in the area: NSGA-II, MOEA/D and MOEA/D-DE.

1 Introduction

Multi-objective evolutionary algorithms (MOEAs) have been found to be very suitable for solving a wide variety of engineering optimization problems, because of their generality, their ease of use and their relatively low susceptibility to the specific features of the search space of the problem to be solved [1]. Nonetheless, they are normally computationally expensive due to several reasons: (1) real-world optimization problems typically involve high-dimensional search spaces and/or a large number of objective functions, (2) they require finding a set of solutions instead of only one, often requiring, in consequence, large population

⋆ The first author acknowledges support from CONACyT and IPN to pursue graduate studies in computer science at CINVESTAV-IPN.
⋆⋆ The second author acknowledges support from CONACyT project no. 103570.
⋆⋆⋆ The third author acknowledges support from CONACyT project No. 79809.

R. Schaefer et al. (Eds.): PPSN XI, Part II, LNCS 6239, pp. 21–30, 2010.

sizes, and (3) frequently, the task of evaluating the objective functions demands high computational costs (e.g., complex computer simulations are required). All these factors decrease the utility of serial MOEAs for its use in real-world engineering Multi-objective Optimization Problems (MOPs). In order to reduce the execution time required to solve these problems two main types of approaches have been normally adopted[1]: (1) Enhance the MOEA's design, namely improving its convergence properties, so that the number of objective function evaluations can be reduced, and, (2) Use of parallel programming techniques, i.e., to adopt a parallel or distributed MOEA.

Based on the above, the major aim of the present work is to develop two different schemes for improving the performance of a pMOEA. The first is designed for improving effectiveness (i.e., for better approximating the Pareto front), while the second is designed for improving efficiency (i.e., for reducing the execution time by using small population sizes in each sub-population). Either approach (or both) can be of interest in solving real-world enginering MOPs. The two proposed schemes are based on the *island paradigm*, and use the multi-objective differential evolution algorithm MODE-LD+SS [2] as its search engine. The proposed schemes are evaluated using standard test functions and performance measures taken from the specialized literature. The results obtained by the two proposed pMOEAs are compared with respect to their serial counterpart and with respect to the NSGA-II [3], MOEA/D [4], and MOEA/D-DE [5].

The remainder of the paper is organized as follows. In Section 2, the most relevant previous related work on island-based pMOEAs is presented. Section 3 is devoted to describe the proposed approach. Then, the experimental setup is presented in Section 4. In Section 5 the obtained results are presented and discussed. Finally, in Section 6 our conclusions and the corresponding future work is highlighted.

2 Previous Related Work

A pMOEA can be useful to solve problems faster, but also for generating novel and more efficient search schemes, i.e., a pMOEA can be more effective than its sequential counterpart, even when executed in a single processor [6]. From the specialized literature, four major pMOEA paradigms are commonly used [7]: (i) "Master-Slave," (ii) "Island," (iii) "Diffusion," and (iv) "hierarchical" or "hybrid". A comprehensive review of these paradigms can be found in [1,7]. This paper focuses on the *Island Model*, which is based on the phenomenon of natural populations evolving independently. In each island, a serial MOEA is executed for a predefined number of generations called *epoch*. At the end of an epoch, communication between neighboring islands is allowed. In this communication, individuals (or copies of them in the case of pollination) can migrate from its

[1] Our discussion here is focused exclusively on MOEAs that use exact objective function values, since fitness approximation schemes and surrogate models can also be used to deal with expensive MOPs.

actual island to a different one according to a predefined *migration topology* which determines the migration path along which individuals can move.

Kamiura et al. [8] presented a pMOEA called MOGADES (Multi-Objective Genetic Algorithm with Distributed Environment Scheme). In this pMOEA, the population is divided into M islands, and in each of them the MOP is converted into a scalar one, i.e., a different weight vector is assigned to each island. The aim of this algorithm is that each island can capture a different region of the Pareto front. One important aspect in this approach is that when migration occurs, the weights for each island are varied. A major drawback for this approach is that a good distribution of solutions cannot be guaranteed as it depends on the dynamics of the evolutionary system, i.e., of the weight vector variation.

Streichert et al. [9] proposed a pMOEA, which combines an island model with the "divide and conquer" principle. This approach partitions the population using a clustering algorithm (k-means), with the aim of assigning to each island, the search task of a particular Pareto front region. In this approach, at each epoch, the sub-populations are gathered by a master process for performing the clustering/distributing process. The individuals in each island are kept within their assigned Pareto front region using zone constraints. The main drawback of this approach is that *a priori* knowledge of the Pareto front shape is needed to define the zone constraints.

Zahaire and Petcu [10] developed the *multi-population APDE* (APDE stands for Adaptive Pareto Differential Evolution). This approach consists of dividing the main population into sub-populations (islands), each of equal size. In each island, a serial version of the APDE is executed with its own set of randomly initialized adaptive parameters, and is evolved for an epoch. Afterwards, a migration process is started. This process is based on a random connection topology, i.e., each individual from each sub-population can be swapped (with a given migration probability) with a randomly selected individual from another randomly selected island.

3 Our Proposed Approach

The first mechanism is as follows. In our proposed parallel algorithm, each island runs an approach called MODE-LD+SS [2], which adopts operators from differential evolution using the DE/RAND/1/bin scheme. Algorithm 1 shows the basic (serial version) pseudo-code of our proposed MODE-LD+SS approach. In Algorithm 1, the solution vectors $\boldsymbol{u_1}, \boldsymbol{u_2}, \boldsymbol{u_3}$ are selected from the current population, only if they are locally nondominated in their neighborhood $\aleph$. Local dominance is defined as follows:

Definition 1. Pareto Local Dominance: Let $\boldsymbol{x}$ be a feasible solution, $\aleph(\boldsymbol{x})$ be a neighborhood structure for $\boldsymbol{x}$ in the decision space, and $\boldsymbol{f}(\boldsymbol{x})$ a vector of objective functions.

- We say that a solution $\boldsymbol{x}$ is locally nondominated with respect to $\aleph(\boldsymbol{x})$ if and only if there is no $\boldsymbol{x}'$ in the neighborhood of $\boldsymbol{x}$ such that $\boldsymbol{f}(\boldsymbol{x}') \prec \boldsymbol{f}(\boldsymbol{x})$

The neighborhood structure is defined as the NB closest individuals to a particular solution. Closeness is measured using the Euclidean distance between solutions. The major aim of using the local dominance concept, as defined above, is to exploit good individuals' genetic information in creating DE trial vectors, and the associated offspring, which might help to improve the MOEA convergence rate toward the Pareto front. From Algorithm 1, it can be noted that this mechanism has a stronger effect during the earlier generations, where the portion of nondominated individuals is low in the global population, and progressively weakens, as the number of nondominated individuals grows during the evolutionary process. This mechanism is automatically switched off, once all the individuals in the population become nondominated, and has the possibility to be switched on, as some individuals become dominated. Aditionally, the diversity of the created offspring can be controled by the local dominance neighborhood size NB. Low values of NB will increase the diversity of the offspring, and viceversa.

The second mechanism that is introduced in MODE-LD+SS is called *envirommental selection based on a scalar function*, and is based on the Tchebycheff scalarization function given by [4]:

$$g(\boldsymbol{x}|\lambda^j, z*) = \max_{1\leq i\leq m} \{\lambda_i^j |f_i(x) - z_i^*|\} \quad (1)$$

In the above equation, $\lambda^j, j = 1, \ldots, N$ represents the weight vectors used to distribute the solutions along the whole Pareto front (see Figure 1(a)). $z*$ corresponds to a reference point, defined in objective function space and determined with the minimum objective values of the population. This reference point is updated at each generation, as the evolution progresses. The procedure *MinimumTchebycheff*$(Q, \lambda^j, z*)$ finds, from the set Q (the combined population consisting on the actual parents and the created offspring), the solution vector that minimizes equation (1) for each weight vector λ^j and the reference point $z*$.

Based on the serial MOEA previously described, we present here two parallelization schemes. The first is designed for improving effectiveness and is called pMODE-LD+SS(A). The second is designed for improving efficiency, and is called pMODE-LD+SS(B). Both of them share the following characteristics:

- Use of a "random pair-islands" bidirectional migration scheme. In this scheme, at each epoch, pairs of islands are randomly selected. Then, the communication is performed between each pair of islands. Migrants from one island are considered as immigrants in the receptor island, and viceversa.
- Use of a pollination scheme, i.e., copies of selected migrants are sent, while the original individuals are retained in their own population.
- The migration policy is based on randomly selected individuals.
- The replacement policy is based on the environmental selection mechanism adopted in the serial version running in each island. In this case, immigrants are added to the receptor island's population, and the environmental selection process is applied to this extended population.

Algorithm 1. MODE-LD+SS

INPUT:
N = Population Size
F = Scaling factor
CR = Crossover Rate
$\lambda[1,\ldots,N]$ = Weight vectors
NB = Neighborhood Size
$GMAX$ = Maximum number of generations
OUTPUT:
PF = Pareto front approximation
Begin
$g \leftarrow 0$
Randomly create P_i^g , $i = 1,\ldots,N$
Evaluate P_i^g , $i = 1,\ldots,N$
while $g < GMAX$ **do**
  $\{LND\} = \{\emptyset\}$
  for $i = 1$ to N **do**
    $DetermineLocalDominance(P_i^g, NB)$
    if P_i^g is locally nondominated **then**
      $\{LND\} \leftarrow \{LND\} \cup P_i^g$
    end if
  end for
  for $i = 1$ to N **do**
    Randomly select $\boldsymbol{u_1}$, $\boldsymbol{u_2}$, and $\boldsymbol{u_3}$ from $\{LND\}$
    $v \leftarrow CreateMutantVector(u_1, u_2, u_3)$
    $P_i^{g+1} \leftarrow Crossover(P_i^g, v)$
    Evaluate P_i^{g+1}
  end for
  $Q \leftarrow P^g \cup P^{g+1}$
  Determine $z*$ for Q
  for $i = 1$ to N **do**
    $P_i^{g+1} \leftarrow MinimumTchebycheff(Q, \lambda^i, z*)$
    $Q \leftarrow Q \backslash P_i^{g+1}$
  end for
  $PF \leftarrow Q$
end while
ReturnPF
End

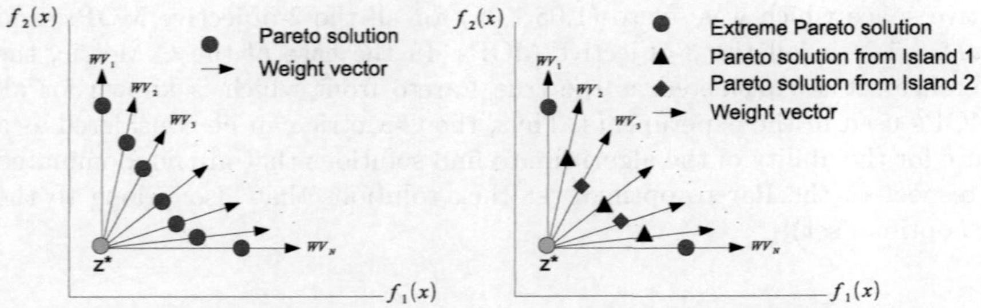

(a) pMODE-LD+SS(A) (b) pMODE-LD+SS(B)

Fig. 1. Weight vectors distribution

The main difference between the two proposed approaches is on the weight vectors distribution used. The pMODE-LD+SS(A) approach can be seen as the serial version of MODE-LD+SS running in p processors and exchanging information among them. For this approach, the same weight vector distribution (see Figure 1(a)) is used in each island. For maintaining diversity of the global population and to evolve each island in an independent manner, different seed values are used in the islands' random numbers generators. In the second case, for the pMODE-LD+SS(B) approach, each island is also instructed to search for the whole Pareto front, but in this case, using a reduced population and different weight vectors sets. It is important to note that all islands contain weight vectors for searching the extreme Pareto solutions. The main idea for the second parallel approach is that the combination of all islands' weight vectors covers the whole Pareto front region. Figure 1(b) illustrates this situation for the case of a bi-objective MOP with two islands participating in the pMOEA.

4 Experimental Setup

In order to validate the two proposed parallel approaches, their results are compared with respect to those generated by their serial counterpart (MODE-LD+SS), and to NSGA-II [3], MOEA/D [4], and MOEA/D-DE [5] which are MOEAs representative of the state-of-the-art in multiobjective evolutionary optimization. Our approaches were validated using nine test problems: five from the ZDT (Zitzler-Deb-Thiele) set with 2 objectives (ZDT1, ZDT2, ZDT3, ZDT4, and ZDT6[2]), and four more from the DTLZ (Deb-Thiele-Laumanns-Zitzler) set with 3 objectives (DTLZ1, DTLZ2, DTLZ3, and DTLZ4). The selected test functions comprise different difficulties such as convex, concave, and disconnected Pareto fronts, as well as problems with multiple fronts. Two performance measures were adopted in order to assess our results: *Hypervolume (Hv)* and *Two Set Coverage (C-Metric)*. A description of these performance measures are omitted here but can be found elsewhere [1]. The Hv measure uses a reference point in the objective space which was set to (1.05,1.05) for all the 2-objective MOPs, and to (5.0,5.0,5.0) in all the 3-objective MOPs. In the case of the C-Metric, the A reference set is considered as the true Pareto front which is known for all the MOPs used in the experiments. Thus, the C-Metric can be considered as a measure for the ability of the algorithm to find solutions that are nondominated with respect to the Pareto optimal set (i.e., solutions that also belong to the Pareto optimal set).

5 Results and Discussion

In this section we present the results obtained by the proposed parallel approaches. As a first step, the serial version of MODE-LD+SS is compared with

[2] ZDT5 is a binary problem and was not included because we adopted real-numbers encoding in our experiments.

respect to NSGA-II, MOEA/D and MOEA/D-DE. Then, the number of islands, epoch and migration rate adopted in the parallel approaches are tuned by means of an empirical study, using ZDT1. Finally, the results obtained with the two parallel approaches are presented and compared to those of the serial version of MODE-LD+SS, and those obtained by NSGA-II, MOEA/D and MOEA/D-DE. These comparisons are based on the average results from 32 independent runs executed by each algorithm and for each MOP.

Comparison of serial versions

Table 1 presents the results of the serial versions for NSGA-II, MOEA/D, MOEA/D-DE and MODE-LD+SS. The population size in all the algorithms was set to N=50 for all the 2-objective MOPs, and 153 for all the 3-objective MOPs the maximum number of generations was set to GMAX = 150 for all problems, except for ZDT4 and DTLZ3, where we used GMAX = 300. The common parameters for NSGA-II, MOEA/D and MOEA/D-DE were: crossover probability $p_c = 1.0$; mutation probability $p_m = 1/\text{NVARS}$ (NVARS correspond to the number of decision variables for each MOP); distribution index for crossover $\eta_c = 15$; and distribution index for mutation $\eta_m = 20$. As for the MOEA/D and MOEA/D-DE the replacing neighborhood size was set as indicated in [4] and [5], respectively. For the MODE-LD+SS algorithm, we used: F = 0.5 for all MOPs; CR = 0.5 for all MOPs except for ZDT4, where CR = 0.3 was used; Neighborhood size NB = 5 for all MOPs except for ZDT4, where we used NB = 1. From the results presented in Table 1, it can be observed that MODE-LD+SS obtains the best results in 6 of 9 MOPs for the Hv measure. It also obtains the best results in 8 of 9 MOPs regarding the C-Metric, which indicates that it converged closer to the true Pareto front.

Parameters for the pMOEA approaches

For any pMOEA approach based on the island model, additional to the parameters required by its serial counterpart, we have to define the number of islands, migration rate, and epoch period. The choice of these parameters has a great influence in the performance of the pMOEA and is problem dependent. For selecting a set of parameters to be used in the present work, ZDT1 was selected to conduct an experimental study for assessing how the parameters affected performance with respect to the serial version. For this study, the following set of parameters was used: epochs = 10, 20, and 50 generations; migration rate MR = 0.1, 0.2, 0.3 and 0.5; and number of islands NI = 4, 6, and 8. All the combinations were tested. The parameters for population size, maximum number of generations, F, CR, and NB were set the same for all the islands, as in the serial version previously described. From the results of this study, and regarding the C-Metric, it was observed that high migration rates with shorter epoch periods produce the best improvements with respect to the serial version. However, this can lead to higher communication costs. From the study, the final set of parameters selected correspond to the following: Number of Islands = 6; epoch = 10 generations, and migration rate = 0.4. This will be used in assessing the two parallel approaches proposed here. Table 2 shows the results of the two proposed parallel approaches.

Table 1. Comparison of Hv and C-Metric measures for the serial versions

Hypervolume measure								
Function	NSGA-II		MOEA-D		MOEA-D-DE		MODE-LD+SS	
	Mean	σ	Mean	σ	Mean	σ	Mean	σ
ZDT1	0.740382	0.003323	0.716729	0.024506	0.583847	0.076507	**0.757395**	0.000397
ZDT2	0.377348	0.070194	0.176615	0.079320	0.082341	0.115367	**0.424895**	0.000331
ZDT3	0.604214	0.003199	0.585094	0.023488	0.277813	0.111381	**0.613846**	0.000307
ZDT4	0.073098	0.122631	**0.730980**	0.016966	0.450990	0.215977	0.349325	0.285549
ZDT6	0.292164	0.020894	0.375312	0.007755	0.239793	0.084688	**0.407638**	0.000009
DTLZ1	124.139600	1.113898	**124.969600**	0.000768	119.402900	7.771898	124.967700	0.000383
DTLZ2	123.972600	0.124088	124.397400	0.001778	124.353700	0.027743	**124.397600**	0.003356
DTLZ3	80.131930	39.091680	124.338100	0.250190	85.976200	54.287600	**124.396900**	0.003004
DTLZ4	123.934300	0.125475	**124.400100**	0.002818	124.387900	0.003452	124.393900	0.002684
C-Metric measure								
Function	NSGA-II		MOEA-D		MOEA-D-DE		MODE-LD+SS	
	Mean	σ	Mean	σ	Mean	σ	Mean	σ
ZDT1	0.994591	0.008785	0.997234	0.007463	1.000000	0.000000	**0.748125**	0.153569
ZDT2	1.000000	0.000000	**0.208557**	0.140626	1.000000	0.000000	0.586492	0.100261
ZDT3	0.931490	0.047844	0.813861	0.121330	1.000000	0.000000	**0.384729**	0.092223
ZDT4	1.000000	0.000000	0.975157	0.088624	1.000000	0.000000	**0.845625**	0.364496
ZDT6	0.975723	0.008476	0.978242	0.001639	0.989831	0.016529	**0.000625**	0.003536
DTLZ1	0.535550	0.134412	0.340389	0.234715	0.807088	0.113710	**0.021434**	0.014189
DTLZ2	0.447368	0.035370	0.211798	0.040208	0.678562	0.053395	**0.171215**	0.009018
DTLZ3	1.000000	0.000000	0.725727	0.179974	0.972366	0.063160	**0.160711**	0.007169
DTLZ4	0.453536	0.058747	0.205555	0.036333	0.554462	0.051949	**0.156578**	0.008628

pMOEA for effectiveness improvement

From Table 2, it can be observed that the approach designed for effectiveness improvement produced better Hv values in 4 of the 9 MOPs (ZDT1, ZDT3, ZDT4, and ZDT6), while improving the C-Metric in 7 of the 9 MOPs (ZDT1, ZDT4, ZDT6, DTLZ1, DTLZ2, DTLZ3, DTLZ4), with respect to the serial version of MODE-LD+SS. One important result to remark from this parallel approach, is its ability to reach the true Pareto front of ZDT4 and ZDT6 in the 32 runs performed, as indicated by the mean and standard deviations for the C-Metric for these two MOPs.

pMOEA for efficiency improvement

For this approach, each island uses a reduced population size of $N = 10$ for the 2-objective MOPs and of $N = 28$ for the 3-objective MOPs. Since we used 6 islands, the global population consists of 60 individuals for the 2-objective MOPs, and of 168 individuals for the 3-objective MOPs. Considering that the global population size grows, the maximum number of generations used in pMODE-LD+SS(B) were reduced accordingly to obtain an equivalent number of objective function evaluations as in the serial version. However, once the islands' populations are gathered and a global environmental selection is performed, the maximum population size reported for this approach is of 50 solutions for the 2-objecive MOPs, and 153 for the 3-objective MOPs. This latter condition is due to the fact that each island searches for the Pareto extreme solutions (there are redundant solutions which are filtered out). The parameters for F, and CR were set the same as in the serial version for all islands. However, due to the reduction in island population size, the parameter NB was set to 1 in all MOPs. In Table 2,

Table 2. Results for MODE-LD+SS(A) and pMODE-LD+SS(B)

Function	pMODE-LD+SS(A)				pMODE-LD+SS(B)				
	Hv		C-Metric		Hv		C-Metric		
	Mean	σ	Mean	σ	Mean	σ	Mean	σ	Speed-up
ZDT1	**0.757675**	0.000115	**0.623750**	0.105517	0.750276	0.002323	0.983219	0.043928	2.9977
ZDT2	0.424363	0.000310	0.766844	0.105424	0.421071	0.001954	0.815243	0.099875	2.6918
ZDT3	**0.614156**	0.000201	0.385278	0.078578	0.608588	0.001898	0.741921	0.122588	2.5434
ZDT4	**0.758770**	0.000006	**0.000000**	0.000000	0.034668	0.119046	1.000000	0.000000	2.3341
ZDT6	**0.407650**	0.000001	**0.000000**	0.000000	0.406966	0.000241	0.002552	0.006860	2.5110
DTLZ1	124.967400	0.000238	**0.011863**	0.004840	**124.970200**	0.000681	0.022575	0.009888	4.9934
DTLZ2	124.391900	0.002976	**0.162893**	0.008310	**124.404200**	0.001968	0.175189	0.008008	4.7269
DTLZ3	124.389000	0.003113	**0.155273**	0.005601	124.015800	1.228651	0.242665	0.247538	4.8357
DTLZ4	124.389500	0.002848	**0.154082**	0.008518	**124.402700**	0.002529	0.149008	0.008031	4.8949

the estimated average parallel Speed-Up measure is reported for all MOPs used. Also from this table, it can be seen that the approach designed for efficiency improvement produced better Hv values in 3 of the MOPs adopted (DTLZ1, DTLZ2, and DTLZ4). By taking a closer look to the results for the Hv metric for ZDT1, ZDT2, ZDT3 and ZDT6, it can be seen that pMODE-LD+SS(B) obtained values very close to those of the serial version (MODE-LD+SS), even when each island was using a small population size.

6 Conclusions and Future Work

We have introduced a new pMOEA, called pMODE-LD+SS. For it, two different parallel schemes were proposed, aiming at improving: (a) effectiveness and (b) efficiency, with respect to its serial version, called MODE-LD+SS. From the results presented in the previous section, the first goal was achieved in 4 and 7 of 9 test problems adopted, when considering the Hv and C-Metric performance measures, respectively. It is worth noting that, in some cases the improvement achieved has been quite significant. Regarding the second goal, and even when each island is using a reduced population size, the second approach is able to obtain better results for the Hv measure in 3 of the 9 test problems adopted. Additionally, in 3 other problems, the values attained are very similar to those obtained by the serial version. From the above, we can conclude that the proposed algorithm has good properties both in terms of effectiveness and efficiency. In the present work, the proposed algorithm was run with parameters derived from empirical tests. However, a thorough statistical analysis is required in order to identify the most appropriate parameters to be adopted, and to relate more closely such parameter values to specific types of test problems. These tasks will be part of our future work. Given the good convergence properties of the proposed algorithm and its ability to improve both effectiveness and efficiency in the test problems adopted, it is also desirable to test them in real-world optimization problems, and that is actually part of our ongoing research. Finally, we also plan to compare our proposed pMOEA with respect to other pMOEAs currently available in the specialized literature.

References

1. Coello Coello, C.A., Lamont, G.B., Van Veldhuizen, D.A.: Evolutionary Algorithms for Solving Multi-Objective Problems, 2nd edn. Springer, New York (2007) ISBN 978-0-387-33254-3
2. Arias Montaño, A., Coello Coello, C.A., Mezura-Montes, E.: MODE-LD+SS: A Novel Differential Evolution Algorithm Incorporating Local Dominance and Scalar Selection Mechanisms for Multi-Objective Optimization. In: 2010 IEEE Congress on Evolutionary Computation (CEC 2010), Barcelona, Spain. IEEE Press, Los Alamitos (2010)
3. Deb, K., Pratap, A., Agarwal, S., Meyarivan, T.: A Fast and Elitist Multiobjective Genetic Algorithm: NSGA–II. IEEE Transactions on Evolutionary Computation 6, 182–197 (2002)
4. Zhang, Q., Li, H.: MOEA/D: A Multiobjective Evolutionary Algorithm Based on Decomposition. IEEE Transactions on Evolutionary Computation 11, 712–731 (2007)
5. Li, H., Zhang, Q.: Multiobjective Optimization Problems With Complicated Pareto Sets, MOEA/D and NSGA-II. IEEE Transactions on Evolutionary Computation 13, 284–302 (2009)
6. Talbi, E.G., Mostaghim, S., Okabe, T., Ishibuchi, H., Rudolph, G., Coello Coello, C.: Parallel approaches for multi-objective optimization. In: Branke, J., Deb, K., Miettinen, K., Slowinski, R. (eds.) Multiobjective Optimization. LNCS, vol. 5252, pp. 349–372. Springer, Heidelberg (2008)
7. Nebro, A., Luna, F., Talbi, E.G., Alba, E.: Parallel Multiobjective Optimization. In: Alba, E. (ed.) Parallel Metaheuristics, pp. 371–394. Wiley-Interscience, New Jersey (2005) ISBN 13-978-0-471-67806-9
8. Kamiura, J., Hiroyasu, T., Miki, M., Watanabe, S.: MOGADES: Multi-objective genetic algorithm with distributed environment scheme. In: Computational Intelligence and Applications (Proceedings of the Second International Workshop on Intelligent Systems Design and Applications: ISDA 2002), pp. 143–148 (2002)
9. Streichert, F., Ulmer, H., Zell, A.: Parallelization of Multi-objective Evolutionary Algorithms Using Clustering Algorithms. In: Coello Coello, C.A., Hernández Aguirre, A., Zitzler, E. (eds.) EMO 2005. LNCS, vol. 3410, pp. 92–107. Springer, Heidelberg (2005)
10. Zaharie, D., Petcu, D.: Adaptive pareto differential evolution and its parallelization. In: Wyrzykowski, R., Dongarra, J., Paprzycki, M., Waśniewski, J. (eds.) PPAM 2004. LNCS, vol. 3019, pp. 261–268. Springer, Heidelberg (2004)

Improved Dynamic Lexicographic Ordering for Multi-Objective Optimisation

Juan Castro-Gutierrez[1], Dario Landa-Silva[1], and José Moreno Pérez[2]

[1] ASAP Research Group, School of Computer Science, University of Nottingham, UK
jpc@cs.nott.ac.uk, dario.landasilva@nottingham.ac.uk
[2] Dpto. de Estadística, I.O. y Computación, Universidad de La Laguna, Spain
jamoreno@ull.es

Abstract. There is a variety of methods for ranking objectives in multi-objective optimization and some are difficult to define because they require information a priori (e.g. establishing weights in a weighted approach or setting the ordering in a lexicographic approach). In many-objective optimization problems, those methods may exhibit poor diversification and intensification performance. We propose the Dynamic Lexicographic Approach (DLA). In this ranking method, the priorities are not fixed, but they change throughout the search process. As a result, the search process is less liable to get stuck in local optima and therefore, DLA offers a wider exploration in the objective space. In this work, DLA is compared to Pareto dominance and lexicographic ordering as ranking methods within a Discrete Particle Swarm Optimization algorithm tackling the Vehicle Routing Problem with Time Windows.

Keywords: Multi-objective Optimization, Swarm Optimization, Combinatorial Optimization, Vehicle Routing Problem.

1 Introduction and Motivation

Multi-objective Optimization (MOO) problems have a number of objectives that are usually in conflict, so improving one objective leads to worsen another. In particular, *many-objective optimization problems* involve the optimization of four or more objectives, presenting a considerable challenge for some solutions methods.

Solution methods for multi-objective optimization differ mainly on the way they rank solutions. Many successful approaches exist to address this issue in problems with few objectives (i.e. less than four). However, these ranking schemes have not exhibited the same performance in *many-objective optimization problems.* Some classical ranking methods like Pareto dominance, use a strict ranking scheme that sometimes fails to discriminate between solutions, as it only accepts improvements in all objectives at the same time. On the other hand, methods like the lexicographic approach impose a more static behavior, as objectives are ranked according to a fixed relative importance.

In previous work [2] we introduced the Dynamic Lexicographic Approach (DLA). This is an alternative dynamic multi-objective ranking approach for

R. Schaefer et al. (Eds.): PPSN XI, Part II, LNCS 6239, pp. 31–40, 2010.

many-objective optimization. DLA offers an intuitive approach to establish a dynamic ranking among objectives. Rather than establishing a fixed priority among the objectives, the decision-maker establishes a preference. This preference is then used with a probability mass function (*pmf*) to generate a vector of priorities that changes dynamically throughout the search process.

We used DLA in [2] as a an additional mechanism to rank the quality of multi-objective solutions in a Particle Swarm Optimization approach applied to tackle the Solomon's instances of the Vehicle Routing Problem with Time Windows (VRPTW) treated as a MOO problem. We use the DLA to select the best leader in the neighborhood of each particle, while still using the traditional Pareto dominance to decide whether to update solutions. This combined approach (*Pareto + DLA*) was compared against the alternative of using Pareto dominance for both tasks (*Pareto + Pareto*): selection of the leader and update of solutions. Results indicated that (*Pareto + DLA*) was better than (*Pareto + Pareto*) in clustered problems *(cxxx)*. But the (*Pareto + DLA*) approach showed poor performance on random *(rxxx)* and random-clustered *(rcxxx)* instances.

This work investigates the role of the probability mass function (*pmf*) and the use of DLA to both select the leader and choose the solution to update. We propose two versions of DLA, one using a greedier function to assign probabilities to each preference and the other using two phases involving two probability functions. For k iterations, one probability mass function is used to encourage intensification. For the remaining iterations, another probability mass function is used to encourage diversification. The two DLA versions are compared to Lexicographic ordering and Pareto dominance on well-known instances of the VRPTW using the hypervolume as performance metric. Our results indicate that the DLA is a valuable ranking alternative method for MOO.

This paper is organized as follows. The basics of multi-objective optimization are given in Section 2. The algorithm for the Dynamic Lexicographic Approach is detailed and exemplified in Section 3 while Section 4 describes the Particle Swarm Optimization method focusing on those elements important to our research. A brief description of the Vehicle Routing Problem with Time Window is given in Section 5. We describe our experiments in Section 6 and discuss results in Section 7. Finally, our contribution and proposed further research are summarized in Section 8.

2 Multi-Objective Optimization (MOO)

In MOO, we aim to solve a problem of the type: *minimize* $\boldsymbol{f}(\boldsymbol{x}) = f_1(\boldsymbol{x})$, $f_2(\boldsymbol{x})$,...,$f_k(\boldsymbol{x})$, subject to: $g_i(\boldsymbol{x}) \leq 0, i = 1, 2, ..., m$ and $h_i(\boldsymbol{x}) = 0,\ i = 1, 2, ..., p$. Where the decision variable vector is $\boldsymbol{x} = [x_1, x_2, ..., x_n]^T$, each objective function f_i is defined in $\Re^n \rightarrow \Re, i = 1, 2, ..., k$ and the constraint functions are g_i and $h_i, i = 1, 2, ..., m, j = 1, 2, ..., p$ which are defined in the same domain as f_i.

Pareto Dominance is based on the concept of dominance. Without loss of generality, we consider a minimization problem. A vector $\boldsymbol{u}$ is said to *dominate* $\boldsymbol{v}$ (denoted by $\boldsymbol{u} \prec \boldsymbol{v}$), iif $\forall i \in (1, ..., k) : u_i \leq v_i \wedge \exists i \in (1, ..., k) : u_i < v_i$.

Moreover, we say that a vector in the feasible region ($\boldsymbol{u} \in \mathcal{F}$) is *Pareto Optimal*, if there is not any other vector in that region ($\boldsymbol{u'} \in \mathcal{F}$), such that $\boldsymbol{u'}$ *dominates* $\boldsymbol{u}$ ($\boldsymbol{u'} \prec \boldsymbol{u}$). We define the *Pareto Front* as the set of vectors in $\Re^n$, such that all elements in the set are *Pareto Optimal*.

Using Pareto dominance in many-objective optimization has an implicit pitfall derived from its manner to discriminate solutions. This method accepts solutions in a linear fashion that inhibits the worsening of objectives, which sometimes provokes the search to get stuck. Recent approaches have extended this ranking method using the concept of *relaxed dominance or* α*-dominance*. This type of dominance, proposed by Kokolo et al. [9], tries to overcome the weakness above explained by allowing the worsening in some objective if there are improvements in others. However, it might be difficult to set the appropriate bounds to get a desirable performance.

Lexicographic ordering is another widely used ranking method in which the decision-maker establishes fixed preferences among objectives. Given a priority order (a, b), the objective whose priority is a is compared in first place and the one with b in second. This way, we can formally describe the lexicographic comparison as: $(a, b) \leq (a', b')$ *iif* $a < a'$ or ($a = a'$ and $b \leq b'$). The same procedure can be easily extended to k priorities. This technique works well when the number of objectives is small. When dealing with a large number of priorities, objectives with low priority will not be likely to be compared and therefore, could be ignored in practice. Moreover, the fixed order, in which the method discriminates vectors, makes the search liable for premature convergence [3].

In addition to Pareto dominance and lexicographic ordering, there exist many other ranking methods to discriminate solutions in multi-objective scenarios. A review on these methods can be found in the work of Ehrgott and Gandibleux [7].

3 Dynamic Lexicographic Approach

The Dynamic Lexicographic Approach (DLA) offers an intuitive approach to establish a dynamic ranking among objectives. The decision-maker establishes preferences among objectives and a function to associate a probability to each preference. These probabilities are used to create different vectors defining a standard lexicographic ordering each time. This approach overcomes the limitation in the number of objectives because DLA does not rule out any preference. Even preferences with a low probability have a chance to appear in the first position of the vector of priorities. Additionally, the continuous change of priorities makes it possible to avoid premature convergence as the exploration is broader.

To clarify how DLA works, we provide an example for $N = 4$ objectives. Lets assume that objective i is assigned preference i, that is, $pref(o_i) = i$ for $i = 0, \ldots, N-1$, where N is the number of objectives. Suppose the decision-maker provides the function $p(i) = 0.8exp(-0.4i)$. Firstly, we evaluate this function for $i = 0, \ldots, N-1$. Since $N = 4$, we calculate the probabilities as $p(0) = 0.8, p(1) = 0.54, p(2) = 0.36$ and $p(3) = 0.24$. Secondly, these probabilities are normalized as $p'(i) = p(i) / \sum_{k=0}^{N-1} p(k)$, obtaining the values $0.41, 0.28, 0.19, 0.12$. In a third step, we split the segment $[0, 1]$ into sub-segments according to the accumulate probability. Therefore, we obtain $[0, 0.41, 0.69, 0.88, 1]$, such that each preference has a sub-interval assigned. The first preference has interval $[0, 0.41)$, the second $[0.41, 0.69)$, the third $[0.68, 0.88)$ and the fourth $[0.88, 1]$. The algorithm goes through the above steps only once. Regarding the generation of priority vectors, roulette-wheel selection is applied using this segment. First, a random number r is generated with uniform distribution in $[0, 1]$. Lets suppose that $r = 0.70$. As r falls in the sub-interval $[0.68, 0.88)$, the third objective will be pushed back in the vector of priority $v \leftarrow 3$. After this operation, we remove the selected sub-interval and the segment is re-scaled, and another random number is generated. After $N - 1$ times repeating this process, we get a priority vector that can be used in a lexicographic approach to discriminate solutions.

The probability mass function (*pmf*) plays a key role in the performance of the DLA. Depending on the shape of this function, different probability values are assigned to each objective producing different lexicographic orderings. Therefore, assuming that the decision-maker establishes decreasing preferences (as in the example above), there are three main types of functions: linear, quadratic and exponential. Linear functions assign probabilities with a constant step among them. Quadratic functions assign a similar probability to those objectives with the highest preferences and zero to those with the lowest. Finally, an exponential function assigns high and distinct probabilities to objectives with high preference and assigns low but non-zero probability to those with low preference.

4 Particle Swarm Optimization (PSO)

PSO is a swarm-based stochastic algorithm proposed by Kennedy and Eberhart [8]. A swarm is formed by a group of particles that moves on the search space. Each particle knows its current position x_i, its best position b_i achieved so far and the best position reached by the swarm so far g. Moreover, each particle i shares its current position with its neighbors n_i. Each particle possesses an independent velocity which is updated taking into account the position of the particle, that of its neighbors and that of the best positioned particles.

The PSO algorithm was originally proposed to tackle continuous optimization problems. For discrete optimization, a number of alternatives have been proposed re-interpreting how particles move. Here, we have adopted the approach proposed by Consoli et al. [4]. This algorithm lacks the use of velocity conceiving the move of particles by using the equation: $x_{i,j} \leftarrow x_i \vee g_j * x_i \vee b_{i,j} * x_i \vee x_j * g_{i,j}$. In this proposal, each particle can accomplish four types of moves. Three involve the action of another particle (*cognitive*) and in the other only the particle is self-involved (*inertial*). Each particle randomly performs just one of these four moves by evolution. This discrete PSO interprets cognitives moves as crossover operations involving the solution of the moving particle and another solution that acts as an attractor. On the other hand, the inertial move is translated into a mutation operation that randomly changes components of the solution. More information and other proposals for PSO can be found in [1].

5 Vehicle Routing Problem with Time Windows

In the Vehicle Routing Problem (VRP) [5], the goal is to determine a least cost route-plan using a fleet of vehicles to serve certain demand from a set of costumers. The VRP can be formally described as follows. Let $G(V, A)$ be a graph where V and A are set of vertices and arcs respectively. The first vertex v_0 is the depot. A fleet of vehicles of identical capacity Q serve certain demand q_i from a number of costumers $V - \{1\}$. Each arc has a cost associated $c_{ij} \geq 0$, such that $i \neq j$. Moreover, a number of constraints are imposed: 1) each costumer $v_i, i > 0$ can be visited only once, 2) all routes must start and end at the depot and 3) the sum of demands in each route cannot exceed the maximum capacity of each delivery vehicle Q.

In the *Vehicle Routing Problem with Time Window* (VRPTW), both the depot and the costumers have a time window (denoted as $[a_i, b_i]$) in which the delivery must occur. While a late arrival is not usually permitted (hard constraint), arriving before the lower time window limit a_i is allowed. It this case, the delivery incurs in a *waiting time* until the customer can be served. The VRPTW also takes into account the actual delivery time. This time is called *service time* and is usually denoted as s_i. A survey with formulations and solution methods for VRP and VRPTW can be found in [10].

6 Experiments

In this section, we describe the settings used in our experiments. In our previous work [2], DLA was compared against Pareto and Lexicographic ranking for selecting the leading particle in the neighborhood of each moving particle. However, this time DLA is presented in two versions for selecting the leaders and also updating the best personal position b_i and the best position achieved by the swarm so far g. The first version of DLA uses a greedier approach to establish the probability for each preference using the *pmf* $p(x) = 0.9 * exp(-x) + 0.05$. The first coefficient (0.9) increases its curvature and the second moves it up by 0.05. We set these coefficients according to some preliminary tests. The second version of the DLA (DLA^2) splits the ranking process into two phases. If the current number of iterations is less or equal than a given k, a probability mass function is used to encourage intensification. Otherwise, a different probability mass function is employed to encourage diversification. To this aim, the first phase only takes into account the first two highest preferences using this *pmf* $p(x) = 0.6 * cos(0.7x + 0.1) + 0.3$ with $x = \{0, 1\}$. For values between $x = 0$ and $x = 1$, this function (equivalent to a quadratic expression in the given range) assigns high and similar probabilities to the two objectives with the highest preference. The coefficients were set, as in the previous *pmf*, using preliminary computational experiments. This intensification phase lasts k iterations. In our experiments k is set to 500 which corresponds to 25% of the total number of iterations that the algorithm runs. The diversification phase runs for the remaining iterations using the function mentioned above $p(x) = 0.9 * exp(-x) + 0.05$, working with the whole set of preferences.

Our efforts focus on comparing different ranking approaches. For this purpose, we implemented a canonical PSO inspired in the Discrete-PSO (DPSO) proposed by Consoli et al. [4]. This approach, as explained in Section 4, allows particles to move in four possible directions depending on which attractor each particle follows. The probability of all attractors were set to 0.25. In our simulations, the swarm was formed by 50 particles evolving for 2000 generations. We used the 100 costumers Solomon's [6] instances of the VRPTW. These instances are divided in three classes: *c1xx* - costumers positioned in clusters, *r1xx* - costumers randomly spread and *rc1xx* - some costumers forming clusters and others randomly positioned. Regarding the DPSO implementation, two operators were used. The crossover operator copies a random route from an attractor to the moving particle. The mutation operator exchanges costumers from one route to another within the route-plan in the solution of the moving particle.

In order to assess the performance of each ranking approach, a number of objectives were considered: *Travel Time* (Z_{tt}) or elapsed time of the route-plan, *Waiting Time* (Z_{wt}) or time the drivers need to wait in case of an early arrival, *Travel Distance* (Z_{td}) or length of the whole route-plan, *Time Window Violation* (Z_{twv}) or sum of lateness of all arrivals, *Number of Time Window Violations* (Z_{ntwv}) or number of customers not served within the appropriate time, *Capacity Violation* (Z_{cv}) or amount of exceeding capacity on vehicles and *Number of Capacity Violations* (Z_{ncv}) or number of vehicles whose capacity is being exceeded. Reducing violations are considered as objectives in this study. In this way, we entitle the decision-maker to decide on the convenience of serving costumers out of their time windows or exceeding the capacity of some vehicles.

In a preliminary study, we tested a number of different combinations of objectives using Pareto dominance. These experiments showed that the best setting for Pareto dominance is to discriminate solutions using (Z_{td}, Z_{twv}). For the Lexicographic approach, a similar study was carried out finding that the best sequence is $(Z_{ntwv}, Z_{td}, Z_{wt}, Z_{tt}, Z_{twv}, Z_{cv}, Z_{ncv})$. Both studies were based on the results of the best extreme values achieved by each combination. To this aim, we took those combinations that made the search process converge faster to feasible regions, analyzing Z_{twv} and Z_{cv}. With respect to the DLA and DLA^2, they both used the same sequence for preferences as the lexicographic for priorities. But, for DLA^2 this sequence was used after 500 generations as explained above.

7 Results

In order to analyse the performance of the proposed DLA variants, we present the results in two modes. Firstly, we depict the approximated non-dominated sets obtained by each strategy using two objectives. Secondly, we show a table containing the normalized average *S-metric* [11] values for each instance family and technique. Three pairs of plots are shown in Figure 1. Each pair shows the approximated non-dominated sets obtained when using Pareto dominance, Lexicographic ordering, DLA and DLA^2 on instances $c101$, $r101$ and $rc101$ using two pairs of objectives. Due to space limitations, for both the plot and the S-metric, we only show the most meaningful combinations of pairs of objectives.

That is, *Time Window Violations* (Z_{twv}) vs *Travel Distance* (Z_{td}) and *Travel Time* (Z_{tt}) vs *Travel Distance* (Z_{td}).

Respect to the Z_{twv} vs Z_{td} comparison, DLA^2 is clearly superior in both intensification and diversification. The approximation sets obtained by DLA^2 on $c101$, $r101$ and $rc101$ have solutions with closer values to the origin, revealing better intensification behavior. Additionally, these solutions, compared to those of other ranking methods, seem to be more spread along both axis, showing better diversification. Moreover, DLA^2 seems to get better results as the difficulty of the instances increases. Analyzing the properties of the Solomon's instances, it is observed that the time windows are much more restrictive on instance sets $r1xx$ and $rc1xx$. Moreover, the geographical location of costumers in these two sets of instances make the problem grow in complexity. This complexity is due to the fact that optimizing the distance does not guarantee to obtain a good set of solutions. This improvement can be seen on the results for instances $r101$ and $rc101$ where the difference in performance between DLA^2 and the other ranking methods is much more noticeable. For this comparison, DLA gets a slightly better performance than the Lexicographic approach. This is because DLA uses a greedy function to assign probabilities to each preference. Therefore, the priority vectors were generated within a short distance with the one used for the Lexicographic ordering. Pareto shows a reasonable performance in instances $r101$ and $rc101$, but it performs poorly on instance $c101$.

In the comparison of Z_{tt} vs Z_{td}, we do not appreciate a good diversification performance in any technique. This is because the Solomon's instances were not created to be used as a multi-objective test case. These instances have symmetrical and identical matrices for distances and times. The variability when comparing these two objectives arises mainly from the waiting and service times. However, this comparison shows the intensification achieved by each ranking technique. DLA^2 is again the best technique obtaining much better results than the others in these three instances. DLA produced approximation sets with more intensification than Lexicographic in these plots as well. Pareto presented a similar behavior to the one exposed for the other objectives comparison, providing in general a poor performance.

Table 1 shows the performance of each ranking approach on each instance set, calculated with the S-metric [11] or hyper-volume. This metric computes

Table 1. Performance of different ranking approaches on Solomon's instances according to the S-metric quality measure calculated over two comparisons: *Time Window Violations* (Z_{twv}) vs *Travel Distance* (Z_{td}) - on the left, and *Travel Time* (Z_{tt}) vs *Travel Distance* (Z_{td}) on the right. Average values and their respective standard deviations are shown for each family of Solomon's instances ($c1xx$, $r1xx$ and $rc1xx$).

	Z_{twv} vs Z_{td}				Z_{tt} vs Z_{td}			
	Pareto	**Lex**	**DLA**	**DLA^2**	**Pareto**	**Lex**	**DLA**	**DLA^2**
c1xx	.20(.17)	.69(.08)	.53(.18)	1(.0)	.50(.14)	.89(.06)	.74(.21)	1(.0)
r1xx	.21(.08)	.28(.12)	.27(.10)	1(.0)	.72(.13)	.72(.09)	.71(.09)	1(.0)
rc1xx	.11(.04)	.25(.10)	.2(.14)	1(.0)	.57(.07)	.65(.88)	.67(.10)	1(.0)

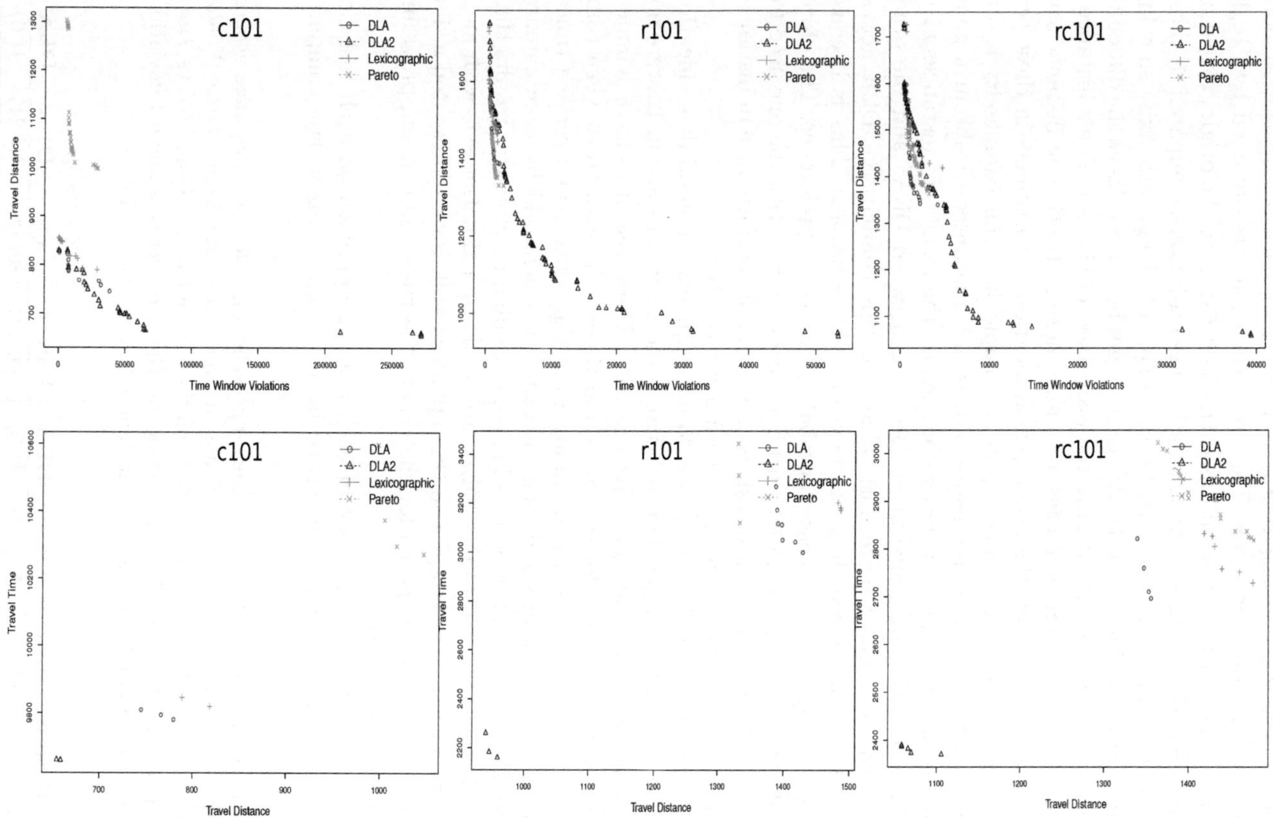

Fig. 1. Approximation sets from different ranking techniques on the Solomon's instances $c101-$ first row, $r101-$ second row and $rc101-$ third row. *Time Window Violations* (Z_{twv}) vs *Travel Distance* (Z_{td}) - on the left, and *Travel Time* (Z_{tt}) vs *Travel Distance* (Z_{td}) on the right.

the hyper-volume of the space limited by the solutions in an approximation set and a reference point. The larger the value of the S-metric, the higher the quality of the approximation set. We calculated the S-metric using the same pairs of comparisons used for the previous graphs. So, these results extend the information conveyed by the plots. All the values were normalized to a value between 0 and 1 according to the highest hyper-volume value obtained in each case. Table 1 is divided in three columns. The first one identifies the instances used. The second and third columns show the S-metric value comparing Z_{twv} against Z_{td} and Z_{tt} against Z_{td} respectively.

For the first comparison, *Time Window Violations* (Z_{twv}) vs *Travel Distance* (Z_{td}), DLA^2 gets the best S-metric value for all instances. On average, it obtains an improvement of 60% over Lexicographic ordering which is the second best ranking method almost in a tie with the DLA. Pareto dominance gives the worst performance with an average of 0.18 across all instances. In the second comparison, *Travel Time* (Z_{tt}) vs *Travel Distance* (Z_{td}), DLA^2 is not the best in only one instance: $c108$. However, it is very close to that value with a distance of only 0.5%. In general, DLA^2 presents an improvement of about 25% over the Lexicographic ordering, which is again the second best ranking technique. Very close to the Lexicographic approach, the DLA gets better results in some instances and worse in others, but on average the former outperforms the later on about 5%. Pareto dominance comes last with an overall performance of 0.60.

The Frieman test was used to analyse the global differences in the S-metric values obtained for each method in both comparisons. There are significant differences at 0.01 significance level. In order to find out where these differences are, we carried out a number of pair-wise comparisons using the four ranking approaches with the Wilcoxon test. According to this test, there is a significant difference between DLA^2 and Lexicographic ordering at 0.01 significance level. The normalized values for the statistic were 4.70 and 4.68, respectively. So, we can safely say that DLA^2 is superior to Lexicographic ordering in this scenario. Similarly, we compared DLA and Lexicographic ordering obtaining the two-way p-values 0.1236 and 0.1285, respectively. This reveals that there is not significant difference between these two ranking approaches at 0.10 significance level. Finally, DLA and Pareto dominance were compared obtaining the two way p-values 0.0007 and 0.0434, respectively. Thus, their results are significantly different at 0.05 significance level.

8 Conclusions

We presented an extended study of a novel approach to rank solutions in multi-objective optimization problems. Our simulations show that Dynamic Lexicographic Approach (DLA) is a valuable technique to discriminate solutions. Specially the variant with double *pmf* (DLA^2) in which the combination of intensification and diversification exhibits much better performance than the other ranking approaches such as Lexicographic ordering and Pareto dominance.

The success of the Dynamic Lexicographic Approach here indicates that it is important to question standard ranking approaches and their suitability for certain problems. This is specially important in those problems where we deal with a large number of objectives. DLA is an alternative easy to implement and intuitive for the decision-maker.

In future research, we will extend these experiments to other MOO problems such as the multi-objective TSP. We are also developing a set of truly multi-objective VRPTW instances based on real-world problems in which the time and distance matrices are not symmetrical nor identical. Moreover, time windows for costumers will be more relaxed than in many other instances in the literature. Another issue that is worth of an in-depth study is the adaptability of the DLA respect to the probability mass function (*pmf*), including setting the parameter k to switch between intensification and diversification.

References

1. Banks, A., Vincent, J., Anyakoha, C.: A review of particle swarm optimization. part II: hybridisation, combinatorial, multicriteria and constrained optimization, and indicative applications. Natural Computing 7(1), 109–124 (2008)
2. Castro-Gutierrez, J., Landa-Silva, D., Moreno Perez, J.: Dynamic lexicographic approach for heuristic multi-objective optimization. In: Proceedings of the Workshop on Intelligent Metaheuristics for Logistic Planning (CAEPIA-TTIA 2009) (Seville (Spain)), pp. 153–163 (2009)
3. Coello, C., Lamont, G., Veldhuizen, D.: Evolutionary algorithms for solving Multi-Objective problems. In: Genetic and Evolutionary Computation, Springer, Heidelberg (2007)
4. Consoli, S., Moreno-Pérez, J.A., Darby-Dowman, K., Mladenović, N.: Discrete particle swarm optimization for the minimum labelling steiner tree problem. Natural Computing 9(1), 29–46 (2010)
5. Dantzig, G., Ramser, J.: The truck dispatching problem. Management Science 6(1), 80–91 (1959)
6. Dorronsoro Díaz, B.: VRP web (2009), `http://neo.lcc.uma.es/radi-aeb/WebVRP/`
7. Ehrgott, M., Gandibleux, X.: Multiple criteria optimization: State of the art annotated bibliographic survey. International Series in Operations Research and Management Science, vol. (52) Springer/Kluwer (2002)
8. Kennedy, J., Eberhart, R.: Particle swarm optimization. In: IEEE International Conference on Neural Networks - Proceedings, vol. 4, pp. 1942–1948 (1995)
9. Kokolo, I., Hajime, K., Shigenobu, K.: Failure of pareto-based MOEAs, does non-dominated really mean near to optimal? In: Proceedings of the 2001 Congress on Evolutionary Computation (CEC 2001), pp. 957–962. IEEE press, Los Alamitos (2001)
10. Laporte, G., Golden, B., Grazia, M., Melissa, A.: The vehicle routing problem: Latest advances and new challenges. Operations Research/Computer Science Interfaces Series, vol. 43. Springer, Heidelberg (2008)
11. Zitzler, E., Brockhoff, D., Thiele, L.: The hypervolume indicator revisited: On the design of pareto-compliant indicators via weighted integration. In: Obayashi, S., Deb, K., Poloni, C., Hiroyasu, T., Murata, T. (eds.) EMO 2007. LNCS, vol. 4403, pp. 862–876. Springer, Heidelberg (2007)

Privacy-Preserving Multi-Objective Evolutionary Algorithms

Daniel Funke and Florian Kerschbaum

SAP Research CEC Karlsruhe, Vincenz-Priessnitz-Str. 1, 76131 Karlsruhe
{daniel.funke;florian.kerschbaum}@sap.com

Abstract. Existing privacy-preserving evolutionary algorithms are limited to specific problems securing only cost function evaluation. This lack of functionality and security prevents their use for many security sensitive business optimization problems, such as our use case in collaborative supply chain management. We present a technique to construct privacy-preserving algorithms that address multi-objective problems and secure the entire algorithm including survivor selection. We improve performance over Yao's protocol for privacy-preserving algorithms and achieve solution quality only slightly inferior to the multi-objective evolutionary algorithm NSGA-II.

1 Introduction

Evolutionary algorithms (EAs) have proven to efficiently find effective solutions for many real-world optimization problems and are therefore widely employed in business practice. Nevertheless in many real-world business problems, such as our use case from collaborative production planning, the data is distributed across a number of parties. A natural objective for each of these parties is to protect their data, in particular if it is sensitive for their well-being, e.g. business secrets such as production costs or capacities.

Privacy-preserving evolutionary algorithms (PPEAs) [1,2] combine the privacy of input data with the effectiveness and efficiency of EA by using ideas from secure computation (SC) [3]. SC is a cryptographic technique that allows a number of parties to jointly compute a function $y = f(\boldsymbol{x})$ on their combined input $\boldsymbol{x}$, such that each party only learns the result y, but nothing else about the input $\boldsymbol{x}$.

The proposals for PPEA in the literature [1,2] suffer from several shortcomings in generality and security. When SC is only applied to cost function evaluation, the selection of individuals from the population reveals significant information about the cost function despite the use of provably secure protocols. Imagine the following situation: Given two individuals differing only in one characteristic, assume one survives and one dies. From this observation one can immediately conclude (even if the cost function was computed privately) that the survivor's characteristic was superior. Due to the many individuals and rounds in EAs this information quickly accumulates.

R. Schaefer et al. (Eds.): PPSN XI, Part II, LNCS 6239, pp. 41–50, 2010.

We improve over this by also securing the generation and selection of individuals in the population. Our protocols are provably secure not only for the cost function evaluation, but for the entire EA. We nevertheless stress that our proposal is not a straight-forward application of the construction by Yao [3]. This would be impractical for complex real-world use case due to communication effort and memory consumption. Instead we use a combination of several techniques which might also be of interest for other complex applications than EA. In order to underpin the practicality of our approach we report the evaluation results of our prototypical implementation.

This paper's contribution is a privacy-preserving multi-objective evolutionary algorithm (PPMOEA) that is more *secure* than previous proposals for PPEA [1,2], that is more *efficient* than general SC [3] and that is almost as *effective* as NSGA-II [4], one of the best known multi-objective evolutionary algorithms (MOEAs) [5].

2 Use Case

Our use case is a finite horizon two-echelon collaborative production planning problem [6], i.e. companies along a two-level supply chain that wish to jointly optimize their production planning for a bounded time period. We limit the two echelons of the supply chain to comprise one party each. On the upstream echelon, a supplier $\mathcal{S}$ provides raw materials to a manufacturer $\mathcal{P}$ on the downstream echelon. Both abide an exclusive relationship, i.e. $\mathcal{P}$ only procures materials from $\mathcal{S}$ and is furthermore $\mathcal{S}$'s single source of revenue. Moreover, we assume periodical shipping with neglected transportation times [6].

A comprehensive description of the use case can be found in the full technical report [7], but due to space constraints we can only highlight some aspects. The use case is characterized by opposing opportunity costs oc_p and capacity consumptions c_p for the supplier's and producer's commodities. This opposition causes tension in the parties' planning objectives, since each party is inclined to act rationally, minimizing only its own cost.

The supplier and producer pursue different objectives – minimizing their own cost – which are opposing due to the differing opportunity costs and capacity consumptions. We thus compose the objective function $F(\mathbf{A})$ as a vector of the supplier's objective function $f_{\mathcal{S}}(\mathbf{A})$ and the producer's one $f_{\mathcal{P}}(\mathbf{A})$.

The producer's fitness function is a sum of the costs for warehousing (excess production), opportunity costs (shortness of production) and penalty costs for exceeding capacity and material supply. The supplier is not constrained by a further upstream party and therefore no costs for excessive material consumption arise and its fitness function is a sum of only three costs. We can show by example that in this scenario local decision making may not lead to the optimum, i.e. if the producer starts with a local production plan then sends the order to the supplier they are likely to waste money. Only a collaborative decision making process as facilitated by our PPMOEA enables finding the global optimum. For details we refer the reader to the technical report [7].

Algorithm 1. Privacy-Preserving MOEA

1: $\mathbb{A} \leftarrow \{\forall i \leq \mu : \mathbf{A}_i \leftarrow \text{random}^{T \times P}\}$
2: **for** $\forall j \leq \eta$ **do**
3: $\quad \mathbb{O} \leftarrow \left\{ \forall i \leq \lambda : \mathbf{O}_i \leftarrow \text{mutate}\left(\mathbf{A}_{\lfloor \frac{i}{\lceil \frac{\lambda}{\mu} \rceil} \rfloor}\right) \right\}$
4: $\quad \mathbb{F} \leftarrow \{\forall \mathbf{O} \in \mathbb{O} : f_i \leftarrow F(\mathbf{O})\}$
5: $\quad \mathbb{A} \leftarrow \begin{cases} \text{select}(\mathbb{O}, \mathbb{F}) & \textit{if non-elitist} \\ \text{select}(\mathbb{O} \cup \mathbb{A}, \mathbb{F}) & \textit{if elitist} \end{cases}$
6: **end for**
7: **return** $\mathbb{A}$

Table 1. Communication complexity ($\mathcal{O}(\cdot)$)

	Yao	PPMOEA
random	$TPm\mu$	0
mutate	$TPm\lambda$	0
F	$T^2Pm^2\lambda$	$TPm^2\lambda$
select	$TPm\lambda \log^2 \lambda$	$\lambda \log^2 \lambda(m + \lambda) + \mu(TPm + \lambda)$
Total per generation	$T^2Pm^2\lambda + TPm\lambda \log^2 \lambda$	$TPm^2\lambda + \lambda^2 \log^2 \lambda + m\lambda \log^2 \lambda$

3 Cryptographic Tools

SC was introduced by Yao [3]. The problem is as follows: Two players, Alice and Bob, both know a joint function $y = f(a, b)$ on their combined input a (Alice) and b (Bob). They are both interested in the result y, but neither is willing to reveal its input. Note that a party may infer information about the other party's input based on y and their own input. This has to be accepted and is excluded from the security definition. Yao [3] constructed a protocol that achieves this for any function f which can be represented as Boolean circuit. Yao's protocol has been implemented [8] using a high-level programming language, but its general construction is too inefficient for complex problems as our PPMOEA. Instead we construct the necessary (optimized) circuits manually and apply further optimizations based on secret sharing and homomorphic encryption [9].

Let s be a secret known to neither Alice nor Bob. Both, Alice and Bob, hold a value (called share) $s^{(A)}$ and $s^{(B)}$, respectively, such that $s = f(s^{(A)}, s^{(B)})$ for some reconstruction function f, e.g. $s = s^{(A)} + s^{(B)} \mod n$ [10]. A secret sharing is perfect if any share $s^{(A)}$ or $s^{(B)}$ does not reveal additional information about the secret s: $Pr(s) = Pr(s|s^{(A)}) = Pr(s|s^{(B)})$.

4 The Privacy-Preserving MOEA

4.1 Algorithm Overview

We follow the conventional structure of an EA as outlined in Algorithm 1. The initial population consists of μ random individuals (operation random). Every individual produces $\lceil \frac{\mu}{\lambda} \rceil$ offspring (operation mutate), perturbing each individual with a small probability p_m by a Gaussian distributed value. We solely use mutation in our use case, but believe that other reproduction schemes including recombination are securely realizable with little additional effort. We then compute the offspring's fitness (operation F) and finally select μ survivors (operation select) using a Pareto-optimal selection algorithm suitable for a multi-objective problem. The evolution iterates for a fixed number η of generations.

Our PPMOEA realizes each operation privacy-preservingly and not only cost function evaluation as previous PPEA. Furthermore all privacy-preserving operations are tied together, such that not even the result of any single operation will be known to any party, but only the result of the entire algorithm, i.e. the optimal production plan, will be revealed to both parties. We emphasize that the parties gain no sensitive information from this production plan, since they need this information to schedule orders and shipments.

In the remainder of the paper we describe the privacy-preserving implementation of the operations, but we start by explaining how we tie the individual operations in Section 4.2, such that no intermediate results are revealed.

4.2 Construction of Secure Protocol

Each operation of the MOEA can be abstractly written as a function

$$y = f(x)$$

e.g. $\mathbb{A} = \text{select}(\mathbb{O}, \mathbb{F})$. Using Yao's algorithm we could construct a privacy-preserving protocol between Alice (Supplier) and Bob (Producer) for any operation, but the results would be revealed and reused as inputs in the next operation. Instead we use secret sharing as follows in order to conceal the results.

We maintain the following invariant as pre- and post-condition of each step: Each variable x or y, i.e. input and output, are distributed as secret shares across Alice and Bob. For our basic data type of integer Alice has $x^{(A)}$ and Bob has $x^{(B)}$, such that $x = x^{(A)} + x^{(B)}$ (we omit the modulus for clarity) [10]. We can now write

$$y^{(A)} + y^{(B)} = f\left(x^{(A)} + x^{(B)}\right)$$

We intend to realize the function f using Yao's protocol which can only implement deterministic functions. However, in order for the security of the secret sharing to hold, the shares need to be chosen randomly. We therefore transform above equation to

$$y^{(B)} = f\left(x^{(A)} + x^{(B)}\right) - y^{(A)}$$

and define this as a new function f'

$$y^{(B)} = f'\left(x^{(A)}, y^{(A)}, x^{(B)}\right)$$

We now implement this function f' using Yao's protocol, such that only Bob will learn the result. In order to obtain the final result Alice and Bob simply exchange shares.

Our secret sharing scheme is linear, and linear operations, such as additions or multiplications with constants, can be performed locally on the shares. Therefore this decomposition into smaller protocols may reduce the communication complexity if used cleverly. Table 1 gives details of the reduced communication complexity in our PPMOEA by application of the decomposition technique. We refer the reader to the full technical report for the security proof [7].

4.3 Operations

Individuals are maintained as secret shares throughout the entire algorithm and each party only possesses a share of every individual, denoted by $\mathbf{A}^{(A)}$ and $\mathbf{A}^{(B)}$. Supplier and producer independently generate the initial population, i.e. μ local production plans. They use these values to initialize their shares for these parts of the individual and simply initialize the shares for the other party's plan with 0. No communication is required for this step and it results in a perfect secret sharing of the random, initial population.

We use mutation for reproduction, due to the high probability that in our use case the combination of two individuals results in prohibitively high penalty costs for excessive material consumption. In our PPMOEA, *each* individual produces $\lceil \frac{\lambda}{\mu} \rceil$ offspring resulting in an ancestor population of λ individuals. We apply Gaussian perturbation to every production quantity with probability p_m. We can use the linearity of the secret sharing scheme in order to realize this operation (almost) without any communication. For each production quantity a party is chosen that applies the perturbation to its share only. The other party does not modify its share. In order to prevent certain active attacks, such as enforcing its locally optimal production plan, we randomly choose the party to perform the perturbation using a pseudo-random function based on a jointly chosen seed.

As noted in the other proposals for PPEA fitness computation is the most sensitive operation. Since it is a simple arithmetic computation, we can implement it straightforwardly using Yao's protocol. Nevertheless in order to speed up the protocol, we employ a number of optimizations over simply implementing $F(\mathbb{A})$ as a circuit.

First note that $F(\mathbb{A})$ can be easily decomposed into λ computations of the fitness $F(\mathbf{A})$ of each individual. The same circuit can be used for each invocation of $F(\mathbf{A})$ and, since all invocations are using independent inputs and outputs, we can run them in parallel. Parallel execution allows computation and communication of different invocations to overlap, maximizing the CPU utilization of each party. Furthermore we use the linearity of the secret sharing in order to reduce the size of the circuit that has to be computed using Yao's protocol. The size of the circuit dominates the communication cost. We use precomputation of sums exploiting the linearity of the secret sharing scheme and term rewriting in order to reduce the size of the circuit for fitness evaluation. In precomputation two secret shares can be added locally resulting in a secret share of the sum without any communication between the parties. If the innermost term of a formula is a sum, it can be precomputed before executing Yao's protocol. Let m denote the bit size of one production quantity. The circuit consists of $12m^2+64m-10$ gates.

For $m = 32$ bits this amounts to 14 326 gates. By term rewriting we reduced the number of expensive ($\mathcal{O}(m^2)$) multiplications from 4 to 2.

The precomputation of the sums does not apply to the capacity penalty cost, since its innermost term is a product of two sensitive values that should be kept private. The resulting circuit's size grows with the number of products P and has $6Pm^2 + 7Pm - 5P + 12m^2 + 32m + 6$ gates. For $m = 32$ bits and $P = 6$ products the circuit has 51 496 gates. We show the impact of the lack of precomputation in our evaluation.

For the fitness computation of the entire population $\mathcal{O}(\lambda T)$ parallel invocations of Yao's protocol with circuits containing $\mathcal{O}(Pm^2)$ gates and $\mathcal{O}(\lambda TP)$ parallel invocations with $\mathcal{O}(m^2)$ gates are required with a total communication complexity of $\mathcal{O}(\lambda TPm^2)$. Due to space limitations we have to refer the reader to the full technical report for details again [7].

Survivor Selection. The privacy-preserving selection of the μ "best" individuals from the population is the final step for every generation. The selected individuals serve as parent population for the subsequent generation. Multi-objective optimization problems (MOPs) require a more complex notion of "better" than single objective ones. Classical methods scalarize the MOP by some (weighted) objective combination scheme, thus requiring knowledge about the relative magnitudes of the parties' fitness functions [11]. This is not desirable in a privacy-preserving setting, since it leaks information, and we therefore embrace Pareto's notion of "better" [12, p.10 et. seq.].

Pareto-sorting is a rather complex task that can even dominate the cost without protection of privacy. The most efficient, non-privacy-preserving algorithms known are $\mathcal{O}(\lambda^2)$ [4]. We therefore choose to implement a traditional sorting algorithm for strictly ordered sets that is only likely to produce a Pareto-sorting. Subsequent to the sorting the non-dominated individuals are then likely occupying the first θ ranks. The first μ individuals are selected for survival; if $\theta > \mu$ optimal individuals will be lost, if $\theta < \mu$ non-optimal ones will survive in order to keep a steady-state population.

Sorting. For a privacy-preserving realization we require a sorting algorithm that is *oblivious* to the outcome of the performed comparisons, i.e. the operations performed after comparing x_i to x_j are exactly the same for the case $x_i < x_j$ and $x_i > x_j$. Sorting networks are an ideal candidate with this property. A sorting network is a sequence of compare-and-exchange (CX) operations to sort any given input $x_1 \dots x_\lambda$. A sorting network is characterized by two properties: *size* refers to the total number of CX operations; *depth* denotes the maximum number of comparators a x_i has to pass through. The most efficient practical sorting networks follow the odd-even merger construction of Batcher [13], resulting in networks of size $\mathcal{O}(\lambda \log^2 \lambda)$ and depth $\mathcal{O}(\log^2 \lambda)$ [13].

A CX operation compares two fitness values x and y (using the Pareto criterion) and if necessary exchanges their position, such that the non-dominated one occupies the lower rank. We decompose the sorting network by secret sharing and implement each CX operation as an invocation of Yao's protocol using the same circuit.

The purpose of privately implementing sorting is to privately implement selection, i.e. the parties should not learn which individuals survive, since this reveals significant information about the private cost functions. Consequently not only the fitness values need to be sorted privately, but also the corresponding individuals. Unfortunately the size (number of bits) of an individual is quite large ($TPm \gg \lambda$), significantly increasing the communication complexity of a CX operation. We therefore realize another optimization by a level of indirection using index variables. An index variable i is a λ-bit string with only one bit $i[\kappa]$ set. At the start of the sorting this bit κ is the index of individual x in the original, unsorted population, hence index variable. After the sorting the lower ranked index variables contain the indices of the "better" individuals.

The index variables are secret shares and we only sort and exchange those shares. This reduces the communication complexity for sorting from $\mathcal{O}(TPm)$ to $\mathcal{O}(\lambda)$, but adds another subsequent selection operation using the index variables and individuals which we describe next. Each CX circuit then has only $66m + 10\lambda + 10$ gates.

The total communication complexity of the sorting network following Batcher's construction is $\mathcal{O}(\lambda \log^2(\lambda)(m + \lambda))$.

Selection. Subsequent to the execution of the sorting network, the indices of the μ best individuals are likely to reside in ranks $1 \ldots \mu$. A separate protocol is required for using an index variable to select the corresponding individual. Implementing this straightforwardly as a circuit in Yao's protocol completely annihilates the advantage from using index variables in the first place. We have developed a novel protocol based on homomorphic encryption with a significantly improved communication complexity. The total communication complexity for this selection is $\mathcal{O}(\mu(\lambda + TPm))$. Due to paper length constraints we have to refer the reader to the full technical report [7].

5 Evaluation

We experimentally evaluated the solution quality and performance of our PPMOEA in comparison to NSGA-II [4].

Solution quality. For evaluation of solution quality we follow the proposals from literature and adopt four of the metrics proposed by Zitzler et al. [5]. Let $\mathbb{F}$ denote the set of the final population's fitness vectors. The average distance of each individual in $\mathbb{F}$ to the true Pareto front of the problem is measured by metric $\mathcal{M}_1^*(\mathbb{F})$. Metric $\mathcal{M}_3^*(\mathbb{F})$ is an indicator for the spread of the identified Pareto-optimal set by computing the maximal distance between any two solutions in $\mathbb{F}$. Another metric for the spread of a population and its distribution along the identified Pareto-front is $\mathcal{S}(\mathbb{F})$. To compare the relative quality of two fitness vector sets $\mathbb{F}'$ and $\mathbb{F}''$, Zitzler et al. [5] propose metric $\mathcal{C}(\mathbb{F}', \mathbb{F}'')$ as the number of individuals in $\mathbb{F}''$ which are dominated by or equal to an individual in $\mathbb{F}'$.

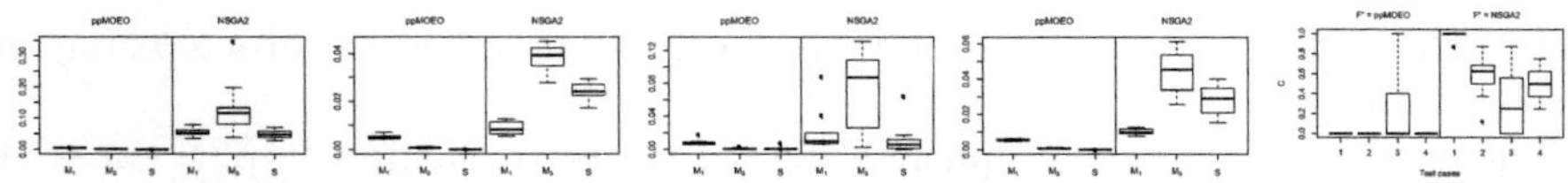

(a) $\eta = 100$, static factors (b) $\eta = 500$, static factors (c) $\eta = 500$, adaptive factors (d) $\eta = 1000$, static factors (e) $\mathcal{C}(\mathbb{F}', \mathbb{F}'')$

Fig. 1. Solution quality comparison of NSGA-II and PPMOEA. Figures (a) through (d) show metrics $\mathcal{M}_1^*(\mathbb{F})$, $\mathcal{M}_3^*(\mathbb{F})$ and $\mathcal{S}(\mathbb{F})$ for the respective test cases. Figure (e) shows the metric $\mathcal{C}(\mathbb{F}', \mathbb{F}'')$ for all four test cases. In all depicted test cases $(\mu, \lambda) = (5; 35)$.

Table 2. Algorithm runtime broken down into preparation, computation, communication and synchronization

	Prep	Comp	Comm	Sync	Other	Total	
Initialization	-	0.02	-	0.02	15.01	**15.08**	**0.49%**
Mutation	-	0.03	-	-	0.01	**0.04**	**0.00%**
Fitness	0.08	308.94	166.72	1 468.84	809.93	**2 754.51**	**89.74%**
Sorting	0.01	100.26	59.48	0.91	98.46	**259.11**	**8.44%**
Selection	-	16.42	18.14	-	6.05	**40.61**	**1.32%**
Total	**0.09**	**425.67**	**244.34**	**1 469.76**	**929.47**	3 069.35	
	0.00%	**13.87%**	**7.96%**	**47.89%**	**30.28%**	*values in [s]*	

In our experiments we varied population sizes (μ and λ) and number of generations (η). Furthermore we ran different variants of the algorithm by using elitist ($\mu + \lambda$) and non-elitist (μ, λ) selection schemes as well as static and adaptive mutation step sizes and penalty factors (σ^2, cp, up). All reported results are the average of 10 runs of each algorithm.[1]

From the entire parameter space we selected four configurations that broadly cover the range of solution quality of our PPMOEA. For details we have to refer the reader to the full technical report [7]. In these four configurations we compared our PPMOEA to NSGA-II, refer to Figure 1. In all four test cases our PPMOEA has better convergence but poorer spread and distribution than NSGA-II. In metric $\mathcal{M}_1^*(\mathbb{F})$ our PPMOEA outperforms NSGA-II by a factor of 3.75, whereas NSGA-II outperforms PPMOEA by a factor of 124.87 and 421.83 in metrics $\mathcal{M}_3^*(\mathbb{F})$ and $\mathcal{S}(\mathbb{F})$. NSGA-II considers the density of the population surrounding an individual in its selection algorithm [4].

Runtime. Our runtime measurements are based on one specific test case and are performed on two servers connected by Gigabit Ethernet, each equipped with four dual-core 2.6GHz CPUs and 16GB of main memory.

Table 2 summarizes the runtime of PPMOEA for one generation with population size $(\mu; \lambda) = (5; 35)$. It requires ≈ 52 minutes. In comparison NSGA-II computes *all* 500 generations in 10.62s; a factor of $\approx 150\,000$ faster. As anticipated privacy protection incurs a very high performance penalty which justifies our optimizations even if they may sacrifice solution quality.

Our decomposition technique enables simple parallelization of several protocol invocations. This is indispensable, since the aggregated CPU time used is ≈ 220 minutes (compared to ≈ 52 minutes runtime).

[1] NSGA-II always uses elitist selection.

Recall that we precompute sums for the partial cost function $\psi_{t,p}(\mathbf{A})$ while not for $C_t(\mathbf{A})$. On average the ψ-circuit requires 2.38s for computation and 1.97s for communication. The C_t-circuit requires 4.48s and 3.74s, respectively. The two circuits differ (almost) only by the P additions in $C_t(\mathbf{A})$ with the ψ-circuit being even slightly larger. Precomputation reduces both measures by $\approx 47\%$.

Almost 90% of the runtime ($\approx$ 45 minutes) is spent on fitness computation which has already been secured in the other proposals for PPEA. Our novel privacy-preserving complementary operations of an EA, particularly secure survivor selection, only contribute the remaining 10%.

6 Related Work

The first PPEAs have been proposed in [2,1]. The former presents a PPEA for combinatorial problems such as traveling salesman problem (TSP). The latter proposes a PPEA for rule discovery in distributed datasets.

Both proposals reveal the result of the cost function (Han and Ng [1] reveal the absolute value while Sakuma and Kobayashi [2] reveal their relative ordering). This allows espionage by inferences from local input and output. Our PPMOEA only reveals the final result.

MOEAs have been studied extensively [12]. State of the art algorithms include NSGA-II [4], SPEA2 [14] and PAES [15]. All use a Pareto-dominance based selection scheme (similar to our PPMOEA), enhanced by some diversity operation (lacking in PPMOEA).

Several implementations of SC exist. Malkhi et al. [8] where the first with a compiler for Yao's protocol. The operations of PPMOEA have been designed based on results from literature [16,17,18].

7 Conclusions

This paper presents a privacy-preserving multi-objective evolutionary algorithm capable of solving distributed multi-objective optimization problems between two parties. We elaborate on a real-world use case from collaborative supply chain management which involves sensitive information, such as mission-critical business secrets.

Our PPMOEA reveals only the optimal solution after the evolution, thus preventing inferences from intermediate results. We introduce several optimizations, including a general decomposition technique for Yao's garbled circuits. As a result the communication complexity is significantly reduced and computation can be parallelized.

Our experimental results show that performance remains the critical parameter. Since cost function evaluation accounts for the majority of the runtime, less complex fitness functions that lead to similar solution quality should be evaluated in future work.

Acknowledgements. The developments presented in this paper were partly funded by the European Commission through the ICT program under Framework 7 grant FP7-213531 to the SecureSCM project.

References

1. Han, S., Ng, W.: Privacy-preserving genetic algorithms for rule discovery. In: Song, I.-Y., Eder, J., Nguyen, T.M. (eds.) DaWaK 2007. LNCS, vol. 4654, pp. 407–417. Springer, Heidelberg (2007)
2. Sakuma, J., Kobayashi, S.: A genetic algorithm for privacy preserving combinatorial optimization. In: Proc. GECCO 2007, pp. 1372–1379. ACM, New York (2007)
3. Yao, A.C.: Protocols for secure computations. In: Proc. IEEE FOCS 1982, Washington, DC, USA, pp. 160–164. IEEE Computer Society, Los Alamitos (1982)
4. Deb, K., Agrawal, S., Pratap, A., Meyarivan, T.: A fast elitist non-dominated sorting genetic algorithm for multi-objective optimization: NSGA-II. In: Deb, K., Rudolph, G., Lutton, E., Merelo, J.J., Schoenauer, M., Schwefel, H.-P., Yao, X. (eds.) PPSN 2000. LNCS, vol. 1917, Springer, Heidelberg (2000)
5. Zitzler, E., Deb, K., Thiele, L.: Comparison of multiobjective evolutionary algorithms: Empirical results. Evolutionary Computation 8(2), 173–195 (2000)
6. Diponegoro, A., Sarker, B.: Finite horizon planning for a production system with permitted shortage and fixed-interval deliveries. Computers and Operations Research 33(8), 2387–2404 (2006)
7. Funke, D., Kerschbaum, F.: Privacy-preserving multi-objective evolutionary algorithms. Cryptology ePrint Archive, Report 2010/326 (2010)
8. Malkhi, D., Nisan, N., Pinkas, B., Sella, Y.: Fairplay - a secure two-party computation system. In: Proc. USENIX Security 2004, pp. 287–302. USENIX Association, Berkeley (2004)
9. Paillier, P.: Public-key cryptosystems based on composite degree residuosity classes. In: Stern, J. (ed.) EUROCRYPT 1999. LNCS, vol. 1592, pp. 223–238. Springer, Heidelberg (1999)
10. Karnin, E., Greene, J., Hellman, M.: On secret sharing systems. IEEE Transactions on Information Theory 29(1), 35–41 (1983)
11. Srinivas, N., Deb, K.: Muiltiobjective optimization using nondominated sorting in genetic algorithms. Evolutionary Computation 2(3), 221–248 (1994)
12. Coello Coello, C., Lamont, G., van Veldhuizen, D.: Evolutionary Algorithms for Solving Multi-Objective Problems, 2nd edn. Springer, Heidelberg (2007)
13. Batcher, K.E.: Sorting networks and their applications. In: Proceedings of the Spring Joint Computer Conference, April 30-May 2, pp. 307–314. ACM, New York (1968)
14. Zitzler, E., Laumanns, M., Thiele, L.: SPEA2: Improving the strength pareto evolutionary algorithm. Technical Report TIK-Report 103, ETH Zurich, Zurich, Switzerland (2001)
15. Knowles, J., Corne, D.: Approximating the nondominated front using the pareto archived evolution strategy. Evolutionary Computation 8(2), 149–172 (2000)
16. Kerschbaum, F.: Practical privacy-preserving benchmarking. In: Proc. IFIP SEC 2008, Boston, MA, USA, pp. 17–31. Springer, Heidelberg (2008)
17. Kerschbaum, F., Dahlmeier, D., Schröpfer, A., Biswas, D.: On the practical importance of communication complexity for secure multi-party computation protocols. In: Proc. ACM SAC 2009, pp. 2008–2015. ACM, New York (2009)
18. Kerschbaum, F., Oertel, N., Weiss Ferreira Chaves, L.: Privacy-preserving computation of benchmarks on item-level data using RFID. In: Proc. ACM WiSec 2010, pp. 105–110. ACM, New York (2010)

Optimizing Delivery Time in Multi-Objective Vehicle Routing Problems with Time Windows

Abel Garcia-Najera and John A. Bullinaria

School of Computer Science, University of Birmingham
Edgbaston, Birmingham B15 2TT, United Kingdom
{A.G.Najera,J.A.Bullinaria}@cs.bham.ac.uk

Abstract. The Vehicle Routing Problem with Time Windows involves finding the lowest-cost set of routes to deliver goods to customers, which have service time windows, using a homogeneous fleet of vehicles with limited capacity. In this paper, we propose and analyze the performance of an improved multi-objective evolutionary algorithm, that simultaneously minimizes the number of routes, the total travel distance, and the delivery time. Empirical results indicate that the simultaneous minimization of all three objectives leads the algorithm to find similar or better results than any combination of only two objectives. These results, although not the best in all respects, are better in some aspects than all previously published approaches, and fully multi-objective comparisons show clear improvement over the popular NSGA-II algorithm.

1 Introduction

The Vehicle Routing Problem (VRP) is one of the most important and widely studied combinatorial optimization problems, with many real-world applications in distribution and transportation logistics. It has several variants that take into account different constraints. The variant *with Time Windows* (VRPTW) is particularly relevant to practical applications, and considers vehicles with limited capacity and specific delivery time windows. Its objective is to obtain the lowest-cost set of routes to deliver demand to customers. Since the problem was originally formulated as a generalization of the Traveling Salesman Problem, cost has primarily been associated with the number of routes and travel distance, but there are several other types of cost [1].

In particular, companies offering transportation services are often more interested in reducing the overall delivery time (or driver salary cost), than the overall distance traveled (or fuel cost), and there are likely to be trade-offs between them. For the standard VRP, if one assumes a constant vehicle velocity, then counting distances and times are equivalent, but that is not true for the VRPTW because of the time wasted due to arriving before delivery windows. The optimization process needs to produce a set of non-dominated solutions that represent the trade-offs between the objectives, rather than a single solution.

Exact methods can be used to find optimal solutions for small instances of the VRPTW, but the computation time required increases considerably for larger

R. Schaefer et al. (Eds.): PPSN XI, Part II, LNCS 6239, pp. 51–60, 2010.

instances [2]. We are therefore interested in using heuristics to solve this problem, in particular, an Evolutionary Algorithm (EA) that automatically generates a whole population of solutions that cover the full range of trade-offs.

There are many past studies that have solved the VRPTW as a single-objective problem using heuristics. Bräysy and Gendreau [3,4] provide an excellent survey of them, and Bräysy et al. [5] focus on evolutionary approaches.

Other studies have considered the bi-objective optimization of the VRPTW, using an EA to minimize the number of vehicles and the travel distance. Tan et al. [7] used the dominance rank scheme to assign fitness to individuals, and designed a problem-specific crossover operator and multi-mode mutation operator. They also considered three local search heuristics. Ombuki et al. [8] used a Pareto ranking method to assign fitness and proposed further crossover and mutation operators. Finally, our own earlier study [9] incorporated a similarity measure in a Bi-objective Evolutionary Algorithm (BiEA) to select parents for the recombination process in a way that preserved a higher population diversity [10], and that enabled a better set of solutions to be obtained.

The work presented here is an improvement of the BiEA proposed in the last-mentioned study, minimizing not only the number of routes and travel distance, but also the delivery time. We analyze the results and compare them with those from previous algorithms, and introduce improved comparisons with the popular NSGA-II algorithm [6] using fully multi-objective performance metrics.

The remainder of this paper is organized as follows: The next section describes formally the VRPTW, and Section 3 reviews the two multi-objective performance metrics that are used to compare algorithms. Our proposed EA for solving the VRPTW as a multi-objective problem is described in Section 4. Then Section 5 presents the results from our algorithm, as well as comparisons with others already published. Finally, we give our conclusions in Section 6.

2 The VRP with Time Windows

Formally, the VRPTW is defined as a set $\mathcal{V} = \{0, \ldots, N\}$ of vertices. Vertices in subset $\mathcal{V}^* = \mathcal{V} \setminus \{0\} = \{1, \ldots, N\}$ are called *customers*. Each customer $i \in \mathcal{V}^*$ is geographically located at coordinates (x_i, y_i), has a demand of goods $g_i > 0$, a time window $[b_i, e_i]$ during which it has to be supplied, and a service time s_i is required to unload its goods. The special vertex 0, called the *depot*, is positioned at (x_0, y_0), has time window $[0, e_0 > \max\{e_i : i \in \mathcal{V}^*\}]$, and demand $g_0 = 0$. It is from the depot that the customers are serviced by a homogeneous fleet of vehicles with capacity $Q \geq \max\{g_i : i \in \mathcal{V}^*\}$. The problem is to design a lowest-cost set of K routes $\mathcal{R} = \{r_1, \ldots, r_K\}$, such that each route begins and ends at the depot, and each customer is serviced by exactly one vehicle.

The travel between vertices i and j has various associated costs, such as the distance d_{ij} and travel time t_{ij}. For the standard benchmark problems to be considered later, one assumes unit velocity and direct travel, so the times and distances are both simply taken to be the Euclidean distances. For real-world problems, however, the distances d_{ij} are unlikely to be Euclidean and the travel times t_{ij} are unlikely to be simply related to the distances. The following will take care to accommodate those possibilities.

Suppose $r_k = \langle u(1,k), \ldots, u(N_k,k) \rangle$ specifies the sequence of N_k customers supplied in the k-th route, where $u(i,k)$ is the i-th customer to be visited in route k, and $u(0,k) = u(N_k+1,k) = 0$ is the depot. Then $\mathcal{V}_k^* = \{u(1,k), \ldots, u(N_k,k)\}$ is the set of customers in route k, and the total travel distance for that route is

$$D_k = \sum_{i=0}^{N_k} d_{u(i,k)u(i+1,k)} \ . \tag{1}$$

In addition to the distances, the times are also needed. Let vehicle k leave the depot at time 0, $a(u(i,k))$ denote its arrival time at the i-th customer, and $l(u(i,k))$ be the time it leaves. The arrival time at the i-th customer is then

$$a(u(i,k)) = l(u(i-1,k)) + t_{u(i-1,k)u(i,k)} \ . \tag{2}$$

Arriving after the end of the customer's time window is simply not allowed. If the vehicle arrives early at the i-th customer's location, it has to wait until the beginning of the customer's time window before it can start unloading, so the departure time $l(u(i,k))$ will be the maximum of the arrival time $a(u(i,k))$ and window opening time $b_{u(i,k)}$, plus the unloading time $s_{u(i,k)}$. Consequently, the waiting time $w(u(i,k))$ involved in serving the i-th customer will be

$$w(u(i,k)) = \max\left(0, b_{u(i,k)} - a(u(i,k))\right) \ . \tag{3}$$

Thus the departure time from the i-th customer in route k can be written as

$$l(u(i,k)) = a(u(i,k)) + w(u(i,k)) + s_{u(i,k)} \ , \tag{4}$$

and the total time required to complete route r_k is the arrival time at the depot

$$T_k = \sum_{i=0}^{N_k} \left(t_{u(i,k)u(i+1,k)} + w(u(i+1,k)) + s_{u(i+1,k)}\right) \ . \tag{5}$$

There are three VRPTW objective functions f_i that we shall concentrate on minimizing in this paper, namely the number of routes or vehicles (f_1), the total travel distance (f_2), and the total delivery time (f_3), computed as follows:

$$f_1(\mathcal{R}) = |\mathcal{R}| = K \ , \quad f_2(\mathcal{R}) = \sum_{k=1}^{K} D_k \ , \quad f_3(\mathcal{R}) = \sum_{k=1}^{K} T_k \ , \tag{6}$$

subject to the demands in each route r_k not exceeding the vehicle capacity, i.e.

$$G_k = \sum_{i \in \mathcal{V}_k^*} g_i \leq Q \ , \tag{7}$$

and no arrival times after the customer's window ends, i.e.

$$a(u(i,k)) \leq e_{u(i,k)} \qquad \forall\, i = 1, \ldots, N_k \ , \ \forall\, k = 1, \ldots, K \ . \tag{8}$$

The simultaneous minimization of all three objectives is not usually possible, so a set of *non-dominated* solutions needs to be obtained, each better than the others on at least one objective. In contrast to single-objective problems, where one can simply compare the best solutions from the various approaches studied, multi-objective problems have to compare whole sets of solutions. For this reason, the definition and use of multi-objective performance metrics is crucial.

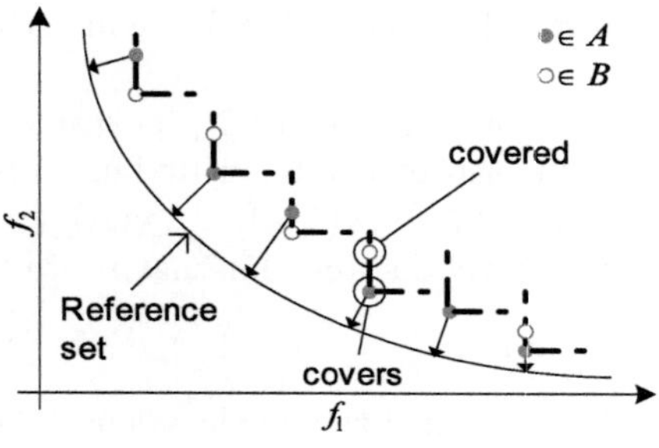

Fig. 1. Representation of $\mathsf{M_C}$ and $\mathsf{M_D}$

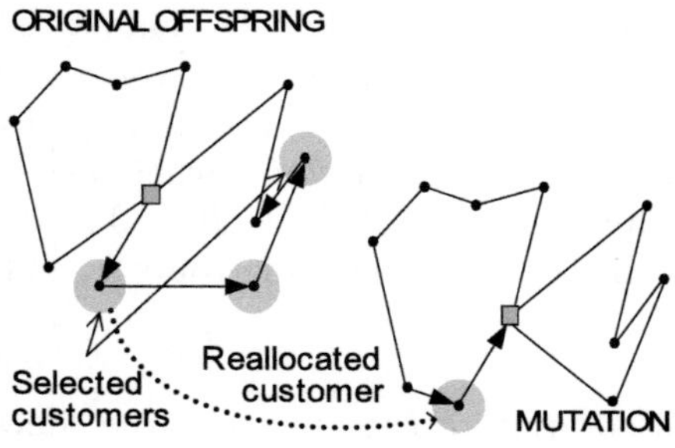

Fig. 2. Reallocation mutation operator

3 Multi-Objective Performance Metrics

Two existing metrics are particularly applicable to the problem at hand.

The coverage metric $\mathsf{M_C}$ [11] compares the number of solutions in set $\mathcal{B}$ that are covered by (i.e. dominated by or equal to) solutions in set $\mathcal{A}$ to the cardinality of $\mathcal{B}$. Thus, $\mathsf{M_C}(\mathcal{A},\mathcal{B})$ maps the ordered pair $(\mathcal{A},\mathcal{B})$ to the interval [0,1], as the fraction of solutions in set $\mathcal{B}$ that are covered by solutions in set $\mathcal{A}$:

$$\mathsf{M_C}(\mathcal{A},\mathcal{B}) = \tfrac{1}{|\mathcal{B}|}\left|\{\mathbf{b} \in \mathcal{B} : \exists\, \mathbf{a} \in \mathcal{A}, \mathbf{a} \preceq \mathbf{b}\}\right| , \tag{9}$$

where the relation $\mathbf{a} \preceq \mathbf{b}$ means that $\mathbf{a}$ is at least as good as $\mathbf{b}$ for all the objectives, and $\mathbf{a}$ is strictly better than $\mathbf{b}$ for at least one objective.

The convergence metric $\mathsf{M_D}$ [12] measures the distance between the approximation set $\mathcal{A}$ and a reference set $\mathcal{R}$. The convergence $\mathsf{M_D}(\mathcal{A},\mathcal{R})$ is defined as

$$\mathsf{M_D}(\mathcal{A},\mathcal{R}) = \tfrac{1}{|\mathcal{A}|}\textstyle\sum_{\mathbf{i}\in\mathcal{A}} d_{\mathbf{i}} , \tag{10}$$

using the smallest normalized Euclidean distance from each point $\mathbf{i} \in \mathcal{A}$ to $\mathcal{R}$

$$d_{\mathbf{i}} = \min_{\mathbf{j}\,\in\,\mathcal{R}} \sqrt{\sum_{k=1}^{F} \left(\frac{f_k(\mathbf{i}) - f_k(\mathbf{j})}{f_k^{\max} - f_k^{\min}} \right)^2} , \tag{11}$$

where $f_k^{\max}$ and $f_k^{\min}$ are the maximum and minimum function values of the k-th objective function in $\mathcal{R}$.

The idea is that the algorithm with the best performance is the one which provides solutions with the largest coverage of the other non-dominated sets and the minimum distance to the reference set. Figure 1 presents a simple example with $\mathsf{M_C}(\mathcal{A},\mathcal{B}) = 3/5$ better than $\mathsf{M_C}(\mathcal{B},\mathcal{A}) = 2/6$, and the average distance from points in set $\mathcal{A}$ to the reference set smaller than that for the points in $\mathcal{B}$. So, the algorithm producing $\mathcal{A}$ is deemed better than that producing $\mathcal{B}$.

4 Multi-Objective EA for VRPTW

Our proposed Multi-Objective EA (MOEA) involves selection, cross-over and mutation as in most EAs, and uses a simple list-based encoding. The main

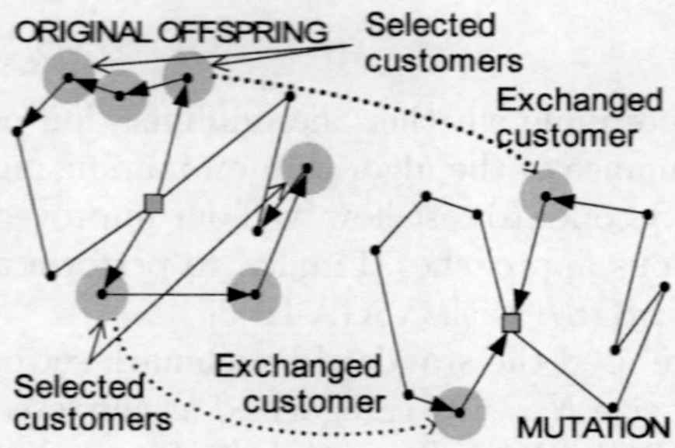

Fig. 3. Exchange mutation operator

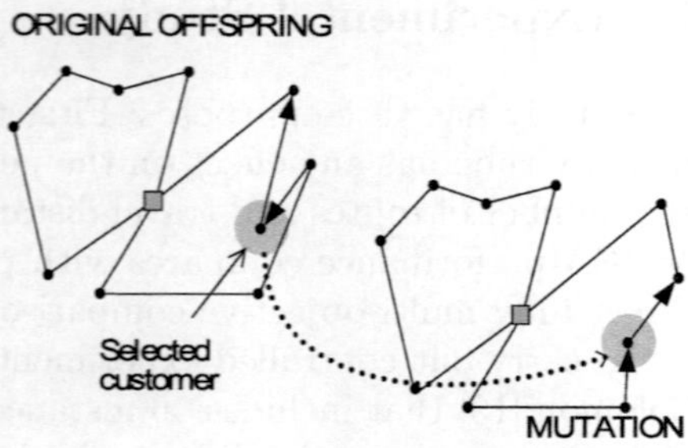

Fig. 4. Reposition mutation operator

novel characteristic is the implementation of a similarity measure to preserve population diversity. This is based on the *Jaccard's similarity coefficient*, which measures how similar two sets are as the ratio between the number of common elements and the total number of elements. We adapted this metric to the VRPTW for our earlier BiEA [9] by treating each solution as a set of arcs, and used it to calculate the average similarity between each solution in the population and every other solution. The MOEA here differs from the BiEA in its improved mutation stage, and in dealing with more than two objectives.

A dominance depth criteria [6] is used to assign fitness to solutions, by which individuals are grouped into non-dominated fronts and the relative depth of the front defines the fitness. Then in the recombination stage, a tournament selection is used to choose the first parent according to fitness as usual, but the second parent is selected on the basis of lowest similarity measure. Afterwards, the recombination process is designed to preserve routes from both parents.

The mutation involves the use of three basic functions: (*i*) `selectRoute()` stochastically selects a route according to the largest ratio between the travel distance and the number of customers, (*ii*) `selectCustomer()` stochastically selects one customer from a specific route according to the longest average length of its inbound and outbound arcs, and (*iii*) `insertCustomers()` tries to insert a set of customers, where the lowest travel distance is obtained, into any specific route, or all existing routes. The last two functions are used by the operators:

- *Reallocation* - Takes a random sequence of customers from a given route and reallocates them to other existing routes, as illustrated in Figure 2.
- *Exchange* - Swaps a sequence of customers between two routes if that is possible, as illustrated in Figure 3.
- *Reposition* - Selects one customer from a specific route and reinserts it into the same route, as illustrated in Figure 4.

Two routes are first chosen using `selectRoute()`. If they are the same, the *Reallocation* operation is performed, otherwise the *Exchange* operator is. Then `selectRoute()` selects another route and the *Reposition* operator is carried out. Finally, the parent and offspring populations are combined and fitness levels assigned, and the individuals with highest fitness form the next generation.

5 Experimental Study

Our study has three purposes: First to determine whether the minimization of delivery time has any effect on the performance of the algorithm on minimizing the number of routes and travel distance. Second, to test how well our improved MOEA performance compares with previous approaches. Finally, to perform a direct fully multi-objective comparison of MOEA with NSGA-II [6].

To carry out controlled experiments, we used the standard benchmark set of Solomon [13] that includes 56 instances of size $N = 100$ categorized as clustered (C1, C2), random (R1, R2), and mixed (RC1, RC2). Tan et al. [7] found that categories C1 and C2 have positively correlating objectives, but the majority of the instances in categories R1, R2, RC1 and RC2 have conflicting objectives.

We ran our algorithm and NSGA-II 30 times for each problem instance. The parameters of our algorithm were set to values that have proved to work well in preliminary testing: population size = 100, number of generations = 500, tournament size = 2, crossover rate = 1.0, and mutation rate = 0.1.

5.1 Minimization of the Delivery Time

The first aim was to analyze the performance of our MOEA with different objective settings, to test the effect of the additional delivery time objective. The number of routes (R), travel distance (D) and delivery time (T) were first set to be minimized in pairs, giving three objective settings (RD, RT and DT), and then all three of them were minimized together (RDT).

To use the coverage metric, for each setting and instance, the outcome set of non-dominated solutions after each of the 30 repetitions was recorded. Then, for each given instance and ordered pair of settings X and Y, $\mathsf{M_C}(\mathrm{X}_i, \mathrm{Y}_j), \forall\, i, j = 1, \ldots, 30$ were computed for the three-objective space, that is 900 $\mathsf{M_C}$ values. The average of these $\mathsf{M_C}$ values over the instances in each category are presented in Table 1, and in brackets are the numbers of instances for which the result is significantly better than the reverse case (determined by a *t-test* at 0.05 significance level). For categories C1 and C2, despite all four settings having a high coverage of each other, DT and RDT have a higher coverage of RD and RT. Otherwise, the coverage of RT, DT and RDT by RD is low ($\leq 14\%$), and the coverage of RD, DT and RDT by RT is nearly zero. The most interesting cases are settings DT and RDT, as their coverage of RD and RT is much higher. Between them, the coverage of DT by RDT is significantly larger than the inverse case in more instances than the inverse case, despite similar average coverages.

To apply the convergence metric, for each instance and objective setting, the overall non-dominated solutions were extracted from the 30 non-dominated sets, and composite non-dominated reference sets $\mathcal{R}$ in the three-objective space were generated. Then, for each setting X, $\mathsf{M_D}(\mathrm{X}_i, \mathcal{R})$ was computed for all $i = 1, \ldots, 30$. The averages of these $\mathsf{M_D}$ values over the instances in each category are presented in Figure 5. It can be seen that, in general, solutions from settings RD and RT are the farthest from the reference set, while those from DT and RDT are the nearest.

Table 1. Coverage metric values, averaged over instance categories, for solutions obtained with MOEA settings RD, RT, DT and RDT. In brackets are the numbers of instances for which the result is significantly better than the reverse case.

Obj.	Covers	C1	C2	R1	R2	RC1	RC2
	RT	0.87 (6)	0.64 (3)	0.04 (6)	0.01 (1)	0.05 (4)	0.02 (2)
RD	DT	0.82 (1)	0.72 (0)	0.08 (1)	0.11 (4)	0.14 (0)	0.08 (0)
	RDT	0.82 (1)	0.72 (1)	0.08 (1)	0.11 (3)	0.13 (0)	0.08 (1)
	RD	0.68 (0)	0.55 (1)	0.01 (2)	0.00 (0)	0.01 (2)	0.01 (0)
RT	DT	0.68 (0)	0.63 (0)	0.03 (0)	0.04 (0)	0.06 (0)	0.02 (0)
	RDT	0.68 (0)	0.62 (0)	0.04 (0)	0.03 (0)	0.07 (0)	0.03 (0)
	RD	0.91 (2)	0.90 (3)	0.31 (11)	0.14 (5)	0.36 (8)	0.21 (6)
DT	RT	0.97 (5)	0.92 (4)	0.49 (12)	0.42 (11)	0.46 (8)	0.48 (8)
	RDT	0.91 (2)	0.88 (2)	0.43 (4)	0.40 (4)	0.42 (2)	0.42 (3)
	RD	0.89 (3)	0.91 (3)	0.32 (11)	0.16 (6)	0.38 (8)	0.23 (7)
RDT	RT	0.97 (6)	0.93 (3)	0.52 (12)	0.44 (11)	0.49 (8)	0.44 (8)
	DT	0.89 (2)	0.89 (1)	0.44 (5)	0.43 (5)	0.46 (6)	0.41 (3)

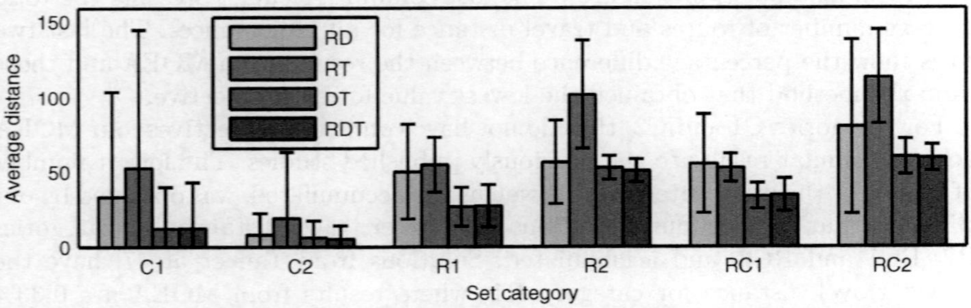

Fig. 5. Convergence metric values, averaged over instance categories, for solutions obtained with MOEA settings RD, RT, DT and RDT

Even though DT and RDT produce similar results for the convergence metric, we can conclude that, considering both metrics, setting the MOEA to minimize all three objectives does lead to better non-dominated solutions. Consequently, the solutions from the RDT case will be used for all the following analyses.

5.2 Comparison with Previous Studies

Although previous studies have considered the VRPTW as a multi-objective problem, they have not presented their results in a multi-objective manner, and have only shown averages of their best results. So the averages of our best results are now compared with those previous studies, before performing proper multi-objective comparisons with NSGA-II. Table 2 presents the average number of routes and travel distances from MOEA, and those from three previous multi-objective studies. For each instance, all the solutions in the resulting non-dominated set are taken from all repetitions, and the average for each objective

Table 2. Numbers of routes (upper figures) and travel distances (lower figures), averaged over categories, for the best solutions found in past studies and by our MOEA

Algorithm	C1	C2	R1	R2	RC1	RC2	Accum.
Tan et al. [7]	10.00	3.00	12.92	3.55	12.38	4.25	441.00
	828.91	590.81	1187.35	951.74	1355.37	1068.26	56293.06
Ombuki et al. [8]	10.00	3.00	13.17	4.55	13.00	5.63	471.00
	828.48	590.60	1204.48	893.03	1384.95	1025.31	55740.33
BiEA [9]	10.00	3.00	12.50	3.18	12.38	4.00	430.00
	830.64	589.86	1191.22	926.97	1349.81	1080.11	56125.35
MOEA	10.00	3.00	12.83	3.82	12.63	4.38	446.00
	828.38	589.86	1191.30	916.32	1349.24	1060.80	55829.68
% diff. R	0.00	0.00	2.67	20.00	2.02	9.38	3.72
% diff. D	0.00	0.00	0.33	2.61	0.00	3.46	0.16

is computed. Then, these are averaged over each category. For each algorithm and category is shown the average number of routes (upper figure) and the average travel distance (lower figure). The last column (Accum.) presents the total average number of routes and travel distance for all 56 instances. The last two rows show the percentage difference between the results from MOEA and those from the method that obtained the lowest value for each objective.

For categories C1 and C2, that do not have conflicting objectives, our MOEA achieved similar results to the previously published studies. The lowest number of routes for the other categories, as well as the accumulated, was obtained by our BiEA [9], but MOEA found solutions with lower travel distances for categories R2, RC1 and RC2, and accumulated. Solutions from Tan et al. [7] have the lowest travel distance for category R1, where results from MOEA are 0.33% higher, and Ombuki et al. [8] obtained the shortest distances for categories R2 and RC2, and accumulated, where results from MOEA are 2.61%, 3.46%, and 0.16% higher, but have smaller numbers of routes. Finally, travel distances from MOEA for category RC1 are the shortest. These results show that, overall, MOEA's performance is comparable to the previously published algorithms.

5.3 Comparison with NSGA-II

Simple averages as in Table 2 are often misleading, so our results are now analyzed using the multi-objective coverage and convergence performance metrics to compare the non-dominated solutions from MOEA with those from NSGA-II [6]. For a fair comparison, the NSGA-II implementation used the same solution representation, with the same crossover and mutation operators, as MOEA. The difference lies in the fact that MOEA makes use of solution similarity, while NSGA-II utilizes the *crowding distance* which does not involve any routing information at all. The coverages $\mathsf{M_C}(\text{MOEA}_i, \text{NSGA-II}_j)$ and $\mathsf{M_C}(\text{NSGA-II}_j, \text{MOEA}_i)$ were determined as described above, and then the convergences $\mathsf{M_D}(\text{MOEA}_i, \mathcal{R})$ and $\mathsf{M_D}(\text{NSGA-II}_i, \mathcal{R})$ were computed using a combined reference set $\mathcal{R}$ of solutions obtained from both algorithms. Table 3 shows the average $\mathsf{M_C}$ (upper

Table 3. Coverage (upper figure) and convergence (lower figure), averaged over instance categories, for solutions obtained with NSGA-II and MOEA. In brackets are the numbers of instances which are significantly better than the other approach.

Algorithm	C1	C2	R1	R2	RC1	RC2
NSGA-II	0.81 (0)	0.87 (2)	0.14 (0)	0.37 (4)	0.13 (0)	0.35 (0)
	25.06 (0)	4.88 (0)	51.56 (0)	49.81 (0)	72.21 (0)	67.17 (0)
MOEA	0.93 (4)	0.88 (2)	0.78 (12)	0.41 (5)	0.80 (8)	0.45 (7)
	12.20 (3)	5.07 (0)	27.89 (12)	50.67 (0)	33.91 (8)	65.86 (1)

figure) and M_D (lower figure) over the instances in each category, and the number of instances significantly better than the other algorithm (in brackets).

Both algorithms show similar coverage for categories C1 and C2, but MOEA has significantly higher coverage of NSGA-II in more instances. For C1, solutions from MOEA are closer to $\mathcal{R}$ on average, and for three instances are significantly better than NSGA-II. For C2, solutions from NSGA-II are are closer to $\mathcal{R}$ on average, but no instances have significant differences. For all instances in categories R1 and RC1, solutions in the non-dominated sets found by MOEA have a significantly higher coverage of those obtained by NSGA-II, and are also significantly nearer to $\mathcal{R}$. For category R2, MOEA has a significantly higher coverage in five instances, and NSGA-II is significantly higher in four, while both show similar distances to $\mathcal{R}$. Finally, for category RC2, MOEA has a significantly higher coverage in seven instances, despite both algorithms having similar average distance to $\mathcal{R}$. Overall then, it can be concluded that our new MOEA, with its similarity-based second parent selection, performs significantly better than NSGA-II, and this is particularly clear for categories R1 and RC1.

6 Conclusions

We have proposed and analyzed the performance of our new Multi-Objective EA (MOEA) for solving the multi-objective VRPTW, which is an improvement of our earlier Bi-objective EA (BiEA) [9] that introduced similarity-based selection to enhance solution diversity. This involved improved mutation operators, improved analysis using fully multi-objective performance metrics, and performance testing for a third objective, namely the delivery time (T), in addition to the previously studied number of routes (R) and travel distance (D).

We tested four different objective settings: first minimizing pairs of objectives (RD, RT and DT), and then all three at once (RDT). The coverage and convergence performance metrics were used to evaluate the algorithm, showing that settings DT and RDT have a higher coverage of RD and RT, and their solutions are closer to a composite reference set, indicating that the minimization of the delivery time improves the algorithm's performance. Moreover, RDT covers DT in more instances, which implies that even better results can be obtained by considering the minimization of all three objectives at the same time.

Averages of the non-dominated sets found by our MOEA with setting RDT were compared with previous studies, and, although not better in all respects, MOEA is better in some, and competitive on average. Significantly, using fully multi-objective coverage and convergence metrics to compare the MOEA against the well-known evolutionary multi-objective optimizer NSGA-II showed that the solutions found by MOEA are better for almost all instance categories.

Given the promising performance of our algorithm, we are now looking at the possibility of optimizing even more objectives (such as route balance [1]), and pursuing the comparison of our results with other multi-criterion optimization methods and further multi-objective performance metrics.

Acknowledgments

The first author acknowledges support from CONACYT (Scholarship 229168).

References

1. Jozefowiez, N., Semet, F., Talbi, E.G.: Multi-objective vehicle routing problems. Eur. J. Oper. Res. 189(2), 293–309 (2008)
2. Desrochers, M., Desrosiers, J., Solomon, M.: A new optimization algorithm for the vehicle routing problem with time windows. Oper. Res. 40(2), 342–354 (1992)
3. Bräysy, O., Gendreau, M.: Vehicle routing problem with time windows, part I: Route construction and local search algorithms. Transport Sci. 39(1), 104–118 (2005)
4. Bräysy, O., Gendreau, M.: Vehicle routing problem with time windows, part II: Metaheuristics. Transport. Sci. 39(1), 119–139 (2005)
5. Bräysy, O., Dullaert, W., Gendreau, M.: Evolutionary algorithms for the vehicle routing problem with time windows. J. Heuristics 10(6), 587–611 (2004)
6. Deb, K., Pratap, A., Agarwal, S., Meyarivan, T.: A fast and elitist multiobjective genetic algorithm: NSGA-II. IEEE T. on Evolut. Comput. 6(2), 182–197 (2002)
7. Tan, K.C., Chew, Y.H., Lee, L.H.: A hybrid multiobjective evolutionary algorithm for solving vehicle routing problem with time windows. Comput. Optim. and Appl. 34(1), 115–151 (2006)
8. Ombuki, B., Ross, B.J., Hanshar, F.: Multi-objective genetic algorithms for vehicle routing problem with time windows. Appd. Intel. 24(1), 17–30 (2006)
9. Garcia-Najera, A., Bullinaria, J.A.: Bi-objective optimization for the vehicle routing problem with time windows: Using route similarity to enhance performance. In: Ehrgott, M., Fonseca, C.M., Gandibleux, X., Hao, J.-K., Sevaux, M. (eds.) EMO 2009. LNCS, vol. 5467, pp. 275–289. Springer, Heidelberg (2009)
10. Garcia-Najera, A.: Preserving population diversity for the multi-objective vehicle routing problem with time windows. In: GECCO (Companion) 2009, pp. 2689–2692. ACM, New York (2009)
11. Zitzler, E., Deb, K., Thiele, L.: Comparison of multiobjective evolutionary algorithms: Empirical results. Evol. Comput. 8(2), 173–195 (2000)
12. Deb, K., Jain, S.: Running performance metrics for evolutionary multi-objective optimization. KanGAL report 2002004, Indian Institute of Technology (2002)
13. Solomon, M.M.: Algorithms for the vehicle routing and scheduling problems with time windows constraints. Oper. Res. 35(2), 254–265 (1987)

Speculative Evaluation in Particle Swarm Optimization

Matthew Gardner, Andrew McNabb, and Kevin Seppi

Department of Computer Science, Brigham Young University

Abstract. Particle swarm optimization (PSO) has previously been parallelized only by adding more particles to the swarm or by parallelizing the evaluation of the objective function. However, some functions are more efficiently optimized with more iterations and fewer particles. Accordingly, we take inspiration from speculative execution performed in modern processors and propose speculative evaluation in PSO (SEPSO). Future positions of the particles are speculated and evaluated in parallel with current positions, performing two iterations of PSO at once.

We also propose another way of making use of these speculative particles, keeping the best position found instead of the position that PSO actually would have taken. We show that for a number of functions, speculative evaluation gives dramatic improvements over adding additional particles to the swarm.

1 Introduction

Particle swarm optimization (PSO) has been found to be a highly robust and effective algorithm for solving many types of optimization problems. For much of the algorithm's history, PSO was run serially on a single machine. However, the world's computing power is increasingly coming from large clusters of processors. In order to efficiently utilize these resources for computationally intensive problems, PSO needs to run in parallel.

Within the last few years, researchers have begun to recognize the need to develop parallel implementations of PSO, publishing many papers on the subject. The methods they have used include various synchronous algorithms [1, 2] and asynchronous algorithms [3, 4, 5]. Parallelizing the evaluation of the objective function can also be done in some cases, though that is not an adaption of the PSO algorithm itself and thus is not the focus of this paper.

These previous parallel techniques distribute the computation needed by the particles in the swarm over the available processors. If more processors are available, these techniques increase the number of particles in the swarm. The number of iterations of PSO that the algorithms can perform is thus inherently limited by the time it takes to evaluate the objective function—additional processors add more particles, but do not make the iterations go any faster.

For many functions there comes a point of diminishing returns with respect to adding particles. Very small swarms do not produce enough exploration to consistently find good values, while large swarms result in more exploration

R. Schaefer et al. (Eds.): PPSN XI, Part II, LNCS 6239, pp. 61–70, 2010.

than is necessary and waste computation. For this reason, previous work has recommended the use of a swarm size of 50 for PSO [6]. Thus, in at least some cases, adding particles indefinitely will not yield an efficient implementation.

In this paper we consider PSO parallelization strategies for clusters of hundreds of processors and functions for which a single evaluation will take at least several seconds. Our purpose is to explore the question of what to do with hundreds of processors when 50 or 100 particles is the ideal swarm size, and simply adding particles yields diminishing returns.

To solve this problem, we propose a method for performing two iterations of PSO at the same time in parallel that we call speculative evaluation. The name comes from an analogy to speculative execution (also known as branch prediction), a technique commonly used in processors. Modern processors, when faced with a branch on which they must wait (e.g., a memory cache miss), guess which way the branch will go and start executing, ensuring that any changes can be undone. If the processor guesses right, execution is much farther ahead than if it had idly waited on the memory reference. If it guesses wrong, execution restarts where it would have been anyway.

We show that the results of standard PSO can be reproduced *exactly*, two iterations at a time, using a speculative approach similar to speculative execution. We prove that the standard PSO equations can be factored such that a set of speculative positions can be found which will *always* include the position computed in the next iteration. By computing the value of the objective function for each of the speculative positions at the same time the algorithm evaluates the objective function for the current position, it is possible to know the objective function values for both the current and the next iteration at the same time. The resulting implementation runs efficiently on large clusters where the number of processors is much larger than a typical or reasonable number of particles, producing better results in less "wall-clock" time.

The balance of this paper is organized as follows. Section 2 describes the particle swarm optimization algorithm. Section 3 describes how speculative evaluation can be done in parallel PSO to perform two iterations at once. In Section 4 and Section 5 we present our results and conclude.

2 Particle Swarm Optimization

Particle swarm optimization was proposed in 1995 by James Kennedy and Russell Eberhart [7]. This social algorithm, inspired by the flocking behavior of birds, is used to quickly and consistently find the global optimum of a given objective function in a multi-dimensional space.

The motion of particles through the search space has three components: an inertial component that gives particles momentum as they move, a cognitive component where particles remember the best solution they have found and are attracted back to that place, and a social component by which particles are attracted to the best solution that any of their neighbors have found.

At each iteration of the algorithm, the position $\boldsymbol{x}_t$ and velocity $\boldsymbol{v}_t$ of each particle are updated as follows:

$$\boldsymbol{v}_{t+1} = \chi\big[\boldsymbol{v}_t + \phi^P U_t^P \otimes (\boldsymbol{x}_t^P - \boldsymbol{x}_t) + \phi^N U_t^N \otimes (\boldsymbol{x}_t^N - \boldsymbol{x}_t)\big] \quad (1)$$

$$\boldsymbol{x}_{t+1} = \boldsymbol{x}_t + \boldsymbol{v}_{t+1} \quad (2)$$

where U_t^P and U_t^N are vectors of independent random numbers drawn from a standard uniform distribution, the $\otimes$ operator is an element-wise vector multiplication, $\boldsymbol{x}^P$ (called personal best) is the best position the current particle has seen, and $\boldsymbol{x}^N$ (called neighborhood best) is the best position the neighbors of the current particle have seen. The parameters ϕ^N, ϕ^P, and χ are given prescribed values required to ensure convergence (2.05, 2.05, and .73, respectively) [8].

Changing the way neighbors are defined, usually called the "topology," has a significant effect on the performance of the algorithm. In the Ring topology, each particle has one neighbor to either side of it; in the Complete topology, every particle is a neighbor to every other particle [6]. In all topologies a particle is also a neighbor to itself in that its own position and value are considered when updating the particle's neighborhood best, $\boldsymbol{x}^N$. Thus with p particles, using the Ring topology each particle with index i has three neighbors: $i-1$, i (itself), and $i+1$. With the Complete topology, each particle has p neighbors.

In this paper we use these topologies as well as a parallel adaptation of the Complete topology, called Random, that has been shown to approximate the behavior of Complete with far less communication [9]. In the Random topology, each particle randomly picks two other particles to share information with at each iteration, along with itself. Thus in both the Ring and the Random topologies, all particles have three neighbors.

3 Speculative Evaluation in PSO

The PSO algorithm can be trivially parallelized by distributing the computation needed for each particle across processors. But for some functions adding particles yields diminishing returns. That is, adding processors does not help reach any given level of fitness appreciably faster. Instead of adding particles, speculative evaluation performs iterations two at a time.

Speculative evaluation is made possible by refactoring PSO such that evaluating the objective function is separate from the rest of the computation. For simplicity, this discussion will describe the case where PSO is performing function minimization using the Ring topology. In this example, each particle has two neighbors, the "right neighbor" and "left neighbor," whose positions are represented as $\boldsymbol{x}^R$ and $\boldsymbol{x}^L$ respectively. Though we will only describe the case of the Ring topology, this method is easily extended to arbitrary topologies.

The refactoring hinges on the idea that once the random coefficients U_t^P and U_t^P are determined, there are only a few possible updates to $\boldsymbol{x}^N$ and $\boldsymbol{x}^P$. For the Ring topology there are 7 possible update cases, identified in Table 1. We label each case with an identifier referring to the source of the update: a minus sign

$(-)$ represents no update, L represents an update to $\boldsymbol{x}^N$ coming from the left neighbor, R represents an update to $\boldsymbol{x}^N$ coming from the right neighbor, and S represents an update to either $\boldsymbol{x}^P$ or $\boldsymbol{x}^N$ coming from the particle itself. As an example, $(S, -)$ refers to the case that the particle finds a new personal best, but neither it nor its neighbors found a position that updated its neighborhood best. In the equations that follow, we refer to an unspecified update case as c, and to the set of cases collectively as $\mathcal{C}$.

Table 1. All possible updates for a particle with two neighbors

Identifier	Source of $\boldsymbol{x}^P$ update	Source of $\boldsymbol{x}^N$ update
$(-,-)$	No update	No update
$(-,L)$	No update	Left Neighbor
$(-,R)$	No update	Right Neighbor
$(S,-)$	Self	No update
(S,L)	Self	Left Neighbor
(S,R)	Self	Right Neighbor
(S,S)	Self	Self

In order to incorporate the determination of which case occurs into the position and velocity update equations, we introduce an indicator function I_{t+1}^c for each case $c \in \mathcal{C}$. When c corresponds to the case taken by PSO, I_{t+1}^c evaluates to 1; otherwise it evaluates to 0. We can then sum over all of the cases, and the indicator function will make all of the terms drop to zero except for the case that actually occurs. For example, the indicator function for the specific case $(S, -)$ can be written as follows:

$$I_{t+1}^{(S,-)}\left(f(\boldsymbol{x}_t), f\left(\boldsymbol{x}_t^L\right), f\left(\boldsymbol{x}_t^R\right), f\left(\boldsymbol{x}_{t-1}^P\right), f\left(\boldsymbol{x}_{t-1}^N\right)\right) = \begin{cases} 1 & \text{if } f(\boldsymbol{x}_t) < f\left(\boldsymbol{x}_{t-1}^P\right) \text{ and } f\left(\boldsymbol{x}_{t-1}^N\right) < \min\left(f(\boldsymbol{x}_t), f\left(\boldsymbol{x}_t^L\right), f\left(\boldsymbol{x}_t^R\right)\right) \\ 0 & \text{otherwise} \end{cases} \tag{3}$$

For each case $c \in \mathcal{C}$, there is also a corresponding velocity update function $\boldsymbol{V}_{t+1}^c$. When the case is known, the specific values of $\boldsymbol{x}_t^P$ and $\boldsymbol{x}_t^N$ may be substituted directly into (1). For example, in case $(S, -)$, $\boldsymbol{x}_t^P = \boldsymbol{x}_t$, as $\boldsymbol{x}^P$ was updated by the particle's current position, and $\boldsymbol{x}_t^N = \boldsymbol{x}_{t-1}^N$, as $\boldsymbol{x}^N$ was not updated at iteration t:

$$\boldsymbol{V}_{t+1}^{(S,-)}\left(\boldsymbol{v}_t, \boldsymbol{x}_t, \boldsymbol{x}_t^L, \boldsymbol{x}_t^R, \boldsymbol{x}_{t-1}^P, \boldsymbol{x}_{t-1}^N, U_t^P, U_t^N\right) = \chi\left[\boldsymbol{v}_t + \phi^P U_t^P \otimes (\boldsymbol{x}_t - \boldsymbol{x}_t) + \phi^N U_t^N \otimes \left(\boldsymbol{x}_{t-1}^N - \boldsymbol{x}_t\right)\right] \tag{4}$$

In the same way we can create notation for the position update function by substituting into (2):

$$\boldsymbol{X}_{t+1}^c\left(\boldsymbol{x}_t, \boldsymbol{v}_t, \boldsymbol{x}_t^L, \boldsymbol{x}_t^R, \boldsymbol{x}_{t-1}^P, \boldsymbol{x}_{t-1}^N, U_t^P, U_t^N\right) = \boldsymbol{x}_t + \boldsymbol{V}_{t+1}^c\left(\boldsymbol{v}_t, \boldsymbol{x}_t, \boldsymbol{x}_t^L, \boldsymbol{x}_t^R, \boldsymbol{x}_{t-1}^P, \boldsymbol{x}_{t-1}^N, U_t^P, U_t^N\right) \tag{5}$$

With this notation we can re-write the original PSO velocity equation (1), introducing our sum over cases with the indicator functions. The velocity (1) and position (2) equations become:

$$\begin{aligned} \boldsymbol{v}_{t+1} = \sum_{c \in \mathcal{C}} \big[I^c_{t+1}\big(f(\boldsymbol{x}_t), f(\boldsymbol{x}^L_t), f(\boldsymbol{x}^R_t), f(\boldsymbol{x}^P_{t-1}), f(\boldsymbol{x}^N_{t-1})\big) \\ \boldsymbol{V}^c_{t+1}\big(\boldsymbol{x}_t, \boldsymbol{v}_t, \boldsymbol{x}^L_t, \boldsymbol{x}^R_t, \boldsymbol{x}^P_{t-1}, \boldsymbol{x}^N_{t-1}, U^P_t, U^N_t\big)\big] \end{aligned} \tag{6}$$

$$\begin{aligned} \boldsymbol{x}_{t+1} = \sum_{c \in \mathcal{C}} \big[I^c_{t+1}\big(f(\boldsymbol{x}_t), f(\boldsymbol{x}^L_t), f(\boldsymbol{x}^R_t), f(\boldsymbol{x}^P_{t-1}), f(\boldsymbol{x}^N_{t-1})\big) \\ \boldsymbol{X}^c_{t+1}\big(\boldsymbol{x}_t, \boldsymbol{v}_t, \boldsymbol{x}^L_t, \boldsymbol{x}^R_t, \boldsymbol{x}^P_{t-1}, \boldsymbol{x}^N_{t-1}, U^P_t, U^N_t\big)\big] \end{aligned} \tag{7}$$

In this form the important point to notice is that there are only 7 values (for this Ring topology) in the set $\{\boldsymbol{X}^c_{t+1} : c \in \mathcal{C}\}$ and that none of them depend upon $f(\boldsymbol{x}_t)$ or any other objective function evaluation at iteration t. Note also that while there are random numbers in the equation, they are assumed fixed once drawn for any particular particle at a specific iteration. Thus PSO has been refactored such that the algorithm can begin computing all 7 of the objective function evaluations potentially needed in iteration $t+1$ *before* $f(\boldsymbol{x}_t)$ is computed. Once the evaluation of $f(\boldsymbol{x}_t)$ is completed for all particles only one of the indicator functions I^c_{t+1} will be set to 1; hence only one of the positions $\boldsymbol{X}^c_{t+1}$ will be kept.

Although this speculative approach computes $f(\boldsymbol{X}^c_{t+1})$ for all $c \in \mathcal{C}$, even those for which $I^c_{t+1} = 0$, these extra computations will be ignored, and might just as well never have been computed. We call the set $\{f(\boldsymbol{X}^c_{t+1}) : c \in \mathcal{C}\}$ "speculative children" because only one of them will be needed.

To see the value of this refactoring, suppose that 800 processors are available, and that the evaluation of the objective function takes one hour. If we only want a swarm of 100 particles, 700 of the processors would be sitting idle for an hour at every iteration, and it would take two hours to run two iterations. If instead we perform speculative evaluation, sending each of the $f(\boldsymbol{X}^c_{t+1})$ to be computed at the same time as $f(\boldsymbol{x}_t)$, we could create a swarm of size 100, each particle with 7 speculative evaluations (700 processors dedicated to speculative evaluation), thus using all 800 processors and performing two iterations in one hour.

In order to do two iterations at once, we use 8 times as many processors as there are particles in the swarm. If these processors were not performing speculative evaluation, they might instead be used for function evaluation needed to support a large swarm. This raises the question of whether a swarm of 100 particles doing twice as many iterations outperforms a swarm of 800 particles. We show in the rest of this paper that in many instances a smaller swarm performing more iterations does in fact outperform a larger swarm. We acknowledge that both intuition and prior research [9] indicate that the optimization of deceptive functions benefits greatly from large and even very large swarm sizes. Thus this work will focus on less deceptive functions.

3.1 Implementation

The number of speculative evaluations needed per particle depends on the number of neighbors each particle has. In a swarm with p particles and n neighbors per particle, $(2n + 1)p$ speculative evaluations are necessary (each additional neighbor adds two rows to Table 1). This dependence on the number of neighbors necessitates a wise choice of topology. The use of the Complete topology, where every particle is a neighbor to every other particle, would require $O(p^2)$ speculative evaluations per iteration. It is much more desirable to have a sparse topology, where $O(np)$ is much smaller than $O(p^2)$. However, some functions are better optimized with the Complete topology and the quick spread of information it entails than with sparse topologies. In such cases, we use the Random topology described in Section 2.

To aid in describing our implementation, we introduce a few terms. We use p_t to denote a particle at iteration t and s_{t+1} to denote one of p_t's speculative children, corresponding to one of the rows in Table 1. n_t is a neighbor of particle p_t. Sets of particles are given by $\mathbf{p}$, $\mathbf{s}$, or $\mathbf{n}$, whereas single particles are simply p, s, or n.

A particle at iteration $t - 1$ that has been moved to iteration t using (1) and (2), but whose position has not yet been evaluated, is denoted as p_t^{-e}. Once its position has been evaluated, but it has still not yet received information from its neighbors, it is denoted as p_t^{-n}. Only when the particle has updated its neighborhood best is it a complete particle at iteration t. It is then simply denoted as p_t.

The outline of the speculative evaluation in PSO (SEPSO) algorithm is given in Algorithm 1.

Algorithm 1. Speculative Evaluation PSO (SEPSO)

1: Move all p_{t-1} to p_t^{-e} using (1) and (2)
2: For each p_t^{-e}, get its neighbors $\mathbf{n}_t^{-e}$ and generate $\mathbf{s}_{t+1}^{-e}$ according to (5).
3: Evaluate all p_t^{-e} and $\mathbf{s}_{t+1}^{-e}$ in parallel
4: Update personal best for each p_t^{-e} and s_{t+1}^{-e}, creating p_t^{-n} and s_{t+1}^{-n}
5: Update neighborhood best for each p_t^{-n}, creating $\mathbf{p}_t$
6: **for each** p_t **do**
7:     Pick s_{t+1}^{-n} from $\mathbf{s}_{t+1}^{-n}$ that matches the branch taken by p_t according to (7).
8:     Pass along personal and neighborhood best values obtained by p_t, making p_{t+1}^{-n}
9: **end for**
10: Update neighborhood best for each p_{t+1}^{-n}, creating $\mathbf{p}_{t+1}$
11: Repeat from Step 1 until finished

3.2 Using All Speculative Evaluations

In performing speculative evaluation as we have described it, $2n + 1$ speculative evaluations are done per particle, while all but one of them are completely ignored. It seems reasonable to try to make use of the information obtained through those evaluations instead of ignoring it. Making use of this information changes the behavior of PSO, instead of reproducing it exactly as the above method explains, but the change turns out to be an improvement in our context.

To make better use of the speculative evaluations, instead of choosing the speculative child that matches that branch that the original PSO would have taken, we take the child that has the best value. The methodology is exactly the same as above except for the process of choosing which speculative child to accept. The only change needed in Algorithm 1 is in step 7, where the s_{t+1}^{-e} with the best value is chosen from $\mathbf{s}_{t+1}^{-e}$ instead of with the matching branch. We call this variant "Pick Best".

4 Results

In our experiments we compared our speculative PSO algorithm to the standard PSO algorithm. At each iteration of the algorithms, we use one processor to perform one function evaluation for one particle, be it speculative or non-speculative. The speculative algorithm actually performs two iterations of PSO at each "iteration," so we instead call each "iteration" a "round of evaluations." For benchmark functions with very fast evaluations this may not be the best use of parallelism in PSO. But for many real-world applications, the objective function takes on the order of at least seconds (or more) to evaluate; in such cases our framework is reasonable.

In each set of experiments we keep the number of processors for each algorithm constant. In all of our experiments SEPSO uses topologies in which each particle has two neighbors in addition to itself. As shown in Table 1, this results in 7 speculative evaluations per particle. With one evaluation needed for the original, non-speculative particle, we have a total of $8p$ evaluations for every two iterations, where p is the number of particles in the speculative swarm. In order to use the same resources for each algorithm, we compare swarms of size p in the speculative algorithms with swarms of size $8p$ in standard PSO.

As discussed in Section 3.1, where the Complete topology would normally be used, we use a Random topology in our speculative algorithm, as Complete leads to an explosion of speculative evaluations. For the standard PSO baseline we have included experiments with the Ring and the Random topologies, both with two neighbors, as well as for the Complete topology. It is important to note however, that in many cases the Complete topology is not a practical alternative in the context of large swarms on large clusters where where the messaging complexity is $O(p^2)$ and can overwhelm the system.

We follow the experimental setup used in [6]. All functions have 20 dimensions, and all results shown are averages over 20 runs. For ease of labeling, we call our speculative algorithm SEPSO, the variant SEPSO Pick Best, and standard PSO just PSO and identify the topology in a suffix, "PSO Ring" for example.

The benchmark function Sphere ($f(\boldsymbol{x}) = \sum_{i=1}^{D} x_i^2$) has no local optima and is most efficiently optimized using a small swarm but many iterations. In our experiments with Sphere, we use 240 processors; thus 30 particles for SEPSO and 240 for PSO. In Figure 1, we can see that SEPSO clearly beats PSO with a Random topology or a Ring topology. SEPSO approaches the performance of PSO with the Complete topology, even though PSO with the Complete topology requires $O(p^2)$ communication. SEPSO Pick Best handily outperforms all other algorithms in this problem.

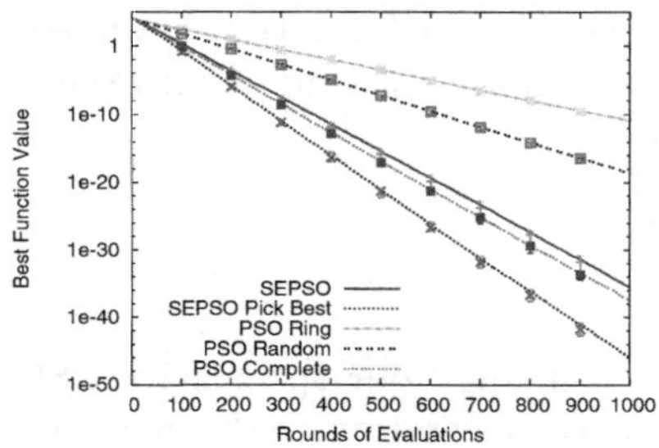

Fig. 1. Function Sphere with a swarm that uses 240 processors per round of evaluations. We show 10th and 90th percentiles every 100 iterations. Note that PSO Complete requires $O(p^2)$ messaging and may not be practical in many cases.

The benchmark function Griewank is defined by the equation $f(\boldsymbol{x}) = \frac{1}{4000} \sum_{i=1}^{D} x_i^2 - \Pi_{i=1}^{D} \cos\left(\frac{x_i}{\sqrt{i}}\right) + 1$. It is best solved in PSO using the Ring topology, as Complete is prone to premature convergence on local optima. Griewank has a global optimum with a value of 0, and sometimes the swarm finds the optimum and sometimes it does not. Instead of showing average function value at each iteration, a more enlightening plot for Griewank shows the percent of runs that have found the global optimum by each iteration.

PSO and SEPSO get caught in local minima with small swarm sizes so we show results in Figure 2 for swarms of size 100 (SEPSO) and 800 (standard) using the Ring topology. Figure 2 shows that SEPSO quickly finds the global optimum, between two and three times faster than running standard PSO.

In Figure 2 we also show results for the Bohachevsky function, defined as $f(\boldsymbol{x}) = \sum_{i=1}^{D}(x_i^2 + 2x_{i+1}^2 - .3\cos(3\pi x_i) - .4\cos(4\pi x_{i+1}) + .7)$. Bohachevsky is a unimodal function best optimized with a Complete swarm. It is similar to Griewank in that there is a global optimum with a value of 0, and swarms either find the optimum or get stuck. Both SEPSO algorithms find the optimum much faster than PSO Random, though only SEPSO Pick Best beats PSO Complete. Also, while the smaller swarm size of SEPSO gets stuck 75% of the time, when using SEPSO Pick Best with the same swarm size, the algorithm finds the optimum every time.

Previous work has shown that the optimization of highly deceptive functions like Rastrigin ($f(\boldsymbol{x}) = \sum_{i=1}^{D} \left(x_i^2 - 10\cos\left(2\pi x_i\right) + 10\right)$) benefit greatly from the addition of particles. Smaller swarms get caught in local optima, up to swarms of at least 4000 particles [9]. Because our speculative algorithms have a significantly

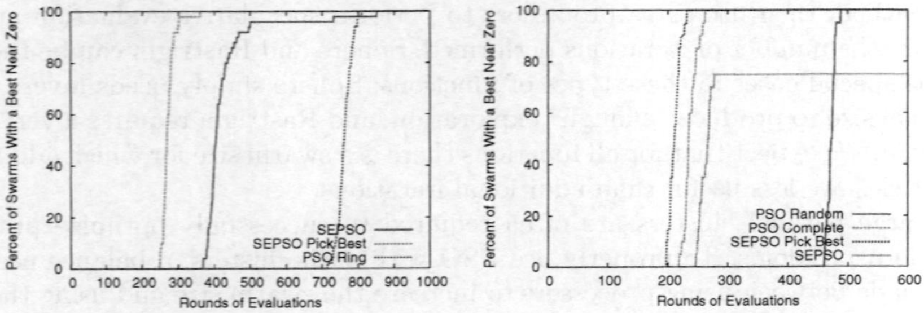

Fig. 2. Function Griewank with a swarm that uses 800 processors per round of evaluations, and function Bohachevsky with a swarm that uses 480 processors per round of evaluations

smaller swarm size, they get stuck at higher values while the larger swarms performing regular PSO continue to improve the best value found. Our experiments with SEPSO on Rastrigin were predicably lack luster, yielding an average value of 31 after 1000 evaluations, as compared to 10 for standard PSO.

5 Conclusions

We have described how parallel implementations of particle swarm optimization can be modified to allow additional processors to increase the number of iterations of the algorithm performed, instead of merely adding more particles to the swarm. Using our modifications, the original PSO algorithm is exactly reproduced two iterations at a time. This technique requires more function evaluations per iteration than regular PSO, but for some functions still performs better when run in a parallel environment. We have also described a method for making use of extra speculative evaluations that performs very well on some functions.

There are some functions for which very little exploration needs to be done; Sphere is an example of such a function. For such functions the best use of processors is to have a small swarm performing speculative evaluation with our Pick Best method, where all speculative evaluations are used.

There are other functions for which it seems there is never enough exploration, such as the Rastrigin function. It has been shown that up to 4000 particles there is no point at which "enough" exploration has been done [9]. With such functions, the smaller swarm size required by speculative evaluation is not able to produce enough exploration to perform better than standard PSO.

Griewank and Bohachevsky are functions between Sphere and Rastrigin. They are deceptive and prone to premature convergence, but by adding particles to the swarm a point is reached where "enough" exploration is done, and the algorithm finds the optimum essentially all of the time. For such functions, the best approach seems to be to increase the swarm size until "enough" exploration is reached, then use extra processors to perform speculative evaluation and increase the number of iterations performed. Sphere and Rastrigin can be thought of as special cases fo these types of functions; Sphere simply needs a very small swarm size to produce "enough" exploration, and Rastrigin requires a very large swarm. We expect that for all functions there is a swarm size for which additional particles are less useful than additional iterations.

Large parallel clusters are often required to successfully optimize practical modern problems. To properly use PSO with such clusters, a balance needs to be made between using processors to increase the swarm size and using them to increase the speed of the algorithm. This work is a first step in that direction.

References

1. Belal, M., El-Ghazawi, T.: Parallel models for particle swarm optimizers. Intl. Journal of Intelligent Computing and Information Sciences 4(1), 100–111 (2004)
2. Schutte, J.F., Reinbolt, J.A., Fregly, B.J., Haftka, R.T., George, A.D.: Parallel global optimization with the particle swarm algorithm. International Journal for Numerical Methods in Engineering 61(13), 2296–2315 (2004)
3. Mostaghim, S., Branke, J., Schmeck, H.: Multi-objective particle swarm optimization on computer grids. Technical Report 502, AIFB Institute, DEC (2006)
4. Venter, G., Sobieszczanski-Sobieski, J.: A parallel particle swarm optimization algorithm accelerated by asynchronous evaluations. In: Proceedings of the 6th World Congresses of Structural and Multidisciplinary Optimization (2005)
5. Koh, B.-I., George, A.D., Haftka, R.T., Fregly, B.J.: Parallel asynchronous particle swarm optimization. International Journal of Numerical Methods in Engineering 67, 578–595 (2006)
6. Bratton, D., Kennedy, J.: Defining a standard for particle swarm optimization. In: Proceedings of the IEEE Swarm Intelligence Symposium, pp. 120–127 (2007)
7. Kennedy, J., Eberhart, R.C.: Particle swarm optimization. In: International Conference on Neural Networks IV, Piscataway, NJ, pp. 1942–1948 (1995)
8. Clerc, M., Kennedy, J.: The particle swarm - explosion, stability, and convergence in a multidimensional complex space. IEEE Transactions on Evolutionary Computation 6(1), 58–73 (2002)
9. McNabb, A., Gardner, M., Seppi, K.: An exploration of topologies and communication in large particle swarms. In: Proceedings of the IEEE Congress on Evolutionary Computation, pp. 712–719 (May 2009)

Towards Directed Open-Ended Search by a Novelty Guided Evolution Strategy

Lars Graening[1], Nikola Aulig[2], and Markus Olhofer[1]

[1] Honda Research Institute Europe GmbH,
Carl-Legien-Straße 30, 63073 Offenbach/Main, Germany
{lars.graening,markus.olhofer}@honda-ri.de
[2] Technical University Darmstadt
Karolinenplatz 5, 64289 Darmstadt
nikola.aulig@web.de

Abstract. In the conceptional phases of design optimization tasks it is required to find new innovative solutions to a given problem. Although evolutionary algorithms are suitable methods to this problem, the search of a wide range of the solution space in order to identify novel concepts is mainly driven by random processes and is therefore a demanding task, especially for high dimensional problems. To improve the exploration of the design space additional criteria are proposed in the presented work which do not evaluate solely the quality of a solution but give an estimation of the probability to find alternative optima. To realize these criteria, concepts of novelty and interestingness are employed. Experiments on test functions show that these novelty guided evolution strategies identify multiple optima and demonstrate a switching between states of exploration and exploitation. With this we are able to provide first steps towards an alternative search algorithm for multi-modal functions and the search during conceptual design phases.

Keywords: Evolutionary algorithm, open-endedness, interestingness, multi-objective optimization, novelty detection, prediction error, niching.

1 Introduction

Evolutionary algorithms have various properties which make them suited to solve complex real world problems. One of these properties is the ability to identify multiple solutions in multi-modal quality functions. The importance of this property lies in the fact that very often conceptually different solutions can be found in real world applications which offer alternative realizations for the design of a system. The selection of the best suited solution can only be done by experts in the related field due to the complexity of the overall problem or due to the non-technical nature of the criteria, for example aesthetic arguments or the necessary distinctiveness to other available solutions used in other products.

In order to enhance this behavior various improvements of the algorithms are described in the literature. A common way is to strengthen the 'exploratory' behavior by increasing the mutation rate. Although this requires a low dimensional

R. Schaefer et al. (Eds.): PPSN XI, Part II, LNCS 6239, pp. 71–80, 2010.

search space in order to find solutions within an acceptable number of generations, successful examples of this approach can be found in the field termed *creative evolutionary search* [1] and in the early phases of design optimisation in which new concepts for a solution have to be identified on simplified models.

Another strategy is followed by *Niching Algorithms* which identify multiple optimal solutions by maintaining diversity within a population [2,3,4]. The simplest niching approach is fitness sharing where the fitness of individuals is reduced if they are located close together within a niching radius. Various improvements of these ideas are available like the dynamic niche sharing which uses a dynamic peak identification (DPI) algorithm to recognize the forming of niches and the fitness is shared among individuals within one niche. The dynamic niching algorithm [2] introduces mating restrictions instead of changing the fitness. While these methods require an a-priori estimation of the size of the niches, the derandomized CMA-ES implements an adaptive niche radius [4]. Although, niching algorithms provides the potential for identifying several distinct optimal solutions, the possible number of optima which can be discovered has to be specified in advance and remains limited by the population size.

An alternative way to guide the search towards new and innovative solutions is the integration of human creativity into the process [5,6] in which a human user is involved in the generation of variations or alternatively in the selection process. Although this process turned out to be very powerful it is limited to problems for which a solution can be found within a small number of evaluations, in which a human operator is able to judge the solutions by their intuition or knowledge of the process.

In this work we outline an alternative method for the determination of optima in a multi-modal fitness landscape which is fundamentally different to existing methods. We propose to guide the search by an additional criterion which directly relates to new and unexplored areas of the search space. This criterion is based on novelty or interestingness measures. Although the concept of novelty exists in the subjective perceptions of individuals and is generally difficult to describe, various attempts to define the concept can be found in different fields of science like psychology, active learning or evolutionary robotics which allow to formulate measures suitable for a numerical calculation.

Silberschatz and Tuzhilin's [7] for example state that interestingness depends on the user who is examining a pattern. They point out that something that is interesting for one user might not be interesting for another one. Schmidhuber [8] argues that if something is too unexpected it appears random and is no longer interesting. Along this line, Saunders [9] refers to the Wundt curve and writes: "... the most interesting experiences are those that are similar-yet-different to those that have been experienced previously". Based on these works it becomes already obvious that novelty as well as interestingness can only be evaluated based past experience. In the field of active learning, Risi et all. [10] evaluates novelty simply by measuring the similarity to existing solutions stored in an archive. Similar to the work of Risi, Lehman at all. [11] defined novelty through sparseness that is evaluated based on already generated solutions. In the domain of developmental robotics, motivated by the concept of intrinsic motivation, Oudeyer and Kaplan [12] provide a comprehensive summary on alternative techniques for quantifying interestingness and novelty. Most of the different attempts share the idea that a model that builds up a compact representation of the search space is used to produce an indicator for novelty or interestingness and allow the identification of parameter regimes which should be sampled. In this work a mechanism for the detection of novelty or interestingness is utilized to actively guide the search to alternative optima in a multi-modal search problem using evolutionary algorithms. The novelty measure provides an additional criteria besides the original quality function. The target is to guide the search by the newly added criteria towards currently unexplored regions of the search space and additionally to start new exploration phases after temporary convergence of the population. It is demonstrated on simple test functions that a combination of both criteria allows the algorithm to guide the population towards alternative local optima after the localization and convergence of the population to a formerly identified optimum. In the next section we describe the novelty metrics used here and their integration into the algorithm in more detail. In section 3 first experimental results are presented where the proposed algorithm is compared to the niching and standard evolutionary algorithm. The paper concludes with a discussion on the results and an outline for future work in section 4.

2 Novelty Guided Evolution Strategy

2.1 Overall Framework

The schematic view of the proposed algorithm is depicted in Fig. 1. After the reproduction, recombination and mutation of the parent population, the fitness is assigned to each individual of the produced offspring population. A second criteria is introduced evaluating the individual's novelty based on a model of the quality function which is adapted by newly evaluated individuals. The selection incorporates at least two criteria, the novelty and the quality function.

2.2 Novelty Evaluation and World Model Adaptation

An individual is said to be novel if it does not meet the expectations derived from the accumulated knowledge about the search space. Therefore, the implemented novelty metric is defined as follows. If the expected quality value, estimated by the world model, differs from the calculated one, a high novelty value is assigned to the individual. As depicted in Fig. 2 the implemented novelty metric is calculated as,

$$\varepsilon(\boldsymbol{x}_i^t) = |f(\boldsymbol{x}_i^t) - \hat{f}(\boldsymbol{x}_i^t)|, \ i \in [1 \dots \lambda] \tag{1}$$

where λ defines the size of the offspring population, t the current generation, $\boldsymbol{x}_i^t$ is the design vector and $f(\boldsymbol{x}_i^t)$, $\hat{f}(\boldsymbol{x}_i^t)$ defines the calculated and the predicted quality function value respectively. $\varepsilon(\boldsymbol{x}_i^t)$ reflects the prediction error of the world

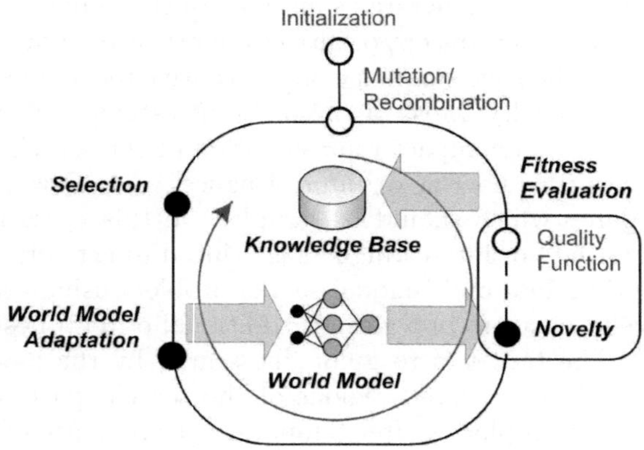

Fig. 1. Flowchart of the novelty guided optimization framework

model and is used for the quantification of the novelty. This formulation of the novelty metric equals the *Predictive novelty motivation* concept described by Oudeyer and Kaplan [12]. Solutions with a high prediction error assign a high novelty value to the individual.

The world model is adopted to predict the quality function value by means of estimating $\hat{f}(\boldsymbol{x}_i^t)$. Since multilayer feed-forward neural networks have successfully been employed as universal function approximators they are implemented for the world model. As already shown by Bishop [13], the neural network model is well suited to estimate the novelty of a solution. Given a pre-defined model structure, the network is updated in each generation $t-1$ using RProp [14], a variation of the back-propagation algorithm, together with cross-validation to prevent over-fitting. Data from generation $t-\gamma$ to generation $t-1$ that is added to the knowledge base during evolution is used for training. The parameter γ controls whether more global or localized model of the search space is generated.

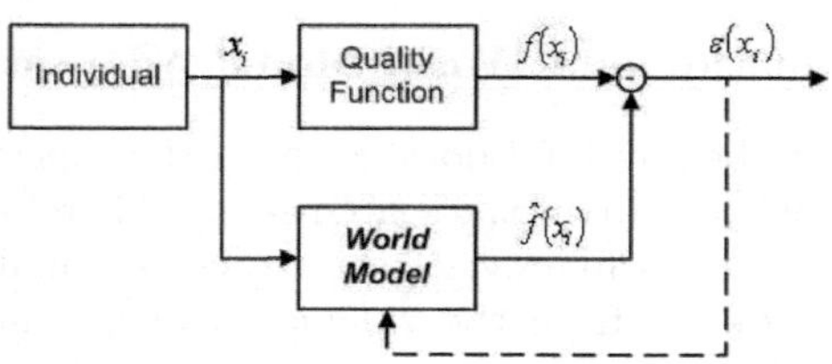

Fig. 2. Estimation of the prediction error which defines the novelty of a solution. The world model builds a compact representation of the search space and is adopted to predict the target function value.

2.3 Selection Strategy

During the fitness evaluation a fitness vector is assigned to each offspring,

$$\boldsymbol{f}(\boldsymbol{x}_i^t) = [f(\boldsymbol{x}_i^t),\ \varepsilon(\boldsymbol{x}_i^t)]^T, \tag{2}$$

containing the actual quality value and the prediction error estimating the novelty. It has to be noted that the additional criteria is dynamic in the sense that each time the world model is updated the estimated novelty value changes for one and the same solution. The multi objective optimisation problem can be transformed into a single objective optimisation by a *linear weighted aggregation.* This method leaves us with the problem of choosing an adequate weight. A high weight on the novelty objective would result in an extensive explorative behavior while a high weight on the actual quality function would result in an intensive exploitation of a single optimal solution. The desired behavior of the evolution strategy is a process that identifies successively several optima in regions with high fitness values but which does not exploit only one optimal solution. Aside the linear weighted aggregation, we employ a Pareto optimal selection criterion. The implemented strategy is derived from the crowded tournament selection suggested in the NSGA-II algorithm [15].

3 Experimental Results

The following experiments target the study of the basic characteristic of the *novelty guided ES.* The proposed evolution strategy is compared to three existing strategies, namely *standard ES*, *dynamic niche sharing* and *open-ended ES.*

3.1 Characteristics of the Novelty Guided Evolution Strategy

In the first experiments, the behavior of the introduced novelty guided ES is studied on a two dimensional multi-modal test function in which the design vector covers two variables, $\boldsymbol{x}_i^t = [x_{1i}^t,\ x_{2i}^t]^T$ or in short $\boldsymbol{x} = [x_1,\ x_2]^T$. The test function is constructed by a superposition of $N_G = 6$ 2D Gauss functions and is mathematically defined as follows:

$$f(\boldsymbol{x}) = -\sum_{j=1}^{N_G} e^{-\frac{1}{2}(\boldsymbol{x}-\boldsymbol{\mu}_j)\boldsymbol{\Sigma}^{-1}\cdot(\boldsymbol{x}-\boldsymbol{\mu}_j)+1}, \tag{3}$$

where $\boldsymbol{\mu}_j$ is the center and $\boldsymbol{\Sigma}^{-1}$ the covariance matrix of the Gauss kernel. Each center $\boldsymbol{\mu}_j$ of the different Gauss functions defines approximately the location of one local optima. The Gauss kernels are inverted, simply to transfer the task from a maximization into a minimization task. The resulting quality function is depicted in Fig. 3 a). All runs were calculated for $N = 500$ generations with a parent population size of $\mu = 20$ and an offspring population size of $\lambda = 100$. The standard evolution strategy is a (μ, λ) strategy with global step size

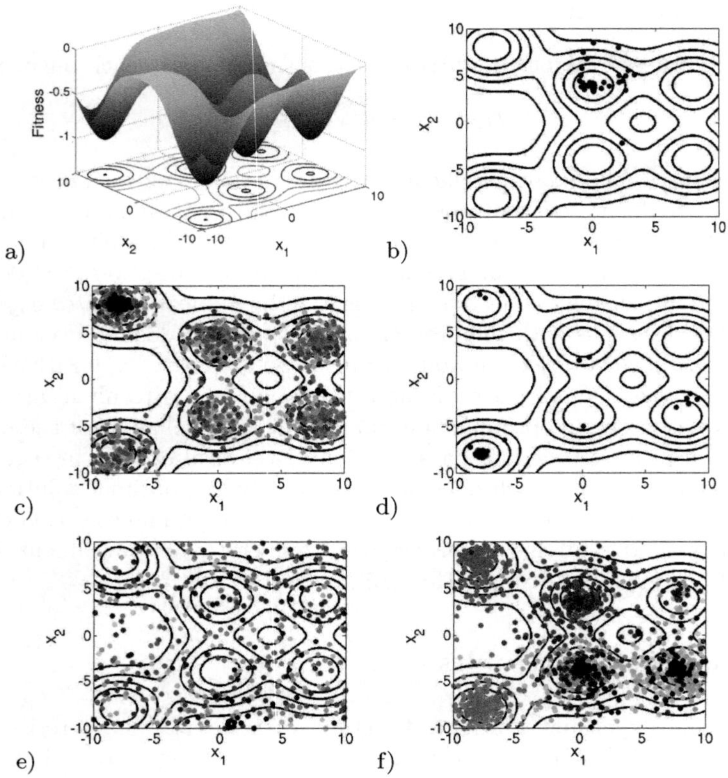

Fig. 3. a) The test function with 6 optima, b) results of the standard ES, c) niching with optimal niching radius, d) niching with an improper niching radius, e) open-ended evolution, targeting the generation of novel designs only and f) the results of the novelty guided ES

control. Recombination as well as mutation are applied to produce the offspring population. For the niching algorithm a niching radius ρ is defined according to Shir [3]. It has to be noted that the calculation of an adequate niching radius requires knowledge about the number of optima, which is usually not available. The world model that is needed for the calculation of the novelty metric is realized using a multi-layer network with 10 hidden neurons and with sigmoidal activation function. The data of the offspring population from 5 generations is used for the training of the network weights. The dominance based ranking with crowding distance is used as selection operator in the novelty guided ES. The results of the different experiments are summarized in Fig. 3. For each experiment, the contour plot of the fitness landscape together with the generated solutions is shown. The fill color of the dots indicate the generation number at which the solution has been produced. Dark indicates early and bright late generations. Fig 3 b) shows the result of a standard ES. The algorithm converges

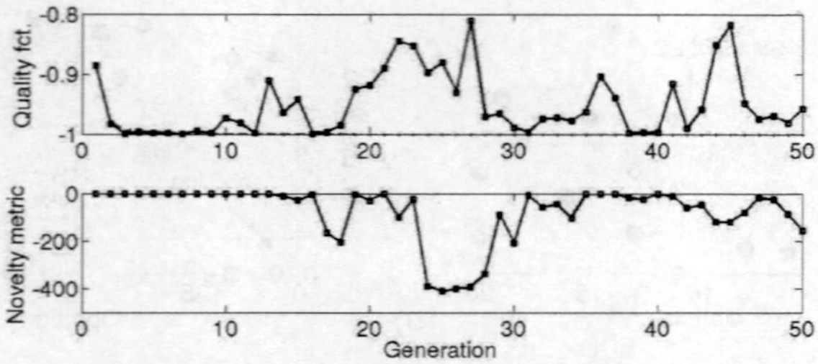

Fig. 4. Progress of quality and novelty in novelty guided ES. The selection of solutions with high degree of novelty allows to escape from optimal solutions.

directly (within about 20 generations) into the next local optimum. After that the algorithm is converged. Fig. 3 c) and d) present the result of the dynamic niching algorithm. While in c) an optimal niching radius has been used, in d) the niching radius is over-estimated (twice the optimal radius) what easily happens if no knowledge about the number of optima is given. In the case of a correct estimation of the niching radius the individuals distribute well between all the optima. If the niching radius is wrong or the population size too small, the niching algorithm might get stuck in a limited number of optimal solutions. In Fig. 3 e) results of a novelty driven evolutionary search are shown in which the search is only based on the novelty criterion neglecting information given by the quality function (open ended evolution). The algorithm does not exploit one of the six optima and diverges towards the boundaries of the search space as expected. Thus, a pure novelty driven optimization is quite inefficient and should be used for the exploration of the search space only. As can be seen from Fig. 3 f) the proposed novelty guided ES is able to locate all optimal solutions. Compared to the niching algorithm the optima are not exploited in parallel but rather sequentially. This sequential exploitation of the optima comes from the interplay between the two objectives, the quality function and the novelty metric. Fig. 4 shows the development of the quality and novelty value of the best offspring in the first 50 generations. In early generations the algorithm starts to exploit a nearby optimal solution exactly as it is done in the standard ES. After about 10 generations the influence of the novelty measure on the selection increases. Novel but worse solutions are selected. This allows the optimizer to escape from one optimum and exploit another one. This interplay between quality and novelty repeats until the algorithm is stopped. Since, a local model is used here, the algorithm visits optima multiple times due to a limited memory of the model.

3.2 Comparative Study on a High Dimensional Test Function

To carry out experiments on higher dimensional quality functions the multivariate Gauss kernel is used for the construction of an n dimensional multi-modal test function. Instead of the superposition of the Gauss kernels the max operator is applied to prevent the shift of the optima from the Gauss center. The n dimensional test function is defined as follows:

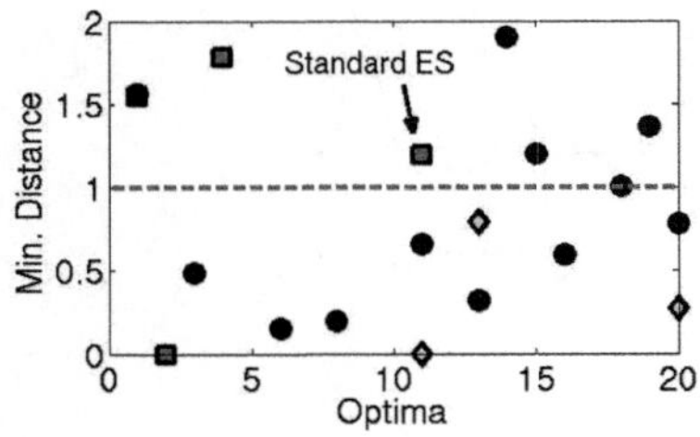

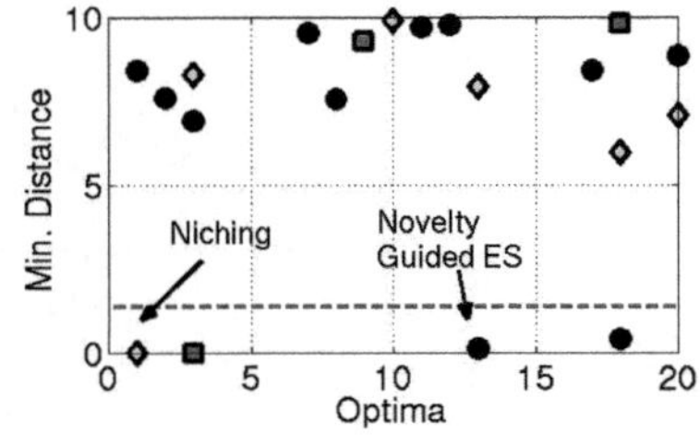

Fig. 5. Illustration of the minimal distance of the produced individuals to each of the 20 optimal solutions (x-axis). Only the closest solutions are shown. a) shows the results on the 5 dimensional and b) on the 10 dimensional fitness landscape.

$$f(\boldsymbol{x}) = - \max_{j \in [1 \ldots N_G]} \{ e^{-\frac{1}{2}(\boldsymbol{x}-\boldsymbol{\mu}_j)\boldsymbol{\Sigma}^{-1} \cdot (\boldsymbol{x}-\boldsymbol{\mu}_j)+1} \}. \quad (4)$$

An $n = 5$ and an $n = 10$ dimensional variant of $f(\boldsymbol{x})$ have been constructed. The number of Gauss kernels and thus the number of optima is set to $N_G = 20$. The Gauss centers are distributed randomly in the search space but remain the same for the different strategies. The minimal distance of all generated individuals to each optima is used to evaluate the strategies. If the distance runs below a threshold of $\tau = 1.0$, an optima is classified as being identified. The novelty guided ES has been compared to niching and standard ES. Since, the open-ended ES does not tend to exploit any of the optimal solutions it is skipped from the comparison in these experiments. Related to the preceding experimental setup, the number of hidden neurons of the world model has been increased to 25. For the niching algorithm, the correct number of optima has been used to estimate the radius ρ. Each optimization has been performed 5 times with different random seeds in order to retrieve a first idea on the reliability of the different strategies.

The results of the experiments are summarized in Fig. 5 and Tab. 1. Fig. 5 visualizes the evaluation of a typical run a) on the 5 and b) on the 10 dimensional quality function. The index of the optima is mapped onto the x-axis while the y-axis shows the distance of the closest solution to each optima. The dotted line indicates the threshold applied for counting the number of approached local optimal solutions. In Tab. 1 the mean $\overline{g}$ and the variance σ^2 of the number of approached optimal solutions over 5 runs is summarized. Again, it can be observed that the standard ES exploits one single optimal solution only independent of

Table 1. Summary of the results on the a) 5 and b) 10 dimensional multi-modal Gauss function. $\overline{g}$, σ^2 are mean and variance of the number of approached optimal solutions.

a)

ES	$\overline{g}$	σ^2
Standard ES	1.4	0.3
Niching	2.2	0.7
Novelty Guided ES	5.6	1.3

b)

ES	$\overline{g}$	σ^2
Standard ES	1.0	0.0
Niching	1.0	0.0
Novelty Guided ES	1.2	0.2

the search space dimension. The distance to the remaining 19 optima remains large. Concerning the number of approached optima, the novelty guided ES outperformes niching on the 5 dimensional test function. The proposed strategy approaches from minimal 4 to about 7 out of 20 optima, while niching reaches only about 3 optima at maximum. As can be seen from Tab. 1 b), none of the strategies is performing well on the 10 dimensional test function. However, the novelty guided ES is at least able to approach 2 optima in one out of the five optimization runs.

4 Discussion

The experiments on the test functions show the general feasibility of the proposed method. Evolutionary Strategies allow, combined with the concept of novelty measures, the determination of multiple optima on a multi-modal quality function. In contrast to other algorithms, novelty guided evolution strategies allow a sequential process of alternating phases of exploration and exploitation on multi-modal quality functions in which the exploration is guided by novelty or interestingness measures instead of randomly sampling the search space. In the presented initial experiments the additional criteria, which is introduced to guide the search to alternative solutions, is based on a novelty measure, purely relying on the prediction error of a model. The generation of purely novel solutions is usually a simple task solved easily by e.g. generating sufficiently large mutations in a unconstrained search space. Preferably, the new direction should be an estimation of the most likely area for new optima or at least an area from which to sample in order to increase the chance to determine a new optima. In this sense the utilization of a novelty measure cannot be the final answer which was already stated in [12]. In order to determine useful search directions, measures of interestingness are required which guide the search towards areas which are interesting in the sense that knowledge about the design space is generated in order to finally determine areas with high probability of high fitness values. Therefore the evaluation of measures based on the learning rate of a model will be the next step to tackle more realistic problems for example in the field of aerodynamic design in which areas of high noise or even chaotic parameter regimes are expected. In general, models of the quality functions are necessary to determine the novelty or the interestingness of design areas. Assuming that approximation models are generally more simple than the original quality function all approximation models can only realize local models, valid in a limited area of the design space. Therefore a second step in our future efforts in the development of the algorithm is the integration and the adaptation of model ensembles in which each single model represent a different area of the global search space. Utilizing model ensembles also avoids the oscillation of the optimization process between to optima which can be observed in the presented results. The reason is that the sampling of one optimum results in an adaptation of the model in a way that old information in the model is removed. Keeping an ensemble of models allows us to avoid the overwriting of former acquired information of the global search space. In this sense the presented work has to be seen as a starting point for the research of novelty guided evolution strategies.

References

1. Bentley, P.J., Corne, D.W. (eds.): Creative Evolutionary Systems. Morgan Kaufmann, San Francisco (2001)
2. Shir, O.M., Bäck, T.: Niching in evolution strategies. In: GECCO 2005: Proceedings of the 2005 Conference on Genetic and Evolutionary Computation, pp. 915–916. ACM, New York (2005)
3. Shir, O.M.: Dynamic niching in evolution strategies with covariance matrix adaptation. In: Proceedings of the 2005 Congress on Evolutionary Computation CEC-2005, Piscataway, pp. 2584–2591. IEEE Press, Los Alamitos (2005)
4. Shir, O.M., Bäck, T.: Niche radius adaptation in the cma-es niching algorithm. In: Parallel Problem Solving from Nature (PPSN IX), Reykjavik, Iceland (2006)
5. Herdy, M.: Evolution strategies with subjective selection. In: PPSN IV: Proceedings of the 4th International Conference on Parallel Problem Solving from Nature, London, UK, pp. 22–31. Springer, Heidelberg (1996)
6. Takagi, H.: Interactive evolutionary computation: Fusion of the capabilities of ec optimization and human evaluation. Proceedings of the IEEE 89, 1275–1290 (2001)
7. Silberschatz, A., Tuzhilin, A.: What makes patterns interesting in knowledge discovery systems. IEEE Transactions on Knowledge and Data Engineering 8, 970–974 (1996)
8. Schmidhuber, J.: What's interesting? Idsia-35-97, IDSIA, Switzerland (1997)
9. Saunders, R.: Curious Design Agents and Artificial Creativity. PhD thesis, Faculty of Architecture, The University of Sydney (2001)
10. Risi, S., Vanderbleek, S.D., Hughes, C.E., Stanley, K.O.: How novelty search escapes the deceptive trap of learning to learn. In: GECCO 2009: Proceedings of the 11th Annual Conference on Genetic and Evolutionary Computation, pp. 153–160. ACM, New York (2009)
11. Lehman, J., Stanley, K.O.: Exploiting open-endedness to solve problems through the search for novelty. In: Proceedings of the Eleventh International Conference on Artificial Life (ALIFE XI). MIT Press, Cambrige (2008)
12. Oudeyer, P.Y., Kaplan, F.: What is intrinsic motivation? a typology of computational approaches. Frontiers in Neurorobotics (2007)
13. Bishop, C.M.: Novelty detection and neural network validation. In: Proc. IEE Conference on Vision and Image Signal Processing, pp. 217–222 (1994)
14. Igel, C., Husken, M.: Improving the rprop learning algorithm. In: Second International Symposium on Neural Computation, pp. 115–121 (2000)
15. Deb, K., Agrawal, S., Pratap, A., Meyarivan, T.: A fast elitist non-dominated sorting genetic algorithm for multi-objective optimization: NSGA-II. In: Parallel Problem Solving from Nature, pp. 849–858. Springer, Heidelberg (2000)

Consultant-Guided Search Algorithms with Local Search for the Traveling Salesman Problem

Serban Iordache

SCOOP Software GmbH, Köln, Germany
siordache@acm.org

Abstract. Consultant-Guided Search (CGS) is a recent metaheuristic for combinatorial optimization problems, which has been successfully applied to the Traveling Salesman Problem (TSP). In experiments without local search, it has been able to outperform some of the best Ant Colony Optimization (ACO) algorithms. However, local search is an important part of any ACO algorithm and a comparison without local search can be misleading. In this paper, we investigate if CGS is still able to compete with ACO when all algorithms are combined with local search. In addition, we propose a new variant of CGS for the TSP, which introduces the concept of confidence in relation to the recommendations made by consultants. Our experimental results show that the solution quality obtained by this new CGS algorithm is comparable with or better than that obtained by Ant Colony System and MAX-MIN Ant System with 3-opt local search.

Keywords: Metaheuristics, combinatorial optimization, swarm intelligence, traveling salesman problem.

1 Introduction

Consultant-Guided Search (CGS) [5] is a swarm intelligence technique based on the direct exchange of information between individuals in a population. It takes inspiration from the way real people make decisions based on advice received from consultants. CGS can be used to solve hard combinatorial optimization problems and it has been first applied to the Traveling Salesman Problem (TSP). Experimental results have shown that the CGS-TSP algorithm introduced in [5] can outperform some of the best Ant Colony Optimization (ACO) [3] algorithms. However, these results may be misleading, because the experiments have been performed without local search and, in practice, ACO algorithms for the TSP are always combined with local search. While most metaheuristics work better when combined with local search, the performance improvement can vary significantly from one algorithm to another. Therefore, the main goal of this paper is to investigate whether CGS is still able to compete with ACO when the algorithms are combined with local search.

Moreover, we propose a variant of the CGS-TSP algorithm, where consultants also indicate the confidence in the recommendations they make. In experiments

R. Schaefer et al. (Eds.): PPSN XI, Part II, LNCS 6239, pp. 81–90, 2010.

with 3-opt local search, we compare the performance of this new algorithm to that of the standard CGS-TSP algorithm, as well as to that of Ant Colony System (ACS) [2] and MAX-MIN Ant System (MMAS) [7].

This paper is organized as follows: to make the paper self-contained, we describe in Section 2 the CGS metaheuristic and the standard CGS-TSP algorithm; in Section 3, we combine CGS with local search, we introduce the new CGS-TSP variant that attaches confidence levels to recommendations and we report our experimental results; in Section 4 we conclude the paper and present future research directions.

2 The CGS Metaheuristic and Its Application to the TSP

In this section, we briefly describe the CGS metaheuristic and the CGS-TSP algorithm. We refer the reader to [5] for a more detailed presentation.

2.1 The Metaheuristic

CGS is a population-based method. An individual of the CGS population is a virtual person, which can simultaneously act both as a client and as a consultant. As a client, a virtual person constructs at each iteration a solution to the problem. As a consultant, a virtual person provides advice to clients, in accordance with its ***strategy***. Usually, at each step of the solution construction, there are several variants a client can choose from. The variant recommended by the consultant has a higher probability to be chosen, but the client may opt for one of the other variants, which will be selected based on some heuristic.

At the beginning of each iteration, a client chooses a consultant based on its ***personal preference*** and on the consultant's ***reputation***. The reputation of a consultant increases with the number of successes achieved by its clients. A client achieves a ***success***, if it constructs a solution better than all solutions found until that point by any client guided by the same consultant. Each time a client achieves a success, the consultant adjusts its strategy in order to reflect the sequence of decisions taken by the client. Because the reputation fades over time, a consultant needs that its clients constantly achieve successes, in order to keep its reputation. If the consultant's reputation sinks below a minimum value, it will take a ***sabbatical leave***, during which it will stop offering advice to clients and it will instead start searching for a new strategy to use in the future.

The pseudocode that formalizes the CGS metaheuristic is shown in Fig. 1. During the initialization phase (lines 2-5), virtual people are created and placed in sabbatical mode. Based on its mode, a virtual person constructs at each iteration either a solution to the problem (line 13) or a consultant strategy (line 9). A local optimization procedure (line 17) may be applied to improve this solution or consultant strategy.

After the construction phase, a virtual person in sabbatical mode checks if it has found a new best-so-far strategy (lines 20-22), while a virtual person in

```
 1. procedure CGSMetaheuristic()
 2.    create the set P of virtual persons
 3.    foreach p ∈ P do
 4.       setSabbaticalMode(p)
 5.    end foreach
 6.    while(termination condition not met) do
 7.       foreach p ∈ P do
 8.          if actionMode[p]=sabbatical then
 9.             currStrategy[p] ← constructStrategy(p)
10.          else
11.             currCons[p] ← chooseConsultant(p)
12.             if currCons[p] ≠ null then
13.                currSol[p] ← constructSolution(p, currCons[p])
14.             end if
15.          end if
16.       end foreach
17.       applyLocalOptimization()                          // optional
18.       foreach p ∈ P do
19.          if actionMode[p]=sabbatical then
20.             if currStrategy[p] better than bestStrategy[p] then
21.                bestStrategy[p] ← currStrategy[p]
22.             end if
23.          else
24.             c ← currCons[p]
25.             if c ≠ null and currSol[p] is better than all solutions
26.                            found by a client of c since last sabbatical then
27.                successCount[c] ← successCount[c]+1
28.                strategy[c] ← adjustStrategy(c, currSol[p])
29.             end if
30.          end if
31.       end foreach
32.       updateReputations()
33.       updateActionModes()
34.    end while
35. end procedure
```

Fig. 1. The CGS Metaheuristic

normal mode checks if it has achieved a success and, if this is the case, it adjusts its strategy accordingly (lines 24-29).

Reputations are updated based on the results obtained by clients (line 32): the reputation of a consultant is incremented each time one of its clients achieves a success and it receives an additional bonus when a client obtains a best-so-far result. Each consultant is ranked based on the best result obtained by any client working under its guidance. For a number of top-ranked consultants, CGS prevents their reputations from sinking below a predefined level.

Finally, the action mode of each virtual person is updated (line 33): consultants whose reputations have sunk below the minimum level are placed in sabbatical mode, while consultants whose sabbatical leave has finished are placed in normal mode.

2.2 The CGS-TSP Algorithm

The CGS-TSP algorithm introduced in [5] is an application of the CGS metaheuristic to the TSP. In this algorithm, the ***strategy*** of a consultant is represented by a tour, which it advertises to its clients. The heuristic used to build new strategies during the ***sabbatical leave*** is based on a pseudorandom proportional rule that strongly favors the nearest city. A virtual person k located at city i moves to a city j according to the following rule:

$$j = \begin{cases} argmin_{l \in \mathcal{N}_i^k} \{d_{il}\} & \text{if } a \leq a_0, \\ J & \text{otherwise.} \end{cases} \tag{1}$$

where: $\mathcal{N}_i^k$ is the feasible neighborhood of person k when being at city i; d_{il} is the distance between cities i and l; a is a random variable uniformly distributed in $[0, 1]$ and $a_0 \in [0, 1]$ is a parameter; J is a random variable selected according to the probability distribution given by formula (2), where β is a parameter.

$$p_{ij}^k = \frac{(1/d_{ij})^\beta}{\sum_{l \in \mathcal{N}_i^k} (1/d_{il})^\beta} \tag{2}$$

Based on the tour advertised by the consultant, a client receives at each step of the solution construction a recommendation regarding the next city to be visited. The client does not always follow the consultant's recommendation. Again, a pseudorandom proportional rule is used to decide which city to visit at the next step:

$$j = \begin{cases} v & \text{if } v \neq null \wedge q \leq q_0, \\ argmin_{l \in \mathcal{N}_i^k} \{d_{il}\} & \text{if } (v = null \vee q > q_0) \wedge b \leq b_0, \\ J & \text{otherwise.} \end{cases} \tag{3}$$

where: v is the city recommended by the consultant for the next step; q is a random variable uniformly distributed in $[0, 1]$ and q_0 ($0 \leq q_0 \leq 1$) is a parameter; $\mathcal{N}_i^k$ is the feasible neighborhood of person k when being at city i; d_{il} is the distance between cities i and l; b is a random variable uniformly distributed in $[0, 1]$ and b_0 is a parameter ($0 \leq b_0 \leq 1$); J is a random variable selected according to the probability distribution given by formula (2).

In CGS-TSP, the ***personal preference*** is given by the inverse of the advertised tour length. It is used by clients in conjunction with the ***reputation***, in order to compute the probability to choose a given consultant k:

$$p_k = \frac{reputation_k^\alpha \, preference_k^\gamma}{\sum_{c \in \mathcal{C}} (reputation_c^\alpha \, preference_c^\gamma)} \tag{4}$$

where $\mathcal{C}$ is the set of all available consultants and α and γ are parameters that determine the influence of reputation and personal preference.

At each iteration, the consultant's k reputation fades at rate r:

$$reputation_k \leftarrow reputation_k(1-r) \tag{5}$$

CGS-TSP adjusts its reputation fading rate according to the total number s_w of successes achieved during the last w iterations by the best $fadingRanks$ consultants, where w and $fadingRanks$ are parameters:

$$r = r_0 \left(1 + \frac{s_w}{\sqrt{1 + \left(\frac{s_w}{f}\right)^2}} \right) \tag{6}$$

The parameter r_0 gives the reputation fading rate for the case where no successes were achieved by the best $fadingRanks$ consultants during the last w iterations. The value of f is computed from the value of another parameter, k_w, which indicates how much greater is the decrease in reputation for a very high number of successes than the decrease in the case when no successes were achieved:

$$f = \left(\frac{1}{r_0} - 1\right)\left(1 - \frac{1}{\sqrt[w]{k_w}}\right) \tag{7}$$

3 CGS with Local Search for the TSP

In this section, we present the details of combining CGS-TSP with local search and we propose a new variant of this algorithm, which introduces the concept of confidence in relation to the recommendations made by consultants. Then, we describe the experimental setting and we compare the results obtained by the competing algorithms considered.

3.1 Applying Local Search to CGS-TSP

Since the CGS metaheuristic provides an optional local optimization step, combining CGS-TSP with local search is a straightforward process. Local search improves the results, but it is a time-consuming procedure. Therefore, significantly fewer iterations can be performed in the same amount of time when the algorithm is combined with local search. For this reason, different parameter settings are appropriate in this case. The standard CGS-TSP algorithm fixes the value of the parameter w to 1000 and the *sabbaticalDuration* to 100 iterations. In our experiments with local search we fix the value of w to 100 and the *sabbaticalDuration* to $500/m$ iterations, where m is the number of virtual persons. For the other parameters with fixed values in the standard CGS-TSP, we preserve the original values when combining the algorithm with local search: $a_0 = 0.9$ and $minReputation = 1$.

3.2 CGS-TSP with Confidence

In addition to the algorithm described in the previous subsection, we propose a variant of the CGS-TSP algorithm, which we refer to as CGS-TSP-C, where each arc in the tour advertised by a consultant has an associated ***strength***. Strengths are updated each time the consultant adjusts its strategy. If an arc in the new advertised tour was also present in the old advertised tour, its strength will be incremented; otherwise, its strength is set to 0. The strength of an arc could be interpreted as the consultant's confidence in recommending this arc to a client. A client is more likely to accept recommendations made with greater confidence. This idea is expressed in CGS-TSP-C by allowing the value of the parameter q_0 to vary in a given range, at each construction step:

$$q_0 = \begin{cases} q_{min} + s \cdot \frac{q_{max} - q_{min}}{s_{max}} & \text{if } s < s_{max}, \\ q_{max} & \text{otherwise.} \end{cases} \tag{8}$$

where s is the strength of the recommended arc and q_{min}, q_{max} and s_{max} are parameters.

3.3 Experimental Setup

To allow a meaningful comparison between heuristics, we have created a software package containing Java implementations of the algorithms considered in our experiments. The software package is available as an open source project at `http://swarmtsp.sourceforge.net/`. At this address, we also provide all configuration files, problem instances and results files for the parameter tuning and for the experiments described in this paper, as well as the outcome of other experiments briefly mentioned in this paper, but not further detailed due to space limitations.

We run a series of experiments in order to compare the performance of CGS-TSP and CGS-TSP-C with that of Ant Colony System (ACS) [2] and MAX-MIN Ant System (MMAS) [7]. We combine all algorithms used in our experiments with 3-opt local search and we use candidate lists of length 20 for all algorithms. Each run is terminated after $n/50$ seconds CPU time, where n is the problem size.

We have tuned the parameters of all algorithms through the ParamILS [4] and F-Race [1] procedures. As training set, we have used 600 generated Euclidean TSP instances, with the number of cities uniformly distributed in the interval $[1000, 2000]$. For CGS-TSP and CGS-TSP-C, the parameter settings are given in Table 1.

The best configuration found for ACS is: $m = 12$, $\rho = 0.6$, $\xi = 0.4$, $q_0 = 0.98$, $\beta = 2$. For MMAS, the best configuration found is: $m = max(10, 44 - 0.175 \cdot n)$, $\rho = 0.15$, $\alpha = 2$, $\beta = 2$. Some of these values differ significantly from the standard values given in [3, p.96], which are: $m = 10$, $\rho = 0.1$, $\xi = 0.1$, $q_0 = 0.98$, $\beta = 2$ for ACS and: $m = 25$, $\rho = 0.2$, $\alpha = 1$, $\beta = 2$ for MMAS. For this reason, for ACS and MMAS, we have performed our experiments using both the tuned and

Table 1. Parameter settings for CGS-TSP and CGS-TSP-C

Parameter	CGS-TSP	CGS-TSP-C	Description
m	$max(3, 21 - n/125)$	$max(3, 16 - n/250)$	number of virtual persons
b_0	0.98	0.95	*see formula* (3)
q_0	0.98	$\begin{cases} q_{min} = 0.8 \\ q_{max} = 0.99 \\ s_{max} = 3 \end{cases}$	*see formulas* (3), (8)
α	7	7	reputation's relative influence
β	12	12	heuristic's relative influence
γ	7	8	result's relative influence
maxReputation	40	50	maximum reputation value
initialReputation	6	3	reputation after sabbatical
bonus	8	6	best-so-far reputation bonus
protectedRanks	$0.7 \cdot m$	$0.6 \cdot m$	protected top consultants
r_0	$3 \cdot 10^{-7}$	$3 \cdot 10^{-6}$	basic reputation fading rate
fadingRanks	2	10	top consultants for fading rate
k_w	3	30	reputation decrease factor

the standard values. As shown in the next subsection, the results obtained using the tuned parameters outperform those obtained with the standard settings.

3.4 Experimental Results

We have applied the algorithms to 27 symmetric instances from the TSPLIB benchmark library, with the number of cities between 654 and 3038. Table 2 reports for each algorithm and TSP instance the best and mean percentage deviations from the optimal solutions over 25 trials. The best mean results for each problem are in boldface. We also report for each problem the p-values of the one-sided Wilcoxon rank sum tests for the null hypothesis (H_0) that there is no difference between the solution quality of CGS-TSP-C and that of the competing ACO algorithm, and for the alternative hypothesis (H_1) that CGS-TSP-C outperforms the considered algorithm. Applying the Bonferroni correction for multiple comparisons, we obtain the adjusted α-level: $0.05/27 = 0.00185$. The p-values in boldface indicate the cases where the null hypothesis is rejected at this significance level.

For a few pairs of algorithms, we use the one-sided Wilcoxon signed rank test to compute the p-values for the null hypothesis (H_0) that there is no difference between the means of the first and the means of the second algorithm considered, and the alternative hypothesis (H_1) that the means of the first algorithm are smaller than the means of the second algorithm considered. The p-values are given in Table 3.

The null hypothesis can be rejected at a high significance level when CGS-TSP-C is compared with the two ACO algorithms, which means that for runs terminated after n/50 seconds CPU time, CGS-TSP-C clearly outperforms both

Table 2. Performance over 25 trials. Runs are terminated after $n/50$ CPU seconds.

Problem instance	ACS		MMAS		CGS-TSP		CGS-TSP-C			
	Best (%)	Mean (%)	Best (%)	Mean (%)	Best (%)	Mean (%)	Best (%)	Mean (%)	p-value (ACS)	p-value (MMAS)
p654	0.000	0.023	0.000	0.062	0.000	**0.009**	0.000	**0.009**	**0.0004**	**0.0000**
d657	0.002	0.167	0.031	0.133	0.002	0.129	0.002	**0.097**	0.0029	0.0326
gr666	0.056	0.159	0.000	**0.066**	0.040	0.069	0.000	0.109	0.0142	0.8059
u724	0.026	0.102	0.045	0.151	0.005	0.128	0.005	**0.101**	0.5932	0.0070
rat783	0.000	0.252	0.023	0.185	0.000	0.215	0.000	**0.147**	0.0157	0.1564
dsj1000	0.038	0.403	0.133	0.316	0.030	0.296	0.000	**0.224**	0.0042	0.1315
pr1002	0.000	0.273	0.034	0.201	0.000	0.189	0.000	**0.162**	0.0203	0.0863
si1032	0.000	**0.023**	0.000	0.035	0.000	0.029	0.000	0.024	0.6031	0.1758
u1060	0.116	0.331	0.104	0.359	0.026	**0.117**	0.026	0.119	**0.0000**	**0.0000**
vm1084	0.000	**0.064**	0.001	0.120	0.000	0.066	0.000	0.071	0.5096	**0.0012**
pcb1173	0.002	0.325	0.021	**0.219**	0.185	0.449	0.002	0.297	0.2456	0.9376
d1291	0.000	0.111	0.000	**0.105**	0.000	0.108	0.000	0.186	0.3690	0.4518
rl1304	0.000	0.196	0.000	0.229	0.000	0.200	0.000	**0.189**	0.5291	0.0243
rl1323	0.077	0.243	0.041	0.245	0.000	**0.131**	0.010	0.152	**0.0005**	**0.0000**
nrw1379	0.088	**0.256**	0.305	0.486	0.083	0.286	0.152	0.264	0.5668	**0.0000**
fl1400	0.020	0.247	0.298	0.668	0.000	0.190	0.000	**0.177**	0.0046	**0.0000**
u1432	0.218	**0.389**	0.250	0.601	0.292	0.481	0.153	0.426	0.8654	**0.0000**
fl1577	0.031	0.293	0.220	0.597	0.004	**0.060**	0.004	0.172	**0.0001**	**0.0000**
d1655	0.006	0.435	0.019	0.325	0.064	0.332	0.000	**0.322**	0.0754	0.4446
vm1748	0.113	0.272	0.154	0.471	0.061	0.191	0.004	**0.159**	**0.0000**	**0.0000**
u1817	0.192	0.480	0.080	**0.295**	0.156	0.386	0.107	0.347	0.0043	0.9048
rl1889	0.345	0.719	0.270	0.545	0.004	0.237	0.000	**0.217**	**0.0000**	**0.0000**
d2103	0.017	0.332	0.040	0.127	0.000	0.345	0.000	**0.047**	**0.0000**	**0.0000**
u2152	0.112	0.426	0.254	0.488	0.115	0.356	0.131	**0.352**	0.0472	**0.0015**
u2319	0.287	**0.372**	0.427	0.599	0.707	0.863	0.742	1.052	1.0000	1.0000
pr2392	0.235	0.507	0.214	0.542	0.019	0.441	0.051	**0.340**	**0.0001**	**0.0012**
pcb3038	0.243	**0.499**	0.671	0.940	0.704	1.056	0.428	0.810	1.0000	**0.0003**

ACS and MMAS. In addition, CGS TSP C outperforms CGS-TSP, which means that the use of confidence in relation to the recommendations made by consultants can lead to better results.

As shown in [5], CGS-TSP clearly outperforms ACS and MMAS in experiments without local search. Combined with 3-opt local search, CGS-TSP still outperforms ACS and MMAS, but only at a moderate significance level. This means that ACS and MMAS benefit more than CGS-TSP from the hybridization with local search. One explanation for this discrepancy could be that the solution construction mechanism of CGS-TSP already bears some resemblance to a local search procedure: a client builds a solution in the neighborhood of the solution promoted by a consultant, while the consultant updates its promoted solution each time one of its clients finds a better one.

Table 3. Performance comparison using the one-sided Wilcoxon signed rank test

First algorithm	Second algorithm	p-value
CGS-TSP-C	ACS	0.00472
CGS-TSP-C	MMAS	0.00148
CGS-TSP-C	CGS-TSP	0.02004
CGS-TSP (without local search)	ACS (without local search)	* 0.00003
CGS-TSP	ACS	0.05020
CGS-TSP (without local search)	MMAS (without local search)	* 0.00269
CGS-TSP	MMAS	0.03051
ACS	ACS (standard settings)	0.00256
MMAS	MMAS (standard settings)	< 0.00001

* p-values taken from [5].

As mentioned in the previous subsection, some of the parameter values found during the tuning phase for ACS and MMAS differ significantly from the values recommended in [3, p.96]. The last two lines of Tab. 3 compare the performance obtained using the tuned parameters and the standard settings. The tuned algorithms clearly outperform the algorithms that use standard settings, thus confirming the effectiveness of the tuning procedures.

Figure 2 shows the development of the mean percentage deviations from the optimum over 25 trials as a function of the CPU time, over 10000 seconds. We consider u1060, for which the CGS algorithms have obtained good results in the previous experiments, and u2319, for which poor results have been obtained. For u1060, the CGS algorithms outperform the ACO algorithms during the entire interval. Although CGS-TSP-C performs best in the initial phases, it is outperformed by CGS-TSP in the long run. In the case of u2319, the CGS algorithms are not able to reach the performance of ACO.

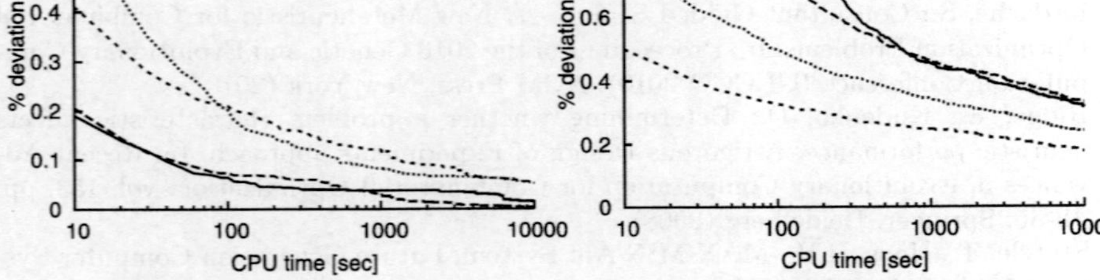

Fig. 2. The development of the mean percentage deviations over 25 trials

4 Conclusions and Future Work

Experimental results indicate that CGS is still able to compete with ACO when the algorithms are combined with local search, although the performance improvement in the case of CGS is not as significant as in the case of ACO. The CGS-TSP-C algorithm proposed in this paper shows that correlating the recommendations with a level of confidence may improve the results. Still, more research is needed in order to determine in which cases CGS-TSP-C should be preferred to CGS-TSP.

Although for our experimental setup the CGS algorithms generally outperform the ACO algorithms, there are a few cases where the performance of CGS is relatively poor. The most striking example in our experiments is the TSP instance u2319. It is therefore worthwhile to investigate which characteristics of the TSP instances influence the performance of the CGS algorithms and how these characteristics relate to the values of the different parameters used by these algorithms. In the case of ACO, it has been shown [6] that an increase in the standard deviation of the cost matrix of TSP instances leads to a decrease in the performance of the algorithm. Like in ACO, the decisions taken in the solution construction phase of CGS algorithms depend on the relative lengths of edges in the TSP. Therefore, we expect that the standard deviation of the cost matrix also affects the performance of these algorithms.

A drawback of CGS algorithms is the large number of parameters to be tuned. We plan to devise a variant of CGS-TSP with only a small number of parameters. One way to achieve this is to identify parameters whose optimal value is not problem specific and to remove these parameters by hard-coding their values in the algorithm.

References

1. Birattari, M., Stützle, T., Paquete, L., Varrentrapp, K.: A racing algorithm for configuring metaheuristics. In: Proceedings of GECCO 2002, pp. 11–18 (2002)
2. Dorigo, M., Gambardella, L.M.: Ant Colony System: A Cooperative Learning Approach to the Traveling Salesman Problem. IEEE Transactions on Evolutionary Computation 1(1), 53–66 (1997)
3. Dorigo, M., Stützle, T.: Ant Colony Optimization. MIT Press, Cambridge (2004)
4. Hutter, F., Hoos, H.H., Leyton-Brown, K., Stützle, T.: ParamILS: An Automatic Algorithm Configuration Framework. Journal of Artificial Intelligence Research (JAIR) 36, 267–306 (2009)
5. Iordache, S.: Consultant-Guided Search - A New Metaheuristic for Combinatorial Optimization Problems. In: Proceedings of the 2010 Genetic and Evolutionary Computation Conference (GECCO 2010). ACM Press, New York (2010)
6. Ridge, E., Kudenko, D.: Determining whether a problem characteristic affects heuristic performance. A rigorous Design of Experiments approach. In: Recent Advances in Evolutionary Computation for Combinatorial Optimization, vol. 153, pp. 21–35. Springer, Heidelberg (2008)
7. Stützle, T., Hoos, H.H.: MAX-MIN Ant System. Future Generation Computer Systems 16(8), 889–914 (2000)

Many-Objective Test Problems to Visually Examine the Behavior of Multiobjective Evolution in a Decision Space

Hisao Ishibuchi, Yasuhiro Hitotsuyanagi, Noritaka Tsukamoto, and Yusuke Nojima

Department of Computer Science and Intelligent Systems, Graduate School of Engineering, Osaka Prefecture University, 1-1 Gakuen-cho, Naka-ku, Sakai, Osaka 599-8531, Japan
{hisaoi@,hitotsu@ci,nori@ci,nojima@}cs.osakafu-u.ac.jp

Abstract. Many-objective optimization is a hot issue in the EMO (evolutionary multiobjective optimization) community. Since almost all solutions in the current population are non-dominated with each other in many-objective EMO algorithms, we may need a different fitness evaluation scheme from the case of two and three objectives. One difficulty in the design of many-objective EMO algorithms is that we cannot visually observe the behavior of multiobjective evolution in the objective space with four or more objectives. In this paper, we propose the use of many-objective test problems in a two- or three-dimensional decision space to visually examine the behavior of multiobjective evolution. Such a visual examination helps us to understand the characteristic features of EMO algorithms for many-objective optimization. Good understanding of existing EMO algorithms may facilitates their modification and the development of new EMO algorithms for many-objective optimization.

Keywords: Evolutionary multiobjective optimization (EMO), many-objective optimization, multiobjective optimization problems, test problems.

1 Introduction

Evolutionary multiobjective optimization (EMO) has been a very active research area in the field of evolutionary computations [3], [5], [26]. A number of EMO algorithms have been proposed and successfully applied to various application tasks [1], [17], [18], [20]. Whereas well-known and frequently-used Pareto dominance-based EMO algorithms such as NSGA-II [6] and SPEA2 [30] work well on two-objective problems, their search ability is often severely degraded by the increase in the number of objectives as pointed out in the literature [4], [9]-[12], [19], [21], [23]-[25], [27], [32].

In the case of many-objective optimization, it is not easy to understand the behavior of multiobjective evolution by EMO algorithms. This is because we cannot visually monitor how a population of solutions is evolved in a high-dimensional objective space. This contrasts to the case of two objectives where we can visually show all solutions at each generation in a two-dimensional objective space in order to examine the move of a population from the initial generation to the final one. Such a visual examination helps us to understand the characteristic features of EMO algorithms such as the convergence-diversity balance and the uniformity of solutions along the Pareto

R. Schaefer et al. (Eds.): PPSN XI, Part II, LNCS 6239, pp. 91–100, 2010.

front. Better understanding of existing EMO algorithms may facilitates their modification and the development of new algorithms for many-objective optimization.

In this paper, we propose the use of many-objective test problems in a two- or three-dimensional decision space to visually examine the behavior of multiobjective evolution. A class of our test problems can be written in the following generic form:

$$\text{Minimize } \mathbf{f}(\mathbf{x}) = (f_1(\mathbf{x}),\ f_2(\mathbf{x}),\ ...,\ f_k(\mathbf{x}))\,, \tag{1}$$

where $\mathbf{x}$ is a two- or three-dimensional decision vector (i.e., a point on a two- or three-dimensional decision space) and $f_i(\mathbf{x})$ is defined by the minimum distance from $\mathbf{x}$ to m points $\mathbf{a}_{i1}$, $\mathbf{a}_{i2}$, ..., $\mathbf{a}_{im}$ in the decision space:

$$f_i(\mathbf{x}) = \min\{\text{dis}(\mathbf{x}, \mathbf{a}_{i1}), \text{dis}(\mathbf{x}, \mathbf{a}_{i2}), ..., \text{dis}(\mathbf{x}, \mathbf{a}_{im})\}\,,\ i = 1, 2, ..., k. \tag{2}$$

In this formulation, dis($\mathbf{x}$, $\mathbf{a}$) is a distance between the two points $\mathbf{x}$ and $\mathbf{a}$. We assume the use of the Euclidean distance throughout this paper.

Since m points ($\mathbf{a}_{i1}$, $\mathbf{a}_{i2}$, ..., $\mathbf{a}_{im}$) are used to define each objective $f_i(\mathbf{x})$, we need km points to define a k-objective test problem. As shown in this paper, we can generate various types of test problems in a two- or three-dimensional decision space using different combinations of those km points. For example, some test problems have small Pareto optimal regions while others have large ones. Some test problems have multiple equivalent Pareto optimal regions while others have disconnected ones. Two examples of our test problems are shown in Fig. 1. Fig. 1 (a) is a four-objective problem with a single rectangular Pareto optimal region (shaded area) while Fig. 1 (b) is a four-objective problem with four equivalent square Pareto optimal regions.

In this paper, first we briefly review related studies on many-objective test problems in Section 2. Next we show some interesting experimental results on our test problems with $m = 1$ (i.e., with a single Pareto optimal region) in Section 3. Then we discuss our test problems with $m > 1$ (i.e., with multiple Pareto optimal regions) and explain their usefulness in Section 4. Finally we conclude this paper in Section 5.

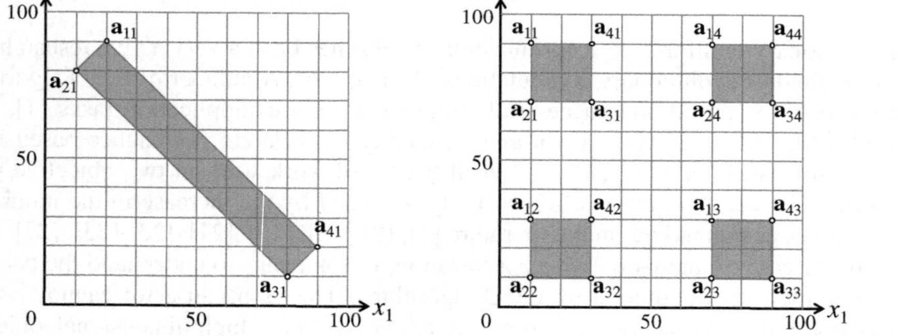

(a) Four-objective problem (m = 1 and k = 4). (b) Four-objective problem (m = 4 and k = 4).

Fig. 1. Two examples of our test problems in Eq. (1) and Eq. (2)

2 Related Studies on Many-Objective Test Problems

One of the most frequently-used test problems in the EMO community is ZDT [29]. This is a set of six two-objective problems (ZDT1 to ZDT6). For many-objective optimization, seven test problems were proposed by Deb et al. [7], which are called DTLZ (DTLZ1 to DTLZ7). The main feature of the DTLZ problems is its scalability: The number of objectives can be arbitrarily specified. Two problems were added by Deb et al. [8]. Nine DTLZ problems have been frequently used in the literature [16].

Multiobjective 0/1 knapsack problems have also been used in many studies since Zitzler & Thiele [31]. They used nine problems with 250, 500 and 750 items and two, three and four objectives. In some studies [15], [16], [23], knapsack problems with more than four objectives have been generated to examine the performance of EMO algorithms for many-objective optimization. Other combinatorial optimization problems with many objectives (e.g., TSP [4], nurse scheduling [25], and job-shop scheduling [4]) have been also used as test problems in the literature [16].

The use of many-objective test problems in a two-dimensional decision space was proposed by Köppen & Yoshida [21]. They used a single regular polygon for problem definition. Thus their test problems can be viewed as a special case of our formulation with $m = 1$ (i.e., with a single Pareto optimal region). Singh et al. [24] used the same test problems as [21] to examine the performance of many-objective EMO algorithms. Some of our experiments in this paper have been motivated by [21] and [24].

On the other hand, Rudolph et al. [22] used two-objective test problems with multiple equivalent Pareto optimal subsets in a two-dimensional decision space. Each Pareto optimal subset was defined by two points as a line (or a curve) in the decision space. Thus their problems can be viewed as a special case of our formulation with $k = 2$ (i.e., our formulation is a general form of their test problems).

3 Results on Test Problems with a Single Pareto Region (m = 1)

As in Köppen & Yoshida [21], our test problems with a single Pareto optimal region (i.e., our test problems with $m = 1$) can be used to examine the distribution of solutions in a decision space for many-objective optimization. In our computational experiments, we used the following four EMO algorithms: NSGA-II [6], SPEA2 [30], MOEA/D [28] with the Tchebycheff (Chebyshev) function, and SMS-EMOA [27]. The first two are well-known and frequently-used Pareto dominance-based EMO algorithms. The other are recently-developed high-performance EMO algorithms with different fitness evaluation schemes: Scalarizing functions are used in MOEA/D for fitness evaluation while the hypervolume measure is used in SMS-EMOA.

First we applied these EMO algorithms to a five-objective problem with five points at the vertices of a regular pentagon using the following setting:

Population size: 200 (NSGA-II, SPEA2) and 210 (MOEA/D),
Total number of examined solutions (Termination conditions): 100,000,
Crossover probability: 1.0 (SBX with $\eta_c = 15$),
Mutation probability: 0.5 (Polynomial mutation with $\eta_m = 20$),

Reference point: Minimum value of each objective (MOEA/D)
Maximum value of each objective × 1.1 (SMS-EMOA).

In MOEA/D, the population size is the same as the number of weight vectors. Due to the combinatorial nature of uniformly distributed weight vectors, the population size cannot be arbitrarily specified (for details, see [28]). We used the closest integer to 200 among the possible values as the population size. The neighborhood size in MOEA/D was specified as 10% of the population size. The same termination condition (i.e., the examination of 100,000 solutions) was used for all algorithms whereas the computation time of SMS-EMOA was much longer than the other algorithms.

In Fig. 2, we show the final population in a single run of each algorithm. All points in the regular pentagon are Pareto optimal solutions. We can observe different characteristic features of each EMO algorithm in Fig. 2.

We can also generate test problems for examining both the convergence and the distribution of solutions. We show an example of such a test problem in Fig. 3 where a four-objective test problem was defined by four vertices of a long and thin rectangle.

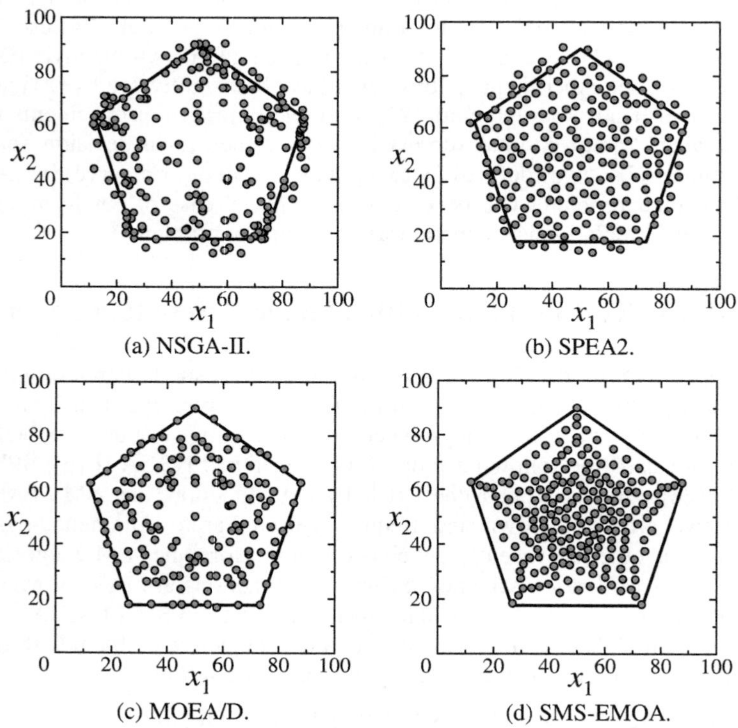

Fig. 2. The final population of a single run of each algorithm on the five-objective problem

In the same manner as in Fig. 2, we applied the EMO algorithms to the four-objective test problem in Fig. 3. Each plot of Fig. 3 shows the final population in a single run of each algorithm. The Pareto optimal region is the inside of the slender rectangle. It looks difficult for NSGA-II and SPEA2 to converge all solutions into the Pareto optimal region (i.e., inside the slender rectangle including the boundary). On the other hand, good distributions of solutions were not obtained by MOEA/D.

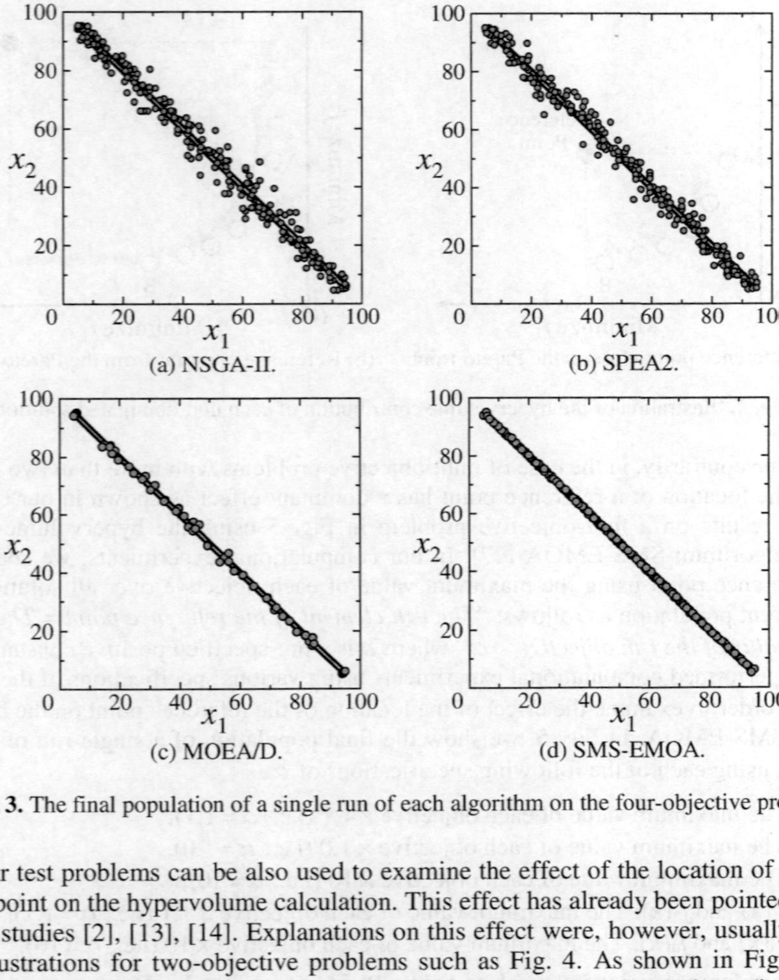

Fig. 3. The final population of a single run of each algorithm on the four-objective problem

Our test problems can be also used to examine the effect of the location of a reference point on the hypervolume calculation. This effect has already been pointed out in some studies [2], [13], [14]. Explanations on this effect were, however, usually based on illustrations for two-objective problems such as Fig. 4. As shown in Fig. 4, the hypervolume contribution of the two extreme non-dominated solutions (i.e., non-dominated solutions with the best value for either objective: Points A and B in Fig. 4) strongly depends on the location of the reference point. The two plots in Fig. 4 show the same non-dominated solution set with different reference points. When the reference point is far from the Pareto front as in Fig. 4 (b), the two extreme solutions A and B have large hypervolume contributions as indicated by the two large shaded rectangles. On the other hand, if the reference point is close to the Pareto front as in Fig. 4 (a), the two extreme solutions A and B have small hypervolume contributions as indicated by the two small shaded rectangles. It should be noted that the hypervolume contribution of each of the other solutions is independent of the location of a reference point. Since the two extreme non-dominated solutions of a two-objective problem usually have the highest fitness values in most EMO algorithms, the location of a reference point has not a large effect on hypervolume-based EMO algorithms.

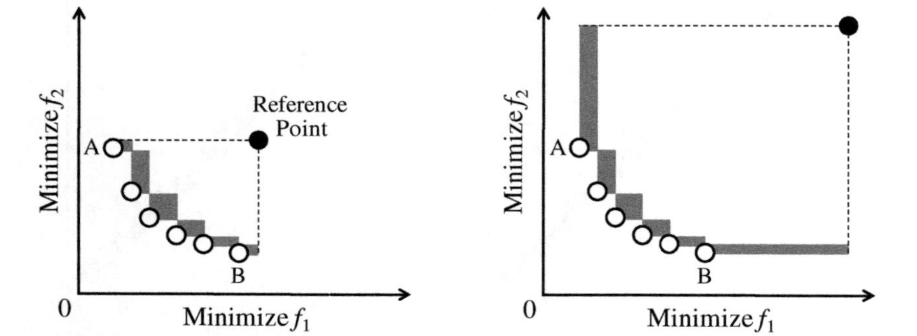

(a) Reference point close to the Pareto front. (b) Reference point far from the Pareto front.

Fig. 4. Illustration of the hypervolume contribution of each non-dominated solution

On the contrarily, in the case of multiobjective problems with more than two objectives, the location of a reference point has a dominant effect as shown in our experimental results on a four-objective problem in Fig. 5 using the hypervolume-based EMO algorithm: SMS-EMOA [27]. In our computational experiments, we specified the reference point using the maximum value of each objective over all solutions in the current population as follows: "*The i-th element of the reference point* = *The maximum value of the i-th objective* × α" where α is a pre-specified positive constant.

We performed computational experiments using various specifications of the value of α in order to examine the effect of the location of the reference point on the behavior of SMS-EMOA. In Fig. 5, we show the final population of a single run of SMS-EMOA using each of the following specifications of α:

(a) The maximum value of each objective × 1.1 (i.e., $\alpha = 1.1$),
(b) The maximum value of each objective × 1.0 (i.e., $\alpha = 1.0$),
(c) The maximum value of each objective × 10 (i.e., $\alpha = 10$),
(d) $f_1(\mathbf{x})$ and $f_2(\mathbf{x})$: The maximum value of each objective × 1.1 (i.e., $\alpha = 1.1$),
$f_3(\mathbf{x})$ and $f_4(\mathbf{x})$: The maximum value of each objective × 10 (i.e., $\alpha = 10$).

When a reference point is too close to the Pareto front, good solution sets were not obtained as shown in Fig. 5 (b). Good result was obtained in Fig. 5 (a) with $\alpha = 1.1$. By increasing the value of α (i.e., by increasing the distance of a reference point to the Pareto front), solutions moved to the lines between two points as shown in Fig. 5 (c). In Fig. 5 (d), $f_3(\mathbf{x})$ and $f_4(\mathbf{x})$ are distances from a solution $\mathbf{x}$ to the right and bottom points, respectively (while $f_1(\mathbf{x})$ and $f_2(\mathbf{x})$ are distances from a solution $\mathbf{x}$ to the left and top points, respectively). Many solutions are along the line between the left and top points (i.e., $\mathbf{a}_{11}$ and $\mathbf{a}_{21}$) for which the smaller value of α was used in our computational experiment in Fig 5 (d).

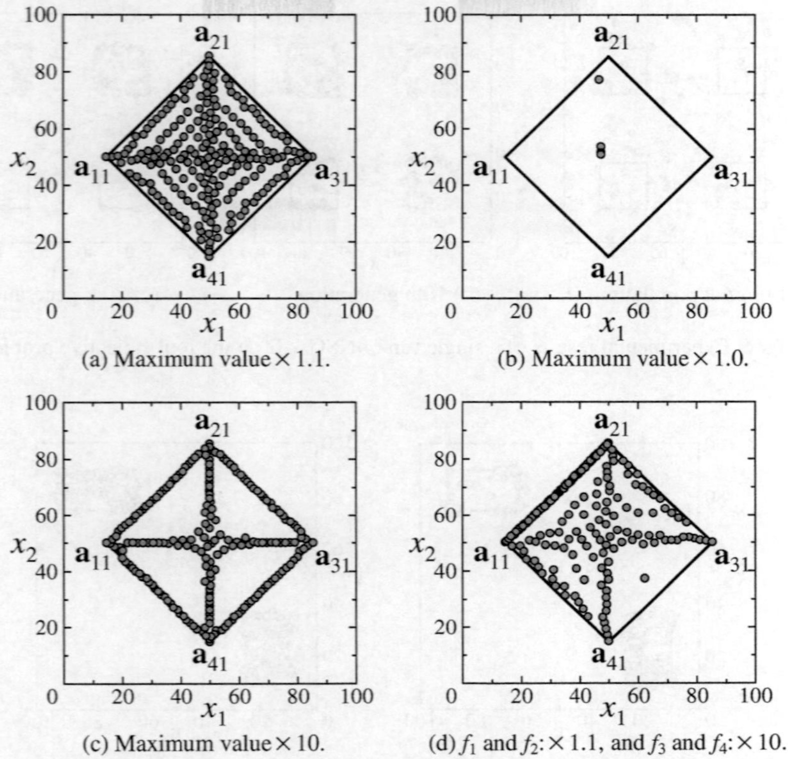

Fig. 5. Experimental results of SMS-EMOA with different specifications of a reference point

4 Results on Test Problems with Multiple Pareto Regions ($m > 1$)

Using multiple polygons with the same shape, we can generate multiobjective problems with multiple equivalent Pareto optimal regions as shown in Fig. 1 (b) in Section 1. In Fig. 6, we show experimental results of a single run of NSGA-II on the four-objective problem in Fig. 1 (b). Fig. 6 shows a randomly generated initial population (a) and two intermediate populations (b) and (c). From the three plots in Fig. 6, we can see that every solution quickly moved to one of the four squares within the first 10 generations. Then they continued to move in the four squares. We performed computational experiments many times. In some runs, solutions converged to one or two squares. In other runs, all the four squares had at least one solution even after 500 generations. That is, final results were totally different in each run.

We can also generate test problems with disconnected Pareto regions by using multiple polygons with different shapes. In Fig. 7, we show experimental results on such a test problem. Each plot of Fig. 7 is the final population in a single run of each algorithm on the four-objective test problem with two rectangles. Since the two rectangles in Fig. 7 are not equivalent, solutions did not converge into one rectangle.

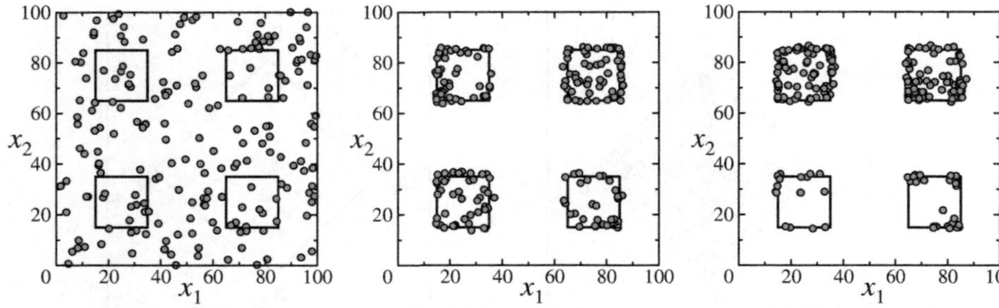

(a) Initial population. (b) 10th generation. (c) 20th generation.

Fig. 6. Experimental results of a single run of NSGA-II on the four-objective problem

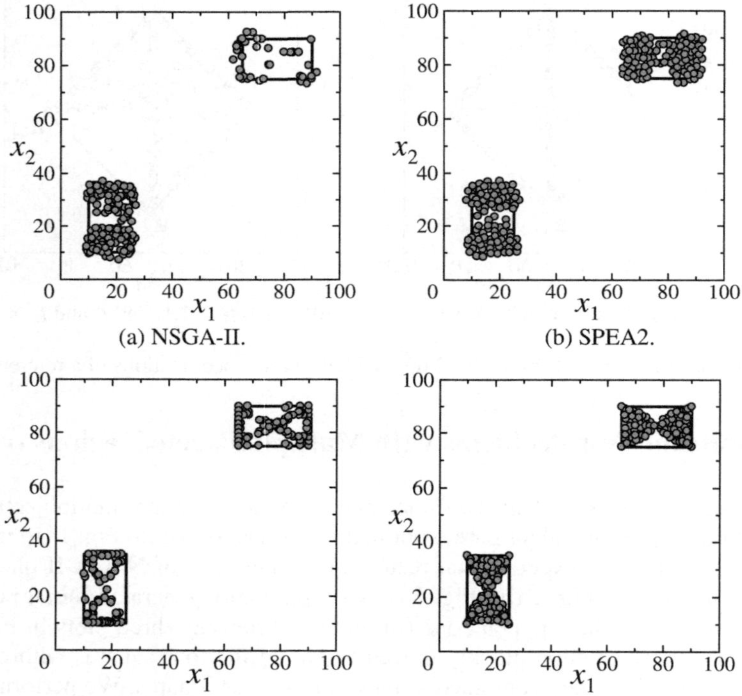

(a) NSGA-II. (b) SPEA2. (c) MOEA/D. (d) SMS-EMOA.

Fig. 7. Results of a single run on a four-objective problem with disconnected Pareto regions

Since the distribution in the decision space has not been taken into account in the design of almost all EMO algorithms, our intention is not to say which EMO algorithm is the best using our test problems but to visually examine the behavior of each EMO algorithm for many-objective optimization problems. However, performance measures for solution sets in the decision space may be an interesting research issue.

5 Conclusions

We proposed the use of many-objective test problems in a two- or three-dimensional decision space in order to visually examine multiobjective evolution for many-objective problems. Our test problems can be viewed as a generalized version of single polygon problems of Köppen & Yoshida [21] and multi-line (or multi-curve) problems of Rudolph et al. [22]. It is the main advantage of our test problems (and test problems in [21], [22]) that we can visually examine multiobjective evolution in the decision space. Whereas we generated test problems in a two-dimensional decision space for visual examination, it is also easy to generate test problems in a high-dimensional space by specifying multiple points with the required dimensionality.

References

1. Abbass, H.A., Alam, S., Bender, A.: MEBRA: Multiobjective Evolutionary-Based Risk Assessment. IEEE Computational Intelligence Magazine 4, 29–36 (2009)
2. Auger, A., Bader, J., Brockhoff, D., Zitzler, E.: Theory of the Hypervolume Indicator: Optimal μ-Distributions and the Choice of the Reference Point. In: Foundations of Genetic Algorithms: FOGA 2009, pp. 87–102 (2009)
3. Coello, C.A.C., Lamont, G.B.: Applications of Multi-Objective Evolutionary Algorithms. World Scientific, Singapore (2004)
4. Corne, D., Knowles, J.: Techniques for Highly Multiobjective Optimization: Some Non-Dominated Points are Better Than Others. In: Proc. of 2007 Genetic and Evolutionary Computation Conference, pp. 773–780 (2007)
5. Deb, K.: Multi-Objective Optimization Using Evolutionary Algorithms. John Wiley & Sons, Chichester (2001)
6. Deb, K., Pratap, A., Agarwal, S., Meyarivan, T.: A Fast and Elitist Multiobjective Genetic Algorithm: NSGA-II. IEEE Trans. on Evolutionary Computation 6, 182–197 (2002)
7. Deb, K., Thiele, L., Laumanns, M., Zitzler, E.: Scalable Multi-Objective Optimization Test Problems. In: Proc. of 2002 IEEE Congress on Evolutionary Computation, pp. 825–830 (2002)
8. Deb, K., Thiele, L., Laumanns, M., Zitzler, E.: Scalable Test Problems for Evolutionary Multiobjective Optimization. In: Abraham, A., Jain, L.C., Goldberg, R. (eds.) Evolutionary Multiobjective Optimization: Theoretical Advances and Applications, pp. 105–145. Springer, Heidelberg (2005)
9. Fleming, P.J., Purshouse, R.C., Lygoe, R.J.: Many-Objective Optimization: An Engineering Design Perspective. In: Coello Coello, C.A., Hernández Aguirre, A., Zitzler, E. (eds.) EMO 2005. LNCS, vol. 3410, pp. 14–32. Springer, Heidelberg (2005)
10. Hughes, E.J.: Evolutionary Many-Objective Optimization: Many Once or One Many? In: Proc. of 2005 IEEE Congress on Evolutionary Computation, pp. 222–227 (2005)
11. Hughes, E.J.: MSOPS-II: A General-Purpose Many-Objective Optimizer. In: Proc. of 2007 IEEE Congress on Evolutionary Computation, pp. 3944–3951 (2007)
12. Ishibuchi, H., Hitotsuyanagi, Y., Nojima, Y.: Scalability of Multiobjective Genetic Local Search to Many-Objective Problems: Knapsack Problem Case Studies. In: Proc. of 2008 IEEE Congress on Evolutionary Computation, pp. 3587–3594 (2008)
13. Ishibuchi, H., Nojima, Y., Doi, T.: Comparison between Single-Objective and Multi-Objective Genetic Algorithms: Performance Comparison and Performance Measures. In: Proc. of 2006 IEEE Congress on Evolutionary Computation, pp. 3959–3966 (2006)
14. Ishibuchi, H., Sakane, Y., Tsukamoto, N., Nojima, Y.: Single-Objective and Multi-Objective Formulations of Solution Selection for Hypervolume Maximization. In: Proc. of 2009 Genetic and Evolutionary Computation Conference, pp. 1831–1832 (2009)

15. Ishibuchi, H., Tsukamoto, N., Hitotsuyanagi, Y., Nojima, Y.: Effectiveness of Scalability Improvement Attempts on the Performance of NSGA-II for Many-Objective Problems. In: Proc. of 2008 Genetic and Evolutionary Computation Conference, pp. 649–656 (2008)
16. Ishibuchi, H., Tsukamoto, N., Nojima, Y.: Evolutionary Many-Objective Optimization: A Short Review. In: Proc. of IEEE Congress on Evolutionary Computation, pp. 2424–2431 (2008)
17. Jeong, S., Hasegawa, S., Shimoyama, K., Obayashi, S.: Development and Investigation of Efficient GA/PSO-HYBRID Algorithm Applicable to Real-World Design Optimization. IEEE Computational Intelligence Magazine 4, 36–44 (2009)
18. Jin, Y., Sendhoff, B.: A Systems Approach to Evolutionary Multiobjective Structural Optimization and Beyond. IEEE Computational Intelligence Magazine 4, 62–76 (2009)
19. Khara, V., Yao, X., Deb, K.: Performance Scaling of Multi-Objective Evolutionary Algorithms. In: Fonseca, C.M., Fleming, P.J., Zitzler, E., Deb, K., Thiele, L. (eds.) EMO 2003. LNCS, vol. 2632, pp. 376–390. Springer, Heidelberg (2003)
20. Knowles, J.: Closed-loop Evolutionary Multiobjective Optimization. IEEE Computational Intelligence Magazine 4, 77–91 (2009)
21. Köppen, M., Yoshida, K.: Substitute Distance Assignments in NSGA-II for Handling Many-Objective Optimization Problems. In: Obayashi, S., Deb, K., Poloni, C., Hiroyasu, T., Murata, T. (eds.) EMO 2007. LNCS, vol. 4403, pp. 727–741. Springer, Heidelberg (2007)
22. Rudolph, G., Naujoks, B., Preuss, M.: Capabilities of EMOA to Detect and Preserve Equivalent Pareto Subsets. In: Obayashi, S., Deb, K., Poloni, C., Hiroyasu, T., Murata, T. (eds.) EMO 2007. LNCS, vol. 4403, pp. 36–50. Springer, Heidelberg (2007)
23. Sato, H., Aguirre, H.E., Tanaka, K.: Controlling Dominance Area of Solutions and Its Impact on the Performance of MOEAs. In: Obayashi, S., Deb, K., Poloni, C., Hiroyasu, T., Murata, T. (eds.) EMO 2007. LNCS, vol. 4403, pp. 5–20. Springer, Heidelberg (2007)
24. Singh, H., Isaacs, A., Ray, T., Smith, W.: A Study on the Performance of Substitute Distance Based Approaches for Evolutionary Many Objective Optimization. In: Li, X., Kirley, M., Zhang, M., Green, D., Ciesielski, V., Abbass, H.A., Michalewicz, Z., Hendtlass, T., Deb, K., Tan, K.C., Branke, J., Shi, Y. (eds.) SEAL 2008. LNCS, vol. 5361, pp. 411–420. Springer, Heidelberg (2008)
25. Sülflow, A., Drechsler, N., Drechsler, R.: Robust Multi-Objective Optimization in High Dimensional Spaces. In: Obayashi, S., Deb, K., Poloni, C., Hiroyasu, T., Murata, T. (eds.) EMO 2007. LNCS, vol. 4403, pp. 715–726. Springer, Heidelberg (2007)
26. Tan, K.C., Khor, E.F., Lee, T.H.: Multiobjective Evolutionary Algorithms and Applications. Springer, Berlin (2005)
27. Wagner, T., Beume, N., Naujoks, B.: Pareto-, Aggregation-, and Indicator-Based Methods in Many-Objective Optimization. In: Obayashi, S., Deb, K., Poloni, C., Hiroyasu, T., Murata, T. (eds.) EMO 2007. LNCS, vol. 4403, pp. 742–756. Springer, Heidelberg (2007)
28. Zhang, Q., Li, H.: MOEA/D: A Multiobjective Evolutionary Algorithm Based on Decomposition. IEEE Trans. on Evolutionary Computation 11, 712–731 (2007)
29. Zitzler, E., Deb, K., Thiele, L.: Comparison of Multiobjective Evolutionary Algorithms: Empirical Results. Evolutionary Computation 8, 125–148 (2000)
30. Zitzler, E., Laumanns, M., Thiele, L.: SPEA2: Improving the Strength Pareto Evolutionary Algorithm. TIK-Report 103, Computer Engineering and Networks Laboratory (TIK), Department of Electrical Engineering, ETH, Zurich (2001)
31. Zitzler, E., Thiele, L.: Multiobjective Evolutionary Algorithms: A Comparative Case Study and the Strength Pareto Approach. IEEE Trans. on Evolutionary Computation 3, 257–271 (1999)
32. Zou, X., Chen, Y., Liu, M., Kang, L.: A New Evolutionary Algorithm for Solving Many-Objective Optimization Problems. IEEE Trans. on Systems, Man, and Cybernetics: Part B - Cybernetics 38, 1402–1412 (2008)

Preference-Based Multi-Objective Particle Swarm Optimization Using Desirabilities

Sanaz Mostaghim, Heike Trautmann, and Olaf Mersmann

Karlsruhe Institute of Technology, Karlsruhe, Germany
TU Dortmund University, Dortmund, Germany
sanaz.mostaghim@kit.edu,
{trautmann,olafm}@statistik.tu-dortmund.de

Abstract. The integration of experts' preferences is an important aspect in multi-objective optimization. Usually, one out of a set of Pareto optimal solutions has to be chosen based on expert knowledge. A combination of multi-objective particle swarm optimization (MOPSO) with the desirability concept is introduced to efficiently focus on desired and relevant regions of the true Pareto front of the optimization problem which facilitates the solution selection process. Desirability functions of the objectives are optimized, and the desirability index is used for selecting the global best particle in each iteration. The resulting MOPSO variant DF-MOPSO in most cases exclusively generates solutions in the desired area of the Pareto front. Approximations of the whole Pareto front result in cases of misspecified desired regions.

Keywords: Particle swarm optimization, MOPSO, desirability function, desirability index, preferences.

1 Introduction

Multi-objective particle swarm optimization (MOPSO) methods are designed to approximate the Pareto-front by a set of diverse solutions [5,18,19,10,20]. Incorporating the user preferences in the algorithm to find such a set in a desired area is of high practical usage.

In this paper, we investigate the influence of desirability function (DF) transformations of the objectives in combination with MOPSO. Less desirable solutions are mapped to lower DF values than highly desired ones where a desired area has to be specified by two points only. The application of DFs allows to avoid using penalty functions for restricting the objective space, and therefore the complete set of individuals contributes to the search process. This is of particular interest as boundary handling methods in MOPSO have a great impact on the solutions. We first employ a typical MOPSO to find a rough approximation of the Pareto-front. After obtaining some solution(s) close to a predefined desired area, the so-called DF-MOPSO approach transforms the objective functions to DFs. In this way, the definitions of personal and global best particles and the domination criterion change. In DF-MOPSO, there is only one global

R. Schaefer et al. (Eds.): PPSN XI, Part II, LNCS 6239, pp. 101–110, 2010.

best particle which guides the population. This is determined as the particle with the highest desirability index (DI) among the particles in the population. DF-MOPSO is tested on several test problems where the desired area is selected in different parts of the objective space.

The state of the art of the methodology is given in Section 1.1. In Section 2, we describe MOPSO and explain DFs and DIs as well as our new approach called DF-MOPSO. The results of the conducted experiments are explained in Section 3 while conclusions are drawn in Section 4.

1.1 State Of The Art

In [3] and [17] overviews of existing approaches are given for evolutionary multi-objective optimization algorithms (EMOA). Jaszkiewicz and Branke [11] especially focus on interactive approaches. Recently, a categorization of existing approaches with respect to the kind of preference specification required by the user has been given in [2]. The usage of DFs for focusing on relevant regions of the Pareto front has already successfully been introduced in combination with NSGA-II [4,21], both for deterministic as well as noisy environments. Objective transformations for preference articulation have been discussed by [13] and [23] in the context of Physical Programming as well, which is a very general but very complex concept for constructing desired ranges of objective functions.

In MOPSO, using an appropriate guidance scheme is essential for focusing on relevant regions. A vast majority of methods investigates this aspect by selecting "global best" or "personal best" particles for guiding the population towards a desired area, e.g. [5,18,19,10,20,1]. Interactive MOPSO manipulate global and personal best particles by replacing or influencing them with selected desired solutions by the user in each iteration. The first guiding MOPSO is studied by Mostaghim and Teich in [16] where they give a predefined set of non-dominated solutions as an input set of global best particles which lead the population to a desired area. Hettenhausen et al. [8] have investigated a tool for visually guiding a MOPSO towards a set of selected solutions. In each generation the user is being asked to select one or more solutions from the current population or the archive of non-dominated solutions. Only the selected solutions are considered as the global best solutions and thereby guide the population of the next iteration towards the desired area. Recently, Wickramasinghe and Li [24] study a MOPSO in which the user defines one or more preferred solution(s) in the objective space without worrying about the feasibilities. The solutions closest to the reference points are selected as the global best solutions in two MOPSO variants.

2 Methodology

2.1 MOPSO

A typical MOPSO contains a population of individuals (usually known as particles) which update their positions in the search space by moving with a certain velocity. The velocity of each particle is influenced by a social impact coming from the population and the individual experience of the particle. We denote a set of n_{pop} particles as a population P_t in the generation t. Each particle i has a position defined by $\boldsymbol{x}^i = \{x_1^i, x_2^i, \cdots, x_n^i\}$ and a velocity defined by $\boldsymbol{v}^i = \{v_1^i, v_2^i, \cdots, v_n^i\}$ in the search space S. In generation $t+1$, the new velocity and position for each particle i is computed by:

$$\begin{aligned} v_{j,t+1}^i &= wv_{j,t}^i + c_1R_1(PB_{j,t}^i - x_{j,t}^i) + c_2R_2(GB_{j,t}^i - x_{j,t}^i) \\ x_{j,t+1}^i &= x_{j,t}^i + v_{j,t+1}^i \end{aligned} \tag{1}$$

where $j = (1, \cdots, n)$, w is called the inertia weight, c_1 and c_2 are two positive constants, and R_1 and R_2 are random values in the range $[0, 1]$. In Equation (1) $\boldsymbol{PB}_t^i$ is the best position that particle i could find so far (personal best particle). It is like a memory for the particle i which gets updated in each generation. One good strategy for selecting $\boldsymbol{PB}_t^i$ is called the newest method [1]. This method compares the new position of the particle with $\boldsymbol{PB}_t^i$. If $\boldsymbol{PB}_t^i$ is dominated by the new position or if they are indifferent to each other, $\boldsymbol{PB}_t^i$ is replaced by the new position. In Equation (1), $\boldsymbol{GB}_t^i$ is the position of the global best particle selected for particle i. The global best particle is selected from the set of non-dominated solutions. In most of the MOPSO methods [18], the non-dominated solutions are stored in an archive and each particle selects its own global best from the archive. There are several strategies to find the global best (SMOPSO [5], Epsilon dominance [19,14], Pareto-dominance Concept [10], Sigma method [15]) which have a high impact on the diversity and convergence of the solutions. However, a random selection of the global best from the archive can be used as a simple strategy for getting close to the optimal front. In addition, diversity preserving methods are applied to MOPSO in order to avoid particles to converge to local optima [25,12]. The well-known turbulence factor [15] randomly reinitializes some solutions in each generation with a predefined probability.

2.2 Desirability Functions and Indices

The desirability concept was designed by Harrington for industrial multiobjective quality optimization [7]. Desirability functions (DF) range in $[0, 1]$ and are based on expert knowledge and preference specification with respect to exemplary objective values. The quality of an individual solution in the corresponding objective increases simultaneously with increasing DF value, which ideally equals one [21,22]. Thus, a mapping to a unitless scale is performed removing dependencies on scaling issues in optimization. Two types of DFs were introduced, one suited for minimization or maximization purposes of an objective (one-sided case), and one for target value optimization (two-sided case). We will concentrate on the former as a monotonic transformation of the objectives is desired which does not change the dominance relation. Harrington's one-sided DF (Figure 1) for an objective Y is defined as follows:

$$d(Y) = \exp(-\exp(-(b_0 + b_1Y))) \tag{2}$$

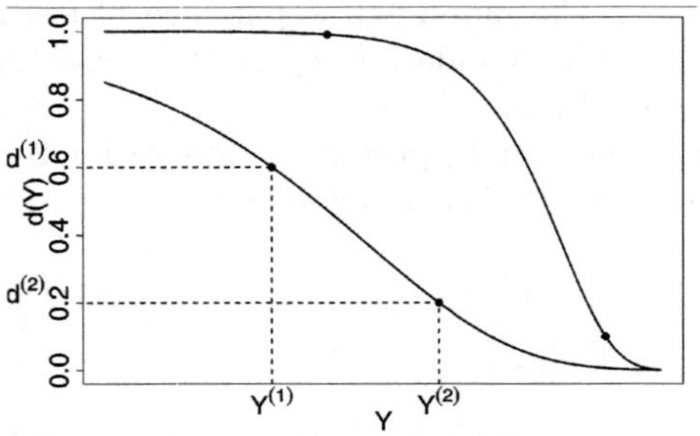

Fig. 1. Two realizations of Harrington's one-sided DF based on the specified 2-tuples depicted as black dots

where two 2-tuples of an objective value and a corresponding desirability are required, denoted by $(Y^{(1)}, d^{(1)})$ and $(Y^{(2)}, d^{(2)})$ (see black dots in Fig. 1) for the computation of the parameters b_0 and b_1. Usually, one nearly optimal and one marginally infeasible objective value are specified reflected by the choices of $d^{(1)} \approx 0.99$ and $d^{(2)} \approx 0.01$. The shape of the function is thus determined by the solutions

$$b_0 = -\log(-\log(d^{(1)})) - b_1 Y^{(1)} \tag{3}$$
$$b_1 = (-\log(-\log(d^{(2)})) + \log(-\log(d^{(1)})))/(Y^{(2)} - Y^{(1)}) \quad \text{to eqns.} \tag{4}$$
$$d^{(i)} = \exp(-\exp(-(b_0 + b_1 Y^{(i)}))),\ i = 1, 2 \tag{5}$$

based on the 2-tuples $(Y^{(1)}, d^{(1)})$ and $(Y^{(2)}, d^{(2)})$ described above. The Desirability Index (DI) reduces the multi-objective optimization problem to a single-objective optimization problem by a scalarizing function $D \colon [0,1]^k \to [0,1]$. Higher DI values correspond to a higher overall quality with respect to the compromise between the preferences encoded in the DFs. The most common used DI is the geometric mean of the DFs $D_{GM} := \left(\prod_{j=1}^{k} d_j(Y_j)\right)^{1/k}$.

2.3 DF-MOPSO

An important property of MOPSO is that the selection of the global and personal best solutions has a great impact on the search mechanism. Suppose that we select one global best for all the population members. After only one iteration, the particles move around that selected solution. In DF-MOPSO, we use this property and find the particle with the highest desirability index value among the population members. Such a particle is the closest to the preferred area so far and therefore is a good candidate for being the only global best GB_t for the population P_t.

Like a typical MPOPSO [18], DF-MOPSO is designed as an iterative method containing an archive A_t for keeping non-dominated solutions (Algorithm 1). Starting from a random population P_t $(t = 1)$ and personal best particles the same as the current positions of the particles PB_t, the DF-MOPSO is run for

Algorithm 1. Desirability MOPSO

```
P_1 := randomPopulation(n_pop);
PB_0 := P_1 ;
dominationCriterion := dominationByObjectives;
for t = 1 to n_gen do
    Evaluate P_t;
    A_t := nonDominatedSolutions(A_t ∪ P_t, dominationCriterion);
    A_t := filterArchive(A_t, dominationCriterion);
    Calculate desirability function values of solutions in P_t;
    D_t := desirabilityIndex(P_t) ;
    if max(D_t) > t_D then
        GB_t := P_t[arg max_{k=1:n_pop} D_t[k]];
        dominationCriterion := dominationByDesirability;
    end
    else
        GB_t := RandomIndividual(A_t);
        dominationCriterion := dominationByObjectives;
    end
    for j = 1 to n_pop do
        PB_t[j] := betterIndividual(P_t[j], PB_{t-1}[j], dominationCriterion);
    end
    P_{t+1} := updatePopulation(P_t, GB_t, PB_t);
end
```

Algorithm 2. filterArchive

```
Data: Archive A
Result: Reduced Archive A''
begin
    if dominationCriterion == dominationByDesirability then
        A' := {};
        for a ∈ A do
            d := desirabilityValues(a);
            if max(d) < t_dmax or min(d) > t_dmin then
                A' := A' ∪ a;
            end
        end
    end
    else
        A' := A ;
    end
    A'' := reduceByClustering(A', n_amax);
end
```

n_{gen} generations. After evaluations, the non-dominated solutions from the population are inserted into the archive and hereby update the archive. If the size of the archive exceeds a certain maximum value (n_{amax}), the archive is filtered and its size is reduced by a clustering method. At this stage the DFs are computed for all the objectives and the DI values (D_{GM}) are calculated. Since it usually takes a few generations for a typical MOPSO to get close to a predefined desired area, we let the DF-MOPSO algorithm run as a classical MOPSO for several generations. Since we only need to get close to the optimal front, it is preferable to select the global best particles randomly from the archive. This typical MOPSO contains epsilon-domination from [14]. As soon as the maximum DI value in the population is larger than a certain threshold t_D we switch to DF-MOPSO.

We transform the objective functions to the Harrington's one sided DF functions (2). Hence, from this point the DF-MOPSO algorithm optimizes the converted functions meaning (1) the definition of domination changes to the transformed functions, (2) the selection of personal best solutions is based on the transformed objective values and (3) the solution with the largest DI value is selected as the only global best for the population. Algorithm 1 describes DF-MOPSO in detail. Algorithm 2 is meant to filter the archive. In case of using the normal domination criterion (classical MOPSO), this algorithm only reduces the size of the archive to the maximum size n_{max}. In case we perform DF-MOPSO, only solutions with DF values between t_{dmin} and t_{dmax} are inserted into the archive.

3 Experiments

We analyze MOPSO and DF-MOPSO on the test functions FF [6], shifted ZDT1 (S-ZDT1) and S-ZDT2 [9] which are known to have higher complexity than the classical ZDT functions. The following parameter settings were determined to be reasonable based on preliminary runs: inertia weight $w = 0.4$, turbulence factor $t_f = 0.1$, DI threshold $t_D = 0.2$, population size 100 and maximum archive size 100.

The user preferences are defined as areas specified by two points for each objective function separately. These are assigned the DF values of 0.01 and 0.99. In the experiments, we first suppose that we know the Pareto front and three preferred areas are defined as *left*, *middle* and *right* parts of the front. With these experiments, we analyze the convergence and the diversity of the solution sets found. Since in reality the Pareto front is usually unknown to the user, we additionally test the DF-MOPSO by specifying (a) an area dominated by the front and (b) an infeasible area dominating the Pareto-front denoted. The (a) and (b) variations are denoted as *upper* and *lower* in the results. Ten runs of MOPSO and DF-MOPSO with 5000 function evaluations (FE) each are conducted for each combination of test function and desired area.

3.1 Results

Figure 2 shows histograms of the nondominated solutions in the final front of all runs for each pair of test function and focused part of the front. The boundaries of the desired areas of the objective functions Y_1 and Y_2, i.e. objectives with DF values in $[0.01, 0.99]$, are visualized by red lines. It becomes obvious that the DF-MOPSO is able to efficiently focus on the desired part of the front. Only a very small proportion of solutions falls outside the specified boundaries.

In case the desired regions are located outside the front, the histograms are replaced by a visualization of the objective space where the desired region, the true Pareto front and the generated nondominated solutions of all runs are plotted. This is a more adequate means to analyse the performance of the DF-MOPSO in these cases as the solutions cannot (*lower*) or should not (*upper*) concentrate in the desired regions for both objectives simultaneously. For S-ZDT1 and S-ZDT2 *lower* and *upper* regions do not prevent the DF-MOPSO from finding the true Pareto front nevertheless. Thus, the algorithm does not get stuck in

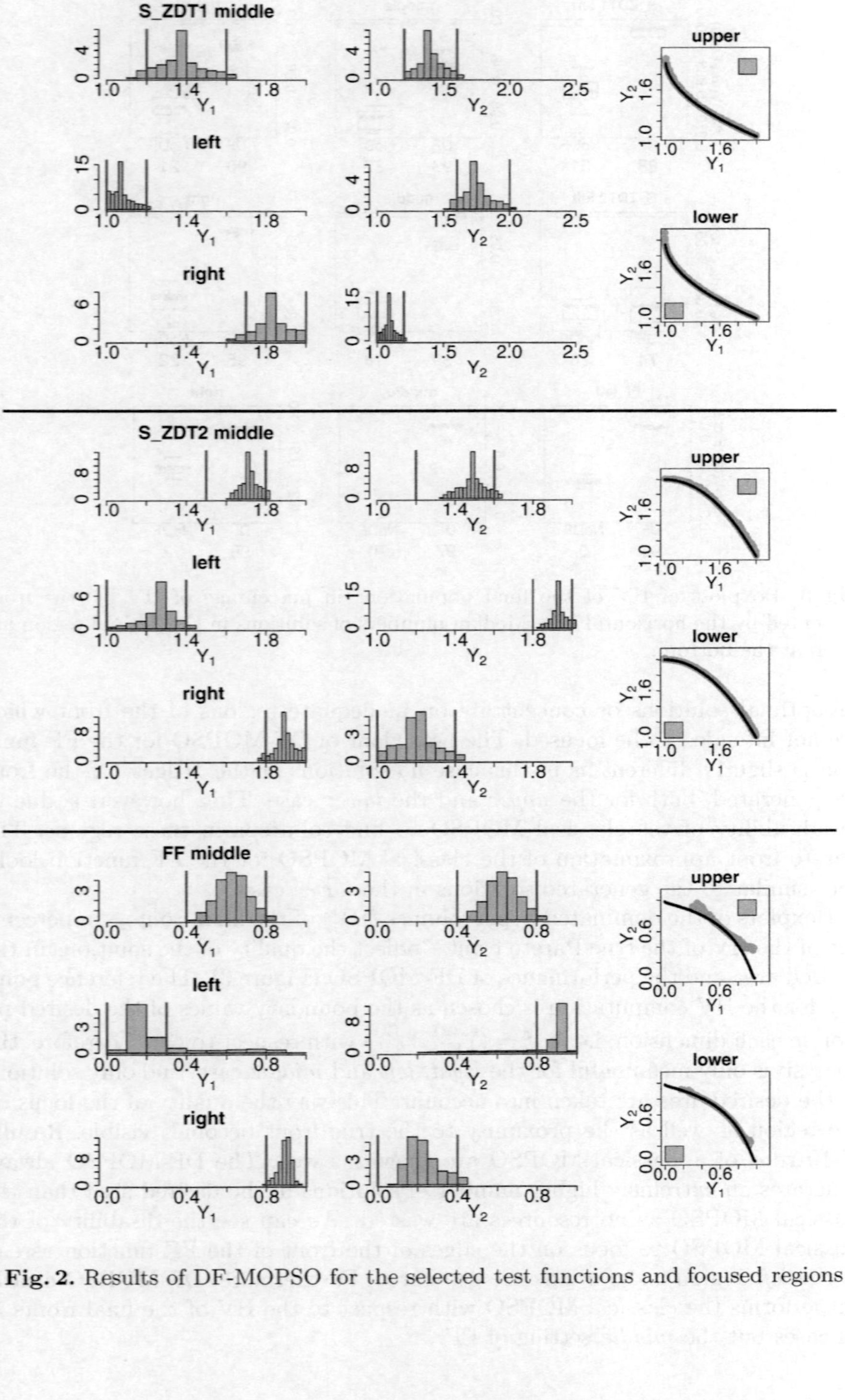

Fig. 2. Results of DF-MOPSO for the selected test functions and focused regions

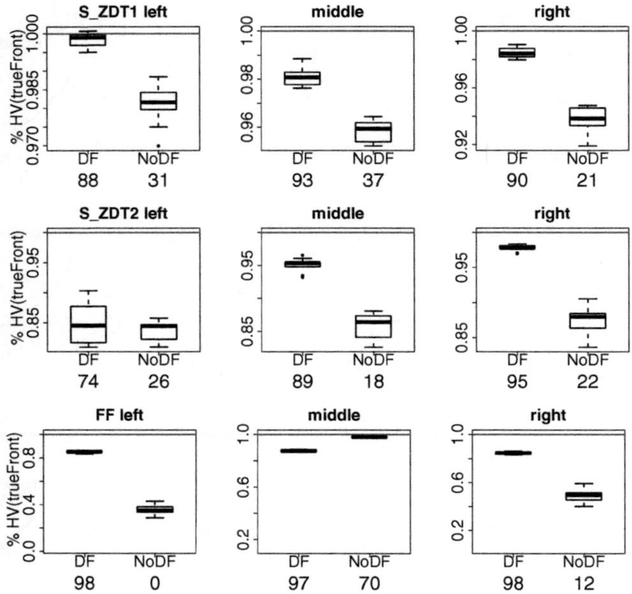

Fig. 3. Boxplots of HV of the final populations in percentage of HV of true front presented by the horizontal line. Median numbers of solutions in the desired region are given at the bottom.

suboptimal solutions or concentrate on inadequate regions of the front which are not intended to be focused. The behaviour of DF-MOPSO for the FF function is slightly different as in this case no solutions at the "edges" of the front are generated, both for the *upper* and the *lower* case. This, however, is due to the disability of the classical MOPSO to find solutions in these regions. The Pareto front approximation of the classical MOPSO for the FF function looks very similar to the generated solutions in the *lower* case.

Boxplots of the dominated hypervolumes (HV) of the final fronts – in percentage of the HV of the true Pareto front – reflect the quality of the solutions in the desired area and the performance of DF-MOPSO (Figure 3). The reference point ref for the HV computation is chosen as the boundary values of the desired region in each dimension, i.e. $ref = (Y_1^{(2)}, Y_2^{(2)})$ with respect to (3). Therefore, the analysis is only meaningful for the *right, left* and *middle* case, and only solutions in the desired area are taken into account. This way the quality of the focus on the region as well as the proximity to the true front becomes visible. Results of 10 runs of a classical MOPSO are shown as well. The DF-MOPSO always generates an extremely higher number of solutions in the desired area than the classical MOPSO as no resources are wasted. We can see the disability of the classical MOPSO to focus on the edges of the front of the FF function as e.g. in the *left* region no solutions are generated. Furthermore, DF-MOPSO clearly outperforms the classical MOPSO with respect to the HV of the final fronts in all cases but the *middle* setting of FF.

4 Conclusions and Outlook

The MOPSO variant DF-MOPSO based on the integration of desirability concepts is designed to generate nondominated solutions in relevant regions of the true Pareto front which extremely facilitates the solution selection process after optimization. Solutions are much denser in this region than in classical MOPSO as no resources are wasted in regions of the objective space which are unimportant to the expert. Experimental results on standard test functions show the ability of DF-MOPSO to efficiently focus on the desired region of the front. Even in case the true front and the desired region do not overlap due to incomplete or insufficient knowledge of the underlying optimization problem the algorithm does not get stuck in suboptimal regions but generates a complete approximation of the true Pareto front.

In future work we will investigate how the results generalize to higher dimensional optimization problems and how a covering of the whole Pareto front can be realized by a parallelization of MOPSO and dynamically setting the focus to different areas of the front by using desirability functions.

Acknowledgements. This work was partly supported by the Collaborative Research Center SFB 823 of the German Research Foundation.

References

1. Branke, J., Mostaghim, S.: About selecting the personal best in multi-objective particle swarm optimization. In: Runarsson, T.P., Beyer, H.-G., Burke, E.K., Merelo-Guervós, J.J., Whitley, L.D., Yao, X. (eds.) PPSN 2006. LNCS, vol. 4193, pp. 523–532. Springer, Heidelberg (2006)
2. Branke, J.: Consideration of partial user preferences in evolutionary multiobjective optimization. In: Multiobjective Optimization, pp. 157–178 (2008)
3. Coello Coello, C.A.: Handling preferences in evolutionary multiobjective optimization: A survey. In: Congress on Evolutionary Computation (CEC), pp. 30–37 (2000)
4. Deb, K., Pratap, A., Agarwal, S.: A fast and elitist multi-objective genetic algorithm: NSGA-II. IEEE Trans. on Evolutionary Computation 6(8) (2002)
5. Durillo, J.J., Garca-Nieto, J., Nebro, A.J., Coello, C.A.C., Luna, F., Alba, E.: Multi-objective particle swarm optimizers: An experimental comparison. In: Ehrgott, M., Fonseca, C.M., Gandibleux, X., Hao, J.-K., Sevaux, M. (eds.) EMO 2009. LNCS, vol. 5467, pp. 495–509. Springer, Heidelberg (2009)
6. Fonseca, C.M., Fleming, J.: An overview of evolutionary algorithms in multiobjective optimization. Evolutionary Computation 3(1), 1–16 (1995)
7. Harrington, J.: The desirability function. Industrial Quality Control 21(10), 494–498 (1965)
8. Hettenhausen, J., Lewis, A., Mostaghim, S.: Interactive multi-objective particle swarm optimisation with heatmap visualisation based user interface. Journal of Engineering Optimization 42(2), 119–139 (2010)
9. Huang, V.L., Qin, A., Deb, K., Suganthan, P., Liang, J., Preuss, M., Huband, S.: Problem definitions for performance assessment on multi-objective optimization algorithms. Technical report, Nanyang Technological University, Singapore (2007)

10. Alvarez-Benitez, J.E., Everson, R.M., Fieldsend, J.E.: A MOPSO algorithm based exclusively on pareto dominance concepts. In: Coello Coello, C.A., Hernández Aguirre, A., Zitzler, E. (eds.) EMO 2005. LNCS, vol. 3410, pp. 459–473. Springer, Heidelberg (2005)
11. Jaszkiewicz, A., Branke, J.: Interactive multiobjective evolutionary algorithms. In: Multiobjective Optimization, pp. 179–193 (2008)
12. Lovberg, M., Krink, T.: Extending particle swarm optimization with self-organized criticality. In: Proceedings of IEEE Conference on Evolutionary Computation, pp. 1588–1593 (2002)
13. Messac, A.: Physical Programming: Effective Optimization for Computational Design. AIAA Journal 34(1), 149–158 (1996)
14. Mostaghim, S., Teich, J.: The role of e-dominance in multi-objective particle swarm optimization methods. In: Proceedings of the Congress on Evolutionary Computation, CEC (2003)
15. Mostaghim, S., Teich, J.: Strategies for finding good local guides in multi-objective particle swarm optimization. In: Swarm Intelligence Symposium, pp. 26–33 (2003)
16. Mostaghim, S., Teich, J.: Covering pareto-optimal fronts by subswarms in multi-objective particle swarm optimization. In: the Proceedings of The Congress on Evolutionary Computation, CEC (2004)
17. Rachmawati, L., Srinivasan, D.: Preference incorporation in multi-objective evolutionary algorithms: A survey. In: Congress on Evolutionary Computation (CEC), pp. 962–968 (2006)
18. Reyes-Sierra, M., Coello, C.C.: Multi-objective particle swarm optimizers: A survey of the state-of-the art. International Journal of Computational Intelligence Research 2(3), 287–308 (2006)
19. Sierra, M.R., Coello, C.A.C.: Improving pso-based multi-objective optimization using crowding, mutation and e-dominance. In: Coello Coello, C.A., Hernández Aguirre, A., Zitzler, E. (eds.) EMO 2005. LNCS, vol. 3410, pp. 505–519. Springer, Heidelberg (2005)
20. Toscano-Pulido, G., Coello, C.A.C., Santana-Quintero, L.V.: Emopso: A multi-objective particle swarm optimizer with emphasis on efficiency. In: Obayashi, S., Deb, K., Poloni, C., Hiroyasu, T., Murata, T. (eds.) EMO 2007. LNCS, vol. 4403, pp. 272–285. Springer, Heidelberg (2007)
21. Trautmann, H., Mehnen, J.: Preference-Based Pareto-Optimization in Certain and Noisy Environments. Engineering Optimization 41, 23–38 (2009)
22. Trautmann, H., Weihs, C.: On the distribution of the desirability index using Harrington's desirability function. Metrika 63(2), 207–213 (2006)
23. Utyuzhnikov, S., Fantini, P., Guenov, M.: Numerical method for generating the entire Pareto frontier in multiobjective optimization. In: Schilling, R., Haase, W., Periaux, J., Baier, H., Bugeda, G. (eds.) Evolutionary and Deterministic Methods for Design, Optimization and Control with Applications to Industrial and Societal Problems, EUROGEN (2005)
24. Wickramasinghe, U., Li, X.: Integrating user preferences with particle swarms for multi-objective optimization. In: Proceedings of the Conference on Genetic and Evolutionary Computation (GECCO), pp. 745–752 (2008)
25. Xie, X.F., Zhang, W.J., Yang, Z.L.: Adaptive particle swarm optimization on individual level. In: Proceedings of the Sixth International Conference on Signal Processing, pp. 1215–1218 (2002)

GPGPU-Compatible Archive Based Stochastic Ranking Evolutionary Algorithm (G-ASREA) for Multi-Objective Optimization

Deepak Sharma[1] and Pierre Collet[2]

[1] LogXlabs Research Center, Paris, France
deepak.sharma@logxlabs.com
[2] FDBT, LSIIT, Université de Strasbourg, France
pierre.collet@unistra.fr

Abstract. In this paper, a GPGPU (general purpose graphics processing unit) compatible Archived based Stochastic Ranking Evolutionary Algorithm (G-ASREA) is proposed, that ranks the population with respect to an archive of non-dominated solutions. It reduces the complexity of the deterministic ranking operator from $O(mn^2)$ to $O(man)^\star$ and further speeds up ranking on GPU.

Experiments compare G-ASREA with a CPU version of ASREA and NSGA-II on ZDT test functions for a wide range of population sizes. The results confirm the gain in ranking complexity by showing that on 10K individuals, G-ASREA ranking is $\approx \times 5000$ faster than NSGA-II and $\approx \times 15$ faster than ASREA.

1 Introduction

In the last two decades, the field of multi-objective optimization (MOO) has attracted researchers and practitioners to solve real world optimization problems that involve multiple objectives with a set of solutions known as *Pareto-optimal* solutions. Many optimization algorithms exist in the literature for MOO problems, but the heuristic population based algorithms (also known as multi-objective evolutionary algorithms (MOEAs)) are most suitable to evolve trade-off solutions in one run [1,2].

Though many MOEAs have been developed, there are only a few dominance-ranking based algorithms that are really effective to solve MOO problems. Mostly, these MOEAs differ in their ranking methods which helps to select and propagate good individuals to the next iteration. According to [2], dominance ranking methods can be categorized by dominance rank, dominance count and dominance depth. MOGA [3] and NPGA [4] use a dominance rank method by checking the number of solutions that dominate an individual. NSGA-II [5] sorts individuals according to dominance depth, using the concept of non-dominated sorting [6]. SPEA2 [7] assigns rank, based on dominance depth and dominance count, where the count of dominated individuals by an individual is used. PESA-II [8] is based

* Where, m = nb. of objectives, n = population size and a = archive size.

R. Schaefer et al. (Eds.): PPSN XI, Part II, LNCS 6239, pp. 111–120, 2010.

on dominance count, and uses an archive of non-dominated solutions. All these dominance-based methods evaluate rank serially in a deterministic way (except NPGA), with a quadratic ranking complexity on the population size ($O(mn^2)$ for NSGA-II, where m is the number of objectives and n the population size).

For solving many objective problems [1,2] or Genetic Programming with more than one objective (error minimization and parcimony [9]), a very large population is often required which takes time to both rank and evaluate. However, the advent of General Purpose Graphic Processing Units (GPGPUs) allow to evaluate very large populations in parallel [10]. In [11], the dominance sorting of NSGA-II has been implemented on GPU, but the sorting of individuals in different fronts is still done on CPU. Although this is the first appreciable effort of parallelizing MOEA on GPUs, this paper does attempt to reduce NSGA-II's $O(mn^2)$ ranking complexity.

In [12], we have developed a MOEA called ASREA (Archive based Stochastic Ranking Evolutionary Algorithm) with an $O(man)$ ranking complexity (where a is the size of an archive, that depends on the number of objectives m) which breaks the $O(mn^2)$ complexity, while yielding improved results over NSGA-II.

The present paper shows how ASREA can be fully parallelized, by performing ranking and function evaluation on the GPU. The details of G-ASREA (GPU-based ASREA) are discussed in section 2 followed by experimental results in section 3, in which ranking time and function evaluation of G-ASREA are compared with serial versions of ASREA and NSGA-II. The paper is concluded in section 4.

2 GPGPU Compatible ASREA (G-ASREA)

In [12], the Archive-based Stochastic Ranking Evolutionary Algorithm (ASREA) was shown to converge to the Pareto-optimal front at least as well as NSGA-II and SPEA2, while reaching the front much faster on two-objectives ZDT functions and three-objectives DTLZ functions.

However, the archive updating rule during stochastic ranking is based on CPU serial operations, making the ranking operator difficult to parallelize. This paper shows how it is possible to address this problem, and make ASREA fully GPU-compatible.

2.1 GPGPU

A GPGPU is a General Purpose Graphics Processing Unit, which is a high-performance multithread many-core processor. GPUs have originally been developed for computer graphics, but today's GPUs are capable of performing parallel data computing with support for accessible programming interfaces and industry-standard languages such as C [13]. nVidia GPUs can be programmed using CUDA, which is a general purpose parallel computing architecture, that allows to use the many cores in NVIDIA GPUs to run other algorithms than graphic rendering algorithms. Interested readers may refer to the CUDA programming guide [14].

2.2 Description of the Algorithm

ASREA is an archive-based MOEA that starts with evaluating a random initial population (**ini_pop**). The distinct non-dominated individuals of **ini_pop** are copied to the archive according to an archive updating algorithm 1 described later in this section. As seen in fig. 1, **ini_pop** is copied to **parent_pop** and a rather standard EA loop starts, by checking if a termination criterion is met.

If it is not, a **child_pop** is created by repeatedly selecting *randomly* two parents from the **parent_pop**, and creating children through a crossover, followed by a mutation (in this paper, a standard SBX crossover is used, followed by a polynomial mutation [15]), after which the **child_pop** is evaluated.

Function evaluation on the GPU: Function evaluation (FE) is one of the most time consuming parts of MOEAs and can be easily parallelized on GPUs. In order to perform FE, a single float array X of size vm is allocated in the global memory of the GPU, where v is the number of variables and m is the number of objectives. The variable set of **child_pop** is then copied to X as shown in fig. 2. At the same time, a single float array OBJ of size mn (where n is the child population size) is also allocated in the global memory of the GPU, in order to store the objective function values of **child_pop**, that are computed in parallel by the single instruction multiple data (SIMD) GPU function evaluation kernel.

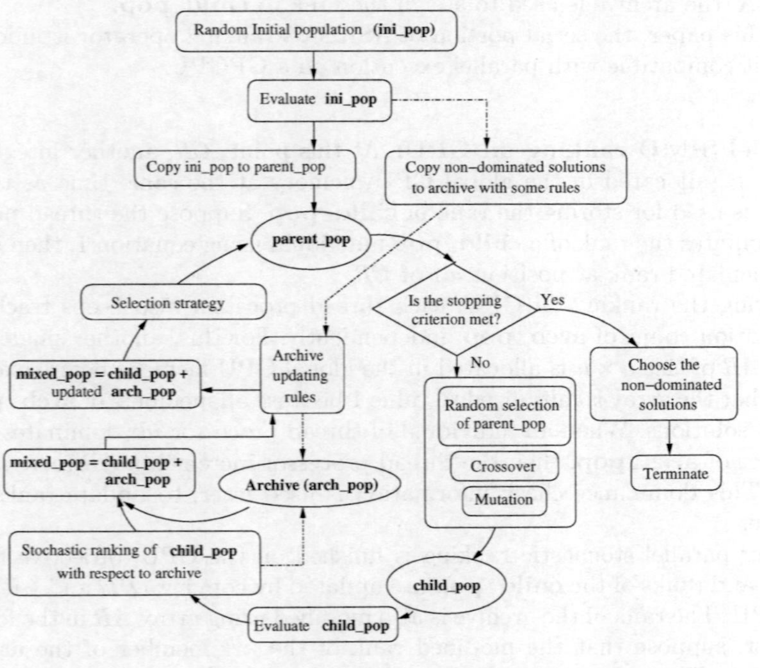

Fig. 1. ASREA flowchart

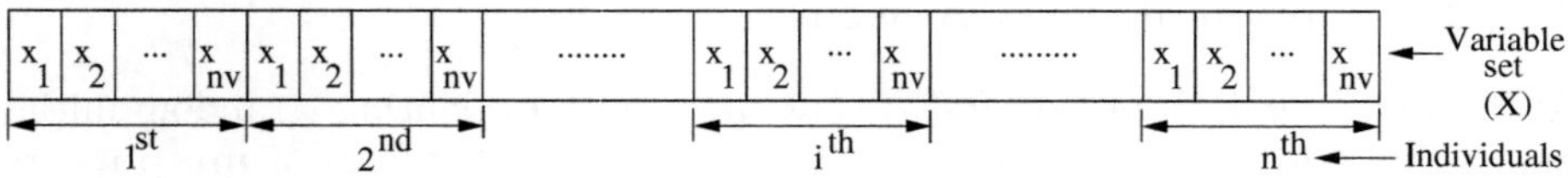

Fig. 2. Single array representation for GPU

Now, comes the time to rank **child_pop**. ASREA's ranking operator [12] reduces the typical ranking complexity of deterministic algorithms such as NSGA-II from $O(mn^2)$ to $O(man)$ (where m is the number of objective, n is the population size and a is the archive size) by comparing individuals of **child_pop** with the members of the archive. The rank of individual (A) of **child_pop** is calculated on the following dominance rank criterion:

$$\text{rank}(A) = 1 + \textit{number of } \textbf{\textit{arch\_pop}} \textit{ members that dominate } A \qquad (1)$$

Note that in ASREA, the lower rank is better, with best rank = 1. The ranking procedure discussed above makes the difference in the working principle of ASREA from other MOEAs, and specially the archived-based ones, in which the archive/parent and current offspring populations are mixed, after which the rank is assigned deterministically [5,7] or every individual of child population is checked for non-domination to become a member of archive [8]. However in ASREA, the archive is used to assign the rank to **child_pop**.

In this paper, the serial portion of ASREA's ranking operator is modified to make it compatible with parallel execution on a GPGPU.

Parallel SIMD ranking on GPU: At this point, *CR*, another integer array of size n (allocated in the global GPU memory at the same time as the *OBJ* array) is used for storing the rank of **child_pop**. Suppose the thread processor *idx* computes the rank of a **child_pop** individual using equation 1, then it stores the calculated rank at position *idx* of *CR*.

During the ranking on GPU, each thread processor also keeps track of the domination count of **arch_pop**, independently. For this, another single integer array *AR* of size $a \times n$ is allocated in the global GPU memory before execution. Note that the array is initialized to value 1 because all members of **arch_pop** are rank 1 solutions. When an individual of thread processor *idx* dominates the kth member of **arch_pop**, then the thread processor increments $AR[(a \times idx) + k]$ by 1. This dominance check information is used later, to update ranks in the archive.

After parallel stochastic ranking is finished on the GPU, objective function values and ranks of the **child_pop** are updated by copying *OBJ* and *CR* back to the CPU. The rank of the archive is also modified using array *AR* in the following manner: suppose that the modified rank of the kth member of the archive is evaluated. Then, for $i = 0$ to $n-1$, the integer value of every $AR[(i \times a) + k]$ is added and finally subtracted by n. If the kth member is still non-dominated,

```
if Number of non-dominated solutions ≤ archive size then
 | Copy distinct non-dominated solutions to archive;
else
 | Copy the extreme solutions of non-dominated front to archive and evaluate
 | crowding distance (CD [5]) of rest of the non-dominated solutions. Fill the
 | remaining archive with the sorted CD-wise individuals in descending order;
end
```

Algorithm 1. Archive updating rules

```
Copy the extreme solutions of updated arch_pop to the parent population.
Fill 20% of parent_pop from the updated arch_pop in the following way:
while (20% of parent_pop is not filled) do
  | Pick randomly two individuals of updated arch_pop;
  | if Crowding distances are different then
  |   | Copy the individual with larger crowding distance into parent_pop;
  | else
  |   | Copy any individual randomly;
  | end
end
Fill rest of parent_pop from child_pop in the following way:
while (rest of parent_pop is not filled) do
  | Pick two individuals randomly from child_pop without replacing them;
  | if Ranks are different then
  |   | Copy the individual with smaller rank into parent_pop;
  | else
  |   | Copy the individual with larger crowding distance into parent_pop;
  | end
end
```

Algorithm 2. Selection strategy to fill the **parent_pop**

then its modified rank is 1. Otherwise, the rank of the kth member depends on the number of **child_pop** individuals who dominated it.

Replacement Strategy. The next step of ASREA is then to update the archive and propagate good individuals to the next generation. First, the ranked **child_pop** and **arch_pop** with modified ranks are mixed together to form **mixed_pop**. Now, the archive is updated from the set of non-dominated solutions ($rank = 1$) of **mixed_pop** as given in algorithm 1.

The next generation (new parent population) is also selected from **mixed_pop** according to the strategy discussed in algorithm 2. Here, 20% of the new parent population is filled from the updated archive with randomly picked members. The rest of **parent_pop** is filled from **child_pop** individuals using tournament selection. The EA loop is then completed and can start again, by checking whether a termination criterion is met (e.g. number of generations) as in fig. 1.

2.3 Salient Features of G-ASREA

The first important feature of G-ASREA comes from its reduced ranking computational complexity $O(man)$ and its implementation on GPGPU for assigning rank to the population by dominance check.

A second important feature of G-ASREA is its inherent ability to preserve diversity in the population. This comes when the not-so-good individuals of **child_pop** may nevertheless obtain a good rank (including rank $= 1$) because they are only compared with the archive of limited size. Had the ranking method been deterministic, then these not-so-good individuals would not have made it into the next generation and it may have been necessary to implement a diversity preserving scheme in order to avoid premature convergence. This is why ASREA's ranking procedure is *stochastic*.

Another feature of G-ASREA is the drastic but subtle selection strategy that is used to propagate good individuals to the next iteration. Finally, G-ASREA also uses the GPU for function evaluation.

3 Experimental Results

In this paper, five Zitzler-Deb-Thiele functions [16] (summarized in appendix A) are chosen to effectively compare G-ASREA over CPU versions of ASREA and NSGA-II on different population sizes. Every MOEA is run for 25 times using different seeds and average computation time of ranking and function evaluations are shown. Note that G-ASREA's computation time includes copy of data from CPU to GPU, computation on GPU and copy back the data from GPU to CPU. All ZDT functions are executed with identical parameters of MOEAs (cf. table 1) on Intel(R) Core(TM)2 Quad CPU Q8200 @ 2.33GHz computer with one of the 2 GPUs of a GTX295 card.

Algorithms ranking complexity (such as $O(mn^2)$ or $O(man)$) correspond to the *worst* case, which may not appear in a real life. Beyond their theoretical complexity, it is therefore important to assess the complexity of algorithms on real world problems, or at least benchmarks that are supposed to exhibit behaviors encountered in real world problems.

Table. 2 shows the average number of ranking comparisons with NSGA-II, ASREA and G-ASREA for ZDT functions over different population sizes (100 to 100,000). Depending on the problems and their Pareto fronts, different comparison counts are observed for a same algorithm (354.10^6 comparisons for NSGA-II with 10,000 individuals on ZDT1, compared to $3,808.10^6$ comparisons for the same algorithm and same population size on ZDT2).

Table 1. Common parameters for all tested algorithms

No. of generations	100		
Individual Xover prob	0.9	Variable Xover prob	0.5
Individual Mut prob	1.0	Variable Mut prob	1/(no. of var)
Distrib crossover index	15	Distrib mutation index	20

Table 2. Real number of ranking comparisons for different population sizes

Problems	ZDT1			ZDT2		
Pop↓ MOEAs→	NSGA-II	ASREA	G-ASREA	NSGA-II	ASREA	G-ASREA
100	1.17 E6	198,463	198,260	968,464	189,059	189,970
1,000	77.8 E6	2.17 E6	2.17 E6	65.2 E6	2.08 E6	2.07 E6
10,000	354 E6	26.2 E6	26.2 E6	3,808 E6	23.9 E6	23.8 E6
100,000	-	413 E6	413 E6	-	346 E6	342 E6

ZDT3			ZDT4			ZDT6		
NSGA-II	ASREA	G-ASREA	NSGA-II	ASREA	G-ASREA	NSGA-II	ASREA	G-ASREA
1.19 E6	198,073	197,940	729,419	181,269	179,710	753,581	191,993	189,120
79.8 E6	2.15 E6	2.15 E6	43.8 E6	1.98 E6	1.98 E6	45.3 E6	2.06 E6	2.05 E6
522 E6	25.5 E6	25.5 E6	2382 E6	22.9 E6	22.9 E6	2395 E6	23.5 E6	23.1 E6
-	392 E6	391 E6	-	330 E6	330 E6	-	372 E6	337 E6

Table 3. Ranking comparison time (in sec.) and speed-up ratio of MOEAs over wide range of population

Problem	ZDT1			Speedup ratio		
Pop↓ MOEAs→	NSGA-II (N)	ASREA (A)	G-ASREA (G)	N/A ratio	N/G ratio	A/G ratio
100	0.000573	0.000101	0.000119	5.6732	4.8151	0.8487
1000	0.037833	0.000999	0.000162	37.8708	233.5370	6.1666
10000	4.304927	0.009971	0.000817	431.7447	5269.1884	12.2044
100000	-	0.098142	0.007011	-	-	13.9983
1000000	-	0.974255	0.065586	-	-	14.8546

Problem	ZDT2			Speedup ratio		
Pop↓ MOEAs→	NSGA-II (N)	ASREA (A)	G-ASREA (G)	N/A ratio	N/G ratio	A/G ratio
100	0.000481	0.000097	0.000116	4.9587	4.1466	0.8362
1000	0.031627	0.000998	0.000160	31.6903	197.6688	6.2375
10000	3.105234	0.009884	0.000813	314.1677	3819.4760	12.1575
100000	-	0.097470	0.006991	-	-	13.9422
1000000	-	0.972147	0.065659	-	-	14.8059

Problem	ZDT3			Speedup ratio		
Pop↓ MOEAs→	NSGA-II (N)	ASREA (A)	G-ASREA (G)	N/A ratio	N/G ratio	A/G ratio
100	0.000574	0.000103	0.000121	5.5728	4.7438	0.8512
1000	0.038939	0.001014	0.000162	38.4013	240.3641	6.2592
10000	4.538463	0.009968	0.000834	455.3032	5441.8021	11.9520
100000	-	0.098616	0.007113	-	-	13.8641
1000000	-	0.992453	0.067140	-	-	14.7818

Problem	ZDT4			Speedup ratio		
Pop↓ MOEAs→	NSGA-II (N)	ASREA (A)	G-ASREA (G)	N/A ratio	N/G ratio	A/G ratio
100	0.000374	0.000094	0.000115	3.9787	3.2521	0.8173
1000	0.020379	0.000952	0.000160	21.4065	127.3687	5.9500
10000	1.159553	0.009277	0.000866	124.9922	1338.9757	10.7124
100000	-	0.092044	0.006947	-	-	13.2492
1000000	-	0.920826	0.066112	-	-	13.9283

Problem	ZDT6			Speedup ratio		
Pop↓ MOEAs→	NSGA-II (N)	ASREA (A)	G-ASREA (G)	N/A ratio	N/G ratio	A/G ratio
100	0.000372	0.000105	0.000120	3.5428	3.1000	0.8750
1000	0.021340	0.001071	0.000161	19.9253	132.5465	6.6521
10000	1.161053	0.009821	0.000829	118.2214	1400.5464	11.8468
100000	-	0.095291	0.006969	-	-	13.6735
1000000	-	0.947676	0.066241	-	-	14.3064

Significant differences can be seen with ASREA and G-ASREA over NSGA-II, which increase drastically with larger populations, confirming the difference shown by their respective $O(man)$ and $O(mn^2)$ complexities (small differences are observed between ASREA and G-ASREA because the algorithms were run with different seeds).

ASREA and G-ASREA are further tested for 100,000 individuals (Genetic Programming typical population sizes) where they show a comparable number of ranking comparisons than NSGA-II for 10,000. NSGA-II was not tested for 100,000 individuals over 25 different runs because it took 19.2 hours for just one single run on a ZDT1 function.

ASREA algorithms are not only frugal on comparisons, but G-ASREA can parallelize the ranking (and evaluation) over the GPU processors, allowing for the speedups shown in Table 3, in which the average time of ranking comparison per generation of NSGA-II, ASREA and G-ASREA are given. The N/A and N/G ratios represent speedup of ASREA and G-ASREA over NSGA-II. N/A speedup ranges from 3.54 to 455.3, whereas N/G speedup ranges between 3.1 to 5,441 for 10,000 individuals.

For 10,000 to 1 million[1] population, G-ASREA shows the speedup between 10 to 15 over CPU-ASREA (cf. table 3). The study [11] also showed the same range of speedup in ranking comparison for dominance check on ZDT functions. However, the speedup is limited to this range for two-objective problems because the dominance check of G-ASREA concerns only parallelization of ranking comparisons on the GPU. Nevertheless, if the number of comparisons increases for many objective optimization, then the speedup for dominance check can further increase.

G-ASREA's ranking operator offers a twofold advantage: first, the speedup comes from the smaller ranking complexity over NSGA-II. Then, another advantage comes by being able to perform ranking comparisons on GPU.

Table 4. Function evaluation time (in seconds) of serial and GPU ASREA

Problems	**ZDT6**		Speedup
Pop↓ MOEAs→	ASREA	G-ASREA	ratio
100	0.00053	0.000119	0.4453
1000	0.00521	0.000161	3.2360
10000	0.004798	0.000816	5.8821
100000	0.04728	0.006921	6.8319
1000000	0.4785	0.06526	7.3322

Function evaluation time can also benefit from being run on the GPU. However, this aspect is already studied in several papers [17,11], so only the computation time for ZDT6 is shown in table 3. The speedup is not impressive because the ZDT6 function is not computationally intensive (speedups of up to ×250 have been obtained on GP function evaluations in [10]).

This paper is not so much concerned with solving the ZDT benchmarks (this can be done with hundreds of individuals only [12]) than to study and minimize the bottleneck induced by the ranking operator that may limit the usage of very large populations in multi-objective optimization problems.

[1] Note that only one run is performed for 1 million individuals on each ZDT function.

4 Conclusions and Future Work

The advent of massively parallel GPGPU cards allows to parallelize the evaluation of very large populations in order to solve complex optimization problems. In MOEAs, the bottleneck then becomes the multi-objective ranking stage. In this paper, a GPGPU compatible stochastic ranking operator is proposed that not only requires fewer comparisons for dominance check, but that can parallelize on the GPU card to assign the rank of population. G-ASREA's benefit in ranking complexity is verified as speedups of $\approx \times 5000$ are observed over dominance sorting of NSGA-II on CPU on a population of 10,000 individuals. However, the speedup was limited to $\approx \times 15$ over ASREA for a large population on two-objective problems because the dominance check of G-ASREA concerns only parallelization of ranking comparisons on the GPU.

A future research work will address G-ASREA's clustering method, in order to make it GPGPU compatible so as to obtain further speedups over ASREA and also, to solve many objective problems for which the current crowding distance clustering operator is not very effective. Further investigation is required on the size of archive for many objective problems.

References

1. Deb, K.: Multi-Objective Optimization using Evolutionary Algorithms, 1st edn. Wiley, Chichester (2001)
2. Coello, C.A.C., Lamont, G.B., Veldhuizen, D.A.V.: Evolutionary Algorithms for Solving Multi-Objective Problems. Springer, New York (2007)
3. Fonseca, C.M., Fleming, P.J.: Genetic algorithms for multi-objective optimization: Formulation, discussion, and generalization. In: Proceedings of the Fifth International Conference on Genetic Algorithms, pp. 416–423 (1993)
4. Horn, J., Nafploitis, N., Goldberg, D.E.: A niched Pareto genetic algorithm for multi-objective optimization. In: Proceedings of the First IEEE Conference on Evolutionary Computation, pp. 82–87 (1994)
5. Deb, K., Agrawal, S., Pratap, A., Meyarivan, T.: A fast and elitist multi-objective genetic algorithm: NSGA-II. IEEE Transactions on Evolutionary Computation 6(2), 182–197 (2002)
6. Goldberg, D.E.: Genetic algorithms in search, optimization and machine learning. Addison-Wesley, Reading (1989)
7. Zitzler, E., Laumanns, M., Thiele, L.: SPEA2: Improving the Strength Pareto Evolutionary Algorithm for Multiobjective Optimization. In: Giannakoglou, K., et al. (eds.) Evolutionary Methods for Design, Optimisation and Control with Application to Industrial Problems (EUROGEN 2001), International Center for Numerical Methods in Engineering (CIMNE), pp. 95–100 (2002)
8. Corne, D.W., Jerram, N.R., Knowles, J.D., Oates, M.J.: PESA-II: Region-based selection in evolutionary multiobjective optimization. In: Proceedings of the Genetic and Evolutionary Computation Conference (GECCOO 2001), pp. 283–290. Morgan Kaufmann, San Francisco (2001)
9. Baumes, L., Blansch, A., Serna, P., Tchougang, A., Lachiche, N., Collet, P., Corma, A.: Using genetic programming for an advanced performance assessment of industrially relevant heterogeneous catalysts. Materials and Manufacturing Processes 24(3) (March 2009)

10. Maitre, O., Querry, S., Lachiche, N., Collet, P.: Easea parallelization of tree-based genetic programming. In: Congress on Evolutionary Computation (CEC 2010) (2010) (to appear)
11. Wong, M.L.: Parallel multi-objective evolutionary algorithms on graphics processing units. In: GECCO 2009: Proceedings of the 11th Annual Conference Companion on Genetic and Evolutionary Computation Conference, pp. 2515–2522. ACM, New York (2009)
12. Sharma, D., Collet, P.: An archived-based stochastic ranking evolutionary algorithm (ASREA) for multi-objective optimization. In: Proceedings of the 12th Annual Conference Genetic and Evolutionary Computation Conference (GECCO 2010), pp. 479–486. ACM, New York (2010)
13. GPGPU: General-purpose computation on graphics hardware, `http://gpgpu.org/`
14. nVidia: nVidia CUDA^{TM} programming guide version 2.3, `http://developer.nvidia.com/object/cuda.html`
15. Deb, K., Agrawal, R.B.: Simulated binary crossover for continuous search space. Complex Systems 9(2), 115–148 (1995)
16. Zitzler, E., Deb, K., Thiele, L.: Comparison of multiobjective evolutionary algorithms: Empirical results. Evolutionary Computation Journal 8(2), 125–148 (2000)
17. Maitre, O., Baumes, L.A., Lachiche, N., Corma, A., Collet, P.: Coarse grain parallelization of evolutionary algorithms on gpgpu cards with easea. In: GECCO 2009: Proceedings of the 11th Annual Conference on Genetic and Evolutionary Computation, pp. 1403–1410. ACM, New York (2009)

A Test Functions

Problem definition [16]: **Minimize** $\Gamma(\mathbf{x}) = (f_1(x_1), f_2(\mathbf{x}))$; *subjected to:* $f_2(\mathbf{x}) = g(x_2, ..., x_m)h(f_1(x_1), g(x_2, ..., x_m))$, where $\mathbf{x} = (x_1, ..., x_m)$

Functions	Properties
ZDT1: $f_1(x_1) = x_1$, $g(x_2, ..., x_m) = 1 + 9.\sum_{i=2}^{m} x_i/(m-1)$, $h(f_1, g) = 1 - \sqrt{f_1/g}$, where $m = 30$, and $x_i \in [0, 1]$	Convex Pareto-Optimal (P-O) front
ZDT2: $f_1(x_1) = x_1$, $g(x_2, ..., x_m) = 1 + 9.\sum_{i=2}^{m} x_i/(m-1)$, $h(f_1, g) = 1 - (f_1/g)^2$, where $m = 30$, and $x_i \in [0, 1]$	Non-convex P-O front
ZDT3: $f_1(x_1) = x_1$, $g(x_2, ..., x_m) = 1 + 9.\sum_{i=2}^{m} x_i/(m-1)$, $h(f_1, g) = 1 - \sqrt{f_1/g} - (f_1/g)\sin(10\pi f_1)$, where $m = 30$, and $x_i \in [0, 1]$	Discontinuous convex P-O front
ZDT4: $f_1(x_1) = x_1$, $g(x_2, ..., x_m) = 1 + 10(m-1) + \sum_{i=2}^{m}(x_i^2 - 10\cos(4\pi x_i))$, $h(f_1, g) = 1 - \sqrt{f_1/g}$, where $m = 10$, $x_1 \in [0, 1]$, and $x_2, ..., x_m \in [-5, 5]$	21^9 local optimal P-O fronts (Multi-modal)
ZDT6: $f_1(x_1) = 1 - \exp(-4x_1)\sin^6(6\pi x_1)$, $g(x_2, ..., x_m) = 1 + 9.(\sum_{i=2}^{m} x_i/(m-1))^{0.25}$, $h(f_1, g) = 1 - (f_1/g)^2$, where $m = 10$, and $x_i \in [0, 1]$	1. Non-uniform distribution of P-O solutions; 2. Lowest density near the front and highest away from the front

Hybrid Directional-Biased Evolutionary Algorithm for Multi-Objective Optimization

Tomohiro Shimada, Masayuki Otani, Hiroyasu Matsushima, Hiroyuki Sato, Kiyohiko Hattori, and Keiki Takadama

The University of Electro-Communication
1-5-1 Chofugaoka, Chofu, Tokyo 182-8585 Japan
{shimada@cas.,masa-o@cas.,matsushima@cas., sato@,hattori@,keiki@}hc.uec.ac.jp

Abstract. This paper proposes the hybrid Indicator-based Directional-biased Evolutionary Algorithm (hIDEA) and verifies its effectiveness through the simulations of the multi-objective 0/1 knapsack problem. Although the conventional Multi-objective Optimization Evolutionary Algorithms (MOEAs) regard the weights of all objective functions as equally, hIDEA biases the weights of the objective functions in order to search not only the center of true Pareto optimal solutions but also near the edges of them. Intensive simulations have revealed that hIDEA is able to search the Pareto optimal solutions widely and accurately including the edge of true ones in comparison with the conventional methods.

Keywords: Multi-objective optimization, evolutionary algorithm, indicator, multi-objective knapsack problem.

1 Introduction

Recently, the several methods of the Multi-objective Optimization Evolutionary Algorithm (MOEA) are proposed, and have been much paid attention on as the methods to solve the multi-objective optimization problem effectively. However, it is easy for the conventional MOEAs (*e.g.*, Elitist Non-Dominated Sorting Genetic Algorithm (NSGA-II)[5]) to find the center of solutions, while it is hard for them to find the edge of the Pareto optimal solutions. This problem on the search causes by considering the objective functions equally when exploring the solutions, and this characteristic becomes serious problem when the whole of the true Pareto optimal solutions should be found as much as possible.

To overcome this problem, this paper proposes the hybrid Indicator-based Directional-biased Evolutionary Algorithm (hIDEA), which is extended from Indicator-Based Evolutionary Algorithm (IBEA)[3] that explores the solutions by employing *Indicator* which evaluates the dominant-subordinate relationship between two individuals to determine the fitness. In detail, hIDEA optimizes the solutions from the viewpoint of one objective function as the single-objective optimization in the beginning of the search, and gradually shifts to the multi-objective optimization. Concretely, hIDEA changes the weight parameter which

R. Schaefer et al. (Eds.): PPSN XI, Part II, LNCS 6239, pp. 121–130, 2010.

is required to calculate Indicator *i.e.*, the weight of the focused objective function comparing with others. This mechanism contributes to find the Pareto optimal solutions in a certain objective function, which becomes the trigger to find the other Pareto optimal solutions. This approach is the completely different from the conventional methods that explores the Pareto optimal solutions by considering all objective functions equally. In particular, hIDEA has the great potential of finding one of the Pareto optimal solutions by focusing on one objective function, while the conventional methods cannot guarantee it. To investigate the effectiveness of hIDEA, this paper compares hIDEA with the conventional methods through the multi-objective 0/1 knapsack problem.

This paper is organized as follows. Section 2 explains the introduction of the multi-objective optimization, and the details of hIDEA are described in Section 3. hIDEA is compared with the conventional methods through the multi-objective 0/1 knapsack problem in Section 4, and Section 5 discusses the results. Finally, our conclusion is given in Section 6.

2 Multi-Objective Optimization

The multi-objective optimization[4] searches the solutions in the trade-off relationship among a number of the objective functions (*e.g.*, if an objective function value is good, the others become bad value). It is formulated as Eq. (1).

$$\begin{cases} \max(\min)\ f_i(x_1, x_2, \cdots, x_n)\ (i = 1, \cdots, k) \\ \text{subject to } g_j(x_1, x_2, \cdots, x_n) \leq 0\ (j = 1, \cdots, m) \end{cases} \tag{1}$$

In Eq. (1), the solutions are optimized from the viewpoint of k objective functions under m conditions using n decision variables. The basic algorithm of MOEA repeats evaluation, selection, crossover and mutation as the evolution of the population (*i.e.*, solutions). In this cycle, MOEA keeps the good solutions in one generation as the archived solutions.

Typical methods of MOEA include NSGA-II (Elitist Non-Dominated Sorting Genetic Algorithm)[5] and IBEA(Indicator-Based Evolutionary Algorithm)[3]. NSGA-II is based on the ranking from the Pareto front by the non-dominated sort and the tournament selection with the crowding distance which selects the solutions with less crowding than others to search widespread solution space. IBEA, on the other hand, evaluates each solution using the fitness calculated by Indicator which treats the solutions in the same Pareto front differently by biasing them in objective function.

3 Hybrid Indicator-Based Directional-Biased Evolutionary Algorithm

3.1 Overview

As the basic mechanism, the hybrid Indicator-based Directional-biased Evolutionary Algorithm (hIDEA) is based on the directional-biased search by changing the weight of objective functions.

3.2 Directional-Biased Search

In the beginning of the search, hIDEA optimizes the solutions from the viewpoint of one objective function as the single-objective optimization and gradually shifts to the multi-objective optimization by changing the weight parameter α required to calculate *Indicator* which shows dominant-subordinate relationship between two solutions, and which is used for calculating the fitness. This mechanism enables hIDEA to find the optimal solution by biasing one objective function as shown in Fig.1, which shows the example of the optimization between two objective functions with the directional-biased search. In this figure, the objective function 1 is represented by f_1, while the objective function 2 is represented by f_2. Concretely, (1) hIDEA optimizes the edge of the solution area as the single objective optimization in the beginning of the search, and then (2) hIDEA gradually explores the solutions toward the other functions.

3.3 Indicator Calculation

Indicator function used in IBEA$_{\epsilon+}$[3] is calculated by Eq.(2).

$$I(A,B) = \min_{\epsilon} \{f_i(A) + \epsilon \geq f_i(B)\} \ (\forall i \in \{1, \cdots, m\}) \tag{2}$$

Indicator $I(A,B)$ is defined as the shortest distance ϵ, *i.e.*, the length required for the individual A to dominate the individual B. Here, smaller $I(A,B)$ is better for individual A. To calculate Indicator biased on one objective function, we improve this equation as shown in Eq.(3).

$$I(A,B) = \min_{\epsilon} \begin{cases} \alpha f_i(A) + \epsilon \geq \alpha f_i(B) \ (\forall i \in \{1, \cdots, m | i \neq j\}) \\ f_j(A) + \epsilon \geq f_j(B) \end{cases} \tag{3}$$

This equation means that hIDEA changes the bias of evaluation for the solutions by changing the weight parameter α. In detail, hIDEA considers only one function f_j (*i.e.*, it does not consider the other objective functions) by setting α to nearly 0 ($\alpha \approx 0$) in the beginning of the search, which executes the single objective optimization for one objective function. For example, Indicator is calculated when $\alpha \approx 0$ as shown in Fig.2. Concretely, in the case of the non-dominant relationship where some objective function values of one individual are

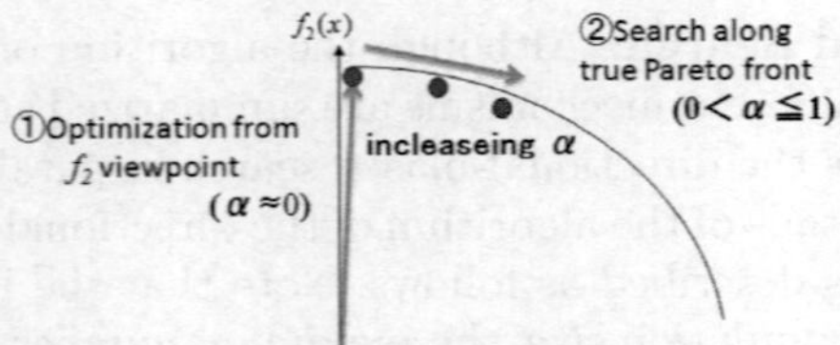

Fig. 1. Overview of directional-biased search

better than the others, Indicator is calculated by only f_2 value as Fig.2(a). In the case of the dominant relationship where all objective function values of one individual are better than the others, on the other hand, Indicator is also calculated by only f_2 as Fig.2(b). Indicator which evaluates how much $\boldsymbol{x}$ is stronger than $\boldsymbol{y}$ is $I(\boldsymbol{x}, \boldsymbol{y})$, while Indicator which evaluates how much $\boldsymbol{y}$ is stronger than $\boldsymbol{x}$ is $I(\boldsymbol{y}, \boldsymbol{x})$. In Fig.2(a), the Indicator of IBEA is calculated by considering relationship between $\boldsymbol{x}$ and $\boldsymbol{y}$, however, the Indicator of hIDEA is calculated by considering relationship between $\boldsymbol{x}'$ and $\boldsymbol{y}'$ since f_1 is not considered (because of $\alpha \approx 0$). Consequently, when the original Indicator is calculated as $I(\boldsymbol{x}, \boldsymbol{y}) = l$ and $I(\boldsymbol{y}, \boldsymbol{x}) = m$, the Indicator of hIDEA is calculated as the shortest distance between $\boldsymbol{x}'$ and $\boldsymbol{y}'$, which are $I(\boldsymbol{x}', \boldsymbol{y}') = -m'$ and $I(\boldsymbol{y}', \boldsymbol{x}') = m'$, meaning that $\boldsymbol{x}$ is better than $\boldsymbol{y}$. This indicates that hIDEA calculates Indicator of each solution by only f_2 in the beginning of the search ($\alpha \approx 0$). Fitness is calculated by following equation used in [3] with weighted Indicator value (*i.e.*, Indicator calculated by using α).

$$F(\boldsymbol{x}) = \sum_{\boldsymbol{y} \in P \setminus \{\boldsymbol{x}\}} -e^{-I(\boldsymbol{x}, \boldsymbol{y})/\kappa} \tag{4}$$

In this equation, P, κ, $\boldsymbol{x}$, and $\boldsymbol{y}$ indicate the *population*, the *fitness scaling factor* which avoids Indicator value from being divergence, the individual $\boldsymbol{x}$, and individual $\boldsymbol{y}$, respectively.

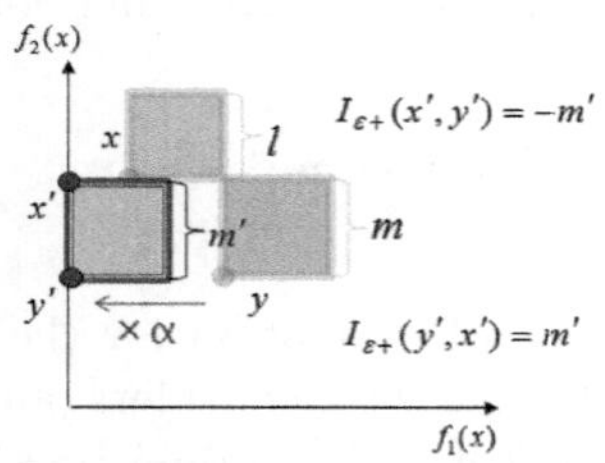

(a) Non-dominant relationship

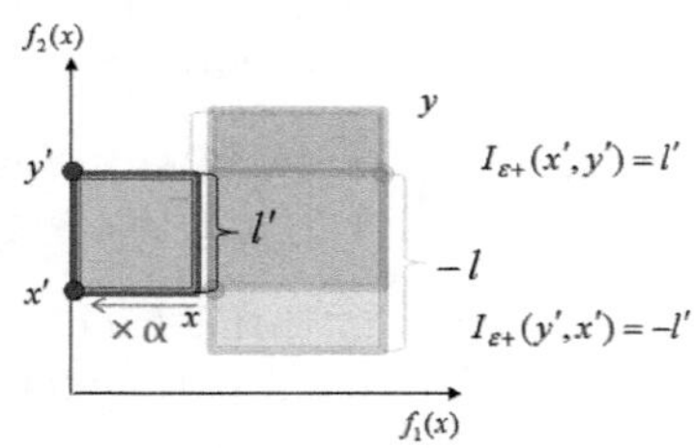

(b) Dominant relationship

Fig. 2. Indicator calculation

3.4 Hybrid Indicator-Based Directional-Biased Search

Directional-biased Search: Although the algorithm of hIDEA is almost the same as IBEA, the different mechanisms are summarized as follows: (1) to update α and (2) to execute the directional-biased search in parallel with the separated populations. The details of the algorithm of the directional-biased search for each objective function is described as follows. Note that the input parameter s, N, and κ indicate the *population size*, the *maximum number of generation*, and the *fitness scaling factor*, respectively, while the output parameter A indicates *Pareto optimal solutions*. Other parameters t, Q, and R indicate the *generation counter*, the *population after operated crossover and mutation operations, population which merges P with Q*, respectively.

Step1. Initialization

Generate an initial population P(size s)Cand set the generation counter t to 0.

Step2. Fitness assignment

1. *Determine upper bound* $\overline{b_i}$ *and lower bound* $\underline{b_i}$ *for each objective function* f_i *D*

$$\begin{aligned}\overline{b_i} &= \max_{\boldsymbol{x}\in P} f_i(\boldsymbol{x})\\ \underline{b_i} &= \min_{\boldsymbol{x}\in P} f_i(\boldsymbol{x})\end{aligned} \tag{5}$$

2. *Scale each objective function to the interval* $[0,1]$*D*

$$f'_i(\boldsymbol{x}) = (f_i(\boldsymbol{x}) - \underline{b_i})/(\overline{b_i} - \underline{b_i}) \tag{6}$$

3. *Calculate Indicator using* $f'(\boldsymbol{x})$ *from Eq.(6) instead of* $f(\boldsymbol{x})$.

$$I(\boldsymbol{x},\boldsymbol{y}) = \min_{\epsilon}\begin{cases}\alpha f'_1(\boldsymbol{x}) + \epsilon \geq \alpha f'_1(\boldsymbol{y})\\ f'_2(\boldsymbol{x}) + \epsilon \geq f'_2(\boldsymbol{y})\end{cases} \tag{7}$$

4. *Determine the maximum absolute indicator value cD*

$$c = \max_{\boldsymbol{x},\boldsymbol{y}\in P} |I(\boldsymbol{x},\boldsymbol{y})| \tag{8}$$

5. *Calculate fitness for all individualD*

$$F(\boldsymbol{x}) = \sum_{\boldsymbol{y}\in P\backslash\{\boldsymbol{x}\}} -e^{-I(\boldsymbol{x},\boldsymbol{y})/(c\cdot\kappa)} \tag{9}$$

Step3. Save archive

Copy initialized population to empty population $Q_t D P_t$ *keeps previous solutionsD*

Step4. Selection

Select parents from $Q_t D$

Step5. Crossover and mutation

Crossover the parents from Q_t *to create their children and mutate the children.*

Step6. Fitness assignment

Merge P_t *with* Q_t *to create* R_t *which size is* $2s$. *Assign fitness to* R_t *as the same as Step2.*

step7. Environmental selction

Until the size of R_t *become s, continue the following steps.*

1. *Select the individual* $\boldsymbol{x}^*$ *having the least fitness F.*
2. *Delete* $\boldsymbol{x}^*$ *from* R_t.

3. *For all individuals $\boldsymbol{x}$ in R_t, renew the fitness F using the following equation. "c" is as same as calculated in Step 2.*

$$F(\boldsymbol{x}) = F(\boldsymbol{x}) + e^{-I(\boldsymbol{x}^*,\boldsymbol{x})/(c\cdot\kappa)} \tag{10}$$

Step8. Save archive
Add 1 to the generation counter t and copy R_t to P_{t+1}. If $t > N$, stop and output Pareto optimal solutions A, which is P_{t+1} as the final Pareto set. If $t < N$, on the other hand, copy P_{t+1} to empty population Q_{t+1}.

Step9. Update α
Update α by increasing it and back to Step4.

Algorithm of hIDEA: The whole algorithm of hIDEA that employs the directional-biased search is explained. hIDEA executes the directional-biased search in parallel for each objective function. The algorithm of hIDEA is described as follows. First, the population is separated as the sub-population equally and executes the directional-biased search for the sub-separated populations in parallel. However, this approach is not enough to achieve the effective search since the searched areas become to be overlapped each other by optimizing from the opposite directions independently as shown in Fig.3 where the red and blue solutions indicate the sub separated populations. In this case, the overlapped areas should be reduced to achieve the effective search. To overcome this problem, each sub separated populations are integrated as one population when the overlapped areas are found in both the sub separated populations. For this purpose, hIDEA changes to execute IBEA when the sub-separated populations are overlapped each other. The algorithm of hIDEA is summarized as follows: As shown in Fig.3, (a)hIDEA executes the directional-biased search for the sub separated populations in each objective function in parallel; (b)when the search areas are overlapped each other, (c)they are integrated as one population and executes IBEA. In all steps from (a) to (c), to keep the diversity of solutions in the population, hIDEA introduces the mechanism to dismiss the duplicate solutions which are have same objective function values for all objectives in the population. The mechanism dismisses the duplicate solutions in the selection process. Only in the case of the number of the duplicate solutions is less than population size, the duplicate solutions are selected as parent solutions in fitness order.

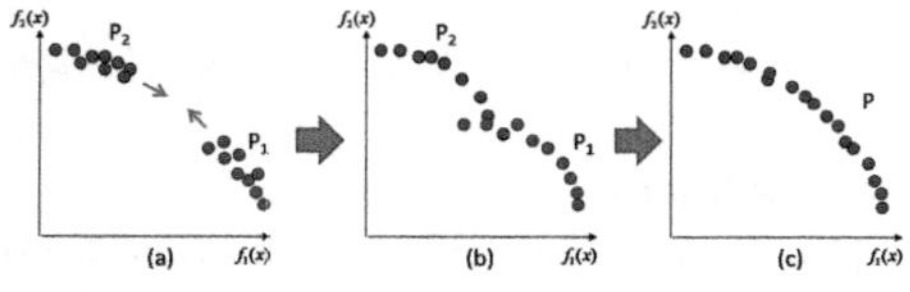

Fig. 3. The image of searching of hIDEA

4 Experiment

4.1 Experiment Design

In order to verify the effectiveness of hIDEA, we compare hIDEA with the conventional methods (*i.e.* NSGA-II and IBEA in this experiment) through multi-objective 0/1 knapsack problem[2] as representative benchmark problem of the multi-objective combinatorial optimization problem. Concretely, we employ the two objective 0/1 knapsack problem (KP500-2) and three objective one (KP100-3). Since hIDEA needs to design the update equation of the parameter α, the following update equations are considered.

$$\begin{aligned} &\text{(a) } \textit{Proportional update}: \quad \alpha = \frac{\text{Generation}}{\text{Max generation}} \\ &\text{(b) } \textit{Exponential update}: \quad \alpha = 2^{\frac{\text{Generation}}{\text{Max generation}}} - 1 \\ &\text{(c) } \qquad \textit{Power update}: \alpha = \left(\frac{\text{Generation}}{\text{Max generation}}\right)^2 \end{aligned} \tag{11}$$

In the above equations, *Generation* and *Max generation* indicate the current generation and the maximum number of generation, respectively. These equations work to start $\alpha \approx 0$ in the beginning of the search and gradually move $\alpha = 1$ according to progress of the search. Since we consider the three equations of updating α, we employ the term hIDEA_P which is hIDEA with Eq.(11)(a) as the proportional update, hIDEA_exp which is hIDEA with Eq.(11)(b) as the exponential update, and hIDEA_pow which is hIDEA with Eq.(11)(c) as the power update. The experiment parameters are summarized as shown in Table 1. The population sizes are set 250 (KP500-2) and 1000 (KP100-3) since the evaluation criteria (*i.e.*, *GD* and *IGD* described later) require enough solutions to evaluate accurately.

Table 1. Experiment Parameters

	NSGA-II	IBEA	IDEA
Max generation	10000		
Crossover rate	1.0		
Crossover type	Two points crossover		
Mutation rate	0.01		
Selection type	Tournament with crowding distance	Tournament (size 2)	
κ	-	0.05	

4.2 Evaluation Criteria

As evaluation criteria, we employ the following criteria: (1) *IGD* (Inverse General Distance)[6] and (2) *GD* (General Distance)[1] to evaluate how accurately the found Pareto optimal solutions are approximated to true ones. In detail, *GD* indicates the averaged distance from the found Pareto optimal solutions (Q) to

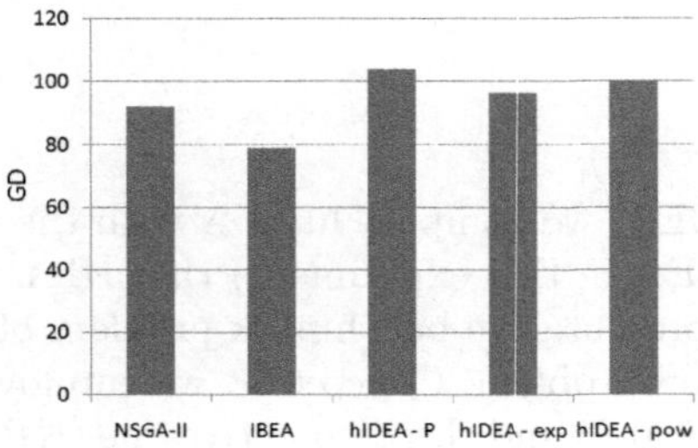

Fig. 4. *GD* (KP500-2)

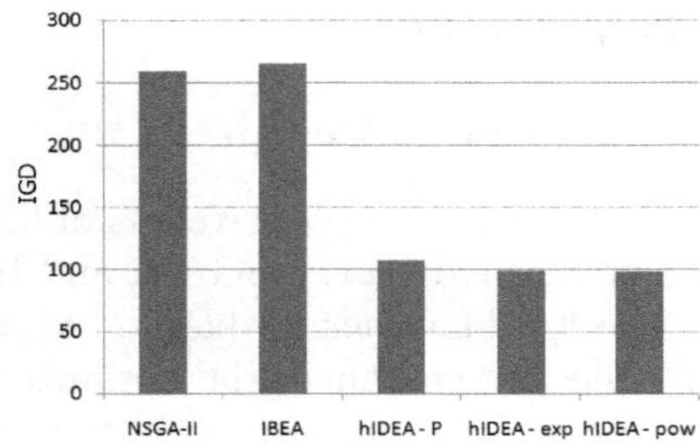

Fig. 5. *IGD* of whole area (KP500-2)

the true Pareto optimal solutions (T), while *IGD* indicates the averaged distance from the true Pareto optimal solutions (T) to the nearest found Pareto optimal solutions (Q). *GD* evaluates how much the result of Pareto optimal solutions are close to true ones, while *IGD* evaluates both of how much the found Pareto optimal solutions are close to true ones and how much the found ones are widespread area. In both *GD* and *IGD*, the less value shows the better performance. The reason why we employ *GD* and *IGD* is that the true Pareto optimal solutions is known in KP500-2 and KP100-3 published in [7]. Concretely, *GD* and *IGD* calculated by Eqs. (12) and (13).

$$GD = \frac{\sum_{i=1}^{|Q|} \delta_i}{|Q|}, \quad \delta_i = \min_{k=1}^{|T|} \sqrt{\sum_{j=1}^{m} (f_j(\boldsymbol{x}_k^{(T)}) - f_j(\boldsymbol{x}_i^{(Q)}))^2} \tag{12}$$

$$IGD = \frac{\sum_{i=1}^{|T|} \delta_i}{|T|}, \quad \delta_i = \min_{k=1}^{|Q|} \sqrt{\sum_{j=1}^{m} (f_j(\boldsymbol{x}_i^{(T)}) - f_j(\boldsymbol{x}_k^{(Q)}))^2} \tag{13}$$

In each method, we calculate both *GD* and *IGD* from every 10 trials.

4.3 Experimental Result

Figs.4 and 5 show the result of KP500-2. From these results, the solutions of hIDEA is more widespread than the conventional methods because *IGD* of hIDEAs are lower than that of the conventional methods. The difference of *IGD* of hIDEA (*i.e.*, hIDEA_P, hIDEA_exp, hIDEA_pow) is few among the different update equations. From the *GD* viewpoint, hIDEA, NSGA-II and IBEA do not show the clear difference. In KP100-3, on the other hand, both value *GD* and *IGD* of hIDEA is the best from Figs.6 and 7. Among three update equations, hIDEA_exp which increases α exponentially is the best in *GD* and *IGD*. These results suggest that hIDEA has an ability of (1) finding more widespread area than NSGA-II and (2) finding the solutions more close to the true Pareto optimal solutions than IBEA.

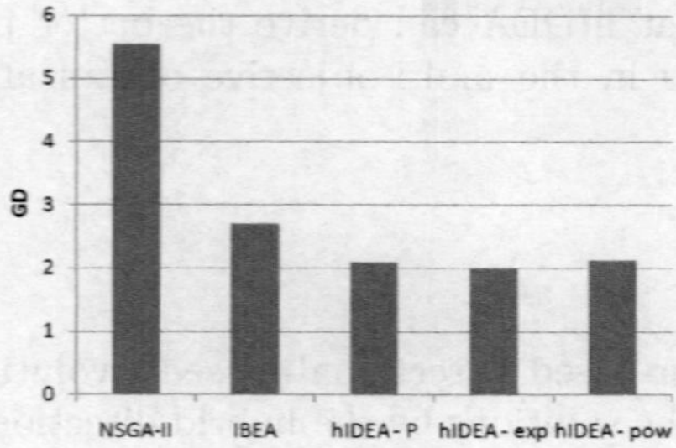

Fig. 6. *GD* (KP100-3)

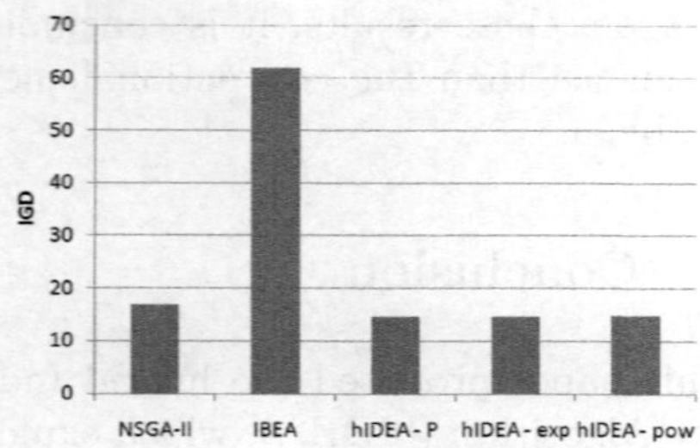

Fig. 7. *IGD* (KP100-3)

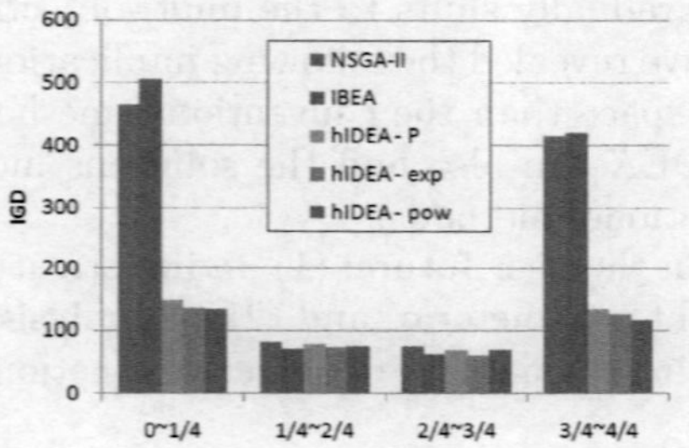

Fig. 8. *IGD* of four separated area in f_1 (KP500-2)

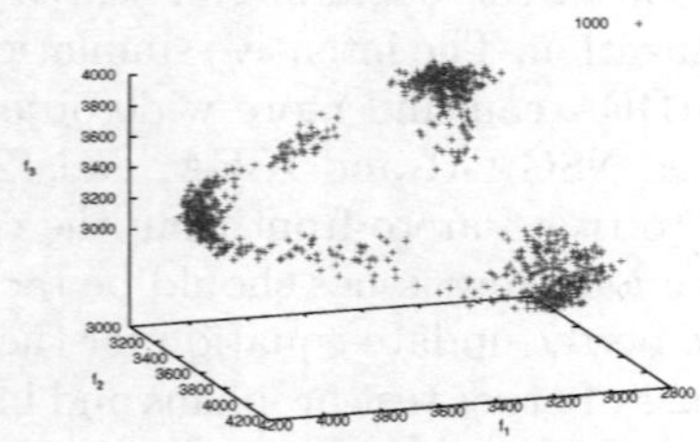

Fig. 9. Plot of hIDEA at 1000 generation (KP100-3)

5 Discussion

In KP500-2, Fig.5 shows that *IGD* of hIDEA is best. This can be also found in Fig.8, which shows the *IGD* of the four separated area in f_1 from the minimum true Pareto optimal solution (15780) to maximum one (20094)[7]. For example, the area[0, 1/4] means the value from 15780 to 16858. This result suggests that hIDEA has the strong ability of search more than the conventional methods in the area of [0, 1/4] and [3/4, 4/4]. This is caused by that hIDEA executes the single objective optimization in the beginning of the search for each sub-population towards the corresponded objective function.

In KP100-3, on the other hand, hIDEA is better than both *GD* and *IGD*, because hIDEA can find the solutions which are widely spread and are close to true Pareto front. As the large search space in KP100-3 which area is larger than KP500-2, the effectiveness of hIDEA in KP100-3 clearly appear in *GD* and *IGD*. Fig. 9 shows the progress of the search at the 1000 generations. This suggests that the populations in the early stage of search are clearly separated into the sub-populations optimized awing to each directional-biased search. The difference of *IGD* among three update equation is not clear, however, the best equation is Eq. (11)(b) which increases α exponentially because of the low *GD* in Figs.4 and 6.

From these results, it is concluded that hIDEA can derive the better performance than the conventional methods in the multi-objective optimization problem.

6 Conclusion

This paper proposed the hybrid Indicator-based Directional-biased Evolutionary Algorithm (hIDEA), which explores the solutions by the hybrid directional-biased search in parallel, and verify its effectiveness through multi-objective the 0/1 knapsack problem (*i.e.* KP500-2 and KP100-3). hIDEA optimizes the solutions from the viewpoint of one objective function as single objective optimization in the beginning of search and gradually shifts to the multi-objective optimization. The intensive simulations have revealed the following implications: (1) hIDEA can find more wide objective space than the conventional methods such as NSGA-II and IBEA; and (2) hIDEA can also find the solutions more close to true Pareto front than the conventional methods.

The following issues should be tackled in the near future: (1) an investigation of the better update equation for the weight parameter α; and (2) on analysis of the case of other test problems and increasing the number of objective functions.

References

1. Veldhuizen, D.A.V., Lamont, G.B.: On Measuring Multiobjective Evolutionary Algorithm Performance. In: Proc. of IEEE Congress on Evolutionary Computation (CEC 2000), vol. 1, pp. 204–211 (2000)
2. Zitzler, E., Thiele, L.: Multiobjective Optimization Using Evolutionary Algorithms - A Comparative Case Study. In: Proc. of the 5th International Conference on Parallel Problem Solving from Nature (PPSN V), pp. 292–304 (1998)
3. Zitzler, E., Künzli, S.: Indicator-based Selection in Multiobjective Search. In: Proc. of the 8th International Conference on Parallel Problem Solving from Nature (PPSN VIII), pp. 832–842 (2004)
4. Deb, K.: Multiobjective Optimization using Evolutionary Algorithms. Wiley, Chichester (2001)
5. Deb, K., Agrawal, S., Pratap, A., Meyarivan, T.: A Fast Elitist Non-Dominated Sorting Genetic Algorithm for Multi-Objective Optimization: NSGA-II, KanGAL report 200001 (2000)
6. Sato, H., Aguirre, H., Tanaka, K.: Local Dominance Using Polar Coordinates to Enhance Multi-objective Evolutionary Algorithms. In: Proc. of IEEE Congress on Evolutionary Computation (CEC 2004), pp. 188–195 (2004)
7. `http://www.tik.ee.ethz.ch/sop/people/zitzler/` (accessed April 1, 2010)

A Framework for Incorporating Trade-Off Information Using Multi-Objective Evolutionary Algorithms

Pradyumn Kumar Shukla, Christian Hirsch, and Hartmut Schmeck

Institute AIFB, Karlsruhe Institute of Technology
Karlsruhe, D-76128, Germany
{pradyumn.shukla,c.hirsch,hartmut.schmeck}@kit.edu

Abstract. Since their inception, multi-objective evolutionary algorithms have been adequately applied in finding a diverse approximation of efficient fronts of multi-objective optimization problems. In contrast, if we look at the rich history of classical multi-objective algorithms, we find that incorporation of user preferences has always been a major thrust of research. In this paper, we provide a general structure for incorporating preference information using multi-objective evolutionary algorithms. This is done in an NSGA-II scheme and by considering trade-off based preferences that come from so called proper Pareto-optimal solutions. We argue that finding proper Pareto-optimal solutions requires a set to compare with and hence, population based approaches should be a natural choice. Moreover, we suggest some practical modifications to the classical notion of proper Pareto-optimality. Computational studies on a number of test problems of varying complexity demonstrate the efficiency of multi-objective evolutionary algorithms in finding the complete preferred region for a large class of complex problems. We also discuss a theoretical justification for our NSGA-II based framework.

1 Introduction

Since the early nineties, multi-objective evolutionary algorithms have been used for finding a well-diverse approximation of the efficient front of a multi-objective optimization problem. Such a knowledge could then be used by the designer/ decision maker (DM) to choose a preferred solution or concentrate on a region of preferred solutions. Although such an approach has its advantages [1], this burdens the DM and is not suitable if the number of objectives is large. In contrast to evolutionary algorithms, if we look at the rich history of classical multi-objective algorithms since the late sixties, we find that incorporation of user preferences has always been a major thrust of research [2].

Working on the strengths of both classical and evolutionary approaches, in this paper, we provide a general structure for incorporating preference information using multi-objective evolutionary algorithms. This is done in an NSGA-II framework and by considering trade-off based preferences that come from proper Pareto-optimal solutions [3]. Although proper Pareto-optimal solutions

R. Schaefer et al. (Eds.): PPSN XI, Part II, LNCS 6239, pp. 131–140, 2010.

have been extensively studied in the classical domain, there is a dearth of studies in the evolutionary domain. In fact, we found just one paper investigating this [4]. This only work dealt with finding a kind of proper Pareto-optimal solutions using a constrained approach, and we found that it fails for multi-modal problems. We argue that finding proper Pareto-optimal solutions requires a set to compare with and hence, population based approaches like multi-objective evolutionary algorithms should be an ideal choice. Moreover, we suggest some practical modifications to the notion of proper Pareto-optimality. We think that none of the classical point-by-point techniques could handle such practical modifications. However, computational studies on a number of test problems of varying complexity demonstrate the efficiency of multi-objective evolutionary algorithms in finding the complete preferred region for a large class of complex problems. In addition, we discuss a theoretical justification for our NSGA-II based framework. Although we only consider preferences coming from the notion of proper Pareto-optimality, the framework that we present is able to handle any kind of preference structure. This paper adequately demonstrates the niche of population based algorithms in finding the preferred region.

The paper is structured as follows. The next section presents various notions of proper Pareto-optimality and some theoretical results that form the background of our algorithm described in Section 3. The fourth section presents extensive simulation results. Conclusions as well as extensions which emanated from this study are presented at the end of this contribution.

2 Theoretical Results

Let $f_1, \ldots, f_m : \mathbb{R}^n \to \mathbb{R}$ and $X \subseteq \mathbb{R}^n$ be given. Consider the following multi-objective optimization problem (MOP):

$$\min \mathbf{f}(\mathbf{x}) := (f_1(\mathbf{x}), f_2(\mathbf{x}), \ldots, f_m(\mathbf{x})) \qquad \text{s.t. } \mathbf{x} \in X.$$

A central optimality notion for the above problem is that of Pareto-optimality.

Definition 1 (Pareto-optimality). *A point* $\mathbf{x}^* \in X$ *is called* Pareto-optimal *if no* $\mathbf{x} \in X$ *exists so that* $f_i(\mathbf{x}) \leq f_i(\mathbf{x}^*)$ *for all* $i = 1, \ldots, m$ *with strict inequality for at least one index* i.

Let X_p denote the set of Pareto-optimal points. A criticism of Pareto-optimality is that it allows unbounded trade-offs. To avoid this, starting with the classical work of Geoffrion [3], various stronger optimality notions have been defined.

Definition 2 (Geoffrion proper Pareto-optimality [3]). *A point* $\mathbf{x}^* \in X$ *is* Geoffrion proper Pareto-optimal *if* $\mathbf{x}^* \in X_p$ *and if there exists a number* $M > 0$ *such that for all* i *and* $\mathbf{x} \in X$ *satisfying* $f_i(\mathbf{x}) < f_i(\mathbf{x}^*)$*, there exists an index* j *such that* $f_j(\mathbf{x}^*) < f_j(\mathbf{x})$ *and moreover* $(f_i(\mathbf{x}^*) - f_i(\mathbf{x}))/(f_j(\mathbf{x}) - f_j(\mathbf{x}^*)) \leq M$.

Definition 3 (M-proper Pareto-optimality [4]). *Let* $M > 0$ *be given. Then, a point* $\mathbf{x}^* \in X$ *is* M-proper Pareto-optimal *if* $\mathbf{x}^* \in X_p$ *and if for all* i *and* $\mathbf{x} \in X$ *satisfying* $f_i(\mathbf{x}) < f_i(\mathbf{x}^*)$*, there exists an index* j *such that* $f_j(\mathbf{x}^*) < f_j(\mathbf{x})$ *and moreover* $(f_i(\mathbf{x}^*) - f_i(\mathbf{x}))/(f_j(\mathbf{x}) - f_j(\mathbf{x}^*)) \leq M$.

Note that Definitions 2 and 3 are not the same as using a trade-off on fixed objectives (like in [5]). We see that trade-offs are inherent in the above definitions of proper Pareto-optimality. However, the DM could require that the constant M in the above two definitions be some (preferred) function of $\mathbf{x}$. With such a preference structure, we present a generalized notion as follows.

Definition 4 ($\mathcal{M}$-proper Pareto-optimality). *Let a function $\mathcal{M} : \mathbb{R}^n \to (0, \infty)$ be given. Then, a point $\mathbf{x}^* \in X$ is called* $\mathcal{M}$-proper Pareto-optimal *if $\mathbf{x}^* \in X_p$ and if for all i and $\mathbf{x} \in X$ satisfying $f_i(\mathbf{x}) < f_i(\mathbf{x}^*)$, there exists an index j such that $f_j(\mathbf{x}^*) < f_j(\mathbf{x})$ and moreover $(f_i(\mathbf{x}^*) - f_i(\mathbf{x}))/(f_j(\mathbf{x}) - f_j(\mathbf{x}^*)) \leq \mathcal{M}(\mathbf{x}^*)$.*

Depending on the choice of the function $\mathcal{M}$ we can obtain other proper Pareto-optimality notions. We obtain the notions in Definitions 2 and 3 by

$$\mathcal{M}(\mathbf{x}^*) := \sup_{\substack{\mathbf{x} \in X \\ f_i(\mathbf{x}) < f_i(\mathbf{x}^*) \\ f_j(\mathbf{x}^*) < f_j(\mathbf{x})}} (f_i(\mathbf{x}^*) - f_i(\mathbf{x}))/(f_j(\mathbf{x}) - f_j(\mathbf{x}^*))$$

and $\mathcal{M}(\mathbf{x}^*) := M$, respectively. Throughout the rest of this paper, we assume that $\mathcal{M}$ is given. Let us denote the set of all $\mathcal{M}$-proper Pareto-optimal solutions as $X_{\mathcal{M}}$.

Lemma 1. *For a given $\mathbf{x}^* \in X$ and a given index i let $\mathcal{S}_i$ denote the system:*

$$\begin{aligned} &-f_i(\mathbf{x}^*) + f_i(\mathbf{x}) < 0, \mathbf{x} \in X, \\ &-f_i(\mathbf{x}^*) + f_i(\mathbf{x}) < \mathcal{M}(\mathbf{x}^*)(f_j(\mathbf{x}^*) - f_j(\mathbf{x})) \quad \forall j \neq i. \end{aligned}$$

Then, $\mathbf{x}^* \in X_{\mathcal{M}}$ if and only if $\mathcal{S}_i$ is inconsistent for every index i.

Proof: If $\mathbf{x}^* \in X_{\mathcal{M}}$ then it is clear that $\mathcal{S}_i$ is inconsistent for every index i. Now, suppose $\mathcal{S}_i$ is inconsistent for every index i. We claim that $\mathbf{x}^* \in X_p$. If $\mathbf{x}^* \notin X_p$, then $\mathbf{x} \in X$ and an index l exist so that $f_l(\mathbf{x}) < f_l(\mathbf{x}^*)$ and $f_k(\mathbf{x}) \leq f_k(\mathbf{x}^*)$ for all $k \neq l$. Moreover, note that by definition $\mathcal{M}(\mathbf{x}^*) > 0$. Thus we see that system $\mathcal{S}_l$ has a solution, a contradiction.

If $\mathbf{x}^* \notin X_{\mathcal{M}}$, then there is an index i and an $\mathbf{x} \in X$ satisfying $-f_i(\mathbf{x}^*) + f_i(\mathbf{x}) < 0$ and $-f_i(\mathbf{x}^*) + f_i(\mathbf{x}) < \mathcal{M}(\mathbf{x}^*)(f_j(\mathbf{x}^*) - f_j(\mathbf{x}))$ for every j such that $-f_j(\mathbf{x}^*) + f_j(\mathbf{x}) > 0$ (such a j exists since $\mathbf{x}^* \in X_p$). For other j's, i.e., for every j such that $-f_j(\mathbf{x}^*) + f_j(\mathbf{x}) \leq 0$, $-f_i(\mathbf{x}^*) + f_i(\mathbf{x}) < \mathcal{M}(\mathbf{x}^*)(f_j(\mathbf{x}^*) - f_j(\mathbf{x}))$ is trivially true. Thus, $\mathcal{S}_i$ is consistent and we arrive at a contradiction. □

Note that in Definition 4, $\mathbf{x} \in X$. However, as shown in the next lemma, when $\mathcal{Y} := \mathrm{f}(X)$ is $\mathbb{R}^m_+$ compact, i.e., the section $(\mathbf{y} - \mathbb{R}^m_+) \cap \mathcal{Y}$ is compact for every $\mathbf{y} \in \mathcal{Y}$, then $\mathbf{x} \in X$ can be replaced by $\mathbf{x} \in X_p$.

Lemma 2. *Suppose that $\mathcal{Y}$ is $\mathbb{R}^m_+$ compact. Then, $\mathbf{x}^* \in X_{\mathcal{M}}$ if $\mathbf{x}^* \in X_p$ and if for all i and $\mathbf{x} \in X_p$ satisfying $f_i(\mathbf{x}) < f_i(\mathbf{x}^*)$, there exists an index j such that $f_j(\mathbf{x}^*) < f_j(\mathbf{x})$ and moreover $(f_i(\mathbf{x}^*) - f_i(\mathbf{x}))/(f_j(\mathbf{x}) - f_j(\mathbf{x}^*)) \leq \mathcal{M}(\mathbf{x}^*)$.*

Proof: Suppose that $\mathbf{x}^*$ satisfies the conditions of the lemma. Then following the proof of Lemma 1, we easily obtain that for all indices i, and for all $\hat{\mathbf{x}} \in X_p$, the system

$$\begin{aligned} &-f_i(\mathbf{x}^*) + f_i(\hat{\mathbf{x}}) < 0 \\ &-f_i(\mathbf{x}^*) + f_i(\hat{\mathbf{x}}) < \mathcal{M}(\mathbf{x}^*)(f_j(\mathbf{x}^*) - f_j(\hat{\mathbf{x}})) \quad \forall j \neq i, \end{aligned}$$

which we denote as $\tilde{\mathcal{S}}_i$, is inconsistent.

Take any $\mathbf{x} \in X \setminus X_p$. Since $\mathcal{Y}$ is $\mathbb{R}^m_+$ compact, there exists a $\hat{\mathbf{x}} \in X_p$ so that $f_i(\hat{\mathbf{x}}) - f_i(\mathbf{x}) \leq 0$ for all $i = 1, 2, \ldots, m$ and $f_k(\hat{\mathbf{x}}) - f_k(\mathbf{x}) < 0$ for some k. Since $\tilde{\mathcal{S}}_i$ is inconsistent for all i, we see that the system

$$\begin{aligned} -f_i(\mathbf{x}^*) + f_i(\hat{\mathbf{x}}) &< f_i(\hat{\mathbf{x}}) - f_i(\mathbf{x}) \\ -f_i(\mathbf{x}^*) + f_i(\hat{\mathbf{x}}) &< \mathcal{M}(\mathbf{x}^*)(f_j(\mathbf{x}^*) - f_j(\hat{\mathbf{x}})) + \mathcal{M}(\mathbf{x}^*)(f_j(\hat{\mathbf{x}}) - f_j(\mathbf{x})) \\ &\quad + f_i(\hat{\mathbf{x}}) - f_i(\mathbf{x}) \quad \forall j \neq i \end{aligned}$$

is inconsistent. This implies that the following system is also inconsistent:

$$\begin{aligned} &-f_i(\mathbf{x}^*) + f_i(\mathbf{x}) < 0 \\ &-f_i(\mathbf{x}^*) + f_i(\mathbf{x}) < \mathcal{M}(\mathbf{x}^*)(f_j(\mathbf{x}^*) - f_j(\mathbf{x})) \quad \forall j \neq i. \end{aligned}$$

Thus, we see that $\tilde{\mathcal{S}}_i$ is inconsistent for all i and for *any* $\mathbf{x} \in X$. Employing Lemma 1, we obtain that $\mathbf{x} \in X_{\mathcal{M}}$. □

Remark 1. It is important to highlight the importance of Lemma 2. This result shows that in order to check if a point is proper Pareto-optimal, it is sufficient to check the boundedness of the trade-offs with the Pareto-optimal points only. For any algorithm, this would drastically reduce the computational effort.

3 A NSGA-II Framework for Incorporating Trade-Offs

If we look at the definition of proper Pareto-optimality, we see the difficulties of applying any point-by-point method. The trade-off information requires working with two points from the set X_p. It is here that a population based algorithm, like an MOEA, could be helpful. Since population based algorithms work with a population, trade-off information is inherently there. It is just a matter of using this trade-off information in a way so as to obtain desired proper Pareto-optimal points. Moreover, MOEAs can be tailored for finding the *complete* set of proper Pareto-optimal points, an issue that has never been examined till now.

Taking a clue from Lemma 2, we could choose a population based MOEA which uses/ computes the non-dominated front. This is because we need at least the non-dominated front (or an approximation) to check for the appropriate trade-offs. We take the NSGA-II algorithm [6] for this purpose. Using Lemma 2, we tailor NSGA-II for finding *any* kind of proper Pareto-optimal points.

Let us assume that the decision maker's (DM's) preferences can be put together in an appropriate $\mathcal{M}$ function. From Lemma 2, we know that for checking

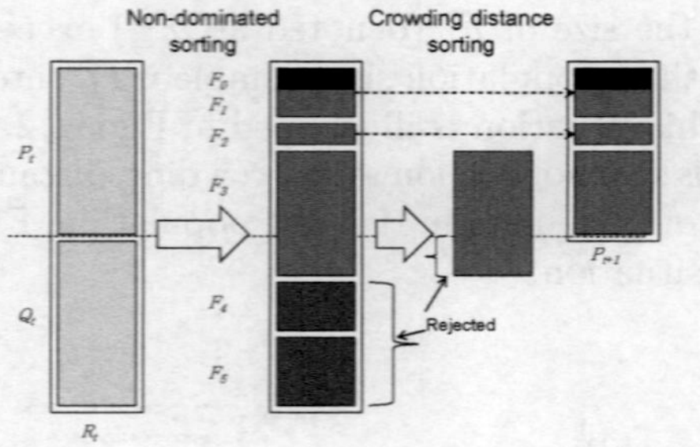

Fig. 1. Schematic showing behavior of pNSGA-II in earlier generations. F_0 is shown in black.

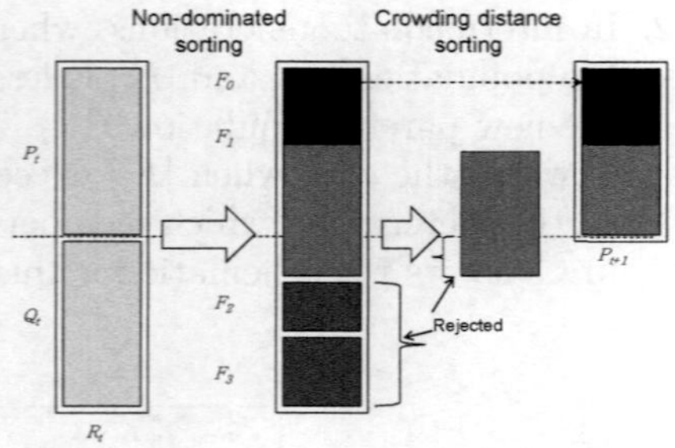

Fig. 2. Schematic showing behavior of pNSGA-II in intermediate generations. F_0 is shown in black.

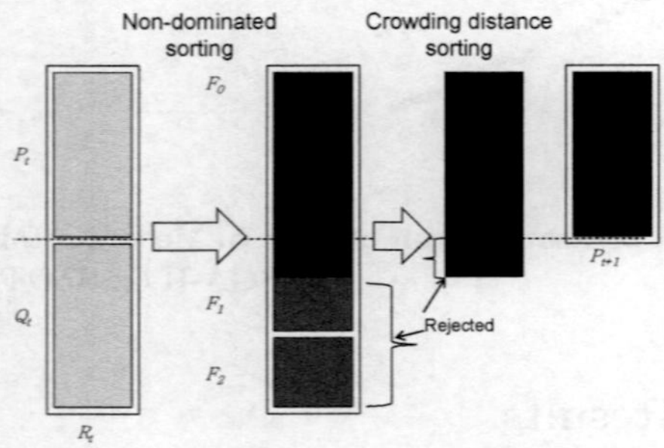

Fig. 3. Schematic showing behavior of pNSGA-II in final generations. F_0 is shown in black. In this case, the size of F_0 exceeds the population size.

the appropriate trade-offs, we need information about the X_p only. This can be introduced in the NSGA-II algorithm, which we call as pNSGA-II, in the non-dominated front at every generation. In the standard NSGA-II, the parent population P_t and the offspring population Q_t are combined and non-dominated fronts $F_1, \ldots, F_r$ of this combined population are found. We do the same in pNSGA-II, however from F_1 (best non-dominated set), we additionally find a set F_0 which satisfies the trade-offs in Definition 4, considering the entire population. So we have one additional (preferred) front. Members of F_0 are assigned a rank of 0 (better than 1, the rank of F_1, as we consider minimization in tournament selection). This extra front changes the search mechanism of NSGA-II:

1. In earlier generations, when there are many fronts, this does not change the new parent population P_{t+1} from NSGA-II. However, due to their better rank, members of F_0 win more tournaments and in this way search is steered towards them. Figure 1 shows the schematic for this situation.

2. In intermediate generations, when the size of F_1 (denoted as $|F_1|$) exceeds the population size and $|F_0|$ is less than population size, complete F_0 enters the new parent population P_{t+1}. This situation is illustrated in Figure 2.
3. Towards the end, when $|F_0|$ exceeds the population size, crowding distance sorting determines which members of F_0 remain in the new population. Figure 3 shows the schematic for this situation.

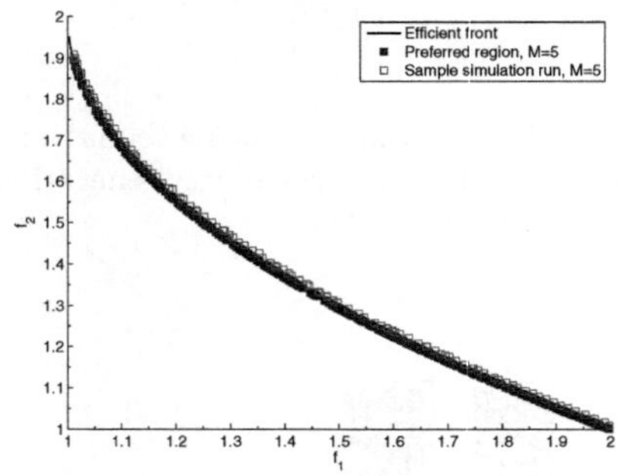

Fig. 4. Preferred front and sample run of pNSGA-II on SZDT1

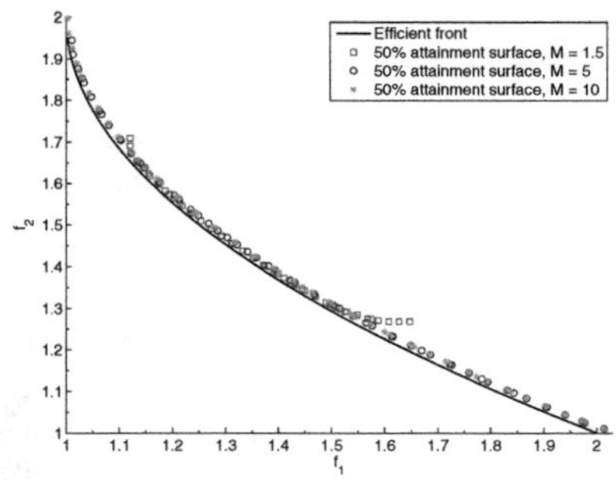

Fig. 5. Median attainment surface plot of pNSGA-II on SZDT1

4 Simulation Results

We test our pNSGA-II on a number of test problems of varying complexity, including two problems from the CEC-2007 competition (SZDT1, SZDT2) [7], four from the ZDT suite (ZDT3, ZDT4, ZDT5, ZDT6) [6], one from the DTLZ family (DTLZ4-3D) [8], and four from the WFG suite (WFG1, WFG2, with both 2 and 3 objectives) [9]. We note that this paper is among the few studies that consider ZDT5, a difficult discrete problem. This additionally shows that the approach is not limited to continuous problems. For all problems solved, we use a population of size 100 and set the maximum number of function evaluations as 20,000 (200 generations). We use a standard real-parameter SBX and polynomial mutation operator with $\eta_c = 15$ and $\eta_m = 20$, respectively [6].

In this paper, we use $\mathcal{M}(\mathbf{x}^*) := M$ with $M = 1.5, 5.0$ and 10.0. Note that these M values restrict the efficient front. This is the preferred efficient front that needs to be found. For all problems we compute a well-distributed approximation of the preferred front (reference set) as follows. Corresponding to the problem, first we generate 10,000 well-diverse points on the efficient front. Then, we calculate the preferred points, i.e., the points that satisfy the M-proper Pareto-optimality criteria (from Lemma 2, we only need to consider X_p). In order to evaluate the results, we use the Inverted generational distance (IGD) and Generational distance (GD) metrics (wrt. the obtained reference set). For statistical evaluation we use the attainment surface based statistical metric [10]. We run each algorithm for 101 times and the median (50%) attainment surface (51st) is plotted.

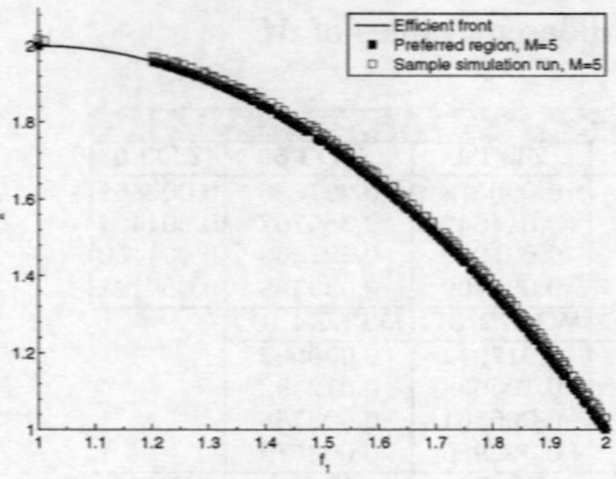

Fig. 6. Preferred front and sample run of pNSGA-II on SZDT2

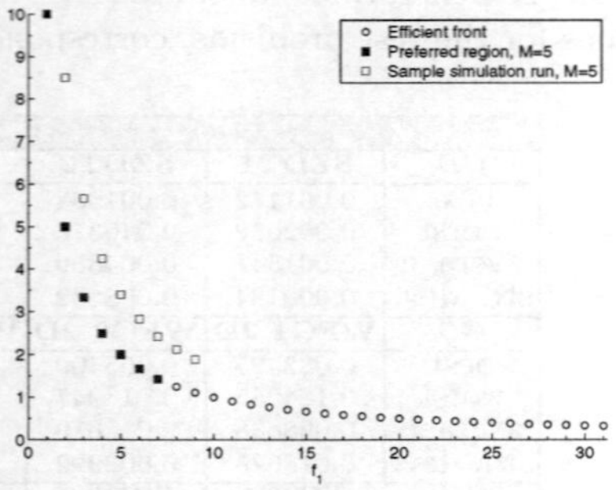

Fig. 7. Preferred front and sample run of pNSGA-II on ZDT5

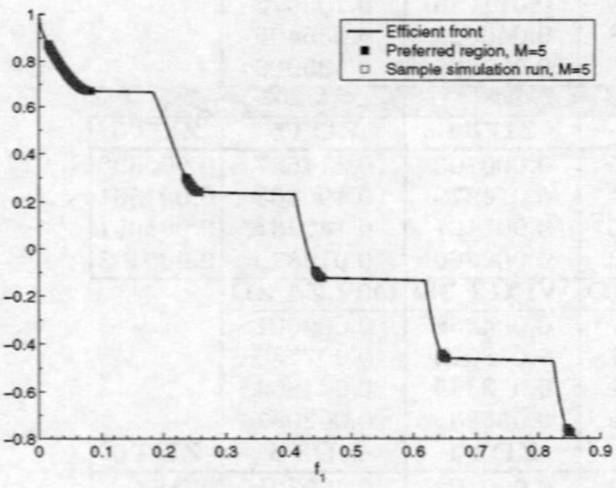

Fig. 8. Preferred front and sample run of pNSGA-II on ZDT3

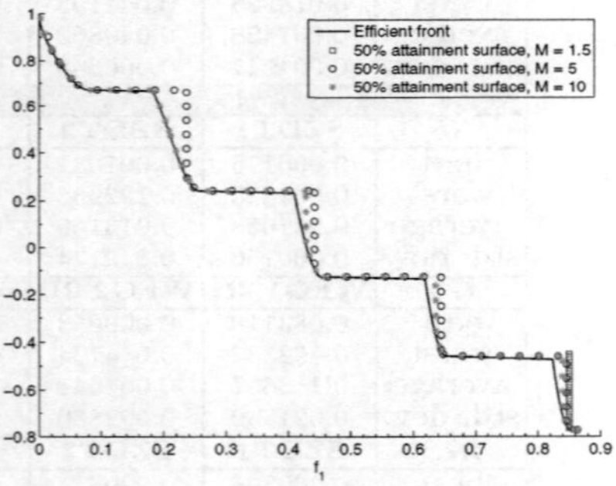

Fig. 9. Median attainment surface plot of pNSGA-II on ZDT3

The source code of pNSGA-II is made available[1]. The data files for all the 101 runs of all the problems are available on request.

Table 1 shows the performance of pNSGA-II on all the eleven test problems after 20,000 function evaluations. From the table we see that pNSGA-II is able to obtain good approximation sets which are close to the preferred efficient front (in terms of GD metric) and perform well wrt. the IGD metric. As the GD metric evaluates convergence and IGD evaluates *both* convergence and diversity, it is expected that the IGD be higher. We see this from most of the values in the table. However, alone the use of IGD metric is not sufficient. For example, we see that for the SZDT2 problem, IGD<GD (considering $M = 5.0$). This happens as the preferred front is now disconnected. The point $(1, 2)^\top$ (minimizer of f_1) is in the preferred part and if there are more population members near this one point, the GD value increases.

[1] `http://www.aifb.kit.edu/web/pNSGA-II/en`

Table 1. Generational distance (GD) and Inverted generational distance (IGD) metric values for the test problems, corresponding to different values of M

			M=10.0			
GD	**SZDT1**	**SZDT2**	**ZDT3**	**ZDT4**	**ZDT5**	**ZDT6**
best	0.001172	0.001683	0.007186	0.000578	0.227636	0.000861
worst	0.002032	0.040376	0.008701	0.014642	0.380797	0.001454
average	0.001541	0.004856	0.007982	0.001230	0.301269	0.001150
std. dev.	0.000184	0.008832	0.000284	0.001409	0.033598	0.000122
GD	**WFG1 2D**	**WFG2 2D**	**WFG1 3D**	**WFG2 3D**	**DTLZ4 3D**	
best	0.082525	0.005400	0.130897	0.017942	0.000002	
worst	0.164055	0.018947	0.143182	0.125658	0.012459	
average	0.098838	0.011619	0.137448	0.066561	0.008756	
std. dev.	0.015025	0.002992	0.002387	0.022980	0.003519	
IGD	**SZDT1**	**SZDT2**	**ZDT3**	**ZDT4**	**ZDT5**	**ZDT6**
best	0.000424	0.000636	0.000271	0.000166	0.499391	0.000625
worst	0.000746	0.020782	0.008399	0.008463	0.623377	0.000791
average	0.000574	0.004272	0.000794	0.001020	0.530842	0.000693
std. dev.	0.000070	0.006320	0.001896	0.001584	0.032058	0.000038
IGD	**WFG1 2D**	**WFG2 2D**	**WFG1 3D**	**WFG2 3D**	**DTLZ4 3D**	
best	0.046000	0.040072	0.103648	0.009658	0.006618	
worst	0.078225	0.044193	0.114329	0.071738	0.105976	
average	0.067428	0.040862	0.108758	0.040323	0.036495	
std. dev.	0.004947	0.000688	0.002145	0.028895	0.035966	
			M=5.0			
GD	**SZDT1**	**SZDT2**	**ZDT3**	**ZDT4**	**ZDT5**	**ZDT6**
best	0.000775	0.001031	0.000095	0.000103	0.114687	0.000806
worst	0.001420	0.192955	0.000179	0.014979	0.190868	0.001361
average	0.001058	0.013165	0.000116	0.001127	0.149261	0.001051
std. dev.	0.000150	0.040144	0.000013	0.002360	0.018870	0.000103
GD	**WFG1 2D**	**WFG2 2D**	**WFG1 3D**	**WFG2 3D**	**DTLZ4 3D**	
best	0.083118	0.000653	0.111507	0.004358	0.000001	
worst	0.192537	0.014764	0.154367	0.018912	0.007230	
average	0.113607	0.004541	0.121821	0.012385	0.004863	
std. dev.	0.021339	0.002860	0.004459	0.005895	0.002087	
IGD	**SZDT1**	**SZDT2**	**ZDT3**	**ZDT4**	**ZDT5**	**ZDT6**
best	0.000395	0.000571	0.000112	0.000162	0.499391	0.000611
worst	0.000681	0.027332	0.008395	0.005952	0.623377	0.000760
average	0.000519	0.006448	0.000789	0.001154	0.535262	0.000665
std. dev.	0.000061	0.008956	0.002239	0.001551	0.033893	0.000029
IGD	**WFG1 2D**	**WFG2 2D**	**WFG1 3D**	**WFG2 3D**	**DTLZ4 3D**	
best	0.057712	0.040111	0.093623	0.068945	0.011257	
worst	0.076654	0.124117	0.104347	0.072751	0.105976	
average	0.069586	0.041821	0.100340	0.070703	0.037346	
std. dev.	0.003761	0.008314	0.001857	0.001091	0.031579	
			M=1.5			
GD	**SZDT1**	**SZDT2**	**ZDT3**	**ZDT4**	**ZDT5**	**ZDT6**
best	0.000495	0.004923	0.000329	0.000055	0.102656	0.001467
worst	0.002592	0.019355	0.003407	0.003387	0.179188	0.003497
average	0.000919	0.011413	0.000928	0.000559	0.147026	0.002140
std. dev.	0.000256	0.003148	0.000367	0.000515	0.019000	0.000370
GD	**WFG1 2D**	**WFG2 2D**	**WFG1 3D**	**WFG2 3D**	**DTLZ4 3D**	
best	0.243957	0.002322	0.307302	0.001694	0.000002	
worst	0.901878	0.055353	0.801592	0.046378	0.074721	
average	0.357252	0.015852	0.483613	0.015292	0.056703	
std. dev.	0.089888	0.009038	0.113862	0.008868	0.021578	
IGD	**SZDT1**	**SZDT2**	**ZDT3**	**ZDT4**	**ZDT5**	**ZDT6**
best	0.006775	0.011828	0.101584	0.005430	0.922765	0.004852
worst	0.011382	0.041212	0.132935	0.009409	1.149415	0.005170
average	0.009299	0.025300	0.131246	0.006695	0.991575	0.005004
std. dev.	0.000879	0.007789	0.006787	0.000564	0.098511	0.000055
IGD	**WFG1 2D**	**WFG2 2D**	**WFG1 3D**	**WFG2 3D**	**DTLZ4 3D**	
best	0.066580	0.210325	0.112330	0.086209	0.058421	
worst	0.107757	0.243042	0.123535	0.097982	0.105977	
average	0.074930	0.213457	0.116712	0.092701	0.069885	
std. dev.	0.010473	0.003167	0.002157	0.002050	0.015424	

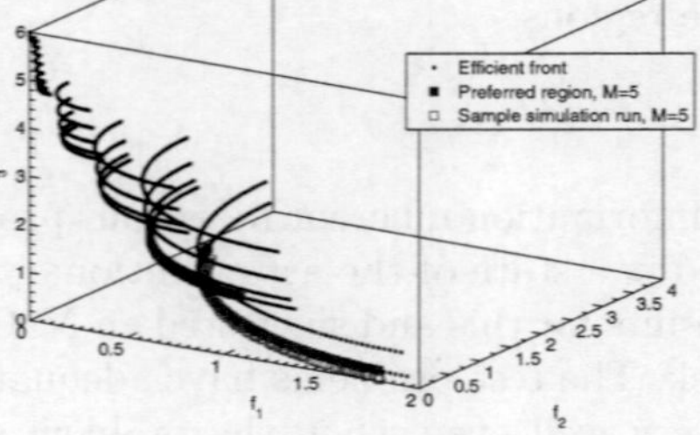

Fig. 10. Preferred front and sample run of pNSGA-II on WFG2 3D

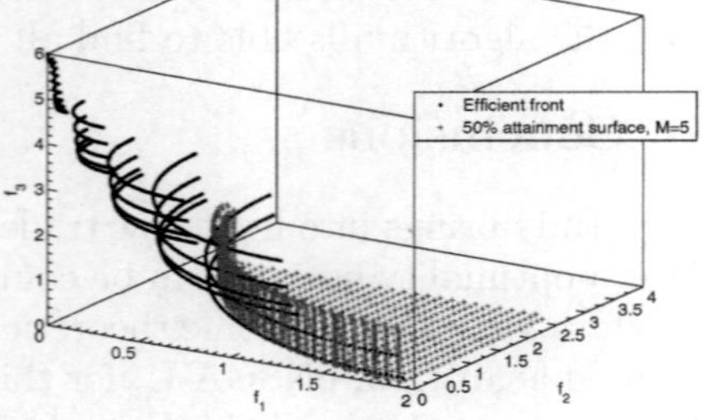

Fig. 11. Median attainment surface plot of pNSGA-II on WFG2 3D

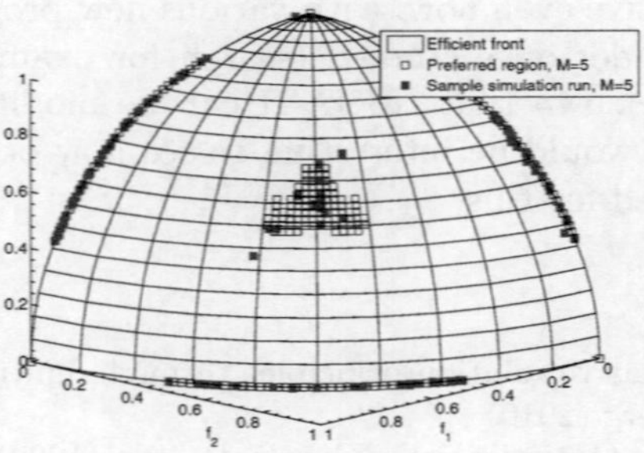

Fig. 12. Preferred front and sample run of pNSGA-II on DTLZ4 3D

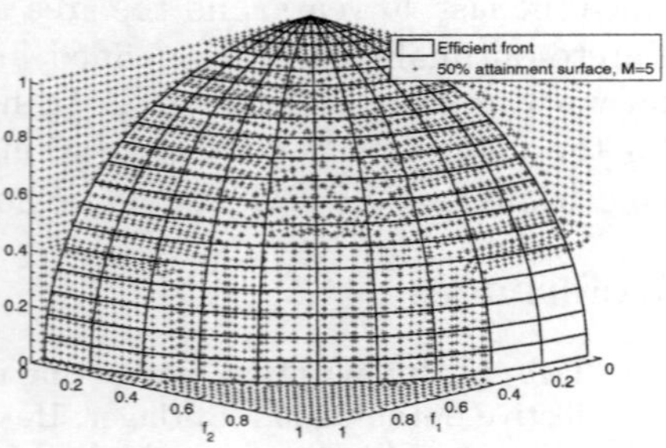

Fig. 13. Median attainment surface plot of pNSGA-II on DTLZ4 3D

Figures 4 and 5 show the performance of the pNSGA-II algorithm on the convex problem SZDT1. It can be seen that pNSGA-II is able to find a well-diverse set of preferred solutions. From Figure 5 we also see how the preferred efficient front is reduced as the value of M decreases. Figures 6 and 7 show the performance of the pNSGA-II algorithm on the non-convex problem SZDT2 and the discrete problem ZDT5. It can be seen that pNSGA-II is able to find a well-diverse set of preferred solutions for the SZDT2 problem. This is not the case in 20,000 function evaluations for ZDT5. However, we see that the obtained sample run approached the preferred region in a way so as not to explore the unnecessary regions of the efficient front. This is a nice feature of pNSGA-II emerging due to the use of front F_0 information in all the generations.

Figures 8 and 9 show the performance of the pNSGA-II algorithm on the disconnected problem ZDT3. We see how here all the *knees* are now part of the preferred front and they are all found by pNGSA-II. This example shows that pNSGA-II could also be used to find the knees by using a low M value. Figures 10, 11, 12 and 13 show the performance of pNSGA-II on three dimensional problems. The M values give rise to disconnected preferred fronts. In general, this happens even if the problem has a connected front (like DTLZ4 3D). However, the algorithm is able to find all these regions.

5 Conclusions

This study brings into light how trade-off information inherent in various proper Pareto-optimality notions can be combined in a state-of-the-art evolutionary algorithm. We presented some theoretical results for this and suggested an NSGA-II based algorithm, pNSGA-II, for this task. The test problems have adequately demonstrated that pNSGA-II performs very well even when the problem size and search space complexity is large. As this is the first study towards finding a well-diverse representation of various proper Pareto-optimal solutions, we hope that the study sheds adequate light in this area and motivates others. Proper Pareto-optimal solutions have been around in the classical literature since the last 40 years, and the area is active even now, with various new proper Pareto-optimality notions defined and older ones studied (search for example on www.ams.org/mathscinet). We have shown that NSGA-II can be modified for the task of finding these solutions. It would be interesting to see how other algorithms, like SPEA-2, could be adapted for this.

References

[1] Deb, K.: Innovization: Extracting innovative solution principles through multiobjective optimization. Springer, Heidelberg (2010)
[2] Miettinen, K.: Nonlinear Multiobjective Optimization. Kluwer, Boston (1999)
[3] Geoffrion, A.M.: Proper efficiency and the theory of vector maximization. Journal of Mathematical Analysis and Applications 22, 618–630 (1968)
[4] Shukla, P.K.: In search of proper pareto-optimal solutions using multi-objective evolutionary algorithms. In: Shi, Y., van Albada, G.D., Dongarra, J., Sloot, P.M.A. (eds.) ICCS 2007. LNCS, vol. 4490, pp. 1013–1020. Springer, Heidelberg (2007)
[5] Branke, J., Kauβler, T., Schmeck, H.: Guidance in evolutionary multi-objective optimization. Advances in Engineering Software 32, 499–507 (2001)
[6] Deb, K.: Multi-objective optimization using evolutionary algorithms. Wiley, Chichester (2001)
[7] Huang, V.L., Qin, A.K., Deb, K., Zitzler, E., Suganthan, P.N., Liang, J.J., Preuss, M., Huband, S.: Problem definitions for performance assessment of multi-objective optimization algorithms. Technical report, NTU, Singapore (2007)
[8] Deb, K., Thiele, L., Laumanns, M., Zitzler, E.: Scalable test problems for evolutionary multi-objective optimization. In: Abraham, A., et al. (eds.) Evolutionary Multiobjective Optimization, pp. 105–145. Springer, London (2005)
[9] Huband, S., Hingston, P., Barone, L., White, L.: A review of multi-objective test problems and a scalable test problem toolkit. IEEE Transactions on Evolutionary Computation 10, 280–294 (2005)
[10] Fonesca, C.M., Fleming, P.J.: On the performance assessment and comparison of stochastic multiobjective optimizers. In: Ebeling, W., Rechenberg, I., Voigt, H.-M., Schwefel, H.-P. (eds.) PPSN 1996. LNCS, vol. 1141, pp. 584–593. Springer, Heidelberg (1996)

Topography-Aware Sensor Deployment Optimization with CMA-ES

Vahab Akbarzadeh, Albert Hung-Ren Ko, Christian Gagné, and Marc Parizeau

Laboratoire de vision et systèmes numériques (LVSN), Département de génie électrique et de génie informatique, Université Laval, Québec (QC), G1V 0A6, Canada

{vahab.akbarzadeh.1,albert.ko.1}@ulaval.ca,
{christian.gagne,marc.parizeau}@gel.ulaval.ca

Abstract. Wireless Sensor Networks (WSN) have been studied intensively for various applications such as monitoring and surveillance. Sensor deployment is an essential part of WSN, because it affects both the cost and capability of the sensor network. However, most deployment schemes proposed so far have been based on over-simplified assumptions, where results may be far from optimal in practice. Our proposal aims at automating and optimizing sensor deployment based on realistic topographic information, and is thus different from previous work in two ways: 1) it takes into account the 3D nature of the environment ; 2) it allows the use of anisotropic sensors. Based on the Covariance Matrix Adaptation Evolution Strategy (CMA-ES), the proposed approach shows good potential for tackling diverse problems in the WSN domain. Preliminary results are given for a mountainous area of North Carolina where coverage is maximized.

1 Introduction

In recent years, Wireless Sensor Networks (WSN) have been studied intensively for various applications such as environmental monitoring and surveillance. A WSN usually consists of numerous wireless devices deployed in a region of interest, each able to collect and process environmental information and communicate with neighbour devices [2, 9, 19].

Sensor deployment is an essential issue in WSN, as it affects how well a region is monitored by sensors. Considering a region monitored by sensors, one of the most critical issues is the region coverage [9, 11, 12, 13, 19, 20]. One goal of a WSN is that each location in a region should be within the sensing range of at least one sensor. An alternative approach is to have a region covered simultaneously by at least K sensors [19, 20].

Many deterministic methods have been explored to address the problem of coverage. It has been shown that covering an area with disks of equal radius can be done in an optimal manner [2, 9, 11]. Similar results have been reported when multiple coverage of the target area is required [2, 12, 19, 20]. Besides, the majority of optimization methods proposed are deterministic, and are generally functions of omnidirectional sensor with a fixed sensing range.

R. Schaefer et al. (Eds.): PPSN XI, Part II, LNCS 6239, pp. 141–150, 2010.

However, most suggested methods are based on over-simplified assumptions [12, 14, 15, 19, 20], and the theoretical perfect coverage shown in these deterministic methods may not hold true in practice for a number of reasons. First, most sensor deployment optimization methods assume that sensors are placed on a 2D plane, without taking terrain into account [2, 9, 11]. Second, most deterministic methods suppose that sensors have omnidirectional sensing capabilities, which is generally not accurate [10]. For instance, antennas have different 3D reception area, depending on factors like orientation, distance, and other environmental factors [10].

The disadvantages of deterministic deployment optimization methods are thus evident, and the 100% coverage that they claim is often over-estimated. This issue is critical because it further complicates the problem of sensor deployment: while WSN seems to satisfy the requirements to achieve full coverage on a target area using a deterministic method, the deployers have no means to ensure that this coverage is truly effective in the real environment. This uncertainty of coverage thus presents a challenge in sensor deployment.

Facing this challenge, we follow a more flexible non-deterministic avenue. Our aim is to achieve automated sensor deployment optimization based on realistic topographic terrain information, and realistic sensor modelling. It differs from traditional deterministic methods in that: 1) deterministic schemes only consider 2D environments and ignore the effects of elevation, whereas our method takes into account the 3D terrain information; 2) deterministic schemes usually assume omnidirectional sensors, whereas our method allows for constraints to be applied on sensors, such as limited sensing angles.

In an effort to tackle this more realistic problem, we opt for an evolutionary algorithm approach. Some prior work has been conducted with such paradigms [3, 16], but using more or less the same over-simplifying assumptions as the deterministic approaches. Among available evolutionary algorithms, we chose the Covariance Matrix Adaptation Evolution Strategy (CMA-ES) [8] for its good performance and stability [6, 7]. The position and orientation of the sensors are encoded inside an individual, and a population of individuals is evolved through generations. At the end of the evolution, the individual with the best coverage is chosen as the final solution. This CMA-ES optimization is linked to a Geographical Information System (GIS) to provide essential environmental data such as elevation of region of interest and obstacles in the area, to compute the fitness of individuals.

The remainder of the paper is organized as follows. The problem statement is presented in the next section (Sec. 2), followed by a presentation of the proposed method (Sec. 3). The experimental protocol and results are then summarized (Sec. 4), concluding the paper with discussions and perspectives (Sec. 5).

2 Problem Statement

The main objective of this proposal is to build a realistic model of the environment and sensor network, and to optimize the sensor deployment accordingly.

The sensing model depends on distance, orientation, and visibility. We first assume that all sensors are positioned at a certain constant height τ above the ground level. The sensor position is thus described by a 3D point $\mathbf{p} = (x, y, z)$, where (x, y) are free parameters, and $z = g(x, y)$ is constrained by the terrain elevation at position (x, y), as defined by a GIS. We further assume that the anisotropic properties of sensors are fully defined by a pan angle θ around the vertical axis (tilt angle is currently assumed null). Given the GIS, a sensor network $N = \{\mathbf{s}_1, \mathbf{s}_2, \ldots, \mathbf{s}_n\}$ of n sensors is thus fully specified by $3n$ free parameters $\mathbf{s}_i = (\mathbf{p}_i, \theta_i)$, $i = 1, 2, \ldots, n$, with $\mathbf{p}_i = (x_i, y_i)$.

Now the coverage $C(\mathbf{s}_i, \mathbf{q})$ of sensor $\mathbf{s}_i$ at point $\mathbf{q}$ in the environment can be defined as a function of distance $d(\mathbf{s}_i, \mathbf{q}) = ||\mathbf{p}_i - \mathbf{q}||$, angle of view $a(\mathbf{s}_i, \mathbf{q}) = \theta_i - \angle(\mathbf{q} - \mathbf{p}_i)$, and visibility $v(\mathbf{s}_i, \mathbf{q})$ from the sensor:

$$C(\mathbf{s}_i, \mathbf{q}) = f[\mu_d(||\mathbf{p}_i - \mathbf{q}||), \mu_a(\theta_i - \angle(\mathbf{q} - \mathbf{p}_i)), v(\mathbf{s}, \mathbf{q})], \tag{1}$$

where $\angle(\mathbf{q} - \mathbf{p}_i)$ is the pan angle of point $\mathbf{q}$ relative to $\mathbf{p}_i$. For $\mathbf{q}$ to be covered by sensor $\mathbf{s}_i$, it needs to be within its sensing range AND its field of view AND must be visible, that is not blocked by any terrain obstacle such as hills. Let $\mu_d, \mu_a \in [0, 1]$ represent the fuzzy membership functions of the first two conditions, then Eq. 1 can be rewritten as:

$$C(\mathbf{s}_i, \mathbf{q}) = \min \left\{ \begin{array}{c} \mu_d(||\mathbf{p}_i - \mathbf{q}||) \\ \mu_a(\theta_i - \angle(\mathbf{q} - \mathbf{p}_i)) \\ v(\mathbf{s}_i, \mathbf{q}) \end{array} \right\}. \tag{2}$$

Function $v(\mathbf{s}_i, \mathbf{q})$ is usually binary. If the line of sight between $\mathbf{s}_i$ and $\mathbf{q}$ is obscured, then we assume that the coverage cannot be met ($v = 0$), otherwise the visibility condition is fully respected ($v = 1$). Fig. 1 illustrates different scenarios that assume rotational topographic symmetry. For real environments, the visibility induces coverage which can produce many more complex shapes. For instance, Fig. 2 gives a concrete example of an environment and how the visibility condition can affect sensor coverage in this environment.

At each position $\mathbf{q} \in \Xi$ of environment Ξ, the coverage for a single sensor is thus the minimum of the three above conditions. Value $C = 1$ means full coverage, and $C = 0$ indicates no coverage. If more than one sensor covers $\mathbf{q}$, then we can compute the local network coverage C_l using:

$$C_l(N, \mathbf{q}) = \max_{i=1,\ldots,n} C(\mathbf{s}_i, \mathbf{q}), \tag{3}$$

and the global coverage C_g using:

$$C_g(N, \Xi) = \frac{1}{|\Xi|} \sum_{\mathbf{q} \in \Xi} C_l(N, \mathbf{q}). \tag{4}$$

Given an environment Ξ, the problem statement is thus to determine the sensor network deployment N that maximizes $C_g(N, \Xi)$.

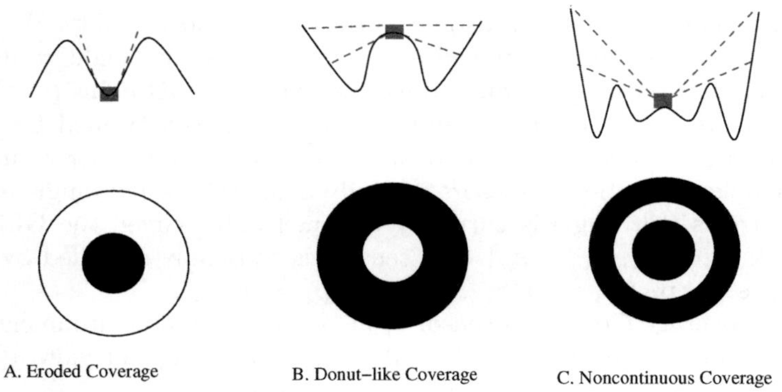

Fig. 1. Examples of visibility induced coverage. The upper row shows different terrain elevations as curves. The small rectangle boxes are sensors. The lower rank shows the true coverage of sensors with those terrain elevations and sensor positions (assuming rotational symmetry).

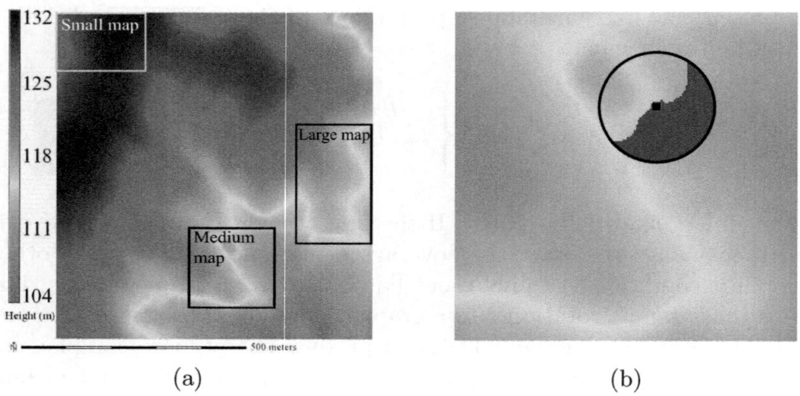

Fig. 2. Impact on coverage of visibility conditions for a given topographical map: (a) elevation is depicted by colour (the subsections of the environment surrounded inside the rectangles will be used in the result section to evaluate the performance of the algorithm); (b) assuming that the sensor is placed in the medium map, the small black square depicts the sensor location, the grey area exposes the induced visibility mask, while the black circle represents the maximum sensing range (assuming an omnidirectional sensor)

3 Methodology

The previous problem statement suggests a straight forward evolutionary algorithm solution. We choose the Evolution Strategy (ES) paradigm and, in particular, the Covariance Matrix Adaptation Evolution Strategy (CMA-ES) algorithm [8]. In our simulations, we attempt to gain insights into these three scenarios:

a) Deterministic deployment with 360° sensors;
b) CMA-ES deployment with 360° sensors;
c) CMA-ES deployment with 90° sensors.

For each of the three scenarios, sensors are positioned at $\tau = 1$ meter above the ground, and coverage is computed using Eq. 4. The fuzzy sets used, μ_d and μ_a, are crisp sets:

$$\mu_d(\delta) = \begin{cases} 1 & \text{if } 0 \leq \delta \leq r_s \\ 0 & \text{otherwise} \end{cases}$$

and either:

$$\mu_a(\theta) = \begin{cases} 1 & \text{if } -180 \leq \theta \leq 180 \\ 0 & \text{otherwise} \end{cases}$$

for 360° sensors, or:

$$\mu_a(\theta) = \begin{cases} 1 & \text{if } -45 \leq \theta \leq 45 \\ 0 & \text{otherwise} \end{cases}$$

for 90° sensors.

The deterministic method has been shown to achieve full coverage on the Cartesian plane [2, 9]. Fig. 3 illustrates this deployment pattern, where sensors are organized in layers of horizontal strips. Assuming sensors with sensing range r_s, they are simply distributed $\sqrt{3}\,r_s$ apart on every strip, and the strips are themselves separated from one another by $\frac{3}{2}r_s$. Furthermore, the strips are interleaved to form a triangular lattice pattern.

To conduct our experiments, we selected a mountainous area in North Carolina. The data come from a raster layer map in the "OSGeo Edu" dataset [18], that stores geo-spatial information about parts of North Carolina State, USA. More specifically, we focus on a portion of the map that covers a small watershed in a rural area near NC capital city, Raleigh. The coordinate system of the map is the NC State Plane (Lambert Conformal Conic projection), metric units and NAD83 geodetic datum. We used three portions of the map for our experiments (See Fig. 4, 5, and 6).

The full GIS data can be read using an open-source GIS software called Geographic Resources Analysis Support System (GRASS) [17], and these data provide crucial information on the terrain information of the target region (See Fig. 2). With the environmental data provided by GRASS, CMA-ES can carry out the optimization task by modifying positions and orientations of deployed sensors. CMA-ES is implemented using Evolutionary Algorithms in Python (EAP) [5], an open source software developed at the Computer Vision and Systems Laboratory of Université Laval.

4 Results

The algorithm's parameters include the number of variables, the population size (μ), the number of offspring (λ), the mutation factor (σ), and the number of

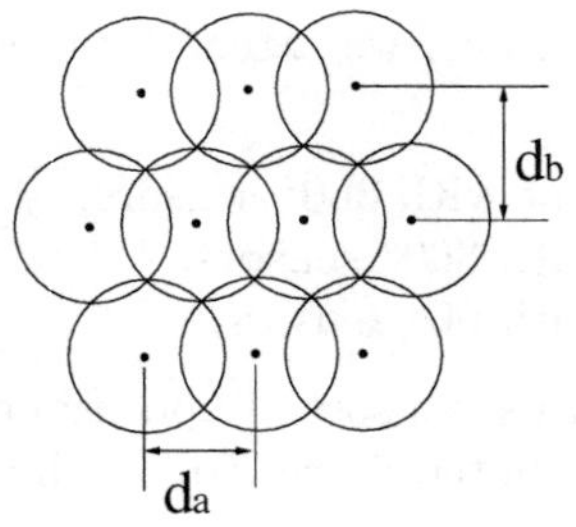

Fig. 3. Pattern of the deterministic method [2, 9] implemented in the paper, where $d_a = \sqrt{3}r_s$, $d_b = \frac{3}{2}r_s$, and r_s is the sensing range for a sensor. Circles are sensor sensing ranges, and dots are sensor positions.

Table 1. Parameters used for the CMA-ES runs

Parameter	Small map 360°	Med. map 360°	Large map 360°	Small map 90°
Dimensionality	24	32	48	144
μ	6	7	7	9
λ	13	14	15	18
σ	0.167	0.167	0.167	0.167
Generations	350	350	350	450

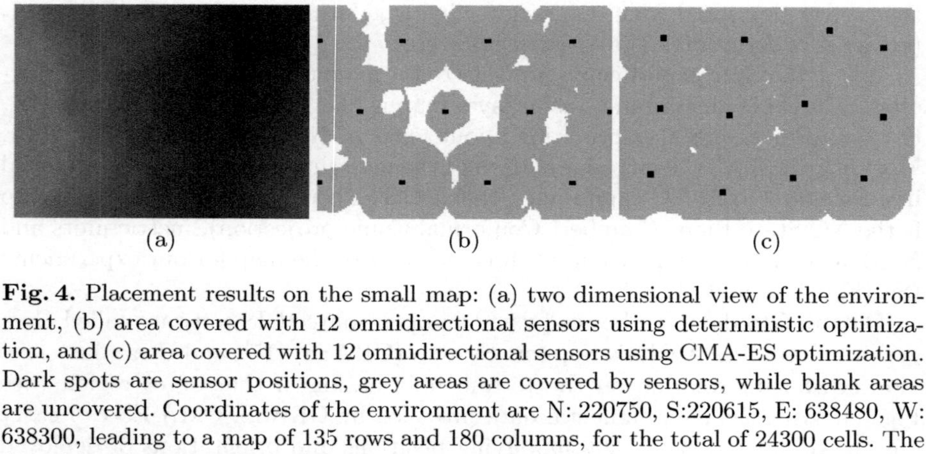

(a) (b) (c)

Fig. 4. Placement results on the small map: (a) two dimensional view of the environment, (b) area covered with 12 omnidirectional sensors using deterministic optimization, and (c) area covered with 12 omnidirectional sensors using CMA-ES optimization. Dark spots are sensor positions, grey areas are covered by sensors, while blank areas are uncovered. Coordinates of the environment are N: 220750, S:220615, E: 638480, W: 638300, leading to a map of 135 rows and 180 columns, for the total of 24300 cells. The data (elevation of terrain) range from 123.9 m to 131.5 m.

generations through which the algorithm runs. Tab. 1 summarizes these values. As for the sensors, we assumed that these are Passive Infrared (PIR) sensors with a sensing range of 30 meters.

We have tested the system on three different portions of the environment. While the deterministic method is supposed to achieve full coverage in each

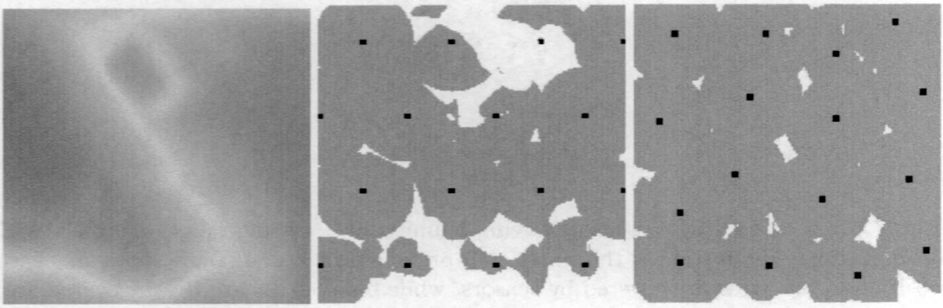

(a) (b) (c)

Fig. 5. Placement results on the medium map: (a) two dimensional view of the environment, (b) area covered with 16 omnidirectional sensors using deterministic optimization, and (c) area covered with 16 omnidirectional sensors using CMA-ES optimization. Dark spots are sensor positions, grey areas are covered by sensors, while blank areas are uncovered. Coordinates of the environment are N: 220250, S:220070, E: 638786, W: 638606, leading to a map of 180 rows and 180 columns, for a total of 32400 cells. The data (elevation of terrain) range from 109.7 m to 120.3 m.

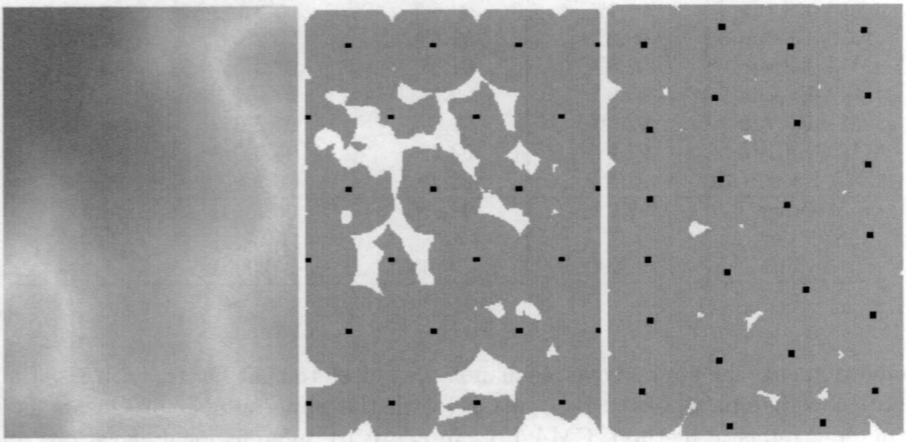

(a) (b) (c)

Fig. 6. Placement results on the large map: (a) two dimensional view of the environment, (b) area covered with 24 omnidirectional sensors using deterministic optimization, and (c) area covered with 24 omnidirectional sensors using CMA-ES optimization. Dark spots are sensor positions, grey areas are covered by sensors, while blank areas are uncovered. Coordinates of the environment are N: 220490, S:220220, E: 639000, W: 638820, leading to a map of 270 rows and 180 columns, for a total of 32400 cells. The data (elevation of terrain) range from 111.5 m to 123.8 m.

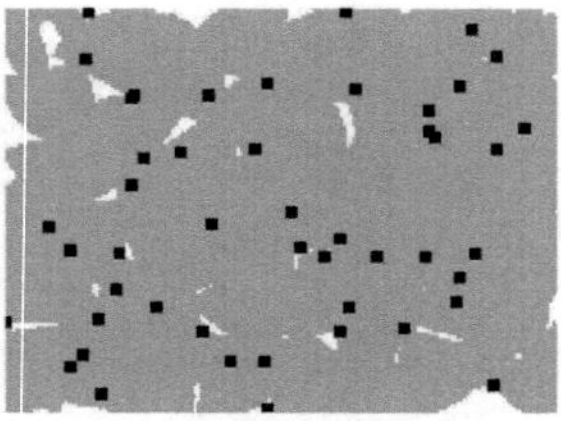

Fig. 7. Area covered by 48 sensors having a limited 90° sensing range, using CMA-ES optimization. The terrain is the small map presented in Fig. 4. Dark spots are sensor positions, grey areas are covered by sensors, while blank areas are the uncovered area.

Table 2. Coverage percentage on the target areas with various numbers of sensors and approaches. The p-value shows the probability of the CMA-ES performance being statistically similar to the deterministic one.

Method	Small map 360°	Med. map 360°	Large map 360°	Small map 90°
Deterministic	85.9%	74.3%	86.1%	–
CMA-ES run 1	98.1%	95.8%	94.5%	96.7%
CMA-ES run 2	98.9%	94.1%	98.0%	94.3%
CMA-ES run 3	95.9%	88.1%	97.1%	94.3%
CMA-ES run 4	95.7%	94.6%	97.5%	95.9%
CMA-ES run 5	98.0%	92.8%	95.2%	94.3%
CMA-ES run 6	97.8%	91.0%	97.3%	93.7%
CMA-ES run 7	95.3%	94.7%	97.9%	96.8%
CMA-ES run 8	97.6%	91.8%	97.2%	96.4%
CMA-ES run 9	97.2%	92.2%	95.6%	90.8%
CMA-ES run 10	97.4%	93.2%	96.2%	95.8%
Average	97.2%	92.8%	96.7%	94.9%
Std. dev.	1.2%	2.2%	1.2%	1.8%
p-value	$p < 0.0001$	$p < 0.0001$	$p < 0.0001$	–

environment, the actual coverage is not even close to that figure, with less than 90% coverage in all cases. By contrast, with the same number of sensors as in the deterministic method, CMA-ES can adapt to different local elevations and thus achieve significantly better coverage (see Tab. 2).

If we add the constraint of sensing angle, the problem is even more challenging. No deterministic method has ever been proposed to solve this type of problem. However, one possible deterministic approach is to deploy four sensors instead of one sensor at the optimal positions, and each sensor will be deployed in a way that four sensors together can have an omnidirectional sensing angle. A clear drawback of this deterministic approach is that we need four times as many sensors to cover the region of interest. What is worse, the coverage is not optimal, as demonstrated before. However, an evolutionary based method such as CMA-ES again has proven its ability to deal with these problem. Using 48 sensors with 90° of sensing angle, CMA-ES demonstrates its capability to adapt to the environment and fine-tune the orientation of each sensor deployed (See Fig. 7).

CMA-ES does not only optimize sensor positions, but also sensor orientations. This capability is critical because a large proportion of existing sensors are not omnidirectional, such as vision sensors, and the need to deploy them in an efficient and optimal manner is thus of great importance.

5 Conclusion

Experimental results on topography-aware sensor deployment with CMA-ES suggest that the proposed method is fully feasible and shows good promise in optimizing sensor deployment. This project is the very first scheme to construct a realistic model for sensor deployment. To our knowledge, no similar initiatives have ever been reported in the literature.

This serves as a starting point to further investigate the use of evolutionary algorithms in sensor deployment optimization. One of our future works is to implement a probabilistic or fuzzy sensing range models rather than traditional disk-like-models [2,11]. Although some probabilistic sensing range models [1,2,9, 11,21] and sensing models with irregular sensing range [4] have been proposed, without any exception they all work on a 2D flat space with omnidirectional sensors. The combinational effects of terrain variations, of constraint sensing angle or irregular sensing range, and probabilistic sensing property of sensors have never been studied. Moreover, another potential future project will be the multi-objective optimizations for sensor deployment, given the multiple concerns such as number of sensors used, energy saving, multiple coverage, and robustness of the network to sensor failures.

Acknowledgment

This work was supported by grant SII-PIV-70 from the GEOIDE Network of Centres of Excellence (Canada). The authors are grateful to Annette Schwerdtfeger for proofreading this manuscript.

References

1. Ahmed, N., Kanhere, S.S., Jha, S.: Probabilistic coverage in wireless sensor networks. In: LCN 2005: Proceedings of the The IEEE Conference on Local Computer Networks 30th Anniversary, pp. 672–681 (2005)
2. Bai, X., Kumar, S., Xuan, D., Yun, Z., Lai, T.H.: Deploying wireless sensors to achieve both coverage and connectivity. In: Proceedings of the 7th ACM International Symposium on Mobile Ad Hoc Networking and Computing, pp. 131–142 (2006)
3. Bhondekar, A.P., Vig, R., Singla, M.L., Ghanshyam, C., Kapur, P.: Genetic algorithm based node placement methodology for wireless sensor networks. In: Proceedings of the International MultiConference of Engineers and Computer Scientists, vol. 1 (2009)

4. Boukerche, A., Fei, X.: A coverage-preserving scheme for wireless sensor network with irregular sensing range. Ad Hoc Network 5(8), 1303–1316 (2007)
5. De Rainville, F.-M., Fortin, F.-A., Gagné, C., Parizeau, M.: Evolutionary Algorithms in Python (2010), `http://deap.googlecode.com`
6. García, S., Molina, D., Lozano, M., Herrera, F.: A study on the use of non-parametric tests for analyzing the evolutionary algorithms' behaviour: a case study on the CEC'2005 Special Session on Real Parameter Optimization. Journal of Heuristics 6(15), 617–644 (2009)
7. Hansen, N.: Compilation of results on the CEC benchmark function set. Technical Report, Institute of Computational Science, ETH Zurich, Switzerland (2005)
8. Hansen, N., Ostermeier, A.: Completely derandomized self-adaptation in evolution strategies. Evolutionary Computation 9(2), 159–195 (2001)
9. Hefeeda, M., Ahmadi, H.: Energy efficient protocol for deterministic and probabilistic coverage in sensor networks. IEEE Transactions on Parallel and Distributed Systems 99, 579–593 (2009)
10. Holland, M.M., Aures, R.G., Heinzelman, W.B.: Experimental investigation of radio performance in wireless sensor network. In: Proceedings of IEEE SECON (2006)
11. Kar, K., Banerjee, S.: Node placement for connected coverage in sensor networks. In: Proceedings of the Workshop on Modeling and Optimization in Mobile, Ad Hoc and Wireless Networks (2003)
12. Kumar, S., Lai, T.H., Balogh, J.: On k-coverage in a mostly sleeping sensor network. Wireless Network 14, 277–294 (2006)
13. Liu, B., Towsley, D.: A study of the coverage of large-scale sensor networks. In: Proceedings of the IEEE International Conference on Mobile Ad-hoc and Sensor Systems, pp. 475–483 (2004)
14. Nickerson, J.V., Olariu, S.: Protecting with sensor networks: Attention and response. In: Proceedings of the 40th Annual Hawaii International Conference on System Sciences (2007)
15. Olariu, S., Nickerson, J.V.: Protecting with sensor networks: Perimeters and axes. In: MILCOM (2005)
16. Seo, J.H., Kim, Y.H., Ryou, H.B., Kang, S.J.: A genetic algorithm for sensor deployment based on two-dimensional operators. In: SAC 2008: Proceedings of the 2008 ACM symposium on Applied computing, pp. 1812–1813 (2008)
17. GRASS Development Team. Geographic resources analysis support system (2007), `http://grass.itc.it`
18. GRASS Development Team. Grass sample data sets (2010) `http://grass.itc.it/download/data.php`
19. Wang, Y.C., Tseng, Y.C.: Distributed deployment schemes for mobile wireless sensor networks to ensure multilevel coverage. IEEE Transactions on Parallel and Distributed System 19(9), 1280–1294 (2008)
20. Zhou, Z., Das, S., Gupta, H.: Connected k-coverage problem in sensor network. In: Proceedings of 13th International Conference on Computer, vol. 1 (2007)
21. Zou, Y., Chakrabarty, K.: A distributed coverage- and connectivity-centric technique for selecting active nodes in wireless sensor networks. IEEE Transaction on Computers 54(8), 978–991 (2005)

Evolutionary Optimization on Problems Subject to Changes of Variables

Richard Allmendinger and Joshua Knowles

University of Manchester, Manchester, Great Britain
allmendr@cs.man.ac.uk, j.knowles@manchester.ac.uk

Abstract. Motivated by an experimental problem involving the identification of effective drug combinations drawn from a non-static drug library, this paper examines evolutionary algorithm strategies for dealing with changes of variables. We consider four standard techniques from dynamic optimization, and propose one new technique. The results show that only little additional diversity needs to be introduced into the population when changing a small number of variables, while changing many variables or optimizing a rugged landscape requires often a restart of the optimization process.

1 Introduction and Motivation

The purpose of this paper is to analyze the performance of evolutionary algorithms (EAs), such as genetic algorithms and evolution strategies, on problems that are subject to changes of variables. Such problems belong to the class of dynamic optimization problems and they are relevant, amongst other applications, to optimization problems in the experimental sciences (see e.g. [1,2,3]).

Our motivation for considering this problem type is an ongoing experimental study concerned with the identification of effective drug combinations using EAs. In this study, a human experimentalist needs to arrange a library of promising or relevant drugs to be considered by an optimizer. In our case, each drug corresponds to a single binary variable indicating whether the drug is included into a drug mixture or not; the drug mixture itself represents an entire solution. In real-world applications of this type, the experimental equipment often allows it to do many experiments (which are here the mixing of drugs) in parallel. In our scenario, a robot is taking care of the mixing of drugs, and this robot is able to mix up to 50 combinations in parallel. During the running experimental optimization, which may take several months, the human experimentalist may decide to replace drugs from the library with new ones because, for instance, the old drugs have a too dominant undesired effect on the efficiency of a drug cocktail or are simply not of interest anymore[1]. Hence, whenever a change of drugs takes place, the corresponding variables are changed as well. Note, a change of drugs does not change the effectivity of cocktails that did not use any of the

[1] The situation where drugs are only added to or removed from an existing library is realistic too but will not be considered in this work.

R. Schaefer et al. (Eds.): PPSN XI, Part II, LNCS 6239, pp. 151–160, 2010.

replaced drugs; i.e. solutions with no 1-bit at any of the replaced variables retain their fitness. Obvious ways of dealing with this are to carry on regardless, or to restart optimization entirely. We wish to see if other strategies fare any better.

Optimizing problems that are subject to changes of variables is similar to other dynamic optimization problems considered for example in [4] as both problem types feature a changing fitness function and thus a changing fitness landscape. An example of a conceptually similar dynamic optimization problem is the dynamic traveling salesman (DTSP) or vehicle routing problem [5], where cities may be replaced with new ones during the optimization. The two main differences between our dynamic combinatorial problem and these existing dynamic problems are that: (i) we do not need to detect changes in the landscape because it is always known when variables are changed (sometimes the exact generation of a change can even be controlled), and (ii) the fitness value remains unchanged for a known subset of solutions (see above). The fact that in our case solutions are evaluated by performing physical experiments, whose number is often limited by resource restrictions, can be considered as another distinction to previous analyses of dynamic problems. The presence of resource limitations requires a quick tracking of new optima. Fortunately, thanks to (i) and (ii), fulfilling this task is usually easier in our scenario than, say, in DTSP.

Work carried out on modifications of EAs enabling them to track optima has looked at three main approaches: diversity control [6], memory-based [7] and multi-population approaches [4]. For example, in [6], the authors study a triggered hypermutation operator and a random immigrants method. While the former method increases the mutation rate drastically whenever the performance degrades, the latter aims at maintaining population diversity throughout the optimization by seeding a part of the population with randomly generated individuals. In [7], the authors suggest the use of diploid representation and dominance in order to memorize and retrieve successful alleles, but this may only be useful when part of the environment/problem can return to a formerly visited state. For a comprehensive review of further modifications please refer to [4].

The remainder of this paper is organized as follows. The next section proposes a modification to EAs specifically designed for problems subject to changes of variables. The test problems we use to compare this approach against four standard strategies from dynamic optimization are described Section 3. An experimental study follows in Section 4, and Section 5 concludes the paper.

2 Strategy to Account for Changes of Variables

The strategy described in this section can be assigned to the class of diversity-control approaches (such as the triggered hypermutation and random immigrants [6]) to account for dynamic environments. Unlike for the majority of strategies belonging to this class, we do not need to answer the question of how to detect environmental changes. Instead, the question is rather how to exploit best the (often expensively obtained and partially intact) information contained in the current population in the generation of a new population after changing variables. The strategy proposed here introduces diversity into the population only in the initial generations upon an environmental change. This is done by using effectively the information stored in the current population and by making use of the fact only bits with value 1 have potentially an effect. The strategy, which we are calling *fair mutation*, is explained in detail in the following.

2.1 Fair Mutation

The idea of this strategy is to allow an optimizer both to continue optimization as normal and at the same time to rapidly explore the space of solutions that *use* any of the new variables. The latter aspect is facilitated by two mechanisms: (i) allowing an optimizer to test each new variable (i.e. its bit being set to 1) within the same number of offspring solutions and (ii) enabling an optimizer to explore solutions with a 1-bit at any of the new variables by applying a raised mutation rate p_r for Δh generations; p_r and Δh are user-defined parameters. Algorithm 1 describes the method $generateOffspringPopulation(Pop, \lambda, V, \Delta t)$, which we are calling to generate the offspring population at each generation. The parameters required by this method are the current population Pop, the number of required offspring solutions λ, the set of changed variables V, and the number of generations Δt passed after the last environmental change.

From Lines 8-9 in Algorithm 1, one can see that in the first Δh generations after changing variables we apply a raised mutation rate (from which 1-bits among the new variables are protected) if an offspring has one or more 1-bits among any of the new variables. Although this step allows an optimizer to perform exploration among solutions that *use* one or more of the new variables, it also runs the risk that an offspring may be perturbed so much that it turns into a poor one. However, when we embed this strategy within a plus (elitist) population update scheme (see Section 4) it achieves both the continued normal optimization of strings that do not contain any of the new variables, and the fair and rapid exploration of each of the subspaces (schemata) that contain a new variable.

3 Testing Environment

We use a *main-and-joint effects (MJ)* problem, which models the real drug mixture problem we are interested in. We present the *MJ* model for the first time in 3.1. We also consider variable changes on *NK* landscapes as a comparison.

3.1 *MJ* Model

The experience of the human experimentalist and a previously performed pilot study suggested that only some drugs (bits) of the library have a positive effect while others have a negative side effect or no effect at all. Moreover, it is believed that interactions among the considered drugs can be described by main and joint effects only while higher order interaction effects are negligible.

Algorithm 1. Fair mutation

Require: p_r (raised mutation rate), Δh (number of generations for which p_r is used)

generateOffspringPopulation($Pop, \lambda, V, \Delta t$){
if $\Delta t = 0$ **then**
  for each variable $k \in V$ **do**
    for $1 \leq i \leq \lfloor \lambda/\text{card}(V) \rfloor$ **do**
      generate an offspring $\boldsymbol{x}_{off}$ as normal; mutate $\boldsymbol{x}_{off}$ using a rate of p_r; set variable k of $\boldsymbol{x}_{off}$ to 1 and add $\boldsymbol{x}_{off}$ to *offPop*; i.e. $offPop = offPop \cup \boldsymbol{x}_{off}$
  // if card(*offPop*) $< \lambda$, then fill *offPop* in the same way as above but instead of setting variable k to 1, set to 1 a variable selected from V at random without replacement
else
  if $0 < \Delta t \leq \Delta h$ **then**
    Generate offspring population as normal but for any offspring containing a 1-bit at any of the variables in V, mutate the other bits using p_r
  else
    Generate offspring population as normal
return *offPop*}

Let M denote the number of randomly selected bits $x_{h(k)}, k = 1, \ldots, M \leq N$ (N is the total number of bits) which are assigned a main effect; $h(k)$ is the variable's index of the kth main effect. And let $J \cdot M$ denote the number of randomly selected distinctive bit pairs $(x_{i(l)}, x_{j(l)}), l = 1, \ldots, J \cdot M, i(l) \neq j(l)$, which are assigned a joint effect; $i(l)$ and $j(l)$ are the indexes of the two variables that have the lth joint effect. In our case, the strengths of all main effects $f_{h(k)}$ and joint effects $g_{i(l),j(l)}$ are drawn uniformly from the interval $[-3, 1]$; i.e. on average, a quarter of all main and joint effects will be positive. Of course, the strength of a main effect $f_{h(k)}$ or joint effect $g_{i(l),j(l)}$ is only considered if the variable $x_{h(k)}$, respectively, both variables $x_{i(l)}$ and $x_{j(l)}$ are 1-bits. To evaluate a candidate solution, we look up all the main and joint effects which are *on* and take the average of the sum of all these values. In other words, the fitness function f to be maximized is defined as

$$f(x_1, ..., x_N) = \frac{1}{N} \left(\sum_{k=1}^{M} f_{h(k)} \cdot x_{h(k)} + \sum_{l=1}^{J \cdot M} g_{i(l),j(l)} \cdot x_{i(l)} \cdot x_{j(l)} \right) . \quad (1)$$

3.2 *NK* Landscapes

The two parameters of *NK* landscapes [8] are the total number of bits N, and the number of bits that interact epistatically at each of the N loci, K. Compared to an *MJ* model, an *NK* landscape assigns a positive effect (drawn usually from $[0, 1]$) to each of the possible 2^{K+1} bit-wise neighborhood configurations; i.e. 0 and 1-bits are treated equally. For fair mutation this means that (i) offspring solutions devoted to a new variable need to be set to 1 and to 0 in equal number and (ii) new variables are always protected from mutation.

3.3 Performing a Change of Variables

On a *NK* landscape, a change of a variable involves to reselect its K neighbors and to reinitialize the corresponding 2^{K+1} effects. Similarly, on a *MJ* model, if the replaced variables had a main and/or joint effect, then these effects are reinitialized and the joint bit reselected. Remember to ensure that the resulting set of $J \cdot M$ bit pairs includes only non-identical bit pairs. In this study, a change of variables is performed every Δg generations whereby always $\#v$ randomly selected variables are changed. That is, while Δg controls the frequency of environmental changes, $\#v$ controls the severity of a change.

4 Experimental Study

We compare the performance of fair mutation on dynamic *NK* landscapes and *MJ* models against four strategies from dynamic optimization: triggered hypermutation, restart of the optimization, reevaluating the current population after changing variables and then carrying on as normal, and a niching algorithm.

Parameter settings: For triggered hypermutation, we use a base mutation rate of $p_m = 0.001$ and a hypermutation rate of $r = 500$. Hypermutation is triggered when there is a decrease in the moving average of the best-of-generation fitness over five generations. These are standard parameter setting combinations for this strategy. The niching algorithm is a deterministic crowding EA as described in [9, p. 140]. At the generation of an environmental change, all strategies reevaluate the current population before they generate the offspring population.

The algorithm on which we test all strategies (except the niching algorithm) is a genetic algorithm (GA) with a $(\mu+\lambda)$ES reproduction scheme; this choice is guided by simplicity but accounts for our belief that elitism is generally useful in this domain. The algorithm also uses binary tournament (with replacement) for parental selection, uniform crossover (crossover probability of 0.7) and bit flip mutation (per-bit mutation rate of $p_m = 1/N$, although triggered hypermutation uses its own mutation rates). We fix the population size of the GA to $\lambda = 50$ and also set $\mu = 50$. Note, initially, triggered hypermutation was implemented in a standard GA. We integrate it into an elitist GA because the performance obtained with that GA proved to be better.

For both test problem models we consider different settings of $\#v$ and Δg. The total number of variables remains thereby fixed at $N = 30$, which is also the size of the drug library in our real-world scenario. The considered *NK* and *MJ* models have the parameters $N = 30$ and $K = 2$, respectively, $M = 30$ and $J = 1$; i.e. all bits have a main effect and 30 bit pairs have a joint effect.

We are mainly interested in the dynamics resulting from changing a small number of variables frequently. On this class of problems, the parameter setting for which fair mutation performed best is $\Delta h = 0$ and $p_r = 1/N$, which will also be used here. The interpretation of this parameter setting is that new variables are protected from mutation only at the generation at which an environmental change occurs, and that the raised mutation rate is the same as the standard mutation rate. Higher values of Δh and p_r are required for more severe environmental changes and different interval ranges of main and joint effects.

Any results shown are average results across 100 independent algorithm runs.

Performance measurement: To measure the performance of the investigated strategies we use three measurements: *accuracy*, *adaptability*, and *average best-of-generation fitness*. Accuracy and adaptability have been proposed in [10] and rate an algorithm's overall performance with a single number. The accuracy value represents the difference between the value of the best solution in the population at the generation just before a change and the optimal value for that variable set, averaged over the entire run. The adaptability value represents the difference between the value of the current best solution of each generation and the optimal value (for the current variables), averaged over the entire run. In both cases the optimal solution is approximated by using the best solution found from running an elitist generational GA for 100 generations, 100 times independently. Obviously, an algorithm is better the smaller its accuracy and adaptability measurements are.

The average best-of-generation fitness is a more standard per-generation measure, suitable for plotting.

4.1 Results

Figure 1-3 show the best-of-generation fitness of all five strategies on *NK* and *MJ* models for three different settings of $\#v$ and Δg: $\#v = 2$ and $\Delta g = 10$ (Figure 1), $\#v = 8$ and $\Delta g = 25$ (Figure 2), and $\#v = 15$ and $\Delta g = 40$ (Figure 3). The accuracy and adaptability measurements of all strategies on these problem instances are shown in Table 1. For comparison reasons, the table includes also the results obtained by random immigrants, using a replacement rate of 30%, and triggered hypermutation using a base mutation rate of $p_m = 1/N$ and $r = 15$.

For each problem instance it is apparent that all strategies perform better on the *MJ* model than on the *NK* landscape, which is due to the higher degree of variable interactions featured by the *NK* landscape. Figure 1 investigates the situation where the environment changes only little but frequently. This is the most realistic and for us most interesting case. Here, we observe that:

- Using a restart policy or triggered hypermutation results in a poor performance. In the case of the restart strategy, the reason are the low frequency and severity of environmental changes. The low base mutation rate combined with the high frequency of changes causes the poor performance of triggered hypermutation, particularly on the *NK* landscape.
- The other three strategies yield similar performances though a slight advantage is apparent for fair mutation, which is closely followed by the reevaluation strategy; this is also confirmed by the accuracy and adaptability measurements. Fair mutation tends to recover significantly faster than the other strategies when there is a severe change in the landscape; see generation 90 on the *NK* landscape and generation 100 on the *MJ* model.

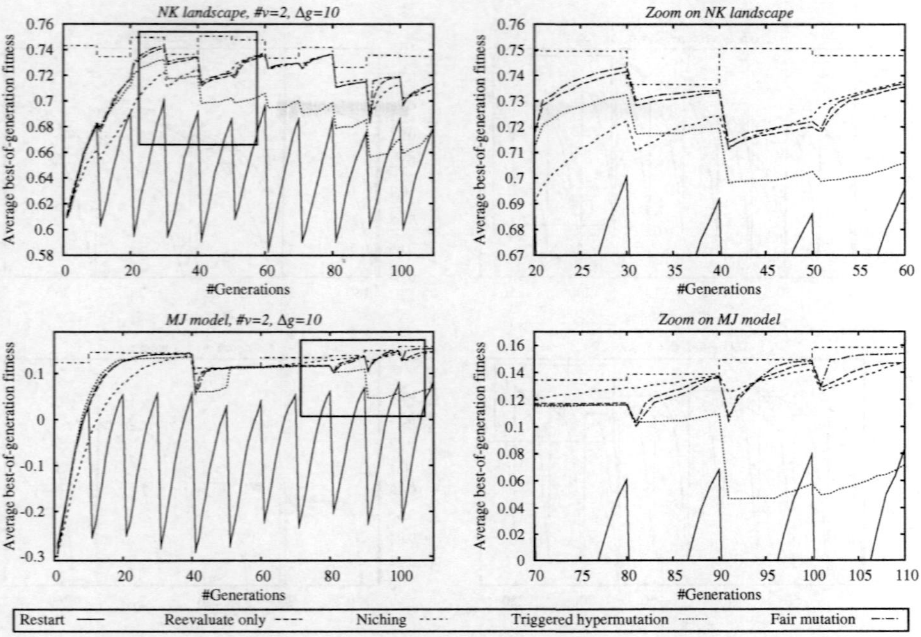

Fig. 1. Shown is the average best-of-generation fitness obtained by various dynamic optimization strategies on a *NK* (top) and *MJ* model (bottom) with $\#v = 2, \Delta g = 10$. The plots on the left hand side zoom into the area indicated by the box in the corresponding left plot. The step-shaped line indicates the (estimated) optimal fitness.

– The niching algorithm maintains a relative high population diversity throughout the search as it searches within several promising niches simultaneously. This slows down the convergence, as can be seen at the beginning of the optimization, but it also increases the probability of finding new optima quicker after undergoing an environmental change; this is for example the case on the *MJ* model at generation 80, or the *NK* landscape at generation 50.

From Figure 2, where more variables are changed less frequently we observe that:

– The restart policy can sometimes outperform the other strategies on the *NK* landscape. This is, for example, the case at generation 50 where the landscape seems to undergo a quite severe change.
– Fair mutation outperforms the other strategies again slightly. The fact that only a small number of offspring solutions can be devoted to each new variable prevents a better performance.
– The deficit in the convergence speed of the niching algorithm is now more apparent. This is because a more severe landscape change reduces the probability that any of the currently occupied niches is close to the new optimal search region; i.e. the population needs to shift its search focus more severely and this takes time when trying to maintain several niches simultaneously.

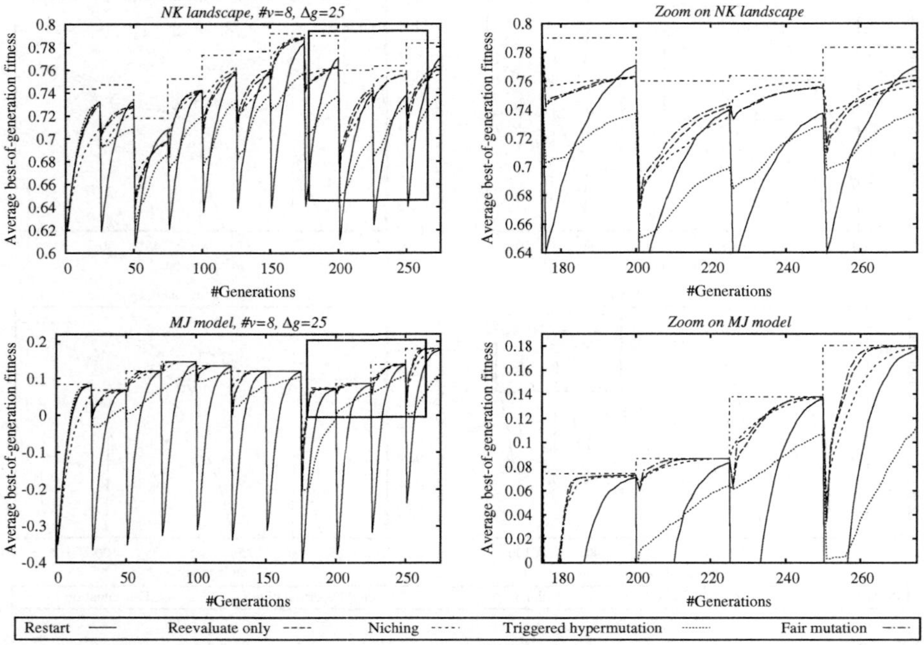

Fig. 2. Shown is the average best-of-generation fitness obtained by various dynamic optimization strategies on a *NK* (top) and *MJ* model (bottom) with $\#v = 8, \Delta g = 25$. The plots on the left hand side zoom into the area indicated by the box in the corresponding left plot. The step-shaped line indicates the (estimated) optimal fitness.

Figure 3 considers the case where half of all solution variables are replaced with new ones every 40 generation. This setting may simulate the case where a human experimentalist wants to test many different drugs without being able (e.g. due to storage or budget limitations) or wanting to change the drug library size. From the figure and the accuracy and adaptability measurements it is apparent that a restart policy significantly outperforms the other strategies on the *NK* landscape. On the *MJ* model, a restart policy yields best results in terms of the accuracy with fair mutation being best in terms of the adaptability. The very small number of offspring solutions devoted to each new variables reduces the *local search* abilities of fair mutation considerably. The niching algorithm performs worst on the *NK* landscape. The reason is again its deficit in the convergence speed combined with the severity of the environmental changes. On the *NK* landscape, triggered hypermutation performs sometimes quite well loosing often only to the restart policy (see e.g. generation 200, 250, and 440).

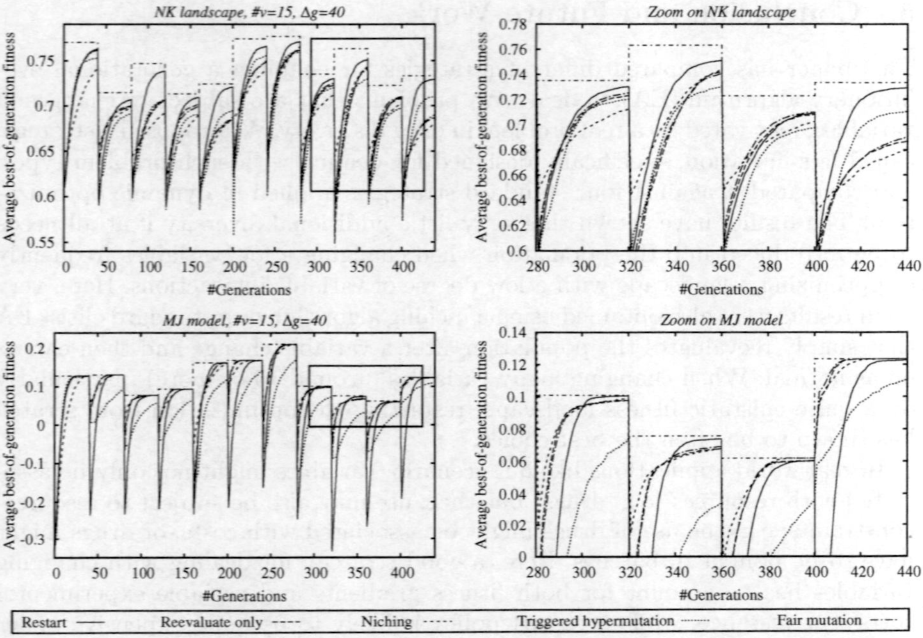

Fig. 3. Shown is the average best-of-generation fitness obtained by various dynamic optimization strategies on a *NK* (top) and *MJ* model (bottom) with $\#v = 15, \Delta g = 40$. The plots on the left hand side zoom into the area indicated by the box in the corresponding left plot. The step-shaped line indicates the (estimated) optimal fitness.

Table 1. Accuracy (Ac.) and adaptability (Ad.) results for various dynamic optimization strategies on different problem instances of dynamic *NK* and *MJ* landscapes. For each problem instance and metric, we highlighted all strategies in bold face that are not significantly worse than any other strategy. The statistical test applied here is the non-parametric Kruskal-Wallis test with a significance level of 5% (2-sided).

		$\#v = 2, \Delta g = 10$		$\#v = 8, \Delta g = 25$		$\#v = 15, \Delta g = 40$	
		NK	*MJ*	*NK*	*MJ*	*NK*	*MJ*
Fair mutation	Ac.	**0.0163**	**0.0155**	**0.0150**	**0.0002**	0.0193	0.0011
	Ad.	**0.0242**	**0.0344**	**0.0298**	**0.0210**	0.0480	**0.0279**
Restart	Ac.	0.0520	0.0815	**0.0151**	0.0022	**0.0076**	**0.0003**
	Ad.	0.0942	0.2133	0.0552	0.1232	**0.0395**	0.0609
Reevaluation	Ac.	**0.0170**	**0.0163**	0.0202	**0.0003**	0.0202	0.0012
	Ad.	**0.0252**	**0.0350**	**0.0302**	**0.0220**	0.0516	**0.0282**
Niching	Ac.	0.0242	0.0290	0.0188	0.0032	0.0264	0.0012
	Ad.	0.0321	0.0471	**0.0314**	0.0271	0.0560	**0.0296**
Tr. hyp. $p_m = 0.001$	Ac.	0.0387	0.0313	0.0376	0.0334	0.0241	0.0290
	Ad.	0.0492	0.0554	0.0586	0.0778	0.0552	0.0833
Tr. hyp. $p_m = 1/N$	Ac.	0.0194	0.0195	**0.0163**	**0.0004**	0.0146	0.0015
	Ad.	0.0299	0.0442	0.0360	0.0354	0.0487	0.0444
Random imm.	Ac.	0.0214	0.0228	0.0209	0.0011	0.0231	0.0019
	Ad.	0.0297	0.0434	0.0357	0.0266	0.0537	0.0337

5 Conclusion and Future Work

This paper has compared different strategies for enabling a generational evolutionary algorithm (EA) to deal with problems that are subject to changes of variables, motivated by a real problem in drug discovery. We proposed a strategy, called fair mutation, specifically designed for dealing with such problem types, and compared it against four standard strategies applied in dynamic optimization. The results have shown that only little additional diversity if at all needs to be introduced into the population when changing a few variables frequently or optimizing a landscape with a low degree of variable interactions. Here, very good results were also obtained using a niching algorithm or a standard elitist EA that simply reevaluates the population after a variable change and then carries on as normal. When changing many variables (around 10 or more), particularly on a quite epistatic fitness landscape, restarting the optimization from scratch has shown to be often the best choice.

In real-world applications like our scenario, variables might not only be associated with resources (e.g. drugs) but their use may also be subject to resource-constraints: e.g. the use of drugs might be associated with costs, or drugs might have to be bought in batches. Here, a good strategy for dealing with changing variables has to account for both fitness gradients and variable experimental costs. For instance, while a restart policy is likely to be quite expensive under these circumstances, a strategy that aims at wasting very little resources might take too long to find fit solutions. Hence, future research into the development of strategies that perform well in the presence of resource-constraints is needed.

References

1. Schwefel, H.-P.: Experimentelle optimierung einer Zweiphasendüse. Bericht 35 des AEG-Forschungsinstitut Berlin zum Projekt MHD-Staustrahlrohr (1986)
2. Rechenberg, I.: Case studies in evolutionary experimentation and computation. Computer Methods in Applied Mechanics and Engineering 186(2-4), 125–140 (2000)
3. Knowles, J.: Closed-Loop Evolutionary Multiobjective Optimization. IEEE Computational Intelligence Magazine 4(3), 77–91 (2009)
4. Branke, J.: Evolutionary Optimization in Dynamic Environments. Kluwer Academic Publishers, Dordrecht (2001)
5. Larsen, A.: The Dynamic Vehicle Routing Problem. PhD thesis, Technical University of Denmark (DTU), Denmark (2001)
6. Cobb, H.G., Grefenstette, J.J.: Genetic algorithms for tracking changing environments. In: Proc. of ICGA, pp. 523–530 (1993)
7. Goldberg, D., Smith, R.E.: Nonstationary function optimization using genetic dominance and diploidy. In: Proc. of ICGA, pp. 59–68 (1987)
8. Kauffman, S.: Adaptation on rugged fitness landscapes. In: Lectures in the Sciences of Complexity, pp. 527–618 (1989)
9. Mahfoud, S.W.: Niching methods for genetic algorithms. PhD thesis, University of Illinois at Urbana-Champaign, USA (1995)
10. Trojanowski, K., Michalewicz, Z.: Searching for optima in non-stationary environments. In: Proc. of IEEE CEC, pp. 1843–1850 (1999)

On-Line Purchasing Strategies for an Evolutionary Algorithm Performing Resource-Constrained Optimization

Richard Allmendinger and Joshua Knowles

University of Manchester, Manchester, Great Britain
allmendr@cs.man.ac.uk, j.knowles@manchester.ac.uk

Abstract. We consider an optimization scenario in which resources are required in order to realize or evaluate candidate solutions. The particular resources required are a function of the solution vectors, and moreover, resources are costly, can be stored only in limited supply, and have a *shelf life*. Since it is not convenient or realistic to arrange for all resources to be available at all times, resources must be purchased on-line in conjunction with the working of the optimizer, here an evolutionary algorithm (EA). We devise three resource-purchasing strategies (for use in an elitist generational EA), and deploy and test them over a number of resource-constraint settings. We find that a just-in-time method is generally effective, but a sliding-window approach is better in the presence of a small budget and little storage space.

1 Introduction

We are currently interested in using evolutionary algorithms (EAs) to optimize a number of things that require us to physically realize and experimentally test candidate solutions in order to evaluate them (similarly to [1,2]), including: combinatory drugs, nucleic acids, food and energy crops, and complex instrumentation equipment. In these contexts, the allele of a particular solution may represent (or require) a *resource* that is needed in order for the solution to be realized and/or tested. It is not always convenient to ensure that all resources are available at all times (because they may be expensive, or need storage facilities, or require people who have other commitments, etc.). So, even though the optimization problem itself may be unconstrained and essentially static, the fitness of a solution $\boldsymbol{x}_t$ may be undefined (or null) at time step t during the optimization. We model the temporary unavailability of resources using what we are calling *ephemeral resource constraints* (ERCs).

In a recent paper [3], we introduced a number of different sub-classes of ERCs, studied how to simulate them and investigated how standard EAs were affected by the key parameters describing the ERCs. An essential finding of the paper was that ERCs can increase drift effects in EAs, or even cause stagnation, due to the fact that the sampling of the search space is limited by the resource availability in addition to the 'normal' drift coming from sampling errors. The reader is also referred to the paper (ibid.) for all material relating the general ERC problem

R. Schaefer et al. (Eds.): PPSN XI, Part II, LNCS 6239, pp. 161–170, 2010.

to some other problem types [4,5,6] and for applications in the literature [7,8,9], which we cannot cover here.

In this paper, we consider a particular ERC, which we call a *commitment composite ERC*. Essentially, with this type of ERC, some part of the solution vector defines a complex sub-part (or *composite*) which must be purchased in advance. Once the composite arrives (after some lag) it can be stored only for a certain length of time, and/or re-used in different experiments only a certain number of times. Thus ordering it represents a sort of *commitment* to using it a number of times (or else it will be wasted).

Here we consider some purchasing strategies to be used in conjunction with an elitist generational EA in order to optimize a problem subject to this sort of ERC, assuming some budget limiting the usage of composites, and/or limiting time. The next section describes the commitment composite ERC in more detail. The three strategies we consider are then given in Section 3, and an experimental study follows (Section 4). Section 5 concludes the paper.

2 Commitment Composite ERCs

A commitment composite ERC occurs when some variables of a candidate solution define a *composite* that requires resources to be locally available (e.g. in a cache) in order for the solution as a whole to be realized and/or evaluated. We can conveniently use the notion of schemata to describe the resource-requiring composite part of a solution. For example, assuming a binary representation of solutions, we would use $H_{\#} = \{**\#\#\#*****\#\#\}$ to state that bit positions 3, 4, 5, 11 and 12 define a composite; we refer to the bit positions denoted by # as the *composite-defining bits*, and the order $o(H_{\#})$ to be the number of composite-defining bits in the schema (we refer to $H_{\#}$ as the *high-level constrained schema*). In the problems tackled here the composite-defining bits are static, and form a part of the ERC problem definition.

When a solution is to be evaluated, we must look at the composite-defining bits in the solution and compare them to a local cache of resources. Each resource in the local cache is indexed by a bit-string of the same length as the order of the schema described above. If there is a match, the solution can be evaluated; if not, the solution may not be evaluated at the current time step.

We define the cache to be made up of a number of storage cells, $\#SC$. Typically, the number of storage cells is smaller than the space of possible resources, which is $2^{o(H_{\#})}$ in a binary search space. A resource available in a storage cell may be used in the evaluation of more than one solution: each resource may be used up to RN (reuse number) times and has a shelf life of SL time steps, and we assume $SL \geq RN$. Finally, the resources available in the cache at time t are a function of previous purchase orders made, and a fixed time lag TL between a purchase being made and it arriving. When resources arrive at a particular time, they are immediately put in a storage cell (and any existing resource in that cell is discarded); which storage cell is selected is defined either at the time of purchase or at the time of arrival (this detail will become clear below).

The corresponding ERC function can be defined by

$$h: \text{if } \left[\bigwedge_{i=1}^{\#SC} \boldsymbol{x_t} \notin H_i(t) \right] \Rightarrow \boldsymbol{x_t} \notin E_t, \tag{1}$$

where, E_t represents the *evaluable search region* at optimization time step t, and $H_i(t), i{=}1, ..., \#SC$, the constraint schema that corresponds at time step t to the composite that is stored in storage cell i. In future, we will denote such a commitment composite ERC by $commCompERC(H_{\#}, \#SC, TL, RN, SL)$.

In this paper we consider the case where time steps refer to function evaluations of single solutions $\boldsymbol{x}$, so does the reuse number RN. Furthermore, to make the constraint more realistic we associate costs of c_{order} and c_{time_step} units with each submitted composite order and time step, respectively. The available budget, which cannot be exceeded, is denoted by C. To simulate the constraint we use the function wrapper, whose functionality is described below. In [3], this wrapper is referred to as the *communication channel* and calling it is identical to calling an objective function f in a standard optimization problem. If the optimizer wants to increment only the time counter without evaluating a solution, it submits what we are calling a *null solution*; note the difference between a null solution and a solution with a fitness value of null, which is is submitted with the purpose of being evaluated but then is not evaluated due to a lack of resources. The variables representing the time counter t, cost counter c, storage cell information SC (i.e. contents, remaining shelf lives and remaining reuse number), and the queue Q of submitted but not arrived composites, are global state variables and visible to the optimizer. For tasks like ordering a composite or *repairing* a solution the optimizer needs to implement a method, which is then called by the channel. Here, repairing means to modify a solution such that it falls into the constrained schema of a composite that is currently in storage.

Once the optimizer submits a solution $\boldsymbol{x}$, the channel works as follows: First, composite orders are collected (given the budget allows it) and added in the order of submission to Q. The storage cell information SC is then updated (i.e. arriving composites are put in the assigned storage cells and expired ones are removed), and the submitted solution checked for evaluability. While an evaluable solution is then evaluated using some objective function f, for a non-evaluable one, an optimizer's repairing method is called. If the optimizer repairs the solution (i.e. the repairing method returns a different solution), then it is evaluated, otherwise it is assigned an objective value of null. After an evaluation we update SC (i.e. decrement the reuse number of the corresponding composite), and increment the time and cost counter. This process is repeated until a budget of C is used up.

3 Online Purchasing Strategies

This section proposes three strategies for dealing with commitment composite ERCs: a just-in-time strategy, a just-in-time strategy with repairing, and a sliding window strategy. In advance, all strategies are designed such that an optimizer has never to deal with solutions that have a null fitness value.

In general, a strategy designed to deal with commitment composite ERCs is composed of three mechanisms, which are concerned with: (i) selecting the composite that is ordered and the point of time at which it is ordered; (ii) selecting the storage cell into which an arrived composite is put, which may mean selecting an existing composite that is to be replaced by a new one; (iii) selecting an available composite to repair a non-evaluable solution (given it is repaired). Each of these mechanisms is described for our strategies in the following.

3.1 Just-in-Time Strategy

Without repairing and any composites in storage, the minimum number of time steps for evaluating all solutions of a population is $Popsize + TL$ (TL time steps are needed for the first composite to arrive; this period is bridged by submitting TL null solutions). This is achieved if orders are processed in contiguous groups organized by the composites they require, e.g. *ccbbbadd...*, where a, b, c, and d shall represent different composites required by solutions. Thus, the first main mechanism of the *just-in-time* (JIT) strategy is to arrange solutions of a population *Pop* into these contiguous groups, and then to make purchase orders so that composites arrive just in time for the scheduled experiment time. This mechanism, which is performed by the method *alignOrdersAndSolutions(Pop)* (see below), prolongs the availability of resources.

When composites are already in storage (we call them *old composites*) savings in purchase orders may be made if those composites are used first. For example, suppose the optimizer requires the composites *ccdadcac*, and composite a is available in one of the storage cells and has 3 uses and 5 time steps of its shelf life remaining. Then, by placing a first, the permutation *aaccccdd* will save us a purchase order since only two a composites are needed. Thus, the second main mechanism of the JIT strategy is to efficiently schedule the evaluation of solutions in a population *Pop* using old composites. This is performed by the method *useUpOldComposites(Pop)* (see below). Notice that, at any given time, JIT (and JIT with repairing) maintain non-identical composites in storage.

During the optimization, the JIT strategy checks at each generation whether old composites can be used in the evaluation of solutions of the current population. If so, the method *useUpOldComposites(Pop)* is applied first and then the method *alignOrdersAndSolutions(Pop)*, otherwise we proceed directly with the method *alignOrdersAndSolutions(Pop)*.

alignOrdersAndSolutions(*Pop*): Recalling that solutions are grouped by the composite they require, this method must just choose the order in which these groups of solutions are to be submitted for evaluation. To obtain a permutation of groups we select groups one by one using roulette wheel selection (without replacement) based on the number of solutions in *Pop* associated with them, and then reverse the order so obtained. This strategy increases the chances that left-over composites at the end of the generation will be useful next generation (because the solutions associated with them are over-represented in the population). Regarding the storage location of arriving composites, the default is to place them in an empty storage cell. If no cell is empty, then the composite replaces the composite that can be used in the fewest evaluations within the subsequent generation; ties between composites are broken by replacing the composite that has the shortest shelf life remaining; further ties are broken randomly.

useUpOldComposites(*Pop*): Our aim when using up old composites is to save as many composite orders as possible. To achieve this when there are several storage cells to consider, we solve the lexicographical optimization problem $\text{lexmax}_{i \in \pi(Z)}(F_1, F_2)$ over the permutation set of $Z = \{\zeta | \zeta \in S \cap C, 1 \leq \theta_\zeta \text{mod} RN \leq \min(\psi_\zeta, \rho_\zeta)\}$, where S is a set of old composites and C the set of composites required by the optimizer to evaluate *Pop*. Associated with each $s \in S$ there is a remaining shelf life ψ_s and a remaining number of reuses ρ_s, and associated with each $c \in C$ there is a required number θ_c. The objective functions to be optimized are $F_1 = \sum_{j \in 1,\dots,|Z|} f_1(\pi_{ij}(Z))$ and $F_2 = \sum_{j \in 1,\dots,|Z|} f_2(\pi_{ij}(Z))$, where π_{ij} is the jth element of permutation i,

$$f_1(\pi_{ij}(Z)) = \begin{cases} 1 & \text{if } \psi_{\pi_{ij}} - \sum_{k=1}^{j-1} \delta_{\pi_{ik}} \geq \theta_{\pi_{ij}} \text{mod} RN \\ 0 & \text{otherwise} \end{cases}, \qquad f_2(\pi_{ij}(Z)) = \delta_{\pi_{ij}}$$

$$\text{and } \delta_{\pi_{ik}} = \begin{cases} \theta_{\pi_{ik}} \text{mod} RN & \text{if } k = 1 \\ \max\left[0, \min\left(\psi_{\pi_{ik}} - \delta_{\pi_{ik-1}}, \theta_{\pi_{ik}} \text{mod} RN\right)\right] & \text{if } k > 1. \end{cases}$$

While F_1 determines the permutations of Z that save the most purchase orders, F_2 breaks ties among these permutations by selecting the permutation(s) that evaluate the most solutions; further ties are broken randomly. Required composites left in storage after evaluating the last solution of the selected permutation, i.e. after $\max(F_2)$ time steps, are used up further in a random fashion; i.e. we first select a composite at random and then select a random (unscheduled) solution from *Pop* that requires this composite, and we repeat this until no future evaluations can be scheduled. Having devised a schedule, we can specify at which time step the first new composite order needs to be submitted (to evaluate the remaining solutions in *Pop*) in order to reduce our waiting time for it to arrive. Subsequently, we know when and how many null solutions need to be submitted.

3.2 Just-in-Time Strategy with Repairing

Although the JIT strategy avoids repairing of solutions and thus allows an optimizer to perform an unconstrained optimization, this may be associated with a waste of up to *Popsize* × $(RN - 1)$ composites per generation (if each composite is used exactly once). To reduce wastage, the *JIT with repairing* (JITR) strategy extends the basic JIT strategy with a repairing procedure. The method *performRepairing(Pop)* (see Algorithm 1) performs the repairing and this method is always called before calling the method *alignOrdersAndSolutions(Pop)*. The idea of the repairing strategy is to repair solutions such that they use a composite that is *nearly* the one required. Solutions to be repaired are identified by first

Algorithm 1. Just-in-time strategy with repairing

Require: w (weighting factor), $\#mIter$ (number of shifting rounds)
1: performRepairing(Pop){
2: filter out solutions from Pop that use up composites entirely
3: **for** $\lceil \frac{|Pop|}{RN} \rceil \leq k \leq \#CompsInPop$ **do**
4: cluster solution in Pop using k-medoids
5: if required, shift solutions between clusters (perform $mIter$ rounds of shifting); retain configuration with the smallest Hamming distance loss and record HDL_k
6: normalize the values HDL_k and k of all configurations, and calculate the scores $HDLN_k \cdot (1 - w) + kN \cdot w$, where $HDLN_k = \frac{HDL_k - \min_k(HDL_k)}{\max_k(HDL_k) - \min_k(HDL_k)}$ and k is normalized in the same fashion
7: repair solutions in Pop according to the configuration with the smallest score; add the filtered-out solutions back to Pop.}

clustering their composites, and then trying to find an assignment of solutions to clusters that minimizes the total Hamming distance of all repairs.

performRepairing(Pop): The first step of this method is to filter out all solutions that use up a composite entirely (without being repaired). If there are more than RN solutions requiring the same composite type, then we filter out a subset of RN solutions among them at random. The remaining solutions take part in the clustering process and are subject to being repaired. The number of clusters k into which we can partition these solutions, or, more appropriate, their required composites, varies between $\lceil \frac{|Pop|}{RN} \rceil \leq k \leq \#CompsInPop$, where $\#CompsInPop$ is the number of non-identical composites in Pop. Let us first describe how we perform the clustering for a given k, and then how we select a particular cluster configuration according to which we repair.

To obtain a partition with k clusters we apply k-medoids on the different composites. The distance between two composites is the Hamming distance between their constraint schemata. The medoid composite of a cluster is the composite that would be used to repair all solutions in that cluster that require a different composite. For diversity reasons, we want to avoid to order a medoid composite more than once, or, in other words, we want to keep the number of solutions within a cluster smaller than the reuse number RN. A simple way to ensure this aspect is to randomly shift solutions from too large clusters to clusters that can accommodate further solutions. We perform $mIter$ shifting rounds in total and select the configuration with the smallest total Hamming distance of all repairs HDL_k to be the best configuration obtained with k clusters.

The cluster configuration according to which we repair is the one with the smallest score $HDLN_k \cdot (1 - w) + kN \cdot w$, where $w \in [0, 1]$ a predetermined weighting factor representing the degree of repairing. Roughly speaking, the larger the value w the more solutions are repaired; note a value of $w = 0$ corresponds to the configuration one would select in the basic JIT strategy. We can now add the solutions that were filtered out again to Pop and process that population further as usual using the method *alignOrdersAndSolutions(Pop)*.

3.3 Sliding Window Strategy

While the JIT and JITR operate in a sequential mode in that they devise a schedule of solution evaluations and purchase orders upon receiving a population from the EA first, the *sliding window* (SW) strategy deals with the working of the algorithm and the purchasing of orders in parallel. More precisely, the strategy submits solutions in the order they are generated, non-evaluable solutions are always repaired, and it aims for the 'most useful' composites to be kept in storage by (i) replenishing composites so that there can never be an empty storage cell and (ii) maintaining composites that were recently requested by the optimizer.

The first aspect, avoiding empty storage cells, is achieved by making purchase orders to fill all storage cells every $\min(SL, RN)$ time steps. The second aspect, maintaining recently requested composites, is achieved by ordering composites from the sliding window, which we define here as a set $\kappa(t)$ containing the most recently requested but unavailable composites; $\kappa(t)$ is limited in size such that $\text{card}(\kappa(t)) \leq WS$, where WS is the *window size*. Here, a requested composite means one that was part of a solution that the EA submitted to the function wrapper for evaluation. In this paper, we simply order the $\#SC$ composites from $\kappa(t)$ that have been added to this set most recently. That is, if we set $WS = \#SC$, then we order always all composites in $\kappa(t)$; note, this means that all existing composites are replaced by new ones. If we cannot order a composite for each storage cell because $|\kappa(t)| < \#SC$, then we order random composites for the remaining storage cells; this is, for example, the case at $t = 0$.

To repair a non-evaluable solution, we use the composite from the storage cell that has the smallest Hamming distance to the actually required composite; ties between equally distant composites are broken randomly. Note, different to JITR, this repairing step is performed within an optimizer's repairing strategy that is called by the communication channel.

4 Experimental Study

Experimental setup: Our strategies are tested on a genetic algorithm with a $(\mu + \lambda)$ES reproduction scheme; this choice is guided by simplicity but accounts for our beliefs that elitism is generally useful in this domain. The algorithm also uses binary tournament selection (with replacement) for parental selection, uniform crossover (crossover probability of 0.7), and bit flip mutation (per-bit mutation rate of $1/l$). We fix the population size to $\lambda = 50$ and also set $\mu = 50$. As the objective function f we consider a MAX-SAT instance, which is a well-known and generally challenging problem. The instance is a uniform random 3-SAT problem instance with $l = 50$ variables and 218 clauses[1], and it is satisfiable. We treat this 3-SAT instance as a MAX-SAT optimization problem, with fitness calculated as the proportion of satisfied clauses. Any results shown are average results across 100 independent algorithm runs on this instance. For SW, we

[1] The instance can be downloaded online at `http://people.cs.ubc.ca/~hoos/SATLIB/benchm.html`; the name of the instance is "uf50-218/uf50-01.cnf".

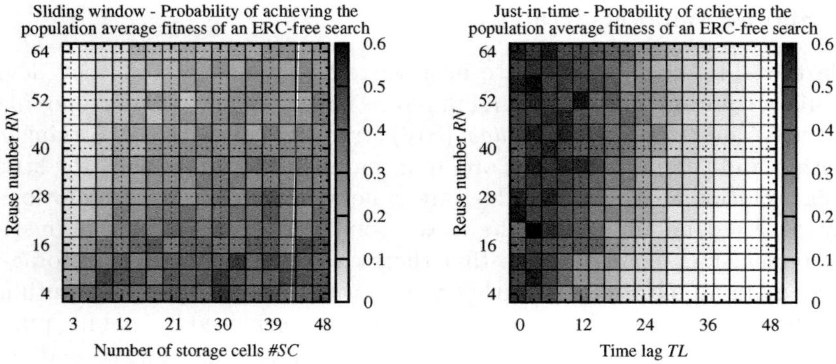

Fig. 1. Shown is the probability of SW (left) and JIT (right) of achieving the population average fitness of our base algorithm obtained in an ERC-free environment after 1500 time steps. For SW this probability is shown as a function of $\#SC$ and RN for the ERC $commCompERC(o(H_{\#}) = 30, \#SC, TL = 10, RN, SL = RN)$, and for JIT it is shown as a function of TL and RN for the ERC $commCompERC(o(H_{\#}) = 10, \#SC = 10, TL, RN, SL = RN)$; cost were set to $c_{order} = 0$, $c_{time\_step} = 1$, $C = 1500$.

found that ordering a composite that first undergoes mutation yields better results. Preliminary experiments revealed a per-bit mutation rate of 5%, which will also be used here, to be promising. For JITR, we set $mIter = 500$ and cool down the weighting factor w stepwise as a function of c as $w = 0.1 \times \max(0, 4 - \lfloor c/250 \rfloor)$.

Experimental results: Figure 1 shows the probability of SW and JIT of achieving the population average fitness of our base algorithm obtained in an ERC-free environment. For SW (left plot), we observe that the performance improves the more storage cells are available and the lower the reuse number is. The reason therefore is that for these settings we order a larger number of potentially different composites more frequently. This in turn reduces the probability of needing to repair and if in case it needs to be repaired it increases the diversity of the repaired solutions among the composite-defining bits. For JIT (right plot), we can see that the performance improves as the time lag decreases and the reuse number increases. As this strategy does not repair and composites are gratis in this experiment, any positive effect on the performance comes from a shorter waiting period between transitions of population generations. While a shorter time lag has a direct positive effect on the waiting period, a greater reuse number allows us to evaluate more solutions from a new population using old composites and thus to compensate a slightly larger time lag.

The performance of JITR obtained at large budgets is not significantly different from the performance of JIT, which is due to the rather conservative cooling scheme of w. However, when looking at the performance during the optimization (see Figure 2 right plot) it is apparent that JITR is able to locate fit solutions at a lower budget than required by JIT (see region $0 < c \leq 600$, $0 \leq c_{time\_step} \leq 0.5$). The weakness of JIT of being too expensive or wasteful in the early stages of

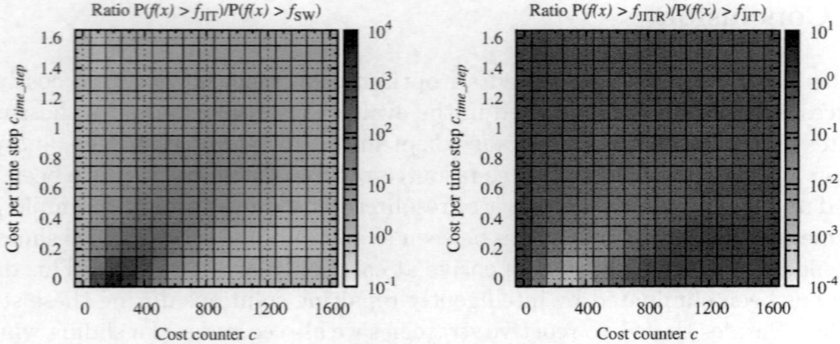

Fig. 2. Shown is the ratio $P(f(x) > f_{\text{JIT}})/P(f(x) > f_{\text{SW}})$ (left) and $P(f(x) > f_{\text{JITR}})/P(f(x) > f_{\text{JIT}})$ (right) as a function of c and c_{time_step} for the ERC $commCompERC(o(H_{\#}) = 10, \#SC = 5, TL = 5, RN = 30, SL = 30)$, $c_{order} = 1$; here, x is a random variable that represents solutions drawn uniformly at random from the search space and f_* the population average fitness obtained with strategy $*$

the optimization is also apparent when comparing it with SW (see Figure 2 left plot). A comparison between JITR and SW is not shown here, but it is easy to see that repairing improves the performance of JIT in early optimization stages.

A comparison between JIT and SW with respect to $o(H_{\#})$, and $\#SC$ for a budget of $C = 1500$ and $C = 3000$ is shown in the left and right plot of Figure 3, respectively; JITR achieved a similar performance to JIT. One can see that, for the lower budget, SW achieves a higher population average fitness than JIT in the range $8 < o(H_{\#}) < 40$, $\#SC < 15$. Compared to JIT, SW is too expensive in the presence of more storage cells. As the budget increases, JIT is able to match the performance of SW and eventually to overtake it, particularly if the order is high and many storage cells are available.

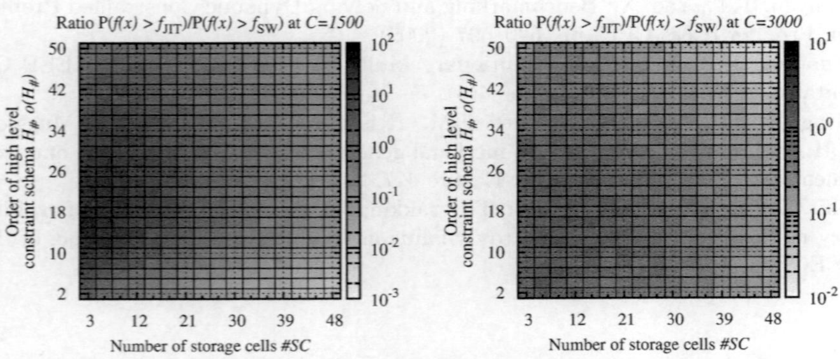

Fig. 3. Shown is the ratio $P(f(x) > f_{\text{JIT}})/P(f(x) > f_{\text{SW}})$ for a budget of $C = 1500$ (left) and $C = 3000$ (right) as a function of the number of $\#SC$ and $o(H_{\#})$ for the ERC $commCompERC(o(H_{\#}), \#SC, TL = 25, RN = 25, SL = 25)$, $c_{order} = c_{time\_step} = 1$

5 Conclusion

In this paper we have considered an optimization scenario in which costly resources (composites) are required in the evaluation of solutions, and these composites had to be ordered in advance, kept in capacity-limited storage, and used within a certain time frame. Three resource-purchasing strategies have been proposed and analyzed over a number of resource-constraint settings. A simple just-in-time (JIT) strategy, which can be seen as the minimal strategy, has shown to be generally effective but too expensive at early optimization stages. This drawback has been eliminated by intelligently repairing solutions during these stages (JITR). Besides these two reactive strategies we also considered a sliding window (SW) strategy, which applied some from of anticipation. This (simple) strategy performs better than JITR and JIT for some constraint settings, although it is costly in the presence of much storage space.

Our current research is looking at other types of resource constraints in evolutionary search, and what strategies can be used to handle them effectively.

References

1. Klockgether, J., Schwefel, H.-P.: Two-Phase Nozzle and Hollow Core Jet Experiments. Engineering Aspects of Magnetohydrodynamics, 141–148 (1970)
2. Rechenberg, I.: Case studies in evolutionary experimentation and computation. Computer Methods in Applied Mechanics and Engineering 186, 125–140 (2000)
3. Allmendinger, R., Knowles, J.: Ephemeral resource constraints in optimization and their effects on evolutionary search. Technical report MLO-20042010, University of Manchester (2010), `http://www.cs.man.ac.uk/~allmendr/publications.html`
4. Bosman, P.A.N., La Poutré, H.: Learning and anticipation in online dynamic optimization with evolutionary algorithms: the stochastic case. In: Proc. of GECCO, pp. 1165–1172 (2007)
5. Branke, J.: Evolutionary Optimization in Dynamic Environments. Kluwer Academic Publishers, Dordrecht (2001)
6. Nguyen, T.T., Yao, X.: Benchmarking and Solving Dynamic Constrained Problems. In: Proc. of IEEE CEC, pp. 690–697 (2009)
7. Knowles, J.: Closed-Loop Evolutionary Multiobjective Optimization. IEEE Computational Intelligence Magazine 4(3), 77–91 (2009)
8. King, R.D., Whelan, K.E., Jones, F.M., Reiser, P.G.K., Bryant, C.H., Muggleton, S.H., Kell, D.B., Oliver, S.G.: Functional genomic hypothesis generation and experimentation by a robot scientist. Nature 427, 247–252 (2004)
9. Shir, O.M., Emmerich, M., Bäck, T., Vrakking, M.J.J.: The application of evolutionary multi-criteria optimization to dynamic molecular alignment. In: Proc. of IEEE ICEC, pp. 4108–4115 (2007)

Parallel Artificial Immune System in Optimization and Identification of Composite Structures

Witold Beluch[1], Tadeusz Burczyński[1,2], and Wacaw Kuś[1]

[1] Department of Strength of Materials and Computational Mechanics, Silesian University of Technology, Gliwice, Poland

[2] Institute of Computer Science, Cracow University of Technology, Cracow, Poland

Abstract. The paper deals with the application of the Artificial Immune System to the optimization and identification of composites. To reduce the computational time parallel computations are performed. Composite structures in form of multilayered laminates are taken into account. Simple and hybrid (with laminas made of different materials) laminates are examined. Different optimization criteria connected with stiffness and modal properties of laminate structures are considered. Continuous and discrete variants of design variables are regarded. The aim of the identification is to find laminate elastic constants on the basis of measurements of state variable values. The Finite Element Method is employed to solve the boundary-value problem for laminates. Numerical examples presenting effectiveness of proposed method are attached.

Keywords: Artificial Immune System, optimization, identification, parallel computing, composite, laminate.

1 Introduction

Optimization and identification tasks are often met in engineering practice. Identification tasks can be typically treated as a kind of optimization. Gradient-based optimization methods are fast and precise, but their application meets many obstacles. In many real problems the calculation of the objective function gradient is not easy or impossible. If the objective function is multimodal, the application of gradient methods can lead to local optima. To avoid the mentioned problems the global optimization methods, like Evolutionary Algorithms or Artificial Immune Systems can be employed. The only information such algorithms require is the objective function value. An application of the Evolutionary Algorithm for optimization and identification of laminates has been presented e.g. in [4] and [2]. In the present paper an Artificial Immune System [13] is used to solve optimization and identification tasks. As the calculation of the objective function is usually the most time-consuming part of engineering computations, the parallel version of Artificial Immune System is introduced. The algorithm has been applied for different composites' optimization and identification problems.

R. Schaefer et al. (Eds.): PPSN XI, Part II, LNCS 6239, pp. 171–180, 2010.

Composites are materials playing an increasing role in modern industry. The main reasons of their popularity are: low specific weight, high specific strength and stiffness properties as well as tailorable characteristics compared to metals. Composite structures can be designed to meet the structural specifications. The most widely used group of composites state layered laminates. They consist of many layers composed of matrix and long fibres, typically placed unidirectionally in each ply. The component materials, stacking sequence, layer thicknesses are generally taken as design variables in laminates optimization. Laminates can be typically treated as thin two-dimensional structures with four independent elastic constants: axial Young modulus E_1, transverse Young modulus E_2, axial-transverse shear modulus G_{12} and axial-transverse Poisson ratio ν_{12}.

There are two goals of the present paper: i) to obtain the optimal properties of the material for different optimization criteria connected with the stiffness and the dynamic behavior of the structure ii) to identify material constants of the laminates on the basis of the measurements.

The laminate stacking sequence optimization is usually a discrete optimization problem due to the manufacturing limitations. Even for a continuous optimization problem and also for an identification problem the objective function is typically multimodal. A commercial FEM software package MSC.Patran/Nastran has been used to solve the direct boundary-value problem for composite structures.

2 Parallel Artificial Immune System

Artificial Immune Systems (AISs) belong to biologically-inspired algorithms. To some extent similar to Evolutionary Algorithms [15], AIS are inspired by a mammal immune system [16]. A natural immune system (NIS) is one of the most complex systems in a human body. It is decentralized and contains distributed groups of dedicated organs, tissues and cells. The NIS is a distributed system capable of learning and adapting to a varying environment. The main role of the NIS is to protect organism against intruders – pathogens. The NIS works on four immune layers: i) skin layer, ii) physiological layer (e.g. temperature, pH), iii) non-specific (innate) immunity and iv) specific immunity [18].

The main elements of the NIS are lymphocytes, mainly B-cells and T-cells. T-cells are formed in the thymus where they are trained to differentiate self cells from non-self cells. Only properly trained T-cells (ignoring self antigen molecules) are released to the blood and lymphatic systems. B-cells, which are produced in the bone marrow, contain antibodies. Antibodies can recognize and neutralize pathogens. Some B-cells become memory cells increasing detection speed of known intruders. To obtain high diversity of antibodies, B-cells undergo a hypermutation process.

The AISs simulate some elements of the specific immunity layer of NIS, like B-cells and T-cells, proliferation, hypermutation and clonal selection. The optimization is treated as a searching for a B-cell containing an antigen best-fitted to a pathogen which represents a global optimum of an objective function [8]. In many engineering optimization problems the calculation of the objective function is the most time-consuming part of the algorithm. It is typically necessary to perform hundreds or even thousands computations of the boundary-value problem. The overall optimization time can be reduced by introducing a parallel approach. In the present paper an in-house parallel AIS (PAIS) is proposed as the global optimization method.

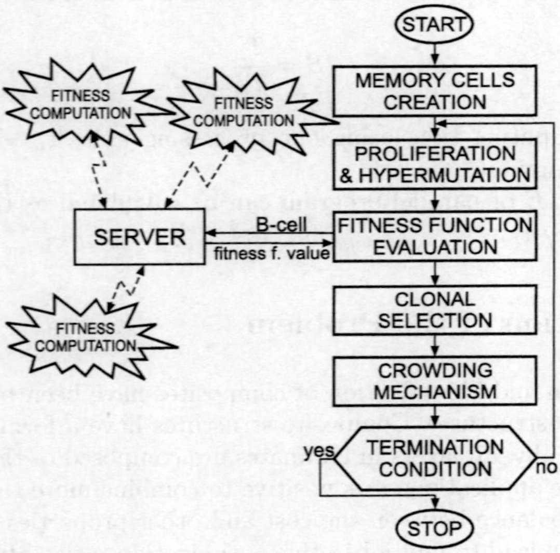

Fig. 1. Parallel Artificial Immune System

The block diagram of PAIS is presented in Fig. 1 ([13],[14]). The PAIS is implemented using Windows system threads. The number of threads during computations is constant. It can be smaller or equal to the number of B-cells. The PAIS is implemented as main (master) thread supported by other threads (workers) for the evaluation of B-cells objective function. The B-cells objective function values are computed by threads by using of round robin method. The number of parallel evaluated B-cells is constrained due to the number of available cores but the memory and disk usage necessary for the objective function solvers have to be taken into account.

The algorithm starts with the random generation of memory cells. The memory cells proliferate – the number of clones depends on the objective function value of the memory cells. The clones undergo gaussian hypermutation and create B-cells. The objective function is calculated for each new B-cell with application of MSC.Nastran Finite Element Method solver. The objective function values may be calculated on more than one processing unit. In the next step the selection procedure takes into account the distance between each memory cell and B-cells and replaces a few memory cells by better B-cells. Similar memory cells are removed by means of the crowding mechanism. The procedure is repeated until the terminal condition is satisfied.

To solve the objective function, the program interface between PAIS and MSC.Nastran had to be developed. On the basis of the design vector values generated by PAIS the input MSC.Nastran file (.bdf) is generated. The result file (.f06) generated by MSC.Nastran is used to calculate the objective function value.

The speed-up of the parallel computations is described as:

$$S = \frac{T_1}{T_n} \tag{1}$$

where: T_1 – computing time using one processing unit, T_n – computing time using n processing unit.

The efficiency E of parallel program can be calculated as the inverse of the speed-up ($E = S/n$).

3 Formulation of the Problem

The optimization and identification of composites have been performed for different composite structures. Composite structures have a form of multi-layered laminates. Typically, all layers in laminates are composed of the same material, however, in some applications it is positive to combine more than one material, e.g. to find the balance between the cost and other properties of the structure. Such attitude can lead to multiobjective optimization tasks. Multiobjective evolutionary optimization of laminates has been presented in [3]. Laminates with laminas (plies) made of different materials are called hybrid ones [1]. The interply hybrid laminates with internal layers built of a weaker/cheaper material and external layers made of a high-stiffness and more expensive material are considered. It is assumed that all considered structures are made of symmetrical laminates – as a result, there is no coupling between shield and bending states in laminate [11].

3.1 Optimization of Composites

Seeking for the optimal properties of composite structures is an important engineering task. An optimal stacking sequence describing fibre angles in particular plies of the composite is looked for. The optimization means a searching for a vector $\mathbf{x}$ which maximizes (or minimizes) the objective function $J(\mathbf{x})$.

In aircraft industry a box-beam sections are popular and used e.g. for helicopter rotor blade structures. Box-beam composite sections resist torsion and bending load efficiently [12]. The analytical solution for box-beam with fixed cross-section is presented in [10]. The presented theory allows predicting the coupling effects (bending-torsion, extending-torsion) in box-beam structures under following assumptions: i) the box-beam is thin-walled; ii) cross-section is fixed and closed; iii) each wall has a plane of symmetry; iv) the loads are: normal longitudinal load, torsion moment about the longitudinal axis and bending moments along two other axis of the box-beam.

The analysis of more complex composite structures is possible due to the development of numerical methods, especially Finite Element Method (FEM). In the present paper an optimization of the box-beam of varying cross-section is considered. The structure has 4 walls composed of an interply hybrid laminate. The aim of an optimization is to find the optimal properties of the composite structures for given criteria. Dynamic and static criteria are taken into account:

1. The maximization of the fundamental eigenfrequency ω_1 of the structure:

$$arg\max\{\omega_1(\mathbf{x}); \mathbf{x} \in \mathbf{D}\} \tag{2}$$

2. The maximization of the distance between external excitation frequency ω_{ex} and the closest eigenfrequency ω_{cl}:

$$arg\max\left\{\left|\omega_{ex}(\mathbf{x}) - \omega_{cl}(\mathbf{x})\right|; \mathbf{x} \in \mathbf{D}\right\} \tag{3}$$

3. The maximization of the total stiffness of the structure. The mean stiffness of the structure can be represented by its total potential energy Π_u [6]:

$$arg\max\{-\Pi_\sigma(\mathbf{x}); \mathbf{x} \in \mathbf{D}\}; \Pi_\sigma = \int_\Omega W(\sigma)d\Omega - \int_{\Gamma_1} \mathbf{p}\mathbf{u}^0 d\Gamma_1 \tag{4}$$

where: Π_σ – total complementary energy of the structure; $W(\sigma)$ – stress potential referred to the unit volume; Ω – a domain occupied by the body; Γ_1 – a part of the boundary on which the function $\mathbf{p}\mathbf{u}^0$ is defined; $\mathbf{p}$ – tractions on the Γ_1; $\mathbf{u}^0$ – displacements on the Γ_1.

3.2 Identification of Composites

In many engineering problems some unknown parameters, like material parameters, shape parameters or boundary conditions must be identified. Identification problems belong to the inverse problems, which are ill-possed ones [5]. To solve such problems it is necessary to collect measurements in form of state fields' values from the considered structure. Then, they are compared with the values of the state fields calculated from the numerical model of the structure.

Composite structures are usually produced in short series or individually. Due to the technological reasons different imperfections resulting non-precise output parameters can occur in composites. As a result, non-destructive methods of composite elastic constants identification should be employed.

The identification can be treated as a the minimization of the functional $J(\mathbf{x})$ with respect to a design variables vector $\mathbf{x}$. In the present paper the following functional is minimized:

$$\min : \left[J(\mathbf{x}) = \frac{1}{N}\sum_{i=1}^{N}(v_i - \hat{v}_i)^2 \right] \tag{5}$$

where: $\mathbf{x}$ – the vector of the design variables, $\hat{v}_i$ – measured values of the state fields, v_i – the same state fields values calculated from a numerical model of the laminate plate, N – the number of measurements.

The design variables vector for a simple laminates can be written as:

$$\mathbf{x} = \{E_1, E_2, G_{12}, \nu_{12}\} \tag{6}$$

The design variables vector for a hybrid laminates consisting of two different materials has a form:

$$\mathbf{x} = \{E_1^1, E_2^1, G_{12}^1, \nu_{12}^1, \rho^1, E_1^2, E_2^2, G_{12}^2, \nu_{12}^2, \rho^2\} \tag{7}$$

where: ρ – material density, superscripts denote the number of material.

The identification procedure can be performed for the data obtained from the structure response to the static load, e.g displacements, which may require many sensors. To reduce the number of sensor points modal analysis methods are employed [17] and the eigenfrequencies values or frequency response data can be treated as a measurements. It is also possible to combine different measurement data which can lead to less ambiguous identification results [7].

4 Numerical Examples

4.1 Numerical Examples: Optimization

A box-beam with varying cross-section is considered (Fig. 2). The wider end of the structure is fixed. Each of 4 walls of the structure is made of the same hybrid, symmetric laminate having identical stacking sequence. The thickness of each ply is h_i=0.2e-3m. External laminas are made of graphite-epoxy material M_e while internal layers are built of glass-epoxy material M_i. The parameters of materials are:

M_e: E_1=141.5GPa, E_2=9.80GPa, G_{12}=5.90GPa, ν_{12}=0.42, ρ=1445.5kg/m^3;
M_i: E_1=38.6GPa, E_2=8.27GPa, G_{12}=4.14GPa, ν_{12}=0.26, ρ=1800.0kg/m^3.

The aim of the optimization is to find an optimal stacking sequence to maximize the fundamental eigenfrequency ω_1 of the structure. It is assumed that the number of plies on each side of the symmetry plane is equal to 7 but external plies angle is fixed and equal to 0, so the number of design variables reduces to 6. The stacking sequence for each wall can be presented as:

$$(0^e/\theta_1^e/\theta_2^i/\theta_3^i/\theta_4^i/\theta_5^i/\theta_6^i)s \tag{8}$$

where subscripts denote a design variable number while superscripts refer to the materials: e – external material (M_e), i – internal material (M_i). Different optimization variants are considered: i) ply angles can vary in a continuous way from the range $-90° \div 90°$; ii) ply angles can vary in a discrete way (every 5°, 15° or 55°) from the same range.

The parameters of the PAIS are: the number of memory cells n_{mc}=5; the number of clones n_{cl} =20; the number of design variables n_m=6; the minimum

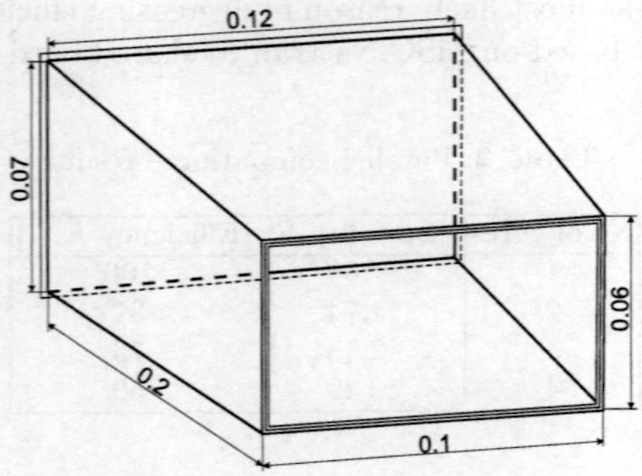

Fig. 2. The box-beam – dimensions and bearing

crowding distance c_{dist}=0.2; the mutation range m_r=0.5; the maximum number of iterations n_i=30. The parameters of PAIS were chosen after some preliminary tests.

The calculations were repeated 30 times for each ply angles variant. The results of the optimization with statistical data in form of the best, the worst and average values of the objective function as well as the standard deviation are collected in Table 1. The best optimization results were obtained for variants with wide search space. The repetitiveness of the algorithm described by the standard deviation is much higher for continuous and 5° variants.

Table 1. The box-beam – optimization results

Variant	ω_1 [Hz]				Stacking sequence for the best solution
	the best	the worst	average	std. dev.	
cont.	900.125	893.649	898.790	1.444	(0/89.9/90/89.5/69.7/-44.4/38.5)s
5°	899.973	896.879	898.663	0.902	(0/90/90/85/-70/45/40)s
15°	898.764	867.803	893.512	10.600	(0/90/90/75/75/-45/45)s
45°	894.589	656.044	855.674	91.896	(0/90/90/90/45/-45/-45)s

The speed-up and efficiency tests of PAIS have been performed. The number of B-cells for each memory cell n_{cl} was reduced to 10. The total number of objective function evaluations in one iteration of PAIS equals 50. The tests have been performed on a dual-processor Intel Xeon E5345 (4 cores each) server station. The PAIS has been implemented by using windows threads. The total number of degrees of freedom (DOFs) in analyzed problem was above 100 000. The eigenvalue problem has been solved during the boundary-value problem computing with the matrix size corresponding to number of DOFs. The computations wall time has been measured for 10 iterations. The computation wall time for iteration on one core was approximately equal to 1 hour. The average wall time from 10 tests has been used for speedup and efficiency evaluation. The results are shown in Table 2. The most likely reason of decreasing efficiency is a concurrent access of the solvers based on MSC.Nastran to shared hard disk.

Table 2. Parallel computing – results

No of cores	Speedup S	Efficiency $E[\%]$
1	1.00	100
2	1.74	87
3	2.18	73
4	2.39	60

4.2 Numerical Examples: Identification

A rectangular symmetrical laminate plate 0.3x0.2m is considered (Fig. 3a). The number of layers, stacking sequence and material density (ρ=1600 kg/m^3) are assumed to be known. The plate consists of 18 plies of the thickness h_i = 0.5e-3m each. Each ply of simple laminate is made of carbon-epoxy material. The stacking sequence of the plate is (0/15/-15/45/-45/90/30/-30/0)s.

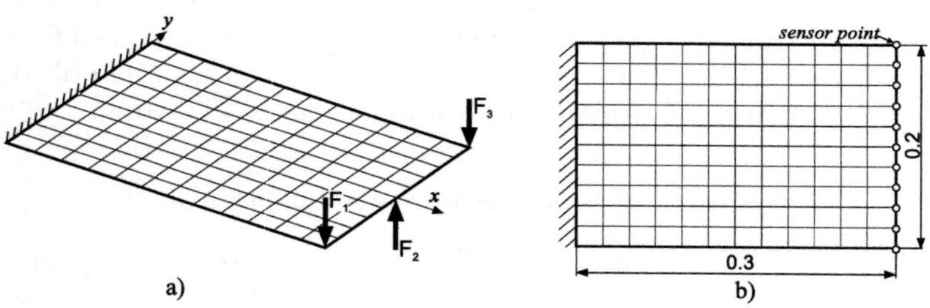

Fig. 3. The laminate plate: a) bearing and loading; b) dimensions and sensors location

The aim is to identify the values of four independent elastic constants: E_1, E_2, G_{12} and ν_{12}. Two measurement data are taken into account: displacements and eigenfrequencies. In the first case, the plate is loaded by 3 concentrated forces F_1=500N, F_2=-1500N and F_3=500N, as presented in Fig. 3a). Displacements are measured in 11 sensor points (Fig. 3b). In the second case first 10 eigenfrequencies of the structure are considered.

The measurements are simulated numerically in both variants. It is assumed that measurements are ideal or measurements are disturbed by a random noise with normal distribution: the expected value $E(x)$ is equal to the non-disturbed value and noise does not exceed 10% [9]. To solve the boundary-value problem the plate is divided into 120 4-node plane finite elements (QUAD4).

The parameters of the PAIS are: the number of memory cells n_{mc}=5; the number of clones n_{cl}=10; the number of design variables n_m=4; the minimum crowding distance c_{dist}=0.2; the mutation range m_r=0.5; the maximum number of iterations n_i=25.

The variable ranges, actual values and identification results for ideal (no noise) and disturbed (with 10% noise) measurements of displacements and eigenfrequencies are collected in Table 3.

Table 3. The laminate plate – identification results

Displacements case						
Constant	Actual value	Range	No noise		Noise 10%	
			Found value	Error [%]	Found value	Error [%]
E_1 [MPa]	1.81e05	1.30e05÷2.20e05	1.811e05	0.02	1.811e05	0.02
E_2 [MPa]	1.03e04	0.80e04÷1.30e04	1.032e04	0.17	9.885e04	4.03
G_{12} [MPa]	7.17e03	5.00e3÷9.00e03	7.155e03	0.21	6.605e03	7.88
ν_{12}	0.28	0.22÷0.32	0.276	1.47	0.287	2.33
Eigenfrequencies case						
Constant	Actual value	Range	No noise		Noise 10%	
			Found value	Error [%]	Found value	Error [%]
E_1 [MPa]	1.81e05	1.30e05÷2.20e05	1.828e05	0.99	1.804e05	0.35
E_2 [MPa]	1.03e04	0.80e04÷1.30e04	1.053e04	2.28	1.102e04	6.95
G_{12} [MPa]	7.17e03	5.00e3÷9.00e03	7.126e03	0.61	7.105e03	0.91
ν_{12}	0.28	0.22÷0.32	0.261	6.77	0.248	11.6

Positive identification results have been obtained for both types of measurements. Slightly better identification results were obtained for displacements as the measurement data. Noisy data reduce the precision of identification procedure, especially for quantities, which does not influence the measurements significantly, like Poisson ratio.

5 Final Conclusions

The application of the Parallel Artificial Immune System for optimization and identification of composite structures has been presented. The aim of the optimization was to find the optimum stacking sequence of laminate structures for static and dynamic optimization criteria. Discrete and continuous variants of the design variables have been considered. The identification with use of PAIS allowed to find composites' elastic constants on the basis of static and dynamic measurement data. The parallelization of calculations allowed significantly reduce computational time. The Finite Element Method commercial software has been used to solve the boundary-value problem for composites. The numerical examples presenting efficiency of the proposed attitude have been enclosed.

Acknowledgments. The research is financed from the Polish science budget resources in the years 2008-2011 as the research project.

References

1. Adali, S., Richte, A., Verijenko, V.E., Summers, E.B.: Optimal design of symmetric hybrid laminates with discrete ply angles for maximum buckling load and minimum cost. Compos. Struct. 32, 409–415 (1995)
2. Beluch, W.: Evolutionary Identification and Optimization of Composite Structures. Mech. Adv. Mater. Struct. 14(8), 677–686 (2007)
3. Beluch, W., Burczyński, T.: Multi-objective optimization of composite structures by means of the evolutionary computations. In: 8th World Congress on Structural and Multidisciplinary Optimization (WCSMO-8), CD-ROM, Lisbon (2009)
4. Beluch, W., Burczyński, T., Kuś, W.: Evolutionary optimization and identification of hybrid laminates. In: Evolutionary Computation and Global Optimization 2006, Warsaw, pp. 39–48 (2006)
5. Bui, H.D.: Inverse Problems in the Mechanics of Materials: An Introduction. CRC Press, Bocca Raton (1994)
6. Burczyński, T.: The Boundary Element Method in Mechanics (in Polish). WNT, Warsaw (1995)
7. Burczyński, T., Beluch, W., Długosz, A., Orantek, P., Nowakowski, M.: Evolutionary methods in inverse problems of engineering mechanics. In: Int. Symposium on Inv. Problems in Eng. Mechanics ISIP 2000, Nagano (2000)
8. de Castro, L.N., Von Zuben, F.J.: Learning and Optimization Using the Clonal Selection Principle. IEEE Transactions on Evolutionary Computation, Special Issue on Artificial Immune Systems 6(3), 239–251 (2002)
9. Dziatkiewicz, G., Fedeliński, P.: Indirect Trefftz method for solving Cauchy problem of linear piezoelecricity. In: Int. Conf. on Comput. Modelling and Advanced Simulations, Bratislava, pp. 29–30 (2009)
10. Ferrero, J.F., Barrau, J.J., Segura, J.M., Sudre, M., Castanie, B.: Analytical theory for an approach calculation of non-balanced composite box beams. Thin-Walled Struct. 39(8), 709–729 (2001)
11. German, J.: Basics of the fibre-reinforced composites' mechanics (in Polish). Publ. of the Cracow University of Technology, Cracow (2001)
12. Kathiravan, R., Ganguli, R.: Strength design of composite beam using gradient and particle swarm optimization. Compos. Struct. 81, 471–479 (2007)
13. Kuś, W., Burczyński, T.: Parallel artificial immune system in optimization of mechanical structures. In: Recent Developments in Artificial Intelligence Methods, Gliwice, pp. 163–166 (2004)
14. Kuś, W., Burczyński, T.: Parallel Bioinspired Algorithms in Optimization of Structures. In: Wyrzykowski, R., Dongarra, J., Karczewski, K., Wasniewski, J. (eds.) PPAM 2007. LNCS, vol. 4967, pp. 1285–1292. Springer, Heidelberg (2008)
15. Michalewicz, Z.: Genetic Algorithms + Data Structures = Evolution Programs. Springer, New York (1992)
16. Ptak, M., Ptak, W.: Basics of Immunology (in Polish). Jagiellonian University Press, Cracow (2000)
17. Uhl, T.: Computer-aided identification of constructional models (in Polish). WNT, Warsaw (1997)
18. Wierzchoń, S.T.: Artificial Immune Systems. Theory and Applications (in Polish). Akademicka Oficyna Wydawnicza EXIT, Warsaw (2001)

Bioreactor Control by Genetic Programming

Dimitris C. Dracopoulos and Riccardo Piccoli

School of Electronics and Computer Science
University of Westminster
London, UK
d.dracopoulos@wmin.ac.uk, riccardo.piccoli@googlemail.com

Abstract. Genetic programming is applied to the problem of bioreactor control. This highly nonlinear problem has previously been suggested as one of the challenging benchmarks to explore new ideas for building automatic controllers. It is shown that the derived control law is successful in a number of test cases.

Keywords: Genetic programming, bioreactor control, nonlinear control.

1 Introduction

The limitations of classical control theory has led to the usage of computational intelligence techniques (neural networks, evolutionary algorithms, fuzzy logic) over the past two decades. Such techniques are also referred to as "intelligent control". Research in intelligent control aims in the derivation of control laws for which no control regimes are known, and the plant operation over large ranges of uncertainty [6]. Uncertainty can be attributed to noise and variations in parameters values, the environment and the inputs.

Traditional adaptive control, also aims in the controller's operation under uncertainty. However, intelligent controllers must be able to operate well, even when the level of uncertainty is substantially greater than that which can be tolerated by traditional adaptive controllers [2,4].

Genetic programming (GP) offers an ideal candidate for the derivation of controllers, since its individuals in a population can be directly mapped to control laws. However, GP has only been applied so far to a small number of challenging control problems.

Bioreactor control has been included in the set of challenging control problems which researchers have to address, in order to explore new ideas for building automatic controllers [1].

This paper describes the application of genetic programming to the control of a bioreactor. The paper is organised as follows. Section 2 provides background information on the bioreactor problem, and details the dynamic system model used for the simulations. Section 3 describes the genetic programming approach for bioreactor control. Finally, Section 4 presents the conclusions for this work, and outlines current work in progress and future research.

R. Schaefer et al. (Eds.): PPSN XI, Part II, LNCS 6239, pp. 181–188, 2010.

2 The Bioreactor Control Problem

Chemical process control offers a set of challenging testbeds for the development of new control algorithms. The problem of bioreactor control was proposed in [5] as such a difficult control benchmark.

In the simplest form, a bioreactor consists of a continuous flow stirred tank reactor (CFSTR) with water containing cells (e.g. yeast or bacteria) which consume nutrients ("substrate"), and produce desired and undesired products and more cells. The cell growth depends only on the nutrient fed to the system.

The objective of the bioreactor control problem is to control the cell mass yield at a prespecified target value. Such a bioreactor is described by the following system of differential equations:

$$\frac{dC_1}{dt} = -C_1 w + C_1(1 - C_2)e^{C_2/\gamma} \tag{1}$$

$$\frac{dC_2}{dt} = -C_2 w + C_1(1 - C_2)e^{C_2/\gamma}\frac{1+\beta}{1+\beta+C_2} \tag{2}$$

where C_1, C_2 are, respectively, the cell mass and the substrate conversion.

The control parameter (output of the controller) is w, the flow rate through the reactor. The constants β and γ are temperature dependent parameters determining the cell growth and nutrient consumption. In all the simulations described in this paper the values of β and γ were kept constant to 0.02 and 0.48 respectively.

Equation (1) describes that the rate of change in the cell mass depends on the amount of cell leaving the tank ($-C_1 w$) and the amount of new cells created $C_1(1 - C_2)e^{C_2/\gamma}$. Similarly, equation (2) describes the change of nutrient based on the rate it is coming out of the reactor, and the rate at which it is metabolised by the cells.

The dynamic system described by equations (1), (2) is very difficult to control for a number of different reasons (despite the fact that it involves only a few variables), and has proved challenging for conventional controllers [5]:

- The equations are highly nonlinear and exhibit limit cycles.
- Optimal behaviour occurs in or near an unstable region.
- The problem exhibits multiplicity: two different values of the control action w can lead to the same desired target value of the cell mass yield.
- Significant delays exist between changes in flow rate and the response in cell concentration.

Based on the values of β and γ aforementioned and used for the experiments described in the paper, a Hopf bifurcation occurs for the system at $w = 0.829$.

3 The GP Approach

A plain GP was applied for the bioreactor problem, based on equations (1), (2). A simulator was built by numerically integrating the equations using the fourth order Runge-Kutta method. The goal was to derive a control law which maintains a desired cell amount in the reactor throughout some period of time T, independent of the initial conditions $C_1[0], C_2[0]$.

Thus, the objective is to minimise the cumulative measure:

$$\sum_{t=0,50,100,\ldots,T} (C_1[t] - C_1^*[t])^2 \tag{3}$$

where $C_1^*[t]$ is the desired cell amount. The summation in equation (3) is across times t which differ by 50 units in value. This is done according to the description of the benchmark in [5]: the sampling interval in simulations is 0.01s and the control interval is 0.5s (50 times as long as the sampling interval). The above implies that a new control action w is calculated by the controller every 0.5s and the rest of the times the previous control action is applied to the plant:

$$w[t] = \begin{cases} \text{controller output}, \; t = 0, 50, 100, \ldots \\ w[t] = w[t-1], \quad \text{all other times } t \end{cases} \tag{4}$$

The controller for which GP searches for, requires three inputs: the current cell mass C_1, the current nutrient amount C_2 and the desired cell amount C_1^*. The single output is the flow rate w. Such a controller is shown in Figure 1.

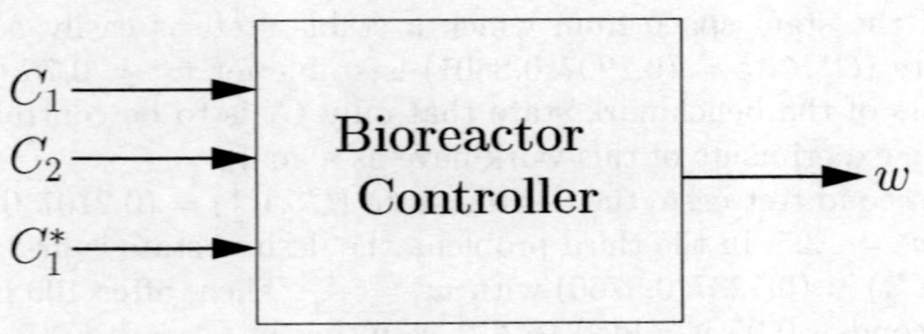

Fig. 1. The controller structure for the bioreactor problem

The fitness function used by the GP runs was:

$$F = \sum_{t=0,50,100,\ldots}^{T} \frac{1}{e^{\sqrt{|C_1 - C_1^*|}}} \tag{5}$$

The above fitness function has to be summed across all fitness cases (different target values of the cell mass C_1^*) used in simulations.

The following terminal and function set were used:

Terminal set: $T = (C_1^*, C_1, C_2, w, R)$
Function set: $F = (+, -, *, /, \ln, \text{IGT-ELSE}, exp)$

where R is the random number generator. / and ln are the protected division and protective natural logarithm respectively as defined by [3], and exp is the exponential function. IGT-ELSE is the *if-greater-than-then-else* operator which tests whether the first argument is greater than the second, and if true it returns the third argument, otherwise it returns its fourth argument.

The GP parameters used, are shown in Table 1.

Table 1. GP Parameters for the Bioreactor control problem

population size	250
crossover probability	0.80
reproduction probability	0.10
mutation probability	0.10
P of function crossover	0.90
maximum initial depth	6
maximum allowed depth	18
generative method	ramped half-and-half
selection method	fitness proportionate

In [5] the benchmark was defined by three test cases, and the results shown here include all of them. The first test case involves starting the bioreactor in a region of the state space from which a stable state is easily achieved. The desired state $(C_1^*, C_2^*) = (0.1207, 0.8801)$ is stable for $w^* = 0.75$ (actually the requirements of the benchmark state that only C_1 is to be controlled and this what all the experiments of this work have as a goal).

For the second test case, the desired state $(C_1^*, C_2^*) = (0.2107, 0.7226)$ is unstable for $w^* = 1.25$. In the third problem, the desired state is first set to stable value $(C_1^*, C_2^*) = (0.1237, 0.8760)$ with $w^* = \frac{1}{1.3}$. Then, after 100 control intervals (50 seconds), 0.05 is added to C_1^*, resulting in $C_1^* = 0.1737$. This change shifts the problem from one of controlling about a stable desired state to one involving an unstable desired state.

The initial conditions for all the experiments require that $C_1[0], C_2[0], w[0]$, are within 10% of the target values [5].

A number of different simulation runs were made, which used both the above test cases (stable and unstable states) and ten other random states as fitness cases (the results presented use one initial condition for each state in training mode) for the individuals control laws evolved. The best individual found in generation 31 (maximum number of generations was 50) in one of the runs is shown (without editing) in Appendix A. This individual consists of 212 nodes and has a depth of 18.

The behaviour of the best individual for the stable state $(C_1^*, C_2^*) = (0.1207, 0.8801)$ can be seen in Figure 2. Note that the data in the figure use a different initial condition for the stable state, than the initial condition used in training. Figure 3 illustrates how the controller derived by GP performs when the target is the unstable state $(C_1^*, C_2^*) = (0.2107, 0.7226)$. Again, the figure uses an initial condition for the unstable state which is different than that used in training mode.

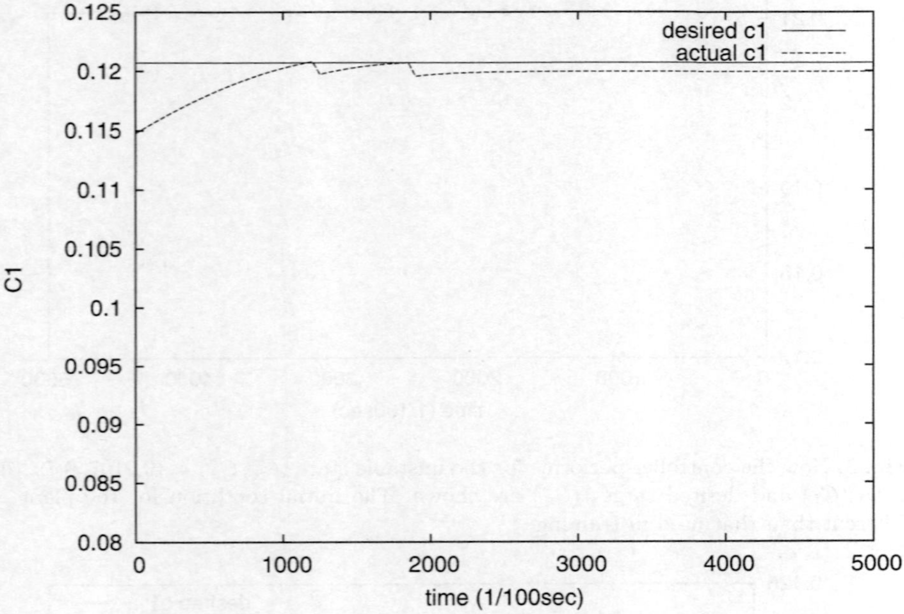

Fig. 2. How the controller performs for the stable state $(C_1^*, C_2^*) = (0.1207, 0.8801)$. Cells (C_1) and desired target (C_1^*) are shown. The initial condition for the plant is different than that used in training.

Following the testing of the best individual against the two fitness cases used in GP runs (although different initial conditions were used in training and testing), its capability to generalise to unseen target states was tested for target state $(C_1^*, C_2^*) = (0.113, 0.8902)$. The results are shown in Figure 4.

It is worth noting that the final state of the plant after applying the GP controller for 50s is $C_1 = 0.11310948323366185$. The results reported in [5] achieved for this case the end state $C_1 = 0.1224$ when a neural network controller was used. From this, it can be clearly seen that the GP controller achieves much better accuracy than the neural network controller results published by the benchmark proposer.

In Figure 5, the derived controller is tested in the third test problem defined in [5], where after 50 seconds the target changes and control moves to a desired state in the unstable region.

The GP controller performed similarly well, when tested with different random initial conditions of the bioreactor dynamic system.

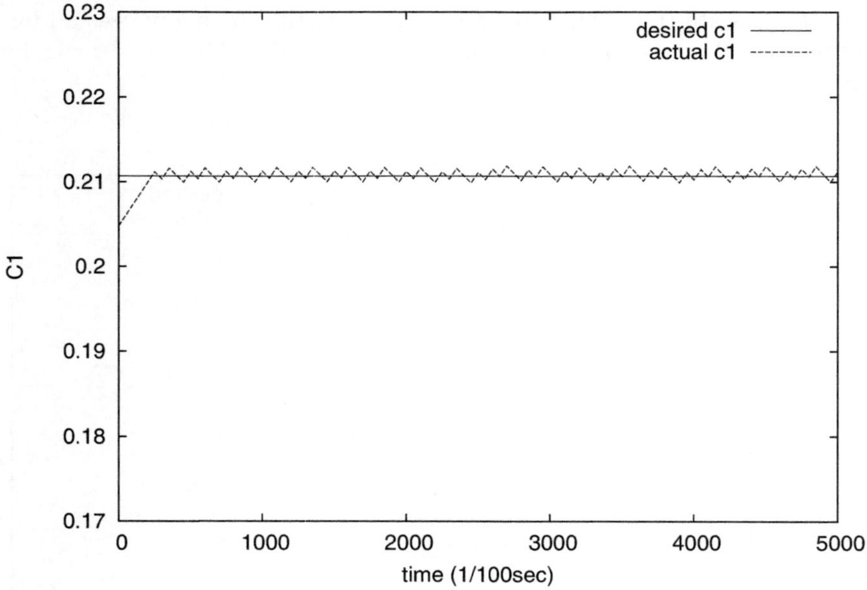

Fig. 3. How the controller performs for the unstable state $(C_1^*, C_2^*) = (0.2107, 0.7226)$. Cells (C_1) and desired target (C_1^*) are shown. The initial condition for the plant is different than that used in training.

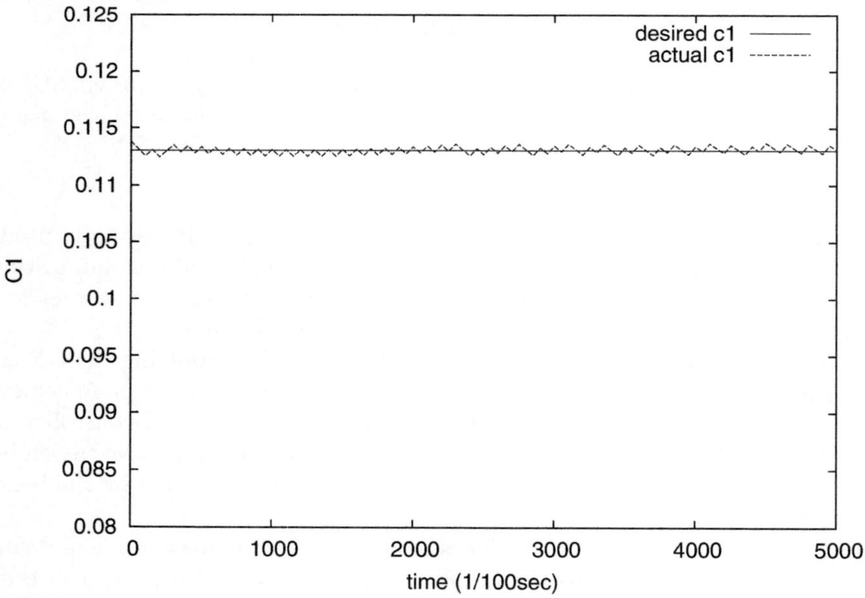

Fig. 4. How the controller performs for the state $(C_1^*, C_2^*) = (0.113, 0.8902)$. Cells (C_1) and desired target (C_1^*) are shown.

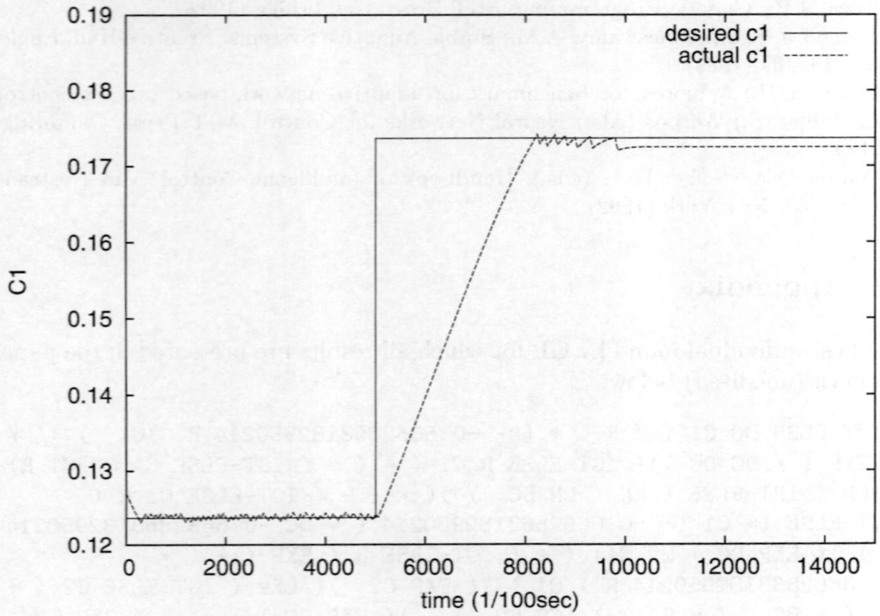

Fig. 5. How the controller performs for a problem where control is started in a stable region and then the target point moves to the unstable region. Initially, the target state is $C_1^* = 0.1237$ and after 50 seconds 0.05 is added to C_1^*, resulting in control about the unstable point $C_1^* = 0.1737$. Cells (C_1) and desired target (C_1^*) are shown.

4 Conclusions and Further Work

The GP approach is tested in a control problem, a chemical bioreactor, which is included in the set of challenging control problems for computational intelligence techniques [6]. A control law which successfully controls the cell mass in the bioreactor was found.

Current work in progress examines whether this individual control law (or other control regimes derived by GP) can control the bioreactor subject to noise.

Future research should compare the results of this approach with the results obtained by more advanced neural networks approaches based on adaptive critics. Additional work could also aim at the investigation of whether any of the derived controllers can be proven to be stable by theory.

References

1. Anderson, C.W., Miller III, W.T.: Challenging control problems. In: Miller, S., Werbos (eds.) Neural Networks for Control, MIT Press, Cambridge (1990)
2. Astrom, K.J., Wittenmark, B.: Adaptive Control, 1st edn. Addison-Wesley, Reading (1989)

3. Koza, J.R.: Genetic Programming. MIT Press, Cambridge (1992)
4. Narendra, K.S., Annaswamy, A.M.: Stable Adaptive Systems. Prentice-Hall, Englewood Cliffs (1989)
5. Ungar, L.H.: A bioreactor benchmark for adaptive network-based process control. In: Miller, S., Werbos (eds.) Neural Networks for Control. MIT Press, Cambridge (1990)
6. White, D.A., Sofge, D.A. (eds.): Handbook of Intelligent Control. Van Nostrand Reinhold, New York (1992)

A Appendix

The best individual found by GP for which all results are presented in the paper is shown (unedited) below:

```
( IGT-ELSE DC C1 ( / R ( * ( - -0.6662863182950214 R  )C1  ) )( /
 ( EXP ( / DC DC ))( IGT-ELSE R C1 ( * ( * ( IGT-ELSE C1 C2 C1 R)
 ( LN ( IGT-ELSE ( LN ( LN DC  ) )( * ( - ( IGT-ELSE C1 R (
 IGT-ELSE (* C1 (+ -0.6662863182950214 ( * DC -0.6662863182950214
 ) ) )( EXP DC ) C2 C1) ( + ( IGT-ELSE R ( EXP ( + ( *
 -0.6662863182950214 R ) C1 ) )( EXP C2  )( EXP ( IGT-ELSE C2 ( +
 C2 ( * DC ( / R R  ) ) )C2 C2  ) ) )( EXP DC ) ) )( - ( EXP ( /
 DC C2  ) )R  ) )( + ( * ( LN ( * DC C1  ) )C2 )( IGT-ELSE (+ R (
 EXP DC))(IGT-ELSE R ( IGT-ELSE ( / ( LN ( LN -0.6662863182950214
 ) ) C1 )( - ( / ( IGT-ELSE R C2 0.8886634785838663 ( + C2 DC  ))
 ( + ( + DC C2 ) C2  ) )( EXP (LN ( EXP DC  ) ) ) )C2 C1  )C1 ( *
 DC C1  ) )C1 R  ) ) )( * ( / -0.6662863182950214 (- C1 C1  ) )(
 IGT-ELSE -0.6662863182950214 ( IGT-ELSE C1 C1 C1 DC  )( IGT-ELSE
 DC (IGT-ELSE DC DC C2 ( + C2 ( LN ( - ( * C2 ( LN C1  ) ) ( +
 -0.6662863182950214 ( EXP DC ) ) ) ) ) )( EXP ( IGT-ELSE ( + (
 LN ( - ( - -0.6662863182950214 C1  )( * DC C1 ) ) )
 -0.6662863182950214 )( + ( * C1 ( EXP DC  ) )DC ) C1
 -0.6662863182950214  ) )( EXP C2  ) )DC  ) )C1  ) ) )R  )( + (
 LN ( LN ( + ( - ( IGT-ELSE ( + R ( EXP DC  ) )( IGT-ELSE R (
 IGT-ELSE ( / ( LN ( LN -0.6662863182950214  ) )C1  )( - C2 DC  )
 C1 C1 ) C1 ( * DC C1  ) )C1 R  )-0.6662863182950214  )DC  ) ) )
 DC  ) ) ) )
```

Solving the One-Commodity Pickup and Delivery Problem Using an Adaptive Hybrid VNS/SA Approach

Manar I. Hosny and Christine L. Mumford

Cardiff School of Computer Science & Informatics
{M.I.Hosny,C.L.Mumford}@cs.cardiff.ac.uk

Abstract. In the One-Commodity Pickup and Delivery Problem (1-PDP), a single commodity type is collected from a set of pickup customers to be delivered to a set of delivery customers, and the origins and destinations of the goods are not paired. We introduce a new adaptive hybrid VNS/SA (Variable Neighborhood Search/Simulated Annealing) approach for solving the 1-PDP. We perform sequences of VNS runs, where neighborhood sizes, within which the search is performed at each run, are adaptable. Experimental results on a large number of benchmark instances indicate that the algorithm outperforms previous heuristics in 90% of the large size test cases. Nevertheless, this comes at the expense of an increased processing time.

1 Introduction

The One-Commodity Pickup and Delivery Problem (1-PDP) is an important problem in transportation and logistics. The problem deals with supplying and collecting one type of commodity from a number of customers, some of them are designated as pickup customers and the others as delivery customers. Each pickup customer provides a certain amount of the commodity, while each delivery customer consumes a certain amount of the same commodity, i.e., goods collected from pickup customers can be delivered to any delivery customer. All customers are served by one vehicle with a limited capacity, and the journey of the vehicle should start and end at a central depot. The depot can supply or consume any additional amount of the commodity that is not supplied or consumed by the customers. Our goal is to find a feasible and minimum cost route for the vehicle, such that all customers are served without violating the vehicle capacity constraint (see [4] for a formal definition of the 1-PDP).

The 1-PDP has many applications in practice. For example, the commodity could be milk that should be collected from farms and delivered to factories with no restriction on the origin and the destination of the product, or it could be money that should be distributed between the branches of a bank [5]. It can also model any logistic situation in which some warehouses have an extra supply of some commodity, while others are in short of the same commodity. A typical situation is when some hospitals need to transfer a certain medicament

R. Schaefer et al. (Eds.): PPSN XI, Part II, LNCS 6239, pp. 189–198, 2010.

to other hospitals, which are in short of this medicament. For example, an H1N1 vaccination could be transferred in urgent epidemic circumstances [7].

In our research we apply an adaptive hybrid VNS/SA (Variable Neighborhood Search/Simulated Annealing) approach to the 1-PDP. The algorithm is distinguished by performing the VNS repeatedly, each time starting from the final solution obtained from the previous VNS run. Also, the algorithm is *adaptive*, in the sense that the maximum neighborhood size allowed in each VNS run is not fixed and depends on the current stage of the search. Early runs are allowed to perform wider jumps in the solution space from the current solution, using large neighborhood sizes. Later runs, on the other hand, are only allowed smaller explorations of the search space, in the vicinity of the current solution, to maintain the solution quality. The stopping criterion for each VNS run is also adaptive and depends on the improvement realized in the current solution. The basic VNS meta-heuristic systematically increases the neighborhood size, within which the search is performed, up to a pre-specified maximum size. In our approach, nevertheless, each VNS run is also terminated when a further increase in the neighborhood size seems not beneficial, even if the maximum neighborhood size was not reached. During each VNS run, an SA acceptance criterion is used to allow the algorithm to escape local optima.

The rest of this paper is organized as follows. Section 2 is a summary of some related work. Section 3 briefly describes the VNS meta-heuristic and highlights previous 1-PDP research that utilizes this approach. Section 4 explains in detail the main components of our proposed heuristic, which we will call an **Adaptive Hybrid VNS/SA (AVNS-SA)** technique for solving the 1-PDP. The complete AVNS-SA algorithm is shown in Section 5. Experimental results are presented in Section 6, and a summary and future directions are given in Section 7.

2 Related Work

Since the 1-PDP is $\mathcal{NP}$*-hard*, exact algorithms are only suitable for small problem sizes. For example, [5] presented a branch-and-cut exact algorithm to solve instances of up to 60 customers. To deal with large size problems, the same authors tried two heuristic approaches in [6]. The first approach is a nearest-neighbor insertion heuristic followed by an improvement phase, using 2-Opt and 3-Opt edge exchanges. The second approach is an incomplete optimization procedure, to find the best solution in a restricted feasible region.

In [4] a heuristic approach, named hybrid GRASP/VND, is proposed. The approach combines two optimization heuristics. The first is called GRASP (Greedy Randomized Adaptive Search Procedure), and is based on a repetition of a construction phase and a local search phase. On the other hand, Variable Neighborhood Descent (VND) is a variant of VNS, which is simply a local search that gradually increases the neighborhood size whenever a local optimum is reached. The hybrid GRASP/VND is basically a GRASP, with an added local search using VND. After the basic GRASP/VND, a further post-optimization phase is performed using *move forward* and *move backward* operators.[1]

On the other hand, a GA approach was introduced in [9] to solve the 1-PDP. The algorithm first starts by creating a population of feasible solutions using a new nearest-neighbor construction heuristic. The initial population is then optimized using a 2-Opt neighborhood move. The most distinguishing feature of the algorithm is a new *pheromone-based* crossover operator, where the selection of the next node to be inserted in the child is based on a probabilistic rule that takes into account the pheromone trail of the edge connecting the last inserted node and the potential new node, such that edges that have proved successful in the past are favored by an increased pheromone value. The offspring are further optimized using a 2-Opt local search. The mutation operator is based on a 3-exchange procedure. The algorithm was tested on the benchmark instances created by [6], producing the best so far results in most test cases.

3 Variable Neighborhood Search (VNS) and Its Application to the 1-PDP

VNS is a relatively new meta-heuristic that was introduced in [2] and [3]. The idea is to generate new solutions that are distant from the incumbent solution, by systematically increasing the neighborhood size within which the search is performed. In addition, a local search is performed on the new solution to reach a local optimum within the current neighborhood. After the local search phase, the new solution replaces the current solution if it is better in quality. The basic steps of the VNS algorithm, as described in [3], are shown below:

- *Initialization*: Select the set of neighborhood structures N_k, $(k = 1, ...k_{max})$; find an initial solution x; choose a stoping condition;
- *Repeat* the following until the stopping condition is met:
  1. Set $k \leftarrow 1$;
  2. Repeat the following steps until $k = k_{max}$:
     (a) *Shaking*: Generate a point x' at random from the kth neighborhood of x $(x' \in N_k(x))$;
     (b) *Local Search*: Apply some local search method with x' as initial solution; denote with x'' the so obtained local optimum;
     (c) *Move or not*: if the local optimum x'' is better than the incumbent, move there $(x \leftarrow x'')$, and continue the search with N_1 $(k \leftarrow 1)$; otherwise set $k \leftarrow k + 1$.

As previously mentioned in Section 2, a variant of VNS,called Variable Neighborhood Decent (VND) has been tried for the 1-PDP as part of the hybrid GRASP/VND heuristic proposed in [4]. The VND heuristic does not include a shaking phase, but only a local search that involves a 2-Opt followed by a 3-Opt move. The algorithm achieved promising results that were better than the

[1] More details about the heuristic in [4] will be given in Section 3.

results obtained by the same authors in their previous heuristics suggested in [6]. However, their algorithm apparently was not fully capable of escaping the trap of local optima. This is evident by the fact that they had to use a post-optimization phase to improve the final result. According to the authors, this post-optimization often made the difference between beating the results obtained by their previous heuristic in [6] or not. The hybrid GRASP/VND heuristic was also outperformed by the GA in [9], in most test cases. A possible shortcoming of this heuristic is the absence of a shaking phase, which should help the diversification of the incumbent solution and allow the succeeding local search phase to escape local optima. In our proposed approach, we try to apply the basic VNS, with both the shaking and the local search, hoping to overcome the limitation of the previous VND attempt on the 1-PDP.

4 The AVNS-SA Heuristic

The main features of our proposed AVNS-SA approach are described below:

The Initial Solution: The construction algorithm we use is the same as the algorithm proposed by [9], which is a nearest-neighbor construction heuristic. However, in [9], they stop the construction process when infeasibility is encountered and try a new starting node. In our approach, we allow violations in the capacity constraint by continuing the construction process despite infeasibility.

The Objective Function: The objective function we use to estimate the quality of a solution S is set to: $F = (NCV(S)+1) \times Dist(S)$, where $NCV(S)$ is the number of capacity violations along the route, and $Dist(S)$ is the total distance of the solution, given the current visiting order of nodes. If there are no capacity violations in the solution, i.e., the solution is feasible, the total distance will be the sole measure of the solution cost.

The Shaking Procedure: The shaking procedure is used for *diversification* of the search in the VNS. We chose as a shaking procedure a displacement of a sequence of nodes with some probability of inverting this sequence. Our VNS algorithm passes the current neighborhood size ($NhSize$) as a parameter to the shaking procedure, which will in turn use this parameter as the *maximum* possible number of nodes that will be displaced. Specifically, the number of nodes to be displaced is a random number between 1 and $NhSize$. So even for large values of $NhSize$, small sequences of nodes could still be displaced, and in fact there is a 'bias' toward such small moves, since they have a chance of being executed in all neighborhood sizes. This is intended to prevent a large disruption of the current solution, and is recommended by some VNS implementations, as in [8].

The local search procedure: This procedure is the tool that VNS uses for *intensification* of the search. In our algorithm, we chose as a local search a simple 2-Opt edge exchange algorithm. Our local search exhaustively tests all possible edge exchanges, and uses *best improvement* as a replacement strategy.

Multiple VNS Runs and the Maximum Neighborhood Size: A sequence of several runs of the VNS procedure is performed to achieve the best result. Each run starts from the final solution obtained in the previous run. The initial maximum neighborhood size ($NhSize_{max}$) value, sent to the first VNS run in the sequence, was set to $NhSize_{max} = 2 \times \sqrt{n}$, where n is the total number of nodes.

However, we realized that during the first VNS run, improvement happens quickly for most $NhSize$ values, even for the large ones among them. Subsequent VNS runs, though, usually respond only to smaller changes in the solution. In other words, smaller neighborhood sizes seem to be more beneficial in subsequent VNS runs, since larger changes seem to cause a disturbance of the current solution and may reduce its quality. Accordingly, after each VNS run, $NhSize_{max}$ was reduced by 1/4 of its value. The reduction is repeated until $NhSize_{max}$ reaches a minimum value of $NhSize_{max}/4$, at which stage no further reduction is performed, and the VNS procedure uses the current $NhSize_{max}$ for all remaining runs. Thus, the algorithm is adaptive in the sense that $NhSize_{max}$ passed to the VNS is not fixed and depends on the current stage of the search.

The VNS could be repeated for a fixed number of iterations, or until no improvement is realized in the current solution for a number of attempts. We chose the second approach, and stopped the repetition of the VNS when no improvement happens in 5 consecutive attempts.

Stopping and Replacement Criteria for Individual VNS Runs: As explained in Section 3, VNS is based on systematically increasing the neighborhood size ($NhSize$), within which a new solution is generated, from 1 up to $NhSize_{max}$. Thus, the shaking and the local search procedures are repeated for all values of $NhSize = 1, 2, 3...NhSize_{max}$. However, we found that in some cases, the current solution may not respond to changes in the neighborhood size and reach a stage of stagnation. Therefore, rather than indiscriminately increasing $NhSize$ up to the pre-specified maximum, we chose to also end each VNS run when the solution has not changed for a certain number of consecutive attempts of increasing $NhSize$. The number of attempts was again chosen to be $NhSize_{max}/4$. Thus, the VNS will be *adaptive* in the sense that it will stop the shaking and the local search cycle, when no benefit seems to be realized from increasing $NhSize$.

In our algorithm, the VNS procedure repeats the shaking and the local search for the ***same*** $NhSize$ for a number of trials. When the number of trials reaches a certain pre-defined limit, the shaking and the local search cycle stops for the current $NhSize$, and the VNS moves on to the next $NhSize$.

The SA Temperature Value: As previously mentioned, we use an SA acceptance criterion within the VNS to replace the current solution. To allow for an adaptive calculation of the SA starting temperature for each instance individually, we adopted the approach in [1]. The temperature is calculated based on the average value of $\Delta cost$, where $\Delta cost$ is the difference in the objective value between some randomly generated solutions for the current problem instance.

Normally, by the end of each complete VNS run, the SA temperature would have reached a small value that should not permit the acceptance of any worse solutions. If we were then to start the next VNS run in the sequence with such small value, there would be no benefit to the SA acceptance criterion, since all worse solutions would be rejected. On the other hand, starting a new VNS run with the initial temperature too high is not advisable, since many worse solutions would be accepted, possibly causing serious loss of solution quality. To achieve a balance between these two situations, the final temperature value reached in the current VNS run is *doubled* before the beginning of the next VNS run.

5 The AVNS-SA Algorithm

To put it all together, Algorithm 1 shows the main Adaptive VNS-SA (AVNS-SA) heuristic, which will invoke the VNSSA procedure (Algorithm 2).

Algorithm 1. Adaptive VNS-SA (AVNS-SA) Algorithm

Find an initial solution (*InitSol*) and calculate the starting SA temperature (*StartTemp*).
$NhSize_{max} \leftarrow 2 \times \sqrt{n}$, where n is the number of nodes
$MaxStagnation \leftarrow NhSize_{max}/4$
$Decrement \leftarrow NhSize_{max}/4$
Initialize *MaxAttempts* to a small number (e.g. 5).
$NoImprovement \leftarrow 0$
repeat
  $NewSol = VNSSA(InitSol, NhSize_{max}, StartTemp, MaxStagnation)$
  if $(NhSize_{max} > Decrement)$ **then**
    $NhSize_{max} \leftarrow NhSize_{max} - Decrement$
  else
    $NhSize_{max} \leftarrow Decrement$
  if (*NewSol* is not better than *InitSol*) **then**
    $NoImprovement{+}{+}$
  else
    $NoImprovement \leftarrow 0$
  $InitSol \leftarrow NewSol$
  $StartTemp \leftarrow StartTemp \times 2$
until (*NoImprovement* reaches *MaxAttempts*)

6 Experimental Results

The algorithm was tested on instances created by [6]. There are 2 types of problem instances. Small instances have a number of customers n in {20, 30, 40, 50, 60}. For these instances, the optimum is known and was obtained using the exact method proposed in [5]. There are also large instances with n in {100, 200,

Algorithm 2. The VNSSA Algorithm

```
Input: InitSol, NhSize_max, StartTemp, MaxStagnation
Output: a new, possibly improved, solution X
k ← 0 { Initialize the current neighborhood size k }
Stagnation ← 0
NumTrials ← LIMIT {LIMIT is the maximum allowed number of trials}
X ← InitSol
repeat
   k++ { Increment the current neighborhood size}
   Trials ← 0
   while (Trials < NumTrials) do
      Shaking(X, XI, k) {Displace a sequence of nodes in X up to a maximum of
      k, with or without inversion. Result stored in XI}
      LocalSearch(XI, XII) {2-Opt applied on XI. Result stored in XII}
      if (Objective(XII) < Objective(X)) then
         X ← XII
      else
         Accept XII using SA acceptance probability
         StartTemp ← StartTemp × α {Decrement current temperature}
         Trials++ {Increment number of trials only when a worse solution is found}
   end while
   if (X did not change in the last iteration (i.e., for the current neighborhood size
   k)) then
      Stagnation++
   else
      Stagnation = 0
until (Stagnation = MaxStagnation) or (k = NhSize_max)
```

300, 400, 500}. For each combination of n and a different vehicle capacity Q in {10, 15, 20, 25, 30, 35, 40, 45, 1000}, 10 problem instances have been created and given the letters {'A' to 'J'}. The data set and the results obtained in [4] can be downloaded from the **Pickup and Delivery Site** of Hernández-Pérez[2]: `http://webpages.ull.es/users/hhperez/PDsite/index.html`

We chose the test cases with the smallest vehicle capacity ($Q = 10$), i.e, the hardest instances. The algorithm was run 5 times on each test case from 20-300 customers. On the other hand, only one run was performed on test cases of 400 and 500 customers, due to time limitation. Also, a number of computers with different specifications were used to run the algorithm. For this reason, the run times we quote here will vary according to the platform.

Results on Problem Sizes 20-60 customers: In this experiment, the algorithm was able to achieve the optimum results at least once in the 5 runs for 39 out of the 50 test cases. The maximum relative difference to the optimum was less than 2% among the 11 cases where the optimum was not found, which was

[2] New best results were obtained by the GA in [9] for vehicle capacity $Q = 10$, but they do not appear in the pickup and delivery site yet.

for test case N50q10D. The processing time ranged on average from 0.66 seconds for 20 customers problems to 47.79 seconds for 60 customers problems.

Results on Problem Sizes 100-500 customers: Table 1 shows the results of the AVNS-SA algorithm on large size problems, from 100 to 500 customers. The table shows the best result achieved and the average result of the 5 runs. The average value is replaced by the best result for problems of size 400 and 500, since the algorithm was run only once on these problems. Finally, the previous best known results are also shown in the table. Most of the best known results were found by the GA in [9]. The best result achieved by the AVNS-SA algorithm is shown in boldface if it was better than the best known result.

Table 1. AVNS-SA Results (100-500 customers)

Name	Best	Avg	Prev-Best	Name	Best	Avg	Prev-Best
N100q10A	**11741**	12175.8	11828	N300q10F	**24042**	24290.6	24826
N100q10B	**13066**	13410.6	13114	N300q10G	**23683**	23945	23868
N100q10C	**13893**	14073.8	13977	N300q10H	**21555**	21824.6	21625
N100q10D	14328	14597	**14253**	N300q10I	**23871**	24110.2	24513
N100q10E	11430	11823.6	**11411**	N300q10J	**22503**	22688.8	22810
N100q10F	11813	11947	**11644**	N400q10A	**30657**	30657	31486
N100q10G	**12025**	12118	12038	N400q10B	**24248**	24248	24262
N100q10H	12821	12844	**12818**	N400q10C	**27853**	27853	28741
N100q10I	**14025**	14278.6	14032	N400q10D	**23750**	23750	24508
N100q10J	13476	13642.8	**13297**	N400q10E	**24798**	24798	25071
N200q10A	17690	17849.2	**17686**	N400q10F	**26625**	26625	26681
N200q10B	**17618**	17887.8	17798	N400q10G	23925	23925	**23891**
N200q10C	16535	16626.6	**16466**	N400q10H	25628	25628	**25348**
N200q10D	**21228**	21594.2	21306	N400q10I	**28262**	28262	28714
N200q10E	**19220**	19485.2	19299	N400q10J	**24847**	24847	26010
N200q10F	**21627**	21677.4	21910	N500q10A	**27904**	27904	28742
N200q10G	**17361**	17634	17712	N500q10B	**26612**	26612	26648
N200q10H	**20953**	21191.4	21276	N500q10C	**30247**	30247	30701
N200q10I	**18020**	18328.2	18380	N500q10D	**29875**	29875	30794
N200q10J	19016	19240.4	**18970**	N500q10E	**29978**	29978	30674
N300q10A	**22940**	23163	23242	N500q10F	**28527**	28527	28882
N300q10B	**22473**	22920.4	22934	N500q10G	**26171**	26171	27107
N300q10C	**21183**	21454	21800	N500q10H	**35805**	35805	36857
N300q10D	**25220**	25500.6	25883	N500q10I	**30247**	30247	30796
N300q10E	**26636**	26934	27367	N500q10J	**30428**	30428	31255

Table 1 shows that the algorithm was able to improve best known results for 50% of the 100 test cases, 70% of the 200 test cases, 100% of the 300 test cases, 80% of the 400 test cases, and finally 100% of the 500 test cases.

The overall average of the 5 runs for all test cases of size 100 is 13091.12, which is only 1% worse than the average result of the GA in [9], having a value of 12954.16. Moreover, our overall average for the 200 test cases is 19151.44, while the average of the heuristic in [9] for the same test cases in 10 runs is 19339.48, i.e., our results account for an improvement of approximately 1%. On the other hand, the overall average of our results for the 300 test cases was 23683.12, with an improvement of more than 2% compared the average of their results for the same test cases, which is 24224.28.

In addition, the average result of the 10 instances of size 500 achieved by our algorithm was 29579.4. This is an improvement of approximately 3% over the average of the best results of [9], having the value 30377.1. These results may also indicate that our algorithm seems to perform even better on larger size problems. The average processing time in this experiment ranged from approximately 542.22 seconds for 100 customers instances to 151103.04 seconds for 500 customers instances.

To further test the robustness of the AVNS-SA algorithm, we performed an additional experiment by running the algorithm on 100-customers problems vehicle capacities of 20 and 40, running the algorithm 5 times on each test case. In this experiment, the algorithm outperformed previous heuristics in 4 out of 10 test cases, for vehicle capacity $Q = 20$, and was able to match the best known result in 6 out of 10 cases, for vehicle capacity $Q = 40$. The average processing time for $Q = 20$ instances was 155.76 seconds, and for $Q = 40$ instances was 146.17 seconds.

Contrary to the exceptional results achieved by our AVNS-SA algorithm, its processing time in general was to a large extent disappointing. For example, the average processing time for 100 customers problems was 542.22 seconds, while the processing time reported by [9] was 21.12 seconds for the same category.

7 Summary and Future Work

In this research, we investigated a new adaptive hybrid VNS/SA approach to the 1-PDP. Adaptation is applied in both the maximum neighborhood size allowed in each VNS run, and in the stopping condition for each VNS run. Searching within smaller neighborhood sizes is preferred in our approach, and larger sizes are only attempted when this looks promising from the search perspective.

Experimental results on a large number of problem instances indicated that our algorithm outperforms previous heuristics in most hard test cases, where the vehicle capacity is smallest. This is especially noticeable for large problem sizes. The algorithm was able to achieve the optimum results for all but few test cases in the small size problems, and was able to achieve new best known results for 90% of the large test cases. The algorithm is also robust enough, since it performs equally well on a wide range of problem instances, e.g. instances with a different vehicle capacity, without the need for any parameter tuning.

These distinguished results, though, come at the expense of computation time. Although we cannot provide an accurate analysis at this stage, because of the use of different processors to run the experiments, we recognize that the running time of the algorithm is rather too long.

In the future, we will continue investigating possible techniques to reduce the run time, for example by reducing the number of VNS runs, or changing the stopping condition for each individual run. Our computational experimentation indicates that some problem instances need fewer than 5 consecutive attempts (without improvement) to reach the best results. However, for other instances, reducing the maximum number of attempts to less than 5 may cause the algorithm to stop prematurely and produce lower quality result. More investigation of the best termination criterion is therefore needed to reduce the overall processing time. Other possible improvement attempts, with respect to the run time, should be oriented towards the local search procedure, since it is the most time consuming part of the algorithm. We can try to reduce the number of calls to this algorithm, or make it optimize only part of the solution rather than whole solution. For example, exchanges can be only restricted to edges with a certain number of closest neighboring nodes.

References

1. Dorband, J., Mumford, C.L., Wang, P.: Developing an ace solution for two-dimensional strip packing. In: 18th International Parallel and Distributed Processing Symposium Workshop on Massively Parallel Processing (2004)
2. Hansen, P., Mladenović, N.: An introduction to variable neighborhood search. In: Voss, S., Martello, S., Osman, I.H., Roucairol, C. (eds.) Meta-heuristics, Advances and Trends in Local Search Paradigms for Optimization, pp. 433–458. Kluwer Academic Publishers, Dordrecht (1998)
3. Hansen, P., Mladenović, N.: Variable neighborhood search: Principles and applications. European Journal of Operational Research 130(3), 449–467 (2001)
4. Hernández-Pérez, H., Rodríguez-Martín, I., Salazar-González, J.-J.: A hybrid GRASP/VND heuristic for the one-commodity pickup-and-delivery traveling salesman problem. Computers & Operations Research 36(5), 1639–1645 (2008)
5. Hernández-Pérez, H., Salazar-González, J.-J.: A branch-and-cut algorithm for a traveling salesman problem with pickup and delivery. Discrete Applied Mathematics 145(1), 126–139 (2004)
6. Hernández-Pérez, H., Salazar-González, J.-J.: Heuristics for the one-commodity pickup-and-delivery traveling salesman problem. Transportation Science 38(2), 245–255 (2004)
7. Martinović, G., Aleksi, I., Baumgartner, A.: Single-commodity vehicle routing problem with pickup and delivery service. Mathematical Problems in Engineering 2008, 1–17 (2008), doi:10.1155/2008/697981
8. Placek, M., Hartl, R.F., Doerner, K.: A variable neighborhood search for the multi depot vehicle routing problem with time windows. Journal of Heuristics 10, 613–627 (2004)
9. Zhao, F., Li, S., Sun, J., Mei, D.: Genetic algorithm for the one-commodity pickup-and-delivery traveling salesman problem. Computers & Industrial Engineering 56(4), 1642–1648 (2008)

Testing the Dinosaur Hypothesis under Empirical Datasets

Michael Kampouridis[1], Shu-Heng Chen[2], and Edward Tsang[1]

[1] School of Computer Science and Electronic Engineering, University of Essex, Wivenhoe Park, CO4 3SQ, UK

[2] AI-Econ Center, Department of Economics, National Cheng Chi University, Taipei, Taiwan 11623

Abstract. In this paper we present the Dinosaur Hypothesis, which states that the behaviour of a market never settles down and that the population of predictors continually co-evolves with this market. To the best of our knowledge, this observation has only been made and tested under artificial datasets, but not with real data. In this work, we attempt to formalize this hypothesis by presenting its main constituents. We also test it with empirical data, under 10 international datasets. Results show that for the majority of the datasets the Dinosaur Hypothesis is not supported.

Keywords: Dinosaur Hypothesis, Genetic Programming.

1 Introduction

The Dinosaur Hypothesis (DH) is inspired by an observation of Arthur [1]. In his work, Arthur and his group conducted the following experiment under the Santa Fe Institute Artificial Stock Market. They first allowed the market evolve for long enough. They then took the most successful agent with his winning predictor[1] out of this continuously evolving market, "froze" him for a while, and then returned the agent back to the market. They found that the early winner could not perform as well as he used to do in the past. His predictors were out of date, which had turned him into a *dinosaur*. This is quite an interesting observation, because it indicates that *any successful predictor or trading strategy can only live for a finite amount of time.*

In addition, Chen and Yeh [3] also tested the existence of this non-stationary market behaviour in their artificial stock market framework; their results verified Arthur's observation. Furthermore, they observed that a dinosaur's performance decreases monotonically.

Based on these observations, Chen [2] suggested a new hypothesis, called the Dinosaur Hypothesis. The DH states that the market behaviour never settles down and that the population of predictors continually co-evolves with this market.

[1] Predictor is the model that the agents use for forecasting purposes. In Arthur's work, predictor is a GP parse tree. In this work, predictors are Genetic Decision Trees (see Sect. 3 for more details). We also refer to them as trading strategies.

R. Schaefer et al. (Eds.): PPSN XI, Part II, LNCS 6239, pp. 199–208, 2010.

In this paper, we first formalize the DH by presenting its main constituents. In addition, motivated by the fact that both Arthur, Chen and Yeh made their observations under an artificial stock market framework, we want to examine whether the same observations hold in the 'real' world. We thus test the hypothesis with empirical data. We run tests for 10 international markets and hence provide a general examination of the plausibility of the DH. Our tests take place under an evolutionary environment, with the use of GP [7]. One goal of our empirical study is to use the DH as a benchmark and examine how well it describes the empirical results which we observe from the various markets.

The rest of this paper is organized as follows: Section 2 elaborates on the DH, and Section 3 briefly presents the GP algorithm that is going to be used for testing the DH. Section 4 then presents the experimental designs, Section 5 addresses the methodology employed to test the DH, and Section 6 presents and discusses the results of our experiments. Finally, Section 7 concludes this paper.

2 The Dinosaur Hypothesis

Based on Arthur's work, we can derive the following statements which form the basic constituents of the DH:

1. The market behaviour never settles down
2. The population of predictors continuously co-evolves with the market

These two statements indicate the non-stationary nature of financial markets and imply that strategies need to evolve and follow the changes in these markets, in order to survive. If they do not co-evolve with the market, their performance deteriorates and makes them ineffective.

However, as we said earlier, these observations were made in an artificial stock market framework. What we thus do in this paper is to test the above statements against our empirical data. We propose the following *Fitness Test*:

The average fitness of the population of predictors from future periods should

1. Not return to the range of fitness of the base period (P1)
2. Decrease continuously, as the testing period moves further away from the base period (P2)

As we can see, there is a population of predictors, which in our framework these are Genetic Decision Trees (GDTs); what we do in this work is to monitor the future performance of these GDTs in terms of their fitness, in accordance with Arthur's and Chen and Yeh's experiments. More details about the testing methodology can be found at Sect. 5.

Statement P1 is quite straightforward and is inspired by Arthur [1]. The term 'range of fitness' is also explained in Sect. 5. Statement P2 is inspired by the observation that Chen and Yeh made [3], regarding the monotonic decrease of a predictor's performance. However, in our framework we do not require the performance decrease to be monotonic. This is because when Chen and Yeh tested for the Dinosaur Hypothesis (they did not explicitly use this term), they only tested it over a period-window of 20 days, which is relatively short, hence easy to achieve monotonic decreasing. Thus, requiring that a predictor's performance decreases monotonically in the long run would be very strict, and indeed hard to achieve. For that reason, statement P2 requires that the performance decrease is continuous, but not monotonic. It should also be mentioned that we are interested in qualitative results, meaning that we want to see how close the real market behaves in comparison with what is described by the DH.

Finally, in order to make the reading of this paper more comprehensive, we present two definitions, inspired by Arthur's work: *Dinosaur*, is a predictor who has performed well in some periods, but then ceased performing well in the periods that followed. This means that his predictor may or may not become effective again. If it does, then it is called a *returning dinosaur*.

3 GP Algorithm

Our simple GP is inspired by a financial forecasting tool, EDDIE [6], which learns and extracts knowledge from a set of data. This set of data is composed of the daily closing price of a stock, a number of attributes and signals. The attributes are indicators commonly used in technical analysis [5]: Moving Average (MA), Trader Break Out (TBR), Filter (FLR), Volatility (Vol), Momentum (Mom), and Momentum Moving Average (MomMA). Each indicator has two different periods, a short- and a long-term one, 12 and 50 days respectively.

The signals are calculated by looking ahead of the closing price for a time horizon of n days, trying to detect if there is an increase of the price by $r\%$. For this set of experiments, n was set to 1 and r to 0. In other words, the GP tries to forecast whether the daily closing price will increase in the following day.

Furthermore, Fig. 1 presents the Backus Naur Form (BNF) (grammar) of the GP. The root of the tree is an If-Then-Else statement. Then the first branch is a Boolean (testing whether a technical indicator is greater than/less than/equal to a value). The 'Then' and 'Else' branches can be a new GDT, or a decision, to buy or not-to-buy (denoted by 1 and 0). Thus, each individual in the population is a GDT and its recommendation is to buy (1) or not-to-buy (0). Each GDT's performance is evaluated by a fitness function presented below.

Depending on what the prediction of the GDT and the signal in the training data is, we can define the following 3 metrics:
Rate of Correctness

$$RC = \frac{TP + TN}{TP + TN + FP + FN} \tag{1}$$

Rate of Missing Chances

$$RMC = \frac{FN}{FN + TP} \tag{2}$$

Rate of Failure

$$RF = \frac{FP}{FP + TP} \tag{3}$$

<Tree> ::= If-then-else <Condition> <Tree> <Tree> | Decision
<Condition> ::= <Condition> "And" <Condition> |
<Condition> "Or" <Condition> |
"Not" <Condition> |
Variable <RelationOperation> Threshold
<Variable> ::= MA_12 | MA_50 | TBR_12 | TBR_50 | FLR_12 |
FLR_50 | Vol_12 | Vol_50 | Mom_12 | Mom_50 |
MomMA_12 | MomMA_50
<RelationOperation> ::= ">" | "<" | "="
Decision is an integer, Positive or Negative implemented
Threshold is a real number

Fig. 1. The Backus Naur Form that the simple GP uses to construct trees

We use these metrics to define the following fitness function:

$$ff = w_1 * RC - w_2 * RMC - w_3 * RF \tag{4}$$

where w_1, w_2 and w_3 are the weights for RC, RMC and RF respectively. The weights are given in order to reflect the preferences of investors. For our experiments, we chose to include GDTs that mainly focus on correctness and reduced failure. Thus these weights have been set to 0.6, 0.1 and 0.3 respectively, and are given in this way in order to reflect the importance of each performance measure for our predictions.

4 Experimental Designs

Experiments were conducted for a period of 17 years (1991-2007) and the data was taken from the daily closing prices of 10 international market indices: CAC 40 (France), DJIA (USA), FTSE 100 (UK), HSI (Hong Kong), NASDAQ (USA), NIKEI 225 (Japan), NYSE (USA), S&P 500 (USA), STI (Singapore) and TAIEX (Taiwan). For each of these markets, we run each experiment for 10 times.

Each year was split into 2 halves (January-June, July-December), so in total, out of the 17 years, we have 34 periods[2]. The GP system was hence executed 34 times. Table 1 presents the GP parameters for our experiments. The behavior of each GDT can be represented by its series of market timing decisions over the entire trading horizon. Thus, the behaviour of each rule is a binary vector of 1s and 0s (buy and not-to-buy). The length or the dimensionality of these vectors is then determined by the length of the trading horizon, which in this study is 6 months, i.e., 125 days long; hence, the market timing vector has 125 dimensions.

Here we should emphasize that the GP was only used for creating and evolving the trading strategies. No validation or testing took place, as it happens in

[2] At this point the length of the period was chosen arbitrarily to 6 months. We leave it to a future research to examine if and how this time horizon can affect our results.

Table 1. GP Parameters. The GP parameters for our experiments are the ones used by Koza [7]. Only the tournament size has been changed (lowered), and the reason for that was because we have observed premature convergence under a larger tournament size. Other than that, the results seem to be insensitive to these parameters.

GP Parameters	
Max Initial Depth	6
Max Depth	17
Generations	50
Population size	500
Tournament size	2
Reproduction probability	0.1
Crossover probability	0.9
Mutation probability	0.01

the traditional GP approach. The reason for this is because we were not using the GP for forecasting purposes; instead, we were interested in using the GP as a *rule inference engine* which would evolve profitable trading strategies for a certain period of time. The GP was thus used for each of the 34 periods to create and evolve trading strategies. After the evolution of the strategies under a specific period, these strategies are not tested against another set. This approach is consistent with the Lo's Adaptive Market Hypothesis [8], as it states that the heuristics of an old environment are not necessarily suited to the new ones. Our no-testing approach is also consistent with the well-tested *overreaction hypothesis* [4], which essentially states that top-ranked portfolios are outperformed by bottom-ranked portfolios during the next period. Thus, after evolving a number of generations (50 in this paper), what stands (survives) at the end (the last generation) is, presumably, a population of financial agents whose market-timing strategies are financially rather successful. This population should, therefore, interest us in spirit of Arthur's adaptive market process; therefore, we use them to test how those competitive strategies perform in the future periods.

5 Testing Methodology

This section presents the testing methodology. But before we do this, let us first present some frequently used terms:

- *Base period*, is the period during which GP was used to create and evolve GDTs that are going to be used for testing the DH
- *Future period(s)*, is a period(s) which follow the base period (in chronological order)

We are interested in observing how the average fitness of the population of GDTs changes throughout time. As we have already seen, we used a simple GP system to generate and evolve trading strategies for each one of the 34 periods.

After this step, we apply this evolved population of GDTs to the future periods' data. In order to better explain this, let us use an example. Let us suppose that the period we trained the GDTs (base period) was the first semester of 1991 (1991a); we can then calculate the average fitness of the population of these trees for this period. From this point on, we will be calling this 'average fitness of the population of GDTs' as *population fitness*. We thus have an indication of how well the population performs during the base period. Then, we apply all evolved GDTs to the data of future periods: second semester of 1991 (1991b), first semester of 1992 (1992a),..., second semester of 2007 (2007b) and calculate the population fitness for each one of these periods. As a result, we can observe how this fitness changes over the future periods.

The same procedure is followed for all periods until 2007a, so that all of them act as a base period. This means that when 1991b is the base period, the GDTs that were created and evolved during 1991b will be applied to all future periods. After 1991b, 1992a takes over as the base period and the same procedure happens again. We do this until 2007a. We obviously cannot do this for 2007b, since there are no data available after this year. The reader should also bear in mind that we only apply the evolved GDTs to future periods; for instance, when the base period is 2000a, we do not apply the GDTs backwards in time, only forwards. We are not interested in looking what happens in the past; we are only interested in observing how the fitness of the GDTs is affected in the future.

Given a base period, the population fitness of all periods is normalized by dividing those population fitnesses by the population fitness in the base period. Hence, each base period has its normalized population fitness equal to 1 and a returning dinosaur is a population of strategies from future periods that has its normalized population fitness 'close to 1'. At this point, we need to define the term 'close to 1'. Strictly speaking, this means that this population's normalized fitness is greater or equal to 1. However, in our opinion, other future periods which do not necessarily satisfy this condition could be considered as returning dinosaurs, too. Let us consider the case of a future period with normalized population fitness very 'close to 1', e.g. 0.99. When this happens, it indicates that there exist those similar market conditions in this future period, as in the base period, so that the dinosaurs can again have high performance. Although this performance may not be exactly equal to 1, we believe that the fact that the normalized population fitness of these strategies (GDTs) is this 'close' to 1, indicates that these GDTs have become successful again, and should thus be considered as returning dinosaurs.

However, defining a specific range of fitnesses for 'close' would be arbitrary; after all, closeness is only a matter of degree. We therefore present in the next section the statistics of fitness observed for each stock market. Besides, as we said in Sect. 2, we are interested in qualitative results; we want to see how close the 10 empirical markets behave in comparison by what is described by the DH.

If DH holds, we should observe two things: firstly, the normalized population fitness of the future periods has decreased and does not return to the range of fitness of the base period (P1), and secondly, this decrease is continuous (P2).

6 Results

6.1 Statement P1

According to P1, the future periods' population fitness will not return to the range of fitness of the base period. As we saw earlier, we test this statement for one period at a time. The subject period forms our base period.

In order to examine how often dinosaurs return, we iterate through each base period and calculate the maximum fitness among its future periods. Let us give an example. If 1991a is the base period, then there is a series of 33 population fitness values for its future periods. We obtain the maximum value among these 33 values, in order to check how close to 1 this future period is. This process is then repeated for 1991b and its 32 future periods, 1992a, and so on, until base period 2007a. We thus end up with a 1×33 vector, which shows the potential returning dinosaur per base period. The graph of this vector is presented in Fig. 2. Each line represents the results on a different dataset and they have been divided in four separate subfigures: CAC40-DJIA-FTSE100 (top-left), HSI-NASDAQ-NIKEI (top-right), NYSE-S&P500 (bottom-left), and STI-TAIEX (bottom-right).

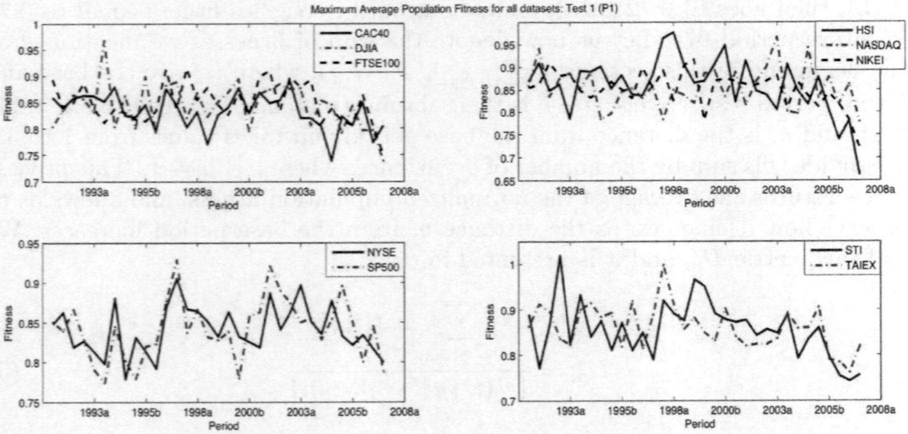

Fig. 2. Fitness Test, P1: The maximum normalized population fitness among all future periods for each base period. Each line represents a single dataset. Results have been divided in 4 subfigures.

What we can see from this figure is that only STI has a base period (1992b) with a maximum normalized population fitness exceeding 1. This indicates returning dinosaurs and goes against statement P1. We cannot observe any more periods that reach the threshold of maximum population fitness greater or equal to 1. Nonetheless, all of our datasets seem to have quite high population fitness values, which many times exceed 0.9 or even 0.95 (e.g. DJIA-1993b, HSI-1998b,

NASDAQ-2003a, TAIEX-1997b). Therefore, although we cannot strictly talk about a returning dinosaur, we should also not neglect the fact that this is an indication that the market environment actually can create conditions that are very similar to the past and as a result, successful strategies from the past do not necessarily have a finite lifetime (as the DH implies), but can again be successful in the future. Thus, our results do not support statement P1.

6.2 Statement P2

To show a continuous decrease in the population fitness, we calculate the sum of the fitness values of all those future periods that are 1 period away from the base period, then the sum of those future periods that are 2 periods away, and so on, up to a period difference between future and base period of 33. In order to do this, we first need to create a table of distances, like the one in Table 2(a). Each row of this table presents the distance of the future periods from their base period. For instance, if 91a is the base (first row), then future period 91b has distance equal to 1, future period 92a has distance equal to 2, and so on. Table 2(b) shows the series of population fitness values for the future periods of each base period. For example, when the base period is 91a (first row), the normalized population fitness starts from 1 in 91a, then drops to 0.66 (91b), then goes to 0.72 (92a), and so on, until it reaches fitness equal to 0.74 in future period 07b. Let us now denote the sum of fitnesses we mentioned at the beginning of this section by $\sum_{|i-j|=m} Fit(i,j)$, where i, j are the base and future period respectively, $|i-j|$ is their absolute distance, as presented in Table 2(a), and m is the distance from the base period and takes values from 1 to 33. We divide this sum by the number of occurrences where $|i-j| = m$. This process hence returns the average of the normalized population fitness, and allows us to observe how it changes, as the distance m from the base period increases. We call this metric D_m and it is presented in (5).

$$D_m = \frac{\sum\limits_{|i-j|=m} Fit(i,j)}{\{\#(i,j), |i-j| = m\}} \quad (5)$$

Let us give an example: if we want to calculate D_{32}, we need to sum up the population fitnesses that have distance $m = 32$. This happens with $Fit(91a, 07a)$ (fitness of GDTs from base period 91a, when applied to future period 07a) and $Fit(91b, 07b)$ (fitness of GDTs from base period 91b, when applied to future period 07b). Therefore D_{32} would be equal to the sum of these two fitness rates divided by 2, as there are only 2 periods that can have $m = 32$.[3] By calculating D_m for all m values, we can have a clear idea of how the average of the population

[3] The distance $m = 32$ can also be found in 07a91a and 07b91b. However, we do not take them into account because, as we said earlier in Sect. 5, we are not interested in applying the evolved GDTs of a base period (here 07a and 07b) backwards in time (91a and 91b, respectively).

Table 2. (a) Distance of future periods from their base period, over the 17 years 1991-2007. The further away we move from a period, a single unit of distance is added. (b) Series of future population fitnesses per base period. Each base period's series is presented as a horizontal line of this table. Fitness values have been normalized, so that the average fitness in the base period is always equal to 1.

(a)

		j					
		91a	91b	92a	92b	...	07b
	91a	0	1	2	3	...	33
	91b	1	0	1	2	...	32
i	92a	2	1	0	1	...	31
	...	...	...	...	...	...	...
	07b	33	32	31	30	...	0

(b)

		j					
		91a	91b	92a	92b	...	07b
	91a	1	0.66	0.72	0.78	...	0.74
	91b		1	0.76	0.72	...	0.70
i	92a			1	0.74	...	0.77
	...	...	...	...	...	...	...
	07b					...	1

fitness changes when we move from periods that are close to the base period (low m), to periods that are further away (high m), and thus observe whether there is a continuous decrease. Figure 3 presents the results for all datasets. Each line represents again a single dataset, similar to that in Fig. 2.

What we observe from this figure is that there are upwards and downwards movements of the D_m metric. This is consistent for all datasets. We do not observe a continuous, or any kind of decrease in general, in the metric. This therefore does not validate the P2 statement.

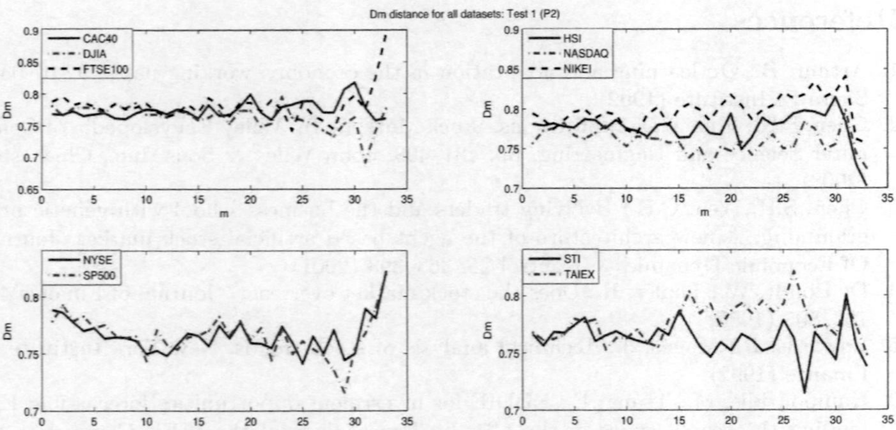

Fig. 3. Fitness Test, P2: D_m values for all m from 1 to 33. Each line represents a single dataset. Results have been divided in 4 subfigures.

7 Conclusion

This paper presented and formalized the Dinosaur Hypothesis. The DH says that the behaviour of a market never settles down and that the strategies in this market continuously co-evolve with it. This was an observation first made by Arthur [1] and later by Chen and Yeh [3]. However, these two works made these observations under an artificial stock market. In this paper, we were interesting in examining whether these observations could also hold in the real world and thus tested the hypothesis with empirical data. For our experiments, we used a *fitness test*, where we created and evolved trading strategies with a GP system. Results showed that 1 of the 10 datasets tested demonstrated the existence of returning dinosaurs; having a returning dinosaur is of course contradicting with statement P1. However, it would not be accurate to say that the remaining 9 datasets fully support P1. This is because all of population strategies have had future periods' average fitness values that are close to the fitness of the base period; in fact, there were many occasions were this fitness was even more than 90% closer to the population fitness of the base period. Therefore, although there is no normalized population fitness among these 9 datasets that reaches 1, we can argue that trading strategies from the past can still be applied to the market and perform satisfactory, even if many years have passed. Markets can thus have a number of 'typical states', where past rules may become useful again. *Returning dinosaurs hence exist.* Finally, regarding statement P2: we did not observe any continuous decrease in the average population fitness of any of the 10 datasets tested, and we can thus argue that P2 is not supported by the empirical data in this work. Overall, we can conclude that the empirical evidence that can support the Dinosaur Hypothesis is quite weak.

References

1. Arthur, B.: On learning and adaptation in the economy, working paper 92-07-038, Santa Fe Institute (1992)
2. Chen, S.H.: Financial Applications: Stock Markets. In: Wiley Encyclopedia of Computer Science and Engineering, pp. 481–498. John Wiley & Sons, Inc., Chichester (2008)
3. Chen, S.H., Yeh, C.H.: Evolving traders and the business school with genetic programming: A new architecture of the agent based artificial stock market. Journal Of Economic Dynamics & Control 25, 363–393 (2001)
4. De Bondt, W., Thaler, R.: Does the stock market overreact? Journal of Finance 40, 793–805 (1985)
5. Edwards, R., Magee, J.: Technical analysis of stock trends. New York Institute of Finance (1992)
6. Kampouridis, M., Tsang, E.: EDDIE for investment opportunities forecasting: Extending the search space of the GP. In: Proceedings of the IEEE Conference on Evolutionary Computation, Barcelona, Spain (2010) (accepted for Publication)
7. Koza, J.: Genetic Programming: on the programming of computers by means of natural selection. MIT Press, Cambridge (1992)
8. Lo, A.: The adaptive market hypothesis: market efficiency from an evolutionary perspective. Journal of Portfolio Management 30, 15–29 (2004)

Fractal Gene Regulatory Networks for Control of Nonlinear Systems

Jean Krohn and Denise Gorse

Department of Computer Science, UCL (University College London)
Gower Street, London WC1E 6BT, UK
{j.krohn,d.gorse}@cs.ucl.ac.uk

Abstract. Gene regulatory networks (GRNs) act as cell controllers; we argue that artificial models of GRNs should therefore make good controllers also. We present the first application of a model GRN to a substantial, well recognised control problem, using the Fractal Gene Regulatory Network model to control a range of versions of the single and jointed pole balancing problem.

Keywords: Gene regulatory network, fractal, genetic algorithm, ALPS, control, pole balancing, inverted pendulum.

1 Introduction

Natural GRNs can be seen as a cell's controller; this is particularly clear in the case of bacteria, but GRNs in multi-cellular organisms also appear as instances of distributed, self-organised control. Natural GRNs moderate the output of the genome via protein substrates whose actions depend on complex three-dimensional shapes only implicit in the genetic code, the mapping between primary and tertiary structures demanding a still incomplete understanding of the mechanisms of protein folding. Fractal GRNs, introduced by Bentley[2][5], aim to exploit an alternative, more tractable, source of complexity in this mapping from genome to control substrate, but with the same aim of producing a system that can display a robust and flexible response to environmental change.

The FGRN system is an evolutionary model of a GRN in which proteins are bitmaps generated from the Mandelbrot set fractal. The system was originally devised as a developmental system but has since been used more widely, for example to approximate the value of π[11] and to compute the square root function[6], and in the early control applications to be described below. FGRNs have been tested for fault-tolerance and have proved more resistant to damage than evolved genetic programming alternatives[6], and have also shown increased robustness and efficiency when evolved beyond explicit fitness requirements[4].

We will evolve FGRNs for direct control and apply them to variations of a well known benchmark control problem, that of balancing a pole on a moving cart (also known as the inverted pendulum problem), studying the ability of the system to balance straight or jointed poles with differing mechanical and physical characteristics.

R. Schaefer et al. (Eds.): PPSN XI, Part II, LNCS 6239, pp. 209–218, 2010.

2 Related Work

2.1 GRNs for Control

Surprisingly, given the role of natural GRNs in biology, artificial GRNs have known little use as controllers and have mainly been applied to relatively simple problems such as thermostat control and the generation of light-following behaviour[12]. FGRNs specifically have so far been evolved to act as a robot controller, guiding a simulated robot to a specific destination while avoiding obstacles, with the final resulting controllers tested successfully on an actual robot [3], and to a grid-world, box-pushing, robot control problem [16]. Dürr et al.[8] did apply a GRN based method to the pole balancing problem (to be described below) but the GRN's role was not control as such but the generation of a neural network based controller.

2.2 The Pole Balancing Problem

Pole balancing is a well-known and well-studied control problem that has been used as a benchmark for the design and test of many controllers[7]. It is usually (and here) defined to be the problem of keeping the angular position θ of the hinged pole within 12° of vertical, and the distance h of the cart on which it is mounted within 2.4m of the centre of the track, using only 'bang-bang' control (a force F of $\pm$10N is applied to the cart at each time step).

The original single pole problem was first solved through reinforcement learning by Barto et al.[1], with Wieland the first to evolve neural networks for the control of the system[15] (single pole, jointed pole, and double pole versions of the problem), and subsequent solutions using various other methods such as genetic programming[13]. More recently, neuroevolution methods of increasing sophistication have used various versions of the pole balancing problem for validation and as a benchmark; Gomez et al. have produced an extensive comparison of the performance of current neuroevolution methods on the single and double pole balancing problems with or without velocities as inputs[9].

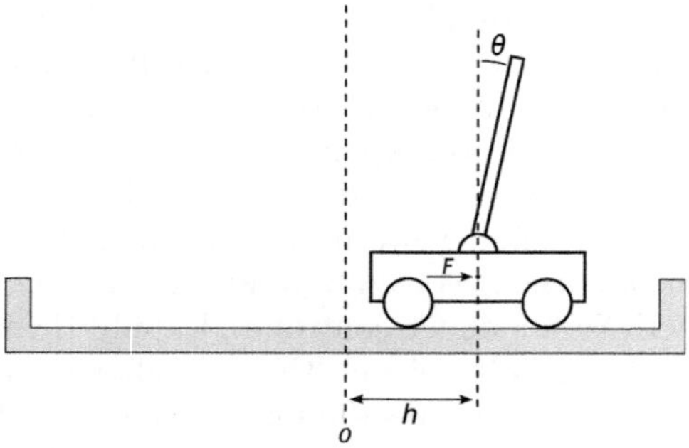

Fig. 1. The pole balancing problem

3 Fractal Gene Regulatory Network (FGRN)

3.1 Fractal Proteins

A fractal protein (x, y, z) is a square subset of the Mandelbrot set with sides of length z and centre coordinates (x, y) that define the real and imaginary parts of a complex number[2]. The colouring from white to black of a sample point within the square represents the speed with which the value at that point falls out of the range $[0, 2]$ upon iteration of the Mandelbrot equation, with black sampling points generating values that remain bounded within a radius of 2. In principle a fractal protein is an object of arbitrarily high complexity, though in practice to limit computational demands the square subsets are usually implemented as 15x15 bitmaps, with the (x, y) value of a pixel being that of the point at its centre.

A fractal protein has an associated concentration that alongside the concentrations of other protein constituents determines the degree to which it can influence the behaviour of the cell containing it. Although a protein's concentration is stored as a single real number it can also be represented as a bitmap for ease of use in the merging and comparison operations described below; Figure 2 shows an example of a fractal protein at 15x15 resolution and its concentration bitmap.

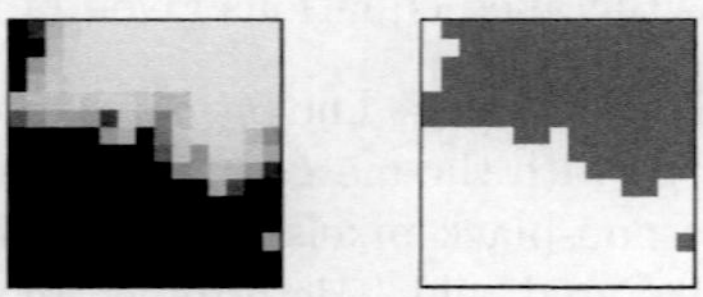

Fig. 2. A protein (left) and associated concentration (right)

3.2 Genetic Representation

A FGRN genome contains a set of genes each of which is defined by nine numbers (two sets of coordinates plus three additional scalars) as shown in Figure 3a. In more detail each gene comprises:

Gene Type

- **Environmental** genes provide potential protein **input** to the system.
- **Receptor** genes act as a **filter**, determining how much of an environmental protein actually enters the cell.
- **Regulatory** genes perform a **control** function, determining the system's response to that part of the environmental protein that has entered the cell.
- **Behavioural** genes, when the merged product of receptor-filtered input and regulatory genes sufficiently matches conditions at the promoter, produce a protein **output** that defines a system behaviour or decision.

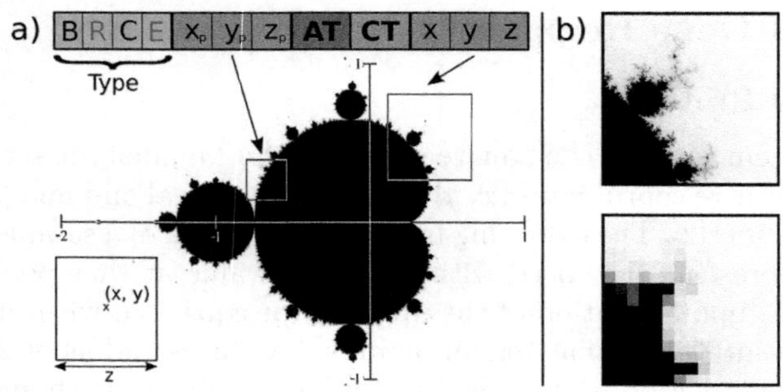

Fig. 3. A gene in detail. a) The type specifies that this gene is expressed as both a **B**ehavioural and a re**C**eptor gene, but not as a **R**egulatory or an **E**nvironmental gene; the promoter (x_p, y_p, z_p) and output (x, y, z) protein coordinates define different fractal portions of the Mandelbrot set. Bottom left: a schema describing the mapping from protein coordinates to fractal portion. b) Top: the fractal portion pointed at by the output coordinates (x, y, z) of the gene. Bottom: the resulting 15x15 bitmap.

Promoter Region. This comprises the promoter protein coordinates (x_p, y_p, z_p) and the real-valued Affinity (AT) and Concentration (CT) Thresholds. This region determines if, and to what degree, the gene will produce an output via the mechanism outlined below(detailed equations given in ref [5]):

- All proteins in the cell are merged. The bitmap associated with the promoter coordinates is compared with the merged product and the summed absolute difference between its non-black pixels and the corresponding pixels in the merged product is calculated, which determines an activation probability for the gene.
- If a gene is determined to be activated the Concentration Threshold is used to regulate the amount by which the concentration of the output protein in the cytoplasm is increased.

Output Protein Coordinates. The coordinates (x, y, z) specify the gene's output protein, which in the case of a behavioural gene will determine a system action.

3.3 Fractal Protein Chemistry

As described in ref [5], the system is run through a number of iterations, comprising successively of setting the environmental protein concentrations to reflect input, filtering using the output of the receptor gene to determine how much of the environmental proteins enters the cell, calculating the merged product of all proteins in the cytoplasm, and comparing the product with conditions at a regulatory or behavioural gene's promoter region to determine protein production or output condition.

In addition to the comparison procedure used by the promoter region there are three further fundamental operations as described below:

Decay. At the end of each iteration the concentration values of the regulatory proteins present in the cytoplasm are decayed and those with values less than a certain threshold have their concentration set to zero (so are no longer considered present in the cell). This process is illustrated in Figure 4a.

Mask. This is the mechanism by which a receptor gene controls the amount of an environmental protein that enters a cell. Black regions of the receptor protein bitmap are treated as opaque and all others as transparent; this means the more similar an environmental protein is to the mask the more of it will be allowed to enter the cell. An example of masking is shown in Figure 4b.

Merge. In principle a merged product is calculated by iterating through the fractal equations for each protein and choosing as winner for each pixel that value that becomes unbounded most quickly. In practice merging can be carried out more simply by comparing the stored values for the bitmaps and choosing at each point the maximum pixel value (pictorially, that closest to white); merged proteins thus tend to be dominated by white regions, as can be seen in Figure 4c. When proteins are merged the concentration at each point becomes that of the winner, producing a 'patchwork' appearance in this example.

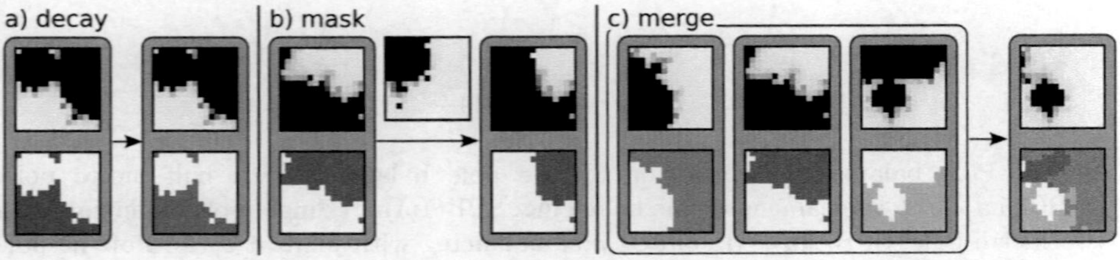

Fig. 4. Protein chemistry. a) Decay: the concentration associated to a protein is reduced. b) Mask: a protein (centre, above arrow) masks another. c) Merge: three proteins and their associated concentrations are merged into one. Note the patchwork nature of the merged protein's concentration.

4 Experiments

In this section we first present the versions of the pole balancing problem studied, and the experimental set-up of the FGRN and associated genetic algorithm, before detailing and discussing the results of these experiments.

4.1 Problem Set-Up

We evolve FGRNs to solve versions of the single and jointed pole balancing problem, defined as follows:

Single Pole Balancing (SPB). The system, modelled with the equations of motion used in ref [14], is used to balance a single unjointed pole with a range of pole lengths: 0.5, 1.0, and 2.0 metres. (The traditional length is one metre.)

Jointed Pole Balancing (JPB). The system, modelled with the equations of motion used in ref [15], is used to balance jointed poles of the same total height (two metres), but with the joint situated at different positions on the pole: one third, half-way, and two thirds of the total height.

Figure 5 illustrates the problems. In all cases the Euler method was used to integrate the equations of motion (integration time step 0.02s for SPB, 0.01s for JPB) with the criterion for success being keeping cart and pole in the required position ranges for 100,000 0.02s time steps.

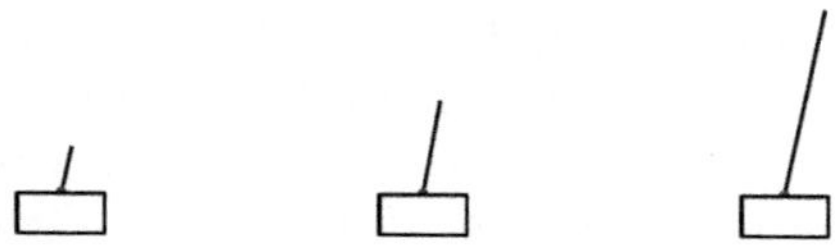

(a) SPB(0.5m) (b) SPB(1m) (c) SPB(2m)

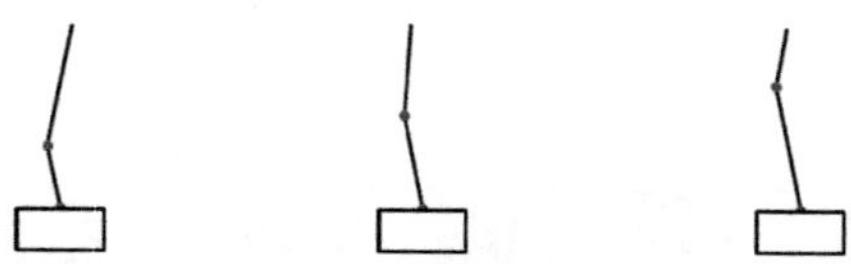

(d) JPB($\frac{2}{3}$: $\frac{4}{3}$) (e) JPB(1 : 1) (f) JPB($\frac{4}{3}$: $\frac{2}{3}$)

Fig. 5. Pole balancing problems. a) Single pole balancing with half metre pole: SPB(0.5m). b) Classical single pole balancing: SPB(1m). c) Single pole balancing with two metre pole: SPB(2m). d) Jointed pole balancing with joint at a third of the pole length: JPB($\frac{2}{3}$: $\frac{4}{3}$). e) With joint in the middle of the pole: JPB(1 : 1). f) With joint at two thirds of the pole length: JPB($\frac{4}{3}$: $\frac{2}{3}$).

4.2 FGRN Set-Up

The initial random FGRN genomes are composed of one behavioural (output) gene, four regulatory (hidden) genes, one receptor (input filter) gene, and as many environmental (input) genes as the problem has inputs (four for SPB, six for JPB). The genome's composition can change throughout a run through the gene duplication/deletion mutation operators.

One aspect of FGRN operation that has not been previously studied is the most appropriate representation of input concentration in situations where the associated physical variables (positions and velocities in this case) take on both positive and negative values. This issue was found to be of considerable significance for the pole balancing problem. Experiments were run with two different input-to-concentration mapping schemes:

[0,1]. The standard FGRN mapping in which the smallest (most negative) inputs are mapped to a concentration value of 0 (total absence of environmental protein) and the largest to 1 (saturation concentration).

[-1,1]. A modified mapping into concentration values $[-1, 1]$, with zero inputs now being represented by zero concentration and negative inputs by negative values. Though negative protein concentrations would not make sense biologically their use is consistent within the FGRN model. This usage is similar to the introduction of negative weights in artificial neural networks, which have no biological counterparts but which are a more flexible and efficient way to represent the action of inhibitory neurons.

4.3 Genetic Algorithms

We will compare the GA used in all previously published FGRN work (here designated FGA) with the more recently introduced ALPS[10] algorithm. It is hoped that the capacity of ALPS to avoid premature convergence will ensure solutions are found more reliably throughout the evolutionary runs. The mutation rate is the same for both GAs (0.02 per gene), and crossover is always applied. Each experiment is run 100 times.

FGA. The set-up here is the standard one used in past FGRN work. The population contains 100 individuals with 80 children generated per generation. The maximum age is set to 10. Parents are selected from the fittest 40 individuals except in 1% of cases in which a parent is selected randomly. In each run, FGA is run for 100 generations.

ALPS. The ALPS GA is set up with a layer size of 20 individuals, an age gap of 10, and a layer age-limit power law ageing scheme. Tournament selection is used in each layer (tournament size of 4 with an elitism of 4). The layer tournament size and elitism value are chosen to match the characteristics of FGA but on a smaller population scale. Each ALPS run continues until the number of fitness evaluations reaches 10,000.

4.4 Results

Figure 6 shows, for each experiment, the number of successful runs as the number of fitness evaluations increases, and Table 1 presents a summary of the results.

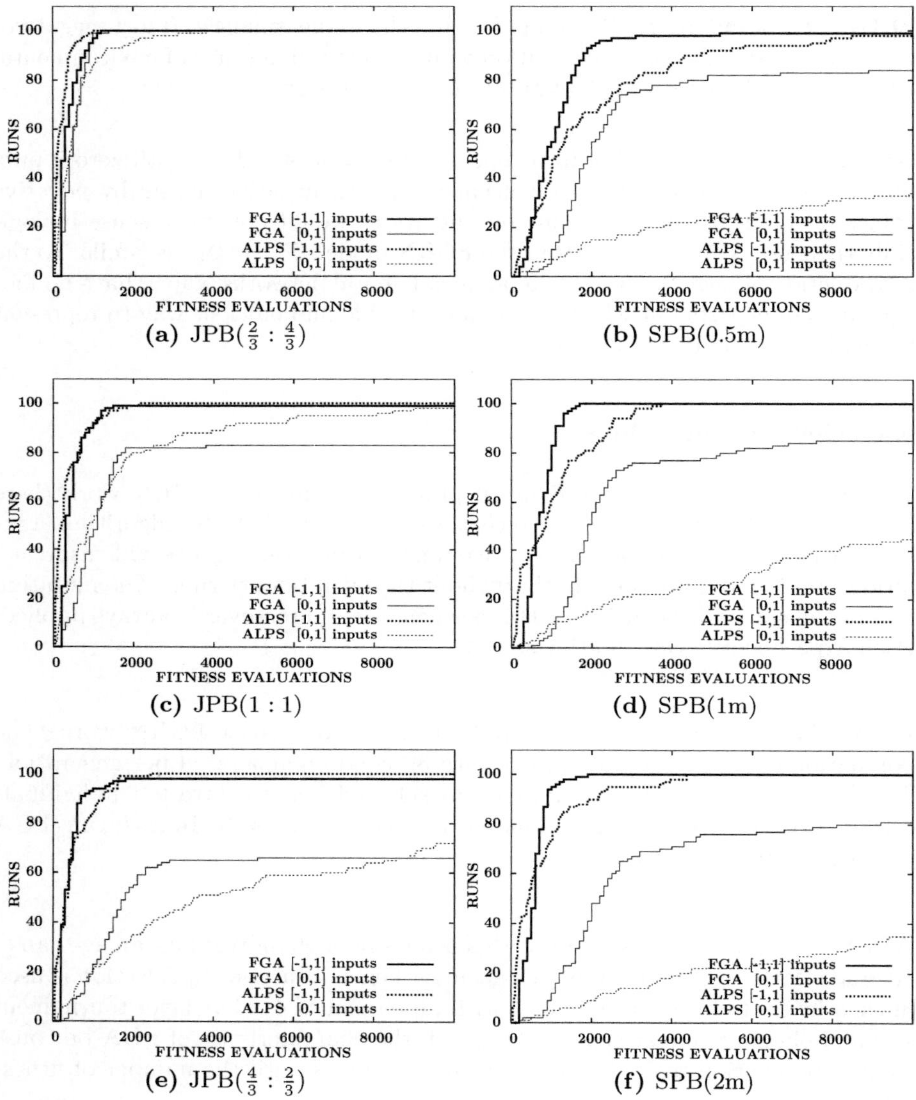

(a) JPB($\frac{2}{3}:\frac{4}{3}$) **(b)** SPB(0.5m)

(c) JPB(1 : 1) **(d)** SPB(1m)

(e) JPB($\frac{4}{3}:\frac{2}{3}$) **(f)** SPB(2m)

Fig. 6. Successful runs (out of a total of 100 runs) as the number of fitness evaluations increases, for each experiment

Table 1. Results: the table shows for each combination of input mapping and GA the percentage of successful runs and the average number of fitness evaluations to success (with the associated standard deviation in parenthesis)

Experiment	[0, 1] inputs				[-1, 1] inputs			
	FGA		ALPS		FGA		ALPS	
SPB(0.5m)	85%	2136(1522)	34%	3629(2842)	99%	1045(696)	98%	1827(1822)
SPB(1m)	85%	2154(1362)	45%	4032(2849)	100%	730(329)	100%	1014(923)
SPB(2m)	81%	2335(1662)	35%	4438(2919)	100%	572(298)	100%	768(949)
JPB($\frac{2}{3}$: $\frac{4}{3}$)	100%	509(256)	100%	580(631)	100%	387(251)	100%	199(192)
JPB(1 : 1)	83%	951(532)	98%	1479(1799)	99%	449(290)	100%	357(418)
JPB($\frac{4}{3}$: $\frac{2}{3}$)	66%	1407(759)	73%	3237(2926)	98%	411(290)	100%	477(480)

5 Discussion

This work constitutes the first application of direct GRN control to a well known, difficult, control problem, and a significant test of the utility of the Fractal Gene Regulatory Network model. The introduction of negative concentrations for inputs was found to substantially increase the reliability with which a solution was found for single and jointed pole balancing ($p < 0.02$) and to reduce the number of fitness evaluations needed to reach a solution ($p < 0.0001$).

The jointed version of the problem appeared easier with a lower pole hinge, possibly because a longer (and therefore heavier) top segment gives the system more latitude before the top segment reaches a critical angle. The increasing number of failures with a higher pole hinge was more pronounced when FGA was used, confirming the capacity of ALPS to avoid premature convergence. Surprisingly, this version also appeared more tractable than single pole balancing, which we speculate might be due to the two extra inputs provided in this case. We intend to study the system's behaviour on more complex variations of the pole balancing problem with and without velocity input, comparing our results with competitor methods, and on other benchmark control problems.

Future work will initially focus on the input representation, given both the notable improvements brought about by the use of negative concentrations and the unexpected result that the inclusion of extra inputs in the jointed pole balancing problem seemed to offset the increased innate difficulty of this task. In the longer term we intend to take further inspiration from the evolutionary mechanisms of natural GRNs, improving the evolvability of FGRNs by allowing the duplication of gene segments such that, for example, a gene's activation might be determined by the combined output of multiple promoter segments.

References

1. Barto, A.G., Sutton, R.S., Anderson, C.W.: Neuronlike Adaptive Elements That Can Solve Difficult Learning Control Problems. IEEE Transactions on Systems, Man, and Cybernetics 13(5), 834–846 (1983)

2. Bentley, P.J.: Evolving Fractal Proteins. In: Cantú-Paz, E. (ed.) GECCO Late Breaking Papers, New York, USA, pp. 23–30. AAAI, Menlo Park (July 2002)
3. Bentley, P.J.: Evolving Fractal Gene Regulatory Networks for Robot Control. In: Banzhaf, W., Ziegler, J., Christaller, T., Dittrich, P., Kim, J.T. (eds.) ECAL 2003. LNCS (LNAI), vol. 2801, pp. 753–762. Springer, Heidelberg (2003)
4. Bentley, P.J.: Evolving beyond perfection: an investigation of the effects of long-term evolution on fractal gene regulatory networks. Biosystems 76(1-3), 291–301 (2004)
5. Bentley, P.J.: Fractal Proteins. Genetic Programming and Evolvable Machines Journal 5, 71–101 (2004)
6. Bentley, P.J.: Evolving Fractal Gene Regulatory Networks for Graceful Degradation of Software. In: Babaoglu, O., Fetzer, C., Jelasity, M., Leonardi, S., Montresor, A., van Moorsel, A., van Steen, M. (eds.) Self-* Properties in Complex Information Systems. Springer Lecture Notes in Computer Science, pp. 21–35. Springer, Heidelberg (2005)
7. Brownlee, J.: The Pole Balancing Problem - A Review of a Benchmark Control Theory Problem. Technical Report 7-01, Swinburne University of Technology (2005)
8. Dürr, P., Mattiussi, C., Floreano, D.: Neuroevolution with Analog Genetic Encoding. In: Runarsson, T.P., Beyer, H.-G., Burke, E.K., Merelo-Guervós, J.J., Whitley, L.D., Yao, X. (eds.) PPSN 2006. LNCS, vol. 4193, pp. 671–680. Springer, Heidelberg (2006)
9. Gomez, F., Schmidhuber, J., Miikkulainen, R.: Accelerated Neural Evolution through Cooperatively Coevolved Synapses. The Journal of Machine Learning Research 9, 937–965 (2008)
10. Hornby, G.S.: ALPS: The Age-Layered Population Structure for Reducing the Problem of Premature Convergence. In: Proceedings of the 8th Annual Conference on Genetic and Evolutionary Computation, pp. 815–822. ACM, New York (2006)
11. Krohn, J., Bentley, P.J., Shayani, H.: The Challenge of Irrationality: Fractal Protein Recipes for PI. In: Genetic and Evolutionary Computation Conference (GECCO 2009), Montreal, Canada (July 2009)
12. Quick, T., Nehaniv, C.L., Dautenhahn, K., Roberts, G.: Evolving Embodied Genetic Regulatory Network-Driven Control Systems. LNCS, pp. 266–277. Springer, Heidelberg (2003)
13. Shimooka, H., Fujimoto, Y.: Generating Equations with Genetic Programming for Control of a Movable Inverted Pendulum. In: McKay, B., Yao, X., Newton, C.S., Kim, J.-H., Furuhashi, T. (eds.) SEAL 1998. LNCS (LNAI), vol. 1585, pp. 179–186. Springer, Heidelberg (1999)
14. Whitley, D., Dominic, S., Das, R., Anderson, C.W.: Genetic Reinforcement Learning for Neurocontrol Problems. Machine Learning 13(2), 259–284 (1993)
15. Wieland, A.P.: Evolving Controls for Unstable Systems. In: Connectionist Models: Proceedings of the 1990 Summer School, pp. 91–102 (1990)
16. Zahadat, P., Katebi, S.D.: Tartarus And Fractal Gene Regulatory Networks With Inputs. Advances in Complex Systems (ACS) 11(06), 803–829 (2008)

An Effective Hybrid Evolutionary Local Search for Orienteering and Team Orienteering Problems with Time Windows

Nacima Labadi[1], Jan Melechovský[1], and Roberto Wolfler Calvo[2]

[1] Institute Charles Delaunay, University of Technology of Troyes, BP 2060, 10010 Troyes Cedex, France
{nacima.labadi,jan.melechovsky}utt.fr

[2] Laboratoire d'Informatique de Paris-Nord, University Paris 13
99, Avenue J.-B. Clement, 93430 Villetaneuse, France
roberto.wolfler@lipn.univ-paris13.fr

Abstract. The orienteering problem (OP) consists in finding an elementary path over a subset of vertices. Each vertex has associated a profit that is collected on the visitor's first visit. The objective is to maximize the collected profit with respect to a limit on the path's length. The team orienteering problem (TOP) is an extension of the OP where a fixed number m of paths must be determined. This paper presents an effective hybrid metaheuristic to solve both the OP and the TOP with time windows. The method combines the greedy randomized adaptive search procedure (GRASP) with the evolutionary local search (ELS). The ELS generates multiple distinct child solutions using a mutation mechanism and a local search. The GRASP provides multiple starting solutions to the ELS. The method is able to improve several best known results on available benchmark instances.

Keywords: Team orienteering problem, Time windows, Evolutionary local search, GRASP.

1 Introduction

The orienteering problem (OP) models for example a game in which competitors must visit a subset of control points in a given area. Once a control point has been visited the competitor ascribes a profit. The profit is counted only at the first visit of the control point. The winner of the game is the competitor who collects the maximum profit and reaches the end point within a prescribed amount of time. Among other practical applications of the problem can be mentioned for example scheduling of traveling salesman visits to the most profitable customers or an intelligent tourist travel guide system. The latter consists in the following problem. A tourist has a time limit to visit some city and thus he wants to visit a subset of tourist sites providing him with maximal satisfaction according to his preferences. Each tourist site is rated by a coefficient of attractiveness. The aim is to determine a route that respects the time limit and maximizes

R. Schaefer et al. (Eds.): PPSN XI, Part II, LNCS 6239, pp. 219–228, 2010.

the total attractiveness of visited sites. To give an example, such a system can implemented in a tourist information board in the airport or in the train station of a city so that it can propose the traveller a schedule of visits with respect to his time availability and personal preferences.

OP belongs to the family of travelling salesman problems with profits (TSPPs). The TSPP is by nature a bi-objective combinatorial optimization problem with two opposite optimization criteria, the first objective forces to extend the route and collect as much profit as possible while the other instigates the traveller to reduce the length of the route. The OP is however formulated as a single-objective optimization problem, in which the route length is stated as a constraint and the sum of the collected profit is maximized.

OP was introduced in [11] and it has been studied extensively since then. A detailed survey and annotated bibliography related to the OP and its variants can be found in a recently published article [13].

In this paper we address the team orienteering problem with time windows (TOPTW). This extension of the classical OP considers additionally a fixed number m of routes to be used in the solution and a time window associated with each customer. To the best of our knowledge only two attempts have been dedicated to the TOPTW: Ant Colony System algorithm (ACS) [7] and Iterated Local Search (ILS) [12].

This paper describes a hybrid metaheuristic composed of a greedy randomized adaptive search procedure (GRASP) and an evolutionary local search (ELS). The ELS generates a number of child solutions at each iteration by means of a procedure of perturbation. These solutions are then improved by a local search procedure. GRASP provides multiple starting solutions to the ELS using a randomized constructive heuristic. The hybrid metaheuristic solves both the OP and the team OP with time windows. A brief description of TOPTW is provided in section 2. The hybrid GRASP-ELS metaheuristic is presented in section 3. Computational results and comparisons to the state-of-the-art methods are given in section 4 and a conclusion ends this paper.

2 Problem Description

Consider a complete graph $G(V, E)$, where $V = \{v_0, v_1, ..., v_n\}$ represents the set of vertices and E the set of edges. The vertex v_0 is reserved for the depot, which is the starting and the ending point of each route. Each vertex v_i $(i = 1, 2, ..., n)$ has associated a positive integer profit p_i while $p_0 = 0$. A service time s_i is spent at each customer vertex v_i, $i \neq 0$. The visit of a vertex v_i can start only within a predefined time window $[e_i, l_i]$. The time window is considered as hard, i.e. the visitor cannot arrive later than the time l_i and in case the visitor arrives earlier than e_i, he must wait before the service can start. For the depot vertex, e_0 represents the earliest time the visitor can leave the starting point and l_0 is the latest possible arrival time to the depot. A nonnegative travel time t_{ij} is associated with each edge $e(i, j) \in E$. Without loss of generality, the service time s_i can be simply added to the arc cost: $\tilde{t}_{ij} = s_i + t_{ij}$. The total travel time of each route is limited by a constant T_{max}. We assume that T_{max} equals the latest arrival time to the depot, i.e. $T_{max} = l_0$. A feasible TOPTW solution must contain exactly m routes starting and ending at vertex v_0 such that each vertex (except v_0) is visited at most once and the limit on each route duration is respected. The objective is to maximize the sum of the collected profit. Mathematical models of the problem can be found in [7,12].

3 GRASP-ELS Algorithm

The proposed method is a greedy randomized adaptive search procedure (GRASP) enhanced by evolutionary local search algorithm (ELS). GRASP was introduced in [4]. It is a simple heuristic which generates independent random solutions (using some randomized heuristic) further improved by a local search procedure.

ELS was originally proposed in [6] for a peer-to-peer problem in telecommunications. The method extends the classical iterated local search (ILS). ILS starts with a solution s obtained by a heuristic and generates several child solutions by applying a perturbation on s. The child solutions are then improved by a local search. ELS additionally generates multiple copies of s and then applies ILS on each copy.

The perturbation procedure is a crucial component in the method. If the perturbation is too strong, the method behaves as a random heuristic and when it is weak, the solutions are quickly trapped in a local optimum. To control the perturbation, a parameter p which ranges from p_{min} to p_{max} is used.

GRASP-ELS hybrid metaheuristic was presented in [8] for the vehicle routing problem. It has been further applied with success to the split delivery capacitated arc routing problem (SDCARP) [1]. Main features of the method is the multi-start solution approach of the GRASP combined with the efficiency of ELS generating multiple child solutions. The framework of GRASP-ELS can be described as follows. At each iteration, GRASP first generates an initial solution $\bar{s}$. A local search procedure is applied to $\bar{s}$ and the resulting solution s enters the ELS mechanism. ELS generates for each copy of s a number of child solutions by performing random mutations and applies a local search. The best child solution s' is recorded and if it improves s then s is replaced with s' for the next iteration of ELS. Otherwise ELS performs the next iteration with s as an initial solution again. Basically the GRASP-ELS is controlled by three parameters determining respectively the number of child solutions nc generated within one ELS iteration, the maximum number of ELS iterations ni and finally the number of starting solutions generated by the GRASP algorithm ns.

3.1 Variable Neighborhood Descent

The local search is organized as a variable neighborhood descent procedure, since it alternates between two neighborhoods. The first one contains the classical routing moves which may involve one or two distinct routes:

1. 2-opt – removes and replaces two arcs in a route and reorders vertices,

2. Or-opt – relocates one vertex,
3. 2-opt* – interchanges two sub-paths between two routes,
4. Exchange – swaps two customers.

These moves aim to reduce route lengths. When $m \geq 2$, all these moves are performed contrarily to the case $m = 1$ where only the moves 2-opt and Or-opt and Exchange are executed. In this neighborhood, feasible moves are tested and the first improving one is retained. Time windows feasibility check is done in $O(1)$ as shown in [5].

The second set of neighborhoods enables to introduce yet unvisited vertices into the current solution. A neighborhood $N_k(s)$, $k = 0, \cdots, k_{max}$, is defined as the number of consecutive vertices k that are removed from one route and replaced by a chain of unrouted vertices. The search starts with $k = 0$, i.e. no vertex is removed, and scans all routes in order to find a move maximizing the difference between the total profit of the inserted vertices and the total profit of the removed ones. Note that only moves with positive difference can be accepted. The best move for the current k found over all routes is performed and the search continues within the same neighborhood until no more improvement can be detected. If no improvement is found with the current value of k, it is incremented by 1 and the search starts again. The procedure stops if $k = k_{max}$ or if an improvement of the solution has been found earlier. Hence the procedure does not explore necessarily all neighborhoods N_k for $k = 0, ..., k_{max}$, since it would require a great computational effort.

The determination of the entering sequence of vertices is NP-hard. It can be formulated as follows: given a time slot T_{ij} between two vertices v_i and v_j, and a set of unvisited vertices $U \subset V$, find such a sequence of vertices $v_{l_1}, v_{l_2}, \cdots, v_{l_q}$ from U that fits the available time slot and maximizes the inserted profit. In addition, the time window of each vertex must be respected. It is in fact equivalent to the single route orienteering problem with time windows; v_i and v_j can be considered respectively as the starting and ending point of a route limited by a time budget T_{ij}. Despite the computational complexity, exact evaluation of the sequence can be practicable for moderate size of k, since the number of feasible sequences is reduced due to the time windows constraints.

Let π denotes a feasible sequence of vertices from U, i.e. π fits the available time slot and the latest arrival for each $v_l \in \pi$ is respected. Furthermore, let S_π be the set of vertices $v \in U \setminus \pi$ that can extend the sequence π such that the insertion remains feasible. Finally, $p(\pi)$ denotes the total profit of π. We define the dominance relation between two feasible sequences π and π': $\pi \succ \pi' \iff$ (i) the endpoints of π and π' are equal; (ii) $p(\pi) > p(\pi')$; (iii) $S_{\pi'} \subseteq S_\pi$.

The evaluation of non-dominated sequences is done with a dynamic programming procedure performing forward and backward search. Similar strategy was used in [9] for solving the OPTW but the authors have used a relaxation of the dominance relation. Our implementation evaluates all non-dominated sequences. The sequence with maximum total profit is returned and if more such sequences exist, the sequence increasing less the route length is preferred.

3.2 Perturbation Procedure

The perturbation can be considered as a random mutation of the current starting solution. The perturbation should ensure the diversification of a solution but it can also deteriorate its quality. Then there is a risk of spending too much computational effort to restore the solution. We have therefore used a simple perturbation mechanism which fulfills both criteria. The procedure first selects randomly a sequence of p consecutive vertices from each route. Then all unvisited vertices are considered to build sequences that can be inserted into the emptied space in each route. The sequence with maximum profit is inserted and the recently modified route is not considered further. The procedure continues until all routes have been modified or no more feasible insertion sequence can be determined. If some route could not be modified at all, the randomly selected sequence of vertices is simply removed. This can happen if p is too small or if almost all vertices are already in the solution.

In order to evaluate the entering sequence efficiently, we propose an alternative heuristic evaluation scheme instead of the dynamic programming procedure. Let v_i be the last vertex of the sequence. Let also A_j be the set of neighbor vertices reachable from v_j without violating their time window. In the following equation q_j denote the cardinality of the set A_j, Δ_{ij} the time increasing if v_j is to be inserted and p_j the profit of v_j. The sequence is extended by a vertex v_j with a maximum ratio $\hat{p}_{ij} = q_j \cdot \frac{p_j}{\Delta_{ij}}$ among all unvisited vertices if the extension is feasible. The evaluation of the entering sequence requires $O(n^2)$.

4 Computational Experiments

The algorithm was coded in Embarcadero Delphi 2010, a Pascal like environment. The computational experiments were carried out on a desktop computer Dell Optiplex GX260 equipped with a Pentium 4 processor; 3 GHz and 1GB of RAM.

We have tested the GRASP-ELS algorithm on two benchmark data sets proposed in [7,12] The first data set was obtained from instances originally designed for the vehicle routing problem with time windows (data sets c/r/rc_100 and c/r/rc_200) by Solomon [10]. The series *100 and *200 are differing in the average size of time windows (wider for c/r/rc_200). The customer vertices are either clustered, randomly distributed or random-clustered in the plane. The second data set (pr01-pr10 and pr11-pr20) was proposed by Cordeau et al [2] for the multiple depot vehicle routing problem. Similarly, the instances in data sets pr11-pr20 have wider time windows on average. The travelling time between two vertices is represented by the Euclidean distance and the profit corresponds to the customer's demand. Solomon's instances contain 56 data files in total, each containing 100 customers. The other set contains 20 data files with the number of customers ranging from 48 to 288. The number of routes ranges from 1 to 4.

Several tests have been performed to determine the parameters setting and to observe the sensitivity. The resulting parameters setting represent the best trade-off between the solutions quality and the computational time. There are seven parameters that must be set up: ns for GRASP iterations, ni for ELS iterations, nc for the number of child solutions, $MaxNoImp$ representing the maximum number of GRASP iterations without improvement of the best solution, the minimal value (p_{min}) and the maximal value (p_{max}) of the perturbation parameter p and the maximum size k_{max} of the neighborhood explored by the local search procedure. The values of ns, ni, nc and $MaxNoImp$ were set differently for the data sets: $(10, 10, 10, 10)$ for the Solomon's clustered data set, $(30, 20, 15, 15)$ for Solomon's random and random clustered data sets and $(30, 15, 10, 10)$ for Cordeau's data sets. Greater values did not lead to much better solutions. For some instances of the Solomon's clustered data set were obtained identical results with lower ni and nc. For the other data sets the reduction of the number of generated solutions caused worse performance on average. The other parameters were identical for all instances: $p_{min} = 1$, $p_{max} = 4$ and $k_{max} = 2$.

Table 1 compares the performance of GRASP-ELS, ILS and ACS. For each method and instance set is reported the average gap to the best known solution and the average computational time. The results of GRASP-ELS and ACS represent the average over five runs with different random seeds. At first glance, GRASP-ELS provides better results than ILS and ACS. ILS is worse on all instance sets, while ACS performed better only on instance set rc100 with $m = 1$ and r100, rc100 with $m = 2$. The average computational times reported for GRASP-ELS are greater than those of ILS, but ACS spends much more effort than GARSP-ELS. Moreover, the processor that we have used is approximately two times slower than the Intel Core 2 with 2.5 GHz used for the ILS. This statement is based on the comparison of various computer systems solving standard linear equation problems presented in [3]. The performance is evaluated on a benchmark problem of a dense system of linear equations given by a matrix of order 100 and 1000. Our Intel Pentium 4 with 3 GHz has achieved 1 571 Mflop/s and 3 650 Mflop/s respectively while for Intel Core 2 (1 core, 2.5 GHz) the values are 2 426 Mflop/s and 7 519 Mflop/s. The processor used for ACS is comparable to ours with the performance 1 470 Mflop/s and 3 654 Mflop/s for the two benchmarks respectively.

Table 2 compares the performance of different components of GRASP-ELS. Each column (I – IV) reports the average gap and computational time per instance set obtained when one component was disabled: column I (GRASP disabled), II (all routing moves disabled), III (ELS disabled), IV (only one child generated). Each test was configured so that approximately the same number of solutions as with the full version was generated. Disabling some components can reduce the computational time, but as expected the obtained solutions are worse.

GRASP-ELS hit or improved several best known solutions. Table 3 summarizes the number of improvements and the number of equal solutions found per instance set. The method improved 141 best known solutions out of 304 tests and it found the best known solution in 118 cases.

Table 1. Summary of results of the four metaheuristics

Instance	GRASP-ELS		ILS		ACS	
Set	Gap(%)	Time(s)	Gap(%)	Time(s)	Gap(%)	Time(s)
m=1						
c100	0.1	4.4	1.1	0.3	0.0	6.3
r100	0.2	9.4	1.9	0.2	0.2	383.4
rc100	0.4	11.2	2.5	0.2	0.0	143.2
c200	0.2	30.0	1.9	1.7	0.2	342.6
r200	-1.2	63.8	1.0	1.7	1.3	1 556.7
rc200	-0.1	111.7	2.3	1.6	0.9	1 544.5
pr01-10	0.4	20.3	4.3	1.8	0.7	1 626.6
pr11-20	-6.7	40.4	1.3	2.0	4.0	887.6
m=2						
c100	0.0	79.7	0.8	1.1	0.1	818.0
r100	0.8	56.3	1.9	0.9	0.3	1 559.4
rc100	1.2	75.5	2.2	0.7	0.9	1 375.8
c200	-0.5	28.1	1.5	3.5	0.7	1 398.1
r200	-2.0	98.1	0.5	2.3	1.5	2 735.1
rc200	-2.1	250.7	1.7	2.2	1.4	2 342.7
pr01-10	-1.5	75.4	3.7	4.8	1.0	1 889.7
pr11-20	-3.1	105.8	3.1	5.2	2.2	2 384.8
m=3						
c100	0.6	113.6	2.3	1.5	0.6	1 043.2
r100	0.3	87.1	1.2	1.7	0.9	1 668.9
rc100	0.2	136.8	1.7	1.1	0.9	1 476.8
c200	-0.8	1.5	1.1	2.2	0.8	1 320.4
r200	-0.1	18.0	0.2	1.4	0.1	1 171.6
rc200	-0.3	102.2	1.0	1.7	0.5	1 509.6
pr01-10	-2.0	131.3	3.7	9.2	1.1	2 163.8
pr11-20	-4.1	216.7	3.9	9.7	1.2	2 349.9
m=4						
c100	0.7	104.6	2.5	2.6	0.8	1 056.1
r100	-0.1	160.3	2.3	2.6	0.8	1 652.5
rc100	-0.1	242.0	2.1	2.0	1.0	1 854.0
c200	0.0	0.0	0.0	1.0	0.0	7.7
r200	0.0	0.5	0.0	0.9	0.0	126.5
rc200	0.0	6.0	0.0	1.1	0.0	566.9
pr01-10	-1.5	200.5	4.4	14.1	0.8	2 447.7
pr11-20	-3.4	280.8	3.1	13.7	0.9	2 583.5

Table 2. Results on Solomon's and Cordeau's instances obtained with different configurations

Instance	I		II		III		IV	
Set	Gap(%)	Time(s)	Gap(%)	Time(s)	Gap(%)	Time(s)	Gap(%)	Time(s)
m=1								
c100	0.6	17.1	0.3	2.1	0.3	4.8	0.9	17.1
r100	1.3	2.3	0.6	13.1	1.3	9.3	0.4	13.2
rc100	3.2	1.3	0.2	5.1	3.2	5.6	0.8	8.1
c200	0.7	28.4	0.4	25.5	0.7	15.5	1.3	31.2
r200	1.5	7.8	1.5	25.8	1.1	32.3	-0.5	67.3
rc200	6.0	5.1	2.5	26.8	5.5	22.1	0.3	47.8
pr01-10	5.0	13.5	1.9	6.0	1.5	44.7	0.4	10.4
pr11-20	-3.1	21.2	-4.1	10.0	-4.3	52.9	-6.2	21.2
m=2								
c100	0.7	83.8	1.1	44.7	1.5	17.2	1.0	99.9
r100	2.3	7.5	2.1	27.3	2.7	12.6	0.9	86.0
rc100	3.5	4.1	3.0	18.4	2.9	8.1	1.9	43.2
c200	-0.4	25.9	0.4	10.4	-0.2	15.0	-0.5	27.5
r200	-1.6	18.6	-0.1	28.1	-1.0	104.2	-1.5	106.6
rc200	-1.4	13.9	0.6	28.1	-0.3	69.8	-0.9	93.0
pr01-10	-0.4	41.9	2.6	13.2	1.5	102.5	-0.5	36.8
pr11-20	-1.5	61.9	2.3	19.6	0.6	185.2	-2.5	57.1
m=3								
c100	1.0	108.8	2.4	44.4	1.6	22.9	1.5	149.0
r100	1.8	15.3	2.4	34.2	2.0	17.2	1.4	64.6
rc100	1.9	8.9	2.8	27.4	3.0	16.2	2.7	29.8
c200	-0.8	1.7	-0.5	4.5	-0.7	4.1	-0.8	1.8
r200	0.0	23.7	0.0	5.6	0.0	16.6	0.0	94.4
rc200	-0.1	13.7	0.1	20.2	0.3	23.6	-0.2	52.6
pr01-10	-1.7	71.2	3.1	27.6	1.7	351.1	-1.9	74.7
pr11-20	-3.7	96.0	2.3	35.4	0.1	280.6	-3.6	104.8
m=4								
c100	0.6	99.8	2.7	52.9	2.5	22.7	0.9	97.5
r100	1.6	22.3	2.8	43.7	-0.2	169.2	0.6	90.1
rc100	2.7	13.2	4.4	47.8	0.3	108.8	2.1	67.5
c200	0.0	0.0	0.0	0.0	0.0	0.0	0.0	0.0
r200	0.0	2.0	0.0	0.4	0.0	0.5	0.0	7.6
rc200	0.0	14.6	0.0	1.3	0.0	3.5	0.0	60.4
pr01-10	-0.8	106.4	5.0	37.7	2.2	208.3	-1.1	116.9
pr11-20	-1.9	136.6	3.1	62.8	0.2	222.4	-3.2	128.7

Table 3. Number of improvements and identical solutions found over the instance sets

	m	1		2		3		4	
Set	# instances	# equal	# impr	# equal	# impr	# equal	# impr	# equal	# impr
c_100	9	9	0	8	1	5	1	4	2
r_100	12	10	0	3	3	3	4	1	8
rc_100	8	6	0	1	2	0	5	2	5
c_200	8	5	2	2	5	2	5	8	0
r_200	11	0	10	1	10	9	2	11	0
rc_200	8	1	7	0	8	3	5	8	0
pr01 - pr10	10	7	1	2	6	2	7	1	7
pr11 - pr20	10	1	9	1	8	1	9	1	9
Sum	76	39	29	18	43	25	38	36	31

5 Conclusion

The implemented GRASP-ELS algorithm has performed well on the TOPTW instances with much less computational effort compared to ACS. The best solutions of many instances have been improved and the average performance of our method is superior to ACS in terms of computational times and to ILS in terms of the quality of obtained results. Still the performance of GRASP-ELS might be improved if much simpler evaluation of newly introduced vertices is used. This could be one stream of future work.

The proposed GRASP-ELS framework can be adapted so that the algorithm handles additional side constraints. This suggests the idea to apply the method to other time constrained routing problems with profits proposed in the literature.

Acknowledgements

The post-doctoral research internship of Jan Melechovský at the University of Technology of Troyes was financed by the region of Champagne-Ardenne.

References

1. Belenguer, J.M., Benavent, E., Labadi, N., Prins, C., Reghioui, M.: Split delivery capacitated arc routing problem: lower bound and metaheuristic. Transportation Science 44, 206–220 (2010)
2. Cordeau, J.F., Gendreau, M., Laporte, G.: A tabu search heuristic for periodic and multi-depot vehicle routing problems. Networks 30(2), 105–119 (1997)
3. Dongarra, J.J.: Performance of various computers using standard linear equations software in a fortran environment. Tech. rep., Electrical Engineering and Computer Science Department, University of Tennessee, Knoxville, TN 37996 - 1301 (2009), `http://www.netlib.org/benchmark/performance.ps`
4. Feo, T.A., Resende, M.G.C.: Greedy randomized adaptive search procedures. Journal of Global Optimization 6, 109–133 (1995)

5. Kindervater, G.A.P., Savelsbergh, M.W.P.: Vehicle routing: handling edge exchanges. In: Aarts, E., Lenstra, J. (eds.) Local search in combinatorial optimization, pp. 311–336. Wiley, Chichester (1997)
6. Merz, P., Wolf, S.: Evolutionary local search for the super-peer selection problem and the p-hub median problem. In: Bartz-Beielstein, T., et al. (eds.) HCI/ICCV 2007. LNCS, vol. 4771, pp. 1–15. Springer, Heidelberg (2007)
7. Montemanni, R., Gambardella, L.M.: Ant colony system for team orienteering problem with time windows. Foundations of Computing and Decision Sciences (34) (2009)
8. Prins, C.: A grasp x evolutionary local search hybrid for the vehicle routing problem. In: Pereira, F., Tavares, J. (eds.) Bio-inspired Algorithms for the Vehicle Routing Problem, vol. 16, pp. 35–53. Springer, Heidelberg (2009)
9. Righini, G., Salani, M.: Decremental state space relaxation strategies and initialization heuristics for solving the orienteering problem with time windows with dynamic programming. Computers & Operations Research 36(4), 1191–1203 (2009)
10. Solomon, M.: Algorithms for the vehicle routing and scheduling problem with time window constraints. Operations Research 35, 254–265 (1987)
11. Tsiligirides, T.: Heuristic methods applied to orienteering. J. Oper. Res. Soc. 35(9), 797–809 (1984)
12. Vansteenwegen, P., Souffriau, W., Berghe, G.V., Oudheusden, D.V.: Iterated local search for the team orienteering problem with time windows. Computers & Operations Research 36(12), 3281–3290 (2009)
13. Vansteenwegen, P., Souffriau, W., Oudheusden, D.V.: The orienteering problem: a survey. European Journal of Operational Research (2010) (accepted manuscript)

Discrete Differential Evolution Algorithm for Solving the Terminal Assignment Problem

Eugénia Moreira Bernardino[1], Anabela Moreira Bernardino[1], Juan Manuel Sánchez-Pérez[2], Juan Antonio Gómez-Pulido[2], and Miguel Angel Vega-Rodríguez[2]

[1] Research Center for Informatics and Communications, Dept. of Computer Science, School of Technology and Management, Polytechnic Institute of Leiria, 2411 Leiria, Portugal
{eugenia.bernardino,anabela.bernardino}@ipleiria.pt

[2] Dept. of Technologies of Computers and Communications, Polytechnic School, University of Extremadura, 10071 CÃ¡ceres, Spain
{sanperez,jangomez,mavega}@unex.es

Abstract. Terminal Assignment is an important problem in telecommunication networks. The main objective is to assign a given collection of terminals to a given collection of concentrators. In this paper, we propose a Discrete Differential Evolution (DDE) algorithm for solving the Terminal Assignment problem. Differential Evolution algorithm is an evolutionary computation algorithm. This method has proved to be of practical success in a variety of problem domains. However, it does not perform well on dealing with Terminal Assignment problem because it uses discrete decision variables. To remedy this, a DDE algorithm is proposed to solve this problem. The results are compared to those given by some existing heuristics. We show that the proposed DDE algorithm is able to achieve feasible solutions to Terminal Assignment instances, improving the results obtained by previous approaches.

Keywords: Communication Networks, Optimisation Algorithms, Discrete Differential Evolution Algorithm, Terminal Assignment Problem.

1 Introduction

In recent years, several combinatorial optimisation problems have arisen in communication networks. This is mainly due to the dramatic growth of communication networks, their increasing complexity and the heterogeneity of the connected equipments. In centralised computer networks, several terminals or workstations are serviced by a central computer. In large networks, concentrators are used to increase the network efficiency. In these networks a set of terminals is connected to a concentrator and each concentrator is connected to the central computer. If the number of concentrators and their locations are known, the problem then is reduced to determine what terminals will be serviced by each concentrator. This is known as the Terminal Assignment (TA) problem. Each concentrator is

R. Schaefer et al. (Eds.): PPSN XI, Part II, LNCS 6239, pp. 229–239, 2010.

limited in the amount of traffic that it can accommodate. For that reason, each terminal must be assigned to one node of the set of concentrators, in such a way that any concentrator does not overstep its capacity [1], [2], [3]. The optimisation goals are to simultaneously produce feasible solutions, minimise the distances between concentrators and terminals assigned to them and maintain a balanced distribution of terminals among concentrators. The TA problem is a NP-Hard combinatorial optimisation problem. Therefore, finding a polynomial time algorithm to solve them to optimality is highly unlikely. In this paper we propose a Discrete Differential Evolution (DDE) algorithm to solve this problem. Our algorithm is based on the DDE algorithm proposed by Pan et al. [4] for solving the permutation flowshop scheduling problem. The DDE algorithm first mutates a target population to produce the mutant population [4]. Then the target population is recombined with the mutant population in order to generate a trial population. Finally, a selection operator is applied to both target and trial populations to determine who will survive for the next generation. A destruction and construction procedure is employed to generate the mutant population. Embedded in the DDE algorithm we use a Local Search (LS) proposed by Bernardino et al. [5], which is used to improve the solutions quality. We compare the performance of DDE with three algorithms: Local Search Genetic Algorithm (LSGA), Tabu Search (TS) algorithm and Hybrid Differential Evolution algorithm with a multiple strategy (MHDE), used in literature.

The paper is structured as follows. In Section 2 we describe the TA problem; in Section 3 we describe the DDE algorithm; in Section 4 we discuss the computational results obtained and, finally, in Section 5 we report about the conclusions.

2 Terminal Assignment Problem

The TA problem involves to determine what terminals will be serviced by each concentrator [1]. In the TA problem a communication network will connect N terminals and each with demand (weight) L_i, via M concentrators and each with capacity C_j. No terminal's demand exceeds the capacity of any concentrator. A terminal site has a fixed and known location CT_i *(x,y)*. A concentrator site has also a fixed and known location CP_j *(x,y)*.

In this work, the solutions are represented using integer vectors. We use the terminal-based representation (see Fig. 1). Each position corresponds to a terminal. The value carried by position i of the chromosome specifies the concentrator that terminal i is to be assigned to.

2	1	2	2	2	3	3	1	3	1

Fig. 1. Terminal-based representation

3 The Proposed DDE Algorithm

Differential Evolution (DE) was introduced by Storn and Price in 1995 [6]. DE is a population-based algorithm using crossover, mutation and selection operators [7]. The crucial idea behind DE is a scheme for generating trial parameter vectors [8], [9]. At the mutation step, distinct individuals (usually three) are selected from population. Mutation adds the weighted difference of two (or more) of the individuals to the third. At the recombination step, new trial individuals are created by combining the mutated individual, with the target individual. Combination takes place according to a strategy (several strategies with different approaches can be found in literature - see [9]). Thereafter, a selection operator is applied to compare the fitness value of both competitive solutions, namely, target and trial solutions to determine who can survive for the next generation. Most of the discrete problems have solutions presented through permutation vectors, while DE usually maintains and evolves floating-point vectors.

In order to apply the DE to solve discrete problems, the most important is to find a suitable encoding scheme, which can transform between floating-point vectors and permutation vectors. DE was first applied to TA problem by Bernardino et al. [5]. They proposed a Hybrid DE (HDE) algorithm and in [10] proposed an improved HDE algorithm with a multiple strategy (MHDE). MHDE combines global and LS by using an evolutionary algorithm to perform exploration while the LS method performs exploitation. MHDE uses the terminal-based representation and not floating-point vectors. After applying the standard equations, the algorithm verifies if the trial solutions contain values outside the allowed range.

In MHDE, if a gene value (concentrator) is outside of the allowed range it is necessary to apply the following transformation:

```
IF concentrator > M THEN          concentrator = concentrator - M
ELSE IF concentrator <=0 THEN     concentrator = concentrator + M
```

This transformation significantly increases algorithm execution time. To solve this problem we use a DDE algorithm based on the algorithm proposed by Pan et al.[4] whose solutions are based on discrete values, which can be applied to all types of combinatorial optimisation problems [11], [12], [13], [14]. The basic model of DDE is different from the DE model.

Main steps of the DDE algorithm:

```
Initialise Parameters
Create initial target Population, P^0
Evaluate target Population P^0
Find Best Solution in P^0, P_g
WHILE stop criterion is not reached
  Create Mutant Population, Pm^t
  Create Trial Population, Pt^t
  Evaluate Trial Population Pt^t
  Make Selection and update target population, P^t
  Find Best solution in P^t, P^t_g
  Apply Local Search to P^t_g
```

Initialisation of parameters
The following parameters, must be defined by the user (1) *ni* - number of individuals; (2) *ms* - maximum number of seconds; (3) *pp* - perturbation probability; (4) *np* - number of perturbations and (5) *pc* - crossover probability.

Initial target Population
The initial target population (P^0) can be created randomly or in a Deterministic Form (DF). DF is based in the Geeedy algorithm proposed by Abuali et al. [15]. This algorithm assigns terminals to the closest feasible concentrator.

Evaluation of solutions
To evaluate how good a potential solution is relative to other potential solutions we use a fitness function. The fitness function is based on: (1) the total number of terminals connected to each concentrator (the objective is to guarantee a balanced distribution of terminals among concentrators); (2) the distance between the concentrators and the terminals assigned to them (the objective is to minimise the distances between concentrators and terminals assigned to them); (3) the penalisation if a solution is not feasible (the objective is to penalise the solutions when the total capacity of one or more concentrators is overloaded). The objective is to minimise the fitness function (equation 4), searching a trade off among the three objectives mentioned.

$$total_c = \sum_{t=1}^{N} \begin{cases} 1 \; if(c(t) = c) \\ 0 \end{cases} \qquad bal_c = \begin{cases} 10 \;\; if(total_c = round(\frac{N}{M}) + 1) \\ 20 * |(round(\frac{N}{M}) + 1 - total_c)| \end{cases} \tag{1}$$

c(t)= concentrator of terminal t, t = terminal, c = concentrator

$$dist_{t,c(t))} = \sqrt{(CP[c(t)].x - CT[t].x)^2 + (CP[c(t)].y - CT[t].y)^2} \tag{2}$$

$$Penalisation = \begin{cases} 0 \quad if(Feasible) \\ 500 \end{cases} \tag{3}$$

$$fitness = 0.9 * \sum_{c=1}^{M} bal_c + 0.1 * \sum_{t=1}^{N} dist_{t,c(t))} + Penalisation \tag{4}$$

Mutant Population
A mutant individual is obtained by perturbing the generation best solution in the target population. The differential variation is achieved in the form of perturbations of the best solution from the previous generation in the target population. To obtain the mutant individual, the following equation is used:

$$Pm_i^t = \begin{cases} DC_{np}(P_g^{t-1}) \quad if(r < pp) \\ MutationClosestConcentrator(P_g^{t-1}) \end{cases} \tag{5}$$

P_g^{t-1} is the best solution from the previous generation in the target population and DC_{np} is the destruction and construction procedure with the destruction size of *np* as a perturbation operator; and *MutationClosestConcentrator* is a simple mutation operator. With this operator, one gene (terminal) is randomly selected and its value (concentrator) is replaced for a new one (the closest feasible concentrator). A uniform random number r is generated between *0* and *1*. If r is less than pp then the *DC* procedure is applied to generate the mutant individual. Otherwise, the best solution from the previous population is perturbed using a simple mutation operator. The *DC* procedure creates a new solution by performing multiple moves whose length is specified as *np*. The algorithm performs *np* perturbations to find a new solution. First the algorithm chooses a random terminal t and searches the closest concentrator. If the concentrator has enough capacity and maintains a balanced distribution of terminals then the terminal t is assigned to the closest concentrator, *closestC*. Otherwise the algorithm generates two random terminals, *t1* and *t2*. The algorithm verifies the concentrators, *c1* and *c2*, assigned to them. If the concentrators have enough capacities and at least one of the concentrators is closest to the terminal that will be assigned, then the algorithm exchanges the terminals, *t1* and *t2*, between the concentrators, *c1* and *c2*. The algorithm repeats this process until at least one exchange is made.

Steps of the DC method:

```
FOR n=1 TO np DO
   t = random(N), closestC=1
   FOR c=1 TO M DO     /*find the closest concentrator*/
       IF distance (t, c) < distance (t, closestC) THEN
             closestC=c
   IF capacityFree(closestC)>=L(t) and mantainBalanced(closestC) THEN
       Assign terminal t to concentrator closestC
   ELSE
        cond=true
        REPEAT
           t1 = random(N), t2 = random(N)
           c1 = solution (t1), c2 = solution (t2)
           IF ( capacityFree(c2) - L(t2)>=L(t1) and capacityFree(c1) -
                L(t1)>=L(t2) ) and (distance(t2,c1)<=distance(t1,c1)
                or distance(t1,c2) <= distance(t2,c2) ) THEN
               Assign t1 to c2 and t2 to c1
               cond = false
        WHILE cond=true
```

Trial Population

Following the perturbation phase, a trial individual is obtained such that:

$$Pt_i^t = \begin{cases} CR(Pm_i^t, P_i^{t-1}) & if(r < pc) \\ Pm_i^t \end{cases} \tag{6}$$

A uniform random number r is generated between *0* and *1*. If r is lower than pc then the crossover operator is applied to generate the trial individual. The crossover operator (CR) adopted was 1-point, widely used in literature. CR produces two children. In this study, we selected one of the children randomly. If r is higher or equal to pc then the trial individual is chosen as: $Pt_i^t = Pm_i^t$ The trial individual is made up either from the perturbation operator or from the crossover operator.

Selection
The selection is based on the survival of the fitness among the trial and target individuals such that:

$$P_i^t = \begin{cases} Pt_i^t \ \ if(fitness(Pt_i^t) < fitness(P_i^{t-1})) \\ P_i^{t-1} \end{cases} \tag{7}$$

Local Search
The LS algorithm applies a partial neighbourhood examination. We generate a neighbour by swapping two terminals between two concentrators *c1* and *c2* (randomly chosen). The algorithm searches for a better solution in the initial set of neighbours. If the better neighbour improves the actual solution then the LS algorithm replaces the actual solution with the better neighbour. Otherwise, the algorithm creates another set of neighbours. In this case, one neighbour results of assigning one terminal of *c1* to *c2* or *c2* to *c1*. The neighbourhood size is *N(c1)*N(c2)* or *N(c1)*N(c2) + N(c1)+N(c2)*.

Steps of the LS algorithm:
```
c1 = random (M), c2 = random (M)
NN = neighbours of ACTUAL-SOL (one neighbour results of interchange
            one terminal of c1 or c2 with one terminal of c2 or c1)
SOLUTION = FindBest (NN)
IF fitness(ACTUAL-SOL) > fitness(SOLUTION) THEN
    NN = neighbours of ACTUAL-SOL (one neighbour results of assign
                          one terminal of c1 to c2 or c2 to c1)
    SOLUTION = FindBest (NN)
    IF fitness(SOLUTION) < fitness(ACTUAL-SOL) THEN
       ACTUAL-SOL = SOLUTION
ELSE
    ACTUAL-SOL = SOLUTION
```

The evaluation process is the most time-consuming step of the algorithm, which is usually the case in many real-life problems. Our LS procedure has some important improvements compared to the LS proposed by Bernardino et al. [5]. After creating a neighbour, the algorithm does not perform a full examination to calculate the new fitness value; it only updates the fitness value based on the modifications that were made to create the neighbour. The running time is reduced considerably.

Termination criterion
The algorithm stops when a maximum number of seconds (ms) is reached.

Further information on DDE can be found in [4], [11], [12], [13], [14].

4 Results

In order to test the performance of our approach, we use a collection of TA instances of different sizes. We take 9 problems from literature [16]. To compare our results we consider the results produced with LSGA, TS and MHDE. GA was first applied to TA by Abuali et al. [15]. GA is widely used in the literature to make comparisons with other algorithms. TS was applied to this problem by Xu et al. [17] and Bernardino et al. [16]. We compare our algorithm with LSGA [18], TS [16] and MHDE [10] algorithms proposed by Bernardino et al. because they (1) used the same test instances; (2) adopted the same fitness function; (3) implemented the algorithms using the same language (C++) and; (4) adopted the same representation (terminal-based).

Table 1 presents the best-obtained results with DDE, LSGA, TS and MHDE. The first column represents the problem number (Prob) and the remaining columns show the results obtained (BestF - Best Fitness, Ts - Run Times). The initial solutions for all algorithms were created using the Greedy algorithm. The algorithms have been executed using a processor Intel Core Duo T2300. The Ts (Run Time) corresponds to the execution time that each algorithm needs to obtain the best feasible solution.

Table 1. Results

Prob	LSGA		TS		MHDE		DDE	
	BestF	Ts	BestF	Ts	BestF	Ts	BestF	Ts
1	65.63	<1s	65.63	<1s	65.63	<1s	65.63	<1s
2	134.65	<1s	134.65	<1s	134.65	<1s	134.65	<1s
3	270.26	<1s	270.26	<1s	270.26	<1s	270.26	<1s
4	286.89	<1s	286.89	<1s	286.89	<1s	286.89	<1s
5	335.09	<1s	335.09	<1s	335.09	<1s	335.09	<1s
6	371.12	1s	371.12	<1s	371.12	<1s	371.12	<1s
7	401.21	1s	401.49	1s	401.21	2s	401.21	**<1s**
8	563.19	7s	563.34	1s	563.19	10s	563.19	**2s**
9	642.83	7s	642.86	2s	642.83	15s	642.83	**3s**

Table 2 presents the average fitnesses and standard deviations. The first column represents the number of the problem (Prob) and the remaining columns show the results obtained (AvgF - Average Fitness, Std - Standard Deviation). To compute the results in table 2 we use $\hat{A}\frac{1}{2}$ second for instances 1-3, *1* second for instances 4-5, *2* seconds for instance 6, *5* seconds for instance 7, *10* seconds for instance 8 and *15* seconds for instance 9.

The suggestions from literature helped us to guide our choice of parameter values for TS [16], LSGA [18] and MHDE [10]. For the TS, we consider a number of elements in the tabu list between *5* and *20*. The parameters of LSGA are set to crossover probability between *0.3* and *0.4*, selection operator="tournament", mutation probability between *0.6* and *0.8*, crossover operator="one-point" and mutation operator="multiple". The parameters of the MHDE algorithm are set

Table 2. Average Fitnesses and Standard Deviations

Prob	LSGA		TS		MHDE		DDE	
	AvgF	Std	AvgF	Std	AvgF	Std	AvgF	Std
1	**65.63**	**0,00**	**65.63**	**0,00**	**65.63**	**0,00**	**65.63**	**0,00**
2	**134.65**	**0,00**	**134.65**	**0,00**	**134.65**	**0,00**	**134.65**	**0,00**
3	270.69	0.23	270.76	0.3	270.75	**0.15**	**270.47**	0.22
4	286.99	0.13	287.93	0.75	287.17	0.14	**286.89**	**0,00**
5	335.99	0.6	335.99	0.59	336.55	0.39	**335.26**	**0.17**
6	371.68	0.24	372.44	0.45	373.19	0.42	**371.38**	**0.22**
7	402.41	0.5	403.25	0.73	403.61	0.33	**401.62**	**0.28**
8	564.94	0.52	564.5	0.54	572.04	0.76	**564.07**	**0.38**
9	646.52	0.84	644.18	0.48	648.46	0.48	**643.96**	**0.46**

to crossover probability between *0.3* and *0.4*, factor F between *0.9* and *1.6* and strategy="Best1Exp". The parameters of DDE are set to crossover probability between *0.1* and *0.3*, perturbation probability between *0.8* and *0.9* and number of perturbations between *[N/10...N/5]*. The MHDE and LSGA were applied to populations of *200* individuals and DDE to populations of *100* individuals. The values presented in table 2 have been computed based on *50* different executions (*50* best executions out of *100* executions) for each test instance.

The four algorithms reach feasible solutions for all test instances. The DDE algorithm can reach the best-known solutions for all instances. MHDE and LSGA can also find the best known solutions but in a higher execution time.

Since we are not trying to dynamically assign terminals to concentrators the running time is not a significant parameter to determine the quality of the algorithms. The differences in terms of execution time are not significant. To establish which is the best algorithm we must observe the average quality of the produced solutions and the standard deviations. As it can be seen in table 2, for larger instances the standard deviations and the average fitnesses for DDE algorithm are smaller. It means that the DDE algorithm is slightly more robust than LSGA, TS and MHDE.

We perform comparisons between all parameters (using the 9 instances) in order to establish the correct parameter setting for the DDE algorithm. We consider the same instance - 7 (a problem with average difficulty) to show the comparisons between parameters. To compute the results we use *1500* iterations.

The better results obtained with DDE use np between $N/20$ and $N/3$, $pp>0.5$, $pc<0.5$ (Fig. 2) and $ni = \{90, 100\}$. These parameters were experimentally found to be good and robust for the problems tested.

In our experiments we use different population sizes. The number of individuals was set to $\{10, 20, ..., 200\}$. We studied the impact on the execution time, the average fitness and the number of best solutions found. A higher number of initial solutions significantly increases algorithm execution time (Fig. 3).

The results show that the best population sizes are *90* and *100*. With these values the algorithm can reach in a reasonable amount of time a reasonable number of good solutions. With a higher number of initial solutions the algorithm can reach a better average fitness but it is more time consuming (Fig. 3).

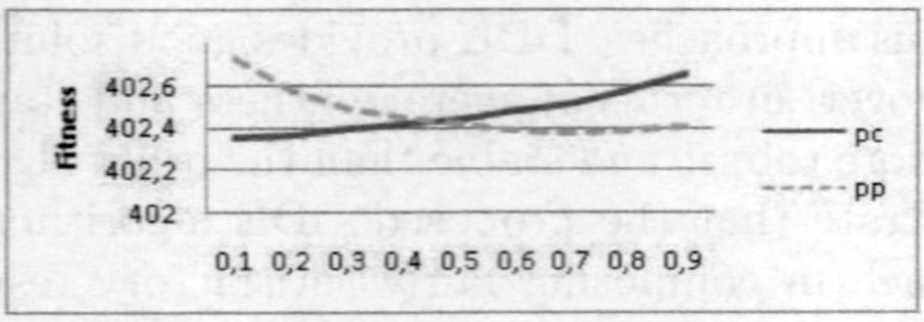

Fig. 2. Influence of parameters, Problem 7

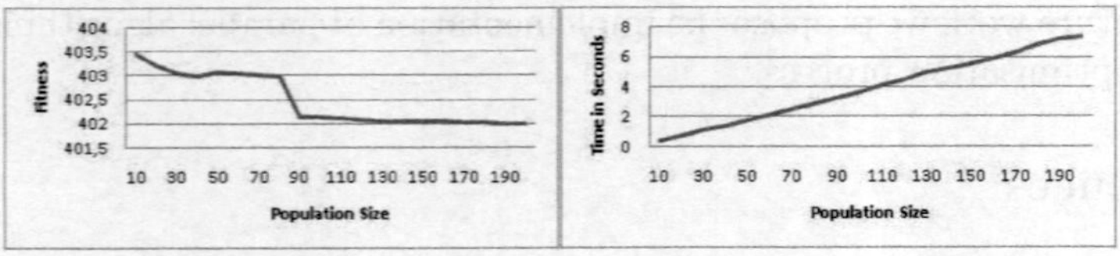

Fig. 3. Number of Individuals, Average Fitness/Execution Time, Problem 7

In our experiments *np* (number of perturbations) was set to {*0, 1, 2, ..., N*} (Fig. 4). In case of a high *np* the resulting permutation tends to be too random, which makes it more difficult to generate new improving solutions. A high *np* has also a significant impact on the execution time (Fig. 4). A small *np* did not allow the system to escape from local minima because the resulting solution was in most cases the same as the starting permutation.

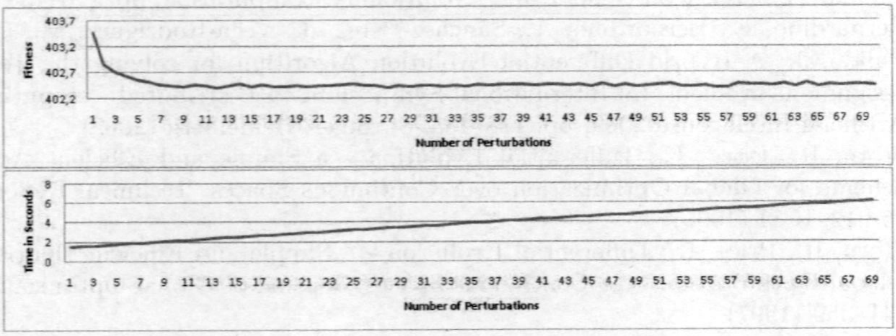

Fig. 4. Number of Perturbations, Average Fitness/Execution Time, Problem 7

In general, experiments have shown that the proposed parameter setting is very robust to small modifications.

5 Conclusions

In this paper we present a DDE algorithm to solve the TA problem. The performance of our algorithm is compared with LSGA, TS and MHDE. The computational results show that DDE performed more strongly, improving the results obtained by previous approaches. DDE provides good solutions in a smaller execution time. Moreover, in terms of average fitness and standard deviation, the DDE also proved more robust and stable than the other algorithms. Experimental results demonstrate that the proposed DDE algorithm is an effective and competitive approach in composing fairly satisfactory results with respect to solution quality and execution time for the TA problem.

In literature the application of DDE for this problem is nonexistent, for that reason this article shows its enforceability in the resolution of this problem.

For future work we propose the implementation of parallel algorithms to speed up the optimisation process.

References

1. Khuri, S., Chiu, T.: Heuristic Algorithms for the Terminal Assignment Problem. In: Proc. of the ACM Symposium on Applied Computing, pp. 247–251 (1997)
2. Salcedo-Sanz, S., Yao, X.: A hybrid Hopfield network-genetic algorithm approach for the terminal assignment problem. IEEE Transaction On Systems, Man and Cybernetics, 2343–2353 (2004)
3. Yao, X., Wang, F., Padmanabhan, K., Salcedo-Sanz, S.: Hybrid evolutionary approaches to terminal assignment in communications networks. In: Recent Advances in Memetic Algorithms and Related Search Technologies, vol. 166, pp. 129–159. Springer, Berlin (2005)
4. Pan, Q.-K., Tasgetiren, M.F., Liang, Y.-C.: A discrete differential evolution algorithm for the permutation flowshop scheduling problem. In: Proceedings of the 9th Annual Conference on Genetic and Evolutionary Computation, pp. 126–133 (2007)
5. Bernardino, E., Bernardino, A., Sánchez-Pérez, J., Vega-Rodríguez, M., Gómez-Pulido, J.: A Hybrid Differential Evolution Algorithm for solving the Terminal assignment problem. In: International Symposium on Distributed Computing and Artificial Intelligence 2009, pp. 178–185. Springer, Heidelberg (2009)
6. Storn, R., Price, K.: Differential Evolution - a Simple and Efficient Adaptive Scheme for Global Optimization over Continuous Spaces. Technical Report TR-95-012, ICSI (1995)
7. Storn, R., Price, K.: Differential Evolution - a Simple and Efficient Heuristic for Global Optimization over Continuous Spaces. Journal of Global Optimization 11, 341–359 (1997)
8. Price, K., Storn, R., Lampinen, J.: Differential Evolution - A Practical Approach to Global Optimization. Springer, Berlin (2005)
9. Differential Evolution, `http://www.icsi.berkeley.edu/~storn/code.html`
10. Bernardino, E., Bernardino, A., Sánchez-Pérez, J., Vega-Rodríguez, M., Gómez-Pulido, J.: A Hybrid Differential Evolution Algorithm with a Multiple Strategy for Solving the Terminal Assignment Problem. In: 6th Hellenic Conference on Artificial Intelligence. Springer, Heidelberg (2010)
11. Tasgetiren, M.F., Pan, Q.-K., Liang, Y.-C.: A discrete differential evolution algorithm for the single machine total weighted tardiness problem with sequence dependent setup times. Computers and Operations Research 36(6), 1900–1915 (2009)
12. Tasgetiren, M.F., Pan, Q.-K., Liang, Y.-C., Suganthan, P.N.: A discrete differential evolution algorithm for the total earliness and tardiness penalties with a common due date on a single machine. In: Proceedings of the 2007 IEEE Symposium on Computational Intelligence in Scheduling (CISched 2007), pp. 271–278 (2007)

13. Tasgetiren, M.F., Pan, Q.-K., Suganthan, P.N., Liang, Y.-C.: A discrete differential evolution algorithm for the no-wait flowshop scheduling problem with total flowtime criterion. In: Proceedings of the 2007 IEEE Symposium on Computational Intelligence in Scheduling (CISched 2007), pp. 251–258 (2007)
14. Tasgetiren, M.F., Pan, Q.-K., Liang, Y.-C.: A discrete differential evolution algorithm for single machine total weighted tardiness problem with sequence dependent setup times. In: IEEE Congress on Evolutionary Computation, pp. 2613–2620 (2008)
15. Abuali, F., Schoenefeld, D., Wainwright, R.: Terminal assignment in a Communications Network Using Genetic Algorithms. In: Proc. of the 22nd Annual ACM Computer Science Conference, pp. 74–81. ACM Press, New York (1994)
16. Bernardino, E., Bernardino, A., Sánchez-Pérez, J., Vega-Rodríguez, M., Gómez-Pulido, J.: Tabu Search vs Hybrid Genetic Algorithm to solve the terminal assignment problem. In: International Conference Applied Computing, pp. 404–409. IADIS Press (2008)
17. Xu, Y., Salcedo-Sanz, S., Yao, X.: Non-standard cost terminal assignment problems using tabu search approach. In: IEEE Conference in Evolutionary Computation, vol. 2, pp. 2302–2306 (2004)
18. Bernardino, E., Bernardino, A., Sánchez-Pérez, J., Vega-Rodríguez, M., Gómez-Pulido, J.: Solving the Terminal Assignment Problem Using a Local Search Genetic Algorithm. In: International Symposium on Distributed Computing and Artificial Intelligence, pp. 225–234. Springer, Heidelberg (2008)

Decentralized Evolutionary Agents Streamlining Logistic Network Design

Stephan Otto and Tobias Bannenberg

Fraunhofer Institute IIS, Nuremberg, Germany
Stephan.Otto@iis.fraunhofer.de, bannents@iis.fraunhofer.de
http://www.iis.fraunhofer.de, http://www.scs.fraunhofer.de

Abstract. We propose a decentralized evolutionary approach for studying autonomous heterogeneous agents interacting in a supply chain. Such logistics networks can be seen as complex networks that need to adapt their internal structure (e.g. transport routes, interactions) as reaction to environmental changes, e.g. the market demand, supplier unavailability or route changes. We model such distributed supply chains as a decentralized multi-agent system in order to draw an analogy to real world scenarios. This paper describes a decentralized evolutionary optimization approach that differs in two ways from traditional EA. First the fitness calculation is replaced by an economic model. Second the entire agent population constructs only one solution. The connections in supply-chains can be seen as a complex network of coexisting but simple interdependent agent strategies producing together the necessary transportation network. We describe how our decentralized approach can be used to solve inherently distributed problems where no central optimization algorithm exist. The simulation results show the applicability of the approach to transport network optimization.

1 Introduction

Logistics processes and systems (e.g. transportation, telecommunication, information) are large and distributed with a huge number of autonomous agents collaborative working in order to fulfill requirements from customers, service providers, organizations and other systems. These systems cannot be fixed in their structure, design and behavior in order to cope with a highly dynamic, complex and unpredictable environment. For inherently distributed systems there cannot be a special agent aware of the whole system. At first elements are limited by how much they can communicate and process [6]. The second reason is information hiding, which means, that not all information can be given to a central control due to intellectual property or security reasons [9,5,12]. Throughout this paper we use the term decentralized system for systems without central control. The aim of this paper is to show a decentralized method to optimize transport networks.

Typically the flow inside logistic networks is organized in processes that describe the sequence of activities in the network. So, according to the process

R. Schaefer et al. (Eds.): PPSN XI, Part II, LNCS 6239, pp. 240–249, 2010.

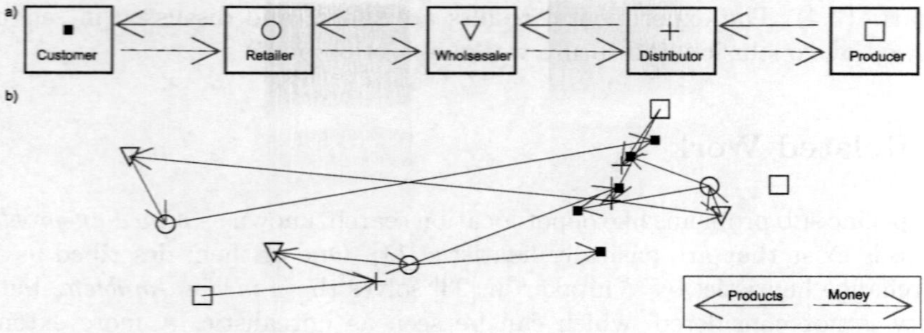

Fig. 1. a) Distribution process from producer to customer, b) Structure of a logistic network according to a)

in figure 1a) there are four steps that determine the flow of goods through the network starting with the production of goods followed by transportation, then to stocking and finally sale to customers. In figure 1b) a possible logistic network consisting of producers, distributors, wholesaler and retailer is shown. An actor may have different capabilities and parameters determining its behavior or strategy (store size, geographic position, etc.) and thus may have an influence on the network. The goal in logistics is to streamline the network structure while improving service to the customer [3]. The optimization of such networks is NP-complete [11].

In this paper we present a multi-agent system that attempts to adapt a distributed system to a distributed problem of network optimization. As an example we use a logistic network based on the process illustrated in Figure 1a) in a decentralized bottom-up approach. Supply chains consist of autonomous and heterogeneous organizations and actors, therefore we use a multi-agent system (MAS) approach where agents represent actors (e.g. producer, retailer, distributor, wholesaler). Holland introduces in [13] flow as a basic property of complex adaptive systems and in natural systems the flow of energy has been studied in a variety of areas [15,4]. Therefore we use evolutionary computation based on an economic model driven by the flow of money. We assume no central element as a manager or coordinator; instead we use an economic bottom-up approach for adaptation in distributed systems by simulation. Decentralization is an inherent characteristic of logistic networks [9,14] and thus we are especially interested in exploring the extend to which a decentralized adaptation can lead to an improved result (from a system designers perspective) and how an EC-enabled and market-based multi-agent system can realize this adaptation appropriately. This paper presents an approach on how distributed systems can be adapted without any central entity.

We start by considering in section 2 decentralized evolutionary approaches and their applications to logistics. Section 3 describes the economic perspective (3.1), formulates the optimization problem to be solved (3.2), describes the evolutionary approach (3.3) and the expected properties related to the given scenario (3.4). The experimental results are shown and discussed in section 4 followed by conclusion and future works in section 5.

2 Related Work

For specific sub problems like depot location search, known as *p-median-problems*, methods exist that are basically heuristic [11]. One of them described as 'fast interchange heuristic' by Whitaker in [23] solves the *p-median-problem*, but capacity is not considered, which can be seen as unrealistic. A more extensive survey is presented in [8,14] whereby only top-down algorithms are considered. First results of an independent market study indicate that most of the available supply chain management software tools support centrally organized networks and lack a collaborative planning and execution support [10] and thus also use a top-down approach with central coordination, which is not applicable to our problem.

So far, very little work has been conducted on evolutionary computation in multi-agent systems concerning logistics. We have excluded all papers not explicitly addressing these three terms. An evolutionary approach using economic agents for studying strategies in electronic markets is presented by [2]. Here trading agents are investigated by studying heterogeneous strategies in large electronic markets. A traditional 'central' evolutionary algorithm is used and the process steps are hard wired in the agent system. This make it inappropriate for the present problem. A decentralized EC-enabled MAS framework is presented in [22] where local selection occurs and has been used in [7,21,16]. In [7,21] Smith and Eymann investigate negotiation strategies in a supply chain for the production of cabinets but concentrate merely on self-organizing coordination effects. Explicit control and optimization is not provided. Further, they consider only coordination effects between agents but not between agents and the environment. Previous EC-enabled MAS approaches use agents that are specific in their function and hence the re-usability of agents in different logistic scenarios or processes is not supported. Further, the use of centralized infrastructure like yellow pages may induce a bottleneck as the agent population, the communication load and the information in the system grows. [16] uses agents that compete for shared resources directly from the environment. Resources are not traded between the agents and subsequently replenished to the environment. In our sample logistic scenario (figure 1) this approach is inadequate, because agents not only interact with the environment but also perform intermediate interactions. Otto describes in [18,19] a decentralized approach but evaluates only the optimization within one step of the supply chain.

To the best of our knowledge, there is no decentralized approach to streamline logistics networks and adapts them by using evolutionary computation. We use flow of money to direct local recombination and selection in a processes based complex system. In this paper we use a robust, modern and scalable state of the art architecture of a multi-agent system without any central control that can run up to several thousands of agents.

3 Approach

This section describes how the simulation model was build and which parameters were taken into consideration. The income possibilities and expenses every agent has are described and the protocols that are used for communication in the multi-agent system are explained in further detail. As well we describe how the steps from the classical, local evolutionary algorithm are mapped onto a supply-chain.

The supply-chain consists of k steps and a number of agents on each step $i = \{0, \ldots, k\}$. The set of agents is denoted with $A = \{a_1, \ldots, a_n\}$ and the subset of agents at i-th step is denoted with A_i. Each agent is located at a position on a two dimensional map (see figure 1). The chain begins with the fixed set of customer agents A_0 which are normal distributed around a point p_0. The other agents (denoted as supply chain agent) representing the steps $s = \{1, ..., k\}$ are initial uniformly distributed over the map. Each customer agent orders one product per round in the subsequent step in the supply chain.

The definition of the multi-agent system is as follows:

Definition 1 (Multi-Agent System (MAS)). *A multi-agent system* $MAS = (A, E)$ *consists of a finite set of agents* $A = \{a_1, ..., a_n\}$ *in environment* E*.*

Since a MAS contains agents, the following definition for an agent is:

Definition 2 (Agent). *An agent* $a = (I, O, f, s, c)$ *consists of a finite set of sensory inputs* $I = \{i_1, \ldots, i_m\}$*, a finite set of effector outputs* $O = \{o_1, \ldots, o_q\}$*, a function* $f : I \rightarrow O$ *which maps sensory inputs to effector outputs, the strategy vector* $s = (s_1, s_2, \ldots, s_r)$ *determining or parameterizing* f*, and the agents current funds* $c \in R$*.*

Without loss of generality the access to specific strategy parameters is denoted by $s_a(param)$ for increased readability. The strategy s_a contains at least the position $s_a(x) = (x_a^1, x_a^2)$, the maximum stock $s_a(stock_{max})$ and the safety stock $s_a(stock_{min})$ of the agent.

3.1 Economic Perspective

In our model we assume a discrete timeline, where all actions take place at consecutive steps. Agents have to pay a tax $\mathcal{T}_a(t)$ to the environment E at every round t for their actions and according to their strategy parameters s_a (e.g. like inventory or backlog). Given the tax, we can calculate the profit π_a an agent a receives at time t by

$$\pi_a(t) = \mathcal{R}_a(t) - \mathcal{P}_a(t) - \mathcal{T}_a(t) \tag{1}$$

where $\mathcal{P}_a(t)$ denotes the payment a has to pay to other agents (e.g. orders placed), $\mathcal{T}_a(t)$ denotes tax turned over to the environment and finally a may have receipts $\mathcal{R}_a(t)$ from the orders of other agents. Based on the profit $\pi_a(t)$, an agent accumulates funds c_a over time expressed by:

$$c_a(t + 1) = c_a(t) + \pi_a(t) \tag{2}$$

where $c_a(t+1)$ denotes the funds of agent a at time $(t+1)$, $c_a(t)$ denotes a's funds at time t and $\pi_a(t)$ is the profit of a at time t. Money cannot be 'created' by the agent, rather it is provided by the customer agents A_0 representing the demand of the market. A_0 provides funds in return to services offered by the set of agents $\{A \setminus A_0\}$.

In our model the flow of money is essential to make the distributed evolutionary algorithm work. The recipts $\mathcal{R}_a(t)$ are the sum of all order payments an agent receives in t:

$$\mathcal{R}_a(t) = \sum_{orders} (p_{buy}(t) + profit) \cdot q_{order}(t) + (\Delta d \cdot p_{\Delta d}) \tag{3}$$

where p_{buy} denotes the buying price, $profit$ the profit per product sold and q_{order} the quantity of products per order. The shipping cost is denoted by the distance Δd multiplied by the price per unit of distance $p_{\Delta d}$. The distance $\Delta d_{1,2}$ is the two dimensional euclidean distance between $a_1 \in A_i$ and its supplying agent $a_2 \in A_j$ on the subsequent step in the supply chain with $i > j$ and $i, j \leq k$. Calculating the order quantity q_{order} is computed by

$$q_{order}(t) = \begin{cases} s_a(stock_{max}) - stock_a(t) - pp_a(t) : stock_a(t) + pp_a(t) < s_a(stock_{min}) \\ 0 : otherwise \end{cases} \tag{4}$$

where $stock_a(t)$ represents the current stock and $pp_a(t)$ the pending products (ordered but not received) at time t.

3.2 Optimization

Based on MAS and the economic perspective the distributed optimization problem is given by:

Definition 3 (Distributed Optimization Problem). *A distributed optimization problem (DOP) is given by:*

$$minimize \quad \sum_{i=0}^{k-1} \sum_{A_i} \Delta d_{i,j}, \; subject \quad MAS$$

where $\Delta d_{i,j}$ denotes the distances between every purchaser i and its supplier j.

The MAS represents a complete supply chain with rational and local agents and their environment. It includes the constraints, e.g. maximum stock $(s_a(stock_{max}))$, current stock $(stock_a(t))$, agents funds (c_a) or the tax. Every agent has exactly one supplier (except the producer at the last step k). On the other hand an agent may have multiple purchasers (except the customers at step 0, see process in figure 1a). Therefore the overall problem is not known to a single agent and can only be processed or solved in a distributed manner.

3.3 Evolutionary Model

Due to the decentralized structure of real-world supply chains a distributed evolutionary algorithm [22,18,19] is used. The various steps of the local EA, like selection and creation of a new individual, had to be mapped onto the multi-agent system. Every agent is performing all evolutionary steps locally. As described within the optimization section 3.2 the solution is distributed over the whole agent population. Thus the novelty of the presented approach differs from traditional EA as the entire population forms only one solution. Every agent contributes a small piece to the entire solution and thus we believe our model is therefore closer to real-world supply chains.

By turning the classical EA design on its head agents act as independent evolutionary entities and can be seen as an artificial life like approach [1]. Local reproduction by an agent includes all steps and operations necessary to produce new offspring, such as local search, recombination and mutation. An agent specific variable θ_a is introduced that serves as a threshold and enables agents to reproduce [18]. Whenever the funds of an agent exceeds this threshold ($c_a \geq \theta_a$) it will reproduce. Before reproduction an agent a chooses a random mate a' and sends out a message to request the strategy. Previous experiments have shown no difference between random selection compared to and other strategies [20]. Upon receiving the mates strategy $s_{a'}$, the recombination and mutation is done local by using the breeder genetic algorithm described in [17] to produce the child strategy s_c. Finally a new agent will be created and given the strategy s_c, half of the funds $\frac{c_a}{2}$ and half of the current stock. Due to the local reproduction trigger the notion of generation number does not exist.

The fitness calculation is done implicitly by following the economic model in section 3.1. Therefore no comparative fitness calculation is necessary. This is an important aspect, as fitness comparison among agents would increase the number of messages and the computation overhead of the overall system.

3.4 Expected Properties

We expect highly adaptive behaviour of the whole system in the following aspects as described in [18,19]: (i) adaptive population size and (ii) Adaptive system strategy as a combination of agents strategies. Similar to eco-systems, economic evolutionary agents tend to stabilize around the so called carrying capacity of the environment. Therefore the presented approach is usefull for dynamic problems as well as one time problems. The carrying capacity is given by the customer agents order behaviour. The emergent behavior of stabilization can be observed in real scenarios [9] where actors enter and leave the market. Further a long term adaptivity is expected on the agents strategy forming the supply-chain. This includes e.g. agent positions in the map and maximum and safety stock values. We expect the positions of the agents to being close to the customer agent positions and thus streamlining the overall transport distance within the supply chain. Even without any central control this effect can be observed.

4 Evaluation

To demonstrate and verify the viability of the developed approach on the supply chain given in figure 1 we have conducted a large number of experiments using different settings and strategies such as mutation rate and *searchrate*. In defined intervals every agent searches for a potential supplier in the previous step of the chain. In order to model decentrality agents can not 'search' for specific agents. Instead they will be connected to a random supplier agent of the specified step. Offers the new supplier a price below the current supplier price (equation 3) the searching agent replaces its current supplier with the new one. Therefore the *searchrate* induces network dynamic and has direct influence on the network structure. This mechanism performs a search for the cheapest supplier with one supplier per agent at each time step t.

The analysis applies to experiments with the following settings: map size is 100×100, customer agents $A_0(|A_0| = 100)$ are normal distributed with $\mathcal{N}(p_0, 5)$, where $p_0 = \{75, 75\}$, all other agents $A_i, i = \{1, \ldots, k\}, |A_i| = 10$ are initial randomly distributed, initial funds of agents $c_a(0) = 140$, initial inventory: 15, buying price for products at the producers site = 10, stock: $s_a(stock_{min}) = 5$ and $s_a(stock_{max}) = 15$, splitting threshold $\theta_a = 15 \cdot s_a(stock_{max}) \cdot p_{buy} \cdot \frac{profit}{p_{buy}+profit}$ (every agent is capable to fill its inventory multiple times dependent to the position in the chain), $profit = 10$, $searchCost = 10$, price per distance unit $p_{\Delta_d} = 0.01$, penalty for one product backlog = 1, $stockCost = 0.5$. Every customer orders one product per round. Every setting was averaged over 50 independent simulation runs with 1500 rounds each.

The evolutionary strategy is realized as a hash map of string value pairs. The values are represented as floats and recombined using the breeder evolutionary algorithm [17]. The search for a new supplier is a simple random connect to a supplier at the previous stage of the supply chain. The agent performs a price request every *searchrate* rounds and replaces its current supplier if the new supplier is cheaper compared to the current one.

Mutation rate

Simulations were computed for mutation rates of 1, 5, 10 and 50 percent, the search rate is 10. Figure 2a) shows the results of this simulations. The overall distance in the supply-chain in every round is plotted for the above mentioned mutation rates. In the beginning the distances are almost equal for all mutation rates. Already after a few rounds, the first search process is carried out and some agents can create child agents whereas others are removed from the agent system because they are running out of money. A high mutation rate causes extreme values for the agents strategy (e.g. position $s_a(x)$, safety stock $s_a(stock_{min})$ and maximum stock $s_a(stock_{max})$). Extreme mutation rates induce a random walk of the solution as shown in figure 2 ($mutationrate = 50\%$). The best performance is shown with mutation rate of 1%.

Search rate

Simulations were computed for search rates of 10, 15 and 20, the mutation rate is set to 1%. In figure 3b) the different results in the supply-chain are plotted. In

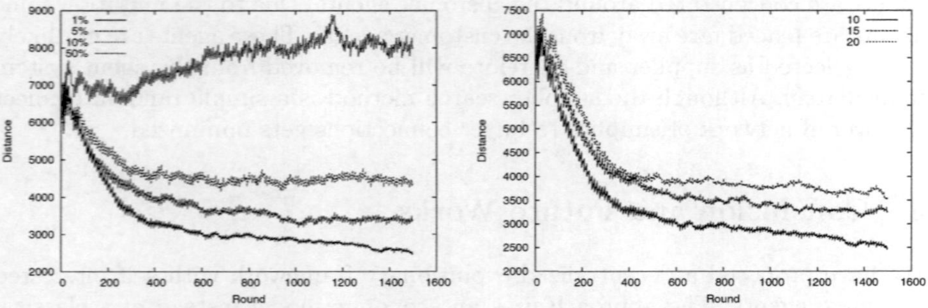

(a) mutation rate (b) search rate

Fig. 2. Different *mutationrates* and *searchrates* and their influence to the overall transport distances

that simulation where the agents look for a more favorable supplier more often (e.g. decreased *searchrate*), agents that are further away from the market agents get removed from the system faster and agents closer to the customer agents get selected more often. The overall dynamic of the supply-chain increases when the *searchrate* decreases.

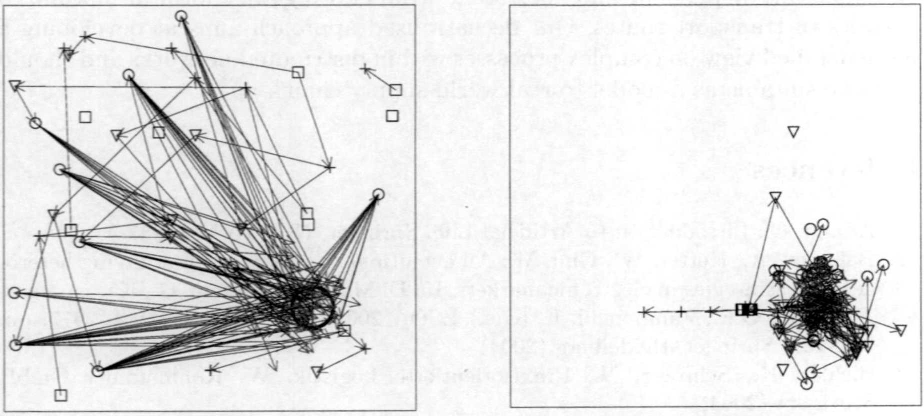

(a) round 1 (b) round 1500

Fig. 3. Snapshots of the supply chain at a) round 1 and b) round 1500

Location of agents

Figure 3 shows two screen shots of the maps taken after round 1 and at the end of the simulation (round 1500). In 3a) the initial random distribution of agents and the direction of the products can be seen. The customers area (mean) is shown in the right bottom corner as a thick black circle. After 1500 rounds the agents are concentrated around the customer agents. Due to the mutation some agents are placed fare away from the customer agents. These agents are not likely to be selected as supplier and therefore will be removed from the agent system immediately. Although the supplier search method is a simple random connect the overall network of supplier to buyer connections gets minimized.

5 Conclusion and Future Works

We have proposed a decentralized evolutionary framework with self interested economic agents. The approach uses an economic model instead of a classical fitness function. This enables the complete distribution of evolutionary steps and their local execution in the agents scope. Another aspect is the distributed solution, which is a combination of partial solutions of all agents. In contrast to conventional EA where each individual contains the complete solution, we spread the solution over the whole population since the problem of real world supply chains is inherently distributed.

The simulation results show the existence of optimized strategies (e.g. positions) throughout the supply chain. Therefore the combination of local strategies form a streamlined global supply chain. In our test cases the agents adapt to the the market, e.g. the position of the customer agents. The tests were performed by varying different values of *mutationrate* and *searchrate*. All the agents strategies are intertwined and produce a network of coexisting strategies. The approach of simple interdependent and coexisting strategies together form an optimized network of transport routes. Our decentralized approach aims at developing a more detailed view on complex processes within distributed networks and should be more suitable as a model to real world supply chains.

References

1. Adami, C.: Introduction to Artificial Life. Springer, New York (1998)
2. Babanov, A., Ketter, W., Gini, M.: An evolutionary approach for studying heterogeneous strategies in electronic markets. In: Di Marzo Serugendo, G., Karageorgos, A., Rana, O.F., Zambonelli, F. (eds.) ESOA 2003. LNCS (LNAI), vol. 2977, pp. 157–168. Springer, Heidelberg (2004)
3. Bichler, K., Schröter, N.: Praxisorientierte Logistik. W. Kohlhammer GmbH Stuttgart (2004)
4. Elton, M.N.C.: The ten-year cycle in numbers of the lynx in canada. Journal of Animal Ecology 11, 215–244 (1942)
5. Cai, W., Turner, S.J., Gan, B.P.: Hierarchical federations: an architecture for information hiding. In: PADS 2001: Proceedings of the Fifteenth Workshop on Parallel and Distributed Simulation, Washington, DC, USA, pp. 67–74. IEEE Computer Society Press, Los Alamitos (2001)
6. Durfee, E.H., Lesser, V.R., Corkill, D.D.: Trends in cooperative distributed problem solving. IEEE Transactions on Knowledge and Data Engineering 1(1), 63–83 (1989)
7. Eymann, T.: AVALANCE - Ein agentenbasierter dezentraler Koordinationsmechanismus für elektronische Märkte. PhD thesis, Universität Freiburg (2000)

8. Graf, H.-W.: Netzstrukturplanung - Ein Ansatz zur Optimierung von Transportnetzen. PhD thesis, Universität Dortmund, Fakultät Maschinenbau, Dortmund (2000)
9. Graudina, V., Grundspenkis, J.: Technologies and multi-agent system architectures for transportation and logistics support: An overview. In: International Conference on Computer Systems and Technologies - CompSysTech, Varna, Bulgaria, pp. IIIA.6-1 – IIIA.6-6(2005)
10. Guenther, S., Laakmann, F.: Efficient evaluation and selection of it-support based on the supply chain management task reference model (2001), `http://www.scm-ctc.de/`
11. Heinrichmeyer, H., Reinholz, A.: Entwicklung eines Bewertungsmodells für die Depotstandortoptimierung bei Servicenetzen. Technical report, Fraunhofer-Institut für Materialfluss und Logistik, Universität Dortmund, Fachbereich Informatik, Lehrstuhl für Systemanalyse (2004)
12. Hohl, F.: A framework to protect mobile agents by using reference states. Technical Report 2000/03, Institut für Parallele und Verteilte Höchstleistungsrechner (IPVR) (March 2000)
13. Holland, J.H.: Hidden Order: How Adaptation Builds Complexity, vol. 9. Addison Wesley Publishing Company, Reading (1996)
14. Jehle, E., Kaczmarek, M.: Organisation der Planung und Steuerung in Supply Chains. Technical report, Universität Dortmund, Lehrstuhl Industriebetriebslehre (2004)
15. Hogan, J.A.: Energy models of motivation: A reconsideration. Applied Animal Behavior Science 53, 89–105 (1997)
16. Menczer, F., Degeratu, M., Street, W.: Efficient and scalable pareto optimization by evolutionary local selection algorithms. Evolutionary Comp. 8(3), 223–247 (2000)
17. Mühlenbein, H.: The breeder genetic algorithm - a provable optimal search algorithm and its application. In: Colloquium on Applications of Genetic Algorithms, vol. 67, IEEE, London (1994)
18. Otto, S.: Ein agentenbasierter evolutionärer Adaptions- und Optimierungsansatz für verteilte Systeme. PhD thesis, Universität Erlangen - Nürnberg (September 2009)
19. Otto, S., Kirn, S.: Adaption in distributed systems: an evolutionary approach. In: Hecht, M.K., et al. (eds.) Proceeding Genetic and Evolutionary Computation Conference (GECCO 2006), vol. 1, pp. 199–206. ACM, New York (2006)
20. Otto, S., Kirn, S.: Evolutionary adaptation in complex systems using the example of a logistics problem. International Transactions on Systems Science and Applications 2(2), 157–166 (2006)
21. Smith, R.E., Bonacina, C., Kearney, P., Eymann, T.: Integrating economics and genetics models in information ecosystems. In: Proceedings of the 2000 Congress on Evolutionary Computation CEC 2000, La Jolla Marriott Hotel La Jolla, California, USA, 6-9, pp. 959–966. IEEE Press, Los Alamitos (2000)
22. Smith, R.E., Taylor, N.: A framework for evolutionary computation in agent-based systems. In: Looney, C., Castaing, J. (eds.) Proceedings of the 1998 International Conference on Intelligent Systems, pp. 221–224. ISCA Press (1998)
23. Whitaker, R.: A fast algorithm for the greedy interchange for large-scale clustering and median location problems. In: INFOR 21, pp. 95–108 (1983)

Testing the Permutation Space Based Geometric Differential Evolution on the Job-Shop Scheduling Problem

Antonin Ponsich[1] and Carlos A. Coello Coello[2]*

[1] Universidad Autónoma Metropolitana, Unidad Azcapotzalco, Dpto. de Sistemas
Av. San Pablo No. 180, Col. Reynosa Tamaulipas. México, D.F. 02200
[2] CINVESTAV-IPN, Departamento de Computación
Av. IPN No. 2508, Col. San Pedro Zacatenco. México, D.F. 07300
antonin@computacion.cs.cinvestav.mx, ccoello@cs.cinvestav.mx

Abstract. From within the variety of research that has been devoted to the adaptation of Differential Evolution to the solution of problems dealing with permutation variables, the Geometric Differential Evolution algorithm appears to be a very promising strategy. This approach is based on a geometric interpretation of the evolutionary operators and has been specifically proposed for combinatorial optimization. Such an approach is adopted in this paper, in order to evaluate its efficiency on a challenging class of combinatorial optimization problems: the Job-Shop Scheduling Problem. This algorithm is implemented and tested on a selection of instances normally adopted in the specialized literature. The results obtained by this approach are compared with respect to those generated by a classical DE implementation (using Random Keys encoding for the decision variables). Our computational experiments reveal that, although Geometric Differential Evolution performs (globally) as well as classical DE, it is not really able to significantly improve its performance.

1 Introduction

The Differential Evolution (DE) technique is an Evolutionary Algorithm proposed by Storn and Price in the mid 1990s [22]. Characterized by a novel mutation operator, this stochastic search technique has been found to be a powerful optimization tool for solving continuous optimization problems [12]. Unlike other methods such as Genetic Algorithms, the canonical DE scheme is based on a floating-point representation of the variables. Thus, the treatment of discrete optimization problems requires an adaptation of its original operators. This latter observation is even more relevant when dealing with permutation-based problems since such problems involve, in addition to the discrete values restriction, inherent constraints of interdependence among variables.

A significant amount of research has been recently devoted to this issue and many techniques have been proposed, particularly focusing on methods for converting real variables into permutations. However, none of all these techniques

* The second author acknowledges support from CONACyT project no. 103570.

R. Schaefer et al. (Eds.): PPSN XI, Part II, LNCS 6239, pp. 250–259, 2010.

has succeeded in avoiding the redundancy in the mapping from real to permutation spaces. As a consequence, the results obtained by DE are not able to satisfactorily compete with those obtained by other metaheuristics when tackling complex permutation-based problems. The other path for adapting DE to problems dealing with permutation variables is the modification of the DE's operators. In 2009, A. Moraglio et al. adapted a geometric framework (previously introduced for Particle Swarm Optimization [13]) for its use with the DE metaheuristic. The differentiation was transformed into a geometric operation, applicable in any space when adopting an appropriate metric. In [14], the authors mainly focused on the consideration of binary spaces and reported interesting results for NK landscapes and Spears-DeJong functions. In a recently released technical report [15], permutation and program spaces are further tackled. The aim of the present paper is thus to evaluate the behavior of such a technique, called *Geometric Differential Evolution* (GDE) for permutation spaces, on a challenging problems class.

As a case study, the Job-Shop Scheduling Problem is chosen, for two main reasons. First, there is no need to emphasize its inherent complexity, already evidenced in many studies and simply justified by the inability for state-of-the-art algorithms to identify optimal solutions of complex instances of this problem. The second reason is a previous work on the JSSP, which analyzed the performance of techniques based on the transformation of real numbers to permutations [20]. This precedent will allow us to compare the two strategies, i.e., transforming either the variable representation mode or the operators. The remainder of this paper is organized as follows. Section 2 presents a short overview on JSSP, while Section 3 is dedicated to the definition of the permutation-based GDE algorithm. The experimental methodology and computational results are presented in Section 4. Finally, some conclusions are drawn in Section 5.

2 Overview of the JSSP

2.1 A Review on Solution Techniques

Scheduling problems, because of their many applications not only in the industrial but also in the service fields [19], have attracted an increasing interest within the Operations Research community. In the manufacturing area, the Job-Shop Scheduling Problem is one of the most complex examples. In this problem, a set of jobs, which all consist of several operations, is processed in a certain order on a set of machines. The processing sequence of each job operations on the machines and the associated processing times are the problem data. The objective is, then, to minimize the completion date of the last scheduled operation.

Because solving exactly the JSSP constitutes a challenging issue, much of the research efforts have focused on the development of efficient methods. Researchers have first concentrated on exact optimization techniques based on the disjunctive graph representation [21]. However, the JSSP is hard for exact algorithms and optimal solutions can be provided for instance sizes up to about 10 machines and 10 jobs. This feature has led to the development of heuristics methods: the Shifting Bottleneck heuristic (particularly the SB-I and SB-II versions presented in [1]) is an example of a heuristic which achieves very good results even for mid-size and large instances.

With the development and enhancement of several metaheuristics, a variety of local search and population-based heuristics have been applied to the JSSP: Simulated Annealing [24], Tabu Search [23], Genetic Algorithms [6], GRASP [2], etc. Aiming at simultaneously exploiting the benefits of several methods, hybrid techniques have also been frequently used: Genetic Local Search [9], Ant Colony Optimization coupled with a Tabu Search approach [10], a Tabu Search/Simulated Annealing hybrid [26]. The two Tabu Search based algorithms proposed by Nowicki and Smutnicki (TSAB [17] and i-TSAB [18]) have provided the best results reported until now for this problem.

2.2 Problem Formulation

The classical JSSP aims at assigning a finite set O of operations to a finite set M of machines ($|M| = m$). Each operation $o_{ij} \in O$ belongs to the sequence of a job j and must be processed on a specific machine i. Conversely, each job $j \in J$ is characterized by a subset of operations $O_j \subset O$ that must be processed according to a defined order. Unlike the Flow-Shop case, the processing sequence differs from one job to another. In most cases, each job j is processed exactly once on each machine i, so that the total number of operations is $|O| = nm$ and the commonly used nomenclature refers to $n \times m$-instances.

The typical objective of the classical JSSP is then to minimize the completion time of the last operation scheduled, namely the makespan, while respecting the following major constraints: (i) one machine cannot simultaneously process more than one job at a time, (ii) preemption is not allowed, (iii) the processing sequence of the operations belonging to a job must be respected (the starting time of any operation is higher than the completion time of its predecessor).

The solution is an assignment of operations to machines on a precise time period and is called a *schedule*. When an appropriate schedule builder is used (see next section), this solution can be formulated as a multi-permutation, i.e., a set of job permutations associated to each machine. The cardinality of all possible solutions to the JSSP is therefore $(n!)^m$.

2.3 Schedule Classes

Attention must be paid to the fact that, for a given multi-permutation, an infinite number of schedules might be built. An important issue therefore concerns the construction of the schedule. Among the feasible schedules, the active schedules are those for which no operations can be brought forward without delaying another operation. It is well known that optimal schedules belong to the *active* class [16]. Another relevant schedule class is the *non-delayed* one, for which no idle time is allowed on a machine if this latter is free and an operation is available for processing. This latter class also constitutes a subset of the active schedules class, but may not contain the optimal solution.

In the following, a schedule builder based on the parameterized Giffler & Thompson's algorithm is used [8]. This technique produces schedules that lie on an region intermediate between the active and non-delayed classes, through the introduction of a parameter δ that controls the number of possible schedules, i.e., the number of solutions in the search space of the considered problem. When δ=0, the built schedule is non-delayed while δ=1 allows the generation of active schedules. This parameter should be tuned for each treated instance.

3 Geometric DE for Multi-permutation Spaces

The aim of this section is not presenting the whole theoretical framework and detailed insights of the geometric adaptation of Differential Evolution to permutation spaces, since such information can be found in [15]. We only provide here the features necessary for a global understanding and for the implementation of permutation-based GDE.

3.1 A Geometric Framework for DE

The basic idea in GDE is a re-interpretation of the evolutionary operators according to a geometric point of view. Considering two vector solutions as points in the space, the crossover of these two parents can be seen, in this sense, as a geometric operation returning a point within the segment defined by the two original solutions. The distance between the offspring vector and both parents is then implicitly determined by the crossover rate. Clearly, in this context, the definition of an appropriate metric associated to the considered seach space is required; but this allows the extension of such concepts to any space endowed with an adequate distance.

In [14], the reformulation of the DE operators is proposed according to this geometric paradigm. Consider the classical DE mutation operator:

$$u_{ij}^G = x_{3j}^G + F(x_{1j}^G - x_{2j}^G), \forall j \in \{1, ..., N\} \tag{1}$$

where u_{ij}^G is the mutated offspring generated for parent x_{ij}^G, while x_{1j}^G, x_{2j}^G, x_{3j}^G are randomly selected individuals in the current population ($x_{1j}^G \neq x_{2j}^G \neq x_{3j}^G \neq x_{ij}^G$) and F ($F \in [0,1]$) is the scaling factor.

Setting $W = \frac{1}{1+F}$, then equation (1) can be written as:

$$W \cdot u_{ij}^G + (1 - W) \cdot x_{2j}^G = W \cdot x_{3j}^G + (1 - W) \cdot x_{1j}^G, \forall j \in \{1, ..., N\} \tag{2}$$

Consider that *U*, *X1*, *X2* and *X3* are the point-wise representations associated to vectors u_{ij}^G, x_{1j}^G, x_{2j}^G and x_{3j}^G, respectively. The differentiation operator can be represented as illustrated in Fig. 1 (taken from [14]): point E is the result of the convex combination of points *X3* and *X1* with weights W and $1 - W$ respectively, and can be constructed since *X1* and *X3* are known. Subsequently, point U can be deduced as the point on the extension ray (ER) going out of *X2* and passing through E (E is the result of the convex combination of points U and $X2$ with weights W and $1 - W$, respectively). Thus, creating algorithmic procedures that translate the convex combination and extension ray operations would allow to design a geometric interpretation of the differentiation operator based on the definition of a metric adapted to the considered space.

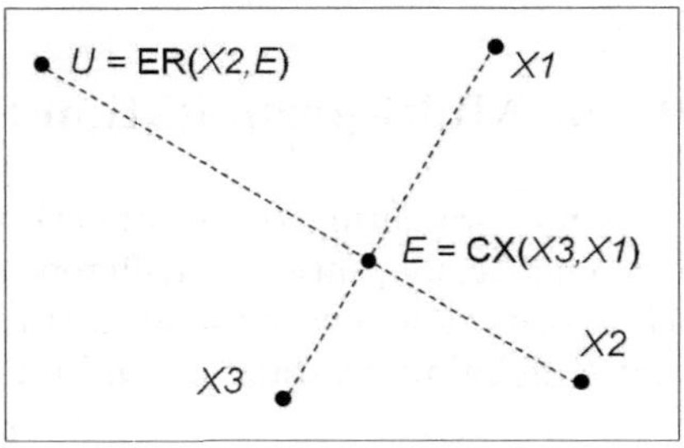

Fig. 1. Construction of U using convex combination and extension ray procedures

Similarly, regarding the crossover procedure, Moraglio et al. prove that the discrete recombination applied to u_{ij}^G and x_{ij}^G (or, according to the geometric paradigm, points U and X_i) produces an offspring that can be seen as the convex combination of points U and X_i with weights C_r and $1-C_r$ respectively, where C_r is the crossover rate used within DE.

3.2 GDE for the JSSP

In order to adapt the GDE framework to the job-shop scheduling problem, a metric must be defined for permutation spaces. The distance between two configurations will subsequently be computed as the sum of the distances between the permutations, in each schedule, associated to every machine. This is proposed in [15]. The swap distance appears to be particularly appropriate for the JSSP: swapping two jobs in a sequence is a neighborhood definition commonly used within many local search methods. The swap distance between two sequences is then equal to the number of swapping mutation steps necessary to transform one permutation into another.

In other words, let us consider that the permutation vectors constitute the nodes of an undirected graph whose edges represent a swapping move between two neighbor permutations (i.e. one permutation is attainable from the other by only one swapping mutation step). The swap distance is thus the shortest path, on this graph, to get from one vector to another. Note that this concept involves, for a permutation having n elements, a total "diameter" of the search space (maximum distance between two permutations) equal to $n - 1$.

Accordingly, for the convex combination procedure, the swap distance between both initial vectors is computed and the new vector is derived in such a way that it lies between the initial strings, at a distance obtained from multiplying each weight by the total swap distance. These distances are interpreted as probabilities in order to construct a mask that determines, for each position, to which parent element the offspring must be equal to. A similar procedure is developed for the extension ray, in such a way that the convex combination of the new vector and one of the initial ones results in the other parent. When the distance between the initial strings is equal to the diameter of the search space, it is impossible to generate an offspring farther away from one of them (outside the segment defined by the initial vectors). Both of these processes are repeated for each permutation in the schedule configuration (i.e., for every machine).

The detailed algorithms for all the above-mentioned procedures (computing swap distances, convex combination and extension ray) are not presented here but are explained in detail in [15].

4 Computational Experiments

4.1 Methodology

The permutation-based GDE algorithm is evaluated on a selection of instances chosen among several sets of JSSP instances commonly used in the specialized literature:

- 3 instances due to [7]: FT06, FT10, FT20.
- 2 instances due to [1]: ABZ5, ABZ6.
- 6 instances due to [3]: ORB01-ORB06.
- 7 instances due to [11]: LA22, LA24, LA25, LA27, LA37, LA38, LA40.
- 4 instances due to [25]: YN1-YN4.

These examples, drawn from the OR-library [5], can be basically divided into three groups, according to their complexity: easy (FT and ABZ), medium (ORB and LA) and difficult instances (YN).

With respect to the GDE parameters tuning, population size NG and generation number NG are set in such a way that the number of objective evaluations $NP \times NG$ is equal to the values commonly reported in the literature (note, however, that this criterion may not always be fair since, in many cases, a simplified objective computation mode is devised in order to shorten computational times). Concerning the crossover rate C_r and the amplification factor F, preliminary computations indicated that the best results are obtained when the former adopts rather high values (between 0.8 and 1) and when the latter adopts values randomly generated between 0.3 and 0.9.

4.2 Standard Random-Keys DE

As mentioned before, the aim of this study is to compare both strategies when adapting DE to problems dealing with permutation variables, i.e., adapting the variable representation mode or the internal DE's operators. So, according to this, the results obtained by the described GDE algorithm are compared here against those of a standard DE.

The DE version used here, fully described in [20], is DE/rand/1/bin, which means that the vectors used within the differentiation operator are randomly selected, only one difference is used and a binary crossover is performed for each variable independently. Besides, the main feature of the algorithm is the encoding procedure. In order to handle permutations, the Random Keys method, initially proposed in [4], was implemented. A real number, bounded between 0 and 1, is used for each operation and operations corresponding to the same machine are sequenced according to the increasing order of their associated variable. For instance, considering 5 jobs with the following variable vector on machine i: [0.41 0.68 0.02 0.85 0.37], the resulting sequencing order is: [2 4 0 1 3]. The Random Keys technique is chosen here because it proved to be more efficient and effective than two others (i.e., evolving dispatching rules and binary matrix priority based on the disjunctive graph) as indicated in the comparative study reported in [20].

Besides, since the differentiation mutation is likely to produce variables lying outside their bounds, a mixed constraint-handling technique is applied according to a given probability P_B (tuned for each instance): (i) setting the variable value to the violated bound; (ii) using the violated bound as a symmetry center to send the considered variable to the feasible side of the boundary.

4.3 Results

For each instance, we performed 20 runs of each method (classical DE and GDE). The comparison is obviously drawn for equal numbers of objective evaluations. The values reported in Table 1 indicate, for each technique, Makespan Relative Errors (with respect to a reference makespan) and standard deviation of the results over the 20 runs. *b-MRE* (respectively, to *m-MRE*) reports the error of the best objective value found over 20 runs f_{Best} (respectively, the mean value of the objective over the 20 runs f_{Mean}). The reference makespan is, typically, a lower bound *LB* (optimal for all instances, except for YN1-YN4). Note that the standard deviations provided in Table 1 are computed according to the makespan values and not to the relative errors. The relative error is computed according to the following formula:

$$b - MRE = 100 \times \frac{f_{Best} - LB}{LB} \tag{3}$$

$$m - MRE = 100 \times \frac{f_{Mean} - LB}{LB} \tag{4}$$

A global observation of Table 1 shows a similarity of both techniques' performance. Regarding the *b-MRE* indicator, both methods provide similar results for 36% of the treated instances. With respect to *m-MRE*, GDE is better in 45% of the considered cases while the Random Keys DE is better for 50%.

However, some differences appear when considering separately each problem category. On the one hand, GDE provides the best results for simple and complex instances and, on the other hand, the Random Keys DE performs better

Table 1. Computational results

Instance	Size	LB/Opt.	Geometric DE			Random Keys DE		
			b-MRE	*m-MRE*	S. Dev.	*b-MRE*	*b-MRE*	S. Dev.
FT06	6×6	55	0	0.18	0.31	0	0.18	0.44
FT10	10×10	930	1.40	2.33	4.50	1.40	1.97	5.07
FT20	20×5	1165	1.12	1.12	0.45	1.29	1.45	0.34
ABZ5	10×10	1234	0.41	0.44	1.47	0.41	0.84	4.87
ABZ6	10×10	943	0.53	0.53	0	0.53	0.73	4.97
	Average		**0.69**	**0.92**	**1.35**	**0.72**	**1.04**	**3.14**
ORB01	10×10	1059	1.04	1.08	1.79	1.04	1.11	2.40
ORB02	10×10	888	0.68	0.92	1.44	0.79	0.99	1.64
ORB03	10×10	1005	1.59	3.86	9.37	1.59	2.54	6.73
ORB04	10×10	1005	1.09	1.94	3.99	1.29	1.92	4.22
ORB05	10×10	887	0.79	1.89	2.05	1.01	1.91	4.36
ORB06	10×10	1010	1.09	1.78	4.70	0.89	1.89	6.38
LA22	15×10	927	3.67	4.95	5.59	1.40	3.67	7.65
LA24	15×10	935	3.21	4.51	4.85	2.14	2.83	3.56
LA25	15×10	977	3.48	5.32	5.83	2.56	3.50	6.12
LA27	20×10	1235	5.67	7.10	7.28	4.78	6.44	8.85
LA37	15×15	1397	3.58	4.82	5.08	2.08	3.57	9.75
LA38	15×15	1254	4.85	6.59	7.96	3.01	3.71	5.30
LA40	15×15	1222	4.09	5.11	5.96	1.96	2.59	5.04
	Average		**2.68**	**3.83**	**5.07**	**1.89**	**2.82**	**5.54**
YN1	20×20	846	12.06	14.02	6.10	13.36	14.59	5.40
YN2	20×20	870	12.64	14.86	7.42	13.86	15.23	6.73
YN3	20×20	840	13.45	14.73	5.47	14.40	15.48	4.60
YN4	20×20	920	14.13	16.01	5.50	14.89	15.49	4.01
	Average		**13.07**	**14.91**	**6.12**	**14.08**	**15.20**	**5.18**

for medium instances. This behavior is further moderated for medium instances since GDE is slightly better for the ORB class while being completely outperformed by the Random Keys DE in the LA class. This comment highlights the fact that understanding why some instances are more difficult for one method than for another still remains as an open question.

It is worth recalling that state-of-the-art algorithms, such as TSAB and *i*-TSAB (complete results available in [17] and [18]) generally perform much better than both of the DE versions considered here (specially on hard instances but also for the simple ones, for which optimal solutions can be systematically determined). This comment must balance the previous observations concerning the quality of the obtained results.

The deceiving results obtained by GDE could be tentatively explained by the extension ray procedure used within the geometrically adapted differentiation. As underlined in [15], this operation consists in generating point U (the mutant, see Fig. 1) beyond point E, which is itself computed as the convex combination of points $X1$ and $X3$. However, in many cases, the distance between $X2$ and E is already equal to the maximum distance between two permutations (called the "search space diameter" in [15]). As a consequence, point U is equal to point E, meaning that the differentiation operator degenerates into the simple convex combination of two parents $X1$ and $X3$: the explorative power of the method is thus significantly damaged. Note that this phenomenon does not occur for smaller alphabets (binary search spaces) because the probability to generate two points separated by the maximum distance is much lower than in the permutation case. Finally, this observation shows that GDE might not provide competitive results for high cardinality alphabets, especially if the variable string size is large.

5 Conclusions

We presented in this study an evaluation of the Geometric Differential Evolution algorithm, adapted to the treatment of problems dealing with permutation variables, recently proposed in [15]. The Job-Shop Scheduling Problem has been adopted because it represents a challenging problem class for which a widely developed bank of instances is available. The results showed that the permutation-based GDE globally performs as well as a classical implementation of DE for this problem class. This may lead to the conclusion that this novel algorithm, based on a modification of the initial DE's operator, is really viable and able to compete with a DE version having a special variables encoding mechanism for the treatment of permutation-based problems. However, GDE is not able to significantly improve the performance of classical DE for non-continuous optimization problems and would be still outperformed by other metaheuristics (particularly Tabu Search in the JSSP framework). Thus, more research is required in order to produce a more competitive DE variant for problems such as JSSP.

References

1. Adams, J., Balas, E., Zawack, D.: The Shifting Bottleneck procedure for job shop scheduling. Management Science 34, 391–401 (1988)
2. Aiex, R.M., Binato, S., Resende, M.G.C.: Parallel GRASP with path-relinking for job shop scheduling. Parallel Computing 29(4), 393–430 (2003)
3. Applegate, D., Cook, W.: A computational study for the job-shop scheduling problem. ORSA Journal on Computing 3(2), 149–156 (1991)
4. Bean, J.: Genetic Algorithms and random keys for sequencing and optimization. ORSA Journal on Computing 6, 154–160 (1994)
5. Beasley, J.E.: Or-library: distributing test problems by electronic mail. Journal of the Operational Research Society 41(11), 1069–1072 (1990), `http://mscmga.ms.ic.ac.uk`
6. Della Croce, F., Tadei, R., Volta, G.: A Genetic Algorithm for the job shop problem. Computers and Operations Research 22(1), 15–24 (1995)
7. Fisher, H., Thompson, G.L.: Probabilistic learning combinations of local job-shop scheduling rules, Industrial Scheduling Edition, pp. 225–251. Prentice Halls, Englewood Cliffs (1963)
8. Giffler, B., Thompson, G.L.: Algorithms for solving production scheduling problems. Operations Research 8(4), 487–503 (1960)

9. Goncalves, J.F., de Magalhaes Mendes, J.J., Resende, M.G.C.: A Hybrid Genetic Algorithm for the job shop scheduling problem. European Journal of Operational Research 167(1), 77–95 (2005)
10. Huang, K.L., Liao, C.J.: Ant Colony Optimization combined with Taboo Search for the job shop scheduling problem. Computers and Operations Research 35(4), 1030–1046 (2008)
11. Lawrence, S.: Resource constrained project scheduling: an experimental investigation of heuristic scheduling techniques (supplement). Technical report, Graduate School of Industrial Administration, Carnegie Mellon University, Pittsburgh (Pennsylvania), USA (1984)
12. Mezura-Montes, E., Velázquez-Reyes, J., Coello Coello, C.A.: Modified Differential Evolution for constrained optimization. In: Proceedings of the 2006 IEEE Congress on Evolutionary Computation (CEC 2006), Vancouver, Canada, pp. 25–32 (2006)
13. Moraglio, A., Di Chio, C., Poli, R.: Geometric Particle Swarm Optimization. In: European Conference on Genetic Programming, pp. 125–136 (2007)
14. Moraglio, A., Togelius, J.: Geometric Differential Evolution. In: Genetic and Evolutionary Computation Conference (GECCO 2009), pp. 1705–1712 (2009)
15. Moraglio, A., Togelius, J., Silva, S.: Geometric Differential Evolution for Combinatorial and Programs Spaces. Technical Report, School of Computing, University of Kent, Canterbury (UK) (March 2010)
16. Morton, T.E., Pentico, D.W.: Heuristic Scheduling Systems: With Applications to Production Systems and Project Management. In: Kocaoglu, D.F. (ed.) Wiley Series in Engineering & Technology Management, New Jersey, USA (1993)
17. Nowicki, E., Smutnicki, C.: A fast Taboo Search algorithm for the job shop problem. Management Science 42(6), 797–813 (1996)
18. Nowicki, E., Smutnicki, C.: An advanced Tabu Search algorithm for the job shop problem. Journal of Scheduling 8(2), 145–159 (2005)
19. Pinedo, M.L.: Planning and scheduling in manufacturing and services. Series in Operations Research. Springer, New-York (2005)
20. Ponsich, A., Castillo Tapia, M.G., Coello Coello, C.A.: Solving permutation problems with Differential Evolution: an application to the jobshop scheduling probem. In: Ninth International Conference on Intelligent System Design and Applications (ISDA 2009), Pisa, Italy (November-December 2009)
21. Roy, B., Sussmann, B.: Les problèmes d'ordonnancement avec contraintes disjonctives. Note DS 9 bis, SEMA, Paris (1964)
22. Storn, R., Price, K.V.: Differential Evolution - a simple and efficient heuristic for global optimization over continuous spaces. Journal of Global Optimization 11, 341–359 (1997)
23. Taillard, E.D.: Parallel Taboo Search techniques for the job shop scheduling problem. ORSA Journal on Computing 6(2), 108–117 (1994)
24. Van Laarhoven, P.J.M., Aarts, E.H.L., Lenstra, J.K.: Job shop scheduling by Simulated Annealing. Operations Research 5(1), 113–125 (1992)
25. Yamada, T., Nakano, R.: A Genetic Algorithm applicable to large-scale job-shop instances. In: Manner, Manderick (eds.) Parallel instance solving from nature, vol. 2, pp. 281–290. North-Holland, Amsterdam (1992)
26. Zhang, C.Y., Li, P., Rao, Y., Guan, Z.: A very fast TS/SA algorithm for the job shop scheduling problem. Computers and Operations Research 35(1), 282–294 (2008)

New Uncertainty Handling Strategies in Multi-objective Evolutionary Optimization

Thomas Voß[1], Heike Trautmann[2], and Christian Igel[1]

[1] Institut für Neuroinformatik, Ruhr-Universität Bochum, Germany
[2] TU Dortmund University, Dortmund, Germany
{thomas.voss,christian.igel}@ini.rub.de,
trautmann@statistik.tu-dortmund.de

Abstract. Since many real-world optimization problems are noisy, vector optimization algorithms that can cope with noise and uncertainty are required. We propose new, robust selection strategies for evolutionary multi-objective optimization in the presence of noise. We apply new measures of uncertainty for estimating the recently introduced Pareto-dominance for uncertain and noisy environments (PDU). The first measure is the inter-quartile range of the outcomes of repeated function evaluations. The second is based on axis-aligned bounding boxes around the upper and lower quantiles of the sampled fitness values in objective space. Experiments on real and artificial problems show promising results.

1 Introduction

Most real-world multi-objective optimization problems (MOPs) are inherently noisy because they rely on noisy measurements or require complex simulations that suffer from noisy conditions as well. Multi-objective evolutionary algorithms relying on indicator-based selection strategies have been successfully applied to various real-world as well as artificial MOPs. However, in their canonical form they do not work well when applied to MOPs with strong noise. By changing the selection scheme and sampling the objective functions for each candidate solution several times, the algorithms' robustness w.r.t. noise and uncertainty can be improved. Recently, an extension of the Pareto-dominance relation that addresses the problem of noisy objective function values has been presented in [1]. This extension originally relies on the convex hull of multiple objective function samples. We propose to consider axis-aligned bounding boxes and per-objective empirical quartiles instead.

In the following, we briefly introduce Pareto-dominance for uncertain and noisy environments (PDU) including our adaptation based on axis-aligned bounding boxes. Thereafter, we describe our empirical evaluation on real-world and artificial fitness functions before we finish with the conclusions and open questions.

R. Schaefer et al. (Eds.): PPSN XI, Part II, LNCS 6239, pp. 260–269, 2010.

2 A Robust Variant of Pareto Dominance

Let the noisy MOP (MNOP) be defined in a very general way without assumptions on the distribution of the objective function values nor on the error structure considered (e.g., the definition is not restricted to multiplicative or additive noise):

Definition 1 (MNOP)

$$\begin{array}{lll} \textit{Min.} & \boldsymbol{y} = (\boldsymbol{f}|\boldsymbol{x}, \boldsymbol{\varepsilon}) = (f_1, \ldots, f_k|\boldsymbol{x}, \boldsymbol{\varepsilon}) & \textit{(MNOP)} \\ \textit{with} & \boldsymbol{y} = (y_1, \ldots, y_k) \in \mathcal{Y}, & \textit{objectives} \\ & (f_i|\boldsymbol{x}, \varepsilon_i) \sim \mathcal{F}_i, \quad i = 1, \ldots, k, & \textit{objective functions} \\ & \mathcal{F}_i, \quad i = 1, \ldots, k, & \textit{distribution functions} \\ & \boldsymbol{x} = (x_1, \ldots, x_n) \in \mathcal{X}, & \textit{decision variables} \\ & \boldsymbol{\varepsilon} = (\varepsilon_1, \ldots, \varepsilon_k), & \textit{error terms.} \end{array}$$

We introduce an approach that handles the inherent uncertainty of the optimization problem by a specific variant of Pareto dominance relying on repeated evaluations of the decision vectors considered. By combining both expectation $E(f_i|\boldsymbol{x}, \varepsilon_i)$ as well as uncertainty $U(f_i|\boldsymbol{x}, \varepsilon_i)$ of the resulting objective realizations an alternative concept of Pareto-dominance (PDU) is set up [2,3]:

Definition 2 (Pareto-Dominance in Uncertain Environments (PDU)). *A solution $\boldsymbol{x}$ of MNOP dominates a solution $\boldsymbol{x}^*$ iff*

$$\begin{aligned} \exists i \in \{1, \ldots, k\} &: E(f_i|\boldsymbol{x}, \varepsilon_i) + U(f_i|\boldsymbol{x}, \varepsilon_i) < E(f_i|\boldsymbol{x}^*, \varepsilon_i) - U(f_i|\boldsymbol{x}^*, \varepsilon_i); \\ \forall j = (1, \ldots, k), j \neq i &: E(f_j|\boldsymbol{x}, \varepsilon_j) + U(f_j|\boldsymbol{x}, \varepsilon_j) \leq E(f_j|\boldsymbol{x}^*, \varepsilon_j) - U(f_j|\boldsymbol{x}^*, \varepsilon_j). \end{aligned}$$

The concept of PDU is based on an uncertainty zone around the expected value at a given solution. Dominance decisions are only made in case the reliability of the objective values is high enough in the sense that the uncertainty zones of two points to be compared do not overlap in any dimension (see Fig. 1). The definition of PDU above implicitly assumes symmetric noise distributions by estimating only a single $U(f_i|\boldsymbol{x}, \varepsilon_i)$ value per objective i at point $\boldsymbol{x}$.

Expectation. As the required expected value $E(f_i|\boldsymbol{x}, \varepsilon_i)$ cannot be observed directly, we estimate the location of the corresponding distribution based on a sample of m evaluations of the objective functions. To get a robust estimate, we consider the median (which is an unbiased estimator of the expectation if the expectation exists and the distribution is symmetric):

$$\begin{aligned} E(f_i|\boldsymbol{x}, \varepsilon_i) &\approx \mathrm{med}\,(f_{im}) \;\forall i \in \{1, ..., k\} \quad \text{with} \\ \mathrm{med}(\boldsymbol{f}_m) &= (\mathrm{med}(f_{1m}|\boldsymbol{x}, \varepsilon_1), \ldots, \mathrm{med}(f_{km}|\boldsymbol{x}, \varepsilon_k)), \\ (f_{im}|\boldsymbol{x}, \varepsilon_i) &= \left(f_i^1, \ldots, f_i^m|\boldsymbol{x}, \varepsilon_i\right), \quad i = 1, \ldots, k. \end{aligned}$$

Theoretically, a sample size of $m \to \infty$ is required. As in practice already moderate sample sizes are unrealistic, we suggest to use racing algorithms (see below) for dynamically adjusting the sample size depending on the amount of variability.

Uncertainty. There are several possibilities for choosing an indicator for the uncertainty of the sampled objective values for a given point in decision space. We will introduce and investigate three different approaches with increasing complexity.

1. One approach, probably the simplest one to handle noisy objectives, is to omit a special uncertainty measure and account for the variability solely by using an estimator for the expected value (i.e., $U(f_i|\boldsymbol{x}, \varepsilon_i) = 0$).
2. The inter-quartile range (IQR) of the objective realizations in each dimension suggests itself as an appropriate estimator for the inherent uncertainty. We refer to a non-parametric estimator because this is in line with using the median as the estimator for the expected value, that is

$$U(f_i|\boldsymbol{x}, \varepsilon_i) = q_{0.75}(f_{im}|\boldsymbol{x}, \varepsilon_i) - q_{0.25}(f_{im}|\boldsymbol{x}, \varepsilon_i) \ ,$$

   where q_α refers to the empirical α-quantile of $\mathcal{F}_i$.
3. Analogous to reverting to convex hulls in [2] we propose the usage of an axis-aligned bounding box (BB, [4]) based on the upper and lower quartiles of the sampled values in $\mathbb{R}^k$. Computing the box has a much lower computational complexity and has the same intuitive interpretation, namely, that with increasing volume of the bounding box the uncertainty of objectives increases simultaneously. The bounding box BBQ is defined as the cartesian product of the k intervals each of which is defined by the lower and upper quartile of the corresponding objective. The average of the distances of $\text{med}(\boldsymbol{f}_m)$ to the closest point on BBQ in each dimension ($\text{BBQ}_j^{\text{closest}}$) is taken as the required uncertainty measure reflecting a deviation to border points (see Fig. 1):

$$U(f_i|\boldsymbol{x}, \varepsilon_i) = \frac{1}{k} \sum_{j=1}^{k} \left| \text{med}(f_{im}) - \text{BBQ}_j^{\text{closest}} \right| \ .$$

Distribution of Function Evaluations. We apply concepts inspired by selection races [5,6,7] for dynamically controlling the sample size per individual such that the number of samples is as low as possible while the single objectives satisfy certain statistical requirements.

3 Preliminary Empirical Evaluation

This section presents the empirical evaluation of the different noise handling approaches on benchmark functions as well as on two real-life optimization problems. We considered the $(\mu + \mu)$-MO-CMA-ES (see [8,9]) and the NSGA-II (see [10]) in our experiments. All experiments have been conducted using the Shark Machine Learning Library [11]. In the following, we first outline the two real-world optimization problems from rapid manufacturing before we describe the experimental setup and the results.

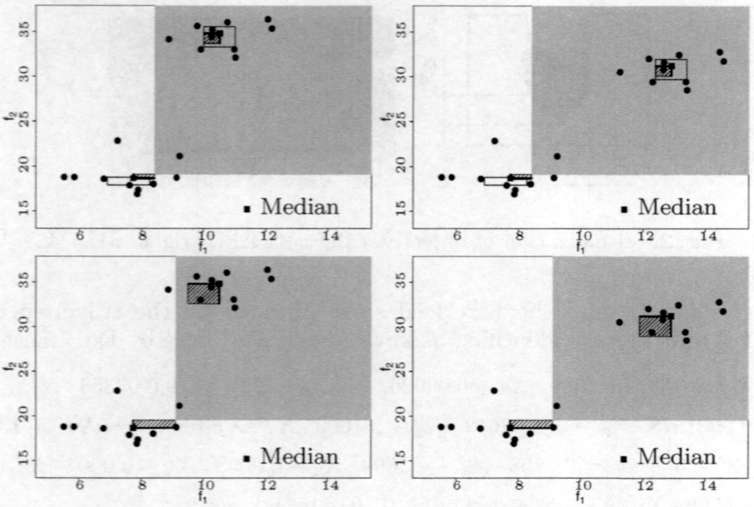

Fig. 1. Examples of PDU. Uncertainty regions are displayed by striped rectangles. The grey rectangle indicates the region dominated by the left solution. Black thick lines visualize distances from the median to the Quartile-BB in each dimension (top, approach 3) or represent the IQR (bottom, approach 2).

Rapid Manufacturing. Ready-to-use additive manufacturing (RUAM) combines welding and grinding by using a single machine while the optimization of the weld bead geometry is the crucial part of an efficient process. Structures are built in a layer-by-layer stepwise procedure which strongly relies on correct wall thickness and shape. Two different RUAM problems are addressed here.

RUAM1 [3]: Liratzis [12] identified depth of penetration P_d and sidewall penetration P_s to be key weld bead geometry quality factors being objectives to be minimized. Travel speed (T, [900,... 1400] mm/min), wire feed speed (W, [8.2,... 13.2] m/min), wire distance from side wall (D, [0.3,..., 1.2] mm) and arc length correction (A_c, [-25,..., 25] %) are the key decision variables of the process. Based on Design of Experiment (DoE) methods the following models were estimated based on coded influence factors in the interval $[0, 1]$ using the variable bounds listed above:

$$\begin{aligned} P_d &= 2.55 + 0.35 \cdot W - 0.23 \cdot T + 0.21 \cdot D + 0.18 \cdot A_c + 0.14 \cdot W \cdot A_c \\ &\quad - 0.25 \cdot T \cdot A_c - 0.073 \cdot A_c^2 + \varepsilon_1,\ \varepsilon_1 \sim \mathcal{N}(0, 0.1661^2), \\ P_s &= 0.60 + 0.038 \cdot W - 0.086 \cdot T - 0.074 \cdot D + 0.13 \cdot A_c + 0.025 \cdot W \cdot T \\ &\quad - 0.024 \cdot W \cdot D - 0.043 \cdot T \cdot D - 0.018 \cdot W^2 - 0.022 \cdot T^2 + \varepsilon_2,\ \varepsilon_2 \sim \mathcal{N}(0, 0.0331^2). \end{aligned}$$

RUAM2 [13]: Another aspect is the maximization of the height (H) and simultaneous target optimization of the width (W) of the weld bead while considering constraints regarding the contact angle (CA) and limitations of the welding machine. The diameter of wire (X_1, [0.8,..., 1.2] mm), wire feed speed (X_2,

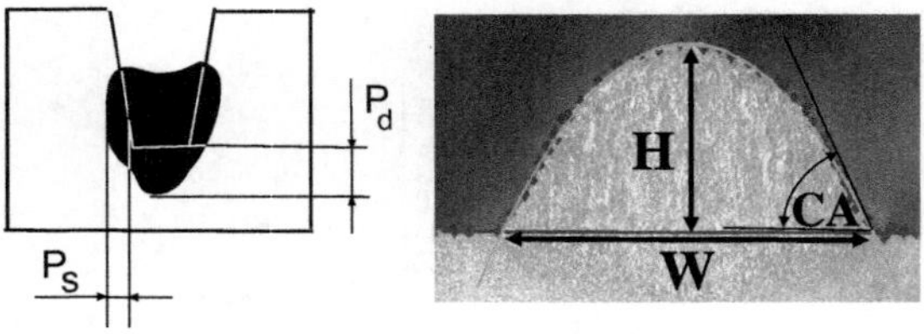

Fig. 2. Visualization of objectives (left: RUAM1, right: RUAM2)

[1.5,...,14] mm/s) and the wire feed speed divided by the travel speed (X_3, [10,...,20] mm/s) were identified as key decision variables by DoE methods:

$$
\begin{aligned}
H &= 3.948 \cdot X_1 - 0.5386 \cdot X_2 + 0.006262 \cdot X_3^2 - 1.123 \cdot X_1^3 + 0.03353 \cdot X_2^2 \\
&\quad - 0.0001578 \cdot X_3^3 - 0.0008612 \cdot X_2^3 + 0.1668 \cdot X_1 \cdot X_2 + \varepsilon_1, \varepsilon_1 \sim \mathcal{N}(0, 0.13^2) \ , \\
W^* &= |W - 5| + \varepsilon_2 = |2.5208 \cdot X_1 - 0.2699 \cdot X_2 + 0.0038 \cdot X_3^2 - 0.0004 \cdot X_2^3 \\
&\quad + 0.6766 \cdot X_1 \cdot X_2 - 5| + \varepsilon_2, \varepsilon_2 \sim \mathcal{N}(0, 0.6116^2) \quad \text{w.r.t.} \\
\text{CA} &= 101.58 \cdot X_2 + 971.94 \cdot X_1 - 74.49 \cdot X_3 - 251.21 \cdot X_2 \cdot X_1 - 0.78 \cdot X_2 \cdot X_3 \\
&\quad - 65.41 \cdot X_1 \cdot X_3 + 0.08 \cdot X_2^2 + 7.67 \cdot X_3^2 + 4.01 \cdot X_2^2 \cdot X_1 + 115.47 \cdot X_2 \cdot X_1^2 \\
&\quad + 0.036 \cdot X_2^2 \cdot X_3 + 28.50 \cdot X_1^2 \cdot X_3 + 0.27 \cdot X_1 \cdot X_3^2 - 0.13 \cdot X_2^3 - 306.33 \cdot X_1^3 \\
&\quad - 0.18 \cdot X_3^3 + 0.12 \cdot X_2 \cdot X_1 \cdot X_3 \leq 90 \ ; X_1 \leq 0.0043 \cdot X_2^2 - 0.1313 \cdot X_2 + 1.7876
\end{aligned}
$$

in addition to the box constraints of the decision variables.

Experimental Setup. We compare the algorithms deploying the respective combinations of noise handling methods on different classes of benchmark functions. We consider the bi-objective, constrained and non-separable WFG6 and WFG8 (see [14]). Moreover, the bi-objective, unconstrained and rotated benchmark function ELLI1 (see [8]) is used for the performance evaluation. We augment the assessment by empirically analyzing the performance of the algorithms on two optimization problems from the domain of rapid manufacturing. For all of the bi-objective functions, we considered different levels of additive, normally-distributed noise (see Table 1). For the selection races, we set δ to 0.0001 according to the suggestions given in [7] and limited the maximum number of repeated fitness function evaluations to $t_{\max} = 15$. The parent and offspring population sizes have been chosen as $\mu = \lambda = 100$. In case of the $(\mu + \mu)$-MO-CMA-ES, we adhered to the parameter setup presented in [9]. For the NSGA-II, we considered the default parameter setup (see [10]). We conducted 50 independent trials and aborted every trial after 500 generations.

Results. The performance of the different noise handling strategies is analyzed by the hypervolume (HV) and the Generalized Spread Indicator (SP) [10] over the course of generations, where the counting of the generations is done independently from the actual number of required fitness reevaluations.

Exemplary visualizations are given in Figs. 3 and 4 for the high noise levels. The objective function WFG8 is omitted as the results are very similar to WFG6.

Table 1. Empirical reference points for computing the unary hypervolume indicator and the per-objective noise levels

	Empirical reference point	Noise levels
WFG6	(2.6131, 5.47336)	$\sigma_{1,2} \in \{0.05, 0.1\}$
WFG8	(2.77644, 5.52929)	$\sigma_{1,2} \in \{0.05, 0.1\}$
ELLI1	(2.21839, 2.14453)	$\sigma_{1,2} \in \{0.1, 1\}$
RUAM1	(3.89213, 1.629089)	$\sigma_1 = 0.1661, \sigma_2 = 0.0331$
RUAM2	(1, 7.28359)	$\sigma_1 = 0.13, \sigma_2 = 0.6116$

As SP has to be minimized, the negative SP is plotted to be visually in line with the HV which has to be maximized.

The results of the algorithms without noise handling suggest that both of them are able to cope with noisy fitness functions to some degree without modifications. This may be attributed to the population-based approach as well as to the rank-based selection scheme employed in both algorithms. However, the performance of the algorithms without noise handling deteriorated quickly for low signal-to-noise ratios.

In all cases most of the variants of both the $(\mu + \mu)$-MO-CMA-ES as well as the NSGA-II employing the PDU performed better with respect to the final HV than the variants of the algorithms without noise-handling (No_NH). The same observations can be made for the lower noise levels showing slightly smaller effects. While the variant without uncertainty measurement (NUT) in some cases seems to outperform the more sophisticated noise handling approaches IQR and BBQUT with regard to HV, the analysis of SP shows that both IQR and BBQUT mostly manage to generate a better spread of solutions in the final front.

In addition, we analyzed the median age of the individuals in the final population. Exemplary results are picked up in Fig. 5. For WFG6, RUAM1 and RUAM2 the results of the NSGA-II show the same tendencies as for the MO-CMA-ES, and the WFG8 results are in line with WFG6. It becomes obvious that the median age of the individuals of both the algorithms without noise handling as well as the NUT variants is considerably higher than for IQR and BBQUT, especially for WFG6 and WFG8 with IQR being dominant for the RUAM. Thus, the approaches No_NH and NUT seem to run the risk of generating biased results with regard to randomly generated solutions which appear to be of high quality but in fact are not, for example they can be even better than the true Pareto front. The composition of the actual fronts lasts much longer than for IQR and BBQUT indicating partial stagnation.

Figure 5 as well lists the median number of nondominated solutions in the final fronts which decreases with increasing noise level. It is noticeable that the number of nondominated solutions for all problems but ELLI1 is much lower than the population size which reveals the increasing complexity of the problems due to the introduced noise.

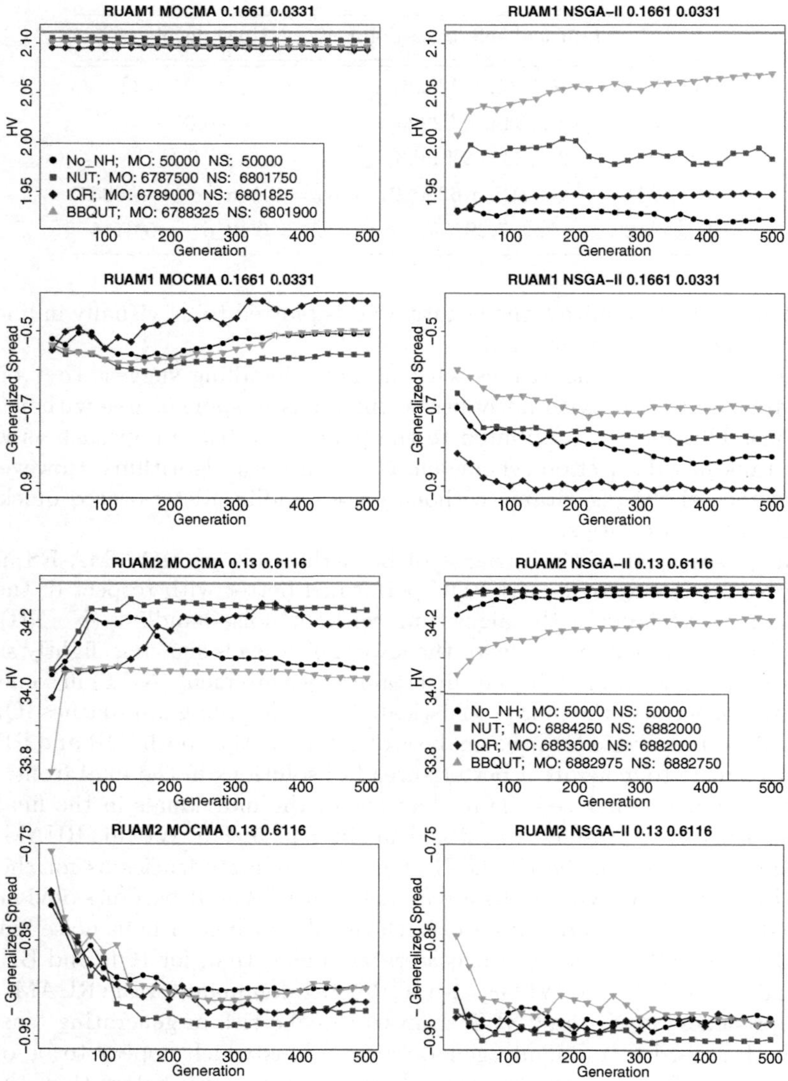

Fig. 3. Evolution of the absolute hypervolume for RUAM1 and RUAM2 over the number of generations. The upper line in the figures indicates the best results for empirically determined Pareto-approximations for the non-noisy case. For each of the noise handling approaches, the average number of fitness function evaluations spent by the $(\mu + \mu)$-MO-CMA-ES (indicated by MO) and the NSGA-II (indicated by NS) are reported.

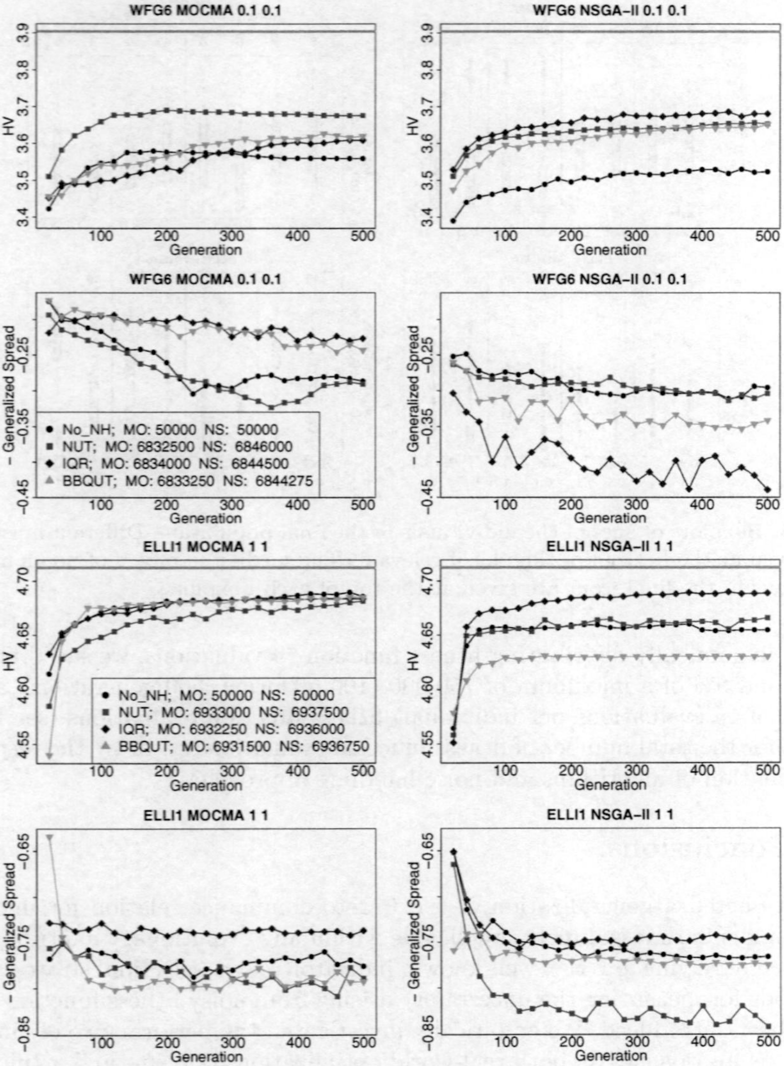

Fig. 4. Evolution of the absolute hypervolume for the fitness functions WFG6 and ELLI1 over the number of generations. All plots refer to the medians of 50 trials. For each of the noise handling approaches, the average number of fitness function evaluations spent by the $(\mu + \mu)$-MO-CMA-ES (indicated by MO) and the NSGA-II (indicated by NS) are given.

The number of non-dominated solutions in the final population is considerably lower than usually observed on standard, non-noisy MOPs. This shows the importance of the first level sorting criterion – here non-dominating sorting using different notions of dominance – for the selection process in the presence of noise and uncertainty.

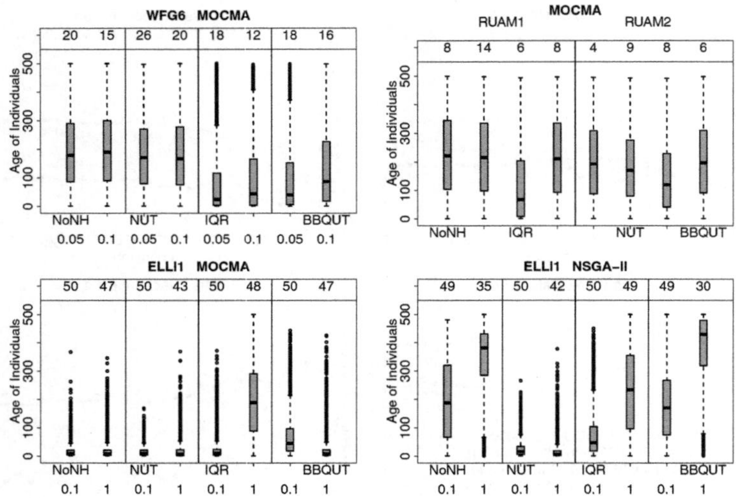

Fig. 5. Boxplots of ages of the individuals in the final population. Different noise levels are given at the bottom of the plot if relevant. The median numbers of nondominated solutions in the final front are given at the top of each boxplot.

By dynamically distributing fitness function reevaluations, we saved between 7.5% and 9% of a maximum of 750,000 (100 offspring, 500 generations, a maximum of 15 evaluations per individual) fitness function evaluations (see Figs. 3 and 4 for the total number of fitness function evaluations spent by the respective combination of algorithms and noise handling approaches).

4 Conclusions

We presented a generalization of the Pareto-dominance relation for uncertain environments that is directly applicable within any evolutionary multi-objective algorithm relying on the well-known indicator-based selection strategy. Two methods for measuring the uncertainty arising from noisy fitness function values have been introduced. We empirically investigated the performance of our noise handling approaches on both real-world optimization problems and artificial fitness functions. Our results suggest that the methods presented here improve the overall robustness of multi-objective evolutionary algorithms relying on the indicator-based selection scheme and thus, enhance the algorithms' performance in the presence of noise.

Nevertheless, the noise handling framework presented here allows for future improvements by considering more sophisticated schemes for measuring the uncertainty. Future work should include an in-depth evaluation on a broad range of fitness functions and consider different noise distributions. Additionally, we will study the integration of selection races and PDU. Further, we will consider non-elitist evolutionary multi-objective algorithms, which may be more appropriate for noisy problems (e.g., see [15]).

References

1. Trautmann, H., Mehnen, J.: Preference-Based Pareto-Optimization in Certain and Noisy Environments. Engineering Optimization 41, 23–38 (2009)
2. Trautmann, H., Mehnen, J., Naujoks, B.: Pareto-dominance in noisy environments. In: Tyrrell, A. (ed.) 2009 IEEE Congress on Evolutionary Computation, pp. 3119–3126. IEEE Press, Los Alamitos (2009)
3. Mehnen, J., Trautmann, H.: Robust Multi-objective Optimisation of Weld Bead Geometry for Additive Manufacturing. In: Teti, R. (ed.) Proceedings of the 6th CIRP International Seminar on Intelligent Computation in Manufacturing Engineering, CIRP ICME 2008 (2008)
4. de Berg, M., Cheong, O., van Kreveld, M., Overmars, M.: Computational Geometry, 3rd edn. Springer, Heidelberg (2008)
5. Maron, O., Moore, A.W.: The racing algorithm: Model selection for lazy learners. Artificial Intelligence Review 11, 193–225 (1997)
6. Audibert, J.Y., Munos, R., Szepesvári, C.: Tuning bandit algorithms in stochastic environments. In: Hutter, M., Servedio, R.A., Takimoto, E. (eds.) ALT 2007. LNCS (LNAI), vol. 4754, pp. 150–165. Springer, Heidelberg (2007)
7. Heidrich-Meisner, V., Igel, C.: Hoeffding and bernstein races for selecting policies in evolutionary direct policy search. In: Bottou, L., Littman, M. (eds.) Proceedings of the International Conference on Machine Learning (ICML 2009), pp. 401–408 (2009)
8. Igel, C., Hansen, N., Roth, S.: Covariance matrix adaptation for multi-objective optimization. Evolutionary Computation 15(1), 1–28 (2006)
9. Voß, T., Hansen, N., Igel, C.: Improved step size adaptation for the MO-CMA-ES. In: Proc. 12th Annual Conference on Genetic and Evolutionary Computation Conference (GECCO). ACM Press, New York (2010)
10. Deb, K., Pratap, A., Agarwal, S., Meyarivan, T.: A fast and elitist multiobjective genetic algorithm: NSGA-II. IEEE Transactions on Evolutionary Computation 6, 182–197 (2002)
11. Igel, C., Glasmachers, T., Heidrich-Meisner, V.: Shark. Journal of Machine Learning Research 9, 993–996 (2008)
12. Liratzis, T.: Tandem Gas Metal Arc Pipeline Welding. PhD thesis, Cranfield University, UK (2007)
13. Wessing, S., Ding, J., Trautmann, H., Mehnen, J., Naujoks, B.: Sequential Parameter Optimization for Multi-Objective Evolutionary Optimization of Additive Layer Manufacturing. In: Teti, R. (ed.) Proceedings of the 7th CIRP International Seminar on Intelligent Computation in Manufacturing Engineering, CIRP ICME 2010 (2010) (accepted)
14. Huband, S., Hingston, P., Barone, L., While, L.: A review of multi-objective test problems and a scalable test problem toolkit. IEEE Transactions on Evolutionary Computation 10, 477–506 (2007)
15. Büche, D., Stoll, P., Dornberger, R., Koumoutsakos, P.: Multiobjective evolutionary algorithm for the optimization of noisy combustion processes. IEEE Transactions on Systems, Man, and Cybernetics 32, 460–473 (2002)

Evolving a Single Scalable Controller for an Octopus Arm with a Variable Number of Segments

Brian G. Woolley and Kenneth O. Stanley

School of Electrical Engineering and Computer Science
University of Central Florida, Orlando, FL 32816
brian.woolley@ieee.org, kstanley@eecs.ucf.edu

Abstract. While traditional approaches to machine learning are sensitive to high-dimensional state and action spaces, this paper demonstrates how an indirectly encoded neurocontroller for a simulated octopus arm leverages regularities and domain geometry to capture underlying motion principles and sidestep the superficial trap of dimensionality. In particular, controllers are evolved for arms with 8, 10, 12, 14, and 16 segments in equivalent time. Furthermore, when transferred *without further training*, solutions evolved on smaller arms retain the fundamental motion model because they simply extend the general kinematic concepts discovered at the original size. Thus this work demonstrates that dimensionality can be a false measure of domain complexity and that indirect encoding makes it possible to shift the focus to the underlying conceptual challenge.

1 Introduction

Whether tackled through neuroevolution or temporal difference-based approaches, in reinforcement learning problems, the number of dimensions in the state and action space is often associated with problem difficulty [6,11,18,19]. Yet the complexity of problems should *not* be determined by the dimensionality of such representations, which are a superficial proxy for the underlying *conceptual* problem. Instead, the problem complexity should correlate to the underlying complexity of the *principle* to be discovered. The argument in this paper is that indirect encoding, which means describing the solution as a *pattern* through a compressed representation [1,9,14,16], is the essential ingredient that will allow reinforcement learning to transcend the superficial aspects of problem dimensionality.

To make this point, the problem domain in this paper is an octopus arm, which is approximated as a structure of interconnected muscles that must act together to create a coordinated behavior. Thus it induces a high-dimensional state space *and* action space (i.e. because each muscle in each segment can be articulated independently). In fact, the high dimensionality of the 10-segment arm provoked previous researchers to dramatically prune the action space by allowing only a small discrete set of pre-coordinated actions [5].

The octopus arm problem is thus an ideal departure for a study on the ability of indirect encoding to transcend such dimensionality. After all, the underlying kinematic control principle is similar regardless of the precise number of segments, muscles and

R. Schaefer et al. (Eds.): PPSN XI, Part II, LNCS 6239, pp. 270–279, 2010.

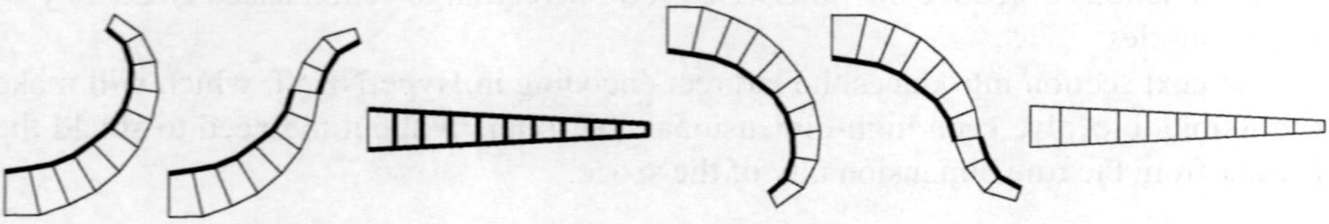

(a) (b) (c) (d) (e) (f)

Fig. 1. Octopus arm actions. While there are theoretically many other combinations of muscle contractions possible, Engel et al. [5] limited their model to these six to make the domain tractable. Line thickness indicates the strength of the contractive force applied.

sensory inputs, suggesting that an approach that is sensitive to its particular dimensionality is missing something fundamental. Thus, to demonstrate this point in this paper, an indirect encoding called *hypercube-based neuroevolution of augmenting topologies* (HyperNEAT) evolves a *description* of how the weights of a neurocontroller relate to each other across the domain geometry irrespective of the arm's precise physical dimensionality [3,8,14]. This approach means that the HyperNEAT controller can actually learn to articulate all the muscles independently without the need to partition the action space up front. Furthermore, as should be the case in learning such problems, the indirect encoding learns equivalently across arms with a variable number of segments. Finally, neurocontrollers trained on arms with eight segments are scaled to a larger number of segments *without further training* and still work because they encode general control principles for the arm.

Thus the major contribution of this work is to demonstrate that indirect encoding is a potentially critical ingredient in reinforcement learning and control problems if they are to focus on the true problem complexity rather than the superficial dimensionality of the state or action space.

2 Background

This section reviews prior work in training multi-segment arms and in indirect encoding of neural networks.

2.1 Reinforcement Learning for Arm Controllers

An interesting study that provides inspiration for this paper demonstrated the ability of Gaussian Process Temporal Differencing (GPTD) [4] to learn value functions in a high-dimensional domain [5]. In it, a control policy is trained for an arm with many degrees-of-freedom (i.e. the octopus arm [20]). GPTD produced value functions for motion trajectories that touch targets at unknown locations within 20 trials.

The details of the original octopus arm experiment are interesting because they set a new standard for high-dimensional control that this paper pushes even further. The 10-segment arm had a state space with 88 dimensions (i.e. position and velocity for each vertex) that map to the six discrete actions shown in figure 1. Engel et al. [5] chose these six actions to reduce the otherwise prohibitively large action space created by so many muscles.

The next section introduces the indirect encoding in HyperNEAT, which will make it possible to evolve such high-dimensional controllers without the need to shield the learner from the true dimensionality of the space.

2.2 Indirect Encoding and HyperNEAT

Neuroevolution, i.e. evolving ANNs, can produce solutions for a broad array of control tasks [6,15,17,19]. Many such methods are based on *direct encodings*, which means each piece of structure in the phenotype is encoded by a single gene, making the discovery of repeating motifs expensive and improbable. Therefore, indirect encodings [1,9,14,16] have become a growing area of interest in evolutionary computation.

One such indirect encoding designed explicitly for neural networks is the Hypercube-based NeuroEvolution of Augmenting Topologies (HyperNEAT) approach [14], which is an indirect extension of the directly-encoded NEAT approach [15,17]. Rather than expressing link weights as distinct and independent parameters in the genome, HyperNEAT allows them to vary across the phenotype in a regular pattern through an encoding called a *compositional pattern producing network* (CPPN) [13].

The idea behind CPPNs is that geometric patterns can be encoded by a *composition of functions* that are chosen to represent common regularities. For example, the Gaussian function is symmetric, so when it is composed with any other function alone, the result is a symmetric pattern. The internal structure of a CPPN is a weighted network, similar to an ANN, that denotes which functions are composed and in what order, which means that instead of evolving ANNs as it normally does, NEAT [15,17] can evolve *CPPNs* that generate connectivity patterns across an ANN.

Formally, CPPNs are *functions* of geometry (i.e. locations in space) that output connectivity patterns whose nodes are situated in n dimensions, where n is the number of dimensions in a Cartesian space. Consider a CPPN that takes four inputs labeled x_1, y_1, x_2 and y_2; this point in four-dimensional space also denotes the connection between the two-dimensional points (x_1, y_1) and (x_2, y_2). The output of the CPPN for that input thereby represents the weight of that connection (figure 2). By querying every pair of points in the space, the CPPN can produce an ANN, wherein each queried point is a neuron position. While CPPNs are themselves networks, the distinction in terminology between CPPN and ANN is important for explicative purposes because in HyperNEAT, CPPNs *encode* ANNs. Because the connection weights are produced as a function of their endpoints, the final structure is produced with *knowledge* of the domain geometry, which is literally depicted geometrically within the constellation of nodes.

As a rule of thumb, nodes are placed in a geometric space called the *substrate* to reflect the geometry of the domain (i.e. the state) [2,8,14]. For example, a visual field can be laid out in two dimensions such that nodes that receive input from adjacent locations in the image are literally adjacent in the network geometry. This way, knowledge of the domain geometry is preserved and exploited by HyperNEAT where regularities (e.g. adjacency, or symmetry, which the CPPN sees) are invisible to traditional encodings. This capability is exploited in the octopus arm substrate introduced next.

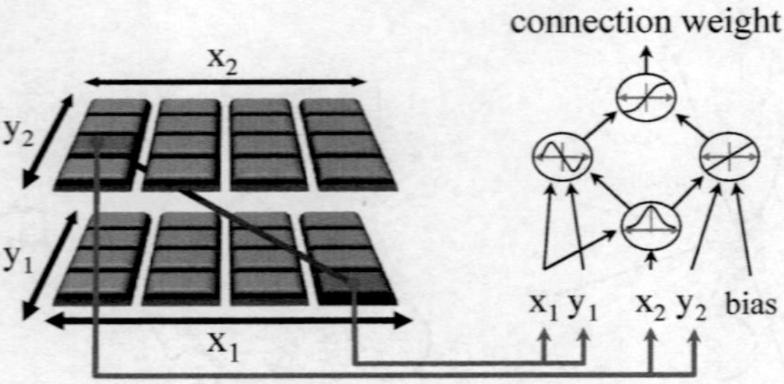

Fig. 2. Encoding the connectivity pattern. The four-dimensional CPPN encodes the connectivity pattern of the substrate through an evolved network of geometric functions. The substrate ANN is generated by querying the CPPN for the value of each potential connection from (x_1, y_1) to (x_2, y_2). In this way, CPPNs capture patterns and regularities in domain geometry.

3 Scalable Neurocontroller for an Octopus Arm

The octopus arm domain formalized in this section is particularly challenging because its state space *and* action space are both high-dimensional. Although Engel et al. [5] reduced dimensionally by choosing only six canonical actions, the aim in this paper is to learn from the full unprocessed action space. Furthermore, unlike any system before, the learned controller will be asked to scale to even larger arms *without further learning*.

The simulation domain in this paper, based on Yekutieli et al. [20], models the kinematics and dynamics of a two-dimensional *muscular hydrostat* (which is the mechanism of the octopus arm [12]) as a chain of quadrilateral polygons with fixed area connected to a fixed base. The model constructs arms based on length (l), width (w), taper (t), mass (m), and number of segments (n). At the vertex of each quadrilateral is a point mass shared by adjacent segments. The *dorsal*, or upper, and *ventral*, or lower, edges of each segment represent longitudinal muscles while the vertical edges between sections represent transverse muscles. The muscles, modeled as spring-joints, are contracted by increasing the spring constant and relaxed by reducing the spring constant.

The fixed size and incompressible nature of the arm are the key features that enable the dynamic motion of the muscular hydrostat. These attributes are modeled by adjusting each segment's internal pressure: as external forces act to compress a segment, pressure increases; conversely, as forces stretch and expand the segment, internal pressure decreases. Thus segments change shape to restore the equilibrium between surface tension and internal pressure. Figure 1 shows the six basic actions utilized in Engel et al. [5] as examples of the model's motion effects. However, in this paper, the ANN will have independent control of all $3n$ muscles in the arm, creating a high-dimensional action space.

Because experiments in this paper involve moving towards a perceived object (unlike Engel et al. [5]), the arm state is defined by sensor inputs that allow the controller to infer the position of each segment relative to the target. Range sensors along the arm provide cues about target position. Each sensor at each segment produces 36 radial distance measurements across the range $[-\pi \ldots \pi]$ (figure 3a), allowing the target to be seen

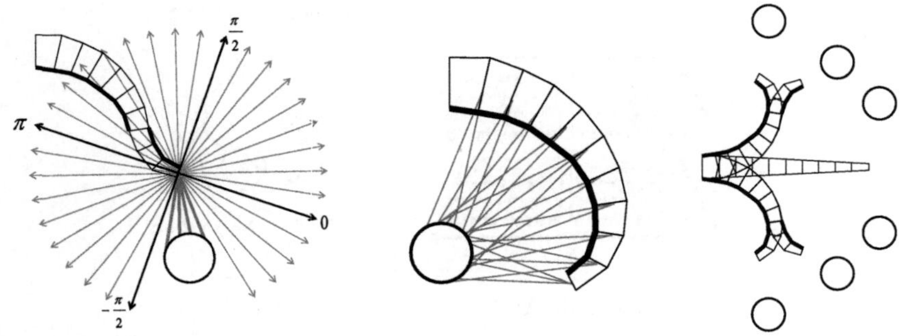

Fig. 3. Target perception and placement. The arm controller perceives the target through range sensors (a) placed at each segment along the arm; the combined effect is shown (b) with non-detecting beams removed for clarity. The training targets in this paper (c) are positioned beyond the reach of the simple actions in figure 1.

by multiple beams simultaneously, especially as the sensor approaches the target or as sensor resolution is increased. Thus the $36n$ total beams also create a high-dimensional input space. Figure 3b illustrates the arm's view of the target with the non-detecting beams removed for clarity.

3.1 Substrate Architecture

The octopus arm substrate (figure 4) closely couples sensing to acting. The input layer accepts sensor data directly and the output layer provides the contractive response for each muscle. Finally, a hidden layer is provided to support nonlinear operation required by the gravity and buoyancy effects acting on the arm.

To represent the sensor array described above, the controller must interpret 36 rangefinder inputs per segment. The arm model is composed of segments that have a necessary order and relationship to the other segments in the arm, i.e. segment 1 connects to segment 2, segment 2 connects to segment 3, etc. Thus, the perception layer is constructed as a two-dimensional sheet with θ as one axis and the arm's proximal-distal (PD) geometry as the other (figure 4, layer A).

To represent the action space, the substrate provides an output for each of the $3n$ muscles in the arm. To take advantage of HyperNEAT's ability to leverage domain geometry, the proximal-distal axis of the sensor layer is mirrored by the output (contractive) layer. Furthermore, note in figure 1 how the dorsal and ventral muscles act together to form coordinated reaching behaviors. This configuration suggests aligning the dorsal, transverse, and ventral muscles along the proximal-distal axis (figure 4, layer C).

By viewing the substrate architecture as a feedforward network spanning from the sensor input layer (A), to the hidden layer (B), to the contractive output layer (C), a CPPN with inputs (x_1, y_1, x_2, y_2) and outputs (AB, BC) provides a complete encoding of the phenotype. Figure 4a illustrates how each of the 7,488 link weights in an eight-segment controller (left) are set by a single CPPN.

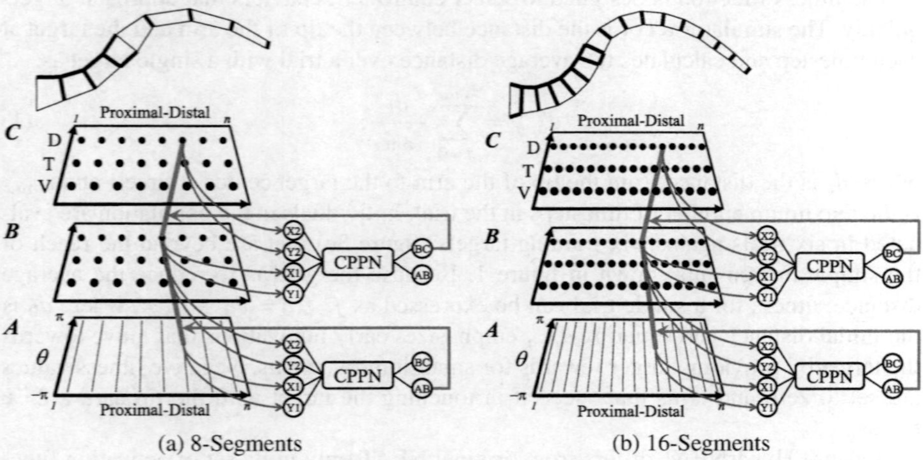

Fig. 4. Scalable neurocontroller architecture. The eight-segment substrate (a) can be scaled to a 16-segment substrate (b) without further training. The substrate at any resolution contains a two-dimensional input layer A that corresponds to the arm's sensory input, a two-dimensional hidden layer B, and a two-dimensional output layer C that controls the musculature. To query connections between A and B, the proximal-distal (PD) axis is the x_1 input and θ is the y_1 input. To query connections between B and C, the x_1 input is also PD, and DTV (Dorsal-Transverse-Ventral) is the y_2 input. Because the CPPN encodes kinematic principles, resampling with the node positions in (b) can produce a similar contractive pattern and arm pose (shown above).

The hope is that the principle that underlies moving a hydrostat is regular across the segments of the arm and therefore can be captured by the CPPN.

3.2 Scaling

By constructing the substrate to reflect the domain geometry (figure 4a), larger arm controllers are generated without further evolution by *requerying the same CPPN* at higher-resolutions (figure 4b). This approach works because adding segments to the arm is analogous to increasing the resolution of the hydrostat model. The CPPN provides a nonlinear interpolation of the behavior policy for each of the $3n$ muscles.

4 Experiment

The first aim of the experiment is to investigate the ability of indirect encoding to facilitate learning to control a hydrostat with dozens of degrees of freedom that are not a priori restricted or pruned in any way. The second aim is to test the ability of an evolved CPPN to generate controllers for higher resolution arms without further training. Both tests can validate that HyperNEAT learns *general principles* of hydrostat control rather than a single solution at a particular dimensionality.

The fitness function is designed to select controller behaviors that approach targets quickly. The simulator records the distance between the tip of the arm and the target at each timestep and calculates the average distance over a trial with a single target as:

$$d_{avg} = \sum_{t=0}^{t_{max}} \frac{d_t}{t_{max}}, \tag{1}$$

where d_t is the distance from the tip of the arm to the target center at time t and t_{max} is the maximum number of timesteps in the trial. Individuals in the population are evaluated in six trials against six training targets (figure 3c) that are beyond the reach of the simple movements shown in figure 1. Because the goal is to reduce the average distance, fitness for a single trial can be expressed as $f_{trial} = d_0 - d_{avg}^2$, where d_0 is the initial distance and squaring d_{avg} emphasizes early innovations that move towards the target by providing larger rewards for small improvements. Negative fitness values are set to zero and arms that succeed in touching the target with the tip earn a 25% bonus.

Because HyperNEAT differs from original NEAT only in its set of activation functions, it uses the same parameters [15]. All experiments were run with a version of the public domain ANJI package [10] augmented to implement HyperNEAT. The population size was 100 and each run lasted 500 generations. The speciation threshold, δ_t, was 0.2 and the compatibility modifier was 0.3. Available CPPN activation functions were sigmoid, Gaussian, sine, and linear functions. Recurrent connections within the CPPN were not enabled. Signed activation was enforced in the CPPN, but the substrate was unsigned, resulting in a node output range of $[-1, 1]$. By convention, a connection was not expressed if the magnitude of its weight is below a minimal threshold of 0.2 [7]. These parameters were found to be robust to moderate variation.

To validate that eight-segment solutions can scale, their evolved CPPNs are requeried to generate controllers for arms with 10, 12, 14, 16, 18, and 20 segments with no further training. It is important to note that these dimensionalities are indeed high because they impact the necessary dimensionality of the corresponding *neurocontroller*, (i.e. an eight-segment controller must set 7,488 connection weights while a 20-segment controller must set 46,800). Also, results cannot be compared directly to controllers trained by Engel et al. [5] because they seek a single target blindly while those in this paper can actively touch targets at multiple locations based on sensory inputs.

5 Results

Figure 5 shows training performance over generations when controllers are separately evolved (i.e. not scaled) for arms with 8, 10, 12, 14, and 16 segments. 20 runs were completed at each resolution. Remarkably, the number of degrees-of-freedom has *no significant effect* on the training curve, suggesting that indirect encoding really is making it possible to focus on learning the underlying *principle* independently of dimensionality.

Across all variants, CPPNs with an average of only 10.1 connections (stdev $= 2.3$) encode substrates with between 7,488 (8 segments) and 29,952 connections (16 segments), demonstrating the considerable compression of the indirect encoding.

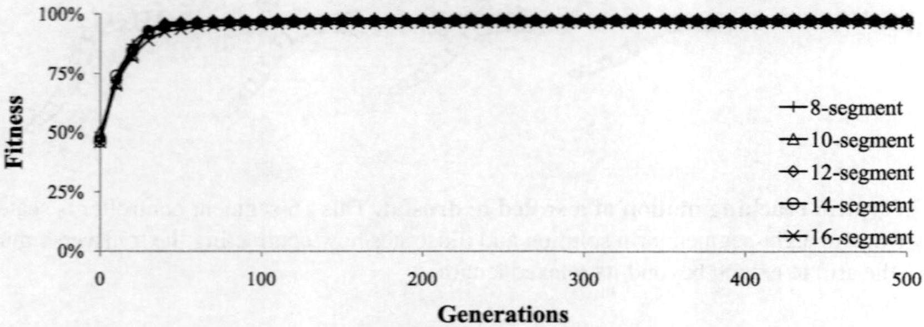

Fig. 5. Training at different arm resolutions. HyperNEAT evolves neurocontrollers for arms with 8, 10, 12, 14 and 16 segments in equivalent time because the CPPN discovers the underlying kinematic patterns. Measurements are averaged over 20 runs.

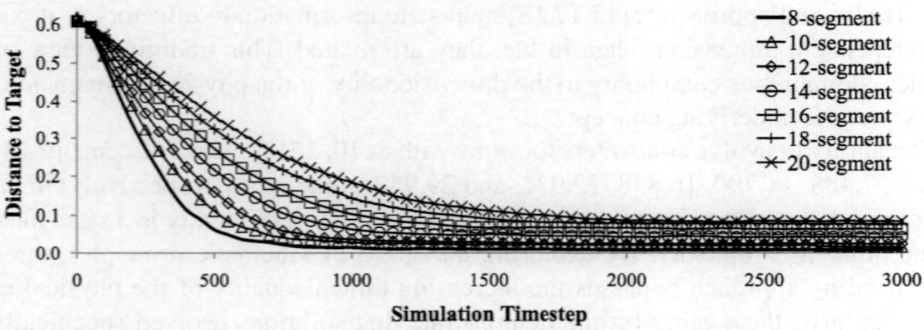

Fig. 6. Scaling solutions to larger arm controllers. The distance from the target surface at each time-step is shown, demonstrating that the ability to move towards the target is preserved as controllers are scaled to support arms with additional segments. Measurements are averaged over 20 runs.

The main scaling result (figure 6) is that the evolved contractive patterns transfer well from controllers trained on eight segments to arms with an increasing number of segments with no additional training. In the figure, the distance from the arm tip to the target surface is graphed over timesteps, demonstrating that controllers maintain the ability to approach targets as the physical structure scales; on average, even the 20-segment (worst) case approaches within 0.084 ($\pm$0.05 at 95% confidence) units of the target surface. It is important to note that the qualitative behavior of the arm at all scales in figure 6 is the same (i.e. they all still approach the target) although the speed of movement slows gradually and emergent physical characteristics begin to render the original solution less effective.

The sequence shown in figure 7 demonstrates a typical scaled reaching behavior. The contractive pattern shown was evolved as an eight-segment arm and applied to a 16-segment arm with no further training. Videos of evolved arms and scaling are available at `http://eplex.cs.ucf.edu/octopusArm`.

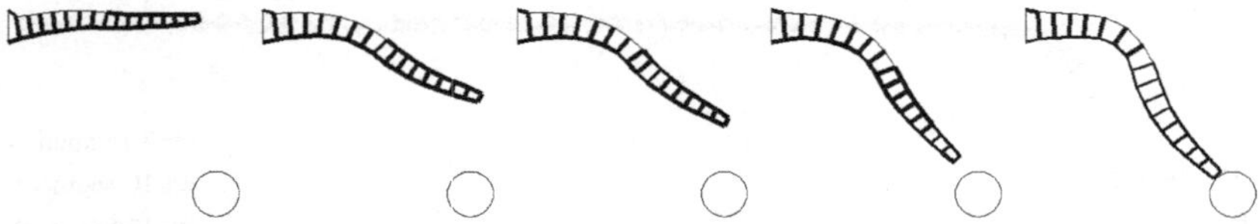

Fig. 7. Typical reaching motion of a scaled hydrostat. This 16-segment controller is scaled directly from an eight-segment arm solution and illustrates how contracting the transverse muscles allows the arm to extend beyond its relaxed length.

6 Discussion and Future Work

Finding solutions to problems in control should be about discovering an underlying principle and not about the number of dimensions in the action or state representation. Traditional approaches [15,17,18] map state information to effectors as if each is an independent dimension when in fact they are related. This traditional view of the problem domain ties complexity to the dimensionality of the physical domain and thus obfuscates the underlying concept.

The ability to evolve controllers for arms with 8, 10, 12, 14, and 16 segments (which contain 7,488, 11,700, 16,848, 22,932, and 29,952 connections, respectively) in equivalent time demonstrates that this physical structure's dimensionality is a false measure of the domain complexity. By exploring the space of kinematic principles, the indirect encoding approach bypasses the increasing dimensionality of the physical structure. Similarly, the scaling results demonstrate that solutions evolved specifically for the eight-segment arm model encompass fundamental kinematic strategies that apply directly to arms with additional segments.

Thus indirect encoding becomes an important consideration for any problem in which state or action dimensionality may be misleading, or for learning scalable control policies. Whether it is a multi-segment arm, a robot hand that can add more fingers, or a centipede with a variable number of legs, indirect encoding shifts the problem away from the precise configuration towards the underlying principle, thereby opening up such problems to machine learning.

7 Conclusions

For many problems, complexity is independent of the number of dimensions. The challenge is to transcend the distraction of superficial dimensionality by preserving meaningful relationships, e.g. geometric principles like order, orientation, and proximity. The octopus arm model is a good platform to test this idea because it can include an increasing number of segments. By discovering an underlying kinematic pattern, the HyperNEAT approach is able to sidestep the increasing dimensionality of the physical structure. Experimental results demonstrate that this approach yields controllers for arms with 8, 10, 12, 14, and 16 segments in equivalent time and that evolved solutions can provide controllers for arms with up to twice as many segments without further training. Thus this paper provides a lesson on the important role of indirect encoding in reinforcement learning.

References

1. Bongard, J.C., Pfeifer, R.: Evolving complete agents using artificial ontogeny. In: Morpho-functional Machines: The New Species (Designing Embodied Intelligence), pp. 237–258. Springer, Heidelberg (2003)
2. Clune, J., Beckmann, B.E., Ofria, C., Pennock, R.T.: Evolving coordinated quadruped gaits with the hyperneat generative encoding. In: Proceedings of the IEEE Congress on Evolutionary Computing Special Section on Evolutionary Robotics. IEEE Press, Los Alamitos (May 2009)
3. D'Ambrosio, D.B., Stanley, K.O.: A novel generative encoding for exploiting neural network sensor and output geometry. In: Proceedings of the 9th annual conference on Genetic and Evolutionary Computation, pp. 974–981. ACM, New York (2007)
4. Engel, Y., Mannor, S., Meir, R.: Bayes meets Bellman: The Gaussian process approach to temporal difference learning. In: Proceedings of the 20th International Conference on Machine Learning, vol. 20, pp. 154–161. AAAI Press, Menlo Park (2003)
5. Engel, Y., Szabo, P., Volkinshtein, D.: Learning to control an octopus arm with gaussian process temporal difference methods. Advances in Neural Information Processing Systems 18, 347–354 (2006)
6. Floreano, D., Dürr, P., Mattiussi, C.: Neuroevolution: from architectures to learning. Evolutionary Intelligence 1(1), 47–62 (2008)
7. Gauci, J., Stanley, K.O.: Generating large-scale neural networks through discovering geometric regularities. In: Proceedings of the 9th Annual Conference on Genetic and Evolutionary Computation, pp. 997–1004. ACM, New York (2007)
8. Gauci, J., Stanley, K.O.: Autonomous evolution of topographic regularities in artificial neural networks. Neural Computation, 38 (2010) (to appear)
9. Hornby, G.S., Pollack, J.B.: Creating high-level components with a generative representation for body-brain evolution. Artificial Life 8(3), 223–246 (2002)
10. James, D., Tucker, P.: ANJI homepage (2004), `http://anji.sourceforge.net/`
11. Kaelbling, L.P., Littman, M.L., Moore, A.W.: Reinforcement learning: A survey. Journal of Artificial Intelligence Research 4(1), 102–138 (1996)
12. Kier, W.M., Smith, K.K.: Tongues, tentacles and trunks: the biomechanics of movement in muscular-hydrostats. Zoological Journal of the Linnean Society 83, 307–324 (1985)
13. Stanley, K.O.: Compositional pattern producing networks: A novel abstraction of development. Genetic Programming and Evolvable Machines 8(2), 131–162 (2007)
14. Stanley, K.O., D'Ambrosio, D.B., Gauci, J.: A hypercube-based indirect encoding for evolving large-scale neural networks. Artificial Life 15(2) (2009)
15. Stanley, K.O., Miikkulainen, R.: Evolving neural networks through augmenting topologies. Evolutionary Computation 10(2), 99–127 (2002)
16. Stanley, K.O., Miikkulainen, R.: A taxonomy for artificial embryogeny. Artificial Life 9(2), 93–130 (2003)
17. Stanley, K.O., Miikkulainen, R.: Competitive coevolution through evolutionary complexification. Journal of Artificial Intelligence Research 21(1), 63–100 (2004)
18. Sutton, R.S., Barto, A.G.: Reinforcement Learning: an introduction. MIT Press, Cambridge (1998)
19. Yao, X.: Evolving artificial neural networks. Proc. of the IEEE 87(9), 1423–1447 (1999)
20. Yekutieli, Y., Sagiv-Zohar, R., Aharonov, R., Engel, Y., Hochner, B., Flash, T.: Dynamic model of the octopus arm. I. Biomechanics of the octopus reaching movement. Journal of Neurophysiology 94(2), 1443–1506 (2005)

An Island Model for the No-Wait Flow Shop Scheduling Problem

Istvan Borgulya

University of Pecs, Faculty of Business and Economics, Hungary

Abstract. In this paper we present an evolutionary algorithm (EA) for the no-wait flow shop scheduling problem. This is a new island model with special master-slave structure. In all islands runs a hybrid steady-state algorithm that uses truncation selection, uses only mutation for the generation of the descendants and it improves the solutions with two local search procedures. The mutation is based on the new, modified version of "the EVL method", that is a memory based method. The algorithm was tested on benchmark problems; its results are similar to or better than the results of the particle swarm optimization algorithms.

Keywords: Scheduling, island model, evolutionary algorithm.

1 Introduction

In the no-wait flow shop scheduling problem (NWFSSP), the operation of each job has to be processed without interruptions between consecutive machines, i.e., when necessary, the start of a job on a given machine must be delayed so that the completion of the operation coincides with the beginning of the operation on the following machine. Two of the most commonly studied criteria in no-wait flow shops are flow time and makespan. In this paper our goal is the makespan.

Some typical applications of this problem are encountered in manufacturing systems, chemical and pharmaceutical processing, metal processing, steel factories, plastic processing and hot rolling industries (e.g. [4], [7]).

The NWFSSP with more than two machines is NP-hard [8], so exact and heuristic algorithms have been introduced. For small problems we can use the branch-and-bound algorithm (e.g. [3]). For larger problems the two most common heuristic groups are the construction methods and the metaheuristics. The construction methods build a feasible solution, i.e. a sequence of n jobs, by successively completing a (partial) solution according to some rule (e.g. [12], [5], [1]). The metaheuristics improve the solutions based on the local search paradigm. We find metaheuristics for the makespan and for the flow time criterions too. Heuristics for the makespan criterion are e.g. the tabu search (TS) (e.g. [6]), simulated annealing (SA) (e.g. [1]). The most popular metaheuristics are the EA versions: genetic algorithm (GA) (e.g. [1]), ant colony optimization (ACO) (e.g. [14]) various hybrids with SA ([1], [15]) and variable neighborhood search (VNS) ([13]). The novel EA technique, named particle swarm optimization (PSO), has been proposed too (e.g. [11], ([7]).

R. Schaefer et al. (Eds.): PPSN XI, Part II, LNCS 6239, pp. 280–289, 2010.

We wish to enrich the field of these heuristic methods with a new EA. To the best of our knowledge, there is no published work addressing the NWFSSP using island model, so we developed a new island model. With this model:

- we want to improve the results of the heuristics for NWFSSP,
- we want to include a better version of a memory based method, called the extended virtual loser (EVL) ([2]).

So in our new algorithm:

- we developed a new island model and
- we prepared a new modified version of the EVL method for the NWFSSP.

The paper is organized in the following way: section 2 describes the problem formulation; section 3 shows the structure of the new island model. Section 4 describes some details of the algorithm: the local search methods used, the modified EVL method with the mutation operator, and the parameter selection. Section 5 shows our computational experience with the new model and compares our model to other heuristics. Section 6 concludes.

2 Problem Formulation

The NWFSSP can be described as follows (based on [11], [7]). Each of n jobs is to be sequentially processed on machines *1,..., m*. The processing time *p(i, j)* of job i on machine j is given. At any time, each machine can process at most one job and each job can be processed on at most one machine. The sequence in which the jobs are to be processed is the same for each machine. The difference between the completion time of the last operation of a job and the start time of its first operation is equal to the sum of the processing times of its operations, and the completion time of a job on a given machine must be equal to the starting time of the job on the next machine. The objective is to find a sequence for the processing of the jobs on the machines that gives the minimum of the maximum completion time, i.e., makespan (C_{max}).

Suppose that the job permutation $\pi=\{\pi_1, \pi_2, \ldots, \pi_n\}$ represents the schedule of jobs to be processed. Let $d(\pi_{j-1}, \pi_j)$ be the minimum delay on the first machine between the start of job π_j and π_{j-1} restricted by the no-wait constraint when the job π_j is directly processed after the job π_{j-1}. The minimum delay can be computed from the following expression:

$$d(\pi_{j-1}, \pi_j) = p(\pi_{j-1}, 1) + \max\left[0, \max_{2\le k\le m}\left\{\sum_{h=2}^{k} p(\pi_{j-1}, h) - \sum_{h=1}^{k-1} p(\pi_j, h)\right\}\right]$$

Then the makespan can be defined as

$$C_{\max}(\pi) = \sum_{j=2}^{n} d(\pi_{j-1}, \pi_j) + \sum_{k=1}^{m} p(\pi_n, k)$$

3 The Island Model for the NWFSSP

3.1 The Structure of the Island Model

Our island model uses a master-slave structure in a way similar to [9], but it organizes the migration differently than in [9]. Our model uses a centralized scheme in which slave processors execute the basic EA on their population and periodically send their best partial results to a master process. The master process stores the partial results in a common migration set (*MS*), and chooses random individuals from the *MS* one after the other for every slave and sends them to the slaves.

Our model can work with *np* parallel processors (e.g. 2, 4, 8 or 16), and the basic EAs run on every parallel processor. The parallel process will be controlled with the *frequ* and *mignumb* parameters. Let EP_i be the evolutionary process on the ith slave processor with P_i population. *frequ* determines the communication frequency: after *frequ* iterations every slave sends migrating individuals into the *MS*. Every EP_i sends number *mignumb* individuals to the *MS* and they select the best individuals for migration. The master process selects randomly *mignumb* individuals from *MS* for every EP_i, and each population P_i get these individuals from *MS*. The master process selects the individuals for P_i randomly excluding the earlier migrant individuals arrived from P_i. Every EP_i replaces its worst individuals in $P_i$with the incoming ones.

With this island model we want to examine the performance of the island model's structure and migration, and to get good results for the NWFSSP. So we simulated the island model on one processor and did not examine the parallel environment characteristics in a network. As cost we computed only the run time that includes the communication time too. Naturally we computed for all parallel processes only one process time that belonged to the longest process.

3.2 The Basic EA

The same basic EA runs on every processor. Let us denote this basic EA by NWFS (No-wait flow shop). It is a hybrid EA that improves the new solutions using local search procedures. It uses a 2-stage algorithm structure where both stages are steady-state hybrid EAs. The first stage is a quick "preparatory" stage which is designated to improve the quality of the initial population.

In the second stage in every generation the algorithm selects a parent by truncation selection and its copy will be the descendant. It uses a mutation based on the EVL method and uses two local search procedures to improve the descendant. As additional operation, the algorithm updates the ECM matrix, the memory of the EVL method and to speed up the convergence it uses the *Delete*, *Filter* and *Restart* procedures.

Let us see first the main steps of NWFS:

Procedure NWFS(*t, itt, kn, itend, ddp*)
Initial population. Initial values of ECM. it=0

/* First stage
Do *itt* times
it = it + 1. A random descendant. Local search. Reinsertion.
In every *kn*th generation: *Update of ECM. Filter, Delete*
od.
/* Second stage
Repeat
it = it + 1. Truncation selection, mutation, local search. Reinsertion.
In every *kn*th generation: *Update of ECM. Filter, Delete, Restart.*
until $it > itend$
end

The parameters of the algorithm:
t - the size of the population,
itt - the number of the generations in the first stage.
kn - the populations are checked in every knth generation.
itend - the maximal number of generations.
ddp - parameter of the Delete procedure.

The operations and the characteristics are the following:
Individuals. An individual is a permutation of the job numbers.
Initial population. The first individual is generated by the NEH heuristic [10], the other individuals of the P population are generated from the first individual with some random swaps.
Fitness function. The fitness function is the objective function C_{max}.
Selection operator. In the first stage the descendants are randomly generated from the first individual with some random swaps. In the second stage the algorithm selects a parent by truncation selection where only the best individuals are selected for parents (from the 0.1 proportion of the population).
Mutation operator. The algorithm uses randomly maximum 10 swaps based on the EVL method (details in section 4.).
Local search. The algorithm uses a sequence of two different local search procedures one after the other. The neighborhood structures used are: insertion and swap (details in section 4.).
Filter. This procedure filters out and deletes the close solutions (based on the Hamming distance, e.g. $d^H(x, x') < 2$).
Restart. If the fittest solution didn't change in the last (e.g. 400) generations, it deletes the weakest solutions (e.g. 90% of the population).
Delete. This deletes the weakest solutions (*ddp*% of the population).
Reinsertion. In the first stage the descendant may replace the most similar one of the former solutions (based on the Hamming distance between the individuals). In the second stage the algorithm selects between the parent and the descendant. (After *Restart*, *Filter* or *Delete* the new descendants only are inserted to the population until the population size is completed).
Stopping criterion. The algorithm is terminated if a pre-determined number of iterations had been performed.

4 Details of the Implementation

In the above description of the algorithm, some heuristic solutions, shortcuts in the local search procedures and the selection of the parameters were not detailed. Let us see them now one-by-one.

4.1 Local Search

In our NWFS algorithm (and the island model, too) one of the most important determinant of the speed is the local search. We have to decide some details to reach a good local search procedure. Researchers ask several questions at a given problem to solve these details e.g.:

- What type of neighborhood structures should be used?
- How can we use different types of neighborhood structures together?
- How often should local search be applied?
- How long should local search be run?

We answered the questions based on the run statistics. We chose 10 difficult test problems from the benchmark set (the *Recxx* problem set from the OR-library) and tried various neighborhood structures, various local search procedures, and various run parameters in the local search procedures to be applied. We computed the run statistics of the various program versions: the average solution values, the average best solution values and the average run times. Based on the best statistics we concluded the following:

- It is appropriate to use only the swap and insert neighborhood structures,
- We can organize two separate 2-opt local search procedures: one with swap and one with insert structures. We can use the procedures one after the other.
- It is appropriate if we use the local search procedure in every knth generation and we apply the local searches for the best descendant of the last kn generations.
- To reduce the run time of the local search procedures we take only the best possible move into account for every individual. So we can use the speed-up methods of [11] to calculate makespan for insert and swap operators. Another possibility to reduce the run time is if we do not repeat the examination of the neighborhood structures if a better solution was found. Based on our examination it is appropriate by larger job numbers (more than 50 jobs).

4.2 The EVL and the Mutation Operator

The principle of the EVL is the following [2]. Let us consider a generic EA, and suppose that each variable of the individual can have k discrete values. Let us notice that ECM (explicit collective memory) is an $l \times k$ matrix that stores and learns the relative frequencies of the different values of the variables. This matrix is updated through the evolution procedure using a few of the worst performing individuals.

Let ECM_{ij}^{gen} be the collected relative frequency of the ith values on the jth position (variable) until the *gen*th generation. We can update the elements of the ECM matrix

$$ECM_{ij}^{gen+1} = (1-\alpha)ECM_{ij}^{gen} + \alpha \Delta ECM_{ij}\,(e.g.\alpha = 0.2)$$

where ΔECM_{ij} is the relative frequency of the ith value on the position j based on the worse individuals of the *gen*th generation and α denotes some relaxation factor. For the probability of mutating the jth variable in individual X we can use the formula

$$pr_j = 1 - \left| \frac{ECM_{X_j j}^{gen}}{\sum_{i=1}^{n} ECM_{ij}^{gen}} - a_j \right|$$

where B is the best individuals and $if\ X_j = B_j\ then\ a_j = 1\ else\ a_j = 0$.

Mutation for NWFS
The generated ECM is based on the worst individuals of the population. The ECM for NWFS is $n\ x\ n$ matrix. Every job has a row and every position of the permutation has a column in the matrix.

The mutation in NWFS is based on the EVL method. Let X be a descendant. The mutation operator first chooses a random j position of X (in the permutation). Next it chooses another job for this position. It chooses with pr_j probability a value; it searches the position of the value and swaps the values of the positions. (The algorithm uses a random number of swaps, maximum 10 swaps in all descendants.) In every *kn*th generation the ECM is updated by using the weakest individuals. In the updating procedure we use 20% of the population.

We observed that the use of the ECM matrix is insufficiently efficient after 200 - 300 generations, the convergence is slow. So we applied two versions of the ECM: a *long term memory* (ECML) and a *short term memory* (ECMS). The ECML is updated only during the first 2000 generations and later it does not change. The ECMS is updated continually but after every 100-150 generations we delete the value of the ECMS, and we begin the update of ECMS with empty matrix.

Using the two memories we have to compute the pr_j probability in another way, too. Now we can compute a probability based on the ECML (let be pl_j) and based on the ECMS (let be ps_j). As new probability we use the $pr_j = pl_j * ps_j$. Using the new probability (and memories) the quality of the results is better to compare with the earlier version.

4.3 Parameter Selection

To achieve a quick and accurate solution we need appropriate parameter values. Studying some of the more complex problems of the benchmark set we analyzed the process of the convergence to find how the parameter values were affecting the convergence, the finding of the global optimum and the speed of the calculation. So we analyzed the population size (t parameter), the frequency of checks (kn parameter), the generation in the first stage (itt parameter) and the ddp - parameter of the *Delete* procedure. Summarizing the results of the analysis, we chose the following parameter settings: the used parameters were the following: $t = 60, itt = 200, kn = 20, ddp = 10\%$. The maximum number of the generations was between 3000 and 7000 depending on the problem. We used these parameters in the island model too and searched appropriate values for the two model parameters. We found that using the $frequ = kn$ and $mignumb = 6$ parameter values we got good results.

5 Experimental Results

Test problems
For the test we chose problems that had already been published in different papers and so we could compare the published results with the results of ours. So we tested the island model with 23 benchmark problems from the OR-Library. Problem dimensions vary from 20 to 100 jobs and 5 to 20 machines. All test instances can be downloaded from the OR-Library (http://mscmga.ms.ic.ac.uk). Every test problem was run ten times. The model was implemented in Visual Basic and run on Intel Core Duo CPU 2.2 GHz with 2 GB RAM under Windows Vista Business (source code: http://netstorage.ktk.pte.hu/˜borgulya/islandm.zip).

Computational experiences with the island model
We applied the island model with different numbers of islands. We used 1, 2, 4, 8 and 16 islands and for every version we computed the average results. The models with 2 or more islands were substantially faster and more accurate than the single island model. Similarly the algorithm with two memories version of the EVL technique was more accurate than the one memory version.

We found the results of best quality with the 4-island and 16-island versions. Table 1 shows their results. In the table we give the problem name (name), the size of the problem (size), the makespan of the RAJ algorithm [12] (*RAJ*), the average relative percentage deviation of the solution (makespan) from the *RAJ* (*ARPD*) in four cases: by the best found solution (*best*), by the worst found solution (*worst*), by the average found solution (*aver.*), by the standard deviation of the solutions (*SD*); and the average run time to the best solutions (*time*) (CPU time in seconds). The bottom row (*average*) shows average values.

We appreciated the result based on the quality of the solutions and on the run time of the algorithm. We found that the quality of the solutions is good as the comparison shows and because the average run time of our method is less than 100 seconds we can say that our run time is acceptable.

Comparative results
We chose several heuristic methods to compare the results of our island model with. We did not find any other parallel model for the NWFS problem, only

heuristics without parallel mechanism. We found the best published results with a TS version and with two PSO versions. So we chose the TS from [6] (*TSG*), the hybrid PSO from [7] (*HPSO*) and the discrete PSO version with VNS from [11] (*DPSO*).

Table 1. The results of the island model

name	RAJ	4-island					16-island				
		ARPD				time	ARPD				time
		best	worst	aver.	SD		best	worst	aver.	SD	
rec01	1590	-4.03	-4.03	-4.03	0.0	0.9	-4.03	-4.03	-4.03	0.0	0.3
rec03	1457	-6.59	-6.59	-6.59	0.0	0.5	-6.59	-6.59	-6.59	0.0	0.2
rec05	1637	-7.70	-7.70	-7.70	0.0	3.1	-7.70	-7.70	-7.70	0.0	1.2
rec07	2119	-3.63	-3.63	-3.63	0.0	5.2	-3.63	-3.63	-3.63	0.0	0.8
rec09	2141	-4.62	-4.62	-4.62	0.0	1.0	-4.62	-4.62	-4.62	0.0	0.7
rec011	1946	-3.34	-3.34	-3.34	0.0	1.4	-3.34	-3.34	-3.34	0.0	0.5
rec013	2709	-6.05	-6.05	-6.05	0.0	11.0	-6.05	-6.05	-6.05	0.0	2.1
rec015	2691	-6.02	-6.02	-6.02	0.0	1.8	-6.02	-6.02	-6.02	0.0	1.1
rec017	2740	-5.58	-5.58	-5.58	0.0	2.2	-5.58	-5.58	-5.58	0.0	1.5
rec019	3157	-9.72	-9.72	-9.72	0.0	4.1	-9.72	-9.72	-9.72	0.0	8.7
rec021	3015	-6.43	-6.43	-6.43	0.0	18.7	-6.43	-6.43	-6.43	0.0	3.8
rec023	3030	-10.89	-10.89	-10.89	0.0	10.89	-10.89	-10.89	-10.89	0.0	8.7
rec025	3835	-6.31	-6.31	-6.31	0.0	16.1	-6.31	-6.21	-6.28	0.0	11.8
rec027	3655	-6.13	-6.05	-6.06	0.0	31.1	-6.13	-6.13	-6.13	0.0	16.8
rec029	3583	-8.15	-7.87	-8.07	0.2	31.8	-8.15	-8.15	-8.15	0.0	7.2
rec031	4631	-6.97	-6.89	-6.93	0.0	44.2	-6.78	-6.56	-6.62	0.1	21.3
rec033	4770	-7.09	-6.94	-7.01	0.1	93.5	-7.25	-6.71	-6.97	0.3	32.6
rec035	4718	-6.80	-6.42	-6.61	0.2	38.0	-6.80	-6.40	-6.65	0.0	62.7
rec037	8929	-9.95	-9.25	-9.71	0.3	421.0	-9.95	-9.25	-9.53	0.4	314.0
rec039	9158	-7.61	-7.23	-7.48	0.2	655.0	-7.64	-6.98	-7.46	0.3	335.0
rec041	9344	-9.25	-9.16	-9.20	0.1	541.0	-9.27	-8.87	-8.95	0.2	398.8
hel1	780	-9.23	-8.33	-8.89	0.5	240.0	-9.10	-8.72	-8.83	0.2	151.0
hel2	189	-5.29	-5.29	-5.29	0.0	1.0	-5.29	-5.29	-5.29	0.0	0.5
Average		-6.84	-6.70	-6.79	0.1	94.5	-6.84	-6.69	-6.76	0.1	60.0

The comparison was encumbered by the use of various programming languages, operating systems and computers. Only one appropriate aspect of comparison could be found, namely the average relative percentage deviation of the solution from the RAJ, so our table of comparison (table 2) is based on the results of comparable accuracies. (The *TSG* was run on Pentium 1GHz, the *HPSO* was run on Pentium IV 2.2 GHz and the *DPSO* was run on Pentium IV 3.0 GHz processors.)

Table 2 shows the comparative results. We give in the table the average *ARPD* values of the best and the average solution for every problem. Some data are missing: the average solutions of *TSG* and the results of *hel1*, *hel2* by *HPSO*. Analyzing the results we concluded that the average results of our models are the best among these methods and concerning the best solutions our models found in four cases new best solutions (they are highlighted with bold characters in table 2). Comparing the best solutions of our model and the *DPSO* we gat in 80% of the test problems the same solutions (probably they are the optimal solutions), in 3 cases the *DPSO* has a little bit better and in 4 cases our model has better solutions.

Table 2. ARPD values of different methods

name	TSG	4-island		16-island		HPSO		DPSO	
	best	best	aver.	best	aver.	best	aver.	best	aver.
rec01	-3.96	-4.03	-4.03	-4.03	-4.03	-3.77	-3.39	-4.03	-4.03
rec03	-6.59	-6.59	-6.59	-6.59	-6.59	-6.59	-6.59	-6.59	-6.59
rec05	-7.64	-7.70	-7.70	-7.70	-7.70	-7.39	-7.15	-7.70	-7.68
rec07	-3.63	-3.63	-3.63	-3.63	-3.63	-3.63	-3.63	-3.63	-3.63
rec09	-4.58	-4.62	-4.62	-4.62	-4.62	-4.58	-4.26	-4.62	-4.62
rec011	-3.34	-3.34	-3.34	-3.34	-3.34	-3.34	-2.30	-3.34	-3.34
rec013	-6.05	-6.05	-6.05	-6.05	-6.05	-6.05	-5.47	-6.05	-6.05
rec015	-6.02	-6.02	-6.02	-6.02	-6.02	-6.02	-5.69	-6.02	-6.02
rec017	-5.58	-5.58	-5.58	-5.58	-5.58	-5.58	-5.42	-5.58	-5.58
rec019	-9.25	-9.72	-9.72	-9.72	-9.72	-9.15	-8.50	-9.72	-9.55
rec021	-6.30	-6.43	-6.43	-6.43	-6.43	-5.70	-5.33	-6.43	-6.33
rec023	-10.73	-10.89	-10.89	-10.89	-10.89	-10.80	-9.72	-10.89	-10.85
rec025	-6.31	-6.31	-6.31	-6.31	-6.28	-5.71	-5.17	-6.31	-6.26
rec027	-6.10	-6.13	-6.06	-6.13	-6.13	-6.13	-5.04	-6.13	-5.89
rec029	-8.28	-8.15	-8.07	-8.15	-8.15	-7.81	-6.93	-8.15	-8.07
rec031	-6.13	**-6.97**	-6.93	-6.78	-6.62	-5.92	-5.20	-6.89	-6.60
rec033	-6.31	-7.09	-7.01	**-7.25**	-6.97	-5.51	-4.08	-7.09	-6.68
rec035	-6.17	**-6.80**	-6.61	**-6.80**	-6.65	-6.02	-5.13	-6.72	-6.53
rec037	-9.49	-9.95	-9.71	-9.95	-9.53	-8.89	-8.20	-10.56	-10.06
rec039	-6.99	-7.61	-7.48	**-7.64**	-7.46	-6.79	-5.64	-7.56	-7.32
rec041	-8.57	-9.25	-9.20	-9.27	-8.95	-7.94	-6.77	-9.28	-9.07
hel1	-8.21	-9.23	-8.89	-9.10	-8.93	-	-	-9.36	-9.17
hel2	-5.29	-5.29	-5.29	-5.29	-5.29	-	-	-5.29	-5.29
Average	-6.59	-6.84	-6.79	-6.84	-6.76	-6.35	-5.65	-6.87	-6.75

6 Summary

In this paper, we present a new island model for the no-wait flow shop problem. The algorithm uses a special master-slave structure. The basic EA in the islands is a hybrid method; it uses a memory based technique for the mutation, the extended virtual loser (called EVL) with two special memories. The paper shows that our algorithm without any sophisticated selection, recombination and mutation operators can solve the no-wait flow shop problem in a simple way. The results of our algorithm have good quality and the run time of our algorithm is acceptable. This new heuristic gets better average result than the earlier heuristics for NWFSSP - with the help of the island-model and the EVL with two memories.

Future efforts can focus on improving the results and the speed of the algorithm with the help of speed up techniques similar to ones in [11].

Acknowledgments. The Hungarian Research Foundation OTKA K 68137 supported the study.

References

1. Aldowaisan, T., Allahverdi, A.: New Heuristics for No-wait Flowshops to Minimize Makespan. Comput. Oper. Res. 30, 1219–1231 (2003)
2. Borgulya, I.: An Algorithm for the Capacitated Vehicle Routing Problem with Route Balancing. Central European Journal of Operations Research 16(4), 331–344 (2008)
3. van Deman, J.M., Baker, K.R.: Minimisation mean flow time in flowshop with no intermediate queues. AIIE Trans. 6, 28–34 (1974)
4. Filho, G.R., Nagano, M.S., Lorena, L.A.N.: Hybrid Evolutionary Algorithm for Flowtime Minimisation in No-wait Flowshop Scheduling. In: Gelbukh, A., Morales, A.F.K. (eds.) MICAI 2007. LNCS (LNAI), vol. 4827, pp. 1099–1109. Springer, Heidelberg (2007)
5. Gangadharan, R., Rajendran, C.: Heuristic Algorithms for Scheduling in the Nowait Flow-shop. Int. J. Prod. Econ. 32, 285–290 (1993)
6. Grabowski, J., Pempera, J.: Some local search algorithms for no-wait flow-scop problem with makespan criterion. Comput. Oper. Res. 32, 2197–2212 (2005)
7. Liu, B., Ling, W., Jin, Y.-H.: An effective hybrid particle swarm optimization for no-wait flow shop scheduling. Int. J. Adv. Manuf. Technol. 31, 1001–1011 (2007)
8. Lin, B.M.T., Cheng, T.C.E.: Batch Scheduling in the No-wait Two-machine Flowshop to Minimize the Makespan. Comput. Oper. Res. 28, 613–624 (2001)
9. Marin, F.J., Trelles-Salazaro, O., Sandoval, F.: Genetic algorithms on LAN-message passing architectures using PVM: Application to the routing problem. In: Davidor, Y., Schwefl, H.P., Männer, R. (eds.) Parallel Problem Solving fron Nature, PPSN III, pp. 534–543. Springer, Berlin (1994)
10. Nawaz, M., Enscore, E., Ham, I.: A heuristic algorithm for the m-machine, n-job flowshop sequencing problem. OMEGA The International Journal of Management Sciences 11, 91–95 (1983)
11. Pan, Q.-K., Tasgetiren, M.F., Liang, Y.-C.: A discrete particle swarm optimization algorithm for the no-wait flowshop scheduling problem. Computers & Operations Research 35, 2807–2839 (2008)
12. Rajendran, C.: A No-wait Flowshop Scheduling Heuristic to Minimize Makespan. J. Oper. Res. Soc. 45, 472–478 (1994)
13. Schuster, C.J., Framinan, J.M.: Approximative procedures for no-wait job shop scheduling. Oper. Res. Lett. 31(3), 308–318 (2003)
14. Shyu, S.J., Lin, B.M.T., Yin, P.-Y.: Application of ant colonyc optimization for no-wait flowshop scheduling problem to minimize the total completion time. Comput. Ind. Eng. 47, 181–193 (2004)
15. Wang, L., Zheng, D.: An effective hybrid optimization strategy for job-shop scheduling problems. Comput. Oper. Res. 28(6), 585–596 (2001)

Environment-Driven Embodied Evolution in a Population of Autonomous Agents

Nicolas Bredeche and Jean-Marc Montanier

TAO - Univ. Paris-Sud, INRIA, CNRS - F-91405 Orsay, France
firstname.name@lri.fr

Abstract. This paper is concerned with a fixed-size population of autonomous agents facing unknown, possibly changing, environments. The motivation is to design an embodied evolutionary algorithm that can cope with the implicit fitness function hidden in the environment so as to provide adaptation in the long run at the level of the population. The proposed algorithm, termed mEDEA, is shown to be both efficient in unknown environment and robust with regards to abrupt, unpredicted, and possibly lethal changes in the environment.

1 Introduction

In this paper, we are interested in a fixed-size population of autonomous physical agents using local communication (e.g. autonomous robots), facing unknown and/or dynamic environments. This class of problems typically arises when the environment remains unknown to the human designer until the population of agents is actually made operational in the real situation [4], or whenever the environment is known to change during operation, without any indication on *when* and *how* these changes will impact survival strategies.

The challenge is to design a distributed online optimization algorithm targeting agent self-adaptation in the long term, that is being able to successfully manage an implicit pressure resulting from environmental particularities *and* algorithmic constraint with regards to the optimization process. Embodied Evolution (EE), as proposed initially in [5], addresses part of this question as it focuses on algorithms for evolutionary optimization of agent behaviors in an on-line, possibly decentralized manner. On the other hand, EE requires an objective function designed from the supervisor, which is unavailable by definition in the problem setting addressed here.

While concepts and methods from EE may be relevant, we can only assume that maximizing the integrity of the agent population as well as maintaining a communication network for exchanging genome are the basic requirements in the present context. To this end, we propose a distributed algorithm for environment-driven self-adaptation based on evolutionary operators that takes into account selection pressure from the environment. The basic assumption behind this algorithm is to consider the strategies as the atomic elements and the population of agents as a distributed resource onto which strategies compete with one another. This approach is better illustrated using the Selfish Gene metaphor [3]: one specific strategy (or set of parameters, or genome) is "successful" if it manages to spread over the population, which implicitly requires to both minimize risk for its "vehicles" (ie. the autonomous agents) and maximizing the number of mating opportunities, though the two may be contradictory.

R. Schaefer et al. (Eds.): PPSN XI, Part II, LNCS 6239, pp. 290–299, 2010.

The general motivation behind the work presented here is to study and provide general evolutionary adaptation algorithms that can utlimately be implemented on real robotic hardware. To this end, the main contribution of this paper is to introduce a new and simple distributed evolutionary adaptation algorithm for use in population of autonomous agents. While it is yet to be applied in a real robotic setup, this paper focuses on an indepth experimental analysis of the robustness of the algorithm with regards to unknown, and changing, environments, under realistic constraints (fixed number of agents, limited sensors and actuators, etc.).

2 Environment-Driven Distributed Evolutionary Adaptation

As stated in the introduction, our objective is to design a distributed online evolutionary algorithm for a fixed population of autonomous physical agents (e.g. autonomous robots), whenever the human engineer fails to provide a proper description of an objective function. As a consequence, the key issue behind Environment-driven Distributed Evolutionary Adaptation (EDEA) relies in the *implicit* nature of the fitness function. However, this implicit fitness may be seen as the result of two possibly conflicting motivations:

- **extrinsic motivation:** agent must cope with environmental constraints in order to maximize survival, which results solely from the interaction between the agent and the environment around (possibly including other agents as well);
- **intrinsic motivation:** set of parameters (ie. "genomes") must spread across the population to survive, which is imposed by the algorithmic nature of the evolutionary process. Therefore, genomes are naturally biased towards producing efficient *mating* behaviors as the larger the number of agents met, the greater the opportunity to survive.

The level of correlation between these two motivations does impact problem complexity to a significant amount: high correlation implies that the two motivations may be treated as one while low correlation implies conflicting objectives. An efficient EDEA algorithm should indeed address this trade-off between extrinsic and intrinsic motivations as the ideal optimal genome should reach the point of equilibrium where genome spread is maximum (e.g. looking for mating opportunities) with regards to survival efficiency (e.g. ensuring energetic autonomy).

These assumptions have also been extensively studied in the field of open-ended artificial evolution, with an emphasis on computational model of evolutionary dynamics [1], including a particular focus on the effect of the environment over the evolutionary adaptation process [6]. However, their application within Embodied Evolution is still an open issue as there is a major difference concerning the working hypothesis as EE is concerned with a fixed number of physically grounded agents that are usually ment to target real world environment (e.g. obstacles, energy constraints, etc.).

2.1 MEDEA: A Minimal EDEA Algorithm

Based on these considerations, we introduce the MEDEA algorithm ("minimal EDEA"), described in table 1. This algorithm describes how evolution is handled on a local basis

Algorithm 1. The MEDEA algorithm

```
genome.randomInitialize()
while forever do
  if genome.notEmpty() then
    agent.load(genome)
  end if
  for iteration = 0 to lifetime do
    if agent.energy > 0 and genome.notEmpty() then
      agent.move()
      broadcast(genome)
    end if
  end for
  genome.empty()
  if genomeList.size > 0 then
    genome = applyVariation(select_random(genomeList))
  end if
  genomeList.empty()
end while
```

and is copied as is within all agents in the population. This algorithm works along with a communication routine, which purpose is to receive incoming genomes and store these in the Imported Genome List for later use.

At a given moment, a given agent is driven by a control architecture which parameters are extracted from an "active" genome, which remains unchanged for a generation. This genome is continuously broadcasted to all agents within (a limited) communication range. This algorithm actually implements several simple, but crucial, features, that can be interpreted from the viewpoint of a traditional evolutionary algorithm structure:

Selection operator: the selection operator is limited to simple random sampling among the list of imported genomes, ie. no selection pressure *on a local individual basis*. However, cumulated local random selection ultimately favor the most widespread genomes *on a global population basis* as such genomes have greater probability to be randomly picked *on average*. In fact, the larger the population and mating opportunities, the more accurate the selection pressure at the level of the population.

Variation operator: the variation operator is assumed to be rather conservative to ensure a continuity during the course of evolution. Generating altered copies of a genome only make sense if there is some continuity in the genome lineage: if no variation is performed, the algorithm shall simply converge on average towards the best existing genome initially in the population. In the following, we assume a gaussian random mutation operator, inspired from Evolution Strategies [2], which conservative behavior can be easily tuned through a σ parameter.

Replacement operator: lastly, replacement of the current active genome to control a given agent is performed by (1) local deletion of the active genome at the end of one generation and (2) randomly selecting a new active genome among the imported genome list (cf. selection operator). On a population level, this implies that surviving genomes are likely to be correlated with efficient mating strategies as a given genome may only survive through (altered) copies of itself in the long run.

The positive or negative impact of environmental variability on genome performance is smoothed by the very definition of the variation operator as newly created genomes are always more or less closely related to their parent. As a consequence, each genome results from a large number of parallel evaluations, both on the spatial scale as closely related copies sharing the same ancester may evaluated in a population, and on the temporal scale, as one genome is also strongly related to its ancestors. Hence, a single genome may get lucky once in a while, but it's highly unlikely that a "family" of closely related genomes manage to survive in the population if there are more efficient competitors.

3 Experimental Setting

This section provides a description of the experimental setting used hereafter as well as implementation details. The motivation is here to design a setting such that it is possible to address several issues regarding evaluation and validation of the proposed algorithm. In particular, robustness of MEDEA with regards to environmental pressure and to sudden environmental changes shall be studied.

3.1 The Problem: Surviving in a Dynamic Unknown Environment

Figure 1 shows the environment used for the experiment: a 2D arena with obstacles, possibly containing food items. The figure also illustrates 100 autonomous mobile agents loosely inspired from the ePuck mobile robot specifications. This environment is used to define two different experimental setups, described hereafter:

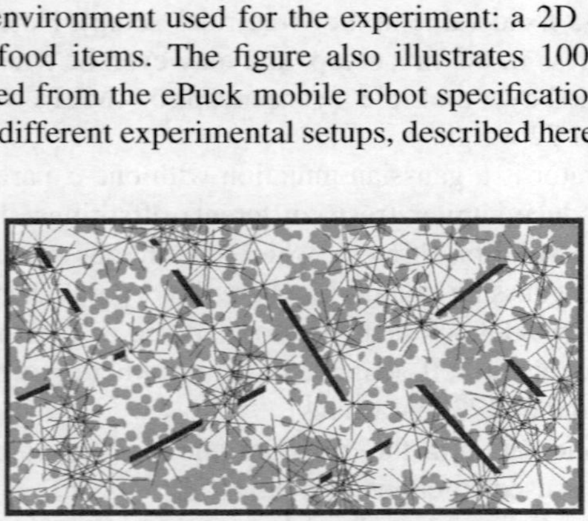

Fig. 1. Snapshot from the simulator with 100 agents. Yellow: food items. Red: agents, modeled after an e-puck robot. Blue: range of proximity sensors (communication range is half this range).

1. **the "free-ride" setup**
   - *Description*: a population of autonomous mobile agents is immersed within an environment with few obstacles. As a consequence, an agent dies only if it was not able to mate with at least one other agent - ie. the current genome is lost for sure as it does not get a chance to survive within any other agents.
   - *Motivation*: this setup makes it possible to evaluate the mechanisms of the MEDEA algorithm as environmental pressure should be limited.

2. **the "energy" setup**
   - *Description*: A set of energy resources ("food items") is spread all over the environment, which can be harvested by the agents. Agents are endowed with an energy level, which depends on harvested food items and power consumption. If the energy level reaches 0, agent dies and genome information is lost. Moreover, harvested food items only "grow" back after a given number of iterations.
   - *Motivation*: In this setup, genomes also compete for agent resources but have to deal with environmental pressure as maximizing mating encounters may not be fully compatible with energy self-sustainability.

The full experimental setup considers starting with the "free-ride" setup, and then suddenly switching to the "energy" setup after a pre-defined fixed number of generations. In the meantime, agents are of course unaware of such a change in the environment and keep on running the same unchanged MEDEA algorithm.

3.2 Representation / Encoding the Problem

Specifications for the autonomous agents are inspired from traditional robotic setup, with 8 proximity sensors dispatched all around the agent body and 2 motor outputs (translational and rotational speeds). Moreover, three additional sensory inputs are considered: the angle and direction towards the nearest food item and the current energy level (which is set to a fixed value in the "free-ride" setup). Note that these additional sensor values are useless in the first setup, and may even be considered as distractors. Each agent is controlled by a multiple layer perceptron (MLP) with 5 hidden neurons, which means a total of 72 weights[1].

The variation operator is a gaussian mutation with one σ parameter: a small (resp. large) σ tends to produce similar (resp. different) offsprings. This is indeed a well known scheme from Evolution Strategy where continuous values are solely mutated using a parameterized gaussian mutation, where the σ parameter may be either fixed, updated according to pre-defined heuristics or evolved as part of the genome. In the scope of this work, we rely on self-adaptive mutation, where σ is part of the genome [2] (ie. the full genome contains 73 real values).

The current implementation of the σ update rule is achieved by introducing α, a σ update value, which is used to either decrease ($\sigma_{new} = \sigma_{old} * (1 - \alpha)$) or increase ($\sigma_{new} = \sigma_{old} * (1 + \alpha)$) the value of σ whenever a genome is transmitted. The idea is that whenever an agent broadcast its own genome, probabilities of transmitting an increased or decreased σ values are equivalent. In the following, α is a predefined value set prior to the experiment so that it is possible to switch from the larger σ value to the smaller in a minimum of approx. 20 iterations.

3.3 Experimental Settings

The whole experiment lasts for 150 generations, switching from the "free-ride" setup to the "energy" setup at generation 75. During the course of evolution, some agents

[1] 11 input neurons ; 5 hidden neurons ; 2 output neurons ; 1 bias neuron. The bias neuron value is fixed to 1.0 and projects onto all hidden and output neurons.

may come to a halt either because they did not meet any other agents, thus failing to import a new genome for use in the next generation, or because they ran out of energy during the "energy" setup (each agent can store a maximum of 400 energy units and consumes 1 unit/step, one generation lasts 400 steps). In the "free-ride" setup, the agents remain still (or "inactive", ie. without genome), waiting for new genomes imported from "active" agents that eventually come into contact. In the "energy" setup though, agents requires an external "human" intervention for refilling energy. Revived agents remains inactive, but are refilled with enough energy to wait until the end of the current generation, listening for new imported genome that may be used for the next generation. While the reviving procedure makes it possible to avoid progressive extinction in the second setup, extinction is nevertheless possible whenever all agents in the population fail to meet any other agents during one generation, whatever the cause (bad exploratory or harvesting strategies). Also, monitoring the number of active agents in a population provides a reliable indicator of the performance of the algorithm as external intervention may be viewed as one important cost to minimize (e.g. minimizing human intervention in a robotic setup). Detailed parameters used for the experiment presented in the next section are given below.

Parameter	Value
arena width and length	$1024 * 530$ inches
"free-ride" setup duration	75 generations
"energy" setup duration	75 generations
lifetime (ie. generation duration)	400 steps per generation
population size	100 agents
proximity sensor range	64 inches
radio broadcast signal	32 inches
agent rotational velocity	30deg/step
agent translational velocity	2 inches/step
genome length	79 real values (78 MLP weights + σ)
variation operator	gaussian mutation with σ parameter
$\sigma_{minValue}$	0.01
$\sigma_{maxValue}$	0.5
$\sigma_{initialValue}$	0.1
α (ie. σ update parameter)	0.35

"energy" setup only:	
food items	2000
food item diameter	10 inches
food item regrow delay	btw 400 and 4000 steps (see text)
energy per food item	100 energy units
agent energy consumption	1 energy unit per step
agent maximum energy level	400 energy units
agent initial energy level	400 energy units

In order to provide a challenging environment, the "energy" setup is designed so that the number of food items in the environment depends on the actual number of active agents. Indeed, a food item grows back whenever harvested, but only after some delay. If the number of active agents is less than half the population size, then $delay_{regrow}$ is set to 400 steps. However, if the number of active agents is between 50 and 100, then the delay linearly increases from 400 steps (fast regrowing) to 4000 steps (slow regrowing, aggressive environment). In the particular setup described here, switching from a possibly efficient population of 100 agents from the "free-ride" setup to the "energy" setup will have a possibly disastrous impact as the number of agents at the beginning of the second setup implies longer regrow delays.

At this point, it is important to note that the motivation behind this experimental setup is both to stress the population for further analysis as well as providing a flexible and challenging experimental settings that could be re-use to evaluate further version and variation over the algorithm presented here. To this end, the source code and parameters for all experiments presented in the following is available on-line in the Evolutionary Robotics Database[2]. On a practical viewpoint, one experiment takes approx. 15 minutes

[2] Evolutionary Robotics Database: `http://www.isir.fr/evorob_db/`

to be performed using one core of a 2.4GHz Intel Core 2 Duo computer. The home-brew agent simulator is programmed in C++ and features basic robotic-inspired agent dynamics with collision.

4 Results and Analysis

The lack of explicit objective function makes it difficult to compare performance during the course of evolution. However, the number of active agents and the average number of imported genomes per generation give a good hint on how the algorithm performs: it can be safely assumed that "efficient" genomes lead to few deaths and many mating opportunities. Moreover, the number of food items harvested gives some indication in the "energy" setup. The four graphs in figure 2 give an a synthetic view of the results over 100 independent runs obtained with MEDEA on the experimental scenario described in the previous section. These graphs compile the average values of selected parameters, or "indicators", over generations: number of active agents, average number of imported genomes per agent, average energy balance per agent, and average σ mutation parameter values.

In both setups, all indicators rise to reach stable average values and some conclusions can be drawn: firstly, both setups show an increase in both mating opportunities (number

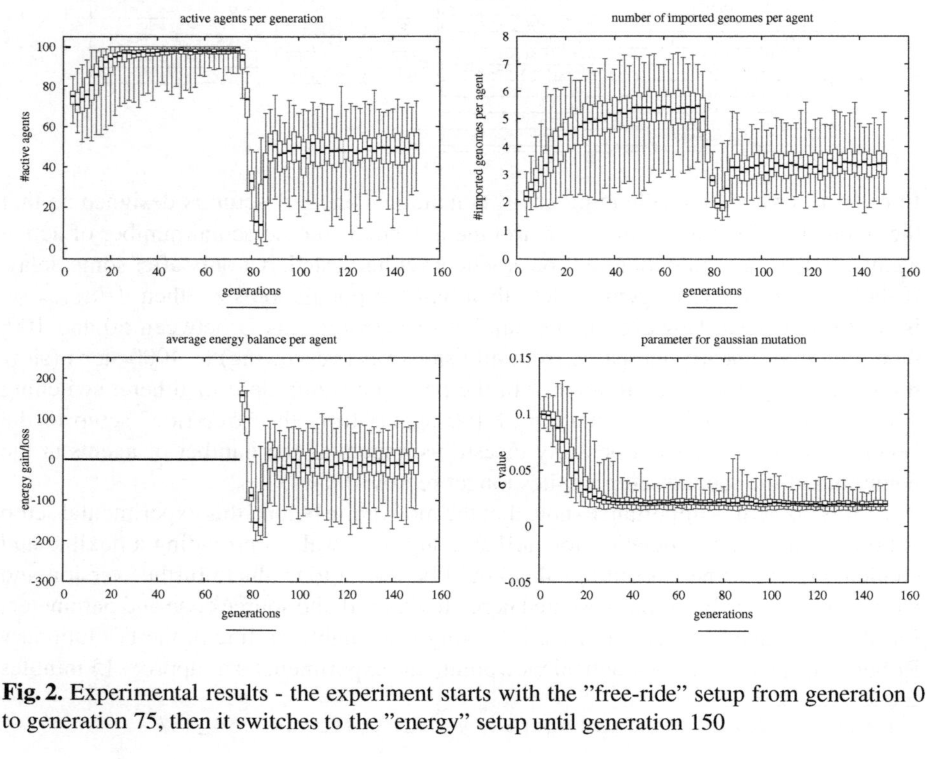

Fig. 2. Experimental results - the experiment starts with the "free-ride" setup from generation 0 to generation 75, then it switches to the "energy" setup until generation 150

of imported genomes) and survival rate (number of active agents). Secondly, switching setups initially leads to a drop for both indicators, followed by a quick recovery through evolutionary adaptation. This interpretation is reinforced by the increasing value of the energy balance which is a key element for the second setup. A notable remark is that the energy balance stays around zero, which is sufficient to guarantee agent survival. This is not a surprise as over-maximizing harvesting may imply a cost with regards to looking for mating partners. On the other hand, the gaussian mutation parameter is not really influenced by the change of environmental setups (except from a slight increase in maximum values). While the results may vary among runs, with a great difference between minimal and maximal values for each indicator, values between the upper and lower quartiles are remarkably close given the noise inherent to this kind of experiment. Indeed, complete extinctions were even observed after switching to the "energy" setup in three of the 100 runs (results not shown).

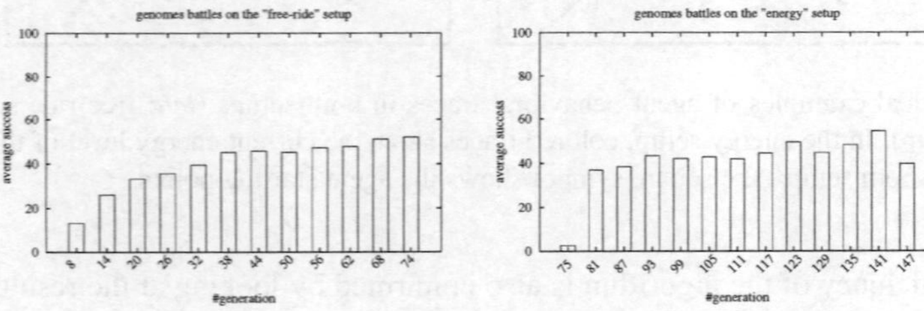

Fig. 3. Genomes battles on both setups (*left*: free-ride setup ; *right*: energy setup). Average success scores for each generation, both histograms are the results of 1000+ battles. See text for details.

However, we must be very cautious with the interpretation of these results. For example, the quality of the equilibrium between maximizing mating opportunity and coping with environmental constraints (ie. avoiding walls, avoiding collisions with other agents, harvesting) is difficult to estimate as such equilibrium may (and appears to) imply sub-optimal values for both related indicators. As a matter of fact, all interpretations provided so far rely on the assumption that values monitored in the experiment are actually correlated with genome survival. In order to support this assumption, a new experimental setup is defined from the results obtained so far: the *post-mortem* battle experiment (or battle experiment, for short). The battle experiment is loosely inspired from competitive coevolution, where each individual competes against a hall-of-fame of the best individuals from every past generations [8], so as to estimate the fitness rank of one individual within all possible (or at least, all available) situations. For the current experiment, one "battle" is achieved by randomly picking up 10 generations from the same setup, and extracting one random genome from each of these generations. Then, each genome is copied into ten different agents, resulting in 100 agents that are immersed in the same setup they were evolving in. Variation is turned off, and evolution is re-launched. After 100 generations of random selection and replacement, the number of copies for each genome is accounted for and used to compute a "survival score". As an example, one genome gets a maximal score if it succeeds in taking control of all the agents. Average results over 1000 battles are given in figure 3. In both setups, genomes from later generations display better survival capability than early genomes. Moreover, battles on the second setup show a very fast recovery after environmental change, possibly to stable, but limited, strategies as the number of active agents is far from the maximum. Also, these histograms lack the misleading artifacts observed in previous graphs regarding the early generations in both setups: genomes from generation 0 do not benefit from uniform sampling of starting location and genomes from generation 75 do not benefit from high initial energy level.

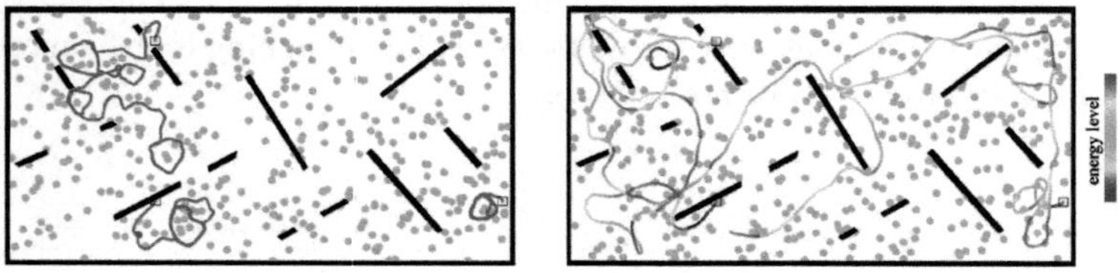

Fig. 4. Typical examples of agent behavioral traces in both setups (*left*: free-ride setup ; *right*: energy setup). In the energy setup, colored traces show the current energy level of the agent (see legend). In both setups, the square symbol shows the agent starting points.

The efficiency of the algorithm is also confirmed by looking at the resulting behavioral strategies. Two examples of behaviors are shown in figure 4, resulting from agent driven by genomes obtained in the late generations of both setups. In the "free-ride" setup, genomes tend to lead to rather conservative behaviors, with obstacle avoidance, but with limited exploratory behavior. On the other hand, genomes from the later generations of the "energy" setup show a different behavioral pattern, favoring long distance travel and few circling around, which is an efficient strategy to avoid being stuck in an exhausted area. Moreover, a closer look at trajectories (including, but not limited to what is shown here) show that agents acquired the ability to drive towards a detected food items under certain conditions, such as favoring safe areas with few obstacles whenever energy level is low.

5 Conclusions and Perspectives

This paper provides a proof-of-concept for the viability of environment-driven distributed evolutionary adaptation in a population of autonomous agents. We have presented the MEDEA algorithm, a particular flavor of Embodied Evolution, tailored to address evolutionary adaptation with implicit fitness. The proposed algorithm was evaluated with regards to our initial motivation and proven to be (1) efficient with regards to providing distributed evolutionary adaptation in unknown environment and (2) robust with regards to unpredicted changes in the environment. Moreover, this algorithm is light-weight and with low complexity, which makes it possible to consider future implementation within hardware/software setups with limited computational capability such as robotic agents.

Many perspectives may be considered from this point, and some are already under investigation. Firstly, the class of problems addressed here is also relevant in the field of embodied Evolutionary Robotics and we are currently working on implementing the MEDEA algorithm within a population of real robots as part of the european Symbrion IP project [4]. Secondly, surviving in aggressive environments requires more complex behavioral patterns, such as self-organization and coordination. Sharing similar concerns, previous works in collective intelligence and reinforcement learning have already stressed the issue of the price of anarchy [9], ie. the cost of efficient selfish behavior with regards to population global welfare. Then again, solving this issue remains an open problem, especially if there is no explicit objective function to decompose. Thirdly, the work presented here targets, and is limited to, providing reliable survival strategies. However, our motivational claim is that one should first aim at a reliable, surviving population before even considering to optimize a pre-defined objective function. Within Evolutionary Robotics, similar ideas have been defended in the very last few years: goal-oriented optimization is often better served with objective that are loosely related with the targeted goal (e.g. maximizing diversity [7]), while objective function with extensive goal description often lead to deceiving fitness landscape, and poor results. Of course, this remains to be extensively studied and demonstrated.

Acknowledgements

This work was made possible by the European Union FET Proactive Intiative: Pervasive Adaptation funding the Symbrion project under grant agreement 216342.

References

1. Bedau, M.A., McCaskill, J.S., Packard, N.H., Rasmussen, S., Adami, C., Green, D.G., Ikegami, T., Kaneko, K., Ray, T.S.: Open problems in artificial life. Artificial Life 6, 363–376 (2000)
2. Beyer, H.-G., Schwefel, H.-P.: Evolution strategies – A comprehensive introduction. Natural Computing 1, 3–52 (2002)
3. Dawkins, R.: The Selfish Gene. Oxford University Press, Oxford (1976)
4. Baele, G., et al.: Open-ended on-board evolutionary robotics for robot swarms. In: Proceedings of the IEEE Conference on Evolutionary Computation, CEC 2009 (2009)
5. Ficici, S., Watson, R., Pollack, J.: Embodied evolution: A response to challenges in evolutionary robotics. In: Wyatt, J.L., Demiris, J. (eds.) Proceedings of the Eighth European Workshop on Learning Robots, pp. 14–22 (1999)
6. Fletcher, J.A., Bedau, M.A., Zwickp, M.: Effect of environmental structure on evolutionary adaptation. In: Artificial Life VI, pp. 189–198. MIT Press, Cambridge (1998)
7. Lehman, J., Stanley, K.O.: Exploiting open-endedness to solve problems through the search fo novelty. In: Proceedings of the Eleventh International Conference on Artificial Life (ALIFE XI), MIT Press, Cambridge (2008)
8. Rosin, C., Belew, R.: New methods for competitive coevolution. Evolutionary Computation (1996)
9. Wolpert, D.H., Tumer, K.: Optimal payoff functions for members of collectives. Advances in Complex Systems 4(2/3), 265–279 (2001)

Large-Scale Global Optimization Using Cooperative Coevolution with Variable Interaction Learning

Wenxiang Chen, Thomas Weise, Zhenyu Yang, and Ke Tang*

Nature Inspired Computation and Applications Laboratory,
School of Computer Science and Technology,
University of Science and Technology of China

Abstract. In recent years, Cooperative Coevolution (CC) was proposed as a promising framework for tackling high-dimensional optimization problems. The main idea of CC-based algorithms is to discover which decision variables, i.e, dimensions, of the search space interact. Non-interacting variables can be optimized as separate problems of lower dimensionality. Interacting variables must be grouped together and optimized jointly. Early research in this area started with simple attempts such as one-dimension based and splitting-in-half methods. Later, more efficient algorithms with new grouping strategies, such as DECC-G and MLCC, were proposed. However, those grouping strategies still cannot sufficiently adapt to different group sizes. In this paper, we propose a new CC framework named Cooperative Coevolution with Variable Interaction Learning (CCVIL), which initially considers all variables as independent and puts each of them into a separate group. Iteratively, it discovers their relations and merges the groups accordingly. The efficiency of the newly proposed framework is evaluated on the set of large-scale optimization benchmarks.

Keywords: Variable Interaction Learning, Large-Scale Optimization, Numerical Optimization, Incremental Group Strategy, Cooperative Coevolution.

1 Introduction

Evolutionary Algorithms (EAs) have been widely applied for solving both numerically and combinational optimization tasks [1]. Finding solutions for a problem usually becomes harder when the number of decision variables increases because of the *curse of dimensionality*. As a consequence, [2] reports that there is a rapid decline in performance of conventional EAs when dealing with large-scale problems. In order to improve the ability to solve high-dimensional optimization tasks, [3] proposes a coevolution approach for combinational optimization problems and [4] further generalizes the approach to a universal framework: Cooperative Coevolution (CC). CC algorithms tackle the curse of dimensionality with a divide-and-conquer method which separates the search space into subspaces of lower dimensionality. They therefore *decompose* the decision vector into groups of variables which can be optimized cooperatively in cycles. After each cycle, the information gained in the separate optimization steps is joined for

* This work is partially supported by two National Natural Science Foundation of China grants (No. 60802036 and No. U0835002).

R. Schaefer et al. (Eds.): PPSN XI, Part II, LNCS 6239, pp. 300–309, 2010.

Algorithm 1: $(subpop, pop, \mathbf{best}) \longleftarrow$ **initializeCC**(NP)

```
for i ⟵ 1 to N do
    subpop_i ⟵ NP_i random one-dimensional samples from a given interval;
pop ⟵ (subpop_1, ..., subpop_N);
best ← arg min f(pop);
return (subpop, pop, best);
```

the next iteration. This approach yields good performance both in benchmark problems and real-world applications [4,5]. Despite this success, for the most important part of CC, the problem decomposition strategy, no satisfying solution has yet been developed. In this paper, we introduce a general, scalable, and highly efficient method for this purpose, called Cooperative Coevolution with Variable Interaction Learning (CCVIL).

The rest of this paper is organized as follows. In the next section, we give a short outline of the cooperative cooperationary idea in general and list related work in the area of problem decomposition in CC. CCVIL is then discussed in detail in Section 3 and experimentally studied by using a set of twenty large-scale benchmark problems in Section 4. It achieves excellent results in these experiments and frequently outperforms two related, state-of-the-art CC techniques as well as the optimizer which it uses internally. We finally conclude our paper in Section 5 where we also give pointers to future work.

2 Cooperative Coevolution

Many cooperative coevolutionary numerical optimization algorithms consist of three basic ingredients: [6]: *1)* A *decomposition* method used to divide the N-dimensional decision vector into groups $G_1 \ldots G_m$ of variables. Each such group is optimized with a separate subpopulation of the corresponding dimension $|G_i| < N$. *2)* In order to evaluate the fitness of the individuals from a certain subpopulation, a representative element from each of the other subpopulations is selected. In this *cooperation* step, a population of complete N-dimensional candidate solutions is constructed by concatenating the representatives to each element of the current subpopulation. *3)* An optimizer is applied to the population for (only) *optimizing* the decision variables in the current group.

In the conventional CC framework, optimizing a group with the corresponding subpopulation is called a *phase*. After finishing a *phase*, CC will turn to optimize the next group and start a new *phase*. One iteration over all groups constitutes a *cycle*. A CC algorithm performs several *cycles*. In Algorithms 1 and 2, we sketch the flow of a simple CC algorithm that treats the problem as completely separable [4].

2.1 Discovering Decision Variable Interactions

The decomposition strategy that identifies interacting decision variables and divides the search space into subspaces of lower dimensionality is the most important component of CC algorithms. A function f is separable according to [7] if Equation 1 holds, i.e., if its global optimum can be reached by successive line search along the axes. Therefore, if a certain function is not separable, there must be interactions between at least two variables in the decision vector. Motivated by this, we provide definition of *interaction*: two decision variables i and j are *interacting* if there is a decision vector $\boldsymbol{x}$ whose i^{th} and j^{th} variable can be substituted with values x'_i and x'_j so that Equation 2 holds.

$$\arg \min_{(x_1,\dots,x_N)} f(x_1,\dots,x_N) = \left(\arg\min_{(x_1)} f(x_1,\cdots),\dots,\arg\min_{(x_N)} f(\cdots,x_N)\right) \tag{1}$$

$$\begin{aligned}\exists \boldsymbol{x}, x'_i, x'_j : \; & (f(x_1,\dots,x_i,\dots,x_j,\dots,x_N) < f(x_1,\dots,x'_i,\dots,x_j,\dots,x_N)) \wedge \\ & (f(x_1,\dots,x_i,\dots,x'_j,\dots,x_N) > f(x_1,\dots,x'_i,\dots,x'_j,\dots,x_N))\end{aligned} \tag{2}$$

The idea behind the decomposition of the decision variables in CC into groups G_1, $G_2,\dots,G_m$ is that the fitness function f can be approximated as a linear combination of *component functions* $f_1, f_2,\dots,f_m$. The domains of the functions f_i have the lower dimensionality $|G_i| < N$ since their results only depend on the variables in the corresponding group G_i. Although it is usually only possible to compute f but not its components f_i, the knowledge that the groups G_i can be optimized separately can speed up the optimization process significantly.

For discovering interactions between variables in the *decomposition* step, a simple method is suggested in [8]: Assume that $\boldsymbol{best}$ would be the vector of the best values for each decision variable discovered so far. After coevolutionary optimizing dimension i, the best individual $\boldsymbol{new}$ in the population *popcc* of the optimizer only focusing on dimension i is extracted as well as a random candidate $\boldsymbol{rand}$ from the global population *pop* (with $\boldsymbol{new} \neq \boldsymbol{rand} \neq \boldsymbol{best}$). Based on these three vectors, two new candidate

$$x_j = \begin{cases} new_i & \text{if } j = i \\ best_j & \text{otherwise} \end{cases} \tag{3}$$

$$x'_j = \begin{cases} new_i & \text{if } j = i \\ rand_k & \text{if } j = k \\ best_j & \text{otherwise} \end{cases} \tag{4}$$

vectors $\boldsymbol{x}$ and $\boldsymbol{x}'$ are created according to Equations 3 and 4 for testing whether dimensions i and k interact. The value at index k of $\boldsymbol{x}$ is better than the kth value of $\boldsymbol{x}'$, since it comes from the vector of best known values $\boldsymbol{best}$ whereas x'_k stems from the

Algorithm 2: $\boldsymbol{best} \longleftarrow$ **basicCC**(*NP*) (as introduced in [4])

(*subpop*, *pop*, $\boldsymbol{best}$) $\longleftarrow$ **initializeCC**(*NP*);
// Decomposition: implicitly performed based on separability assumption
while *stoping criterion not met* **do** // Optimization: Start a new cycle
  for $i \longleftarrow 1$ **to** N **do** // Start a new phase
    for $j \longleftarrow 1$ **to** NP_i **do** // Collaboration
      $popcc_j \longleftarrow (best_1,\dots,best_{i-1}, subpop_{i,j}, best_{i+1},\dots,best_N)$
    $(popcc, \boldsymbol{new}) \longleftarrow$ **optimizer**$(popcc, i)$// Optimize the ith subcomponent
    $subpop_i \longleftarrow popcc_i$;
    $best_i \leftarrow new_i$;
return $\boldsymbol{best}$;

random population member $\boldsymbol{rand}$. Optimizing dimension i should not influence this relation and $f(\boldsymbol{x}) \leq f(\boldsymbol{x}')$ would hold if the variables were separable. Therefore, if $f(\boldsymbol{x}') < f(\boldsymbol{x})$, i.e., $\boldsymbol{x}'$ is better than $\boldsymbol{x}$, there likely is an interaction between dimension i and k [8].

2.2 Related Work

In the early stage of the development of CC, researchers mainly adopted two simple decomposition methods: one-dimension based and splitting-in-half methods [4,9]. These two methods do not take the interaction between components into consideration. Thus, they cannot solve problems consisting of non-trivial variable interactions.

In order to mitigate this problem, a multi-level pyramidal genetic algorithm is utilized in [10] to better deal with multiple-choice scheduling. In the area of numerical optimization, Yang et al. proposed two effective CC-based algorithms, Differential Evolution using Cooperative Coevolution with adaptive grouping strategy (DECC-G) [6] and Multilevel Cooperative Coevolution (MLCC) [11]. DECC-G uses a constant group size, for instance 100, and randomly decomposes the high-dimensional variable vector into several such groups. These are then optimized with a certain EA. DECC-G with a small group size works properly for separable problems while highly nonseparable problems can better be solved with large group size. The problem of DECC-G is that the best group size is not known in advance. In order to overcome it, MLCC adopts a multilevel strategy for decomposition. It maintains a decomposer pool from which decomposers with different group sizes are selected depending on the problem under investigation and the stage of the evolution. For nonseparable problems, MLCC tends to select the decomposers with large group sizes. In the opposite case, MLCC prefers to choose the decomposers with smaller group sizes. However, determining a good pool of decomposers is hard in practice since the interaction between variables is usually not known beforehand. This, in turn, would to lead a waste of function evaluations.

By using the technique of learning variable interactions used in [8] and outlined here in Section 2.1 CC can become more adaptable. Nevertheless, the way it is used in [8] suffers severe shortcomings. For example, the maximum group size is limited to two, which rarely is the case real-world problems. Moreover, the *choice* of the dimensions i and k for interaction investigation (see Equations 3 and 4), frequently leads to the detection of non-existing interactions [8]. In [8], it is possible that dimension i and k are not optimized in successive phases. Thus, changes in other dimensions of $\boldsymbol{best}$ may take place in the mean time which violates the condition of Equation 2.

CCVIL overcomes the problem of the DECC-G and the MLCC algorithm, i.e., the group sizing, by using the interaction learning method of [8]. The choice of the dimensions i and k for interaction detection, however, is conducted in a way which prevents the discovery of non-existing interactions.

3 Cooperative Coevolution with Variable Interaction Learning

We propose a novel CC-framework called *Cooperative Coevolution with Variable Interaction Learning* (CCVIL) with incremental group sizes for solving large-scale optimization problems of separable, partially-nonseparable, and nonseparable character.

Algorithm 3: *groupInfo* $\longleftarrow$ Learning Stage of CCVIL

$K \longleftarrow 0$;
groupInfo $\longleftarrow \{\{1\},\{2\},\ldots,\{N\}\}$// initially assume full separability
repeat // start a new learning cycle
    $\Pi \longleftarrow$ random permutation of dimension indices $\{1,2,\ldots,N\}$;
    $K \longleftarrow K+1$;
    $lastIndex \longleftarrow 0$;
    $(subpop, pop, \mathbf{best}) \longleftarrow$ **initializeCC**$(3,3,\ldots,3)$// use $NP_i = 3\ \forall i \in 1..N$
    for $i = 1$ **to** N **do** // start a new learning phase, i.e., tackle next dimension
        if $lastIndex \neq 0$ **then**
            $G_1 \longleftarrow$ **find**$(groupInfo, \Pi_i)$ // find the group containing Π_i
            $G_2 \longleftarrow$ **find**$(groupInfo, lastIndex)$ // find group of last optimized variable
        if $i = 1 \vee (G_1 \neq G_2)$ **then**
            for $j = 1$ **to** NP **do**
                $\mathbf{popcc}_j \longleftarrow (\mathbf{best}_1, \ldots, \mathbf{best}_{\Pi_i - 1}, \mathbf{subpop}_{\Pi_i, j}, \mathbf{best}_{\Pi_i+1}, \ldots)$
            $(popcc, \mathbf{new}) \longleftarrow$ **optimizer**$(popcc, \Pi_i)$ // any optimizer, we used JADE
            $\mathbf{subpop}_{\Pi_i} \longleftarrow \mathbf{popcc}_{\Pi_i}$;
            $\mathbf{best}_{\Pi_i} \leftarrow \mathbf{new}_{\Pi_i}$;
            if $lastIndex \neq 0$ **then**
                Compose $\boldsymbol{x}$ and $\boldsymbol{x}'$ according to Equations 3 and 4;
                if $f(\boldsymbol{x}) < f(\boldsymbol{x}')$ **then** // interaction between dim. i and $lastIndex$?
                    groupInfo $\longleftarrow ((groupInfo \setminus \{G_1\}) \setminus \{G_2\}) \cup (\{G_1 \cup G_2\})$;
            $lastIndex \longleftarrow \Pi_i$ // *only* test successively optimized dimensions
until $(|groupInfo| = 1) \vee [(K > \check{K}) \wedge (|groupInfo| = N)] \vee (K > \hat{K})$;
return *groupInfo* // return the set of mutually separable groups of interacting variables

Instead of setting the group sizes as a constant or defining a set of values from which to choose them, we allow the optimization process to adapt them by learning the interaction between the decision variables. The whole procedure of CCVIL is divided into two stages, the *learning stage* and *optimization stage* executed once in exactly this sequence. Both stages are divided into cycles and phases, similar to the conventional CC framework (see Section 2).

3.1 Learning Stage

The *learning stage* of CCVIL optimizes one dimension after the other, similar to the simple CC approach given in Algorithm 2. While doing this, it only tests the currently and the previously optimized dimension for interaction. Since only these two dimensions changed between the application of the interaction detection mechanism of [8], Equation 2 can never be violated and only becomes true for real interactions. Therefore, the flaw of detecting non-existent interactions is avoided. Before each learning cycle, the order of visiting the dimensions is randomly permutated so that each two dimension have the same chance to be processed in a row. The details of the learning stage are presented in Algorithm 3.

The efficiency of CCVIL becomes clear from a stochastic point of view. The probability of placing any two (possibly interacting) dimensions i and j in an N-dimensional problem adjacently in one random permutation Π is $2/N$. The probability $p_{capt}(K)$ that this happens in at least once in K learning cycles then is given in Equation 5.

$$p_{capt}(K) = 1 - (1 - 2/N)^K \tag{5}$$

In a 1000-dimensional problem, the probability of putting any two (possibly interacting) variables adjacently in a random permutation and examining interaction between them during $K = 500$ cycles is already $p_{capt}(K) = 0.6325$ and raises to 0.7984 for 800 cycles. Given a limited number of fitness function evaluations, as many learning cycles as possible should thus be performed. Therefore, the population size and generation limit of the internal optimizer should be as small as possible during the learning stage (3 and 1, respectively, in this work). CCVIL issues an independent restart for each cycle to prevent possible loss of population diversity during the learning stage.

We furthermore set a lower and an upper threshold for the number of learning cycles: $\check{K}$ and $\hat{K}$. If CCVIL cannot detect interactions between any two variables in the first $\check{K}$ cycles of learning stage, it assumes that the problem is fully separable or that the interactions are rather weak and immediately switches to the optimization stage. Additionally, a transition to the optimization stage is enforced after $\hat{K}$ cycles even if not all interactions have been discovered in order to limit the runtime used for learning. As default settings, we recommend 10 for $\check{K}$ and to set $\hat{K}$ to the number of cycles needed to achieve 80% for $p_{capt}(\hat{K})$ (see Equation 5). This number can be computed by $\hat{K} \geq \log(1 - 0.8)/\log(1 - 2/N)$. However, $\hat{K}$ should not lead to a consumption of more than 60% of the function evaluations in the learning stage.

3.2 Optimization Stage

The user of our CC framework is completely free in the choice of the optimizer to be used for the variables grouped together. During both, the interaction learning and the optimization stage of CCVIL, we apply JADE for this purpose. JADE is an enhanced variant of Differential Evolution (DE) [12,13] with improved speed and reliability compared with plain DE [14]. In each phase of the optimization stage of CCVIL, the optimizer is applied to the complete subspace defined by one group $G_i \in \mathit{groupInfo}$ (whereas the remaining variables of the candidate vectors are constant and correspond to the representatives from the other groups).

During the learning stage, CCVIL may divide the variables into groups of different sizes. In the optimization step, the population size NP_i and number of generations Gen_i granted to the optimizer for processing the group G_i should depend on its size. As a rule of thumb, we set $Gen_i = \min\{|G_i| + 5, 500\}$. Whereas the population size NP_i of JADE is set as small as possible during the interaction learning, in the optimization stage, we apply an adaptive strategy. The initial value here is $NP_i = |G_i| + 10$, which is sufficient for unimodal problems. After the population loses its diversity and ceases to improve, an independent restart with thrice the population size is performed as suggested in [15]. This is done when the *relative* fitness improvement of the current cycle compared to the previous on is below 10^{-2}, i.e., $(f_{K-1} - f_K)/f_K < 10^{-2}$.

Table 1. The main characteristics of the 20 benchmark functions [7] used in this paper

	Function	Sep	Multi modal	Groups real	Groups found
f_1	Shifted Elliptic Function	Yes	No	1000	1000
f_2	Shifted Rastrigin's Function	Yes	Yes	1000	1000
f_3	Shifted Ackley's Function	Yes	Yes	1000	969
f_4	Single-group Shifted 50-rotated Elliptic Function	No	No	951	963
f_5	Single-group Shifted 50-rotated Rastrigin's Function	No	Yes	951	952
f_6	Single-group Shifted 50-rotated Ackley's Function	No	Yes	951	921
f_7	Single-group Shifted 50-dimensional Schwefel's	No	No	951	952
f_8	Single-group Shifted 50-dimensional Rosenbrock's	No	Yes	951	1000
f_9	10-group Shifted 50-rotated Elliptic Function	No	No	510	627
f_{10}	10-group Shifted 50-rotated Rastrigin Function	No	Yes	510	516
f_{11}	10-group Shifted 50-rotated Ackley Function	No	Yes	510	501
f_{12}	10-group Shifted 50-dimensional Schwefel's	No	No	510	522
f_{13}	10-group Shifted 50-dimensional Rosenbrock's	No	Yes	510	1000
f_{14}	20-group Shifted 50-rotated Elliptic Function	No	No	20	232
f_{15}	20-group Shifted 50-rotated Rastrigin's Function	No	Yes	20	37
f_{16}	20-group Shifted 50-rotated Ackley Function	No	Yes	20	39
f_{17}	20-group Shifted 50-rotated Schwefel's Function	No	No	20	42
f_{18}	20-group Shifted 50-rotated Rosenbrock's Function	No	Yes	20	1000
f_{19}	Shifted Schwefel's Function 1.2	No	No	1	1
f_{20}	Shifted Rosenbrock's Function	No	Yes	1	1000

Furthermore, the difficulties in solving the component functions may vary a lot. Therefore, it is reasonable to stop optimizing converged groups, when no improvements can be achieved for a certain number of cycles (in the context of this work, we set this threshold to five).

4 Experimental Studies

4.1 Experimental Setup

For benchmarking CCVIL, we choose the set of twenty 1000-dimensional functions provided by in [7]. These functions represent high-dimensional problems with different degrees of variable interactions. This makes them especially suitable to test the ability of our algorithm which was designed for tackling this kind of tasks. We compared CCVIL to DECC-G [6], MLCC [11], and JADE [14]. In the experiments, we grant three million fitness function evaluations to each algorithm run. We fixed the population size in JADE to 1000 and used default parameter settings for DECC-G and MLCC are obtained from the related publications [6,11]. For each algorithm and benchmark function, 25 independent runs were performed.

4.2 Benchmark Functions and Learned Groups

In Table 1, we list the characteristics of the benchmark functions applied in our experiments. The column *Sep* denotes functions which are separable according to Equation 1.

Most of the non-separable functions consist of mutually separable groups of interacting variables. f_{10} is the sum of ten rotated instances of Rastrigin's function applied to groups of 50 decision variables each and one non-rotated instance of the same function applied to the remaining 500 decision variables. Since the plain Rastrigin's function is separable (and the rotated version is not), an ideal interaction learning stage would thus discover 500 separable groups of size 1 and plus 10 groups of size 50, as listed in column *Groups (real)*. f_{18} is composed of 20 rotated Rosenbrock's functions, each applied to 50 decision variables, leading to 20 "real" groups.

In the last column of Table 1, we noted the median of the number of groups discovered by CCVIL during the 25 runs. From these results, we can clearly see that CCVIL most often is able to represent the interactions between the variables correctly. Due to the limitation $\hat{K}$ imposed on the runtime of the learning stage, it may not discover all interactions, i.e., may not merge all interacting groups, and thus, yield a slightly higher number of groups.

Furthermore, we notice that, for most of the benchmark functions related to Ackley's function (f_3, f_6, f_{11}), CCVIL combines more variables than expected. The reason for this is that Ackley's function is separable according to Equation 1, but not *additively separable* [16], i.e., it cannot be divided into an exact arithmetic sum of component functions, as pointed out in [17]. Therefore, our algorithm correctly picks up interactions since it aims at representing the fitness function as linear combination of component functions.

The decision variables of Rosenbrock's function, although entirely nonseparable, exhibit only a very weak interaction. Our algorithm discovers that functions f_8, f_{13}, f_{18} and f_{20} can best be treated as separable problems, i.e., problems with 1000 independent decision variables.

4.3 Comparison with Other CC-Based Algorithms and JADE

In Table 2, we provide the results of the comparison of our algorithm with DECC-G, MLCC, and plain JADE. We list the mean results of the 25 runs for each benchmark as well as the standard deviations. The columns named R denote the outcomes of a two-tailed Mann-Whitney U test [18] with 5% significance level. A W in column R_1 stands for statistically significant win of CCVIL against both, DECC-G and MLCC, a L a loss, and "–" means that no significant difference was found. Column R_2 represents the same comparison between CCVIL and the native JADE (used as optimizer in CCVIL).

Table 2 shows that CCVIL is clearly superior to DECC-G and MLCC. It outperforms them in 15 benchmarks and only loses on f_3, the separable but not *additively* separable Ackley function. CCVIL also wins against its internal optimizer JADE alone in ten out of 20 benchmark functions and loses in six. Especially for separable and highly non-separable functions, CCVIL proofs to be advantageous. The performance of CCVIL is worse than its JADE in most single-group non-separable functions. The reason is the structure of the benchmark set [7] where a large factor (1 million) is put in front of the nonseparable component function. If even a single interaction is not discovered by CCVIL, it will perform worse than JADE which treats the fitness function as completely nonseparable. This fact leads us to the conclusion that CCVIL can achieve much better results if the learning phase can proceed sufficiently long to discover all interactions.

Table 2. Comparison with other CC-based algorithms and plain JADE

	CCVIL		DECC-G		MLCC		R_1	Naive JADE		R_2
	Mean	Std Dev	Mean	Std Dev	Mean	Std Dev		Mean	Std Dev	
f_1	1.55e-17	7.75e-17	2.93e-07	8.62e-08	1.53e-27	7.66e-27	–	1.57e+04	1.38e+04	W
f_2	6.71e-09	2.31e-08	1.31e+03	3.24e+01	5.55e-01	2.20e+00	W	7.66e+03	9.67e+01	W
f_3	7.52e-11	6.58e-11	1.39e+00	9.59e-02	9.86e-13	3.69e-12	L	4.52e+00	2.41e-01	W
f_4	9.62e+12	3.43e+12	5.00e+12	3.38e+12	1.70e+13	5.38e+12	W	6.14e+09	3.81e+09	L
f_5	1.76e+08	6.47e+07	2.63e+08	8.44e+07	3.84e+08	6.93e+07	W	1.35e+08	1.21e+07	L
f_6	2.94e+05	6.09e+05	4.96e+06	8.02e+05	1.62e+07	4.97e+06	W	1.94e+01	1.79e-02	–
f_7	8.00e+08	2.48e+09	1.63e+08	1.38e+08	6.89e+05	7.36e+05	–	2.99e+01	3.30e+01	–
f_8	6.50e+07	3.07e+07	6.44e+07	2.89e+07	4.38e+07	3.45e+07	–	1.19e+04	4.92e+03	L
f_9	6.66e+07	1.60e+07	3.21e+08	3.39e+07	1.23e+08	1.33e+07	W	2.70e+07	2.08e+06	L
f_{10}	1.28e+03	7.95e+01	1.06e+04	2.93e+02	3.43e+03	8.72e+02	W	8.50e+03	2.30e+02	W
f_{11}	3.48e+00	1.91e+00	2.34e+01	1.79e+00	1.98e+02	6.45e-01	W	9.29e+01	9.66e+00	W
f_{12}	8.95e+03	5.39e+03	8.93e+04	6.90e+03	3.48e+04	4.91e+03	W	6.21e+03	1.34e+03	–
f_{13}	5.72e+02	2.55e+02	5.12e+03	3.95e+03	2.08e+03	7.26e+02	W	1.87e+03	1.11e+03	W
f_{14}	1.74e+08	2.68e+07	8.08e+08	6.06e+07	3.16e+08	2.78e+07	W	1.00e+08	8.84e+06	L
f_{15}	2.65e+03	9.34e+01	1.22e+04	9.10e+02	7.10e+03	1.34e+03	W	3.65e+03	1.09e+03	W
f_{16}	7.18e+00	2.23e+00	7.66e+01	8.14e+00	3.77e+02	4.71e+01	W	2.09e+02	2.01e+01	W
f_{17}	2.13e+04	9.16e+03	2.87e+05	1.97e+04	1.59e+05	1.43e+04	W	7.78e+04	5.87e+03	W
f_{18}	1.33e+04	1.00e+04	2.46e+04	1.05e+04	7.09e+03	4.77e+03	–	3.71e+03	9.58e+02	L
f_{19}	3.52e+05	2.04e+04	1.11e+06	5.00e+04	1.36e+06	7.31e+04	W	3.48e+05	1.67e+04	–
f_{20}	1.11e+03	3.04e+02	4.06e+03	3.66e+02	2.05e+03	1.79e+02	W	2.06e+03	2.01e+02	W

To find a good strategy to distribute runtime between the learning and the optimization stage of CCVIL is thus an interesting point for future research.

5 Conclusions and Future Work

In this paper, we introduced a novel CC framework called Cooperative Coevolution with Variable Interaction Learning, or CCVIL for short. In the related work study, we showed that currently no efficient method for finding groups of interacting variables in CC exists. CCVIL fills this gap with a two-stage approach: In a learning step, the interactions between the decision variables of a (potentially very high-dimensional) search space are detected. In the second stage, these groups are optimized according to the traditional CC model. We showed in an experimental study based on twenty 1000-dimensional benchmark functions with different degrees of variable interaction and group sizes that CCVIL can outperform two very efficient, state-of-the-art CC approaches as well as its internal optimizer JADE.

The experiments also showed the drawback of CCVIL: finding the optimal distribution of runtime between the learning and the optimization stage is an open question. Here, we could only provide some simple rules-of-thumb, but this problem surely will be subject of our future work. In order to reduce the overall learning time, we will furthermore explore generating the permutations Π in our algorithm according to a deterministic scheme instead of creating them randomly. Additionally, we wish to experiment with possibly more efficient optimizers such as CMA-ES [19] as internal optimizers.

References

1. Sarker, R., Mohammadian, M., Yao, X.: Evolutionary Optimization. Kluwer Academic Publishers, Norwell (2002)
2. van den Bergh, F., Engelbrecht, A.: A cooperative approach to particle swarm optimization. IEEE Transactions on Evolutionary Computation 8(3), 225–239 (2004)
3. Husbands, P., Mill, F.: Simulated co-evolution as the mechanism for emergent planning and scheduling. In: 4th Intl. Conf. on Genetic Algorithms, pp. 264–270. Morgan Kaufmann, San Francisco (1991)
4. Potter, M.A., De Jong, K.A.: Cooperative coevolution: architecture for evolving coadapted subcomponents. Evolutionary Computation 8(1), 1–29 (2000)
5. Yong, C.H., Miikkulainen, R.: Cooperative coevolution of multi-agent systems. Technical Report AI01-287, University of Texas at Austin, Austin, TX, USA (2001)
6. Yang, Z., Tang, K., Yao, X.: Large scale evolutionary optimization using cooperative coevolution. Information Sciences 178(15), 2985–2999 (2008)
7. Tang, K., Li, X., Suganthan, P.N., Yang, Z., Weise, T.: Benchmark functions for the CEC'2010 special session and competition on large scale global optimization. In: TR, NICAL, USTC, Hefei, Anhui, China (2009), `http://nical.ustc.edu.cn/cec10ss.php`
8. Weicker, K., Weicker, N.: On the improvement of coevolutionary optimizers by learning variable interdependencies. In: IEEE CEC, pp. 1627–1632. IEEE Press, Los Alamitos (1999)
9. Potter, M.A., De Jong, K.A.: A cooperative coevolutionary approach to function optimization. In: 3rd Conf. on Parallel Problem Solving from Nature, vol. 2, pp. 249–257 (1994)
10. Aickelin, U.: A Pyramidal Evolutionary Algorithm with Different Inter-Agent Partnering Strategies for Scheduling Problems. In: GECCO Late-Breaking Papers, pp. 1–8
11. Yang, Z., Tang, K., Yao, X.: Multilevel cooperative coevolution for large scale optimization. In: IEEE Congress on Evolutionary Computation, pp. 1663–1670. IEEE Press, Los Alamitos (2008)
12. Price, K., Storn, R., Lampinen, J.A.: Differential Evolution: A Practical Approach to Global Optimization. Springer, Heidelberg (2005)
13. Chakraborty, U.K. (ed.): Advances in Differential Evolution. Springer, Berlin (2008)
14. Zhang, J., Sanderson, A.C.: JADE: Adaptive differential evolution with optional external archive. IEEE Transactions on Evolutionary Computation 13(5), 945–958 (2009)
15. Auger, A., Hansen, N.: A restart CMA evolution strategy with increasing population size. In: IEEE Congress on Evolutionary Computation, vol. 2. IEEE Press, Los Alamitos (2005)
16. Streeter, M.J.: Upper bounds on the time and space complexity of optimizing additively separable functions. In: GECCO, pp. 186–197. Springer, Heidelberg (2004)
17. Hansen, N., Kern, S.: Evaluating the CMA evolution strategy on multimodal test functions. In: Parallel Problem Solving from Nature – PPSN VIII, pp. 282–291. Springer, Heidelberg (2004)
18. Mann, H.B., Whitney, D.R.: On a test of whether one of two random variables is stochastically larger than the other. The Annals of Mathematical Statistics 18(1), 50–60 (1947)
19. Hansen, N., Müller, S.D., Koumoutsakos, P.: Reducing the time complexity of the derandomized evolution strategy with covariance matrix adaptation (CMA-ES). Evolutionary Computation 11(1), 1–18 (2003)

EvoShelf: A System for Managing and Exploring Evolutionary Data

Timothy Davison[1], Sebastian von Mammen[1], and Christian Jacob[1,2]

[1] Dept. of Computer Science, Faculty of Science
[2] Dept. of Biochemistry & Molecular Biology, Faculty of Medicine
University of Calgary, Canada
{tbdaviso,s.vonmammen,cjacob}@ucalgary.ca

Abstract. Systems that utilize evolutionary computation produce large amounts of data. Quite often, this data has a convenient visual representation. However, managing and visualizing evolutionary data can be a difficult and onerous task. By employing techniques used in photo management software, we have produced a system that helps to organize and visualize evolutionary data while retaining a complete record of a simulation. By means of a simple plugin architecture this system can be extended to import data produced by arbitrary evolutionary systems. We present the system's architecture, its features, and we provide a comprehensive example, highlighting its advantages in applied research.

1 Introduction

Evolutionary systems produce large amounts of data. Beyond the obvious data (such as the genotype and phenotype of an individual), there is a considerable amount of meta-data produced as well. Such data includes the hereditary data, fitness values, and other attributes of the evolutionary computation approach being employed.

It is common to manage experimental data by means of a file-system browser, such as the Finder in Mac OS X, and Windows Explorer in Microsoft Windows. Searching or organizing individuals according to various criteria is a laborious task in such systems. Consider a system that organizes the individuals produced by an experiment into sub-directories by generation, giving each individual its own file containing its genotype, and phenotype, along with meta-data such as fitness, or genealogy. Filtering these individuals by fitness value would be a difficult task with either file-system browser.

An evolutionary system may employ an interface of its own for browsing the data that it produces. In this case, the visualization procedures and the management of the genotype/phenotype data are typically implemented specifically for the one evolutionary system. However, the universality of evolutionary algorithmic approaches renders generic visualization and data management techniques valuable across various application domains.

In a way, the situation is very similar to managing individual (digitized) image and music collections. Such libraries can easily consist of thousands of items. A

R. Schaefer et al. (Eds.): PPSN XI, Part II, LNCS 6239, pp. 310–319, 2010.

number of applications have made the organization and management of such data much easier for the end user [1,2,4,9]. We propose a system that can deal with evolutionary data with the same ease of use and flexibility as provided by these mainstream media management applications.

EvoShelf is an extensible system that allows for the importing and exploration of evolutionary data from evolutionary systems. It solves the challenge of organizing imported metadata by providing a navigable, and searchable image-based browser that uses interface design elements from Apple's photo and music management software iPhoto [1], and iTunes [2]. Furthermore, it provides a plugin framework for building additional import modules and visualizations. Both navigation and visualization are optimized for real-time interaction.

In Section 2 we explore the topic of visualizing data in evolutionary systems. Section 3 presents the design of the *EvoShelf* system and its graphical user interface. It also touches upon some of the example visualization techniques included in *EvoShelf*, as well as details about its plugin architecture. In Section 4 we use *EvoShelf* in coordination with an existing evolutionary system for semi-interactive evolutionary computing, and for analyzing the results produced by that system. We will conclude in Section 5 with a summary of our work, along with possible directions in which to take it in the future.

2 Related Work

We briefly outline the data management and user-interface approach of various software systems that inspired the *EvoShelf* visualization and management system. Secondly, we outline various techniques that have been developed for visually supporting computational evolutionary experiments.

2.1 Digital Media Libraries

The framework presented in this article was mainly inspired by iPhoto, Apple's mainstream photo management application [1]. It is capable of organizing and browsing thousands of photos. Despite the large amounts of information that it is capable of presenting to the user, it maintains a very simple and intuitive interface. It consists of two primary views, an organizer view, and an image browsing view (Figure 1(a)). Multiple images, up to and including an entire library of photos, are displayed in the browsing view. The organizer view is used to filter this view into subsets of photos, such as those represented by a photo album containing the user's favorite photos. As photos are imported into the system they are grouped into events. Pictures taken during a certain period of time might have all been taken during a vacation and the respective group of photos could be labeled after the location of the recreational stay.

iTunes is another application from Apple Inc. that manages a large amount of data in a similar fashion to iPhoto. Unlike iPhoto, whose interface is focused on visualizing and managing photos, iTunes is targeted towards playing music and organizing large digital music collections. Visual cover art often decorates individual music files, but the iTunes library is mainly organized by sorting and searching through textual meta-data such as artist name, music category, or album name (Figure 1(b)). Together, iPhoto and iTunes suggest an interface that combines visualization and meta-data management techniques that could be very powerful for organizing evolutionary data.

(a) (b)

Fig. 1. The user interfaces of the media management applications (a) iPhoto and (b) iTunes

2.2 Evolutionary Visualization Techniques

Various data visualization techniques have been presented in the context of evolutionary computing. On the one hand, individuals can be compared at a glance based on their multi-dimensional genotypes, independent of the respective interpretation or phenotype. On the other hand, methods of visualization have been developed that capture characteristics of whole populations, allowing one to visually track the evolutionary process.

Pohlheim, for instance, presented a toolkit of convergence diagrams, 3D line plots, and 2D image plots, to visualize the evolution of fitness values and other individual attributes [10]. Hart and Ross introduced a tree-based visualization to trace the ancestry of the best individual produced by an evolutionary run [5]. Daida et al. unfold genetic ancestry onto concentric circles on a 2D plane to create a compact and highly scalable visualization of hereditary processes [3]. Wu et al. represent genotypes as sequences of color coded stripes whose colours correspond to different genes [12]. Keim et al. designed a system to visualize search queries on a (relational) database [7]. Data items that match the query most closely are arranged in the center of a spiral arrangement. This visualization technique can be used to relate individuals in an evolutionary system in arbitrary ways, e.g. by comparing fitnesses or individual attributes. In [8], Khemka and Jacob have closely investigated the possibilities to visualize population-based optimization processes at various levels of scale—from the individual to sets of experiments. They provide an easily adaptable user interface with various interactive manipulators to explore optimization processes across these scales.

3 The *EvoShelf* System

The interface of *EvoShelf* is divided into three window panes (Figure 2). The organization view on the left-hand side is used for selecting and grouping imported experimentation data (Section 3.1). The user's selection is shown in the browser view in the center pane. An inspector view (right-hand side) shows further details about an individual or an experiment. In addition to importing and inspecting functions, the toolbar at the top of the window gives access to built-in visualization methods which are explained in Section 3.2. Typically, a user of *EvoShelf* writes a plugin to import and visualize data for his respective evolutionary system, if it does not already exist. We provide details about plugins in Section 3.3.

EvoShelf makes use of lazy fetching of data. That is, images and attributes of an individual are only loaded when they are needed. When individuals go off screen, their data is unloaded. In this way, we have manipulated data sets with over 40,000 individuals. Conceivably, *EvoShelf* can work with even larger datasets. To further increase the scalability of *EvoShelf*, high resolution images of individuals are loaded on demand—if no zoom is required, a low resolution image is displayed instead.

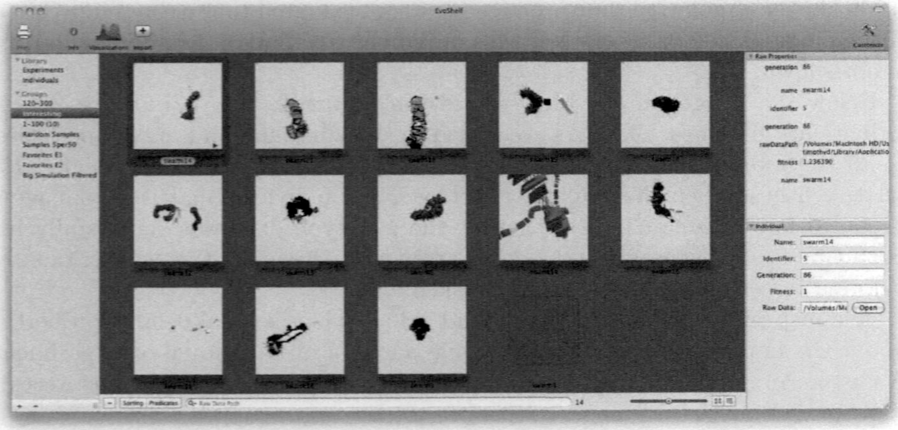

Fig. 2. The graphical user interface of *EvoShelf*

3.1 Individuals, Experiments, and Groups

The organizational view to the left of Figure 2 is divided into a library section and a groups section. In the library section, the user can select either *Individuals* or *Experiments*. In particular, the *Individuals* selection displays the images of all the individuals in the library, whereas *Experiments* shows representative thumbnails of all the imported experiments. The user can browse through the set of individuals of any experiment by hovering with the mouse over its thumbnail. The individuals of an experiment are revealed when the user double clicks on the experiment.

The data in the browser view can be sorted or filtered by the experiments' and individuals' attributes. Once the user has formed a suitable selection he can save his selection in a group, which would be equivalent to photo albums or playlists in [1,2] respectively. In Figure 2, a group labelled *Interesting* is selected, which hosts individuals from multiple experiments that the authors found of interest. Groups can be organized hierarchically. That is, one can form groups containing groups. When such a group is selected a union is formed from all of the individuals contained within the subgroups.

The controls at the bottom of the interface allow the user to remove individuals from a group or from the library, to sort individuals, to search for individuals according to arbitrary attributes (such as fitness value or generation), to scale the size of the images displayed, and to change the display mode. One display mode shows individuals as a collection of images, another one lists them in tabular format. The latter view is convenient for sorting and searching through individuals based upon numeric or textual attributes.

The specimen in the upper left corner is selected in the browser view in Figure 2. The image representing the individual was generated by the evolutionary system used as a test run for *EvoShelf* (see Section 4). A play button (a right pointing arrow) is projected on top of the specimen's description. It allows one to re-run the simulation that produced and/or evaluated the selected individual. If the plugin does not support re-runs of individual simulations, the play button is not shown.

Below the images in the browser view in Figure 2, blue bars represent the individuals' fitnesses. The bars are scaled to the minimum and maximum fitness of all the individuals currently displayed in the browser view. The higher the relative fitness, then the brighter and longer the individual bar is. No image is provided for *Swarm35* indicating that the genotype data was successfully imported but no image was found—in the given case, the simulation was terminated before a screenshot would have been taken.

The inspector view on the right hand side displays several default properties about the imported data, such as the file name of an individual or experiment. A custom interface for the inspector can be defined via the plugin architecture (Section 3.3).

3.2 Built-in Visualization Techniques

EvoShelf employs two basic built-in visualization techniques: star plots of name-value pairs [8] and FitnessRiver, a derivative of the ThemeRiver^{TM} method, which integrates local numeric values with global trends [6].

A star plot in *EvoShelf* visualizes a set of name-value pairs as a series of radially arranged line segments (Figure 3(a)). The length of a line segment is representative of an attribute's value and it is normalized to the attribute's maximum value in respect to the selected individuals. An attribute's line segment will consistently appear at the same location in a star plot to render individuals comparable.

The ThemeRiver^{TM} visualization method produces a stream diagram that is read from left to right. Currents in the stream represent individual themes that occur, grow and decay over time. Currents are visually differentiated from each other by way of colour, and those colours may be reused for non-adjacent currents. Instead of separating equivalent attributes into individual currents of a stream diagram, our FitnessRiver visualization method stacks the fitness values of individuals on top of each other. The width of a current is proportional to the fitness of the corresponding individual. Different colours are used to distinguish between individuals. Discontinuing currents indicate the removal of an individual from the evolutionary process. In the FitnessRiver visualization the x-axis represents the sequence of generations. A flat baseline is used so that the user has a greater sense of the progression of the fitness evolution (Figure 3(b)).

In Figure 3(b) we can see a large jump in the overall fitness at about the middle generation. When we look closely, we see that there are a few very successful individuals in the previous generation. We can see how these individuals likely contributed to the next generation. Furthermore, the majority of individuals in the new generation have noticeably more fitness than those in the previous generation.

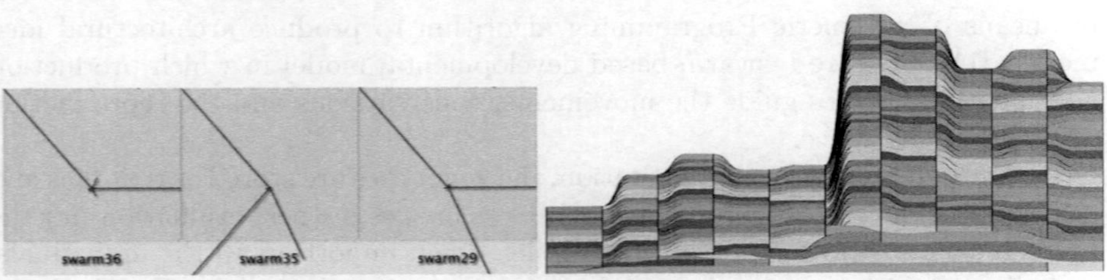

(a) (b)

Fig. 3. (a) Individuals are comparable based on their star plots. (b) The FitnessRiver visualization shows the evolution of local and the global fitness. It is an adaptation of ThemeRiver^{TM} [6].

3.3 Plugins

A user can define additional import modules, visualization modules, data models, and finally custom inspector views for custom data models[1]. A few basic

[1] *EvoShelf* plugins are written in Objective-C and should follow Apple's Cocoa API `http://developer.apple.com/cocoa/`.

classes are provided for these modules and models that serve as plugin templates. The importing process, including control over import dialogue windows, can be adapted and alternative visualization modules can be subclassed from the *EvoShelf* visualization view controller class.

The default data model (Figure 4) is well suited for generational systems such as evolutionary algorithms (EA) but also supports other heuristic computation concepts, such as particle swarm optimization (PSO). For instance, each step in a PSO simulation could correspond to a generation in an EA. *EvoShelf* can be adapted for other evolutionary systems by extending the default data model to add new attributes, or relationships. For instance, the set of attributes of the classes *EVExperiment* and *EVIndividual* can be adapted to match a given evolutionary system. The new attributes automatically determine the searching and sorting options in *EvoShelf*, as well as the information provided by the inspector view. In case a more elaborate inspector view is desired, an interface constructed in Apple's WYSIWG Interface Builder application can be loaded.

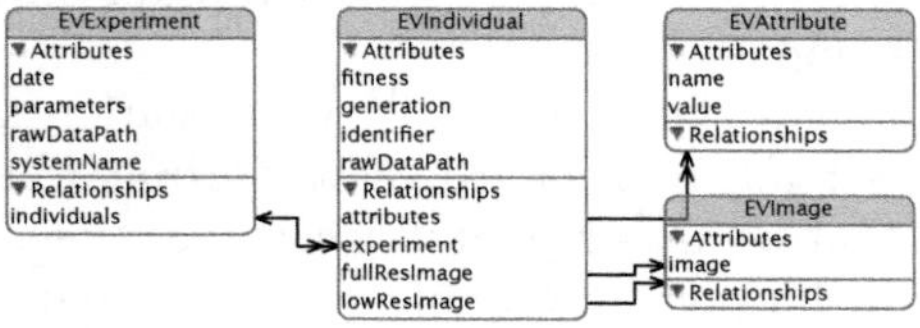

Fig. 4. The default data model for importing and managing *EvoShelf* data

4 Example Scenario

In this section, we explore the use of *EvoShelf* with a preexisting evolutionary system. In the evolutionary system of choice, Swarm Grammars (SGs) are bred by means of a Genetic Programming algorithm to produce architectural idea models [11]. SGs are a swarm-based developmental model in which production and interaction rules guide the movements, constructions and the reproduction of agents in 3D space.

In a subdirectory for each generation, the genotypes are stored as text files and snapshots of the corresponding phenotypes as images. Fitness evaluations for the individuals are stored in an additional file. When importing all the individuals, including their image representations and their meta-data into *EvoShelf*, the original directory structure is automatically copied into *EvoShelf*'s database.

Figure 5(a) shows a set of interesting SG specimens. We want to emphasize that due to the partially very low fitness values of the 2nd, 4th and 20th swarms from the top, we would have very likely not have inspected these phenotypes without relying on *EvoShelf*'s visual browsing functionality. Based on these undervalued, interesting phenotypes, we were able to improve the fitness function that drives the SG evolution by repositioning the geometrical focus of the fitness evaluation in respect to the SGs' constructions to better suit the favored ones. We also used *EvoShelf* for a semi-interactive evolutionary process by repeatedly selecting and exporting interesting individuals, modifying the fitness function and parameters to the GA, breeding their offspring for a fixed number of generations and importing the outcome (Figure 5).

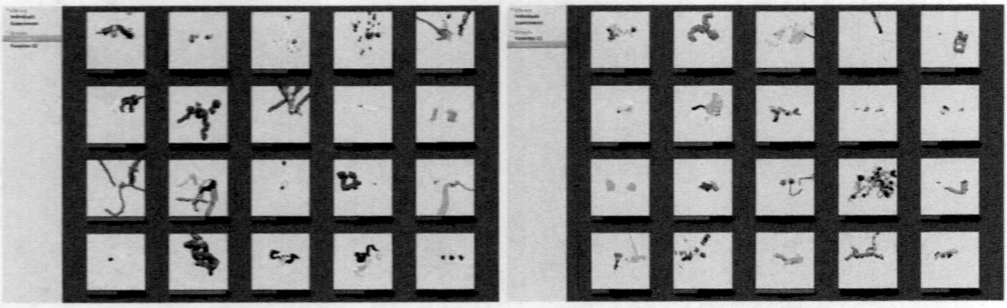

(a) (b)

Fig. 5. (a) 20 interesting individuals are selected from an experiment and served as the initial generation for a (b) follow-up experiment

We discovered that the SG GP evolution usually converged prematurely after at most several hundred iterations. Figure 6 shows the FitnessRiver plot over 300 generations. Overall 20,000 individuals were computed and imported into *EvoShelf*. We noticed that the overall fitness of our individuals had stagnated by the 100*th* generation (there is a very slight improvement in fitness past this point). Figure 7 confirmed our assumption of over-fitting: Up to the fitness stagnation at around generation 100, we randomly chose and plotted one of the ten best individuals every ten generations. For the period afterwards, we plotted one of the ten best individuals at random every 20 generations. And indeed, the phenotype images in combination with the star plots reveal a one-sided development, most easily recognizable by the inverted T-shaped star plots. Upon closer investigation, this similarity corresponds to the deployed amounts of two out of three construction elements provided to the SG agents (rods and layers), and the amount of construction elements that were placed outside of the intended target area. As the latter construction elements reduce the fitness of an individual, their increase might explain the fitness fluctuation as observed in Figure 6.

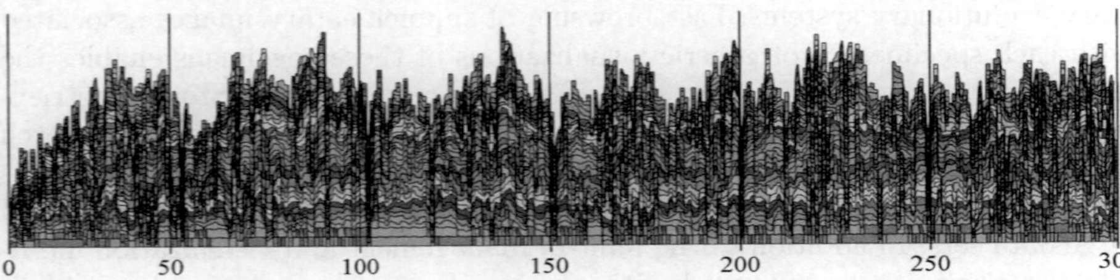

Fig. 6. The FitnessRiver plot shows stagnating and fluctuating fitness development after about 100 generations. The vertical lines denotes each 50*th* generation.

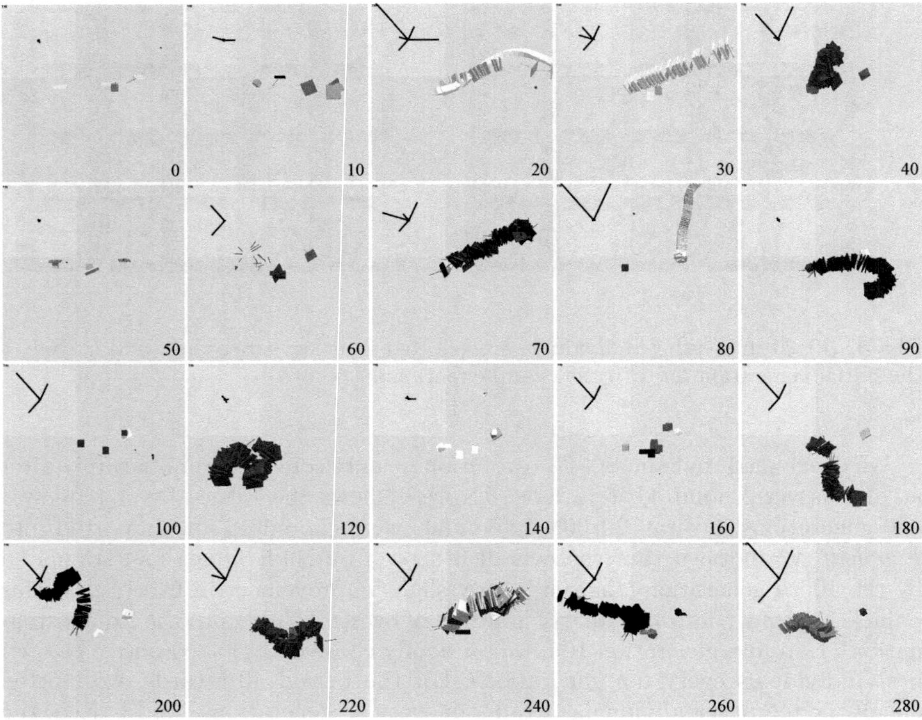

Fig. 7. First, for every ten generations, then (2*nd* half) for every 20 generations, a star plot and phenotype of a randomly selected individual is shown

5 Summary and Future Work

We presented *EvoShelf*, an easy-to-use application in the same vein as mainstream media-browsers for managing the experimental data produced by arbitrary evolutionary systems. Fast browsing of supplementary images associated with each specimen or of generic visualizations of those specimens enables the user to retrace and interactively explore vast amounts of data. Storing, retrieving, and ordering experimental data is facilitated by a simple yet powerful search function that considers a specimen's attributes and meta-data (generation, fitness, etc). Hierarchical grouping structures further facilitates the management of large data sets. In addition to the built-in management and visualization methods, *EvoShelf*-can be extended with plugins that implement new import, visualization or inspection functionalities. According programming templates are provided that can be easily adjusted or majorly extended, depending on the user's demands.

We applied *EvoShelf* to an evolutionary application that breeds Swarm Grammars to generate architectural idea models [11]. Due to the convenient and fast browsing functionality of *EvoShelf*, we have been able to identify specimens that received low fitness values despite their appeal. As a consequence, *EvoShelf* helped us to adjust the fitness function of the SG GP system to better suit our expectations. By means of the visualization techniques that come with *EvoShelf*, Fitness-River and star plots, we have been able to track and investigate an over-fitting process in our evolutionary runs. Finally, by using the selection and storing capabilities of *EvoShelf*, we have also been able to introduce interactivity into an otherwise autonomous evolutionary process.

In the future, we would like to add more visualizations, and to improve the current ones. For instance, it should be possible to automatically overlay different individual-based visualizations as we have done in Figure 7. The star plot visualization should be extended to improve its readability—possibly by using different coloring schemes, or line strengths. Overall, we found it would be useful to automatically associate representative specimens with global trends, as attempted by the combination of Figures 6 and 7 or by revealing details as one mouses over a visualization. The FitnessRiver visualization could also use new visual queues to track the application of genetic operators and the course of inheritance. We would also like to explore the possibility of using *EvoShelf* with running systems, for controlling systems that use interactive evolution as found in [8], and to directly execute and manage experiments. Finally, we would like to release the system as open source software.

References

1. Apple Inc. Apple - iPhoto (April 2010), `http://www.apple.com/ilife/iphoto/`
2. Apple Inc. Apple - iTunes (April 2010), `http://www.apple.com/itunes/`
3. Daida, J., Hilss, A., Ward, D., Long, S.: Visualizing tree structures in genetic programming. Genetic Programming and Evolvable Machines (January 2005)
4. Google. Picasa photo editing (April 2010), `http://picasa.google.com/`
5. Hart, E., Ross, P.: Gavel-a new tool for genetic algorithm visualization. Evolutionary Computation (January 2001)
6. Havre, S., Hetzler, B., Nowell, L.: Themeriver tm: In search of trends, patterns, and relationships. IEEE Transactions on Visualization and Computer Graphics (January 2002)
7. Keim, D., Kriegel, H.: Visdb: Database exploration using multidimensional visualization. IEEE Computer Graphics and Applications (January 1994)
8. Khemka, N., Jacob, C.: Visplore: a toolkit to explore particle swarms by visual inspection. In: GECCO 2009: Proceedings of the 11th Annual Conference on Genetic and Evolutionary computation (2009)
9. Nullsoft. Winamp media player (April 2010), `http://www.winamp.com/`
10. Pohlheim, H.: Visualization of evolutionary algorithms - set of standard techniques and multidimensional visualization. In: Proceedings of the Genetic and Evolutionary Computation Conference, GECCO (1999)
11. von Mammen, S., Jacob, C.: Evolutionary swarm design of architectural idea models. In: GECCO 2008: Proceedings of the 10th annual Conference on Genetic and Evolutionary Computation (July 2008)
12. Wu, A., Jong, K., Burke, D., Grefenstette, J., Ramsey, C.: Visual analysis of evolutionary algorithms. In: Proceedings of the 1999 Congress on Evolutionary Computation, CEC 1999 (1999)

Differential Evolution Algorithms with Cellular Populations

Bernabé Dorronsoro and Pascal Bouvry

Faculty of Science, Technology and Communication,
University of Luxembourg
{bernabe.dorronsoro, pascal.bouvry}@uni.lu

Abstract. Differential Evolution (DE) algorithms are efficient Evolutionary Algorithms (EAs) for the continuous optimization domain. There exist a large number of DE variants in the literature. In this paper, we analyze the effect of adding a cellular structure to the population of some of the most outstanding existing ones. The original algorithms will be compared versus their equivalent versions with cellular population both in terms of accuracy and convergence speed. As a result, we conclude that the cellular versions of the algorithms perform, in general, better than the equivalent state-of-the-art ones in the two considered issues.

1 Introduction

Differential Evolution (DE) algorithms are population based metaheuristics. They were initially proposed in [1] as a variant of Evolutionary Algorithms (EAs) [2–4] for continuous optimization problems. The main difference between DE and other kind of EAs is in the way the population is evolved. In DE, instead of the classical recombination and mutation operators typically applied in EAs, a *mutant* vector is generated, and then it is recombined with the evolving solution. There exist many different variants for generating the mutant vector, and they usually lie in modifying the variables of the *reference vector* (one of the parents) with weighted differences of the same variables of other solutions in the population (two or more additional parents).

It is well known in the literature of EAs that structuring (or decentralizing) the population is generally useful for better guiding the search of the algorithm. The reason is that a structured population provides the algorithm with better exploration/exploitation capabilities for exploring the search space [5–9, 15]. As it is shown in Fig. 1, the two main ways for decentralizing the population are the distributed [15] and cellular [5] schemes.

On the one hand, distributed EAs (dEAs) are composed of several small semi-isolated subpopulations (called *islands*) that are independently evolved by separate EAs. With some given frequency, the different islands exchange some information (e.g., the best local solution found so far) with other neighboring ones.

On the other hand, in the case of cellular EAs (cEAs) individuals are arranged into a toroidal lattice of very small tightly connected subpopulations (typically

R. Schaefer et al. (Eds.): PPSN XI, Part II, LNCS 6239, pp. 320–330, 2010.

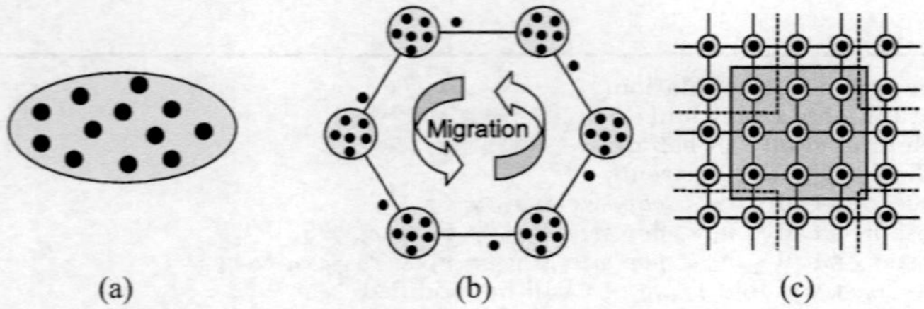

Fig. 1. Panmictic (a), distributed (b) and cellular (c) population topologies

composed by one single solution). In this model, only those solutions that are next to each other in the population mesh are allowed to interact during the evolution process. As an example, we show in Fig. 1 the solutions that could influence in the generation of the new solution at the center position in the grid when using the C9 neighborhood (shadowed), composed by its 8 closest solutions, and itself. We also show the toroidal effect of the lattice in the neighborhood of the top left solution (dashed line).

Despite that the use of decentralized populations is well extended and studied in many EA families, like Genetic Algorithms (GAs) [5, 6, 9], Particle Swarm Optimization (PSO) [10, 11], Genetic Programming (GP) [12], or Estimation of Distribution Algorithms (EDAs) [13], among others [14, 15], there exist only a few works dealing with decentralized population DE algorithms. A review of them can be found in [16].

The main contribution of this work is to propose and evaluate four new designs based on state-of-the-art DE algorithms by adding a cellular structure into their populations. The cellular population used allows the algorithms to keep the diversity of solutions for longer, enhancing their performance in general. The 8 algorithms (4 cellular plus the 4 original ones) are compared in terms of the quality of solutions and the convergence speed on a wide benchmark of complex problems, well known in the literature.

The structure of this paper is detailed next. Section 2 describes the DE algorithms studied. Later, cellular EAs are introduced in Section 3. Section 4 summarizes our experiments, describing the benchmark used, the configuration of the algorithms, and the results obtained. Finally, we conclude this work in Section 5.

2 Differential Evolution Algorithms

This section presents the description for the different DE algorithms we consider in this paper. They are a canonical DE and three state-of-the-art DE algorithms, namely DEGL [17], JADE [18], and SaDE [19].

The canonical DE algorithm we consider in this work follows the design described by Storn and Price in [1]. It implements a *panmictic* (i.e., non-decentralized) population. Its pseudocode is given in Fig. 2. The algorithm starts by randomly generating the solutions composing the initial population. Then, it iterates

```
pop = generate_initial_population()
while ! termination_condition() do
    for each individual i in pop do
        // Choose the three parents
        choose r0 s.t. 0 ≤ r0 < pop.size & r0 ≠ i
        choose r1 s.t. 0 ≤ r1 < pop.size & r1 ≠ r0 & r1 ≠ i
        choose r2 s.t. 0 ≤ r2 < pop.size & r2 ≠ r1 & r2 ≠ r0 & r2 ≠ i
        // At least variable j_rand of i will be modified
        choose j_rand s.t. 0 ≤ j_rand < i.numberOfVariables
        // Generate alternative solution
        v_i = generate_mutant_vector(x_r0, x_r1, x_r2)
        u_i = recombine(x_i, v_i)
        // Keep the best solution of the two ones
        if f(u_i) ≤ f(x_i) then
            pop[ind] = u_i
        end if
    end for
end while
```

Fig. 2. Pseudocode of a classical DE

until the termination condition is reached. This termination condition is usually either finding the optimal solution or performing a maximum number of generations. In every iteration (also called generation), the algorithm sequentially considers all the individuals to be updated. This process lies in randomly choosing three other solutions from the whole population (x_{r_0}, x_{r_1}, and x_{r_2}) and using them to generate a mutant vector. This mutant vector will later be used to modify the current solution in the recombination step to generate a new one, that will replace the current solution with some policy (in this work, the replacement is done if the new solution is better or equal to the current one). Therefore, the new generated solutions are immediately inserted into the population, so they can interact with solutions from their parents generation.

There are many mechanisms to generate the mutant vector proposed in the literature. In this work, we consider a simple one, proposed by Storn and Price [1], and described in (1). According to this operator, the mutant vector v_i is generated by adding to the reference vector (one of the three parents) the difference of the other two parents, weighted by parameter F.

$$v_i = x_{r_0} + F \cdot (x_{r_1} - x_{r_2}) \ . \tag{1}$$

This mutant vector is then used to modify the current solution x_i in the recombination process, thus generating a new one. The most common approach for that is choosing every variable j, of either the mutant vector v_i or the current solution x_i, with some given probability CR, as shown in (2). As it can be seen, we force that at least one variable j_{rand} (randomly chosen) is adopted from the mutant vector for the new solution u_i.

$$u_{i,j} = \begin{cases} v_{i,j} & \text{if } rand(0,1) \leq CR \text{ or } j = j_{rand} \\ x_{i,j} & \text{otherwise} \end{cases} \ . \tag{2}$$

Finally, the fittest solution between the current and the new ones will remain in the population, as (3) defines.

$$x_i = \begin{cases} u_i & \text{if } f(u_i) \leq f(x_i) \\ x_i & \text{otherwise} \end{cases} . \tag{3}$$

The described DE algorithm is usually known in the literature as DE/rand/1/bin, because the base vector x_{r_0} is chosen randomly, one vector difference $x_{r_1} - x_{r_2}$ is added to it to generate the mutant vector, and the new solution variables $u_{i,j}$ are taken from x_i and v_i using a binomial distribution.

In order to avoid a strong dependency on the control parameters of DE (CR and F), we adopt a self-adaptive strategy that automatically chooses *good* values for the two parameters [20]. It lies in initially assigning random CR and F values (in the intervals $[0.0, 1.0]$ and $[0.1, 1.0]$, respectively) to every solution. Then, when a new solution is created, it inherits these values from the currently evolving one. Additionally, they are modified with probability 0.1 with random values in the same intervals considered for the initialization —as shown in (4), where $rand_1$ to $rand_4$ are random values in the interval $[0, 1]$.

$$F = \begin{cases} 0.1 + 0.9 * rand_1 & \text{if } rand_2 \leq 0.1 \\ F & \text{otherwise} \end{cases} ; \quad CR = \begin{cases} rand_3 & \text{if } rand_4 \leq 0.1 \\ CR & \text{otherwise} \end{cases} . \tag{4}$$

DEGL. DEGL [17] is a DE algorithm with structured population. The solutions in the population are arranged in a toroidal ring. A neighborhood is defined in it in such a way that every individual can only interact with a number of the next and previous ones in the ring. In this algorithm, a combination of two mutation strategies is used to generate the new solutions. One of them is using the best solution in the neighborhood, and the other one takes information from the best global solution in the population.

JADE. JADE [18] is an algorithm implementing the new mutation strategy DE/current-to-$p$best. It is similar to the classical current-to-best one, but with the difference that a randomly chosen solution from the top $100 \cdot p\%$ individuals (the authors use $p = 0.05$ in their experiments in [18]) is selected instead of the best overall one in the population. The authors propose a second version of JADE with archive, which is keeping, for every position in the population, a list (or archive) containing the last S individuals that were in this position. We borrow this second version for our comparison, because it was concluded to be better in the original paper.

SaDE. SaDE [19] is a self-adaptive DE that automatically adapts the probability of applying four different mutation strategies (DE/rand/1/bin, DE/rand-to-best/2/bin, DE/rand/2/bin, and DE/current-to-rand/1) depending on the success rate of each one in the past generations. The success rate of every strategy is defined in terms of the number of new solutions they provide to the population in every generation. This way, the authors are able to combine into one single DE algorithm several mutation strategies with different features that perform

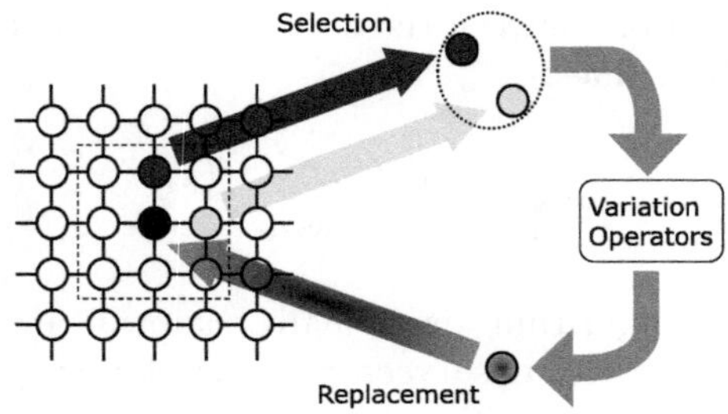

Fig. 3. In cellular GAs, individuals are only allowed to interact with their neighbors during the breeding loop

differently on the distinct problems: some of them are more explorative, and some others are more exploitative.

3 Cellular Evolutionary Algorithms

Cellular Evolutionary Algorithms (cEAs) [5, 9] are a kind of EA with structured population, meaning that individuals can only interact with a subset of the individuals in the whole population. As it was mentioned in Section 1, the use of decentralized populations in EAs usually leads to more accurate algorithms with respect to EAs with panmictic populations. Additionally, the few existing works comparing cellular and distributed EAs usually conclude that cEAs outperform dEAs, specially when dealing with complex problems [5, 8].

In cEAs, individuals are spread in a (usually) two dimensional toroidal mesh, and they are only allowed to interact with their neighbors. The degree of isolation of individuals is given in terms of the neighborhood size: the smaller the neighborhood the higher the isolation of individuals, and thus the slower their genetic information will be spread through the grid. This information spread is guaranteed thanks to the overlapping among neighborhoods.

We graphically show in Fig. 3 the different steps performed during the breeding loop in cEAs for every individual. As it can be seen, the parents are chosen from the neighborhood of the current solution, and then the variation operators are applied. After that, the newly generated individual is inserted back into the population (following a given replacement policy) immediately after its creation, and thus it can interact in the breeding loop of its neighbors, even when they most probably belong to previous generations.

4 Computational Experiments

We summarize in this section the experiments performed for this work. The benchmark we used is presented in Section 4.1, and the configurations of the algorithms are described in Section 4.2. Finally, our results are analyzed in Section 4.3.

4.1 Test Problems

We have selected for this work the benchmark that was built for the special session on continuous optimization in CEC 2005 [22]. It is composed of 25 complex functions with different features. Functions $F1$ to $F5$ are unimodal, while the others are multimodal. Among the latter ones, problems $F6$ to $F12$ are basic multimodal functions, while $F13$ and $F14$ are expanded multimodal functions, and $F15$ to $F25$ are hybrid compositions of functions.

In our experiments, we have considered the highest problem dimensionality this benchmark affords: 50 variables.

4.2 Configuration of Algorithms

The parameters for the DE algorithms we are considering in this paper are discussed in this section. The basic DE algorithm we study implements the DE/rand/1/bin scheme, it has a population of 50 solutions, and the mutation strategy is the standard one already described in Section 2. The three parents are selected by binary tournament, so for every parent we randomly select two candidate solutions and choose the best one as the parent. The new solutions replace the current ones in the population if they have better or equal fitness value, and the F and CR parameters are self-adaptive as presented in Section 2. Finally, the sopping condition of all the algorithms is either to find the optimal solution (with an error $< 4.9E-324$) or to perform 500,000 fitness function evaluations.

For the configuration of the compared state-of-the-art algorithms, we have taken the values proposed in their original papers. However, we would like to clarify some of these parameters that are not clearly set in the original papers (i.e., the authors recommend either some intervals for these values or different alternatives). The population of the algorithms is set to 50 solutions, with the exception of DEGL, which has a population of 500 solutions arranged in a toroidal grid with neighborhood size 50. For the weight factor used to combine the local and global mutation strategies in DEGL, we have chosen the self-adaptive one proposed in [17]. Regarding JADE, we randomly select one solution from the best 5% solutions in the population for the DE/current-to-$p$best mutation used, and the archive size for every position in the population was set to the population size. Finally, in the case of SaDE, the learning period was set to 10 generations.

In the case of the cellular version of the algorithms, the population is composed by 7×7 solutions, arranged in a 2 dimensional toroidal lattice. Additionally, the selection of the parents is limited to the C9 neighborhood of the considered solution. All the other parameters are the same as for the original versions of the algorithms.

4.3 Results

We proceed now to discuss our results. Table 1 shows the average results obtained by the different algorithms after 100 independent runs. We put the percentage of runs in which the problems were solved to optimality in parenthesis when

Table 1. Average solutions found by the different algorithms

	ssDE	cDE	SaDE	cSaDE	DEGL	cDEGL	JADE	cJADE
F1	2.50E−14(56%) ±2.84E−14	1.14E−15(98%) ±8.00E−15	1.02E−14(82%) ±2.20E−14	**0.0(100%) ±0.0**	2.92E−12 ±1.20E−12	8.70E−14(32%) ±7.16E−14	5.68E−14 ±0.0	9.44E−14 ±2.71E−14
F2	4.61E1 ±7.63E1	6.10 ±5.49	3.33E3 ±9.54E3	6.92E−1 ±4.93E−1	2.95E2 ±6.26E1	**9.83E−14(43%) ±9.31E−14**	5.92E−13 ±2.81E−13	1.02E−1 ±2.38E−1
F3	8.47E6 ±4.96E6	8.25E6 ±4.10E6	9.01E6 ±1.81E7	1.14E6 ±4.58E5	7.46E6 ±1.36E6	**6.40E4 ±2.81E4**	9.30E4 ±9.04E4	2.80E6 ±1.07E6
F4	3.05E3 ±1.70E3	4.06E3 ±2.07E3	6.49E3 ±2.83E3	4.35E3 ±2.12E3	1.20E3 ±2.79E2	**4.82E1 ±3.71E2**	3.66E4 ±1.43E4	4.38E2 ±3.28E2
F5	3.59E3 ±7.84E2	4.59E3 ±8.37E2	1.12E4 ±1.47E3	8.28E3 ±1.24E3	**3.32E2 ±1.16E2**	3.46E3 ±6.58E2	3.70E3 ±7.16E2	3.61E3 ±1.00E3
F6	4.55E1 ±3.15E1	4.54E1 ±3.46E1	2.04E2 ±1.86E2	1.02E2 ±5.79E1	3.07E1 ±1.34E1	**2.28 ±2.36**	2.49 ±1.10E1	2.89E1 ±2.77E1
F7	6.20E3 ±6.20E−13	**6.30E−3(1%) ±1.02E−2**	6.20E3 ±4.44E−12	6.20E3 ±6.60E−13	6.20E3 ±2.62E−9	6.20E3 ±1.14E−12	6.20E3 ±7.20E−13	6.20E3 ±4.52
F8	2.11E1 ±3.97E−2	2.11E1 ±3.49E−2	2.08E1 ±3.83E−1	**2.07E1 ±4.78E−1**	2.11E1 ±3.55E−2	2.11E1 ±3.86E−2	2.13E1 ±4.11E−2	2.11E1 ±3.22E−2
F9	1.13E1 ±4.75	1.87E1 ±6.94	9.66E−15(83%) ±2.15E−14	**9.95E−3(99%) ±9.95E−2**	3.26E2 ±1.43E1	1.12E2 ±2.36E1	5.63E−14(1%) ±5.68E−15	4.55E1 ±2.15E1
F10	1.86E2 ±1.14E2	**1.28E2 ±7.54E1**	2.43E2 ±3.54E1	2.75E2 ±4.45E1	3.44E2 ±1.63E1	1.32E2 ±3.15E1	2.64E2 ±1.15E2	3.28E2 ±2.02E1
F11	5.82E1 ±1.04E1	5.74E1 ±5.84	4.19E1 ±4.74	4.10E1 ±4.63	7.27E1 ±1.52	**3.37E1 ±5.22**	6.87E1 ±4.39	6.56E1 ±3.90
F12	2.67E4 ±2.36E4	3.90E4 ±2.47E4	3.58E4 ±2.47E4	2.15E4 ±1.23E4	**9.50E3 ±7.95E3**	2.56E4 ±2.19E4	2.01E4 ±1.93E4	2.66E4 ±2.12E4
F13	5.07 ±3.59	4.42 ±1.88	2.12 ±2.91E−1	**1.87 ±2.03E−1**	3.00E1 ±1.18	7.14 ±1.65	7.80 ±9.39E−1	2.21E1 ±1.49
F14	2.30E1 ±2.1950E−1	2.28E1 ±2.7841E−1	2.18E1 ±6.1830E−1	2.16E1 ±5.3498E−1	2.29E1 ±1.6824E−1	**2.15E1 ±5.8852E−1**	2.37E1 ±2.1198E−1	2.24E1 ±2.8563E−1
F15	**3.00E2 ±9.7326E1**	3.08E2 ±9.2724E1	3.06E2 ±1.5229E2	3.62E2(1%) ±1.0166E2	3.33E2 ±9.4317E1	3.82E2 ±5.6059E1	3.19E2(2%) ±9.2035E1	3.56E2 ±7.2295E1
F16	1.50E2 ±1.0232E2	**1.21E2 ±7.4178E1**	1.57E2 ±5.7039E1	1.25E2 ±1.7855E1	2.54E2 ±2.4010E1	1.60E2 ±1.1217E2	2.29E2 ±1.1575E2	3.01E2 ±7.5831E1
F17	2.47E2 ±8.5184E1	2.20E2 ±6.6414E1	**1.21E2 ±2.1617E1**	1.35E2 ±2.4640E1	2.83E2 ±2.1617E1	1.74E2 ±1.2916E2	3.76E2 ±6.9206E1	3.26E2 ±6.7814E1
F18	9.20E2 ±1.2770E1	9.26E2 ±5.3856	9.87E2 ±3.1410E1	9.92E2 ±1.7562E1	**9.12E2 ±1.0894**	9.55E2 ±1.8367E1	9.36E2 ±1.7285E1	9.30E2 ±1.1939E1
F19	9.22E2 ±3.4133	9.26E2 ±5.5953	9.86E2 ±4.0976E1	9.94E2 ±1.8461E1	**9.12E2 ±9.5186E−1**	9.55E2 ±2.3280E1	9.35E2 ±1.6130E1	9.31E2 ±1.1520E1
F20	9.22E2 ±4.2001	9.26E2 ±5.3533	9.88E2 ±3.1503E1	9.94E2 ±1.8026E1	**9.12E2 ±9.1633E−1**	9.54E2 ±2.3415E1	9.36E2 ±1.6098E1	9.31E2 ±1.1512E1
F21	6.62E2 ±2.3726E2	**5.95E2 ±1.9804E2**	7.65E2 ±3.3996E2	7.29E2 ±3.2749E2	7.16E2 ±2.4986E2	7.64E2 ±3.2608E2	6.86E2 ±2.4168E2	1.02E3 ±8.8117E1
F22	9.17E2 ±1.5901E1	9.22E2 ±1.7524E1	9.67E2 ±1.7584E1	9.85E2 ±1.8126E1	**9.17E2 ±3.9815**	9.64E2 ±2.3627E1	9.65E2 ±2.4867E1	9.46E2 ±2.3938E1
F23	7.48E2 ±2.3645E2	**6.07E2 ±1.6539E2**	7.05E2 ±2.8793E2	6.70E2 ±2.6362E2	7.65E2 ±2.3657E2	8.55E2 ±2.3527E2	6.85E2 ±2.2106E2	1.01E3 ±9.7561E1
F24	2.00E2 ±1.6815E−12	**2.00E2 ±8.1842E−13**	1.02E3 ±3.8723E2	1.18E3 ±1.7456E2	2.00E2 ±1.5836E−12	3.75E2 ±3.7456E2	9.91E2 ±1.8249E2	1.03E3 ±1.0708E1
F25	1.38E3 ±1.1813E1	**2.21E2 ±1.8361**	1.41E3 ±1.7135E1	1.42E3 ±1.1230E1	1.40E3 ±4.3774	1.37E3 ±7.7264	1.40E3 ±7.3243	1.40E3 ±6.6738

appropriate. Notice that the initial population was the same for all the algorithms in the different runs (of course, distinct initial populations were used in every run). In the case of DEGL, 50 solutions were the same as for the other algorithms, while the rest were randomly generated. The best overall result for every problem is emphasized in **bold font**, and the grey background means that the algorithm was significantly better (with 95% confidence) than its equivalent one with/without the cellular model according to the Wilcoxon matched-pairs signed ranks test [23]. The results of this test are shown in Table 2, where p-value ≤ 0.05 means statistical difference on the comparisons, and high values of R^+ are favoring the cellular versions of the algorithms (these p-values are emphasized with **bold font**), while high R^- values favor the original ones.

We can see that the cellular algorithms outperform, in general, their equivalent original ones for all the studied DE variants in the considered problems. The exception is JADE, since it performs better than cJADE in 13 problems (most

Table 2. Results of Wilcoxon matched-pairs signed ranks test. Panmictic vs. cellular algorithms

	DE		SaDE		DEGL		JADE	
	$(R^+, R-)$	p-value	$(R^+, R-)$	p-value	$(R^+, R-)$	p-value	$(R^+, R-)$	p-value
$F1$	(1034, 47)	**7.21E − 8**	(171, 0)	**2.14E − 4**	(5050, 0)	**4.01E − 18**	(0, 2211)	$1.69E-12$
$F2$	(4886, 164)	**4.86E − 16**	(5050, 0)	**4.01E − 18**	(5050, 0)	**4.01E − 18**	(0, 5050)	$4.01E-18$
$F3$	(2516, 2534)	0.98	(4434, 616)	**5.34E − 11**	(5050, 0)	**4.01E − 18**	(0, 5050)	$4.01E-18$
$F4$	(1505, 3545)	$4.56E-4$	(4041, 1009)	**1.88E − 7**	(4950, 100)	**7.76E − 17**	(5050, 0)	**4.01E − 18**
$F5$	(487, 4563)	$2.48E-12$	(4992, 58)	**2.27E − 17**	(0, 5050)	$4.01E-18$	(2663, 2387)	0.64
$F6$	(2809, 2241)	0.33	(3866, 1184)	**4.05E − 6**	(5050, 0)	**4.01E − 18**	(198, 4852)	$1.26E-15$
$F7$	(5050, 0)	**4.01E − 18**	(3630, 835)	**1.38E − 7**	(5050, 0)	**4.01E − 18**	(105, 2380)	$2.87E-11$
$F8$	(2509, 2541)	0.96	(3053, 1997)	0.07	(3168, 1882)	**0.03**	(5050, 0)	**4.01E − 18**
$F9$	(430.5, 4519.5)	$9.8E-13$	(153, 18)	**3.52E − 3**	(5050, 0)	**4.01E − 18**	(0, 5050)	$4.01E-18$
$F10$	(3589, 1461)	**2.56E − 4**	(1156, 3894)	$2.54E-6$	(5050, 0)	**4.01E − 18**	(1027, 4023)	$2.63E-7$
$F11$	(3002, 2048)	0.10	(2936, 2114)	0.16	(5050, 0)	**4.01E − 18**	(3951, 1099)	**9.53E − 7**
$F12$	(1412, 3638)	$1.30E-4$	(3764, 1286)	**2.06E − 5**	(744, 4306)	$9.28E-10$	(1786, 3264)	0.01
$F13$	(2238, 2812)	0.33	(4473, 577)	**2.15E − 11**	(5050, 0)	**4.01E − 18**	(0, 5050)	$4.01E-18$
$F14$	(3626, 1424)	**1.55E − 4**	(3005, 2045)	0.10	(5050, 0)	**4.01E − 18**	(5050, 0)	**4.01E − 18**
$F15$	(1098.5, 1602.5)	0.17	(717, 1428)	0.02	(1454, 3496)	$3.68E-4$	(1189, 3861)	$4.40E-6$
$F16$	(3102, 1948)	**0.05**	(4151, 899)	**2.29E − 8**	(4147, 903)	**2.48E − 8**	(1026, 3924)	$4.30E-7$
$F17$	(3406, 1644)	**2.47E − 3**	(1441, 3609)	$1.95E-4$	(4518, 532)	**7.39E − 12**	(4001, 1049)	**3.92E − 7**
$F18$	(644, 4406)	$1.01E-10$	(2202, 2848)	0.27	(0, 5050)	$4.01E-18$	(2984, 2066)	0.12
$F19$	(590, 4460)	$2.92E-11$	(2242, 2808)	0.33	(100, 4950)	$7.76E-17$	(2835, 2215)	0.29
$F20$	(554, 4496)	$1.25E-11$	(2207, 2843)	0.28	(100, 4950)	$7.76E-17$	(2869, 2181)	0.24
$F21$	(1460.5, 1095.5)	0.30	(2049, 1521)	0.24	(1692, 3358)	$4.21E-3$	(228, 4822)	$2.91E-15$
$F22$	(1646, 3404)	$2.52E-3$	(777, 4273)	$1.88E-9$	(12, 5038)	$5.76E-18$	(4020, 1030)	**2.77E − 7**
$F23$	(3615, 1435)	**1.80E − 4**	(3594, 1456)	**2.39E − 4**	(1989, 3061)	0.07	(186, 4864)	$9.04E-16$
$F24$	(1783.5, 46.5)	**1.67E − 10**	(1109, 3941)	$1.14E-6$	(125, 2503)	$2.58E-11$	(3344, 1706)	**4.90E − 3**
$F25$	(5050, 0)	**4.01E − 18**	(1056, 3994)	$4.44E-7$	(5049, 1)	**4.13E − 18**	(3720, 1330)	**4.01E − 5**

of them unimodal and basic multimodal functions), being worse for 8 functions. We think that the reason is the archive of solutions JADE is keeping in every location of the population, from which the algorithm can get one solution as one of the parents during the evolution process. This archive is keeping the last 100 individuals that were placed in this location during the evolution. Therefore, the effect is that diversity is highly increased, thus slowing down the convergence speed of the population. When combining this technique with the cellular population, which is slowing down the convergence speed too, then for some problems the algorithm cannot run for enough number of iterations to converge to good solutions. In other words, the use of the archive and the cellular topology in cJADE provides the algorithm with too explorative capabilities.

If we pay attention to the algorithms finding the best results for every problem (figures in **bold font** in Table 1), we see that at least one of the DE variants implementing the cellular population finds the best results for 17 problems. Therefore, only for the other 8 functions one of the standard panmictic DEs was found to be the best one.

To better study the behavior of the different algorithms, we present in Fig. 4 the evolution of the best solution in the population during the run. The results plotted are averaged over 100 runs for every generation. We can see how, in general, the new algorithms implementing the cellular population are converging faster than their equivalent original panmictic versions. It stands out the good behavior of cDEGL versus DEGL, since it performs faster convergence for all the functions shown. We can see also how the cellular population helps JADE algorithm to scape from local optimal solutions in which it gets stuck very early (see, for example, plots for $F6$ and $F10$).

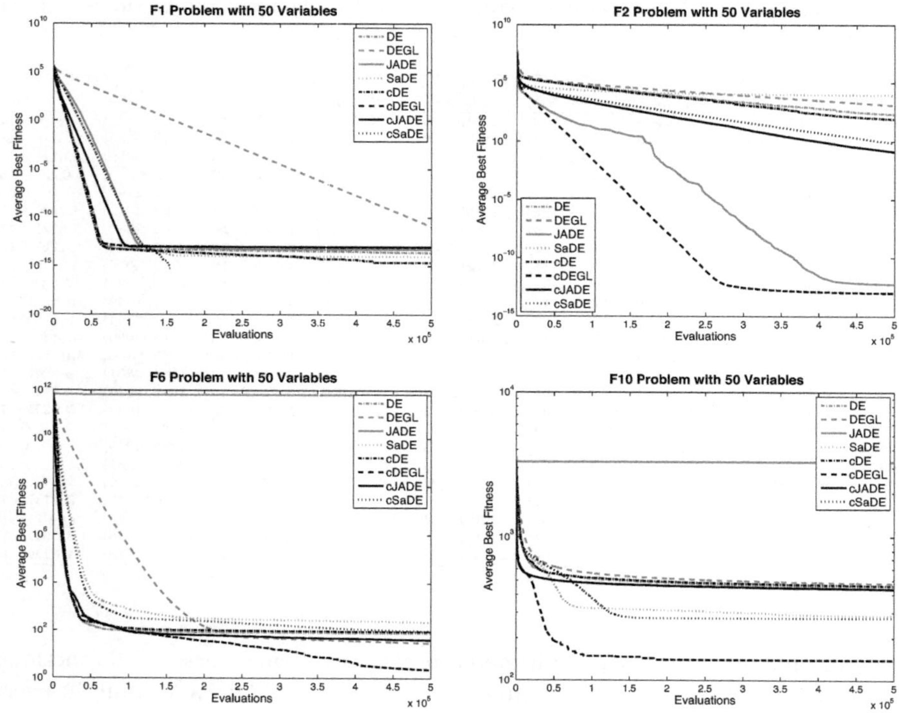

Fig. 4. Evolution of the best solution in the population during the run for some of the considered problems (averaged over 100 runs)

5 Conclusions and Future Work

We propose in this paper to enhance the behavior of four state-of-the-art DE algorithms (a canonical one, DEGL, JADE, and SaDE) by adding a cellular topology structure into their populations. This topology restricts the interactions among individuals to only close ones, therefore slowing down the spread of good solutions through the population. The effect is that different parts of the population will hopefully explore distinct areas of the search space. The algorithms were compared in terms of accuracy and convergence speed. As a main result, we can summarize that the new algorithms with cellular populations outperform, in general, the original ones for the canonical, DEGL, and SaDE variants. The exception is JADE algorithm, because it outperforms cJADE in average.

As future extensions to this work, we plan to make a deeper study by considering different population topologies, other benchmark functions, and more DE variants.

References

1. Storn, R., Price, K.: Differential evolution—a simple and efficient heuristic for global optimization over continuous spaces. J. of Glob. Opt. 11(4), 341–359 (1997)
2. Bäck, T.: Evolutionary Algorithms in Theory and Practice: Evolution Strategies, Evolutionary Programming, Genetic Algorithms. Oxford University Press, Oxford (1996)
3. Bäck, T., Fogel, D., Michalewicz, Z. (eds.): Handbook of Evolutionary Computation. Oxford University Press, Oxford (1997)
4. De Jong, K.: Evolutionary Computation. A Unified Approach. MIT Press, Cambridge (2006)
5. Alba, E., Dorronsoro, B.: Cellular Genetic Algorithms. Springer, Heidelberg (2008)
6. Alba, E., Tomassini, M.: Parallelism and evolutionary algorithms. IEEE Transactions on Evolutionary Computation 6(5), 443–462 (2002)
7. Alba, E., Troya, J.: Improving flexibility and efficiency by adding parallelism to genetic algorithms. Statistics and Computing 12(2), 91–114 (2002)
8. Luque, G., Alba, E., Dorronsoro, B.: Parallel Genetic Algorithms. In: Parallel Metaheuristics: A New Class of Algorithms, pp. 107–125. John Wiley & Sons, Chichester (2005)
9. Tomassini, M.: Spatially Structured Evolutionary Algorithms: Artificial Evolution in Space and Time. Natural Computing Series. Springer, Heidelberg (2005)
10. Kennedy, J., Mendes, R.: Population structure and particle swarm performance. In: CEC, pp. 1671–1676. IEEE Press, Los Alamitos (2002)
11. Janson, S., Middendorf, M.: A hierarchical particle swarm optimizer and its adaptive variant. IEEE Systems, Man and Cybernetics - Part B 35(6), 1272–1282 (2005)
12. Folino, G., Pizzuti, C., Spezzano, G.: A scalable cellular implementation of parallel genetic programming. IEEE Trans. on Evolutionary Comp. 7(1), 37–53 (2003)
13. Alba, E., Madera, J., Dorronsoro, B., Ochoa, A., Soto, M.: Theory and practice of cellular UMDA for discrete optimization. In: PPSN-IX, pp. 242–251. Springer, Heidelberg (2006)
14. Alba, E.: Parallel Metaheuristics: A New Class of Algorithms. Wiley, Chichester (2005)
15. Talbi, E.G.: Parallel Combinatorial Optimization. John Wiley & Sons, Chichester (2006)
16. Dorronsoro, B., Bouvry, P.: Studying the effects of several population management schemes in differential evolution. IEEE Trans. on Ev. Comp. (2010) (submitted)
17. Das, S., Abraham, A., Chakraborty, U.K., Konar, A.: Differential evolution using a neighborhood-based mutation operator. IEEE TEC 13(3), 526–553 (2009)
18. Zhang, J., Sanderson, A.C.: JADE: Adaptive differential evolution with optional external archive. IEEE Trans. on Evolutionary Computation 13(5), 945–958 (2009)
19. Qin, A.K., Huang, V.L., Suganthan, P.N.: Differential evolution algorithm with strategy adaptation for global numerical optimization. IEEE Transactions on Evolutionary Computation 13(2), 398–417 (2009)
20. Brest, J., Greiner, S., Bošković, B., Mernik, M., Žumer, V.: Self-adapting control parameters in differential evolution: A comparative study on numerical benchmark problems. IEEE Transactions on Evolutionary Computation 10(6), 646–657 (2006)
21. Alba, E., Dorronsoro, B., Giacobini, M., Tomassini, M.: 7, Decentralized Cellular Evolutionary Algorithms. In: Handbook of Bioinspired Algorithms and Applications, pp. 103–120. CRC Press, Boca Raton (2006)

22. Suganthan, P.N., Hansen, N., Liang, J.J., Deb, K., Chen, Y.P., Auger, A., Tiwari, S.: Problem definitions and evaluation criteria for the CEC 2005 special session on real-parameter optimization. Technical report, Nanyang Technological University, Singapore, and Kanpur Genetic Algorithms Laboratory, IIT Kanpur (2005)
23. García, S., Molina, D., Lozano, M., Herrera, F.: A study on the use of non-parametric tests for analyzing the evolutionary algorithms' behaviour: A case study on the CEC 2005 special session on real parameter optimization. Journal of Heuristics 15, 617–644 (2009)

Flocking in Stationary and Non-stationary Environments: A Novel Communication Strategy for Heading Alignment

Eliseo Ferrante, Ali Emre Turgut, Nithin Mathews, Mauro Birattari, and Marco Dorigo

IRIDIA, CoDE, Université Libre de Bruxelles, Brussels, Belgium

Abstract. We propose a novel communication strategy inspired by explicit signaling mechanisms seen in vertebrates, in order to improve performance of self-organized flocking for a swarm of mobile robots. The communication strategy is used to make the robots match each other's headings. The task of the robots is to coordinately move towards a common goal direction, which might stay fixed or change over time.

We perform simulation-based experiments in which we evaluate the accuracy of flocking with respect to a given goal direction. In our settings, only some of the robots are informed about the goal direction. Experiments are conducted in stationary and non-stationary environments. In the stationary environment, the goal direction and the informed robots do not change during the experiment. In the non-stationary environment, the goal direction and the informed robots are changed over time. In both environments, the proposed strategy scales well with respect to the swarm size and is robust with respect to noise.

1 Introduction

In nature, we observe different activities performed by animals living in groups. Such activities require collective decision-making even when few individuals have the needed information. This information then spreads in the group according to different mechanisms, depending on the species. In some insect and fish species, information is transferred implicitly without any signaling mechanism [1,2]. Conversely, other species utilize explicit signaling mechanisms such as vocalization. For example, mountain gorillas switch between some daily activities (resting to travelling/feeding) very rapidly using vocalization [3]. In a honeybee swarm, when scouts agree on a new nest site, they fly rapidly towards the nest while signaling the right direction to the rest of the swarm [4].

Coordinated motion of animals is an example of activities which require collective decision-making. For example, flocks of birds and schools of fish move and maneuver coherently as if they were a super-organism [5]. Recent work in biology showed that a group of animals can be guided to particular locations (food sources) even if only a minority of the group is aware of the location [1].

R. Schaefer et al. (Eds.): PPSN XI, Part II, LNCS 6239, pp. 331–340, 2010.

The first studies of flocking in swarm robotics were inspired by Reynolds' seminal work [6], one of the first regarding flocking outside biology. Reynolds obtained a realistic computer animation of a flock of birds through three simple concurrent behaviors: *separation* (avoiding collisions), *cohesion* (staying close to neighbors) and *alignment* (heading in the same direction as neighbors). These behaviors are all based on local sensing and local decision rules. Separation and cohesion control, denoted in the rest of the paper as *proximal control*, is the aggregate behavior controlling the relative distance between individuals.

In Reynolds' work, in order to perform alignment, each individual is assumed to sense the velocity of its neighbors without noise. This assumption is unrealistic in robotics. Subsequent works have tried to relax this assumption. These works can be organized into three different categories. The first category does not feature an explicit alignment behavior. Instead, it tries to obtain alignment intrinsically via other behaviors such as homing [7], leader-following [8] and light-following [9]. The second category achieves alignment by resorting to the emulation of a heading sensor [10] or to the estimation of the heading of neighbors [11]. In the third and last category, heading information is spread within the swarm via local communication [12]. Turgut et al. [13] proposed an algorithm belonging to the third category. A robot measures its heading with respect to the North using a compass and broadcasts it periodically so that the heading is sensed "virtually" by its neighbors. With this method, Turgut et al. [13] achieved self-organized flocking in a random direction.

Çelikkanat et al. [14], inspired by the implicit decision-making mechanisms of some animal species [15,2], extended the flocking behavior proposed by Turgut et al. [13] by providing a goal direction to some of the robots ("informed" robots). They observed that a large swarm can be guided by only a few robots, which is in accordance with theoretical results [1].

In this paper, inspired by both biological [3,4] and swarm robotics ideas [14], we propose a new method for heading alignment. The proposed strategy that we call *information-aware* communication, uses only local communication, which means that robots can communicate only with their neighbors within a given range and in their line of sight. Our system is composed of informed and uninformed robots: informed robots relay the goal direction and uninformed ones just send the average of the messages they receive from their neighbors. We compare the novel *information-aware* communication with *heading* communication, where all of the robots always send the measured heading information to their neighbors by still relying on local communication only. The latter was used by Turgut et al. and Çelikkanat et al. [13,14].

We present the results of two sets of experiments. One is conducted in a stationary environment, in which the goal direction and the informed indivuals do not change during the experiment. The other is conducted in a non-stationary environment, in which both the goal direction and the informed robots are changed at regular time intervals during the experiment.

2 Methodology

The methodology we use to design the flocking behavior is based on artificial physics [9]. At each control step, a virtual force vector is computed as:

$$\mathbf{f} = \alpha\mathbf{p} + \beta\mathbf{h} + \gamma\mathbf{g},$$

where $\mathbf{p}$ is the proximal control vector; $\mathbf{h}$ is the heading alignment vector; $\mathbf{g}$ is the vector that indicates the goal direction. The vectors $\mathbf{p}$ and $\mathbf{h}$ are each calculated by a behavior (explained in Section 2.1 and Section 2.2 respectively). The goal direction vector $\mathbf{g}$ is available to some robots, whereas for the others $\mathbf{g} = \mathbf{0}$. The weights α, β and γ define the relative contribution of the different force components. In this paper, we do not tune these parameters for obtaining optimal performance, but we set them to $\alpha = 1$, $\beta = 5$ and $\gamma = 10$ to reflect our prior knowledge on the relative importance of the three components.

2.1 Proximal Control Behavior

The proximal control behavior assumes that a robot perceives the relative position (range and bearing) of its neighbors in close proximity. This is realized using LEDs and an omni-directional camera as in [12]. Let k denote the number of robots perceived at a given time, d_i and ϕ_i denote the range and bearing measurements of the ith robot, respectively. The virtual force $\mathbf{p}$ is given by:

$$\mathbf{p} = \sum_{i=1}^{k} p_i e^{j\phi_i},$$

where $p_i e^{j\phi_i}$ are vectors expressed in polar coordinates. p_i is calculated as a function of d_i using a force function $p_i(d_i)$ as in [16]. p_i is repulsive when d_i is smaller than the desired distance (D) and it is attractive when d_i is greater than D. The function is:

$$p(d_i) = -\frac{2D^2}{d_i^3} + \frac{2}{d_i},$$

2.2 Heading Alignment Behavior

The heading alignment behavior assumes that, using an onboard light sensor, a robot r measures its heading (θ_r) with respect to the common reference frame represented by a light source. The robot receives an angle θ_i from its ith neighbor. The value sent by each neighbor depends on which communication strategy is used, as explained in the following. Each received angle is transformed into robot r body-fixed reference frame[1]. Having received k angles from its k neighbors,

[1] We define two reference frames. One is the reference frame common to all of the robots, and the other is the body-fixed reference frame specific to each robot. The body-fixed reference frame is fixed to the center of a robot: its $x-axis$ is coincident with the rotation axis of the wheels and its $y-axis$ points to the front of the robot.

robot r calculates the average heading vector as:

$$\mathbf{h} = \frac{\sum_{i=1}^{k} e^{j\theta_i}}{\| \sum_{i=1}^{k} e^{j\theta_i} \|},$$

where $\| \cdot \|$ denotes the norm of a vector.

The proposed communication strategy, that we call *information-aware* communication strategy, is explained in the following.

Information-aware communication: This communication strategy assumes that robots are aware of whether they are provided with the goal direction $\mathbf{g}$ or not, that is whether they are *informed* or *non-informed.* This awareness mechanism is implemented by measuring the length of the goal direction vector $\mathbf{g}$: the robot considers itself *non-informed* if $\mathbf{g} = \mathbf{0}$ and *informed* if $\mathbf{g} \neq \mathbf{0}$. Each robot then communicates the following information: if it is *non-informed*, it sends $\angle\mathbf{h}$ ($\angle\cdot$ denotes the angle of a vector) to its immediate neighbors; otherwise, it sends $\angle\mathbf{g}$. The intuition behind this strategy is the following: if the robot is *non-informed*, it should facilitate the diffusion of the information originating from the *informed* robots; if it is *informed*, it should then directly propagate the information about the goal location to its immediate neighbors. The information then eventually reaches the entire swarm thanks to the uninformed robots. Note that the awareness of each robot is only used to determine which information should be sent, and is never directly communicated to neighboring robots (i.e., each robot never knows if a message is received from an *informed* or a *non-informed* robot).

As a baseline comparison, we implemented another communication strategy that we call *heading* communication strategy. This strategy is similar to the one used in [13].

Heading communication: This communication strategy consists in the local communication of the robot's own current heading θ measured with respect to the common reference frame. In the original work [13], this strategy was used to simulate robots that are able to measure the heading of their neighbors when the actual measurement is not physically possible.

2.3 Motion Control

The computed virtual force vector $\mathbf{f}$ is mapped into rotational speed of the wheels. First, using Newton's second law of motion, the target velocity $\mathbf{u}_{target}$ is computed:

$$\mathbf{u}_{target} = \mathbf{u}^t + \frac{\mathbf{f}\Delta t}{m},$$

where Δt is the control-step size, m is the mass of the robot and $\mathbf{u}^t$ is the current velocity of the robot. The target velocity $\mathbf{u}_{target}$ cannot be followed directly by the robot due to its non-holonomic constraints. Thus, it is mapped into the robot's forward velocity $\mathbf{u}^{t+1}$, which points in the direction of the $y - axis$ and has magnitude $u = \|\mathbf{u}^{t+1}\|$ set to:

$$u = \begin{cases} (\frac{\mathbf{u}_{target}}{\|\mathbf{u}_{target}\|} \cdot \frac{\mathbf{u}^t}{\|\mathbf{u}^t\|}) u_{max}, & \text{if } \mathbf{u}_{target} \cdot \mathbf{u}^t \geq 0; \\ 0, & \text{otherwise.} \end{cases},$$

where u_{max} is set to 0.036 m/s.

The angular velocity ω of the robot is determined by a proportional controller that calculates the deviation of the desired angle from the current heading of the robot: $\omega = K_p(\angle\mathbf{u}_{target} - \angle\mathbf{u}^t)$, where K_p is a proportionality constant whose value is set to 0.5 s^{-1}. Finally, the rotation speeds of the left (N_L) and right (N_R) motors are set to:

$$N_L = \left(u + \frac{\omega}{2} l\right) \frac{1}{r}; N_R = \left(u - \frac{\omega}{2} l\right) \frac{1}{r},$$

where l is the distance between the wheels and r is their radius.

3 Experiments

In this section, we first introduce the metrics and the experimental setup used to evaluate the proposed methodology. We then present the results in a stationary and a non-stationary environment.

3.1 Metrics

In flocking, we are interested in having a group of robots that move compactly, coherently, within their sensory range and without collisions. Furthermore, the group should be aligned towards a common direction (in our case the goal direction) and move towards that direction. In this paper, we use two metrics as in [14]: order and accuracy.

Order: The order metric ψ is used to measure the angular order of the robots. $\psi \approx 1$ when the group has a common heading and $\psi \ll 1$ when each robot is pointing in a random, different direction. The order is defined as:

$$\psi = \frac{1}{N}\|\bar{\mathbf{a}}\| = \frac{1}{N}\|\sum_{i=1}^{N} e^{j\theta_i}\|,$$

where N is the total number of robots in the experiment, and $\bar{\mathbf{a}}$ is the vectorial sum of the measured headings of the N robots.

Accuracy: The accuracy metric δ is used to measure how accurately close to the target direction (dependent on the task) robots are moving. $\delta \approx 1$ when robots are perfectly aligned (which corresponds also to a high value for the order metric $\psi \approx 1$) towards the correct direction of motion. As in [1], accuracy can be defined as:

$$\delta = 1 - \frac{\sqrt{2\left(1 - \psi cos(\angle\bar{\mathbf{a}} - \angle\mathbf{g})\right)}}{2},$$

where $\angle\bar{\mathbf{a}}$ is the direction of $\bar{\mathbf{a}}$ and $\angle\mathbf{g}$ is the goal direction with respect to the common reference frame.

3.2 The Task and the Experimental Setup

In our experiments, N mobile robots are placed at random positions and with random orientations in an empty arena of 5 meters $\times$ 5 meters. Each robot is a realistic simulation of a foot-bot, in development for the Swarmanoid project[2]. We utilized the following sensors and actuators: i) A light sensor, that is able to perceive a noisy light gradient around the robot. It is used to measure θ_r, the orientation of robot r with respect to a common light source. ii) A range and bearing communication system, with which a robot can send a message to other robots that are within 2 meters and in line of sight [17]. iii) Two wheels actuators, that are used to control independently the left and right wheels speed of the robot. iv) 24 LEDs and a camera, which are used to detect distance and bearing from other robots in the proximal control behavior (see Section 2.1). We conducted two sets of experiments.

Stationary environment: In a stationary environment, a proportion ρ of randomly selected robots are given the information about the goal direction $\mathbf{g}$. All the other robots remain uninformed for the entire duration of the simulation. In every run, we randomly choose $\mathbf{g}$, as well as the selection of robots that are informed. The duration of one run is 100 simulated seconds.

Non-stationary environment: A non-stationary environment consists of four stationary phases of equal length. The proportion of informed robots ρ is kept fixed during the entire run. However, at the beginning of every stationary phase, the informed robots are reselected at random. Also, the goal direction $\mathbf{g}$ changes randomly from one stationary phase to the next one. The duration of one run is 250 simulated seconds.

In the stationary environment, we study the effect of changing the swarm size N and the proportion of informed robots ρ. In the non-stationary environment, we also study the effect of noise in the heading alignment vector $\mathbf{h}$. We modeled this noise as a uniformly distributed random variable controlled by a scaling parameter $\sigma \in [0, 1]$, which is used to add noise to θ_r, the robot's measured heading: $\tilde{\theta_r} = \theta_r + \mathcal{U}(-\sigma 2\pi, +\sigma 2\pi)$. For each experimental setting, we execute 100 runs and we report the average results.

3.3 Results in Stationary Environments

The effect of varying the swarm size N is shown in Figure 1a. The information-aware communication strategy outperforms the classical heading communication strategy, in the sense that it achieves higher accuracy both with small (10) and large (100) swarms. In both strategies, the convergence speed is higher with the smaller swarm size. This can be explained by the fact that smaller swarms have smaller inertia than larger swarms. However, in the heading communication strategy, the accuracy level reached with a larger swarm is higher than the one

[2] `http://www.swarmanoid.org`

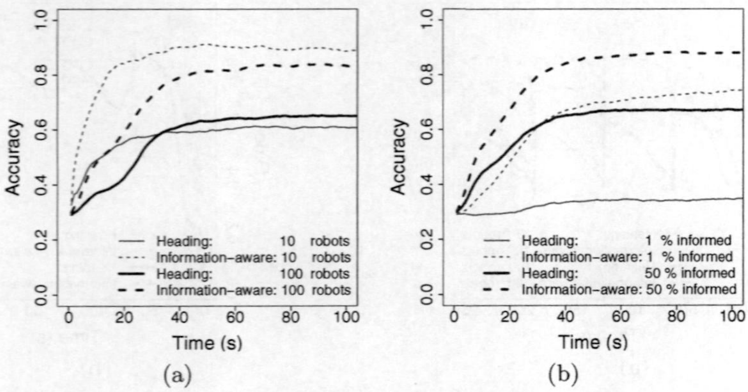

Fig. 1. Information-aware communication against heading communication in a stationary environment: (a) effect of varying swarm size N with fixed $\rho = 0.25$ and $\sigma = 0$; (b) effect of varying the percentage of informed robots ρ with fixed $N = 100$ and $\sigma = 0$

achieved with a smaller swarm. This result is consistent with the findings of [1], which state that the needed proportion of informed robots to achieve a given level of accuracy becomes smaller for increasing swarm sizes.

The effect of varying the proportion of informed robots ρ is shown in Figure 1b. Also in this case, the information-aware communication strategy outperforms the classical heading communication. In both strategies, more informed robots corresponds to higher accuracy. With a proportion of 1% informed robots (corresponding to 1 robot), the classical alignment communication strategy cannot achieve an increasing accuracy over time, which means that at the end of the simulation robots are randomly oriented as they were at the beginning. Differently, the proposed information-aware communication strategy can cope also with a very small proportion of informed robots.

3.4 Results in Non-stationary Environments

The effect of varying the swarm size N is shown in Figure 2a. The same trends observed in the stationary environment are also present here: The information-aware communication strategy always outperforms the classical alignment communication. However, here we observe an interesting phenomenon: The accuracy convergence speed of the information-aware communication strategy in the initial stationary phase is lower than the one in the subsequent three stationary phases. This can be explained by the order shown in Figure 2b. The initial order of the system is very low, and gets higher and higher towards the end of the first stationary phase. When the change in the environment occurs, order decreases but does not reach the initial, very low values for any of the techniques. This means that the swarm is able to make a transition from an ordered state to another ordered state without a detrimental impact on the order itself, which in turn corresponds to faster adaptation to changes in the environment.

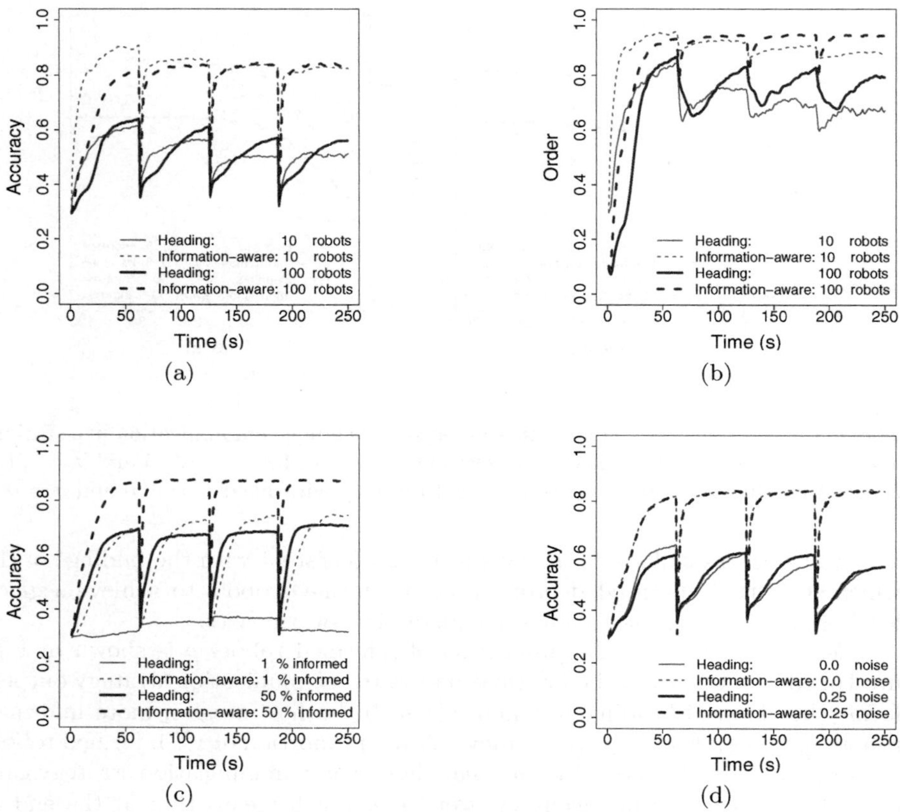

Fig. 2. Information-aware communication againts heading communication in a non-stationary environment: effect on the accuracy (a) and on the order (b) of varying swarm size N with fixed $\rho = 0.25$ and $\sigma = 0$, (c) effect of varying the percentage of informed robots ρ with fixed $N = 100$ and $\sigma = 0$; (d) effect of adding noise $\sigma = 0.25$ with fixed $N = 100$ and $\rho = 0.25$.

The effect of varying the proportion of informed robots ρ is shown in Figure 2c. The same trends observed in the stationary case apply in this case. Hence, the information-aware communication strategy scales well to the non-stationary case even when only 1% of the robots (in this case only one robot) are informed.

Finally, the effect of adding noise σ in the alignment is shown in Figure 2d. As we can see, noise has a non-significant impact on the accuracy of the information-aware communication strategy, which still continues to outperform the classical alignment communication strategy.

4 Conclusions and Future Work

In this paper, we proposed a communication strategy, called information-aware communication, for heading alignment in self-organized flocking. In the system, a swarm of robots is decomposed into informed and uninformed robots: informed robots possess information about a goal direction, whereas uninformed robots do not. The proposed strategy works as follows: each robot communicates the goal direction when it is informed, whereas it communicates the average of its neighbors messages when it is uninformed. We compared this strategy with the heading communication strategy, similar to those used in the literature, in which each robot communicates directly its own heading.

We conducted two sets of experiments. The first set is executed in a stationary environment, where the goal direction and the informed robots do not change over time. The second set is executed in a non-stationary environment, where both the goal direction and the informed robots change over time. Results show that the proposed information-aware communication strategy always outperforms the heading communication strategy. Furthermore, the proposed approach is also robust against noise in the alignment, and it achieves high accuracy values even if only few robots in a large swarm are informed.

The presented work can be extended in many ways. First, a more complete scenario may include multiple sources of different information which can be conflicting, and perceived by different robots in different ways. In this scenario, we may need the swarm to be coherent while heading towards a common direction, or split to different sub-groups, depending on the application. To cope with this scenario, we might need to include some measure of information quality, perceived by each robot, which can be used to determine a new communication strategy. Secondly, we plan to port the presented work into real robots. We believe that the set of assumptions made in this work are all compatible with future real robots experiments, with the exception of the one in which we assume that all robots are able to perceive a common environmental cue. As a matter of fact, the foot-bot robots that we will use are not equipped with a compass, and the light sensor cannot be used for such an aim because the perception of light is not uniform in the swarm due to the robots shadowing each other or to their varying distance from the light source. To avoid the need of a common environmental cue, we plan to extend the information-aware communication strategy to incorporate the capabilities of a situated communication mechanism, in which also the range and bearing of the sender is available and has a meaning.

Acknowledgments. This work was supported by the *SWARMANOID* project funded by the Future and Emerging Technologies programme (IST-FET) of the European Commission (grant IST-022888). M. Birattari and M. Dorigo acknowledge support from the F.R.S.-FNRS of the French Community of Belgium.

References

1. Couzin, I.D., Krause, J., Franks, N.R., Levin, S.A.: Effective leadership and decision-making in animal groups on the move. Nature 433, 513–516 (2005)
2. King, A.J., Johnson, D.D.P., Van Vugt, M.: The origins and evolution of leadership. Current biology 19(19), 911–916 (2009)

3. Stewart, K.J., Harcourt, A.H.: Gorillas' vocalizations during rest periods: signals of impending departure? Behaviour 130(1), 29–40 (1994)
4. Beekman, M., Fathke, R., Seeley, T.: How does an informed minority of scouts guide a honeybee swarm as it flies to its new home? Animal Behaviour 71(1), 161–171 (2006)
5. Camazine, S., Franks, N.R., Sneyd, J., Bonabeau, E., Deneubourg, J.L., Theraulaz, G.: Self-Organization in Biological Systems. Princeton University Press, Princeton (2001)
6. Reynolds, C.: Flocks, herds and schools: A distributed behavioral model. In: Stone, M.C. (ed.) SIGGRAPH 1987: Proc. of the 14th Annual Conference on Computer graphics and Interactive Techniques, pp. 25–34. ACM, New York (1987)
7. Matarić, M.J.: Interaction and Intelligent Behavior. PhD thesis, MIT, MA (1994)
8. Kelly, I., Keating, D.: Flocking by the fusion of sonar and active infrared sensors on physical autonomous robots. In: Proc. of the Third Int. Conf. on Mechatronics and Machine Vision in Practice, pp. 14–17 (1996)
9. Spears, W.M., Spears, D.F., Hamann, J.C., Heil, R.: Distributed, physiscs-based control of swarms of vehicles. Autonomous Robots 17, 137–162 (2004)
10. Holland, O., Woods, J., Nardi, R., Clark, A.: Beyond swarm intelligence: the ultraswarm. In: Proc. of the IEEE Swarm Intelligence Symposium, Piscataway, NJ, pp. 217–224. IEEE Press, Los Alamitos (2005)
11. Hayes, A., Dormiani-Tabatabaei, P.: Self-organized flocking with agent failure: Off-line optimization and demonstration with real robots. In: Proc. of the IEEE Int. Conf. on Robotics and Automation, Piscataway, NJ, pp. 3900–3905. IEEE Press, Los Alamitos (2002)
12. Campo, A., Nouyan, S., Birattari, M., Groß, R., Dorigo, M.: Negotiation of goal direction for cooperative transport. In: Dorigo, M., Gambardella, L.M., Birattari, M., Martinoli, A., Poli, R., Stützle, T. (eds.) ANTS 2006. LNCS, vol. 4150, pp. 191–202. Springer, Heidelberg (2006)
13. Turgut, A.E., Çelikkanat, H., Gökçe, F., Şahin, E.: Self-organized flocking in mobile robot swarms. Swarm Intelligence 2(2), 97–120 (2008)
14. Çelikkanat, H., Turgut, A.E., Şahin, E.: Guiding a robot flock via informed robots. In: Asama, Kurokawa, Ota, Sekiyama (eds.) Distributed Autonomous Robotic Systems (DARS 2008), Berlin, Germany, pp. 215–225. Springer, Heidelberg (2008)
15. Reebs, S.G.: Can a minority of informed leaders determine the foraging movements of a fish shoal? Animal behaviour 59, 403–409 (2000)
16. Tanner, H.G., Jadbabaie, A., Pappas, G.J.: Stable flocking of mobile agents part II: Dynamic topology. In: Proc. of the 42nd IEEE Conference on Decision and Control, Piscataway, NJ, vol. 2, pp. 2016–2021. IEEE Press, Los Alamitos (2003)
17. Roberts, J., Stirling, T., Zufferey, J., Floreano, D.: 2.5d infrared range and bearing system for collective robotics. In: Papanikolopoulos, N., Sugano, S., Chiaverini, S., Meng, M. (eds.) IEEE/RSJ International Conference on Intelligent Robots and Systems, Piscataway, NJ. IEEE Press, Los Alamitos (2009)

Evolution of XPath Lists for Document Data Selection*

Pablo García-Sánchez, Juan J. Merelo Guervós, Pedro Ángel Castillo, Jesús González, Juan L. Jiménez Laredo, Antonio M. Mora García, and Maria I. García Arenas

Dept. of Computer Architecture and Technology, University of Granada, Spain
pgarcia@atc.ugr.es

Abstract. XML has became a standard for structured data, and very often transformations from one specific format to another are needed. XSLT stylesheets are programs designed for this purpose, and they use XPath expressions to select sets of nodes within the document. In this paper a new version of an evolutionary algorithm that creates XSLT from examples is presented, improving on previously obtained results by testing a new individual representation with a new set of operators, based mainly on evolution of XPaths with a fixed XSLT program structure. The experiments show that this new representation, and a lower set of operators, yield better results in less generations that in our previous version.

1 Introduction

Since the Information Technology industry has settled on different Extensible Markup Language (XML) dialects as information exchange format, there is a business need for programs that transform from one XML set of tags to another, extracting information or combining it in many possible ways; a typical example of this transformation could be the extraction of news headlines from an on-line newspaper that uses XHTML.

XSLT stylesheets (XML Stylesheet Language for Transformations) [1], also called *logicsheets*, are programs designed for this purpose: applied to an XML document, they produce another. There are other possible solutions: programs written in any language that work with text as input and output (using, for instance, regular expressions) or SAX filters [2], processing each tag in an XML document in a different way, and not needing to load into memory the whole XML document. However, these solutions require programming in external languages, while XSLT is a part of the XML set of standards (in fact, XSLT logicsheets are XML documents). This is why XSLT is one of the most common ways of specifying document transformations. XSLT applies XPath expressions [3] to select nodes from the source document. XPath (XML Path Language) is a language created to construct expressions which proccess an XML document to find and select specific parts of the hierarchical structure of XML. For example,

* Supported by projects AmIVital (CENIT2007-1010) and EvOrq (TIC-3903).

R. Schaefer et al. (Eds.): PPSN XI, Part II, LNCS 6239, pp. 341–350, 2010.

the XPath route `/rss/item[10]/date` selects the `date` node of the 10th `item` node of the XML document whose root is `rss`.

The work needed for logicsheet creation grows dramatically with the number of input and output formats: for n input and m output formats, $n \times m$ transformations will be needed. Considering that each conversion is a hand-written program and the initial and final formats can vary with certain frequency, any automation of the process means a considerable saving of effort.

So, the problem is to find the XSLT logicsheet that, from one input XML document, is able to obtain an output XML document which contains exclusively the information desired from the first one. This information may be sorted in any possible way (maybe in a different order to the input document). In this work, an Evolutionary Algorithm (EA) to solve this problem is presented. The logicsheet will be evolved using evolutionary operators that take into account the structure of the program and its components.

Thus, XSLT provides a general mechanism for the association of patterns in the source XML document to the application of format rules to these elements, but in order to simplify the search space for the evolutionary algorithm, just three instructions will be considered in this paper: template, which selects the XML fragments that will be included when the element in its match attribute is found; apply-templates, which is used to select the elements to which the transformation is going to be applied, and delegate control to the corresponding templates; and copy-of, which includes the text representation of the nodes of the input XML into the output file (that is, copy all node contents and tags), so the output file will be a complete XML instead a list of content, as we did in our previous work [4]. XPath expressions will also be used to select particular elements and sets of them.

In [5], we published an initial set of XSLT evolution experiments, testing different document structures, operators and fitness. In this paper we will try to improve those results with a more simpler XSLT structure: a list of copy-of nodes with evolvable XPaths. To compare with the previous work we are going to use the same examples as before: a set of input and output XML documents with a tree structure. These documents are composed of several nodes, which makes it easier to compare them with each other. As before, the output XML will be a complete XML document with a (possibly sorted in a different way) list of nodes present in the original document.

The rest of the paper is structured as follows: the state of the art is presented in Section 2. Section 3 describes the solution presented in this work, with the novel elements introduced. Experiments with the automatic generation of XSLT stylesheets for different examples are described in Section 4, and finally the conclusions and lines of future work are presented in Section 5.

2 State of the Art

To our knowledge, there are few works related to the application of genetic programming techniques to the automatic generation of XSLT logicsheets; one of them, by Scott Martens [6], presents a technique to find XSLT stylesheets that transform an XML file into HTML by using genetic programming. Martens works on simple XML documents and uses the UNIX *diff* command as the basis for its fitness function. He concludes that genetic programming is useful to obtain solutions to simple examples of the problem, but it needs unreasonable execution times for complex examples and might not be a suitable method to solve this kind of problems.

Schmidt and Waltermann [7] addressed the problem taking into account that XSLT is a functional language, and using functional language program generation techniques on it, in what they call *inductive synthesis*. Firstly, they create a non-recursive program, and then, by identifying recurrent parts, convert it into a recursive program; this is a generalization of the technique used to generate programs in other programming languages such as LISP [8], and used thoroughly since the eighties [9].

A few other authors have approached the general problem of generating XML document transformations, knowing the original and target structure of the documents, as represented by its DTD (Document Type Definition): [10,11,12,13,14] have proposed the semi-automatic generation of transformations for XML documents, but user input is needed to define the label association. There are also freeware programs that perform transformations on documents from a XSchema to another one. However, they must know both XSchemata in advance, and are not able to accomplish general transformations on well formed XML documents from examples. In some cases, the transformation is carried out in two steps: first the input document is compiled to a generic representation, which is then mapped to the new, required, representation [15,12].

Some other papers try to map the tree structure of input and target documents [16], but this must be a supervised procedure, in the same way as it was proposed in [12].

In our previous work [5], we presented an evolutionary algorithm to obtain an XSLT that extracts information represented in a output XML from an input XML. Several XSLT structures and operators were presented and studied. The main drawback of the proposed method in that work is the difficulty of managing a complete XSLT tree structure since it requires more processing time and larger sets of operators to manage all possibilities of the trees and routes that are formed.

3 Methodology

The EA described in this section evolves simpler XSLT stylesheets, which are generated using a set of operators, and evaluated using a fitness function which is related to the difference between generated XML and output XML associated to the example. The way the algorithm works is shown in Figure 1. The solution has been programmed using JEO [17], an evolutionary algorithm library developed at the University of Granada as a part of the DREAM project [18], which is available from `http://www.dr-ea-m.org`.

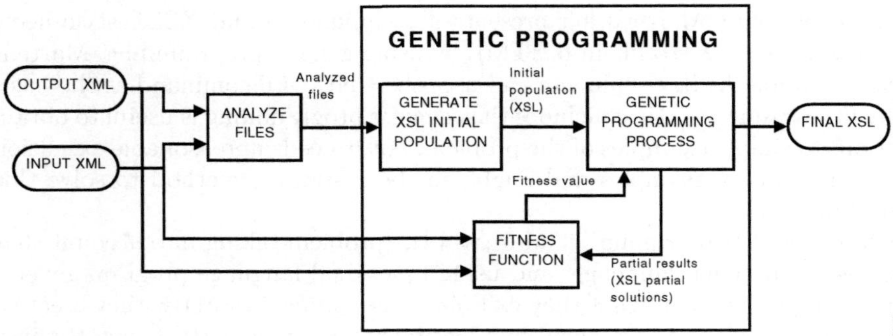

[htb]

Fig. 1. This figure shows the algorithm workflow. Each individual of the population is a XSLT stylesheet, whose fitness is computed comparing the XML generated by the stylesheet (using the input XML) with the output XML.

In our previous work we used a specific XSLT structure, which was selected from [4]. An example of this structure is shown in Figure 2.

However, this structure has some inconveniences, such as complexity of processing, managing and evolution. So, a new XSLT structure is proposed in this paper (see Figure 3). This new structure is not a recursive stylesheet, it is only a list of XPaths in `copy-of` nodes. The bennefits of this kind of structure are quite obvious: recursive processing is not needed and less management operators are used.

Thus, in our EA the individual representation is basically a list of XPath routes, instead of a full tree. This reduces the evolution process and also the complexity since there is a reduction in the number of operators, which improves the fitness calculation, and the XSLT processing from the input to the output file.

There is a smaller set of operators (just 8 instead of 14). The new operators may be classified in two different types: the first one, are those which modify the XPath routes in the attributes of the XSLT instructions `copy-of`); and the other are the operators used to add or remove routes to the list. In order to ensure the existence of the elements (tags) added to the XPath expressions and XSLT instruction attributes, every time one of them is needed it is randomly selected from the input file.

The first class of operators, those that alter the attributes, can be divided in two kinds: the operators that modify the list of elements of a route and those that modify a filter:

- **XpathMutatorAddElement:** Adds a valid element at the end of the XPath route.
  I.e.: `/rss/item` → `/rss/item/date`
- **XpathMutatorRemoveElement:** This mutator removes the last element of a route.
  I.e.: `/rss/item/date` → `/rss/item`

```
<?xml version="1.0"?>
<xsl:stylesheet version="1.0"
  xmlns:xsl="http://www.w3.org/1999/XSL/Transform">
  <xsl:output indent="no" method="xml"/>
  <xsl:template match="/">
    <rss>
      <xsl:apply-templates select="/rss"/>
    </rss>
  </xsl:template>
  <xsl:template match="/rss">
    <xsl:copy-of select="div[3]"/>
    <xsl:copy-of select="div[1]/h3[5]"/>
    <xsl:copy-of select="h1"/>
  </xsl:template>
</xsl:stylesheet>
```

Fig. 2. Example of a final XSLT generated by our previous algorithm. This logicsheet, applied to the input XML document, produces an XML document that equals the desired XML output document. Since it is a tree, it is more complex to build and manage.

On the other hand, the filter operator modifies the cardinality of a filter:

- **XpathMutatorModifyFilter:** Modifies the cardinality of a valid filter (that is, only the elements with possible cardinalities can be modified and the new cardinality is a valid number).
  I.e.: `/rss/item[10]/date` → `/rss/item[3]`
- **XpathMutatorAddFilter:** Adds a filter in a valid position
  I.e.: `/rss/item/date` → `/rss/item[2]/date`
- **XpathMutatorRemoveFilter:** Removes an existing filter
  I.e.: `/rss/item[3]/date` → `/rss/item/date`

Finally the XSL operators modify the list of different XPaths:

- **XSLAddXPath:** Removes an existing XPath.
  I.e.:`/rss/item[10]` → `/rss/item[10]`
  `/rss/title`
- **XSLRemoveXPath:** Adds a valid XPath to the list.
  I.e.: `/rss/item[10]` → `/rss/item[10]`
  `/rss/title`

However, these operators are not enough to perform a smooth search; sometimes the XSLT search converges into a bad solutions area, when we want to select ordered but alternated items from a node. So, as we proved in our previous work, it is necessary to add a new operator to increase the diversification of this solution. This operator, *XSLSplitNode*, expands a random `copy-of` node into a list of complete `copy-of` with all cardinalities (as shown in Figure 4). The result

```
<?xml version="1.0"?>
<xsl:stylesheet version="1.0" xmlns:xsl="http://www.w3.org/1999/XSL/Transform">
 <xsl:output indent="no" method="xml"/>
 <xsl:template match="/">
  <rss>
   <xsl:copy-of select="/rss/channel/title"/>
   <xsl:copy-of select="/rss/channel/item[10]/category[3]"/>
   <xsl:copy-of select="/rss/channel/item[3]"/>
   <xsl:copy-of select="/rss/channel/item[5]"/>
  </rss>
 </xsl:template>
</xsl:stylesheet>
```

Fig. 3. Example of a final XSLT generated by the algorithm. This logicsheet, applied to the input XML document, also produces an XML document equals to the desired XML output document, but it is simpler than the shown in Figure 3.

```
<xsl:template match="book">
    <xsl:copy-of select="chapter"/>
</xsl:template>
```

```
<xsl:template match="book">
  <xsl:copy-of select="chapter[1]"/>
  <xsl:copy-of select="chapter[2]"/>
  <xsl:copy-of select="chapter[3]"/>
  <xsl:copy-of select="chapter[4]"/>
</xsl:template>
```

Fig. 4. The left template is transformed into right template applying the split mutator. The number of chapters in the input XML were 4 (this is known by the algorithm when it process the input XML at the beginning of execution).

of applying the new XSLT and the previous is the same, but it is easier for genetic operators to modify the list of `copy-of` than the generic one (modifying, adding or removing XPath and/or tags).

The crossover has also been simplified, just a one-point crossover of the list of routes, instead of a crossover which must comply all the recursive tree structure restrictions.

The fitness is computed as the XML difference between the desired and the obtained output, that is, the difference in nodes between the desired T and the actual document X. This difference breaks down in insertions (nodes in X but not in T) and deletions (nodes in T but not in X). We will leverage this vectorial structure of fitness so that evolution can profit from it: instead of using a single aggregative function, as we did in previous papers [4], fitness is now a vector that includes the number of node deletions and additions needed to obtain the target output from the obtained output. The XSLT stylesheet is correct only if the number of deletions and additions is 0. So, the fitness is minimized by comparing individuals as follows: An individual is considered better than another

- if the number of deletions is smaller,
- if the number of additions is smaller, being the number of deletions the same.

Separating and prioritizing the number of deletions helps to guide the evolution, by trying to find first a stylesheet that includes all the elements in the target document. As we proved in [5], we do not consider the length of the generated XSLT, in order to minimize selective pressure, helping in that way the evolution of existing XPaths. This aids to keep the solutions having the same number of deletions and insertions but larger size caused by the use of the operator *XSLSplitNode* (expanding the copy-of tags).

4 Experiments and Results

In order to test the value of the algorithm we have performed several experiments with 7 different XML input and output files. The algorithm has been executed 30 times for each input XML. The same input file was used for several experiments: a RSS feed from a weblog and a XHTML file[1].

The computer used to perform the experiments is a Centrino Core Duo at 1.83 GHz, 2 GB RAM, and the Java Runtime Environment 1.6.0.01; the XML and XSLT processors were the default ones included in the JRE standard library. The population size was 128 individuals for all runs, generated using the input XML as information source. The termination criteria was set to 300 generations or until a solution was found, and selection was performed considering a 5-Tournament. 30 experiments were run, with different random seeds, for each input document. The mutation operator chooses one of the operators (commented in Section 3) with the same probability. The crossover and mutation probability have been set to 0.25 and 0.5, after several experimental runs.

We compare the results with the best structure found in our previous work. The breakdown of results per input file is shown in Table 1.

The new structure, in general, yielded better results than previously. The algorithm was able to find an adequate XSLT stylesheet within the pre-assigned number of generations in most cases, using a more restricted set of operators than before.

Examples 1, 2, 3 and 6 are complete and ordered lists of elements of one or several nodes, whose solution is simple, since the algorithm can easily create a logicsheet that extracts all the children of a specific node. Example 7 takes specific and repeated elements from distinct nodes, which makes it more difficult, because different expressions are needed to extract each one of them and the generated logicsheet is more complex. Finally, examples 4 and 5 focus into portions of ordered and unordered fragments of an XML section (namely the 3rd and 6th chapters of a book), so the population converges into solutions with all the elements of a node (selecting all chapters of a book) due to the way the fitness works. As can be seen in Figure 5, the evolution process starts priorizing the individuals with all nodes present (that is, 0 nodes deleted). After this, the algorithm tries to minimize the nodes not present in the output (*Inserted* in

[1] All input and output files and programs used in this experiment are available from our Subversion repository: `http://tinyurl.com/6nxv8c`

Table 1. Number of times, out of 30 experiments, a solution is found (TF, Times Finished) within the predefined number of generations and the average generations/standard deviation to find an optimal solution (or reach 300 generations) of the XSLT tree structure and the XPath list structure.

Example	TF XSLT Tree	TF XPath List	Gens. XSLT Tree	Gens. XPath List
1	25	28	66.33 ± 106.85	74.23 ± 71.53
2	30	30	1.1 ± 1.29	8.1 ± 1.67
3	30	30	3.83 ± 2.90	17.6 ± 4.81
4	24	30	71.68 ± 107.24	28 ± 7.07
5	29	30	36.03 ± 50.19	27.7 ± 7.05
6	30	30	19.0 ± 11.12	36.7 ± 8.84
7	13	14	214.0 ± 112.90	231.17 ± 90.52

Figure 5). Since the length of the generated XSLT is not used to calculate the fitness, this value increases in the period in which best individuals are not found.

When a solution was found, the number of generations and time taken to find it also varies, as shown in Table 1. In general, the exploration/exploitation balance seems to be biased towards exploration. Being such a vast and rough search space makes that, after a few initial generations that create stylesheets with a small difference from the target, mutations are the main operator at work.

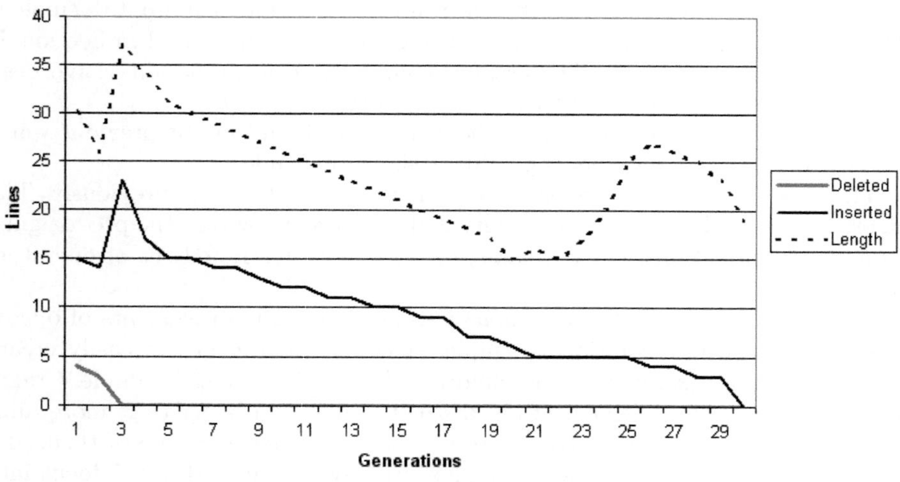

Fig. 5. Evolution of the best fitness in an execution of example 4. The length of the XSLT is not considered in the fitness calculation.

5 Conclusions and Future Work

In this paper we present a new version of an evolutionary algorithm designed to search the XSLT logicsheet that is able to make a particular transformation from an input XML document into a desired output one. One of the advantages of this application is that resulting logicsheets can be used directly in a production environment, without the interaction of a human operator. It tackles a real-world problem found in many organizations and it is open source software, available from `http://tinyurl.com/5lwjcn`.

The performed experiments show that the usage of a simpler structure produces better results, using less operators. Also we have found that the search space is particularly rough, with mutations in general leading to huge changes in fitness. The hierarchical fitness used is probably the cause of having a big loss of diversity at the beginning of the evolutionary search, leading to the need of a higher level of explorations later during the algorithm run. This problem will have to be approached considering explicit diversity-preservation mechanisms, or by using a multiobjective evolutionary algorithm, instead of the one used now. A deeper understanding of how different operator rates affect the results will also be useful; for the time being, operator rate tuning has been very shallow, and geared towards obtaining the result. In addition, obtained results can be used as a baseline for future versions of the algorithm, or other algorithms for the same problem. At any rate, unlike what was mentioned in the pioneering paper [6], solutions can be found effectively and efficiently.

However, there are some other issues that should have to be addressed in future papers. Firstly, using the DTD (associated to an XML file) as a source of information for conversions between XML documents and for restrictions of the possible variations or adding different labels in the XSLT to allow the building of different kinds of documents such as HTML or WML. Another topic to adress is testing evolution with other type of tools, such as a chain of SAX filters. Obviously, testing different kinds and increasingly complex sets of documents, and using several input and desired output documents at the same time, to test the generalization capability of the procedure. Finally, to tackle difficult problems from the point of view of a human operator: in general, the XSLT stylesheets found here could have been programmed by a knowledgeable person in around an hour, but in some cases, input/output mapping would not be so obvious at first sight. This will mean, in general, increase also the XSLT statements used in the stylesheet, and also in general, adding new types of operators.

References

1. Clark, J.: XSL transformations (XSLT), version 1.0, W3C recommendation (November 16, 1999), `http://www.w3.org/TR/xslt.html`
2. Wikipedia: Simple API for XML — Wikipedia, the free encyclopedia (2007) [Online; accessed March 21, 2007]
3. Clark, J., DeRose, S., et al.: XML Path Language (XPath) Version 1.0. W3C Recommendation 16 (1999)
4. García-Sánchez, P., Guervós, J.J.M., Sevilla, J.P., Laredo, J.L.J., Mora, A.M., Valdivieso, P.A.C.: Automatic generation of xslt stylesheets using evolutionary algorithms. In: Genetic and Evolutionary Computation Conference, GECCO 2008, Proceedings, pp. 1701–1702 (2008)

5. García-Sánchez, P., Guervós, J.J.M., Laredo, J.L.J., Mora, A., Castillo, P.A.: Evolving xslt stylesheets for document transformation. In: Rudolph, G., Jansen, T., Lucas, S., Poloni, C., Beume, N. (eds.) PPSN 2008. LNCS, vol. 5199, pp. 1021–1030. Springer, Heidelberg (2008)
6. Martens, S.: Automatic creation of XML document conversion scripts by genetic programming. In: Genetic Algorithms and Genetic Programming at Stanford, p. 269 (2000)
7. Schmid, U., Waltermann, J.: Automatic synthesis of XSL-transformations from example documents. In: Hamza, M. (ed.) IASTED International Conference on Artificial Intelligence and Applications, pp. 252–257 (2004)
8. Biermann, A.: The inference of regular LISP programs from examples. IEEE Transactions on Systems, Man and Cybernetics 8(8), 585–600 (1978)
9. Biermann, A.W., Guiho, G. (eds.): Computer Program Synthesis Methodologies. Reidel, Dordrecht (1983)
10. Leinonen, P.: Automating XML document structure transformations. In: Proceedings of the 2003 ACM Symposium on Document Engineering, pp. 26–28 (2003)
11. Kuikka, E., Leinonen, P., Penttonen, M.: Towards automating of document structure transformations. In: Proceedings of the 2002 ACM Symposium on Document Engineering, pp. 103–110 (2002)
12. Chidlovskii, B., Fuselier, J.: Supervised learning for the legacy document conversion. In: DocEng 2004: Proceedings of the 2004 ACM Symposium on Document Engineering, pp. 220–228. ACM, New York (2004)
13. Suzuki, N., Fukushima, Y.: An XML document transformation algorithm inferred from an edit script between DTDS. In: ADC 2008: Proceedings of the Nineteenth Conference on Australasian Database, pp. 175–184. Australian Computer Society, Inc., Australia (2007)
14. Chuang, T.R., Lin, J.L.: On modular transformation of structural content. In: DocEng 2004: Proceedings of the 2004 ACM symposium on Document Engineering, pp. 201–210. ACM, New York (2004)
15. Soares, L.F.G., Rodrigues, R.F., de Resende Costa, R.M.: Automatic building of frameworks for processing XML documents. In: WebMedia 2006: Proceedings of the 12th Brazilian Symposium on Multimedia and the Web, pp. 118–127. ACM Press, New York (2006)
16. Shin, D.H., Lee, K.H.: Towards the faster transformation of XML documents. J. Inf. Sci. 32(3), 261–276 (2006)
17. Arenas, M.G., Dolin, B., Merelo-Guervós, J.J., Castillo, P.A., de Viana, I.F., Schoenauer, M.: JEO: Java Evolving Objects. In: Proceedings of the Genetic and Evolutionary Computation Conference, p. 991 (2002)
18. Arenas, M., Collet, P., Eiben, A., Jelasity, M., Merelo, J.J., Paechter, B., Preuß, M., Schoenauer, M.: A framework for distributed evolutionary algorithms. In: Guervós, J.J.M., Adamidis, P.A., Beyer, H.-G., Fernández-Villacañas, J.-L., Schwefel, H.-P. (eds.) PPSN 2002. LNCS, vol. 2439, pp. 665–675. Springer, Heidelberg (2002)

PMF: A Multicore-Enabled Framework for the Construction of Metaheuristics for Single and Multiobjective Optimization

Deon Garrett

Department of Computer Science,
University of Memphis,
Memphis, TN, USA
deong@acm.org

Abstract. This paper describes the design and implementation of the Parallel Metaheuristics Framework (PMF), a C++ framework for the construction of single and multiobjective metaheuristics utilizing Intel's Threading Building Blocks library to allow easy parallelization of computationally intensive algorithms. The framework demonstrates a generic approach to the construction of metaheuristics, striving to provide a general representation of core operators and concepts, thus allowing users to more easily tailor the system for novel problems. The paper describes the overall implementation of the framework, demonstrates a case study for implementing a simple metaheuristic within the system, and discusses a range of possible enhancements.

1 Introduction

Many metaheuristics for optimization problems, and almost all aimed at multiobjective optimization, exhibit a large degree of potential parallelism that can be exploited to provide significant gains in execution speed. Historically, most researchers have not focused on this potential gain. One prominent reason is simply that such researchers are accustomed to running many separate trials of each algorithm in order to report statistically significant results, and thus they already have a free form of parallelism in the form of simply running multiple independent trials at once.

However, in recent years, increases in processing power on the desktop have come almost exclusively in the form of additional processing cores, while the processing power of a single core has largely stagnated. This has led to an increase in interest in technologies and techniques that enable programmers and users to take fuller advantage of the parallel processing power on almost every desktop and laptop. It is generally expected that this trend will only accelerate reaching more and more cores in the next few years.

In 2007, Intel announced the release of their open source Threading Building Blocks (TBB) package. TBB provides a high level, task-oriented approach to developing software to better utilize the many processing cores available in modern CPUs. This work describes the design and implementation of a framework

R. Schaefer et al. (Eds.): PPSN XI, Part II, LNCS 6239, pp. 351–360, 2010.

for constructing parallel metaheuristics called, appropriately, the Parallel Metaheuristics Framework (PMF) which uses TBB to enable researchers to build metaheuristics that take full advantage of the power of modern desktop and server class processors.

The remainder of this document is arranged as follows. Section 2 briefly describes several existing packages and compares them to PMF with respect to their design goals. In Section 3, the guiding principles driving the development of PMF are put forth. Sections 4 and 5 briefly describe Intel's TBB library with respect to its usage in PMF. Section 6 details a case study: building a simple genetic algorithm in PMF. Finally, Section 7 details areas of ongoing work with PMF.

2 Previous Work

There are a number of other high-quality open source frameworks for experimenting with different flavors of metaheuristics, each with their own focus and strengths. In this section, some of the most popular packages are briefly described.

ECJ [1], is one of the most complete packages available for research into evolutionary computation. It includes particularly strong support for genetic programming, but all manners of evolutionary algorithms are supported. ECJ provides excellent support for parallel execution of the algorithms as well. However, the coverage of multiobjective optimization in ECJ is somewhat sporadic, and the focus of ECJ is firmly on evolutionary methods, as opposed to general metaheuristics including various forms of local search.

In the C++ world, ParadisEO [2] is a very good option for general purpose metaheuristic research. It includes not only very complete support for numerous multiobjective techniques, but also allows inclusion of several local search and other methaheuristic algorithms. Furthermore, in the form of ParadisEO-MOEO [3], the framework has been extended to provide support for parallel execution of the search algorithms. However, much of the focus of the parallelism in Paradiseo-PEO is intended to model parallel metaheuristics such as Island model Genetic Algorithms rather than conventional algorithms executed in parallel.

3 The Parallel Metaheuristics Framework

PMF aims to serve a different role. By focusing exclusively on multicore parallelism as opposed to distributed multiprocessing, PMF aims to put a thinner layer of complexity atop the well-known terminology that researchers and practitioners are already familiar with. As well, TBB provides a fairly simple concurrency model which is used consistently throughout the framework, and it should be easy for researchers to adapt their own algorithms using this framework. It is the view of the author that there is a large opportunity, due to increases in desktop computing power, for persons outside the normal academic research environment to employ metaheuristics to solve important problems, using only the desktop class computers already available. PMF attempts to fill this niche by providing a simple configuration mechanism by which existing algorithms may be evaluated, and by providing a simpler model of multicore parallelism that can be exploited by practitioners without specialized training in concurrent programming.

PMF is thus intended to fill a need within the metaheuristics community for an accessible package providing relatively simple support for constructing metaheuristics that take advantage of modern multicore processors, particularly with respect to multiobjective optimization using hybridizations of multiple techniques. The combination of, for example, multiobjective evolutionary algorithms with directed local search methods has proven popular and effective on a number of real world problems. PMF makes the process of building and experimenting this type of algorithm in particular very simple. It requires very little awareness on the part of the user of issues such as locking and synchronization which can often obscure the details of the algorithms. The use of Intel's Threading Building Blocks package to abstract away as much of the parallelization code as possible allows the user to concentrate on combining the ready-made modules in PMF with their own custom code as simply as possible.

PMF adheres to other guiding principles as well. One such principle is that all configuration information must be available at runtime in a simply expressed form, and must be easily managed from within the code. All configuration is thus performed through simple text files. In the author's view, XML is too cumbersome for the generally simple requirements of algorithm configuration. Instead, PMF attempts to provide easy ways for the programmer to retrieve configuration values in a completely type-safe way using template specializations. Figure 3 shows an example of a typical configuration file used with PMF.

There are a few things to note about this sample file. All options are specified generally as keyword/value pairs, separated by either a colon (":") or an

```
# configuration file for running NSGA-II on
# an instance of the multiobjective QAP
algorithm = nsga2
encoding = permutation
problem = qap
problem_data = /path/to/data/file.dat
population_size: 100
terminator: evaluation_limit
max_evaluations: 100000
metric: evaluation_count
metric: generation_count
metric: hypervolume
hypervolume_reference_point: 5000000 5000000
trials: 10
```

Fig. 1. Sample PMF configuration file

equals ("="). Comments are allowed, and are considered to run from any "#" character to the end of the line. The value portion of the configuration lines may contain multiple values, denoting a list or vector, as in the case of `hypervolume_-reference_point`. In addition, keywords may be repeated, in which case PMF forms a list of values associated with that keyword in the order in which they appear in the configuration file.

This system provides a great deal of flexibility while retaining a very simple syntax. From the point of view of the programmer, a single (template) function is provided with several specializations to handle retrieving values from the user-specified configuration. Typically, all that is required is to declare a variable of the desired type to hold the specified value, then pass the variable to the `parameter_value` function, along with the keyword in question.

Figure 3 shows example code reading a value from the configuration file, and demonstrates a number of concepts. In PMF, all accepted keywords are defined in a dedicated namespace so as to serve as a convenient source of in-code documentation. Thus the expression, `keywords::PROCESSORS` refers to a string defined in that namespace to denote the parameter name used to specify the number of processors available. The second argument is the variable which will be used to store the retrieved value. Note that the `parameter_value` function is a template which provides instantiations for all supported data types (most integral and floating point types, `bool`, and standard template library lists and vectors of supported types). PMF attempts whenever possible to maintain strong and strict type safety; it encourages the use of unsigned types when appropriate, const correctness, etc, and the specification of `num_processors` as unsigned is an example of this goal. Finally, the third parameter indicates that specification of this value is optional. In the case of the number of processors, the default behavior is to allow the TBB library to apportion threads as it likes.

```
unsigned int num_processors;
configuration::parameter_value(keywords::PROCESSORS,
                                num_processors, false);
```

Fig. 2. Code to read a value from the configuration file

Out of the box, PMF includes support for a large group of optimization algorithms and techniques. In single objective optimization, the broad class of generational and steady-state evolutionary algorithms are supported, including many popular operators for selection, crossover, mutation, and replacement or environment selection. Several local search operators are also available, and may be embedded either within a population based approach such as one of the evolutionary algorithms or within one of a number of local search containers that implement strategies such as random restarts, simulated annealing, and tabu search algorithms.

Multiobjective optimization is supported through the NSGA-II, SPEA2, and ϵ-MOEA algorithms. Again, each supports the incorporation of local search techniques using any of the supported implementations. In addition, multiple options are available concerning how the local search operators adapt the multiple objectives of the problem into a single-objective function suitable for hill climbing behavior. In addition, several multiobjective local search strategies are supported.

PMF treats all problems as multiobjective, but provides comparators that can single out the first or any specific objective or scalarize the objectives to yield a single valued function. These mechanisms allow several simple means to provide classical single objective optimization.

4 Overview of TBB

Threading Building Blocks, or TBB, is Intel's attempt at a C++ library that abstracts much of the tedium of developing parallel versions of certain algorithms [4]. As evidenced by the name, TBB is a thread-oriented library, as opposed to one aimed at distributed memory parallelism such as MPI. TBB provides template functions implementing parallel algorithms such as `parallel_for`, `parallel_while`, and `parallel_reduce`, and also more fine-grained concepts related to parallelism such as locks and atomic types. To truly derive the most benefit from TBB, it is expected that the programmer will do as much work as possible within the specified parallel algorithms, resorting to manual locking and synchronization only when necessary.

As computers have continued to increase in performance, metaheuristics and optimization in general have become available to a wider audience. Many practitioners who could benefit from optimization will be limited to desktop class hardware, and without the support infrastructure to help manage the complexity of massively parallel clusters. Multiprocessing is notoriously difficult, and without this support infrastructure, the problems of writing correct multithreaded code for the desktop will likely be difficult to overcome for many smaller organizations. TBB aims to help solve this problem by removing as much manual intervention as possible from the process of writing multithreaded code. As such, it is hoped the PMF can provide a base on which new problems and algorithms may be built while requiring a minimum of explicit thread management. In practice, this means expressing metaheuristics in terms of the building blocks provided by the TBB library.

5 Evolutionary Algorithms and the Reduce Operator

In PMF, most forms of evolutionary algorithms are expressed as instances of the `parallel_reduce` function. In functional programming languages, `reduce` is a well-known function which combines a sequence of values into a single value through repeated application of a specified operator. For example, given the a hypothetical functional language, an expression such as

```
reduce + 0 [1,2,3,4,5]
```

generates an expression such as

```
(((((0+1)+2)+3)+4)+5)
```

Note that the initial zero in the expression comes from the second argument passed to the `reduce` function. In essence, `reduce` simply applies the given operator to the given primer argument and the head of the specified list. It then applies the operator to the just computed return value and the next element from the list, continuing in this fashion until the list has been exhausted, returning the result of the final application of the given operator.

One important thing to notice about the expansion of the `reduce` function is that, at least for the case of addition, the operator imposing the reduction is associative, and thus we are free to perform the calculations in any equivalent grouping. One such grouping might be expressed as

```
f [1,2,3,4,5] = ((0+1)+2)+((3+4)+5)
```

Expressed in this manner, we see an obvious opportunity for parallelism. A dual-core machine could perform the first "mini-reduction" on one core while simultaneously performing the second on the other core. With no communication overhead, the speedup could be perfectly linear.

The only remaining step is to specify how the algorithm is to combine the results of the two distinct reductions to obtain the correct solution. In the case of our simple addition, the answer is trivial. However, we are free to use any method of combining the results of the independent parallel reductions to obtain the desired solution.

At this point, it is worthwhile to step back and consider a canonical evolutionary algorithm. The basic steps of such an algorithm are shown in Listing 5. Consider the body of the while loop. The basic form is that we perform some initial setup (creating the child population), enter a for loop in which the evolutionary operators are repeatedly applied to generate the offspring population, and then we consult the replacement operator to combine the parent and child populations into the next generation's parent population.

In TBB terms, we could express the inner for loop as a straightforward instance of `parallel_for`, but each running thread would then need to coordinate access to the shared child population. Another option is thus to formulate the algorithm in terms of a `parallel_reduce`. Unlike the simple example above involving addition over a list of numbers, in this case two different operators are needed. The first must apply a single iteration of the inner GA loop, producing two offspring which are inserted into the child population local to that thread. The second operator is needed to combine all the thread-local child populations back into a single population that can be passed to the *Replacement* function to produce the next generation.

Because `parallel_reduce` (and all the other algorithms in TBB) are implemented as template functions, we must arrange some way to pass the "block" of code implementing the GA loop into the function. TBB, like the STL, makes

Algorithm 1. The Canonical Evolutionary Algorithm

```
Generate initial population P of size n
while not done do
    Create child population P'
    for i ← 1, n/2 do
        (p1, p2) ← Selection(P)
        (c1, c2) ← Crossover(p1,p2)
        c1 ← Mutate(c1)
        c2 ← Mutate(c2)
        P' ← P' ∪ {c1, c2}
    end for
    P ← Replacement(P, P')
end while
```

heavy use of functors for this purpose. A functor in C++ is simply a class that provides `operator()`, thus allowing instances of the class to be called as functions. The basic model in TBB is thus to encapsulate all logic that we wish to be executed in parallel inside of a C++ functor. Any shared resources must be available inside the functor, and if the resource must be modified, we must arrange for a new copy to be allocated inside each copy of the functor. The copied resources may then be built independently in each thread, and TBB will call the functor's `join` method after all threads have completed, at which time the shared resource can be combined back into a single unified object.

This model of parallelism is highly adapted for the case of generational parallelism, i.e., algorithms which partition the work done during each generation across multiple cores. PMF has a strong focus toward hybrid algorithms, and this model of parallelism is especially useful in algorithms that embed time consuming local search algorithms inside a larger generational algorithm. Parallelization of the evaluation function, for example, is certainly possible, but PMF does not currently provide special support to address this issue. In the next section, we will demonstrate the process of building algorithms in TBB by walking through the implementation of the canonical genetic algorithm.

6 Case Study: The Simple GA in PMF

Implementing a new optimizer in PMF consists primarily of two steps. First, we must define the high level skeleton of the algorithm. Ideally, we need to express the algorithm in terms of repeated applications of some inner loop, the iterations of which may be performed in parallel. For the classic simple genetic algorithm, this task is fairly easy. As shown in Listing 5, this algorithm consists of a number of iterations of a loop which performs the defined genetic operators to fill a child population which becomes the input to the next iteration of the loop. We can thus parallelize the execution of the genetic operators. Using the `parallel_reduce`

```
void simple_ga::run()
{
    while(!terminate())
    {
        population offspring(m_pop.size());
        // note we pass pointers to ourself, our population,
        // and copies of pointers to each genetic operator
        simple_ga_loop loop(this,&m_pop,m_prob,m_sel_op,m_cross_op,
                            m_mut_op);
        // for simplicity, only show TBB's automatic method of
        // apportioning threads
        parallel_reduce(blocked_range<size_t>(0, m_pop.size()/2),
                        loop,auto_partitioner());
        // after all threads have been joined, copy the final
        // child population created by joining each thread
        copy(loop.offspring().begin(),loop.offspring().end(),
             back_inserter(offspring));
        m_rep_op->merge(m_pop, offspring, m_pop);
        iteration_completed(m_pop);
    }
}
```

Fig. 3. The simple GA control class. (from `simple_ga.cpp`).

template, we can then join the resulting partial child populations into a single unit in preparation for the next iteration of the loop. Figure 6 shows the code required to set up the basic skeleton of the algorithm. (Note that for brevity, the declaration and initialization of the genetic operators has been elided.)

The basic structure of this method is very simple. It creates a new child population, then initializes a functor object to perform the generation of offspring in parallel. Because a `parallel_reduce` is used, TBB will call the functor's join method before returning control to the main loop, and this method will be used to construct a single child population. That child population is then copied into the local population and merged using whatever replacement operator the user has specified (through the `m_rep_op` variable).

The work of generating the child population is performed inside the functor, shown in Figure 6 below. Note that instead of looping over the entire parent population, TBB splits the iteration into disjoint blocks represented by the TBB type `blocked_range`. Inside the functor, we then iterate over the range passed to us by TBB. The important part to notice in the loop functor is that `m_offspring` is not a pointer to a shared resource. Rather, each copy of the functor creates its own local population. In the `join` method of the functor, which TBB will repeatedly call until only one such object remains, the offspring are combined into a single population. It is this unified population which is accessed by the outer loop when all threads have finished.

```
void simple_ga_loop::operator()(const blocked_range<size_t>& r)
{
    m_sel_op->set_population(m_pop);
    for(size_t i=r.begin(); i!=r.end(); ++i)
    {
        candidate p1 = m_sel_op->apply();
        candidate p2 = m_sel_op->apply();
        candidate c1 = p1;
        candidate c2 = p2;
        m_cross_op->apply(p1, p2, c1, c2);
        m_mut_op->apply(c1);
        m_mut_op->apply(c2);
        m_prob->evaluate(c1);
        m_ga->evaluation_completed(c1);
        m_prob->evaluate(c2);
        m_ga->evaluation_completed(c2);
        m_offspring.add(c1);
        m_offspring.add(c2);
    }
}

void simple_ga_loop::join(simple_ga_loop& that)
{
    for(unsigned int i=0; i<that.m_offspring.size(); ++i)
        m_offspring.add(that.m_offspring[i]);
}
```

Fig. 4. The simple GA TBB loop functor. (from `simple_ga.cpp`).

7 Conclusions and Future Work

PMF provides a ready-to-use system that implements a number of useful optimization algorithms in a form suited for efficient execution on today's multicore hardware. In addition, it provides a reasonable simple means by which new problems and optimization techniques may be incorporated using only knowledge of standard C++ constructs.

However, there are definite areas in which improvements can be made. While PMF performs very well, there is some overhead in the mechanism by which it supports arbitrary encodings. An earlier version of the software relied on extremely heavy use of C++ template metaprogramming techniques and type traits to automatically infer at compile time the correct type of numerous parameters. The result was that there was a statistically significant increase in performance in the earlier system. However, the usability of the earlier system suffered due to the drawbacks inherent in extensive use of template metaprogramming in C++. It remains an open question if there is a way to reclaim some of this performance without incurring the most severe of these penalties.

Another issue in PMF is that the focus on generational parallelism means that the framework provides little effective support for certain classes of algorithms. A principled and consistent mechanism for providing additional forms of parallelism is needed to allow maximum flexibility in extending PMF.

PMF has an extremely flexible method of providing performance metrics and termination criteria, but by nature, this type of information demands global access. Particularly for researchers making comparisons between algorithms, it is important that, for example, if we specify a maximum number of fitness function evaluations, that the algorithm halt in precisely that allotment of evaluations. However, this means that each thread must have shared access to these global pieces of information. TBB provides atomic variables and locking primitives, and PMF uses those to ensure data consistency, but this imposes a slight performance hit that becomes more significant as more and more cores are enabled. Thus a further avenue for improvement is to implement a generic and customizable mechanism by which the trade-off between absolute accuracy in the metrics and the inherent contention of resources can be managed by the user.

Finally, there are dozens of compelling search and optimization techniques available that can be incorporated into PMF. The current focus of PMF is on hybrid metaheuristics for multiobjective optimization, which implies a stronger representation in the areas of multiobjective evolutionary algorithms and local search techniques. However, additional paradigms are needed to provide a more complete package for researchers and practititions to build upon.

PMF is available on Github (*http://github.com/˜deong/pmf*) under a non-restrictive BSD license. For length requirements, this paper omitted detailed performance measures. The reader is encouraged to download the PMF package and experiment with the built-in algorithms and problems to get a complete understanding of the performance characteristics of the framework.

References

1. Luke, S., Panait, L., Balan, G., et al.: Ecj 19: A java-based evolutionary computation research system
2. Cahon, S., Melab, N., Talbi, E.G.: ParadisEO: A framework for the reusable design of parallel and distribued metaheuristics. Journal of Heuristics 10(3), 357–380 (2004)
3. Liefooghe, A., Jourdan, L., Talbi, E.G.: A unified model for evolutionary multiobjective optimization and its implementation in a general purpose software framework: ParadisEO-MOEO. Technical Report Research Report RR-6906, INRIA (2009)
4. Reinders, J.: Intel Threading Building Blocks. O'Reilly, Sebastopol (2007)

Parallel Evolutionary Approach of Compaction Problem Using MapReduce

Doina Logofătu[1] and Dumitru Dumitrescu[2]

[1] University of Applied Sciences, Lothstr. 64, 80336, Munich, Germany
[2] Babeş-Bolyai University, Str. M. Kogălniceanu 1, 400084, Cluj-Napoca, Romania

Abstract. A parallel evolutionary approach of Compaction Problem is introduced using MapReduce. This problem is of interest for VLSI testing and bioinformatics. The overall cost of a VLSI circuit's testing depends on the length of its test sequence; therefore the reduction of this sequence, keeping the coverage, will lead to a reduction of used resources in the testing process. The problem of finding minimal test sets is NP-hard. We introduce a distributed evolutionary algorithm (MapReduce Parallel Evolutionary Algorithm−MRPEA) and compare it with two greedy approaches. The proposed algorithms are evaluated on randomly generated five-valued benchmarks that are scalable in size. The MapReduce paradigm offers the possibility to distribute and scale large amount of data. Experiments show the efficiency of the proposed parallel approach. The project, containing the Hadoop implementation can be found at: *http://sourceforge.net/projects/dcpsolver/* [10].

Keywords: Data Compaction, Static Test, Parallel Algorithm, Evolution Strategies, Greedy, Discrete Optimization, Apache Hadoop, MapReduce, Statistical Tests.

1 Introduction

Compaction refers usually to the reduction of a sequence of data with no loss of information. In our case, there is no need to perform the inverse operation (decompactation/decompression) for obtaining the initial data. We need to compact a data sequence by maintaining the property of coverage (the compacted data must cover the initial one). As a real-world problem, there is the need to compact vectors of tests for the VLSI, so that the faults possibilities are completely covered [2,3]. Unlike dynamic compaction, static compaction does not require any modifications to the test generation procedure. Since dynamic compaction is based on heuristics and does not achieve the minimum test length, static compaction is useful even after dynamic compaction has been used, for further reducing the length of the test sequence [4]. There is a dual motivation for studying test compaction. First, by reducing test sequence length, the memory requirements during test application and the test application time are reduced. Second, the extent of test compaction possible for deterministic test sequences indicates that test pattern generators spend a significant amount of time generating test vectors that are not necessary. The compacted test sequences provide

R. Schaefer et al. (Eds.): PPSN XI, Part II, LNCS 6239, pp. 361–370, 2010.

a target for more efficient deterministic test generators. The scope is to minimize the amount of data to test a circuit. On the other hand, in biology, there is the need to reconstruct DNA-sequences, based on the fact that compacted sets of sequences have to be minimal. These DNA-sequences contain only the characters 'A',' 'C', 'G', 'T',' X', where 'X' could initially have been any of the DNA-Acids (A–adenine, C–cytosine, G–guanine, T–thymine), so it plays the same "Don't care" role [7].

2 Problem Description

Consider a set of test sequences $T = S_1, S_2, \ldots, S_n$ detecting (covering) the set of faults $F = \{f_1, f_2, \ldots, f_m\}$ of a circuit. Every test sequence $S_i = \{v_1, v_2, \ldots, v_{L_i}\}$, $i = 1, \ldots, n$ is an ordered set of L_i test vectors $v_1, v_2, \ldots, v_{L_i}$ where L_i is the length of S_i. A fault f_i within a sequence S_j has the detection cost d_{ij} equal to the number of vectors from the beginning of the sequence until f_i becomes detected in S_j. The test compaction problem is to find a collection of subsequences, i.e. subsets of vector sequences, so that all faults in F are covered, and the test length of the collection is a minimum. This problem can be reformulated as a set covering problem, which is NP-complete.

Definition 1 (Compaction Set (CS)). *Every test is seen as a string which contains symbols from the set $CS = \{$'0', '1', 'U', 'Z', 'X'$\}$.*

Definition 2 (Compatible symbols and merge-operation). *Two symbols c_1 and c_2 are compatible if they are the same or if one of them is 'X'. We will denote this relation with $\cong$ and its negation with $\ncong$. 'X' is the "Don't Care" character. If two symbols are the same, then the merged character is one of them, otherwise it is the one different from 'X'. The compatibility/merge-relation is presented in Table 1 ('*' means no compatibility).*

Table 1. Compatibility/merge-relation

$\cong$	'0'	'1'	'U'	'Z'	'X'
'0'	'0'	'*'	'*'	'*'	'0'
'1'	'*'	'1'	'*'	'*'	'1'
'U'	'*'	'*'	'U'	'*'	'U'
'Z'	'*'	'*'	'*'	'Z'	'Z'
'X'	'0'	'1'	'U'	'Z'	'X'

Definition 3 (Compatible tests and merge operation). *Two given tests are compatible if they have the same length and all corresponding symbols (on the same position) in pairs are compatible. The merged test is obtained by substituting every symbol sequentially with the merged symbol from the corresponding position in the two given strings. We will denote this relation also with $\cong$ and its negation with $\ncong$.*

Example 1. The tests $t_1 = 10ZX0XU$ and $t_2 = X0Z10UU$ are compatible because $t_1(i) \cong t_2(i)$ for all $i = 1, \ldots, 7$ and $\text{Merge}(t_1, t_2) = 10Z10UU$.

Definition 4 (Coverage Set). *For two given sets of tests $S_1 = \{t_{11}, t_{12}, \ldots, t_{1i}\}$ and $S_2 = \{t_{21}, t_{22}, \ldots, t_{2j}\}$, S_2 is a coverage set of S_1 if for every test t in S_1 there is a compatible test in S_2. An example is given in Figure 1.*

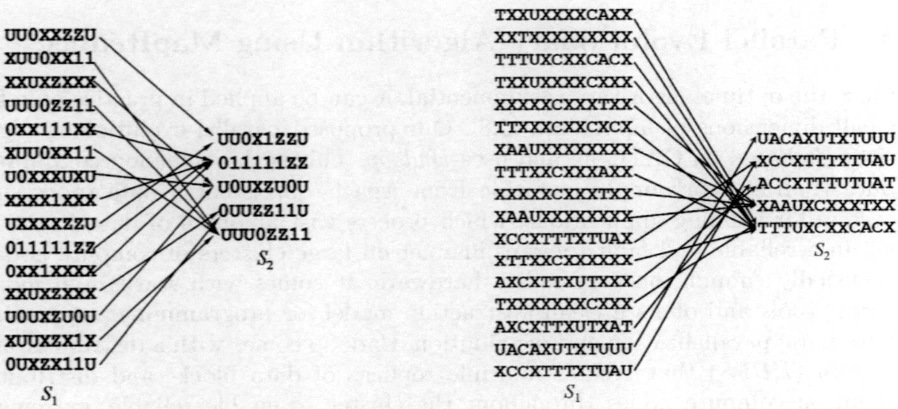

Fig. 1. Examples for VLSI and bioinformatics: S_2 is a coverage set for S_1. a) S_1 contains 15 tests, each having a length of 8. The coverage set contains 5 tests, i.e. its dimension is 33.33% from dimension of S_1. b) S_2 is a coverage set for S_1. S_1 contains 17 tests, each having a length of 11. The coverage set contains 5 tests, i.e. its dimension is 29.41% from dimension of S_1.

Definition 5 (Test Compaction Problem (TCP)). *Given is a set of tests $S_1 = \{t_{11}, t_{12}, \ldots, t_{1i}\}$, every test t_{1k}, for $k = 1, \ldots, i$, having the same length. Find a coverage set $S_2 = \{t_{21}, t_{22}, \ldots, t_{2j}\}$ of S_1, such that the cardinality of S_2 is minimal over all coverage sets of S_1.*

TCP can be generalized by considering any alphabet which contains a “Don't Care” symbol. We name this generalization *Data Compaction Problem (DCP)*.

3 Recent Work

The optimal method and two greedy variants, GRBT (**G**reedy **B**inary **T**ree) and GRNV (**G**reedy **N**aive), are proposed in [5,8]. The transformation is made based on a single sequence, where the start sequence is the given one. At each step, two compatible sequences are merged, if they exist. Experiments show much better GRBT results versus GRNV for small compaction rates, e.g. around 20%—40% compactions from the initial sequence. Also the execution times are better;

therefore GRBT is the method of choice in such cases. A genetic approach, GATC (Genetic Algorithm for Test Compaction), based of cloned individuals and a mutation operator, is introduced in [6]. Here, the fitness function is the total number of "Don't care" in the sequence. It provides some better results than GRBT. As expected, the smaller the compaction rate, the better the quality results. The biological background and many-sided experiments are presented in [7], especially the behaviour of both greedy algorithms for different compaction rates.

4 Parallel Evolutionary Algorithm Using MapReduce

Since the optimal algorithm is exponential, it can be applied in practice only for small dimensions of input data [5,8]. Our proposed parallel evolutionary algorithm is based on GATC [6] and uses *Hadoop*. This is the OpenSource MapReduce [1] framework implementation from Apache [9], a batch data processing system for running applications, which process vast amounts of data in parallel, in a reliable and fault-tolerant manner on large clusters of compute nodes, eventually running on commodity hardware. It comes with status and monitoring tools and offers a clean abstraction model for programming, supporting automatic parallelization and distribution. Hadoop comes with a distributed file system (*HDFS*) that creates multiple replicas of data blocks and distributes them on compute nodes throughout the cluster to enable reliable, extremely rapid computations. The compute and storage nodes are typically the same. This allows the framework to schedule tasks effectively on the nodes where data is already present, resulting in very high aggregate rate across the cluster. The framework consists of a single master *JobTracker* and one slave *TaskTracker* per compute node. The master is responsible for scheduling the tasks for the map- and reduce-operations on the slaves, monitoring them and reexecuting the failed tasks. The slaves execute the tasks, as directed by the master.

The applications specify the input/output locations, supply *map* and *reduce* functions and eventually invariant (contextual) data. These comprise the *job configuration.* The Hadoop *job client* then submits the job (Java byte code packed in a jar-archive) and configuration to the *JobTracker*, which then distributes them to the slaves, schedules the map-/reduce- tasks, and monitors them, providing status and diagnostic information to the *job client.*

A MapReduce *job* splits the input data into independent chunks (splits), which are then processed by the *map tasks* in a completely parallel manner. The framework sorts the *maps* outputs and forwards them as input to the *reduce tasks.*

The parallel evolutionary algorithm runs for a given number of generations. For each generation, it generates a random number of permutations of the "current" set of sequences, splits the set of permutations in several subsets (*map* operation) and executes GATC [6] on each of them during the *reduce* operation. Afterwards, it collects the best individuals (permutations) resulted on each Reducer node in the cluster and concatenates them. The result constitutes the new set of sequences for the next generation. The pseudocode is presented in Figure 1.

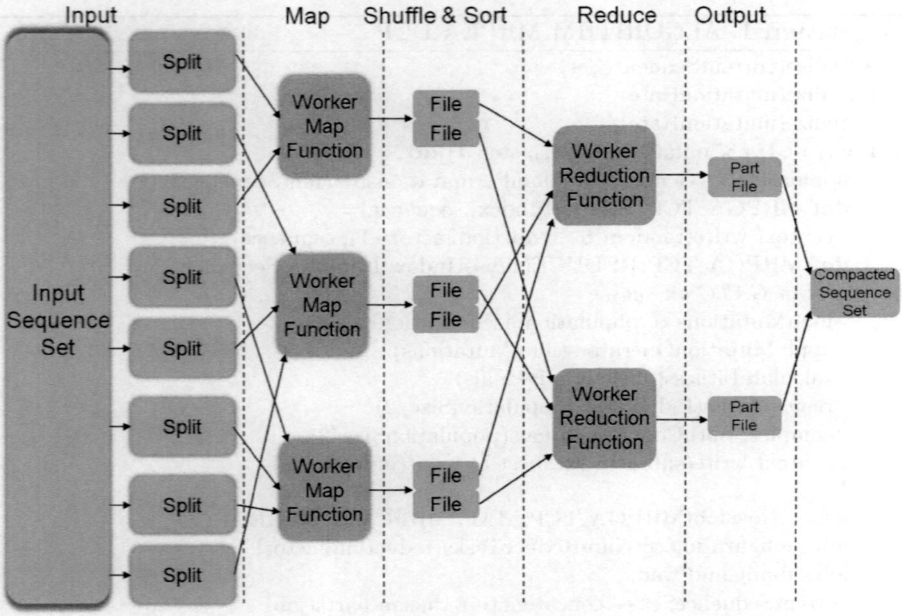

Fig. 2. MapReduce Dataflow

The map operation associates for each generated permutation a subset index between 0 and *reductionFactor*−1 randomly, as output key. This index represents the partition for reduction. In the *Shuffle&Sort* phase, the sequences get sorted after their indices, and each *reducer* node receives and executes GATC in one call over its subset of sequences having the same index.

5 Implementation Details

The Hadoop application for DCP is available as the Open Source project dcpSolver on sourceforge.net[12]. The Java binary application and its dependent libraries are archived in dcpSolver.jar. The archive's entry point (*dcpsolver.DcpSolverDriver*) is a Hadoop program driver that registers the DcpSolver's application (*dcpsolver.DcpSolverJob*). The start command is *hadoop jar dcpSolver.jar*, and it gets the name of the DcpSolver's application: *dcp*.

DcpSolver uses Apache Commons CLI2 for parsing the command line parameters. The command:

hadoop jar dcpSolver.jar dcp -h

displays the help for all available options. If there is the need to build the dcpSolver, one can either check out the Java sources with subversion from

https://dcpsolver.svn.sourceforge.net/svnroot/dcpsolver

Algorithm 1. ALGORITHM_MRPEA_DCP

```
initialize(currentSequenceSet)
initialize(mutationRate)
initialize(mutationRate)
for (i ← 1; i ≤ numGenerations; step 1) do
  population ← GenerateRandomPermutations(currentSequenceSet)
  def MRPGA_TCP_MAP(seqIndex, sequence)=
    context.write(random(0.. reductionFactor−1), sequence)
  def MRPGA_TCP_REDUCE(subsetIndex, Iterable<Sequence>)={
   //run GATC on subset
    numMutations ← populationSize·mutationRate
    applyMutationOperators(numMutations)
    calculateFitness(allNewIndividuals)
    removeWorstIndividuals(populationSize/2)
    completeWithCopyIndividuals(populationSize/2)
    context.write(subsetIndex, best_element(individuals))
  }
  job ← NewJob(MRPGA_TCP_MAP, MRPGA_TCP_REDUCE)
  job.configuration.setNumReduceTasks(reductionFactor)
  job.submit_and_wait
  currentSequenceSet ← concatenate_reducers_parts(job)
end for
return currentSequenceSet
```

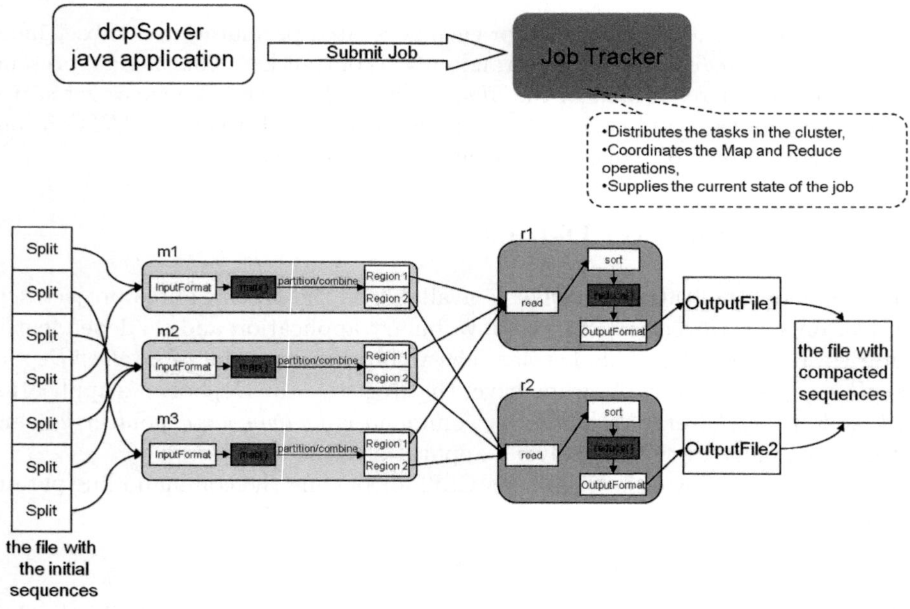

Fig. 3. Implementation of dcpSolver with Java and Hadoop

or browse them at

http://dcpsolver.svn.sourceforge.net/viewvc/dcpsolver/.

For compiling the Java code there are needed the Java libraries Apache Commons Math Version 2.0, Apache Commons CLI2 and JUnit 4 (for test units).

6 Experimental Results and Statistical Tests

We generated sets of data artificially with different parameters. We varied the dimension of the sequences set (n), the size of the sequences (k), the compaction factor (cf) - i.e. the approximately expected percentage of the compacted sequence, and the maximum number of random generated clones (e.g. 100, 200).

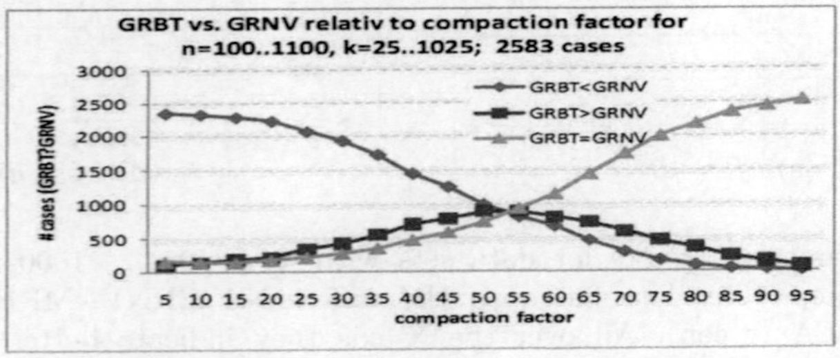

Fig. 4. Evolution of a number of cases with GRBT<GRNV, GRBT>GRNV and GRBT=GRNV for a set of 2583x19 experiments with $100 \leq n \leq 1100$, $25 \leq k \leq 1025$, and compaction factor $= 5, 10, 15, \ldots, 95$

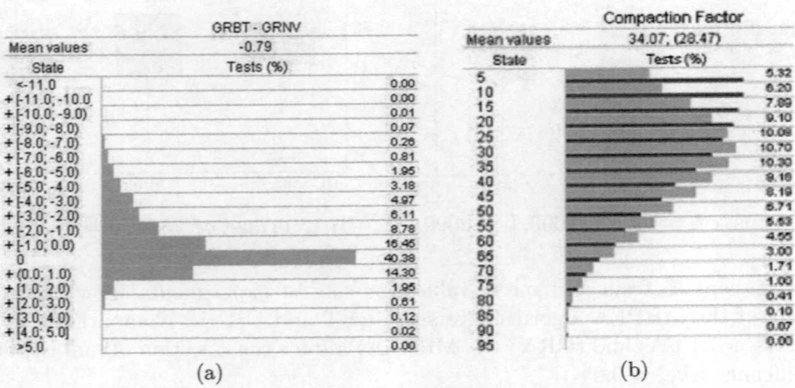

Fig. 5. a) Distribution on intervals of the difference GRBT−GRNV within 49077 experiments with $100 \leq n \leq 1100$, $25 \leq k \leq 1025$, and compaction factor $= 5, 10, 15, \ldots, 95$; b) Distribution of the percentage of results within 9915 experiments with GRBT<GRNV and GRBT execution time smaller as the GRNVś one, the reference (blue) is the set of tests where GRBT provides better results as GRNV

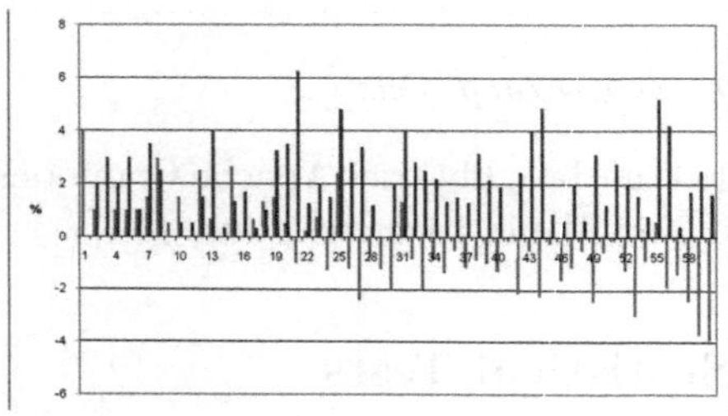

(a) $cf = 25$

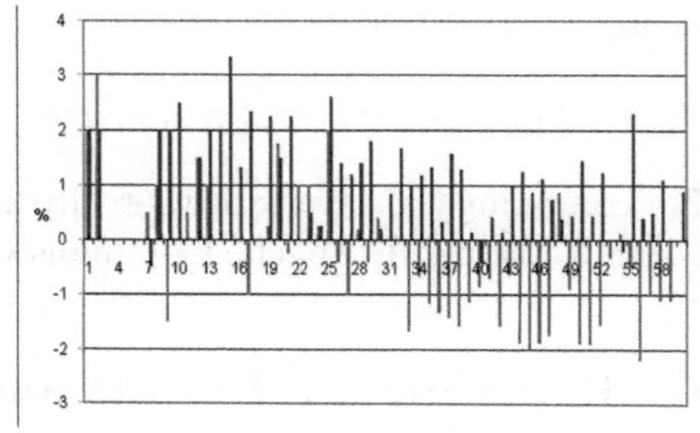

(b) $cf = 35$

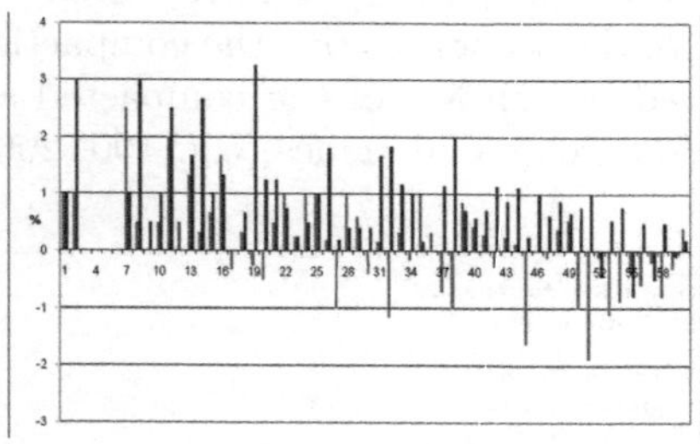

(c) $cf = 45$

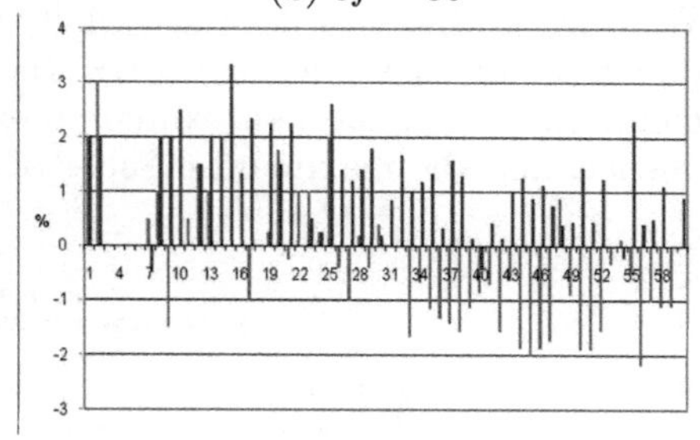

(d) $cf = 55$

Fig. 6. Experimental results for data sets with $n = 100, \ldots, 1000$, step 100, $k = 50, \ldots, 550$, step 100. The bars are the differences GRNV−MRPEA (red) and GRBT−MRPEA (green). All over the X-axis they indicate better results for the MRPEA approach. MRPEA is in the most cases better than GRNV. The bigger the compaction factor, the better the quality of MRPEA results versus GRBT.

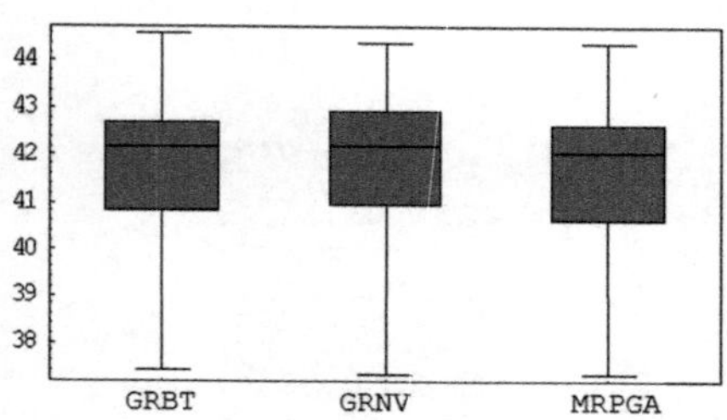

(a) 50 runs, $cf = 45, n = 1000, k = 2000$

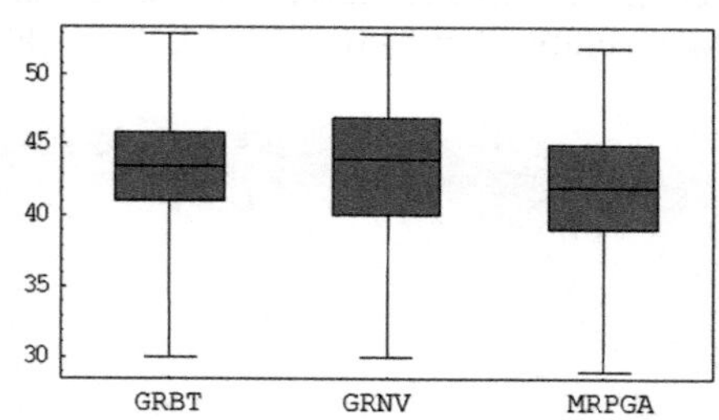

(b) 150 runs, $cf = 45, n = 100, k = 200$

Fig. 7. Student T-Tests and mean values for sets of experiments show the quality of results of the MRPEA algorithm versus GRBT and GRNV. Comparisons in pairs (GRBT vs. MRPEA) and (GRNV vs. MRPEA) induce the rejection of null hypothesis at significance level = 0.05.

First we ran the two greedy algorithms GRBT and GRNV [5] on 49077 various artificial generated instances with $100 \leq n \leq 1100$, $25 \leq k \leq 1025$, and compaction factor $= 5, 10, 15, \ldots, 95$. The GRBT is better the smaller the compaction factor is. For a compaction factor bigger than 50%, the GRNV is sensibly better.

In Figure 5 we show the mean values for the difference GRBT−GRNV (as percentage of compacted sequence) for all experiments.

Table 2. Selective results for the experiments from Fig. 6, (a) and (c)

n	k	$cf = 25$			$cf = 45$		
		MRPEA(%)	GRBT(%)	GRNV(%)	MRPEA(%)	GRBT(%)	GRNV(%)
100	50	**31.00**	35.00	**31.00**	**33.00**	34.00	34.00
100	550	**26.00**	27.00	27.00	**41.00**	**41.00**	**41.00**
200	50	**28.50**	30.00	32.00	**45.50**	48.00	46.50
200	550	**23.50**	26.00	25.00	**47.50**	48.00	**47.50**
300	50	**26.00**	26.67	30.00	**45.00**	46.33	46.67
300	550	**22.67**	24.00	23.67	**40.33**	40.67	41.00
400	50	**29.00**	30.50	32.25	42.25	**42.00**	45.50
400	550	23.25	**22.00**	24.75	40.75	41.75	**41.25**
500	50	**21.60**	24.00	26.40	**47.20**	48.20	48.20
500	550	23.40	**21.40**	25.40	37.80	**37.40**	38.20
600	50	**26.33**	27.67	30.33	**48.00**	48.17	49.67
600	550	23.17	**22.67**	24.67	**40.00**	40.33	**40.00**
700	50	31.71	**30.57**	33.00	48.71	**48.00**	49.86
700	550	27.29	**25.14**	29.71	42.00	**41.71**	43.14
800	50	32.88	**32.38**	36.88	**47.25**	47.50	48.13
800	550	26.38	**25.88**	27.00	**45.00**	45.38	45.88
900	50	28.89	**26.44**	32.00	**49.00**	49.56	49.67
900	550	27.00	**26.11**	27.78	40.56	**39.67**	41.33
1000	50	**28.20**	28.80	33.40	49.20	48.80	**48.40**
1000	550	25.80	**21.90**	27.40	**42.70**	43.10	42.90

The experiments were done on a Hadoop cluster with 2 machines. The first (confi-gured both as Hadoop master and slave) had 4 CPUs Dual Core AMD Opteron(tm) 280 2.4 GHz and 16G RAM. The second (configured as slave) had 4 CPUs Intel(R) Xeon(TM) CPU 2.80GHz and 12G RAM.

7 Conclusions and Future Work

After a formal description of a compaction problem, we described an efficient distributed implementation using the MapReduce paradigm−MRPEA_DCP. Experiments run on artificially generated five-valued benchmarks showed the efficiency of proposed approach, especially for specific traits of data input. The benchmarks are scalable in size, number of symbols and compaction factor, i.e. the approximate expected percentage of the compacted sequence toward the initial one. This one turns to be an important measure of the input data regarding the applied method. In the experiments, first the most efficient proposed greedy approaches for the problem – GRBT_DCP and GRNV_DCP – are compared, and some useful statistics are provided. An application was written for the distributed algorithm using the Hadoop implementation for MapReduce [1,9]. This application, together with some benchmarks and statistics, is available as Open Source project on sourceforge.net: *http://sourceforge.net/projects/dcpsolver/*. Advantages of the Hadoop implementation are, among others the scalability, the robustness, the fault toleration, the possibility to work on commodity hardware, status reporting and monitoring, and performing the computation where the data is. Experiments have to be done for larger data sets, testing different distributions for the input sequences, as well some alternative variants for the map or reduce phases. Applying the proposed algorithms on real-data, e.g. from VLSI or bioinformatics, could bring some further knowledge regarding their efficiency and applicability in practice. Further, the problem could be studied as a multi-valued logic one, eventually with alternative relations for the symbols.

References

1. Dean, J., Ghemawat, S.: MapReduce: simplified data processing on large clusters. Communications of the ACM 51(1), 107–113 (2008)
2. De Micheli, G.: Synthesis and Optimization of Digital Circuits. McGraw-Hill, Inc., New York (1994)
3. Drechsler, R.: Evolutionary Algorithms for VLSI CAD. Kluwer Academic Publishers, Dordrecht (1998)
4. El-Maleh, A., Osais, Y.: Test vector decomposition based static compaction algorithms for combinatorial circuits. ACM Trans. Des. Autom. Electron. Syst. (TODAES) 8, 430–459 (2003)
5. Logofătu, D., Drechsler, R.: Comparative Study by Solving the Test Compaction Problem. In: Proceedings of the 38th International Symposium on Multiple Valued Logic, pp. 44–49 (2008)
6. Logofătu, D.: Static Test Compaction for VLSI Tests: An Evolutionary Approach. Advances in Electrical and Computer Engineering 8(2), 48–53 (2008)
7. Logofătu, D.: DNA Sequence Vectors and Their Compaction. In: AIP Conf. Proceedings of the 1st International Conference on Bio-Inspired Computational Methods Used for Solving Difficult Problems: Development of Intelligent and Complex Systems, vol. 1117(1), pp. 29–39 (2008)
8. Logofătu, D.: Algorithmen und Problemlösungen mit C++, 2nd edn., pp. 402–411. Vieweg+Teubner, Wiesbaden (2010)
9. Apache Hadoop (2009), `http://hadoop.apache.org`
10. Open Source Project, dopiSolver (2010), `http://sourceforge.net/projects/dcpsolver/`

Ant Colony Optimization with Immigrants Schemes in Dynamic Environments

Michalis Mavrovouniotis[1] and Shengxiang Yang[2]

[1] Department of Computer Science, University of Leicester, University Road, Leicester LE1 7RH, United Kingdom
mm251@mcs.le.ac.uk

[2] Department of Information Systems and Computing, Brunel University, Uxbridge, Middlesex UB8 3PH, United Kingdom
shengxiang.yang@brunel.ac.uk

Abstract. In recent years, there has been a growing interest in addressing dynamic optimization problems (DOPs) using evolutionary algorithms (EAs). Several approaches have been developed for EAs to increase the diversity of the population and enhance the performance of the algorithm for DOPs. Among these approaches, immigrants schemes have been found beneficial for EAs for DOPs. In this paper, random, elitism-based, and hybrid immigrants schemes are applied to ant colony optimization (ACO) for the dynamic travelling salesman problem (DTSP). The experimental results show that random immigrants are beneficial for ACO in fast changing environments, whereas elitism-based immigrants are beneficial for ACO in slowly changing environments. The ACO algorithm with hybrid immigrants scheme combines the merits of the random and elitism-based immigrants schemes. Moreover, the results show that the proposed algorithms outperform compared approaches in almost all dynamic test cases and that immigrant schemes efficiently improve the performance of ACO algorithms in DTSP.

Keywords: Ant Colony Optimization, Immigrants Schemes, Dynamic Optimization.

1 Introduction

Ant colony optimization (ACO) algorithms emulate the behaviour of real ant colonies when they search for food from their nest to food sources. Ants communicate using their pheromone trails in order to complete this task as efficiently as possible. ACO algorithms have proved to be able to solve different optimization problems in real-world applications [2,3]. Traditionally, researchers have been focused on stationary optimization problems, where their environment remains fixed during the execution of the algorithm. However, many real-world applications have dynamic environments. The problem then becomes more challenging since the optimum needs to be tracked when dynamic changes occur [12].

Traditional ACO algorithms have been designed for stationary optimization problems [3], and may not be sufficient anymore for DOPs. This is due to the fact

R. Schaefer et al. (Eds.): PPSN XI, Part II, LNCS 6239, pp. 371–380, 2010.

that the pheromone trails of the previous environment will not make sense for a new environment, after a change occurs. A simple way to address this problem is to re-initialize the pheromone trails and consider every change as the arrival of a new problem instance which needs to be solved from scratch. Unfortunately, this restart strategy is computationally expensive and usually not efficient.

Recently, developing ACO algorithms for DOPs has attracted a lot of attention since they can be useful for real-world applications. Thus, more specialized strategies have been proposed to maintain the high quality of output efficiently, which include local and global restart strategies [8], pheromone manipulation schemes to maintain or increase diversity [4], and memory-based approaches [6,9]. These methods have been applied on the dynamic travelling salesman problem (DTSP) due to its importance for many real-world applications. One of the most efficient and well-studied methods is the memory-based version of ACO, known as the population-based ACO (P-ACO) algorithm [7]. It has a different framework from a traditional ACO algorithm since it maintains a population list (memory), which stores the best ant of every iteration, and is used to generate the pheromone trails. Taking a closer look at P-ACO, we see that it has the characteristics of a genetic algorithm (GA) [11] because of the memory. Thus, it inherits the disadvantage of a GA when a dynamic change may affect the individual on the genotypic level, which needs to be repaired. Often, the repair procedure is computationally expensive.

As we have seen on many GAs, immigrants schemes are advantageous when applied to DOPs [14,15,17]. Immigrants schemes enable the algorithm to maintain the diversity of the population to a certain level, by introducing new individuals into the current population. In this paper, we apply immigrants schemes into P-ACO. However, instead of using a long-term memory as in P-ACO, we use a short-term memory, where all the new ants replace the old ones to form a new population. Later on, a percentage of the worst ants are replaced by immigrants. We introduce three types of immigrants, which are traditional random, elitism-based, and hybrid immigrants. The experimental results show that immigrant schemes enhance the performance of ACO into DOPs. However, different immigrants schemes are advantageous under different environmental conditions.

The rest of the paper is organized as follows. Section 2 describes the standard ACO and P-ACO algorithms. Moreover, it describes how they are applied to the DTSP. Section 3 describes our proposed approaches where we apply immigrants schemes into P-ACO. Section 4 describes the experiments carried out by comparing our proposed approaches with P-ACO. Finally, Section 5 concludes this paper with directions for future work.

2 ACO for Dynamic Environments

2.1 Standard ACO

The traditional ACO algorithm consists of a population of μ ants, where each ant consists of two modes, the forward mode and the backward mode. Initially, all ants are placed on a randomly selected city for a TSP and all pheromone trails are initialized with an equal amount of pheromone. All ants on their forward mode choose the next city based on pheromones and some heuristic information using a probabilistic decision rule, which is defined as follows:

$$p_{ij}^k = \frac{[\tau_{ij}]^\alpha [\eta_{ij}]^\beta}{\sum_{l \in N_i^k} [\tau_{il}]^\alpha [\eta_{il}]^\beta}, \text{if } j \in N_i^k \tag{1}$$

where τ_{ij} is the existing pheromone trail between city i and city j, η_{ij} is the heuristic information available a priory, which is defined as $1/d_{ij}$ and d_{ij} is the distance between the cities. N_i^k denotes the neighbourhood of cities of ant k when being on city i. α and β are the two parameters that determine the relative influence of pheromone trail and heuristic information, respectively.

Later on, all ants proceed to their backward mode by retracing their solutions and deposit pheromone according to their solution quality on the corresponding trails. However, before adding any pheromone, a constant amount of pheromone is deduced from all trails due to the pheromone evaporation, which is defined as:

$$\tau_{ij} \leftarrow (1-\rho)\,\tau_{ij}, \forall\ (i,j), \tag{2}$$

where $0 < \rho \leq 1$ is the rate of evaporation. Reducing the pheromone values enables the algorithm to forget bad decisions made in previous iterations [3]. After evaporation, all ants deposit pheromone to the corresponding trails of their tour as follows:

$$\tau_{ij} \leftarrow \tau_{ij} + \Delta\tau_{ij}^k, \forall\ (i,j) \in T^k, \tag{3}$$

where $\Delta\tau_{ij}^k = 1/C^k$ is the amount of pheromone that ant k deposits and C^k is the cost of the tour T^k.

2.2 Population-Based ACO

The P-ACO algorithm is the memory-based version of an ACO algorithm, which was first applied on the stationary TSP [7]. It differs from the standard ACO algorithm described above since it follows a different framework. Generally, the algorithm maintains a population of solutions, which is used to update pheromone trails without any evaporation.

The initial phase and the first iterations of the P-ACO algorithm work in the same way as with the standard ACO algorithm. The pheromone trails are initialized with an equal amount of pheromone and the population list of a size K is empty. For the first K iterations, the iteration best ant deposits a constant amount of pheromone using Eq. (3) where $\Delta\tau_{ij}^k = (\tau_{max} - \tau_{init})/K$. Here, τ_{max} and τ_{init} denote the maximum and initial pheromone amount, respectively. This positive update procedure is performed whenever an ant enters the population list. On iteration $K+1$, the ant that has entered the population list first needs to be removed in order to make room for the new one, and thus, a negative update to its pheromone trails is done, as follows:

$$\tau_{ij} \leftarrow \tau_{ij} - \Delta\tau_{ij}^{k}, \forall\ (i,j) \in T^{k}, \tag{4}$$

where $\Delta\tau_{ij}^{k}$ is defined as in the positive update above.

This strategy is based on the *Age* of ants. However, other strategies have also been proposed by researchers, such as *Quality* and *Prob* [6]. From the experimental results in [6], the default *Age* strategy is more consistent and performs better than the others, since the other strategies have more chances to maintain identical ants into the population list, which leads the algorithm to the stagnation behaviour. This is due to the fact that high levels of pheromone will be generated into a single trail and dominates the search space. Moreover, we have seen the importance of keeping the pheromone trails into a certain level from the Max-Min Ant System (MMAS) [13], which is one of the state-of-the-art ACO algorithms on stationary problems.

2.3 Response to Dynamic Changes

Theoretically, ACO algorithms can adapt to dynamic changes since they are inspired from nature, which is a continuous changing environment [12]. In practice, they can adapt by transferring knowledge from past environments [1]. So far, the description of ACO algorithms above has been made assuming stationary environments. Considering the DTSP, ACO needs to be modified in order to adapt to environmental changes efficiently.

The dynamics of adding/deleting a city affects both the genotypic and, usually, the phenotypic level of the ant. Therefore, considering that the solutions are affected by the change in iteration n, the pheromone trails will not make sense in iteration $n+1$. For the ACO algorithms that follow the traditional framework, it is vital to re-initialize the pheromone trails after a dynamic change, which acts as a restart of the algorithm.

For the P-ACO approach, the solutions stored in the population list are repaired and the pheromone trails are re-generated accordingly. This strategy is called *KeepElitst* [9] and uses two greedy heuristics to repair the genotype of the population: 1) the offended cities are removed from the solutions; and 2) the new cities are placed individually in a greedy fashion where they cause the minimum increase on the phenotype.

3 Incorporating Immigrants Schemes to ACO Algorithms

When addressing DOPs, traditional ACO algorithms cannot adapt well to the environmental changes once the ants reach the stagnation behaviour, where they follow the same path. The algorithm loses its adaption capability since it does not maintain diversity within the population. A good start has been made with the P-ACO algorithm with the use of memory, which maintains a certain level of diversity and enables ACO algorithms to be more efficient for DOPs. However, it is a long-term memory and the solutions need to be repaired once a city is added or removed. Usually, the repair procedures requires prior knowledge of the problem and is computationally expensive.

As mentioned above, the application of immigrants schemes has been found efficient for GAs for DOPs. The principle is to introduce new individuals into the current population by replacing a percentage of individuals in the population [5]. The percentage should be relatively small because a high percentage may lead the algorithm into a too high level of diversity. High diversity does not always mean good performance on DOPs, because it may lead the algorithm into randomization [15,17].

In this paper, we apply immigrants schemes into the P-ACO algorithm to maintain a certain level of diversity in the population and enhance its dynamic performance. However, we use a short-term memory where the ants of the current iteration replace the ants of the old iteration. Moreover, a percentage of immigrants replace the worst ants of the current population.

The advantages of using a short-term memory are closely related to the survival of ants in a dynamic environment, where no ant can survive in more than one iteration. This way, there is no need to use any repair algorithm (apart from the best ant of the previous iteration for the elitism-based immigrant scheme) because the changes do not affect the ants stored. Furthermore, there is one main concern that involves immigrants schemes, i.e., how to generate immigrants.

3.1 Random Immigrants ACO

The random immigrants ACO (RIACO) algorithm uses an immigrants scheme where ants are generated randomly, and replace the worst ones of the current population stored in the short-term memory every iteration. It is believed that "the continuous adaption of such algorithms makes sense only when the environmental changes of a problem are small to medium" [12]. This is due to the fact that the old environment has more chance to be similar with the new one. After a change occurs, transferring knowledge from the old environment may provide a good solution efficiently.

Considering this argument, RIACO may be suitable when changes are not slight since it provides diversity without considering any knowledge from the old environment. Moreover, it may be suitable in fast changing environments where information from the past may not be useful, since the algorithm does not have sufficient time to converge onto a good solution in order to gain knowledge.

3.2 Elitism-Based Immigrants ACO

The elitism-based immigrants ACO (EIACO) algorithm uses an immigrants scheme where ants are generated by mutating the best ant of the previous iteration. These immigrants also replace the worst ones in the short-term memory every iteration as in RIACO. This immigrants scheme transfers knowledge from old environments and, thus, may be advantageous when changes are small to medium. Furthermore, it may be suitable in slowly changing environments since it needs sufficient time to locate a good optimum which can be useful to the new environment since the global optimum may be similar.

The mutation of the best ant is carried out using the inversion operator, where two cities are randomly selected and the sub-tour between them is reversed. However, there are two types of inversion: 1) the simple one is as explained above; and 2) the adaptive one is based on the inver-over operator [10]. The adaptive inversion is much more efficient than the simple one, since several inversions are carried out under some criteria. This type of inversion has adaptive characteristics which may be more suitable for DOPs.

3.3 Hybrid Immigrants ACO

The hybrid immigrant ACO (HIACO) algorithm uses an immigrants scheme that combines both random and elitism-based immigrants. The replacing policy is the same as in RIACO and EIACO algorithms. However, half of the immigrants are randomly generated and the other half are generated by mutating the best ant. HIACO attempts to combine the merits of both RIACO and EIACO, where one is good on slowly and slightly changing environments and the other on fast and significantly changing environments. Therefore, HIACO may possibly be suitable under all environmental conditions.

4 Simulation Experiments

4.1 Experimental Setup

In the experiments, we compare RIACO, EIACO, and HIACO with P-ACO with its best population update policy, that is, *Age*. All the algorithms have been applied on the `kroA200` problem instance, obtained from TSPLIB[1], which consists of 200 cities. The dynamic environment was generated by taking away half of its cities and constructing a "spare pool" of cities before running the algorithms. Every f iterations, a percentage of m cities were randomly chosen from the spare pool and exchanged with a percentage of m random ones from the actual instance (the other half cities). This way, the size C of the problem instance remains the same through the whole run.

The parameters f and m indicate the frequency and magnitude of dynamic changes, respectively. The f parameter is defined as the number of iterations between two environmental changes. The m parameter is defined as the percentage of selected cities from the spare pool that replaces other cities from the actual instance. The common parameters used for the algorithms were set according to the guidelines in [3, pp. 71] as follows: $\alpha = 1$ and $\beta = 2$ for Eq. (1), and $\tau_{init} = 1/(C-1)$. For P-ACO, K was set to 3 and τ_{max} was set to 1.0 as in [6,7]. For all three proposed algorithms, K was set to 25, in which we replace 6 ants with immigrants ($\approx 25\%$ of K). Moreover, μ was set to 25 ants for all

[1] Available on `http://comopt.ifi.uni-heidelberg.de/software/TSPLIB95/`

Table 1. In the first section, values in bold indicate the best results of the overall offline performance. In the second section, "$s-$" or "$s+$" means that the first algorithm is significant better or significantly worse than the second algorithm, respectively, whereas "$\sim$" indicates no significant difference between algorithms.

Algorithms & Instances	kroA200			
$f = 20, m \Rightarrow$	10%	25%	50%	75%
P-ACO	27339.89	28497.20	29072.95	29290.05
RIACO	25798.46	26016.24	26029.56	25975.49
EIACO	25822.68	26001.00	26018.47	25996.12
HIACO	**25752.20**	**25985.19**	**25961.28**	**25907.79**
$f = 100, m \Rightarrow$	10%	25%	50%	75%
P-ACO	**24284.40**	25010.38	25359.90	25394.80
RIACO	24513.54	24799.98	24903.43	24852.92
EIACO	24455.14	24688.14	**24749.48**	**24682.73**
HIACO	24421.26	**24604.94**	24784.14	24683.38
t-Test Results				
$f = 20, m \Rightarrow$	10%	25%	50%	75%
P-ACO⇔RIACO	$s+$	$s+$	$s+$	$s+$
P-ACO⇔EIACO	$s+$	$s+$	$s+$	$s+$
P-ACO⇔HIACO	$s+$	$s+$	$s+$	$s+$
RIACO⇔EIACO	$\sim$	$\sim$	$\sim$	$\sim$
RIACO⇔HIACO	$\sim$	$\sim$	$s+$	$s+$
EIACO⇔HIACO	$s+$	$\sim$	$s+$	$s+$
$f = 100, m \Rightarrow$	10%	25%	50%	75%
P-ACO⇔RIACO	$s-$	$s+$	$s+$	$s+$
P-ACO⇔EIACO	$s-$	$s+$	$s+$	$s+$
P-ACO⇔HIACO	$s-$	$s+$	$s+$	$s+$
RIACO⇔EIACO	$\sim$	$\sim$	$s+$	$s+$
RIACO⇔HIACO	$\sim$	$\sim$	$s+$	$s+$
EIACO⇔HIACO	$\sim$	$\sim$	$\sim$	$\sim$

algorithms in order to have the same number of evaluations in each iteration, that is, 25 evaluations per iteration.

For each algorithm on a DTSP instance, $N = 30$ independent runs were executed on the same random environmental changes. The algorithms were executed for $G = 1000$ iterations and the overall offline performance is calculated as follows:

$$\boldsymbol{P}_{offline} = \frac{1}{G}\sum_{i=1}^{G}\left(\frac{1}{N}\sum_{j=1}^{N} P_{ij}^{*}\right) \tag{5}$$

where P_{ij}^{*} defines the best ant after a change of iteration i of run j [12]. Our implementation closely follows the guidelines of the ACOTSP[2] framework.

The value of f was set to 20 and 100, indicating environmental changes of high and low frequencies, respectively. The percentage of m was set to 10, 25, 50, and

[2] Available on `http://www.aco-metaheuristic.org/aco-code`

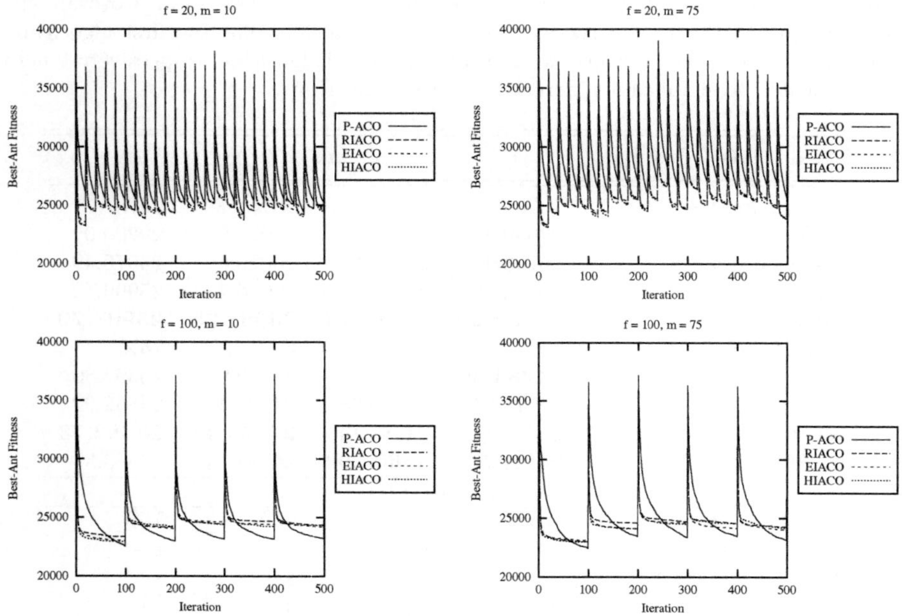

Fig. 1. Overall offline performance for different dynamic test problems

75, indicating the degree of environmental changes from small, to medium, to large, respectively. As a result, eight dynamic environments, i.e., 2 values of $f \times$ 4 values of m, were generated from the stationary TSP instance to systematically analyze the adaption and searching capability of each algorithm on the DTSP.

4.2 Experimental Results

The experimental results regarding the offline performance of the algorithms with the corresponding statistical results of two-tailed t-test with 58 degrees of freedom at a 0.05 level of significance are presented in Table 1. Moreover, to better understand the dynamic behaviour of the algorithm, the results are plotted in Fig. 1 for the first 500 iterations with $f = 20$, $m = 10$ and $m = 75$, and $f = 100$, $m = 10$ and $m = 75$. From the experimental results, several observations can be made by comparing the behaviour of the algorithms.

First, RIACO, EIACO, and HIACO significantly outperform the P-ACO algorithm in almost all test cases. On cases where the frequency is short the P-ACO algorithm is not able to maintain a population list of useful solutions because it has slow convergence. This can be observed from Fig. 1, where under large frequencies P-ACO converges slowly to a better optimum than other algorithms. However, when the magnitude of changes is small with a large frequency, it is significant better than the other algorithms. On the other hand, RIACO, EIACO and HIACO are able to provide a good solution faster after a change since they gain more diversity by incorporating immigrants to the population.

Second, RIACO performs slightly better than EIACO on cases where the frequency is small, as expected. This is because EIACO needs to converge to a good optimum in order to be effective. This task needs sufficient time as with the P-ACO algorithm. Recall that in EIACO we use an adaptive inversion, which provides more exploration than the simple inversion. On the other hand, EIACO performs significant better than RIACO in almost all slowly changing environments since it has sufficient time to locate a good solution.

Third, HIACO improves the performance of EIACO and RIACO on cases where the frequency is small. Incorporating random and elitism-based immigrants, diversity is achieved with random ones and the guidance on promising areas in the search space is achieved by the elitism-based ones. As a result, diversity is controlled more since RIACO may generate high levels of diversity and become ineffective due to the lose of useful solutions found during past iterations. However, HIACO is not improving on cases where the change frequency is large, but it keeps the merits of EIACO since they are not significant different.

5 Conclusions

Different types of immigrants schemes have been successfully applied to EAs to address DOPs efficiently. In this paper, we apply random, elitism-based, and hybrid immigrants schemes into ACO for the DTSP, resulting in the RIACO, EIACO, and HIACO algorithm, respectively. The difference of these algorithms lies in the way immigrant ants are generated. The immigrant ants are generated randomly for RIACO and are generated by mutating the best ant of the previous iteration for EIACO, respectively. For HIACO, half of the immigrant ants are generated randomly and the other half are generated using the elitism-based scheme. All immigrants replace the worst ants of the population on every iteration in order to gain sufficient diversity within the population, which can be useful for the DTSP.

Comparing with P-ACO, an existing ACO framework developed for DOPs, on different cases of dynamic environments, the following concluding remarks can be drawn. First, immigrants schemes are advantageous for ACO algorithms. Second, the performance of EIACO is significant better than RIACO in slowly changing environments. Third, the performance of RIACO is slightly better than EIACO on most fast changing environments, while the performance of HIACO is significant better than both of them. Forth, the performance of HIACO on slowly changing environments is competitive with EIACO. Finally, P-ACO may be a sufficient choice in very slowly and slightly changing environments, or in cyclic environments since it is a memory-based approach [16].

For further work, it would be interesting to compare the algorithms on other dynamic environmental cases, i.e., cyclic environments where past environments reappear, and investigate the effect of other parameters or strategies within the proposed algorithms, i.e., which ants should immigrants replace.

Acknowledgement

This work was supported by the Engineering and Physical Sciences Research Council (EPSRC) of UK under Grant EP/E060722/1.

References

1. Bonabeau, E., Dorigo, M., Theraulaz, G.: Swarm Intelligence: From Natural to Artificial Systems. Oxford University Press, New York (1999)
2. Dorigo, M., Maniezzo, V., Colorni, A.: Ant system: optimization by a colony of cooperating agents. IEEE Trans. on Syst., Man and Cybern., Part B: Cybern. 26(1), 29–41 (1996)
3. Dorigo, M., Stüzle, T.: Ant Colony Optimization. The MIT Press, London (2004)
4. Eyckelhof, C.J., Snoek, M.: Ant Systems for a Dynamic TSP. In: ANTS 2002, Proc. of the 3rd Int. Workshop on Ant Algorithms, pp. 88–99 (2002)
5. Grefenestette, J.J.: Genetic algorithms for changing environments. In: Proc. of the 2nd Int. Conf. on Parallel Problem Solving from Nature, pp. 137–144 (1992)
6. Guntsch, M., Middendorf, M.: Applying population based ACO to dynamic optimization problems. In: Dorigo, M., Di Caro, G.A., Sampels, M. (eds.) Ant Algorithms 2002. LNCS, vol. 2463, pp. 111–122. Springer, Heidelberg (2002)
7. Guntsch, M., Middendorf, M.: A population based approach for ACO. In: EvoWorkshops 2002: Appl. of Evol. Comput., pp. 72–81 (2002)
8. Guntsch, M., Middendorf, M.: Pheromone modification strategies for ant algorithms applied to dynamic TSP. In: EvoWorkshops 2001: Appl. of Evol. Comput., pp. 213–222 (2001)
9. Guntsch, M., Middendorf, M., Schmeck, H.: An ant colony optimization approach to dynamic TSP. In: Proc. of the 2001 Gen. and Evol. Comput. Conf., pp. 860–867 (2001)
10. Guo, T., Michalewicz, Z.: Inver-over operator for the TSP. In: Proc. of the 5th Int. Conf. on Parallel Problem Solving from Nature, pp. 803–812 (1998)
11. Holland, J.: Adaption in Natural and Artificial Systems. University of Michigan Press, Ann Arbor (1975)
12. Jin, Y., Branke, J.: Evolutionary optimization in uncertain environments - a survey. IEEE Trans. on Evol. Comput. 9(3), 303–317 (2005)
13. Stüzle, T., Hoos, H.: The MAX-MIN ant system and local search for the traveling salesman problem. In: Proc. of the 1997 IEEE Int. Conf. on Evol. Comput., pp. 309–314 (1997)
14. Yang, S.: Genetic algorithms with memory and elitism based immigrants in dynamic environments. Evol. Comput. 16(3), 385–416 (2008)
15. Yang, S.: Genetic algorithms with elitism based immigrants for changing optimization problems. In: Giacobini, M. (ed.) EvoWorkshops 2007. LNCS, vol. 4448, pp. 627–636. Springer, Heidelberg (2007)
16. Yang, S.: Memory-based immigrants for genetic algorithms in dynamic environments. In: Proc. of the 2005 Genetic and Evol. Conf., vol. 2, pp. 1115–1122 (2005)
17. Yu, X., Tang, K., Chen, T., Yao, X.: Emprical analysis of evolutionary algorithms with immigrants schemes for dynamic optimization. Memetic Comput. 1(1), 3–24 (2009)

Secret Key Specification for a Variable-Length Cryptographic Cellular Automata Model

Gina M.B. Oliveira, Luiz G.A. Martins,
Giordano B. Ferreira, and Leonardo S. Alt

Faculdade de Computação, Universidade Federal de Uberlândia – UFU
Av. João Naves de Ávila, 2121- Campus Santa Mônica, Bloco B, sala 1B60
CEP: 38400-902 Uberlândia, MG, Brazil
Phone: +55 (34) 3239-4144
{gina,gustavo}@facom.ufu.br, {giordanobsf,leonardoaltt}@gmail.com

Abstract. Reverse algorithm was previously evaluated as encryption method concluding that its simple adoption is unviable, since it does not assurance the pre-image existence. Variable-Length Encryption Method (VLE) was proposed where a alternative algorithm with extra bits is adopted when pre-image computation is not possible. If an adequate secret key is used with VLE it is expected that the final ciphertext length is close to plaintext size. Several CA static parameters were calculated for a set formed by all radius 2 right-toggle rules. A database was generated associating rules performance in VLE ciphering with its parameters. A genetic algorithm-based data mining was performed to discover an adequate key specification based on CA parameters. Using such specification, ciphertext length is short, encryption process returns high entropy and VLE has a good protection against differential cryptanalysis.

Keywords: Cellular Automata, cryptography, pre-image computation, genetic algorithm, data mining.

1 Introduction

Wolfram was the first to suggest the use of CA in cryptography [1]. Several studies on this topic have been accomplished [2–11]. The CA-based cryptographic models can be divided into three classes: *(i)* models that use CA to generate pseudo-random binary sequences, which are used as cryptographic keys, but the effective ciphering process is made by another function [1–4]; *(ii)* models based on additive, non-homogeneous and reversible CA, that use algebraic properties of this kind of rules to generate automata of maximum and/or known cycle [5, 6]; and *(iii)* models based on irreversible CA, which uses the backward interaction of cellular automata in the ciphering process and the forward interaction to decipher [7–11], as the model discussed here.

CA backward interaction is also known as pre-image computation and an efficient reverse algorithm was proposed in [12]. The application of the reverse algorithm as a ciphering algorithm was early investigated in [9]. Using a spatial

R. Schaefer et al. (Eds.): PPSN XI, Part II, LNCS 6239, pp. 381–390, 2010.

entropy measure to evaluate the ciphering quality, the main conclusion in [9] is that the simple adoption of the reverse algorithm is not possible since rules with 100% guarantee of pre-image existence are not appropriate for ciphering because they do not exhibit a chaotic dynamics. A new approach has been emerged from this previous study: it alternates the original reverse algorithm and a variation that uses extra bits in encryption when the pre-image computation fails. Since it is expected that in practice few failures happen, the ciphertext length will be close to the plaintext. This method was evaluated in [11] where it was named Variable-Length Encryption Method (VLE). Only small samples of radius 2 and radius 3 rules were used in [11]. A more representative set formed by all radius 2 right-toggle rules, totalizing 65536 keys (2^{16}), was used to evaluate VLE in [10]. These rules represent about 50% of the possible secrete keys in radius 2 space.

CA rules used as secret keys must be properly specified to reduce the probability of failure occurrence during pre-image computation. Using small samples of radius 2 and 3 rules applied to cipher a sample of lattices, it was shown in [9] that the joint use of symmetry (S) and Z parameters could lead us to a good specification of rules with low probability to fail in pre-image computation. However, based on an exhaustive analysis of the rule set formed by all radius 2 right-toggle rules in [10] it was verified that although the majority of right-toggle rules are suitable to be used with VLE, there are some undesirable behavior rules in this set that most be avoided as secrete keys.

In the present work, we better investigate the secret key specification. We employed an analysis based on several CA static parameters to capture the pattern associated to underperforming rules. Using a genetic algorithm to mine this pattern [13], we were able to find a good specification of rules to be used as secret keys. Such specification had shown to be good to filter the complete set of radius 2 rules and to specify radius 3 rules.

2 Cellular Automata Applied in Cryptography

Cellular automata (CA) are discrete complex systems that possess both a dynamic and a computational nature. A cellular automaton consists of two parts: the cellular space and the transition rule. Cellular space is a regular lattice of N cells subjected to boundary conditions. A state is associated to each cell at time t. The transition rule Φ yields the next state for each cell as a function of its neighbourhood. At each time step, all cells synchronously update their states according to Φ. For 1D CA, the neighbourhood size m is usually written as $m = 2r + 1$, where r is CA radius. In binary-state CA, the transition rule is given by a state transition table which lists each possible neighbourhood together with its output bit, that is, the updated value for the state of the central cell.

CA dynamics is associated with its transition rule Φ represented by its output bits $(b_1, b_2, ..., b_N)$. In order to help forecast the dynamic behavior of CA, several parameters have been proposed [14], some of them are: *(i)* Z derived from the pre-image computation algorithm and it is composed by Z_{left} and Z_{right}. [12];

(ii) S is the symmetry level of the output bits in a rule transition [9]; and *(iii) Neighborhood dominance (ND)* verifies whether the new value of the centre cell "follows" the state that appears the most in the neighborhood [14].

Cellular automata are particularly well suited for cryptographic application. Since CA rule is simple, local and discrete, it can be executed in easily-constructed parallel hardware at fast speeds. Gutowitz proposed a cryptographic model based on backward evolution of irreversible CA [7]. A toggle rule with radius R is used as the secret key in his model. A CA toggle rule is sensible in respect to a specific neighborhood cell - any modification of the state on this cell necessarily provokes a modification on the new state of the central cell. A pre-image of an arbitrary lattice of size N is calculated adding R extra bits in each side of the lattice [7]. Plaintext is the initial lattice and P pre-images are successively calculated. The ciphertext is the last pre-image obtained and its size is given by $N + 2RP$. Such increment is pointed as the major flaw in the model. Reverse algorithm was proposed in [12] for a generic pre-image computation using a periodic boundary CA. Such algorithm was evaluated as encryption method in [9]. However, its usage has the disadvantage that there is no guarantee of pre-image existence for any given lattice and any given rule.

The major challenge to apply the reverse algorithm as a viable cipher method was to find a manner to guarantee the existence of at least one pre-image for any plaintext. Parameters that seemed more relevant to this problem [9] were the components of Z known as Z_{right} and Z_{left} and the symmetry level (S). It was observed that one component of Z, Z_{right} or Z_{left}, must be equal to 1 and the other one must be different of 1. Moreover, S equal to 1 also must to be avoided. An observation is that although it was possible to specify rules with a high probability to find at least one pre-image for any lattice and with a good perturbation spread, even the better rules evaluated can fail during pre-image computation. The major conclusion of the analysis in [9] is that the simple adoption of the reverse algorithm is not viable because rules with 100% guarantee of pre-image existence are not appropriate for ciphering; they are not chaotic. A method based on reverse algorithm adopting a contour procedure when pre-image computation fails was proposed in [9].

3 Variable Length Encryption Method

Since the main conclusion is that the simple adoption of the reverse algorithm is not possible, an alternative method was proposed in [9] and pruned in [11]. This method is based on the original reverse algorithm adopting an alternative procedure to apply when the pre-image computation fails. The alternative procedure adds extra bits only when the pre-image is not possible to calculate, guarantying the possibility to cipher any plaintext. It is expected that with an appropriate secret key specification, there is a low probability to this failure occurrence and, as consequence, rarely use this alternative procedure. For practical reasons, it can be better to limit the method to operate with only toggle rules. Ciphering process is defined by the computation of P consecutive pre-images starting from a lattice of size N corresponding to the plaintext. The secret key is a CA rule Φ with radius R. Suppose that ciphering process started pre-images computation using the reverse algorithm and it fails in the Kth pre-image ($K \leq P$). It uses the alternative algorithm with extra bits to calculate the Kth pre-image, which will have $N + 2R$ cells. Ciphering process returns again using reverse algorithm to calculate the remaining pre-images. If all the subsequent pre-images computation succeeds the final ciphertext will have a size of $N+2R$. If the process fails in F pre-images the final lattice will have $N + 2FR$. Starting from a lattice of N cells, the ciphertext size after P pre-images computation will be between N and $N + 2PR$. Therefore, it is a variable-length encryption model, named VLE.

Using VLE one has the guarantee that ciphering is possible even if the plaintext is a Garden-of-Eden state. However a short length ciphertext depends on the secret key specification. Starting from the information presented in [9], which uses Z_{left}, Z_{right} and S, and by using small rule sets the expectative about the usage of the VLE method had been confirmed [11]: it has a good quality of ciphering entropy and the ciphertext length is close to the original block size. However, a lot of open questions remain since experiments in [11] were performed based on very limited samples of rules. In [10], a more exhaustive analysis was conducted using the complete set of radius 2 right-toggle rules. This set is composed by 65536 (2^{16}) rules, being that all of them have $Z_{left} = 1$. These rules represent about 50% of the possible secrete keys in radius 2 space, if we impose the restriction of using only toggle rules for faster encryption. VLE-based environment was employed to cipher a hundred 256-bits plaintexts using each right-toggle rule. The number of consecutive pre-images was not fixed in [10] aiming to discover a good value of P to be applied with radius 2 rules and 256-bits plaintexts. P was dynamically defined and the pre-image computation stops when ciphering achieves a good entropy level. The investigation in [10] suggests that $P = 32$ is a good value to be applied in such instance. Moreover, it was pointed that the usage of parameters Z and S initially investigated in [9] and [11] are not enough to filter all underperforming rules and a larger number of static parameters must be investigated.

4 Experiments

An initial analysis was performed using the complete set of radius 2 right-toggle rules. VLE-based environment was employed to cipher a hundred 256-bits plaintexts using each right-toggle rule, by calculating P consecutive pre-image steps. These experiments are similar to those presented in [10] except for the fact that here we employed a fixed and predefined number of pre-image steps (P). Initially, we employed $P = 32$ as suggested by results in [10]. However, it was possible to observe that this number of steps was not enough to spread perturbations when using a considerable portion of rules of the entire set. Thus, different values of P were considered in this analysis: 32, 40, 48, 56 and 64. Using $P = 64$, the best result related to perturbation spread was obtained as expected but additionally the number of fails increased (when compared with $P = 32$) in a considerable portion of rules. Considering the trade-off between perturbation spread and number of fails we conclude that the best evaluated value of P is 48. We present here the results obtained using VLE to cipher a hundred 256-bits plaintexts by calculating 48 consecutive pre-image steps. Based on an exhaustive analysis of radius 2 space we evaluated effects of using as secret keys the complete set of rules in which the unique restriction is the right-toggle property ($Z_{left} = 1$ is a consequence). This set is formed by 65536 rules.

Applying VLE method, any rule with $Z_{left} = 1$ is able to complete ciphering process starting from any initial lattice, for any P. However, the final length of the ciphertext can be between N and $N + 2PR$. We want to evaluate if the expected final length is in fact close to N. For this, we calculated the mean length of the ciphertext (L_{mean}) and the mean number of failures occurred during ciphering process (F_{mean}). Mean results were computed considering the application of all rules to cipher a sample of 100 lattices of 256 bits and we obtained $L_{mean} = 257.90$ and $F_{mean} = 0.476$. F_{mean} is below 0.5, which returns a mean ciphertext size very close to the original size (256 bits).

Differential cryptanalysis is based on the analysis of some pairs of ciphertexts generated after similar plaintexts. Sen *et al.* (2001) analyzed the security of their CA cryptosystem named CAC to resist to differential attack comparing it with DES and AES cryptosystems [15]. Several pairs of plaintext (X, X') were used, which differ one of the other by a fixed and small difference D. Each pair (X, X') was used to generate a pair of ciphertexts (Y, Y') which differ one of the other by a difference D' obtained by Y XOR Y' operations. For each pair (Y, Y'), the number of 1s in D' is counted, which corresponds to the number of different bits between Y and Y'. Finally, the standard deviation of this measure is calculated over all the analyzed pairs. As higher is the standard deviation in D', as higher is the probability of the ciphertext to be broken by differential cryptanalysis. An algorithm with standard deviation below 10% is said to be protected against differential cryptanalysis [15]. In our tests, the difference D between two plaintexts X and X' was fixed in only one bit in any arbitrary position over the lattice. The computation of D' to each pair (Y, Y') was performed to obtain the standard deviation (σ). Besides the σ computation, the difference D' was also used to compute a second measure related to ciphering quality. The goal of this measure is to verify if D' does not keep any pattern which eventually could help a cryptanalyst. Spatial entropy [9] was calculated on D' to evaluate the existence of some undesirable regularity on this difference. Entropy above 0.75 indicates a random difference enough to expect that ciphertexts Y and Y' do not maintain any similarity, even so they started from similar plaintexts. Entropy below 0.5 indicates a strong pattern in D' [9]. Entropy values between 0.5 and 0.75 had been considered fuzzy, since it cannot guarantee the existence of an ordered or random pattern. Therefore, if any cryptography method is applied to similar texts returning an $E_{mean} > 0.75$, it indicates that ciphering adds a high level of entropy during the process, a necessary characteristic in any encryption method. We obtained $\sigma_{mean} = 3.83\%$ and $E_{mean} = 0.876$ for the complete set of right-toggle rules. Since σ_{mean} is below 10%, the proposed CA cryptographic model can be considered secure in relation to differential cryptanalysis. E_{mean}

is above 0.85 indicating that rules were able to add a high entropy in ciphering, using $P = 48$.

Therefore, considering only the mean values of the complete set analyzed, it returned excellent values on all the measures. However, as pointed in [10], the analysis of the worst performing rules in the set indicates the existence of secrete keys not appropriate for ciphering purpose. Considering only the 1000 worst performing rules in each metric we obtained $F_{mean} = 17.645$, indicating that these rules returning ciphertexts with size superior to 320 bits in average. Fmean represents the mean value size for each rule considering all the 100 lattices used to test it. However, if we consider the worst result in such lattices, we can find ciphertexts with a considerable size: we calculated a new metric F_{max}, the mean of the maximum ciphertext size obtained considering the 1000 worst rules, and obtained $F_{max} = 25.540$. This metric highlights the existence of several secret keys returning at least one ciphertext with size superior to 350 bits. We observed that there are around 750 rules in the complete set of right-toggle rules that returns ciphertext lengths between 352 and 448 bits in the worst case and there are about 400 rules above 350 bits in average. An analysis of the worst rules related to entropy values can also be done, similar to the previous ciphertext length analysis. Considering only the 1000 worst performing rules, the mean entropy in D' was obtained: $E_{mean} = 0.254$, indicating that these rules does not perform an actual encryption of the plaintexts in average. About 1020 rules returned E_{mean} below 0.5. E_{min} was also calculated representing the worst entropy found for each rule. We obtained an average of $E_{min} = 0.036$, considering the 1000 worst rules. These results indicate that there are lattices not encrypted. The most probable behavior is that the rule only shifts the initial lattices, not performing an actual encryption of these plaintexts. Although this behavior is a minor occurrence considering the entire set of CA rules it cannot be allowed in a secure cryptosystem. Considering the entire rule set, about 3250 rules returned entropy below 0.5 for at least one pair Y and Y'. Concluding, there are undesirable behavior rules in the complete set analyzed that must be avoided as secrete keys. The entire rule space formed by all radius 2 right-toggle rules is not appropriate to be applied as secret keys in VLE method. Approximate 4000 rules (6% of the key space) must be avoided: 3200 due to low entropy and 800 due to long ciphertext length.

Specifications was proposed in [9] and [11] trying to filter such undesirable behavior keys, in which S and the components Z_{left} and Z_{right} were used. However, experiments in [10] using the complete set of right-toggle rules had evidenced that their application is not effective as supposed. Aiming to better understand the relation between CA static parameters and underperforming rules, an analysis using more than Z and S parameters was conducted here. A series of CA parameters was calculated for each rule trying to identify a pattern to filter underperforming rules of radius 2 right-toggle rules set. Nine parameters were used in this analysis: Z_{right}, S, BWLR_Symmetry ($BWLR$), LR_Symmetry (LR), Absolute Activity (AA), Neighborhood Dominance (ND), Sensitivity (μ), Activity Propagation (AP) [14] and Core Entropy (CE) [10]. All these parameters were calculated to each one of the 65536 radius 2 rules. A database was elaborated in which each register corresponds to one right-toggle rule and the fields are composed by the values of the nine parameters calculated for each rule, and the values of the metrics calculated when the rule is applied to cipher 100 plaintexts: F_{mean} and E_{mean}. As a pattern associating parameters with the worst performing rules in ciphering was not possible to recognize by a simple visual inspection, we decided to apply a data mining process. A standard GA was elaborated with this goal based on the model described in [13]. Individual is composed by I genes, where I is the number of CA static parameters analyzed; that is, nine fields in our experiments. It is illustrated in Figure 1. The i-th gene is subdivided into three fields: weight (W_i), operator (O_i) and value (V_i).

Each gene corresponds to one condition in the antecedent part (IF) and the individual as a whole is the rule antecedent. The weight field is an integer between 0 and 10. This field determines the insertion of the correspondent gene in the rule antecedent. If this value is lesser than a boundary-value this gene will not appear in the rule, otherwise the gene appears. In this work, the value 7 was used as the boundary-value. The operator field can be $<$ (minor), $\geq$ (larger or equal) or $\neq$ (different). The value field is a floating-point number that can vary between the 0 and 1, because each parameter was normalized. To establish the consequent part of the rule, we first analyze the fields F_{mean}, and E_{mean} of each register of the database to characterize the underperformed rules in specific classes. Field *Class* was added to the database with the classification of each register in one of these classes: (1) 886 rules with low mean entropy (< 0.5); (2) 750 rules with large mean ciphertext length (> 300 bits); and (3) 63,900 reminiscent rules. The individual in Figure 1 represents only the antecedent part of the rule (IF). The consequent part is always in the format *THEN Class* $=$ *C*, being that *C* can be 1, 2 or 3. However, it is omitted in individual's representation. Conversely, it is a fixed execution parameter of GA. Thus, if the GA is executed with $C = 1$, all the rules of population represent classification rules in the format: IF *Antecedent* THEN *Class is "Low entropy"*. All registers are considered either *Class 1* or *Not Class 1*. Fitness quantifies the quality of the rule associated to each individual. Individual fitness is given by a weighted sum between *Sensitivity* and *Specificity* indicators [13]. Stochastic tournament with Tour $= 3$ is used as the matting selection method. Two-point crossover is applied and a specific mutation operator is used to each type of gene field with a rate of 30%. The next generation is formed by selecting the best individuals in populations. Each GA experiment was formed by 100 runs, using a population of 100 individuals, which were evolved by 100 generations. The classification rule that was indeed intended to be mined is Class 3, because it represents appropriate rules to be used in cryptography. However, the other two classification rules (classes 1 and 2) were also important to achieve our goal, because they better characterize low entropy rules and long ciphertext ones, given us some important information to prune the rules returned by GA for Class 3. After several executions and post-processing pruning procedures we have found the rule following:

IF ($S \neq 1$ AND $ND \leq 0.57$ AND $CE > 0.65$ AND $Z_{right} \neq 1$) THEN $Class = 3$

This rule employs 4 of the 9 CA parameters to characterize adequate secrete keys. We applied it as a filter in the complete radius 2 right-toggle CA rule set to remove all secret keys that non attending its conditions. The filtered set has 51,495 right-toggle rules. It was named *Subset*. By using the environment implemented based on VLE, we employed them to cipher a hundred 256-bits plaintexts, by calculating 48 consecutive pre-image steps, as performed to the complete rule set. Considering all the remaining rules in each set, Table 1 and Figure 2 show the results obtained with *Subset* rules. In such tables and figures, the results of the complete set (2^{16} rules) is also presented, named as *Fullset*. Table 1 shows the mean values obtained with each set considering all the rules in each one. Considering the totality of the rules, *Subset* results are better than those obtained using the entire radius 2 set, but the differences are not so expressive. However, when the worst performing rules are analyzed the advantage of such filters is evidenced. While 1020 rules in *Fullset* returned mean entropy below 0.5, only 28 rules with such inappropriate low entropy remains in *Subset*. Considering the minimum entropy in each rule applied over 100 ciphertexts the benefice of filter rule application is highlighted: 3243 rules in *Fullset* presents an entropy below 0.5 in difference D' in at least one pair Y and Y'. The number of such rules downs to 327 in *Subset*. This improvement can also be recognized by the mean values of the 500 worst performing rules in Table 1 (E_{mean} and E_{min}) and in the curves of Figure 2. An analysis of ciphertext length also evidences the application of filtered rules: while 758 rules presents an average of ciphertext length greater than 280 bits, only 217 rules with a ciphertext length above 280 bits remains in *Subset*. Besides, 835 rules in *Fullset* returns a maximum ciphertext length greater or equal to 300 bits, while there are only 291 such rules in *Subset*. Therefore we conclude that rule filter is a good specification for CA rules.

Gene $_1$			Gene $_2$				Gene $_9$		
Z_{right}			S			...	CE		
W_1	O_1	V_1	W_2	O_2	V_2		W_9	O_9	V_9

Fig. 1. Individual representation

5 Final Remarks

A cryptographic model based on toggle CA rules as secret keys are better investigated here. This method alternates during ciphering process the employment of the original reverse algorithm [12] with a variation, which adds extra bits when a pre-image is calculated. This approach is a variable-length encryption method named VLE. The average of standard deviation found based on more

Table 1. Mean values of the complete rule sets and the 500 worst performing rules

Set	Number of Rules	Complete rule sets				500 worst performing rules			
		L_{mean}	F_{mean}	E_{mean}	σ_{mean}	F_{mean}	F_{max}	E_{mean}	E_{min}
Fullset	65536	257.903	0.476	0.876	3.83%	17.620	33.212	0.125	0.036
Subset	51495	257.410	0.352	0.889	3.69%	7.227	18.514	0.787	0.53

than 6 million of tests is 3.83%, showing this method is robust to a differential cryptanalysis-like attack being much lower than the upper bound limit suggested in [15]: 10%. Comparing with the results in [15] the superiority of VLE is clear: 12%, 7% and 5% returned by DES, AES and CAC respectively. Besides, the absence of an ordered pattern when ciphering two very similar plaintexts was evidenced by the mean entropy found: 0.876. VLE guarantees that ciphering is always possible with a short length ciphertext expectative: 257.9 bits next to 256 bits of the plaintext. The properly key specification was deeper investigated in the present work using all the 65,536 radius 2 right-toggle rules. It became clear that there are some rules in this set inappropriate to be used as secret keys: 1.5% of them returning long plaintexts and 5% of them exhibiting a more dangerous behavior, which is not able to encrypt at least one plaintext. Therefore, some kind of restriction is needed.

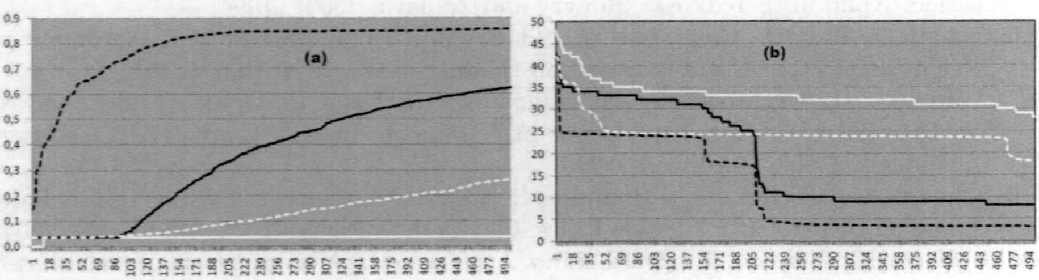

Fig. 2. 500 worst performing rules in fullset (white lines) and subset (black lines): (a) mean D' entropy (dashed line) and minimum D' entropy (continuous line). (b) mean ciphertext length (dashed line) and maximum ciphertext length (continuous line).

We employed an analysis based on several CA static parameters and a genetic algorithm to mine this information. Therefore, we were able to find a good specification of rules to be used as valid secret keys. This specification had shown to be good to filter the complete set of radius 2 rules. A final test was performed using radius 3 rules with $Z_{right} = 1$ aiming to evaluate if the filter rule can also be applied to radius 3 rule space. Since an exhaustive procedure was not possible in this rule space a new GA was implemented to generate 1,000 right-toggle radius 3 rules attending the *ND_filter* rule. We employed them using the VLE environment to cipher a hundred 512-bits plaintexts, by calculating 16 consecutive pre-image steps. Considering 1000 radius 3 rules, we obtained the following average values: $L_{mean} = 514.507$, $F_{mean} = 0.627$, $E_{mean} = 0.895$ and

$\sigma_{mean} = 3.42\%$. They are satisfactory mean values for criptography. Moreover, no rule returns low entropy in any ciphered plaintext.

Acknowledgements

GMBO thanks CNPq and FAPEMIG support.

References

1. Wolfram, S.: Cryptography with cellular automata. In: Williams, H.C. (ed.) CRYPTO 1985. LNCS, vol. 218, pp. 429–432. Springer, Heidelberg (1986)
2. Tomassini, M., Perrenoud, M.: Stream Ciphers with One and Two-Dimensional Cellular Automata. In: Deb, K., Rudolph, G., Lutton, E., Merelo, J.J., Schoenauer, M., Schwefel, H.-P., Yao, X. (eds.) PPSN 2000. LNCS, vol. 1917, pp. 722–731. Springer, Heidelberg (2000)
3. Seredynski, F., Bouvry, P., Zomaya, A.Y.: Secret key cryptography with cellular automata. In: Workshop on Nature Inspired Distributed Computing (2003)
4. Benkiniouar, M., Benmohamed, M.:: Cellular Automata for Cryptosystem. In: Proceedings of IEEE Conference Information and Communication Technologies: From Theory to Applications, pp. 423-424 (2004)
5. Kari, J.: Cryptosystem based on reversible cellular automata. Personal communication. Apud in (Seredynski, Bouvry and Zomaya, 2003) (1994)
6. Nandi, S., Kar, B., Chaudhuri, P.: Theory and Applications of CA Automata in Cryptography. IEEE Transactions on Computers 43, 1346–1357 (1994)
7. Gutowitz, H.: Cryptography with Dynamical Systems. In: Goles, E., Boccara, N. (eds.) Cellular Automata and Cooperative Phenomena, vol. 1, pp. 237–274. Kluwer Acapdemic Press, Dordrecht (1995)
8. Oliveira, G., Coelho, A., e Monteiro, L.: Cellular Automata Cryptographic Model Based on Bi-Directional Toggle Rules. I. J. Modern Physics C 15, 1061–1068 (2004)
9. Oliveira, G., Macêdo, H., Branquinho, A., Lima, M.: A cryptographic model based on the pre-image computation of cellular automata. In: Automata-2008: Theory and Applications of Cellular Automata, pp. 139–155. Luniver Press (2008)
10. Oliveira, G.M.B., Martins, L.G.A., Alt, L.S., Ferreira, G.B.: Exhaustive Evaluation of Radius 2 Toggle Rules for a Variable-Length Cryptographic CA-Based Model. In: Int. Conf. on Cellular Automata for Research and Industry, Ascoli Piceno (2010)
11. Oliveira, G.M.B., Martins, L.G.A., Alt, L.S., Ferreira, G.B.:: Investigating a Cellular Automata-Based Cryptographic Model with a Variable-Length Ciphertext. In: CSC 2010, - International Conference on Scientific Computing, Las Vegas (2010)
12. Wuensche, A., Lesser, M.: Global Dynamics of Cellular Automata. Addison-Wesley, New Mexico (1992) ISBN: 0-201-55740-1
13. Fidelis, M., Lopes, H., Freitas, A.: Discovery comprehensible classification rules with a genetic algorithm. In: C.Evolutionary Computation, CEC 2000, USA (2000)
14. Oliveira, G., de Oliveira, P., e Omar, N.: Definition and applications of a five-parameter characterization of 1D cellular automata rule space. Artificial Life 7(3), 277–301 (2001)
15. Sen, S., Shaw, C., Chowdhuri, D., Ganguly, N., Chaudhuri, P.: Cellular Automata based Cryptosystem (CAC). In: Deng, R.H., Qing, S., Bao, F., Zhou, J. (eds.) ICICS 2002. LNCS, vol. 2513, pp. 303–314. Springer, Heidelberg (2002)

Variable Neighborhood Search and Ant Colony Optimization for the Rooted Delay-Constrained Minimum Spanning Tree Problem

Mario Ruthmair and Günther R. Raidl

Institute of Computer Graphics and Algorithms
Vienna University of Technology, Vienna, Austria
{ruthmair,raidl}@ads.tuwien.ac.at
http://www.ads.tuwien.ac.at

Abstract. The rooted delay-constrained minimum spanning tree problem is an NP-hard combinatorial optimization problem arising for example in the design of centralized broadcasting networks where quality of service constraints are of concern. We present two new approaches to solve this problem heuristically following the concepts of ant colony optimization (ACO) and variable neighborhood search (VNS). The ACO uses a fast construction heuristic based on node delays and local improvement exploiting two different neighborhood structures. The VNS employs the same neighborhood structures but additionally applies various kinds of shaking moves. Experimental results indicate that both metaheuristics outperform existing approaches whereas the ACO produces mostly the best solutions.

1 Introduction

When designing a communication network with a central server broadcasting information to all the participants of the network, some applications, such as video conferences, require a limitation of the maximal delay from the server to each client. Beside this delay-constraint minimizing the cost of establishing the network is in most cases an important design criterion. In another example we consider a package shipment organization with a central depot guaranteeing its customers a delivery within a specified time horizon. Naturally the organization aims at minimizing the transportation costs but at the same time has to hold its promise of being in time.

These network design problems can be modeled using a combinatorial optimization problem called *rooted delay-constrained minimum spanning tree (RDCMST) problem*. The objective is to find a minimum cost spanning tree of a given graph with the additional constraint that the sum of delays along the paths from a specified root node to any other node must not exceed a given delay-bound.

More formally, we are given an undirected graph $G = (V, E)$ with a set V of n nodes, a set E of m edges, a cost function $c : E \rightarrow \mathbb{R}_0^+$, a delay function

R. Schaefer et al. (Eds.): PPSN XI, Part II, LNCS 6239, pp. 391–400, 2010.

$d : E \rightarrow \mathbb{R}^+$, a fixed root node $s \in V$, and a delay-bound $B > 0$. An optimal solution to the RDCMST problem is a spanning tree $T = (V, E')$, $E' \subseteq E$, with minimum cost $c(T) = \sum_{e \in E'} c(e)$, satisfying the constraints:

$$\sum_{e \in P(s,v)} d(e) \leq B, \ \forall v \in V,$$

where $P(s, v)$ denotes the unique path from root s to node v.

The RDCMST problem is $\mathcal{NP}$-hard because already the special case with $d(e) = 1$, $\forall e \in E$, called *hop-constrained minimum spanning tree problem*, is $\mathcal{NP}$-hard [1].

The rest of the paper is organized as follows: In Section 2 existing approaches are discussed, Section 3 introduces some helpful preprocessing steps to reduce the problem size, Section 4 describes a metaheuristic method based on general variable neighborhood search, Section 5 presents a second approach based on ant colony optimization, Section 6 discusses test results, and Section 7 concludes the article.

2 Previous Work

Exact approaches to the RDCMST problem have been examined by Gouveia et al. in [2] based on the concept of constrained shortest paths utilized in column generation and Lagrangian relaxation methods. A flow-based reformulation of the problem on layered acyclic graphs is applied to reduce the RDCMST problem to the well-known and intensively examined *minimum Steiner tree problem* [3]. Since the size of the layered graph and therefore the efficiency of the according model heavily depends on the delay-bound B this approach can in practice only be used for instances with a reasonably small number of possible discrete edge delay values. Furthermore all these methods can only solve small instances with significantly less than 100 nodes to proven optimality in reasonable time when considering complete graphs.

A constructive heuristic approach based on Prim's algorithm to find a minimum spanning tree [4] is described by Salama et al. in [5]. A general problem of this heuristic especially on Euclidean instances is the fact that the nodes in the close surrounding of the root node are typically connected rather cheaply, but at the same time delay is "wasted", and many distant nodes can later only be linked by rather expensive edges. The stricter the delay-bound the more this drawback will affect the costs negatively. This fact led Ruthmair et al. [6] to a more de-centralized approach by applying the basic concept of Kruskal's minimum spanning tree algorithm [7] to the RDCMST problem. Two metaheuristics based on GRASP and variable neighborhood descent improve the constructed solution in [6].

There are many recent publications dedicated to the *rooted delay-constrained minimum Steiner tree problem* which is a generalization of the RDCMST problem. In this variant only a subset of the nodes has to be reached within the given delay-bound, the other nodes can optionally be used as intermediate (Steiner) nodes. Several metaheuristics have been applied to this variant, such as GRASP [8,9], path-relinking [10] and variable neighborhood descent [11]. Also exact methods based on Integer Linear Programming have been explored, e.g. Leggieri et al. [12] describe a node-based formulation using lifted Miller-Tucker-Zemlin inequalities.

3 Preprocessing

The following rules are applied in a preprocessing phase in order to possibly reduce the instance graph without affecting optimal solutions.

3.1 Infeasible Edges

In the following cases an edge $e = (i, j) \in E$ cannot be part of a feasible solution so discarding it safely reduces the search space.
(a) Obviously all edges $e \in E$ having a delay $d(e)$ higher than the bound B can be discarded immediately.
(b) Edges $e = (i, j) \in E$ which would exceed the bound in all possible trees can also be removed from the graph [12] if satisfying these conditions:

$$d_{\min}(s, i) + d(e) > B \quad \wedge \quad d_{\min}(s, j) + d(e) > B,$$

whereas minimum delays $d_{\min}(s, v) := \min_{P(s,v)} \sum_{e \in P(s,v)} d(e)$, $\forall v \in V$, are calculated a priori by Dijkstra's shortest path algorithm [13] applied on the delays. Both preprocessing tests can be done on all edges E in time $\mathcal{O}(m + n \log n)$ including the calculation of d_{min} values.

3.2 Suboptimal Edges

Suboptimal edges can be part of a feasible solution but may not appear in an optimal solution, so discarding them safely prunes the solution space.

Comparison to Root Edges. Applying the arc elimination test presented in [14] to the RDCMST problem results in the removal of edge $e = (i, j) \in E$ if edges (s, i) and (s, j) exist and the following conditions hold:

$$c(s, j) \leq c(e),\ d(s, j) \leq d_{min}(s, i) + d(e) \quad \wedge$$

$$c(s, i) \leq c(e),\ d(s, i) \leq d_{min}(s, j) + d(e)$$

This preprocessing rule can be helpful for rather dense or complete graphs but in sparse graphs only few edges are typically discarded, mainly because of the small out-degree of the root node. Searching those edges takes time $\mathcal{O}(m)$ if d_{min} values are already known.

Extension to Arbitrary Triangles. Considering any triangle in the original graph consisting of edges $e_1 = (v_1, v_2)$, $e_2 = (v_2, v_3)$, $e_3 = (v_3, v_1) \in E$, edge e_1 cannot appear in an optimal solution if:

$$c(e_1) > c(e_2) + c(e_3) \ \wedge \ d(e_1) \geq d(e_2) + d(e_3)$$

If only the second part holds and $c(e_1) = c(e_2) + c(e_3)$, then edge e_1 can be part of an optimal solution but there has to be at least one other optimal solution with $e_1 \notin E'$ and we therefore can also remove it. This preprocessing step can be done in time $\mathcal{O}(mn)$.

Extension to Arbitrary Paths. The last preprocessing step can be further extended in the following way: Edge $e = (i, j)$ cannot appear in an optimal solution if there exists a path $P(i, j) \in E \backslash \{e\}$ satisfying the following conditions:

$$c(e) > \sum_{e' \in P(i,j)} c(e') \quad \wedge \quad d(e) \geq \sum_{e' \in P(i,j)} d(e')$$

Similarly if the second part is met and $c(e) = \sum_{e' \in P(i,j)} c(e')$, then edge e can be part of an optimal solution but there has to be at least one other optimal solution with $e \notin E'$. Applying this preprocessing test reduces to solving the constrained shortest path problem. This problem is $\mathcal{NP}$-hard but approximable by an FPTAS which implicates the existence of an exact pseudo-polynomial algorithm. An efficient dynamic programming approach customized to the RDCMST problem was presented by Gouveia et al. [2]. Nevertheless finding a delay-constrained shortest path runs in time $\mathcal{O}(mB)$ which extends to $\mathcal{O}(m^2 B)$ for the complete preprocessing test.

Especially the last two preprocessing steps must be used with caution. When having a limited runtime (as in our tests in Section 6) the preprocessing phase could dominate or even use up the whole time for large instances. So preprocessing can also be counterproductive and finally worsen solution quality.

4 General Variable Neighborhood Search

For constructing a feasible starting solution we use the Kruskal-based heuristic from [6]. To improve the quality of this solution we apply the metaheuristic framework *general variable neighborhood search* (GVNS) as introduced by Hansen et al. [15]. In each iteration the so far best solution is perturbed by shaking and then improved by an embedded *variable neighborhood descent* (VND).

4.1 Variable Neighborhood Descent

The VND used here is the one introduced in [6]. It performs a local search switching between two neighborhood structures: *Edge-Replace* and *Component-Renew*.

A move in the Edge-Replace neighborhood removes an edge and connects the resulting two components in the cheapest feasible way. A neighborhood search is done by next improvement considering the edges in decreasing cost order until a local optimum is reached, running in time $\mathcal{O}(nm)$.

A Component-Renew move also deletes an edge, but completely dissolves the component which is now separated from the root node; it then re-adds the individual nodes by applying Prim's algorithm [4] respecting feasibility. In some cases not all nodes can be added due to the delay-bound. These remaining nodes are again joined to the root component by shortest delay paths, dissolving created cycles. A neighborhood search is done in a similar way following a next improvement strategy considering the edges in decreasing cost order until a local optimum is reached, running in time $\mathcal{O}(n^3)$.

4.2 Shaking

Three different kinds of shaking moves are used in the GVNS framework. The algorithm always chooses one at random with equal probabilities:
(a) Replacing a random edge by another feasible randomly chosen edge not violating the delay constraint and tree structure.
(b) Adding the shortest delay path to a random vertex (see Section 3.1).
(c) Adding the delay-constrained least cost path to a random vertex (see Section 3.2).

To ensure feasibility of the solution two issues have to be considered: Firstly the last two shaking moves could possibly cause cycles which have to be dissolved, and secondly the delay-bound in the third move can be set at most to the global delay-bound reduced by the maximal delay in the subtree of the randomly chosen vertex. The number of shaking moves performed in one iteration is $\lceil |V| * sr \rceil$, where $sr \in (0, 1]$ is called shaking rate. This shaking rate is either set to a fixed value or dynamically changed in the search process. In the latter case sr is initialized with 0.01, increased by 0.01 when no better solution could be found in an iteration, limited from above by 0.3, and reset again to 0.01 in case of an improvement.

5 Ant Colony Optimization

We apply the concept of the $\mathcal{MAX}-\mathcal{MIN}$ Ant System (MMAS) from Stützle et al. [16] to our problem implementing the following key features:
(a) Only a single ant is allowed to deposit pheromones at the end of an iteration, either the best ant of the iteration or the globally best one, focusing the search to the best solutions found.
(b) Pheromone values are limited to an interval $[\tau_{min}, \tau_{max}]$ preventing a stagnation of the search.
(c) The initial pheromone values are set to τ_{max} leading to a high diversification at the beginning of the search.

Combining all these features provides both intensification and diversification throughout the search process, which is essential for a well-performing metaheuristic.

5.1 Pheromone Values

Pheromone values $\tau_{v,d} \in [\tau_{\min}, \tau_{\max}]$ are defined for each node $v \in V$ in combination with any delay $d \in (0, B]$ it might have. Here, d denotes the sum of delays of all edges on the path from the root to node v. The root s by definition always has a node delay 0 and therefore has no pheromone values associated.

Notice that in general delays are real values and therefore pheromone values do not form a classical finite matrix. However, when considering one special instance graph the number of feasible node delays is finite (but possibly very large). In an implementation one has to consider efficient techniques for handling large sparse matrices [17].

Limits $\tau_{\min}$ and $\tau_{\max}$ are initialized and updated according to [16] using parameter $p_{\text{best}} = 0.00005$. Preliminary tests have shown that these parameter settings work well in general.

5.2 Solution Construction

The method for constructing a solution based on the pheromone values is inspired by the level-based construction heuristic introduced in [18] and runs in time $\mathcal{O}(nB + n^2)$:
(a) For each node a delay value is selected with a probability proportional to the according pheromone value.
(b) All nodes are then sorted by these delay values in ascending order.
(c) The nodes are added in the specified order to the existing tree – initialized with the root node – always choosing the cheapest possible edge without causing a node delay higher than the selected delay. If there is no edge satisfying this constraint, the shortest delay path to the problematic node is added, overriding the given order but guaranteeing a feasible solution.

5.3 Local Improvement

After its construction, a solution is improved either by the VND or by a local search in one of the two neighborhoods Edge-Replace or Component-Renew (see Section 4.1) depending on the instance size. In the latter case the neighborhood Edge-Replace is chosen with probability 0.8 because of its usually higher performance.

5.4 Depositing Pheromones

After each ant constructed and improved a solution the pheromone values are updated. Here the mixed strategy suggested in [16] is used: At the beginning

Table 1. Comparison of GRASP+VND, GVNS and MMAS on random instance sets with 500 and 1000 nodes (B: delay-bound, $\overline{c}$: average final objective values, σ: standard deviations; CPU time limit: 300 seconds; best results are printed bold)

		B	6		20		50		100	
		$\alpha/sr/p$	$\overline{c}$	σ	$\overline{c}$	σ	$\overline{c}$	σ	$\overline{c}$	σ
R500	G+V	0.25	8997.3	672	2048.3	87	942.1	37	616.3	14
	GVNS	dynamic	8703.0	620	1961.5	88	901.1	35	601.1	14
		0.05	8701.1	617	1947.2	88	897.7	35	601.1	15
		0.1	8691.7	618	1938.9	89	893.7	34	599.3	14
		0.15	**8691.6**	618	1942.7	88	894.0	34	599.2	14
		0.2	8696.1	618	1947.8	90	896.4	35	599.8	14
	MMAS	0.6	8727.5	616	1937.4	85	891.4	34	598.0	13
		0.7	8726.4	614	1935.1	85	889.5	34	597.1	12
		0.8	8723.4	612	1932.1	84	**887.5**	34	**596.8**	13
		0.9	8722.4	613	**1930.8**	82	891.2	36	602.2	14
		0.95	8720.4	610	1941.3	83	914.9	40	612.2	14
R1000	G+V	0.25	9775.3	487	2473.0	76	1290.3	31	1026.8	9
	GVNS	dynamic	9497.9	486	2377.7	81	1257.4	33	1020.0	7
		0.05	9397.2	476	2346.9	80	1253.4	31	1020.1	7
		0.1	9412.1	480	2353.6	78	1252.5	31	**1019.4**	7
		0.15	9455.2	487	2365.8	80	1254.7	31	1019.6	7
		0.2	9488.3	485	2374.3	80	1257.0	32	1019.9	7
	MMAS	0.5	9385.3	485	2323.4	75	1243.7	28	1021.3	8
		0.6	9378.4	483	2312.4	73	1239.3	27	1020.9	8
		0.7	9376.4	481	**2308.8**	73	**1238.2**	27	1022.1	10
		0.8	9369.1	478	2309.9	74	1241.3	31	1028.8	14
		0.9	**9367.7**	477	2320.9	76	1281.3	46	1042.8	14

of the search only the best ant of the current iteration is allowed to deposit its pheromones. Later in the search process intensification has higher priority in order to concentrate on the surrounding of the so far best solution. This leads to the following update strategy: The more iterations have been performed, the higher the frequency of reinforcing the pheromone trail of the so far best solution instead of the iteration best one. More precisely, an instance-dependent number of iterations $I = \frac{50000}{|V|}$ is defined. In iterations $[1, I]$ only the iteration best solution deposits pheromones, in $(I, 2I]$ the so far best solution is chosen every fifth iteration, in $(2I, 3I]$ every third iteration, in $(3I, 6I]$ every other iteration and in $(6I, \infty)$ every time. A predefined pheromone decay coefficient p controls the evaporation and enforcement of the pheromone values.

6 Experimental Results

All tests have been executed on a multicore system consisting of Intel Xeon E5540 processors with 2.53 GHz and about 3 GB RAM per core.

Since there are no existing benchmark sets in literature with appropriate graph sizes, we tested our approaches to self-made instance sets called R500 and R1000, now available at `http://www.ads.tuwien.ac.at/~marior/instances/`, each

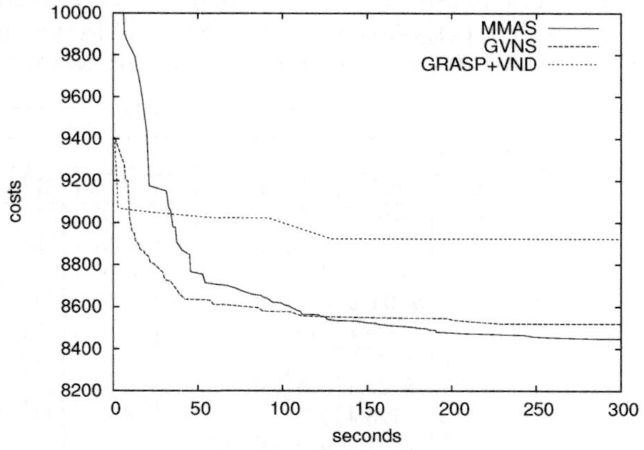

Fig. 1. Typical run characteristics of all three heuristics applied on a R1000 instance with $B = 6$ with a time limit of 300 seconds

containing 30 complete instances with 500 and 1000 nodes, respectively, random integer edge costs and delays uniformly distributed in $[1, 99]$. Depending on the chosen delay-bound B more or less edges can be discarded a priori if $d(e) > B$ (see Section 3.1).

Each result presented in Table 1 is derived from 30 runs with a CPU time limit of 300 seconds for each of the 30 instances. Three metaheuristics are included in the comparison: GRASP+VND from [6], which so far was the best heuristic, and the new GVNS and MMAS (with 5 ants).

All preprocessing steps except the search for alternate constrained paths (see Section 3.2) have been applied before starting the search reducing the complexity of the instances significantly. When having higher time limits the omitted preprocessing step might be advantageous.

The GRASP+VND approach was outperformed by almost all GVNS and MMAS runs almost independent of the parameter settings, which might be explained by the fact that the GRASP+VND has no memory. It "forgets" the solutions of past iterations and therefore cannot build on already obtained information. The MMAS mostly produces the best solutions probably because it has the most effective memory of all three methods in terms of the pheromone values containing the information of many solutions in one data structure.

A matter of high importance when using a MMAS is the fact that it can typically only exhibit its full effectiveness on a rather high number of iterations because of the longer exploration phase in the beginning [16], see Figure 1. When considering the R1000 instances, a full VND improvement of each constructed solution takes much time which leads to a smaller number of iterations within the time limit. The pheromone values therefore have not enough time to converge and the solutions are constructed rather randomly. So a faster local search instead of the VND produces in general worse solutions in a single iteration but yields higher quality in the end due to the higher number of iterations. For the smaller R500 instances it is better to use the VND improvement since it is fast enough to allow a sufficient number of iterations before time is running out.

When considering strict delay-bounds finding feasible solutions is more difficult and therefore the search gets caught in a local optimum more easily. Rather big changes have to be made to catapult the solution to another basin of attraction. Small changes like replacing a single edge to decrease the costs are often not possible because of a lack of residual delay in the nodes. So choosing the wrong way in the beginning of the search has more impact on the quality of the finally best solution than in cases with looser bounds. This fact can be observed in Table 1 independent of the heuristic method: The stricter the bound the higher the standard deviations.

A too small pheromone decay coefficient in the MMAS causes a fast convergence of the pheromone values and thus disregards diversification; a too high p-value has the opposite effect: the exploration phase lasts too long especially when having a tight time limit.

Relating to the MMAS results the following observation can be made: When tightening the delay-bound the p-value has to be increased to obtain better results. This behavior is another consequence of the facts mentioned above: When having strict bounds the quality and structure of the produced solutions varies much more and by using a higher p-value a single bad solution has not that much influence on the pheromone values because of the smoother evaporation of already deposited pheromones.

Generally speaking, the whole parameter setting of the MMAS heavily depends on the predefined target runtime. Here the small number of five ants speeds up the evolution of the pheromone values which is necessary for the time limit of 300 seconds. Regarding the final results it is not disadvantageous that possibly worse solutions are allowed to update pheromone values.

Applying Wilcoxon signed-rank tests with confidence level 0.95 to the results for $\alpha = 0.25$ (GRASP), $sr = 0.1/0.05$ (GVNS on R500/R1000) and $p = 0.8/0.7$ (MMAS on R500/R1000) yields the following error probabilities P: GVNS and MMAS produce better results than GRASP with $P = 2.2 \cdot 10^{-16}$ for all bounds. MMAS performs better than GVNS on R500 instances with $P = 1$ ($B = 6$), $P = 2.6 \cdot 10^{-13}$ ($B = 20$), $P = 2.2 \cdot 10^{-16}$ ($B = 50, 100$) and on R1000 instances with $P = 7 \cdot 10^{-16}$ ($B = 6$), $P = 2.2 \cdot 10^{-16}$ ($B = 20, 50$), $P = 1$ ($B = 100$).

7 Conclusion

We presented two metaheuristic approaches for solving the rooted delay-constrained minimum spanning tree problem, both outperforming existing approaches. The GVNS benefits from sophisticated neighborhood structures and various shaking moves to quickly converge to high quality solutions, whereas the MMAS with its balanced mix of diversification and intensification built on a fast and diverse solution construction extended with efficient local improvement methods could even exceed the results of the GVNS in most cases.

References

1. Dahl, G., Gouveia, L., Requejo, C.: On formulations and methods for the hop-constrained minimum spanning tree problem. In: Handbook of Optimization in Telecommunications, pp. 493–515. Springer Science + Business Media (2006)
2. Gouveia, L., Paias, A., Sharma, D.: Modeling and Solving the Rooted Distance-Constrained Minimum Spanning Tree Problem. Computers and Operations Research 35(2), 600–613 (2008)
3. Gilbert, E.N., Pollak, H.O.: Steiner minimal trees. SIAM Journal on Applied Mathematics 16(1), 1–29 (1968)
4. Prim, R.C.: Shortest connection networks and some generalizations. Bell System Technical Journal 36, 1389–1401 (1957)
5. Salama, H.F., Reeves, D.S., Viniotis, Y.: The Delay-Constrained Minimum Spanning Tree Problem. In: Blum, C., Roli, A., Sampels, M. (eds.) Proceedings of the 2nd IEEE Symposium on Computers and Communications, pp. 699–703 (1997)
6. Ruthmair, M., Raidl, G.R.: A Kruskal-Based Heuristic for the Rooted Delay-Constrained Minimum Spanning Tree Problem. In: Moreno-Díaz, R., Pichler, F., Quesada-Arencibia, A. (eds.) EUROCAST 2009. LNCS, vol. 5717, pp. 713–720. Springer, Heidelberg (2009)
7. Kruskal, J.B.: On the shortest spanning subtree of a graph and the traveling salesman problem. Proceedings of the American Mathematics Society 7(1), 48–50 (1956)
8. Skorin-Kapov, N., Kos, M.: A GRASP heuristic for the delay-constrained multicast routing problem. Telecommunication Systems 32(1), 55–69 (2006)
9. Xu, Y., Qu, R.: A GRASP approach for the Delay-constrained Multicast routing problem. In: Proceedings of the 4th Multidisplinary International Scheduling Conference (MISTA4), Dublin, Ireland, pp. 93–104 (2009)
10. Ghaboosi, N., Haghighat, A.T.: A Path Relinking Approach for Delay-Constrained Least-Cost Multicast Routing Problem. In: 19th IEEE International Conference on Tools with Artificial Intelligence, pp. 383–390 (2007)
11. Qu, R., Xu, Y., Kendall, G.: A Variable Neighborhood Descent Search Algorithm for Delay-Constrained Least-Cost Multicast Routing. In: Proceedings of Learning and Intelligent OptimizatioN (LION3), pp. 15–29. Springer, Heidelberg (2009)
12. Leggieri, V., Haouari, M., Triki, C.: The Steiner Tree Problem with Delays: A Tight Compact Formulation and Reduction Procedures. Technical report, Universita del Salento, Lecce, Italy (2010)
13. Dijkstra, E.W.: A note on two problems in connexion with graphs. Numerische Mathematik 1(1), 269–271 (1959)
14. Gouveia, L.: Multicommodity flow models for spanning trees with hop constraints. European Journal of Operational Research 95, 178–190 (1996)
15. Hansen, P., Mladenović, N.: Variable neighborhood search: Principles and applications. European Journal of Operational Research 130(3), 449–467 (2001)
16. Stützle, T., Hoos, H.: MAX–MIN ant system. Future Generation Computer Systems 16, 889–914 (2000)
17. Duff, I., Erisman, A., Reid, J.: Direct methods for sparse matrices. Oxford University Press, USA (1989)
18. Gruber, M., van Hemert, J., Raidl, G.R.: Neighborhood Searches for the Bounded Diameter Minimum Spanning Tree Problem Embedded in a VNS, EA, and ACO. In: Proceedings of the Genetic and Evolutionary Computation Conference, vol. 2, pp. 1187–1194. ACM, New York (2006)

Adaptive Modularization of the MAPK Signaling Pathway Using the Multiagent Paradigm

Abbas Sarraf Shirazi[1], Sebastian von Mammen[1], and Christian Jacob[1,2]

[1] Dept. of Computer Science, Faculty of Science
[2] Dept. of Biochemistry & Molecular Biology, Faculty of Medicine
University of Calgary, Canada
asarrafs@ucalgary.ca, s.vonmammen@ucalgary.ca, cjacob@ucalgary.ca

Abstract. We utilize an agent-based approach to model the MAPK signaling pathway, in which we capture both individual and group behaviour of the biological entities inside the system. In an effort to adaptively reduce complexity of interactions among the simulated agents, we propose a bottom-up approach to find and group similar agents into a single module which will result in a reduction in the complexity of the system. Our proposed adaptive method of grouping and ungrouping captures the dynamics of the system by identifying and breaking modules adaptively as the simulation proceeds. Experimental results on our simulated MAPK signaling pathway show that our proposed method can be used to identify modules in both stable and periodic systems.

1 Introduction

A signaling pathway is a process by which a cell transfers information from its external receptors to a target inside [1]. It usually consists of a cascade of biochemical reactions carried out by enzymes. From a software engineering point of view, a signaling pathway description is similar to a UML diagram describing which components interact in the cascade. Playing a key role within the cell cycle, the Mitogen-Activated Protein Kinase (MAPK) pathway is one of the most documented signaling pathways in the literature.The MAPK pathway creates responses to extracellular stimuli and regulates cellular activities, such as gene expression, mitosis, differentiation, etc [2].

A multiagent system (MAS) can be composed of a number of agents interacting with their neighbours as well as their environment. This paradigm is a promising approach to model a biological system in which there are different entities that interact locally [3,4,5,6]. One of the key challenges associated with multiagent modeling is its high computational cost. Therefore, there should be a mechanism for efficient usage of computational resources. Modularization is such a mechanism in which the average behaviour of similar processes is learned, thus creating a higher-level algorithmic representation, which is then used instead of the original, more elementary processes. However, in cases when the behaviour

R. Schaefer et al. (Eds.): PPSN XI, Part II, LNCS 6239, pp. 401–410, 2010.

of agents changes over time, a static modularization cannot be used. Instead, the model should have the ability of adaptive modularization, in order to properly reflect the dynamics of the underlying system.

The goal of this work is to propose such a modularization method and demonstrate its effectiveness by example of the MAPK signaling pathway. To achieve this objective, there are various issues to be addressed. The first issue is how to group different agents into a module and learn their behaviour. Another issue is whether and how to break or integrate modules whenever the dynamics of the system is changed. Furthermore, the transition between different states of the model should be seamless. This research will shed light on how to build a smooth transition between various models in a complex and multiscale model and therefore will serve as the first step to multiscale modeling of biological systems using multiagent systems.

The remainder of this paper is organized as follows. Section 2 reviews related work in the field of multiagent modeling of biological systems and multiscale modeling. Section 3 presents the details of our proposed method. Section 4 reports on the experiments conducted to demonstrate the performance of the proposed method. Finally, concluding remarks are presented in Section 5.

2 Related Work

Amigoni and Schiaffonati [1] present a thorough analysis of multiagent-based simulation of biological systems. In particular, they discuss three different multiagent approaches to model the MAPK signaling pathway. The first approach [2], models every chemical reaction as agents, while the approach proposed in [7] defines a multiagent system in which each intracellular component is an agent that uses a blackboard mechanism to interact with other agents in the system. The third approach [8], models each molecular entity as an agent. In this model, a reaction is implemented as messages communicated among the agents.

A modularization approach for the MAPK signaling pathway is presented in [9]. It works by finding the node with the maximum number of neighbours in the biological interaction network. Further expansion of this node into a subgraph is called a module. To this end, it is assumed that the graph of the network is known beforehand and a static graph analysis is performed. Despite being a static and intuitive algorithm, it serves as a starting point for the modularization part of this research toward a multiscale model.

Another approach which is proposed by Papin *et al.* [10] tries to find modules in an unbiased fashion using mathematically based definitions. These authors reviewed three different approaches to calculate correlated reaction sets (Co-Sets). Co-Sets are groups of reactions in a network whose functional states are similar. Network-based pathways methods like elementary modes [11] and extreme pathways [12] aim to optimize a flux-balance equation by finding sets of similar nodes. Another method is referred to as the flux coupling finder, which also minimizes or maximizes the ratio between all pair-wise combinations of nodes [13]. Finally, a correlation coefficient is defined between each pair of nodes in the network based on their reaction fluxes [14].

As the technology advances, the prospect of making a multiscale model becomes more prominent. In [15], different issues and trends in multiscale modeling of complex biological systems are addressed. In [16], a software framework for multiscale model integration and simulation is proposed; however, no specific modeling techniques are described. There are a few physical multiscale models, e.g. CPM [17], and Synergetics [18]. However, as of yet, there is no universally adopted theoretical or computational framework for the assembly of multiscale biological models [19].

Bassingthwaighte *et al.* identify a systems approach for developing multiscale models which includes six steps [20]: (1) defining the model as its highest level of resolution, (2) designing reduced-form modules, (3) determining the range of validity of the reduced form modules, (4) monitoring the variables of the system, (5) replacing higher resolution models with reduced form modules, and finally, (6) validating the performance of the multiscale model against available real data. They further identify issues that must be addressed by any attempt to multiscale modeling. Examples of these issues are parameter identification of closed-loop systems, the identification of input-output delays, and the imposition of known constraints. Their work is among very few attempts to identify challenges ahead of multiscale modeling from a computer science perspective.

3 Adaptive Modularization in a Multiagent Environment

Modularization is the process of identifying modules within a network that are functionally similar. By replacing the behaviour of individual nodes with the behaviour of their enclosing module, the complexity of the network will be reduced. This way, a large network can be efficiently analyzed using a reduced number of nodes. Modularization is usually a static process in which modules are found before the simulation starts. Furthermore, most modularization approaches assume that the agent interaction graph is completely known as a whole. This assumption is restrictive, especially in the case of extended networks where the number of nodes is very large. Furthermore, analyzing the global graph is not scalable, since with the introduction of each new node the analysis must be performed again. As a result, we propose that the multiagent paradigm can be used to tackle the problem of scalability and also complexity of large graphs.

A multiagent system usually has no top-down control unit, which operates on the whole system. Agents cooperate or compete autonomously to perform various tasks. Contrary to traditional systems, a MAS agent only knows about its local interactions. Consequently, agents can form their local directed graph of interaction. Agents can cooperate and share their information (in this case, their interaction graph) with other agents. This way, they can form groups or modules in a bottom-up fashion.

In our proposed approach, we aim to find, integrate and break modules dynamically as the simulation proceeds. Based on the system dynamics, we expect our algorithm to find different modules that act together over a period of time. To this end, we must address several issues as described in [20]. How and when

Algorithm 1. Module Indentification

```
m = current_module;
Module new_module;
Queue q;
q.Enqueue(m);
new_module.Add(m);
while !q.empty() do
  Module head = q.Dequeue();
  for all Agent s in head do
    for all Agent t in s.Neighbours()
    do
      if |ρ_st| ≥ τ_edge then
        new_module.Add(t);
        q.Enqueue(t);
      end if
    end for
  end for
end while
return new_module;
```

Algorithm 2. Validity Monitoring

```
m = current_module;
needToBreak = false;

for all Agent s in m do
  for all Agent t in s.Neighbours() do
    if |ρ_st − ρ'_st)| > τ_valid then
      needToBreak = true;
      break;
    end if
  end for
end for

if needToBreak then
  simulation.remove(m);
  for all Agent s in m do
    simulation.add(s);
  end for
end if
```

to integrate nodes to form a module, how to learn the behaviour of a module, and how to monitor the validity of modules are among the issues that we address in this section.

3.1 Creating Modules

In our system, agents are associated with an interaction graph as well as an interaction history for all their neighbours. The weight of an edge in their interaction graph is equal to their correlation coefficient with their neighbour. A correlation coefficient between two statistical variables indicates their linear dependancy. A zero correlation coefficient means that two variables are independent, while +1 or -1 shows highly correlated variables. The more two variables are correlated, the more similar their function is. In case there is a series of n measurements of agents s and t in the form of s_i and t_i, where $i = 1, 2, ..., N$, their correlation coefficient (ρ_{st}) is defined as follows:

$$\rho_{st} = \frac{\sum_{i=1}^{N}(s_i - \bar{s})(t_i - \bar{t})}{(n-1)\sigma_s\sigma_t} \tag{1}$$

where $\bar{s}$ and $\bar{t}$ are the mean values, and σ_s and σ_t are standard deviations of s and t, respectively.

Having a local weighted graph, each agent then periodically checks if its correlation coefficient with each neighbour is greater than some threshold (τ_{edge}). If so, they form an initial module and repeat this process to identify a cluster

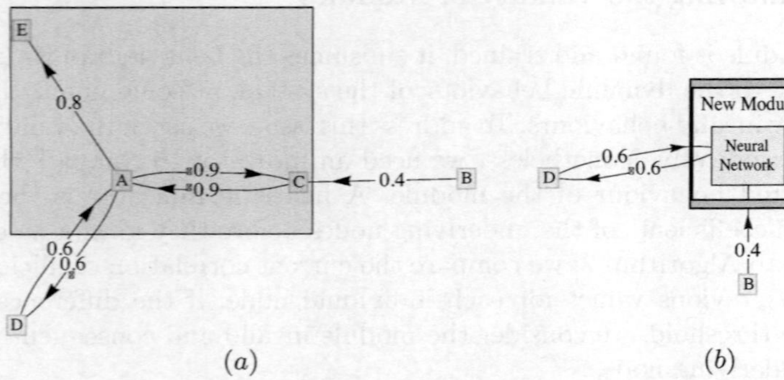

Fig. 1. Example of an interaction graph. The edges denote the correlation coefficients. (a) *Agent A*, *Agent C*, and *Agent E* form a module, (b) The new neighbours of this module are *Agent B* and *Agent D*.

of agents that are highly correlated (Algorithm 1). Fig. 1 shows an example in which *Agent A* finds *Agent C* and *Agent E*, and they form a module. The set of new neighbours is the union of all neighbours of the underlying nodes. Having formed such a module, the next step is to train this new module, so that it learns and imitates the group behaviour of its underlying nodes.

3.2 Learning the Group Behaviour

A module has to subsume the behaviour of its underlying nodes by abstracting from their behaviour. In other words, the new module has to replace its associated nodes and produce the same outputs as if there were individual agents in the system. The learning algorithm can employ neural networks, time series, or any other function approximation algorithm. No matter what learning algorithm is used, each node has to have an interaction history to be used during the learning phase. In our approach, we used a three-layer feed-forward neural network with back propagation learning algorithm [21] to train the network. This way, we also have control over the speed of learning.

The structure of the neural network is determined by its inputs, the number of nodes in the hidden layer, and the outputs. Since in our model, agents are not aware of their dependent agents (in fact, they only know about their outgoing edges), the output of the network should simply be all of the underlying nodes (in the example of Fig. 1, outputs are *Agent A*, *Agent C*, and *Agent E*). Regarding the input to the network, there are different design choices. The first one would be to assign external incoming edges and ignore internal connections (*Agent D* in Fig. 1). An alternative approach is to consider internal nodes as well. This way, the neural network has more meaningful sets of data to be trained with. As for the number of nodes in the hidden layer, we follow a simple rule-of-thumb and assign it to be the number of inputs + 2.

3.3 Monitoring the Validity of Modules

Once a module is found and trained, it subsumes the behaviour of its underlying nodes. Due to the dynamic behaviour of the system, at some point, the module might show invalid behaviours. To address this issue, we check the validity of each module periodically. Nonetheless, we need an indicator to compare the current and expected behaviour of the module. A heuristic indicator is the previous correlation coefficients of the underlying nodes before they form a module (ρ'_{st}). According to Algorithm 2, we compare the current correlation coefficients of the module to previous values for each individual node, if the difference is larger than some threshold, we consider the module invalid and consequently break it into its underlying nodes.

4 Experiments on the MAPK Signaling Pathway

Our proposed adaptive modularization approach can be employed in any system where there are different agents interacting locally. Signal transduction pathways are such ideal candidates, as for most of them there is quantized data available. In general, a signal transduction pathway starts with an external stimulus in a cascade of biochemical processes, which in turn results in a change of state in a cell. In the MAPK signaling pathway [22], a hypothetical enzyme E1 stimulates the cell and results in an increase in production of MAPK-PP enzyme (Fig. 2(a)). In another model [23], a negative feedback loop causes sustained oscillations in the production of MAPK-PP (Fig. 2(b)).

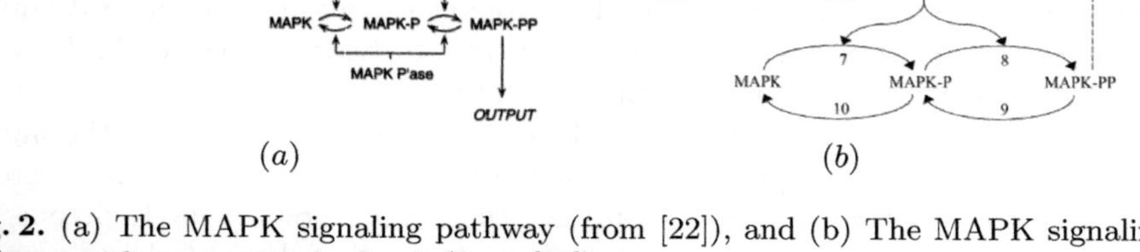

Fig. 2. (a) The MAPK signaling pathway (from [22]), and (b) The MAPK signaling pathway with a negative feedback (from [23])

4.1 Agent-Based Model of the MAPK Signaling Pathway

Contrary to the differential equation-based approach discussed above, in our model each substance is considered to be an independent entity which is loosely

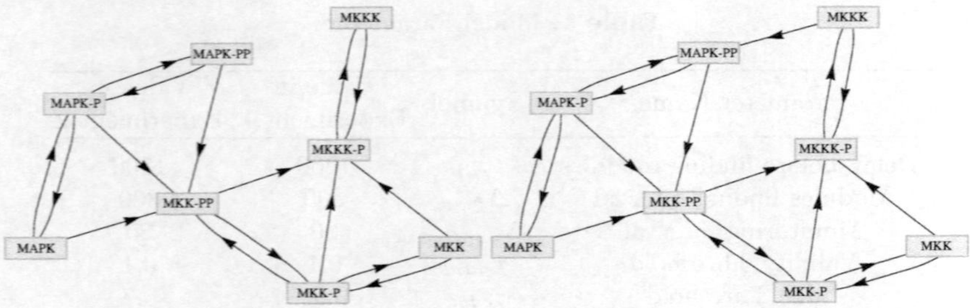

(a) (b)

Fig. 3. Agent graphs for the MAPK signaling pathways of Fig. 2

defined as an agent. For each agent, the interaction graph defines its relations with the substances that appear in its update formula[1]. Fig. 3 shows the complete interaction graph for the signaling pathways of Fig. 2.

To validate the performance of the adaptive modularization, we conducted a series of experiments on both MAPK models. Essentially, there are five parameters in our algorithm which are summarized in Table 1. We let the system run in its normal mode for some time (t_{wait}) and then start looking for modules within a time interval (Δ_{find}). t_{wait} is important in that the system has to reach a rather stable condition before the modularization algorithm starts to work. We keep monitoring the system also in predefined intervals ($\Delta_{monitor}$). A module is valid as long as its correlation coefficients with its neighbours do not vary too much with regards to those of individual agents (τ_{valid}). Finally, to integrate nodes and find modules, the value of an edge in the interaction graph should be greater than some threshold (τ_{edge}). τ_{valid} and τ_{edge} have been found through trial and error. A more detailed exploration of the parameter spaces will be undertaken in our future work.

Fig. 4(a) shows the result of applying our approach to the first model (Fig. 2(a)) in terms of the number of modules. Initially, each agent is its own module. The identification of modules starts after $t = 1200$. The process of construction and deconstruction of modules results in the emergence of a periodic pattern. The reason is that whenever a module is broken, all of its underlying nodes start to work as individual agents again. Naturally, when a module contains a larger number of nodes, the probability of that module to become invalid is higher. In other words, since there is no hierarchical learning, after an all-encompassing single module is created and it breaks, there are again eight individual modules (one for each agent) in the system. As this modularization/demodularization process continues, a periodic pattern appears as illustrated in Fig. 4(a). Fig. 4(b) shows that the final concentration successfully resembles that of the PDE solver.

[1] The complete set of update equations can be found in [22] and [23].

Table 1. Model Parameters

Parameter Name	Symbol	Value in Experiment 1	Value in Experiment 2
Delay before finding modules	t_{wait}	1200	1500
Modules finding interval	Δ_{find}	300	300
Monitoring interval	$\Delta_{monitor}$	20	20
Validity Threshold	τ_{valid}	0.1	0.1
Edge Threshold	τ_{edge}	0.95	0.7

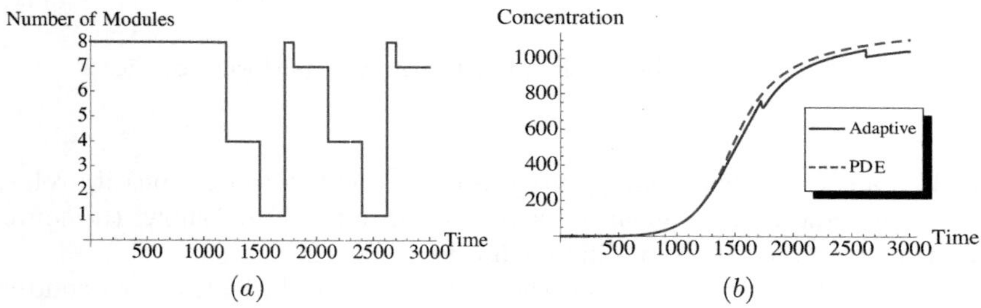

Fig. 4. Adaptive modularization results for the first MAPK pathway model of Fig. 2(a). (a) Number of agents, (b) Concentration of MAPK-PP.

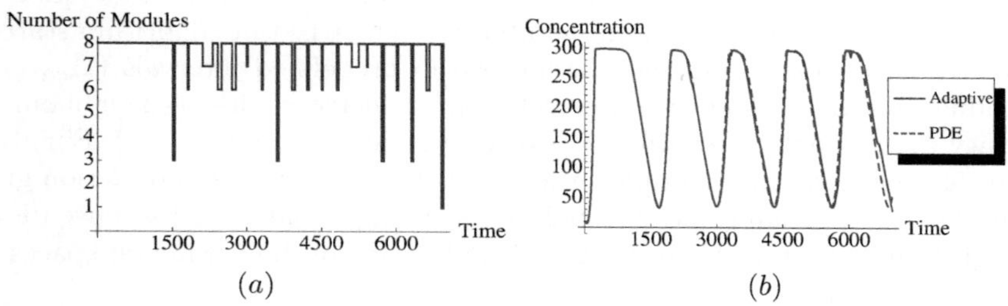

Fig. 5. Adaptive modularization results for the second MAPK pathway model of Fig. 2(b). (a) Number of agents, (b) Concentration of MAPK-PP.

Fig. 5 shows the result of adaptive modularization for the second MAPK pathway. Since this model is periodic, the adaptive modularization algorithm successively finds, trains, and breaks modules over time. The number of spikes in Fig. 5(a) shows that the validity period of a composite module is not long enough. The reason is that the correlation coefficient is a linear indicator which varies from -1 to +1 over a periodic signal. This variation makes a module in a periodic system invalid. This result suggests that we have to look for other parameters when we have a nonlinear system with feedback. Nevertheless, it is still more reliable than if modules were broken at random.

5 Conclusion and Future Works

In this paper, we introduced a bottom-up method to reduce the complexity of a multiagent system simulating the MAPK signaling pathway by adaptive modularization and demodularization. Although we have shown that this approach works very well for this specific example, we believe that our module composition and decomposition algorithm can be applied to a wide range of other multiagent systems. In particular, individual agents share their interaction graph to build a higher-level module which subsumes their behaviour. After a new module is formed, it learns the behaviour of its underlying nodes using a feed-forward neural network. To monitor the validity of a module, the values of any edge in its interaction graph is checked – at defined intervals – and compared against the previous values of its nodes. A module is broken if the difference between the current and previous value of an edge is greater than some threshold.

We use correlation coefficients to determine the edge value in the agent's interaction graphs. Although this indicator is mathematically sound, it does not capture the nonlinear dependance between agents. Looking for other nonlinear indicators seems to be a promising approach. This work is among very few attempts to find an algorithmic framework to address the complexity reduction in an agent-based system and can serve as the first step to address the reduction of complexity in highly complex systems.

References

1. Amigoni, F., Schiaffonati, V.: Multiagent-based simulation in biology a critical analysis. Model-Based Reasoning in Science, Technology, and Medicine 64, 179–191 (2007)
2. Desmeulles, G., Querrec, G., Redou, P., Kerdelo, S., Misery, L., Rodin, V., Tisseau, J.: The virtual reality applied to biology understanding: The in virtuo experimentation. Expert Syst. Appl. 30(1), 82–92 (2006)
3. Hoar, R., Penner, J., Jacob, C.: Transcription and evolution of a virtual bacteria culture. In: IEEE Congress on Evolutionary Computation, Canberra, Australia. IEEE Press, Los Alamitos (2003)
4. Jacob, C., Burleigh, I.: Genetic programming inside a cell. In: Yu, T., Riolo, R.L., Worzel, B. (eds.) Genetic Programming Theory and Practice III, pp. 191–206. Springer, Heidelberg (2006)
5. Jacob, C., Barbasiewicz, A., Tsui, G.: Swarms and genes: Exploring λ-switch gene regulation through swarm intelligence. In: CEC 2006, IEEE Congress on Evolutionary Computation, Vancouver, BC, Canada (2006)
6. Jacob, C., Burleigh, I.: Biomolecular swarms: An agent-based model of the lactose operon. Natural Computing 3(4), 361–376 (2004)
7. Gonzalez, P.P., Cardenas, M., Camacho, D., Franyuti, A., Rosas, O., Lagunez-Otero, J.: Cellulat: an agent-based intracellular signalling model. Biosystems 68(2-3), 171–185 (2003)
8. Khan, S., Makkena, R., McGeary, F., Decker, K., Gillis, W., Schmidt, C.: A multi-agent system for the quantitative simulation of biological networks, pp. 385–392 (2003)

9. Nayak, L., De, R.K.: An algorithm for modularization of mapk and calcium signaling pathways: comparative analysis among different species. J. Biomed. Inform. 40(6), 726–749 (2007)
10. Papin, J.A., Reed, J.L., Palsson, B.O.: Hierarchical thinking in network biology: the unbiased modularization of biochemical networks. Trends Biochem. Sci. 29(12), 641–647 (2004)
11. Schuster, S., Dandekar, T., Fell, D.A.: Detection of elementary flux modes in biochemical networks: a promising tool for pathway analysis and metabolic engineering. Trends Biotechnol. 17(2), 53–60 (1999)
12. Schilling, C.H., Letscher, D., Palsson, B.O.: Theory for the systemic definition of metabolic pathways and their use in interpreting metabolic function from a pathway-oriented perspective. J. Theor. Biol. 203(3), 229–248 (2000)
13. Burgard, A.P., Nikolaev, E.V., Schilling, C.H., Maranas, C.D.: Flux coupling analysis of genome-scale metabolic network reconstructions. Genome. Res. 14(2), 301–312 (2004)
14. Price, N.D., Schellenberger, J., Palsson, B.O.: Uniform sampling of steady-state flux spaces: means to design experiments and to interpret enzymopathies. Biophys. J. 87(4), 2172–2186 (2004)
15. Martins, M., Ferreira Jr., S.C., Vilela, M.: Multiscale models for biological systems. Current Opinion in Colloid & Interface Science 15(1-2), 18–23 (2010)
16. Erson, E.Z., Cavuşoğlu, M.C.: A software framework for multiscale and multilevel physiological model integration and simulation. In: Conf. Proc. IEEE Eng. Med. Biol. Soc., vol. 2008, pp. 5449–5453 (2008)
17. Merks, R.M., Glazier, J.A.: A cell-centered approach to developmental biology. Physica A: Statistical Mechanics and its Applications 352(1), 113–130 (2005)
18. Haken, H.: Synergetics: an introduction: monequilibrium phase transitions and self-organization in physics, chemistry and biology / Hermann Haken. Springer, Berlin (1977)
19. Walker, D.C., Southgate, J.: The virtual cell–a candidate co-ordinator for 'middle-out' modelling of biological systems. Briefings in Bioinformatics 10(4), 450–461 (2009)
20. Bassingthwaighte, J., Chizeck, H., Atlas, L.: Strategies and tactics in multiscale modeling of cell-to-organ systems. Proceedings of the IEEE 94(4), 819–831 (2006)
21. Haykin, S.: Neural Networks: A Comprehensive Foundation. Macmillan, New York (1994)
22. Huang, C.Y., Ferrell, J.E.: Ultrasensitivity in the mitogen-activated protein kinase cascade. Proc. Natl. Acad. Sci. USA 93(19), 10078–10083 (1996)
23. Kholodenko, B.N.: Negative feedback and ultrasensitivity can bring about oscillations in the mitogen-activated protein kinase cascades. Eur. J. Biochem. 267(6), 1583–1588 (2000)

Experimental Comparison of Methods to Handle Boundary Constraints in Differential Evolution

Jarosław Arabas[1], Adam Szczepankiewicz[2], and Tomasz Wroniak[2]

[1] Institute of Electronic Systems,
Warsaw University of Technology, Poland
jarabas@elka.pw.edu.pl
[2] Faculty of Electronics and Information Technology,
Warsaw University of Technology, Poland

Abstract. In this paper we show that the technique of handling boundary constraints has a significant influence on the efficiency of the Differential Evolution method. We study the effects of applying several such techniques taken from the literature. The comparison is based on experiments performed for a standard DE/rand/1/bin strategy using the CEC2005 benchmark. The paper reports the results of experiments and provides their simple statistical analysis. Among several constraint handling methods, a winning approach is to repeat the differential mutation by resampling the population until a feasible mutant is obtained. Coupling the aforementioned method with a simple DE/rand/1/bin strategy allows to achieve results that outperform in many cases results of almost all other methods tested during the CEC2005 competition, including the original DE/rand/1/bin strategy.

1 Introduction

Differential Evolution (DE) [1, 2] has proved an efficient and a very simple algorithm from the Evolutionary Computation (EC) family. In particular, the DE scheme overcomes one of the basic difficulties that are met when tuning an Evolutionary Algorithm (EA), namely, the issue of a proper setting of the mutation range. In DE, mutation is based on differences between chromosomes contained in the population, and, since their distribution is influenced by the shape of the fitness function, the distribution of mutants reflects that shape as well. This effect has been called by Storn a "contour matching property".

The basic DE scheme has been introduced for an unconstrained optimization task, whereas in most of the engineering problems, there may appear additional constraints that any solution must satisfy. In general, handling constraints in EC is a problem itself, and a lot of research has been conducted to elaborate efficient methods of solving it. In this contribution we focus on applying DE in optimization problems with boundary constraints, where each coordinate of the solution must fit into the range between a lower and an upper bound. This form of constraints allows for very easy check for feasibility, and it is also very easy to define constraint handling strategies that are based on repairing. For

R. Schaefer et al. (Eds.): PPSN XI, Part II, LNCS 6239, pp. 411–420, 2010.

these reasons, the choice of a technique to handle boundary constraints does not appear to be a serious problem, and maybe this explains why we were unable to find a systematic study that addresses the issue of boundary constraints handling methods. We give an experimental evidence that this issue cannot be treated too easily, since it may have a surprisingly big influence on the DE effectiveness.

In the research papers devoted to DE we can find several techniques to handle boundary constraints. Price *et al.* [1] suggest that random reinitialization (that is, replacing an infeasible solution by a randomly initialized one) is "the most unbiased approach". This method has been also used e.g. in [3, 4]. Price *et al.* [1] defined also a "bounce back" strategy, where an infeasible solution y, generated by mutating a feasible solution x, is replaced by a new feasible solution located on a line between x and y. This approach has been applied e.g. in [5, 6]. Another approach to repairing infeasible solutions is to reflect them back from the bound [7]. Yet another possibility [8] is to project infeasible solutions on bounds, which consists in changing each parameter that exceeds a boundary value to a new value which equals the boundary.

Some optimization tasks, e.g. digital filter design, are periodic in nature. Then, search can be constrained to parameter values from a certain base interval covering a single period, which are treated feasible. Values outside that range can be shifted by an integer multiple of the interval length to fit it the feasible area [9]. We shall call this strategy a wrapping approach.

The motivation for this paper was given by studying the results of the CEC2005 competition of optimization algorithms [10]. In the CEC2005 proceedings, 18 papers took part in the competition. Among the submitted papers, some authors did not report details of handling constraints, and those who reported used various techniques. In the group of algorithms that performed quite well, one can find a very basic DE method [7]. In this paper we investigate if the results of this method can be improved by changing the method of handling boundary constraints.

2 Differential Evolution Algorithm

The outline of the DE algorithm is depicted in Fig. 1. With P^t we denote the population in the generation number t, and P_i^t stands for the i-th chromosome in the population P^t; finally, with $P_{i,j}^t$ and x_j we denote the j-th coordinate of the chromosome $P_{i,j}^t$ and x, respectively. All chromosomes are n-dimensional vectors of real numbers, and μ is the size of the population P^t.

The algorithm minimizes the fitness function $f : R^n \to R$. We assume that for each dimension, a pair of numbers l_i, u_i is defined which are lower and upper bound of the feasible area in the i-th dimension. Thus, the feasible area is defined by two vectors composed of lower bound values l and upper bound values u.

In our study we considered a simple DE/rand/1/bin strategy which implements DE steps in the following way. In the initialization phase, population P^0 is filled with chromosomes that are generated with uniform distribution in the feasible area $[l, u]$. The selection rule returns a chromosome $x = P_i^t$ where the

```
algorithm DE with constraint handling
P^0 ← initialize(l, u)
repeat until stop condition met
    for i ∈ 1...μ
        x ← selection (P^t)
        v ← differential mutation(x, P^t)
        if v is feasible then w ← v
                else w ← repair(v, x, P^t)
        y ← crossover(P_i^t, w)
        P_i^{t+1} ← replacement(P_i^t, y)
    end for
    t ← t + 1
end repeat
```

Fig. 1. Outline of a basic DE algorithm

index i is a random variable with uniform distribution in $\{1, ..., \mu\}$. Differential mutation uses two chromosomes P_j^t and P_k^t whose indexes are also random variables with uniform distribution in $\{1, ..., \mu\}$. In the result, a chromosome v is generated according to the formula

$$v = x + F(P_j^t - P_k^t) \tag{1}$$

where F is a scaling factor defined by the user. When a constrained optimization problem is considered, chromosome v may become infeasible, therefore a repairing strategy is applied that generates a new feasible chromosome w instead of v. In this paper we study several versions of the repair procedure, and those versions are discussed in the next subsection. The new chromosome w, which resulted from repairing the chromosome v, undergoes crossover with the chromosome P_i^t. In our study we focus on the binary crossover which is defined as follows

$$y_j = \begin{cases} w_j & \text{with probability } CR \\ P_{i,j}^t & \text{with probability } 1 - CR \end{cases} \tag{2}$$

where CR is a user-defined parameter. In the replacement step, i-th chromosome for the next generation results from the tournament of chromosomes P_i^t and y:

$$P_i^{t+1} = \begin{cases} y & f(y) < f(P_i^t) \\ P_i^t & \text{otherwise} \end{cases} \tag{3}$$

Considered Constraint Handling Methods. In the presented study we compare the following methods of handling boundary constraints which we have found in the literature.

- *Conservatism:* If the differential mutation resulted in an infeasible chromosome, it is rejected and $w = x$. This strategy is equivalent to the assumption that the infeasible chromosome will not be repaired, but it will be rejected in the replacement phase.

- *Reinitialization:* Chromosome w is generated with the uniform distribution in the feasible area.
- *Reflection:* Chromosome w is generated by reflecting coordinate values from the exceeded boundary values

$$w_i = \begin{cases} v_i & l_i \leq v_i \leq u_i \\ 2u_i - v_i & v_i > u_i \\ 2l_i - v_i & v_i < l_i \end{cases} \tag{4}$$

  If the resulting chromosome is still infeasible, it is reflected again, and the procedure is repeated until feasibility is obtained.
- *Projection:* All coordinate values that exceed bounds are trimmed to the boundary values

$$w_i = \begin{cases} v_i & l_i \leq v_i \leq u_i \\ u_i & v_i > u_i \\ l_i & v_i < l_i \end{cases} \tag{5}$$

- *Wrapping:* All coordinate values that exceed the admissible range are shifted by an integer multiple of the range such that they become feasible

$$w_i = \begin{cases} v_i & l_i \leq v_i \leq u_i \\ v_i + k_i(u_i - l_i) & \text{otherwise} \end{cases} \tag{6}$$

  where k_i is an integer number that guarantees feasibility for the i-th dimension.
- *Resampling:* Selection of a random chromosome from P^t and its differential mutation is repeated until a feasible chromosome is obtained:

$$w = \text{differential mutation}(\text{selection}(P^t), P^t) \tag{7}$$

  This means that chromosomes x, P_j^t and P_k^t are selected anew by picking them with the uniform distribution from the population P^t.

3 Experiments and Results

Conditions of testing. We performed an experimental comparison using the suite of fitness functions that have been defined for the CEC2005 competition. Detailed description of functions and conditions of testing are defined in [11], therefore we provide only a brief information about the benchmark.

The benchmark is a compilation of 25 minimization problems that can be divided into four groups (in brackets we refer to function numbers assumed originally in the benchmark definition): 1) unimodal functions (f_1 up to f_5), 2) basic multimodal functions (f_6 up to f_{12}) which include functions by Ackley,

Griewank, Rastrigin, Rosenbrock, Schwefel, and Weierstrass, 3) expanded functions (f_{13}, f_{14}) which are combinations of Griewank, Rosenbrock and Shaffer functions, 4) composition functions (f_{15} up to f_{25}) which resulted from combining functions from groups 1) – 3). Most of the problems from that benchmark are defined for a shifted, rotated, and scaled coordinate system. With two exceptions (f_7 and f_{25}), all other problems have boundary constraints. Some of the benchmark functions have their global minimum on the boundary. Tests are performed for 10-dimensional and 30-dimensional problems.

For each optimization problem, the algorithm is run 25 times, and each time it returns the best value found in that run. For each benchmark function its global minimum is known, so the difference between the fitness value of the best chromosome and the true global minimum can be treated as the error of a single run. Thus, for 25 independent runs one obtains populations of 25 error values whose statistics are reported. For a better readability and compactness, in our study we decided to reduce the suggested set of reported error statistics to the mean value and standard deviation. They are computed for populations of solutions obtained by each of 25 independent runs after $n \cdot 10000$ evaluations of the fitness function.

All experiments were performed using a plain DE/rand/1/bin strategy assuming the following parameter values: population size $\mu = 10n$, scaling factor $F = 0.8$, binary crossover rate $CR = 0.9$. We applied these settings and performed testing with each of the aforementioned constraint handling techniques.

Results. Results obtained by DE/rand/1/bin with various constraint handling methods are reported in Tab. 1, 2 for all bounded CEC2005 problems. To facilitate the interpretation of results, for each test function we use the mean error values to assign ranks to all constraint handling methods under comparison. Then, for each constraint handling method we average its ranks over all test functions and we report these averaged rank values.

In addition to the mean value and standard deviation of the fitness function, we indicate the percentage of chromosomes that have been generated outside the admissible area and required repairing. We provide the mean values of the percentage of repaired chromosomes for each constraint handling method (averaged over all test functions). This indicates how effectively a constraint handling method avoids generation of infeasible chromosomes. We also report for each function how frequently a constraint handling method was used (averaging over all constraint handling methods), which informs about the importance of a proper choice of a constraint handling method for that particular function.

A general conclusion is that the choice of the constraint handling technique may significantly influence the final result. Still, for 10-dimensional problems it is not clear which strategy is the most efficient. For some functions, like f_1, f_2, f_{13}, f_{14}, application of any constraint handling technique yields similar results. Projection and reflection work well when the global minimum is located on bounds or close to them, which is the case of functions f_5,$f_{18} - f_{20}$. Reinitialization appears extremely effective for the Schwefel 1.2 problem (f_{12}). Resampling and conservatism are clearly winning strategies only for the

Table 1. Mean and standard deviation of results yielded by DE/rand/1/bin after 100000 fitness function evaluations for 10-dimensional problems from CEC2005

		projection	reflection	resampling	conservative	wrapping	reinitialize	% repaired
f_1	mean	7.89E-09	8.14E-09	7.73E-09	8.59E-09	**7.59E-09**	8.37E-09	8.1
	std	1.85E-09	1.26E-09	1.59E-09	9.57E-10	1.50E-09	1.10E-09	
f_2	mean	8.14E-09	8.65E-09	8.24E-09	8.15E-09	**7.73E-09**	8.52E-09	11.6
	std	1.62E-09	1.05E-09	1.61E-09	1.41E-09	1.39E-09	1.24E-09	
f_3	mean	1.61E-08	3.58E-08	**8.58E-09**	1.35E-07	5.33E-08	5.18E-08	15.3
	std	8.23E-09	3.81E-08	1.10E-09	1.21E-07	3.41E-08	4.10E-08	
f_4	mean	**8.18E-09**	8.21E-09	8.75E-09	8.24E-09	8.28E-09	**8.18E-09**	11.4
	std	1.34E-09	1.48E-09	1.12E-09	1.31E-09	1.26E-09	1.48E-09	
f_5	mean	**1.23E-08**	4.91E-05	3.59E-06	1.63E+00	1.33E+00	1.39E+00	67.9
	std	8.72E-09	3.88E-05	1.84E-06	6.88E-01	4.92E-02	6.01E-01	
f_6	mean	3.81E-06	1.59E-01	**1.69E-06**	1.74E-05	4.66E-06	1.59E-01	7.75
	std	6.10E-06	7.81E-01	2.00E-06	1.84E-05	3.76E-06	7.81E-01	
f_8	mean	**2.05E+01**	**2.05E+01**	2.06E+01	2.07E+01	**2.05E+01**	**2.05E+01**	15.0
	std	7.25E-02	7.35E-02	1.55E-01	1.22E-01	9.82E-02	9.41E-02	
f_9	mean	2.75E+01	3.34E+01	2.46E+01	2.45E+01	**2.32E+01**	2.52E+01	13.3
	std	1.94E+01	1.30E+01	1.32E+01	1.66E+01	1.71E+01	1.45E+01	
f_{10}	mean	2-19E+01	3.39E+01	2.42E+01	2.43E+01	2.49E+01	**1.95E+01**	11.7
	std	1.36E+01	1.45E+01	1.61E+01	1.60E+01	1.66E+01	1.55E+01	
f_{11}	mean	6.58E+00	7.35E+00	**4.12E+00**	4.71E+00	8.79E+00	8.11E+00	16.4
	std	2.78E+00	2.30E+00	3.55E+00	3.79E+00	1.38E+00	2.90E+00	
f_{12}	mean	1.04E+02	5.22E+00	6.23E+01	1.73E+02	5.54E+01	**4.00E-01**	13.3
	std	4.05E+02	8.11E+00	3.05E+02	4.61E+02	2.64E+02	1.96E+00	
f_{13}	mean	**2.76E+00**	2.77E+00	2.97E+00	3.00E+00	2.85E+00	2.79E+00	3.6
	std	1.21E+00	1.22E+00	1.11E+00	9.62E-01	1.12E+00	1.22E+00	
f_{14}	mean	3.80E+00	3.69E+00	3.78E+00	**3.61E+00**	3.94E+00	3.84E+00	17.5
	std	6.37E-01	5.20E-01	4.98E-01	6.86E-01	1.39E-01	4.36E-01	
f_{15}	mean	3.84E+02	3.02E+02	2.08E+02	2.59E+02	2.05E+02	**1.88E+02**	11.6
	std	9.23E+01	1.31E+02	1.05E+02	1.09E+02	1.13E+02	9.87E+01	
f_{16}	mean	1.55E+02	1.53E+02	1.47E+02	**1.35E+02**	1.48E+02	1.50E+02	13.0
	std	3.52E+01	3.16E+01	3.54E+01	2.90E+01	3.08E+01	2.95E+01	
f_{17}	mean	1.69E+02	1.78E+02	1.56E+02	1.60E+02	**1.55E+02**	1.61E+02	16.1
	std	2.81E+01	2.60E+01	2.62E+01	2.41E+01	2.35E+01	3.51E+01	
f_{18}	mean	**3.82E+02**	4.00E+02	8.00E+02	7.40E+02	8.00E+02	8.00E+02	12.0
	std	1.71E+02	2.00E+02	2.64E-11	1.62E+02	7.61E-10	7.15E-10	
f_{19}	mean	3.63E+02	**3.40E+02**	7.80E+02	7.40E+02	7.40E+02	8.00E+02	11.6
	std	1.50E+02	1.36E+02	9.80E+01	1.62E+02	1.62E+02	4.80E-10	
f_{20}	mean	**3.52E+02**	3.60E+02	8.00E+02	7.45E+02	7.64E+02	8.00E+02	11.5
	std	1.44E+02	1.62E+02	2.16E-11	1.66E+02	1.38E+02	1.35E-09	
f_{21}	mean	5.00E+02	5.00E+02	**4.36E+02**	4.52E+02	4.92E+02	4.92E+02	7.9
	std	8.28E-12	3.20E-12	9.33E+01	8.54E+01	3.92E+01	3.92E+01	
f_{22}	mean	7.83E+02	7.04E+02	6.80E+02	6.84E+02	**6.27E+02**	7.43E+02	12.0
	std	2.20E+01	1.76E+02	1.93E+02	1.92E+02	2.24E+02	1.31E+02	
f_{23}	mean	**5.59E+02**	**5.59E+02**	5.98E+02	6.34E+02	5.85E+02	5.72E+02	8.1
	std	1.44E-12	1.68E-12	1.05E+02	1.01E+02	5.93E+01	4.39E+01	
f_{24}	mean	2.14E+02	**2.00E+02**	**2.00E+02**	**2.00E+02**	**2.00E+02**	**2.00E+02**	8.5
	std	6.64E+01	3.15E-12	3.00E-12	9.25E-11	2.86E-11	3.94E-11	
avg. rank		2.87	3.35	2.74	3.39	3.00	3.43	
% repaired		15.5	12.4	14.7	15.0	13.6	13.7	

Weierstrass function f_{11}. Despite of local differences, average percentage of repaired chromosomes stays on roughly similar level for all methods under comparison.

In contrast to the 10-dimensional case, for 30-dimensional problems a clear picture is obtained. Resampling appears the most effective constraint handling strategy. Projection and reflection seem to be the methods of the second choice. Reinitialization performs surprisingly bad. As for the average percentage of infeasible chromosomes, its clear minimum value is recorded for the reinitialization, and for the other methods under comparison, this value is roughly similar. Thus we state a hypothesis that the good performance of the resampling strategy results from its efficiency in avoiding generation of infeasible chromosomes.

Resampling method is the only one where the repairing cost cannot be guaranteed. To estimate it, for each test function we computed the average number of trial chromosomes that are generated to repair a single infeasible chromosome. For 10-dimensional problems, the average repairing cost ranged from 2.3 to 6.8 (the mean value was 3.3), and for 30-dimensional problems, it ranged from 3.8 to 29.3 (the mean value was 6.7).

Comparison with other CEC2005 competitors. Interpretation of the presented results becomes easier after looking at Tab. 3 which contains a comparison of the results between DE/rand/1/bin with constraint handling by resampling and the algorithms that were found superior in the summary of the CEC2005 competition [10]. Those algorithms are: DE variants — plain DE/rand/1/bin[7] (treated as a reference method), DE-535[12] (modified DE/rand/1/bin, scaling factor generated randomly) L-SaDE[13] (DE with adaptive scaling factor and local search incorporated); Memetic EA — BLX-GL50[14], BLX-MA[15], SPC-PNX[16]; Steady state EA — K-PCX[17]; Coevolutionary EA — CoEVO[18]; simple Particle Swarm Optimization DMS-L-PSO[19]; Estimation of Distribution — EDA[20] and CMA-ES using two restart strategies — G-CMA-ES[21], L-CMA-ES[22]. For L-SaDE and DMS-L-PSO, no results have been reported for $n = 30$, and for DE-520, only results for $f_1 - f_{14}$ are available.

Each cell of the table, which corresponds to a function f_i and the competing algorithm number j, indicates if the mean error achieved by DE/rand/1/bin with resampling was smaller (symbol '+') or greater (symbol '–') than the mean error achieved by the algorithm number j for the function f_i. Since the compared methods are stochastic, we applied a generalized t-Student's test of equal means. This generalization, called Welch test, is applicable for samples driven from normal distributions with different standard deviations. Symbols '–' or '+' appear when the difference was statistically significant, i.e., when the probability of accepting the null hypothesis that both mean values were equal was smaller than 0.05. If the null hypothesis could not be rejected, it is indicated with '·'.

Analysis of Tab.3 indicates that, for several test functions, the combination of the DE with the resampling method could have allowed the DE/rand/1/bin to perform significantly better than the DE version with reinitialization [7], which might have allowed to even outperform the winners of the CEC2005 competition for some functions. This effect is clearly visible in 30 dimensions on multimodal

Table 2. Mean and standard deviation of results yielded by DE/rand/1/bin after 300000 fitness function evaluations for 30-dimensional problems from CEC2005

		projection	reflection	resampling	conservative	wrapping	reinitialize	% repaired
f_1	mean	6.82E+01	6.30E+01	**1.53E+01**	7.07E+02	3.00E+02	3.63E+02	50.0
	std	2.61E+01	2.06E+01	7.12E+00	1.98E+02	9.75E+01	1.01E+02	
f_2	mean	1.61E+02	2.06E+02	**1.57E+01**	1.23E+03	5.49E+02	6.37E+02	60.7
	std	7.11E+01	5.59E+01	6.65E+00	6.20E+02	1.70E+02	2.29E+02	
f_3	mean	4.96E+05	6.04E+05	**5.97E+03**	3.93E+06	1.48E+06	1.53E+06	76.8
	std	1.93E+05	1.78E+05	2.68E+03	1.75E+06	4.96E+05	4.44E+05	
f_4	mean	5.57E+02	7.09E+02	**7.61E+01**	4.18E+03	1.65E+03	1.97E+03	67.3
	std	1.93E+02	2.33E+02	3.98E+01	2.54E+03	5.30E+02	5.87E+02	
f_5	mean	3.93E+03	4.43E+03	**9.09E+02**	1.17E+04	8.13E+03	9.80E+03	88.1
	std	8.23E+02	5.49E+02	4.99E+02	1.50E+03	9.92E+02	1.37E+03	
f_6	mean	5.14E+04	3.97E+04	**1.16E+04**	3.90E+06	7.66E+05	1.67E+06	51.4
	std	3.26E+04	2.41E+04	9.02E+03	2.38E+06	5.11E+05	1.11E+06	
f_8	mean	2.10E+01	2.10E+01	2.11E+01	2.11E+01	**2.09E+01**	2.10E+01	46.3
	std	5.22E-02	4.26E-02	2.26E-01	2.18E-01	1.08E-01	5.22E-02	
f_9	mean	1.16E+02	1.30E+02	**8.97E+01**	1.84E+02	1.59E+02	1.68E+02	67.3
	std	3.36E+01	4.19E+01	1.14E+01	2.94E+01	2.91E+01	3.35E+01	
f_{10}	mean	1.41E+02	1.48E+02	**9.91E+01**	2.41E+02	1.98E+02	2.17E+02	70.5
	std	2.58E+01	3.52E+01	4.24E+01	1.60E+01	3.02E+01	1.76E+01	
f_{11}	mean	3.98E+01	3.74E+01	**2.20E+01**	2.56E+01	3.87E+01	3.72E+01	54.2
	std	2.73E+00	5.29E+00	5.89E+00	2.50E+00	3.72E+00	5.59E+00	
f_{12}	mean	1.00E+05	4.79E+04	**1.02E+04**	7.65E+04	8.85E+04	6.34E+04	65.7
	std	4.45E+04	2.12E+04	6.64E+03	3.00E+04	2.96E+04	2.58E+04	
f_{13}	mean	1.85E+01	1.82E+01	**1.57E+01**	1.92E+01	1.77E+01	1.80E+01	30.0
	std	1.07E+00	1.31E+00	1.94E+00	1.44E+00	1.41E+00	1.61E+00	
f_{14}	mean	1.35E+01	1.34E+01	**1.31E+01**	**1.31E+01**	1.36E+01	1.36E+01	55.9
	std	2.46E-01	4.19E-01	9.91E-01	6.40E-01	2.37E-01	2.07E-01	
f_{15}	mean	4.48E+02	4.50E+02	**4.30E+02**	5.11E+02	4.87E+02	4.84E+02	60.9
	std	2.59E+01	2.06E+01	3.56E+01	3.22E+01	1.94E+01	4.42E+00	
f_{16}	mean	1.65E+02	1.70E+02	**1.36E+02**	3.01E+02	2.46E+02	2.60E+02	76.3
	std	3.13E+01	3.76E+01	2.83E+01	4.05E+01	1.95E+01	1.99E+01	
f_{17}	mean	2-47E+02	2.56E+02	**1.89E+02**	3.36E+02	2.97E+02	2.99E+02	85.8
	std	1.87E+01	2.59E+01	5.95E+01	5.34E+01	2.10E+01	1.92E+01	
f_{18}	mean	9.22E+02	9.33E+02	**8.61E+02**	9.93E+02	9.62E+02	9.68E+02	90.6
	std	4.52E+00	3.76E+00	5.34E+01	2.80E+01	1.70E+01	2.75E+01	
f_{19}	mean	9.22E+02	9.33E+02	**8.44E+02**	9.95E+02	9.49E+02	9.63E+02	89.9
	std	4.52E+00	5.23E+00	5.29E+01	2.40E+01	2.65E+01	2.90E+01	
f_{20}	mean	9.22E+02	9.32E+02	**8.39E+02**	9.94E+02	9.53E+02	9.66E+02	89.8
	std	4.51E+00	5.71E+00	5.13E+01	2.41E+01	2.19E+01	2.70E+01	
f_{21}	mean	5.25E+02	5.09E+02	**5.04E+02**	6.76E+02	5.79E+02	6.15E+02	56.1
	std	3.28E+01	3.15E+00	1.96E+00	6.92E+01	3.44E+01	5.10E+01	
f_{22}	mean	**8.88E+02**	9.26E+02	9.29E+02	1.02E+03	9.94E+02	9.96E+02	88.0
	std	1.01E+01	1.12E+01	1.80E+01	1.47E+01	1.26E+01	1.50E+01	
f_{23}	mean	5.50E+02	5.40E+02	**5.35E+02**	7.52E+02	5.97E+02	6.73E+02	57.3
	std	1.09E+01	5.35E+00	2.99E+00	1.24E+02	2.57E+01	1.28E+02	
f_{24}	mean	5.62E+02	2.20E+02	**2.05E+02**	5.45E+02	3.40E+02	4.01E+02	61.9
	std	3.61E+02	5.20E+00	1.72E+00	1.16E+02	5.19E+01	7.65E+01	
avg. rank		2.74	2.65	1.09	5.30	4.09	4.48	
% repaired		73.8	66.2	51.7	71.8	68.6	68.5	

composition functions, where [7] reported rather poor results. For unimodal and basic multimodal functions, DE/1/rand/bin with any constraint handling technique is relatively ineffective. In our opinion this indicates a poor performance in exploitation which can be overcome by intelligent restarting and/or by the hybridization with some local optimization method, like in several other methods that took part in the competition [13–16, 21, 22].

Table 3. Comparison of the DE/rand/1/bin with resampling versus leading optimization methods from CEC2005

$n = 10$	1	2	3	4	5	6	8	9	10	11	12	13	14	15	16	17	18	19	20	21	22	23	24
DE/rand/1/bin	−	−	+	−	−	−	−	−	+	·	+	−	−	·	+	·	−	−	−	+	·	·	·
DE.535	−	·	+	+	+	+	−	−	−	−	·	−	−	·	−	−	−	·	−	+	+	+	−
L.SaDE	−	−	+	·	+	−	−	−	−	·	·	−	−	−	−	−	·	·	−	·	·	·	·
BLX.GL50	·	·	+	·	−	−	−	−	−	−	·	−	−	+	−	−	−	−	−	+	·	·	·
BLX.MA	+	+	+	+	+	+	−	−	−	·	·	−	−	+	−	−	+	·	·	+	·	+	−
SPC.PNX	+	+	+	+	−	+	+	−	−	−	·	−	−	·	−	−	−	−	−	+	·	·	·
K.PCX	+	+	+	·	+	·	−	−	−	+	·	−	−	+	−	−	·	·	·	+	·	+	−
CoEVO	+	·	·	·	+	+	−	·	·	+	+	−	·	+	+	+	+	·	+	+	+	+	−
DMS.L.PSO	−	−	−	+	−	−	−	−	−	·	·	−	−	−	−	−	·	·	+	+	·	+	·
EDA	−	−	−	−	+	−	−	+	+	+	·	·	·	+	+	+	−	−	−	+	+	·	·
G.CMA.ES	−	−	−	−	−	−	−	−	−	−	·	−	−	·	−	−	−	−	−	+	·	·	·
L.CMA.ES	−	−	−	·	−	−	−	+	+	·	·	−	·	·	−	+	−	−	−	·	·	+	−

$n = 30$	1	2	3	4	5	6	8	9	10	11	12	13	14	15	16	17	18	19	20	21	22	23	24
DE/rand/1/bin	−	−	+	−	−	−	−	−	−	+	·	−	·	+	+	+	+	+	+	+	+	+	−
DE.535	−	+	+	+	+	−	·	−	−	·	−	−	·	−	+	−	·	·	+	+	−	+	−
BLX.GL50	−	−	−	−	−	−	−	−	−	·	·	−	−	−	−	−	+	+	+	−	−	+	−
BLX.MA	−	−	+	−	+	−	−	−	·	+	−	−	−	−	+	+	·	+	+	−	−	+	−
SPC.PNX	−	−	+	−	+	−	−	−	−	−	·	−	·	−	−	−	+	+	+	−	−	·	−
K.PCX	−	−	−	+	+	−	−	−	−	+	−	−	+	+	−	·	−	·	·	+	+	+	−
CoEVO	−	−	−	+	+	−	−	+	+	+	+	−	·	·	+	+	+	+	+	+	+	+	−
EDA	+	+	+	+	+	+	−	+	+	+	+	+	+	+	+	+	+	+	+	·	−	+	−
G.CMA.ES	−	−	−	·	−	−	−	−	−	−	·	−	·	−	−	+	+	+	+	−	−	·	−
L.CMA.ES	−	−	−	+	−	−	−	+	+	−	·	−	+	−	−	+	+	+	+	−	−	·	−

4 Summary

We showed that the results obtained by the DE/rand/1/bin algorithm were significantly influenced by the method to handle boundary constraints. A strategy that repeats the differential mutation until a feasible solution is found appeared to be a winning one, and in particular, it appeared superior to reinitialization which is nowadays quite commonly used in DE. This observation might indicate yet another possibility to improve efficiency of the DE by choosing an appropriate boundary constraint handling technique.

References

1. Price, K., et al.: Differential Evolution: A Practical Approach to Global Optimization. Springer, Heidelberg (2005)
2. Neri, F., Tirronen, V.: Recent advances in Differential Evolution: a survey and experimental analysis. Artificial Intelligence Rev. 33(1-2), 61–106 (2010)
3. Qin, A.K., et al.: Differential Evolution algorithm with strategy adaptation for global numerical optimization. IEEE Trans. Evolutionary Computation 13(2), 398–417 (2009)

4. Liu, J., Lampinen, J.: A Fuzzy Adaptive Differential Evolution algorithm. Soft Computing 9(6), 448–462 (2005)
5. Zhang, J., Sanderson, A.C.: JADE: adaptive Differential Evolution with optional external archive. IEEE Trans. Evolutionary Computation 13(5), 945–958 (2009)
6. Doumpos, M., et al.: An evolutionary approach to construction of outranking models for multicriteria classification: The case of the ELECTRE TRI method. Eur. J. of Operational Research 199(2), 496–505 (2009)
7. Rönkkönen, J., et al.: Real-parameter optimization with differential evolution. In: CEC 2005. IEEE, Los Alamitos (2005)
8. Brest, J., et al.: Self-adapting control parameters in differential evolution: A comparative study on numerical benchmark problems. IEEE Trans. Evolutionary Computation 10(6), 646–657 (2006)
9. Karabogal, N., Cetinkayal, B.: Design of digital FIR filters using Differential Evolution algorithm. Circuits, Systems, Signal Processing 25(5), 649–660 (2006)
10. Hansen, N.: Compilation of results on the 2005 CEC benchmark function set (2005), `http://www.ntu.edu.sg/home/epnsugan/index_files/CEC-05/compareresults.pdf`
11. Suganthan, P.N., et al.: Problem definitions and evaluation criteria for the CEC 2005 special session on real-parameter optimization. Technical report, Nanyang Tech. Univ. (2005)
12. Bui, L.T., et al.: Comparing two versions of differential evolution in real parameter optimization. In: CEC 2005. IEEE, Los Alamitos (2005)
13. Qin, A., Suganthan, P.: Self-adaptive differential evolution algorithm for numerical optimization. In: CEC 2005. IEEE, Los Alamitos (2005)
14. Martines, C.G., Lozano, M.: Hybrid real-coded genetic algorithms with female and male differentiation. In: CEC 2005. IEEE, Los Alamitos (2005)
15. Molina, D., et al.: Adaptive local search parameters for real-coded memetic algorithms. In: CEC 2005. IEEE, Los Alamitos (2005)
16. Ballester, P., et al.: Real-parameter optimization performance study on the CEC-2005 benchmark with SPC-PNX. In: CEC 2005. IEEE, Los Alamitos (2005)
17. Sinha, A., et al.: A population-based, steady-state procedure for real-parameter optimization. In: CEC 2005. IEEE, Los Alamitos (2005)
18. Posik, P.: Real parameter optimization using mutation step co-evolution. In: CEC 2005. IEEE, Los Alamitos (2005)
19. Liang, J., Suganthan, P.: Dynamic multi-swarm particle swarm optimizer with local search. In: CEC 2005. IEEE, Los Alamitos (2005)
20. Yuan, B., Gallagher, M.: Experimental results for the special session on real-parameter optimization at CEC 2005: A simple, continuous EDA. In: CEC 2005. IEEE, Los Alamitos (2005)
21. Auger, A., et al.: A restart CMA evolution strategy with increasing population size. In: CEC 2005. IEEE, Los Alamitos (2005)
22. Auger, A., et al.: Performance evaluation of an advanced local search evolutionary algorithm. In: CEC 2005. IEEE, Los Alamitos (2005)

Entropy-Driven Evolutionary Approaches to the Mastermind Problem

Carlos Cotta[1], Juan J. Merelo Guervós[2],
Antonio M. Mora García[2], and Thomas Philip Runarsson[3]

[1] ETSI Informática, Universidad de Málaga, Campus de Teatinos,
29071 Málaga, Spain
ccottap@lcc.uma.es

[2] Dept. of Architecture and Computer Technology, ETSIIT, University of Granada
jj@merelo.net, amorag@geneura.ugr.es

[3] School of Engineering and Natural Sciences, University of Iceland
tpr@hi.is

Abstract. Mastermind is a well-known board game in which one player must discover a hidden color combination set up by an opponent, using the hints the latter provides (the number of places –or pegs– correctly guessed, and the number of colors rightly guessed but out of place in each move). This game has attracted much theoretical attention, since it constitutes a very interesting example of dynamically-constrained combinatorial problem, in which the set of feasible solutions changes with each combination played. We present an evolutionary approach to this problem whose main features are the seeded initialization of the population using feasible solutions discovered in the previous move, and the use of an entropy-based criterion to discern among feasible solutions. This criterion is aimed at maximizing the information that will be returned by the opponent upon playing a combination. Three variants of this approach, respectively based on the use of a single population and two cooperating or competing subpopulations are considered. It is shown that these variants achieve the playing level of previous state-of-the-art evolutionary approaches using much lower computational effort (as measured by the number of evaluations required).

1 Introduction

Mastermind is a board game that has enjoyed world-wide popularity since the 70s, when its current design was put forward (antecedents can be traced back to traditional puzzles such as *bulls and cows* or AB [1] though). Roughly speaking, Mastermind is a two-player code-breaking game, or in some sense a puzzle, since one of the players –the *codemaker* (CM)– has no other role in the game than providing a hidden combination, and automatically providing hints on how close the other player –the *codebreaker* (CB)– has come to guess this combination. More precisely, the functioning of the game is as follows:

– The CM sets a length ℓ combination of κ symbols. Therefore, the CB is faced with κ^ℓ candidates to be the hidden combination. This combination

R. Schaefer et al. (Eds.): PPSN XI, Part II, LNCS 6239, pp. 421–431, 2010.

is typically represented by an array of pegs of different colors. We will use uppercase letters to denote these colors.
- The CB tries to guess this secret code by producing a combination with the same length and using the same set of symbols as the secret code. As a response to this move, the CM provides information on the number of symbols guessed in the right position (black pegs in the physical board game), and the number of symbols in an incorrect position (white pegs).
- The CB uses this information to produce a new combination, that is assessed in the same way. If he correctly guesses the hidden combination in at most N attempts, the CB wins. Otherwise, the CM takes the game.

There exist other variants of the game in which more information is provided (i.e., which –rather than how many– symbols are correctly guessed and/or are out of place), additional constraints are enforced on the hidden combination (e.g., not allowing repeated symbols in the secret code, as in *bulls and cows*, thus reducing the search space), or even allowing the CM to change the code during the game in a way that is compatible with previous moves (this variant is aimed to exploit any bias the CB may have when selecting the next combination played, such as preferring certain symbols over others).

As mentioned before, the game is asymmetrical in the sense that the CM has no freedom of move after setting up the hidden combination (at least in static variants of the game in which the hidden combination does not change), and therefore the CB does not have to outsmart the CM, but he has to put his own analytical skills at play to determine the course of action given the information available at each step. The resulting combinatorial problem is enormously interesting, as it relates to other generally called *oracle* problems such as circuit and program testing, differential cryptanalysis, uniquely identifying a person from queries to a genetic database [2] and other puzzle-like games and problems – check also [3]. It is also a complex problem, which has been shown to be NP-complete under different formulations [4,5], for which several issues remain open, e.g., what is the lowest average number of guesses needed to solve the problem for any given κ and ℓ. Associated to this, there arises the issue of coming up with an efficient mechanism for finding these guesses in any particular case. To this end, several attempts have been made to use evolutionary algorithms (EAs) for exploring the space of feasible (meaning here compatible with the available information) combinations, e.g., [6,7,8].

Most evolutionary approaches presented for this problem are based on providing the CB with a set of potential combinations among which he has to select his next move. This decision-making process is very important, since although all potential candidates may be compatible with the information available, the outcome of the move can be very different, ranging from minimal reductions in the set of potential solutions, to a major pruning of the search space. Several metrics have been defined for this purpose. We consider here the use of an entropy-based criterion, which is further introduced in the fitness function to guide the search, and provide a variable-size, high-quality set of potential candidates. We show that this approach, combined with the seeded initialization of the population can result in a large reduction in the computational effort of the EA, with respect to other evolutionary approaches described in the literature.

2 Background

This section provides a brief overview of the problem, presents its classical formulation, and discusses how it has been tackled in the literature.

2.1 Formulation

As mentioned in Sect. 1, a Mastermind problem instance is characterized by two parameters, namely the number κ of colors and the number ℓ of pegs. Let $\mathbb{N}_\kappa = \{1, 2, \cdots \kappa\}$ be the set of symbols used to denote the colors. Subsequently, any combination, either the hidden one or one played by the CB, is a string $c \in \mathbb{N}_\kappa^\ell$. Whenever the CB plays a combination c_p, a *response* $h(c_p, c_h) \in \mathbb{N}^2$ is obtained from the CM, where c_h is the hidden combination. A response $\langle b, w \rangle$ indicates that the c_p matches c_h in b positions, and there exist other w symbols in c_p present in c_h but in different positions.

A central notion in the context of the game is that of consistency. A combination c is *consistent* with a played combination c_p if, and only if, $h(c, c_p) = h(c_p, c_h)$, i.e., if c has as many black and white pegs with respect to the c_p as c_p has with respect to the hidden combination. Intuitively, this captures the fact that c might be a potential candidate to be the hidden combination in light of the outcome of playing c_p. We can easily extend this notion and denote a combination c as consistent (or feasible) if, and only if, it is consistent with all combinations played so far, i.e.,

$$h(c, c_p^i) = h(c_p^i, c_h) \; 1 \leqslant i \leqslant n \qquad (1)$$

where n is the number of combinations played so far, and c_p^i is the i−th combination played. Any consistent combination is a candidate solution. It is straightforward to see that the number of feasible solutions decreases with each guess made by the CB (provided he always plays feasible solutions). By the same token, the feasibility of a candidate solution is a transient property that can be irreversibly lost upon obtaining further information from the CM. This turns out to be a central feature in the strategies devised to play Mastermind, as shown next.

2.2 Game Strategy

Most naive approaches to Mastermind play a consistent combination as soon as one is found. An example of such an approach within an evolutionary context was proposed by Merelo *et al.* [3]. While this can constitute a simple and fast strategy early in the game (when there are many feasible combinations left), and is ensured to find eventually the hidden combination, it is a poorly performing strategy in terms on the number of attempts the CB needs to win the game. Indeed, unless some bias is introduced in the way solutions are searched, this strategy reduces to random approach, as solutions found (and played) are a random sample of the space of consistent guesses. It is also highly inefficient in the last stages of the game (when there are very few –maybe just one– feasible solutions), since a large part of the feasible space has to be examined to find a feasible solution. We will return to this latter issue later on.

One of the reasons behind the poor performance of these naive strategies was anticipated in Sect. 2.1: the feasibility of a solution can vanish upon receiving feedback from the CM. Thus, a sensible strategy would try to take this into account when selecting the next move. More precisely, the CB would like to play a feasible combination that –were it not the hidden one– left him in the best position in the next move. This leads to a generic framework for defining Mastermind strategies in which (1) a procedure for finding a large set (even a complete set) Φ of feasible combinations is firstly used, and (2) a decision-making procedure to select which combination $c \in \Phi$ will be played is then used.

Regarding the decision-making procedure, its purpose must be to minimize the losses of the CB, putting him in the best position for the next move. This is substantiated in reducing the number of feasible solutions as much as possible, and hence that the set of available options in the next step is minimal. However, it is obvious that the reduction in the number of feasible solutions attainable by a certain guess depends on the hidden combination which is unknown. Therefore, any strategy based on this principle must rely on heuristics. To be precise, let us consider the concept of *partitions* (also called Hash Collision Groups, HCG [1]) defined as follows:

1. Let Φ be the set of feasible solutions available.
2. Let Ξ be a $|\Phi| \times (\ell+1) \times (\ell+1)$ all-zero matrix.
3. **for each** $\{c_i, c_j\} \subseteq \Phi$ **do**
   (a) $\langle b, w\rangle \leftarrow h(c_1, c_2)$.
   (b) $\Xi_{ibw} \leftarrow \Xi_{ibw} + 1$
   (c) $\Xi_{jbw} \leftarrow \Xi_{jbw} + 1$

Notice that each combination c_i has a partition matrix $\Xi_{i[\cdot,\cdot]}$, indicating how it relates to the rest of feasible solutions. Notice also that after playing combination c_i and obtaining response $\langle b, w\rangle$, entry Ξ_{ibw} indicates how many solutions in Φ remain feasible. Most approaches to Mastermind use the information in this matrix to select the next combination played, typically using some kind of worst-case or average-case reasoning. This idea was introduced by Knuth [9], whose algorithm tries to minimize the worst case by following the strategy of minimizing the worst expected set size, i.e., let $\eta(c_i, \Xi) = \max\{\Xi_{ibw} \mid b, w \leqslant \ell\}$, then $c_p = \min^{-1}\{\eta(c_i, \Xi) \mid c_i \in \Phi\}$ (ties are broken by lexicographical order). Using a complete minimax search Knuth shows that a maximum of 5 guesses (4.478 on average) are needed to solve the game using this strategy for $\kappa = 6$, $\ell = 4$.

2.3 Related Work

The path leading to the most successful heuristic (non-evolutionary) strategies to date include minimization of the *worst case* [9] or *average case* [10], or maximization of *entropy* [11,12] or *number of partitions* [13]. We defer to next section a more detailed description of the use of entropy, which has been the strategy chosen in this work.

EAs that try to solve this problem have also historically proceeded more or less in the same way. After using naive strategies that played the first combination found [6], using suboptimal strategies with the objective of avoiding the search to be stuck [7], or even playing the best guess each generation in a policy that resulted in a fast and very bad solution to the puzzle [14,15]. However, it was not until recently when Berghman *et al.* [8] adopted the method of partitions to an EA. The strategy they apply is similar the *expected size* strategy.

The use of the information in Ξ can be considered as a form of look-ahead, which is computationally expensive and requires the availability of set Φ. Notice however that if no look-ahead is used to guide the search, any other way of ranking solutions (i.e., any mechanism that analyze solutions on an individual basis) might find solutions that were slightly better than random, but not more. In any case, it has been shown [16] that in order to get the benefit of using look-ahead methods, Φ need not be the full set of feasible solutions at a certain step: a fraction of around one sixth is enough to find solutions that are statistically indistinguishable from the best solutions found. This was statistically established, and then tested in an EA termed EvoRank [17], where the *most-partitions* strategy was used. Solutions were quite competitive, being significantly better than random search and also similar to the results obtained by Berghman *et al.*, but using an smaller set size and a computationally simpler strategy.

The strategy presented in this paper provides two fundamental steps beyond this previous work. On one hand, the use of an entropy-based fitness function and guided initialization reduces the computational effort required by the algorithm. On the other hand, it is shown that not fixing in advance the size of the set Φ (and thus letting the algorithm use limited computational resources to find a set as large as possible but without any lower bound on its size) does not penalize performance. This indicates the EA can be run on a fixed computational budget, rather than until completing a large enough set Φ.

3 Entropy-Driven Approaches

The general framework used in this work is depicted in Algorithm 1. An EA plays the role of the CB, selecting a combination to be played at each step on the basis of (i) past guesses and their outcomes, and (ii) the set of feasible solutions found in the previous step. In the first step no previous information is available, and all combinations are potentially the hidden one with equal probability. Hence, a fixed guess is made in this case.

Algorithm 1. Outline of the evolutionary Mastermind approach

```
typedef Combination: vector[1..ℓ] of ℕ_κ;
procedure MASTERMIND (in: c_h: Combination, out: guesses, evals: ℕ);
var c: Combination;
var b, w, e: ℕ;
var P: List[⟨Combination, ℕ²⟩] ;                          // game history
var F: List[Combination] ;                       // known feasible solutions
evals ← 0; guesses ← 0; P ← [] ;                         // initialize game
repeat
    guesses ← guesses + 1;
    if guesses = 1 then                                   // initial guess
        c ← INITIALGUESS(ℓ, κ);
        F ← [];
    else
        RUNEA (↓ P, ↕ F, ↑ c, ↑ e) ;                          // run the EA
        evals ← evals + e ;  // update cummulative number of evaluations
    endif
    ⟨b, w⟩ ← h(c, c_h) ;                       // current guess is evaluated
    played.ADD(⟨c, ⟨b, w⟩⟩) ;                   // game history is updated
until b ≠ ℓ ;
```

The EA handles a population of candidate combinations. This population is initialized using the feasible solutions found in the previous step. More precisely, random sampling with replacement is used to select the initial member of the population from the set of known feasible solutions. Notice that this set is probably not exhaustive, since the EA is not guaranteed to have found all feasible solutions in the previous move. Furthermore, many of these feasible solutions will no longer be feasible in light of the outcome of the last move. In any case, using this set as a seed provides the EA with valuable information to focus the search towards the new feasible region of the search space, which will likely intersect with the known feasible set (and ideally would be a subset of the latter).

Solutions can be manipulated using standard reproductive operators for recombination and mutation. The fitness function is responsible for determining the feasibility of a given combination given the currently available information. Every feasible solution discovered during this run of the EA is kept in a secondary population for post-processing after the run finishes. This post-processing is aimed to select the single combination that will be played in the next move. The procedure used for this purpose is analogous to that used for fitness assignment, so let us firstly describe the latter.

As mentioned in Sect. 2.3, entropy is used as a quality indicator to optimize the status of the game after the guessed combination is played. To be precise, the fitness function firstly scans the population to divide the population in two groups: feasible solutions and infeasible ones. Infeasible solutions are assigned a fitness value (to be maximized) that indicates its closeness to feasibility. This is done by computing the Manhattan distance between the score $h(c, c_p^i)$ the solution obtains against the i-th combination played, and the score $h(c_p^i, c_h)$ the latter obtained against the hidden combination. This distance is summed over all previously played combinations, normalized by dividing by the maximum possible distance, and deducted from 1.0. This way, the closer (from below) the fitness is to 1.0, the closer to feasibility the solution is.

Feasible solutions have thus a fitness equal to 1.0, and receive a bonus based on entropy. To this end, the partition matrix Ξ is computed on the basis of the feasible solutions contained in the current population. Subsequently, the entropy $H_i(\Xi)$ of each combination is computed as:

$$H_i(\Xi) = - \sum_{0 \leqslant b,w \leqslant \ell} p_i(b,w|\Xi) \log p_i(b,w|\Xi) \tag{2}$$

where $p_i(b,w|\Xi) = \Xi_{ibw} / \sum_{0 \leqslant b',w' \leqslant \ell} \Xi_{ib'w'}$. The underlying the idea of using this entropy measure is to reward feasible solutions tending to produce a uniform partition of the feasible search space, hence minimizing the worst-case scenario. This procedure can be also regarded as a way of maximizing the average information returned by the CM once the combination is played. The actual selection of this combination is done after the EA terminates, using the procedure described above on the set of feasible solutions discovered in the run.

Three variants of the EA described above have been considered, each one using a different population management strategy. The first one is an EA that follows the procedure above on a single population. We will denote this algorithm as eGA. Secondly, we have considered a version of the algorithm in which the population is divided in two equal-size tiers A and B; partitions (and subsequently entropy) are computed in a crossed way, i.e., determining how solutions in tier A compare to solutions in tier B and vice versa. The rationale is to preclude (or at least hinder) the appearance of endosymbiotic relationship (the population evolving to maximize entropy internally, but not globally). This co-evolving variant is denoted as eGACo. Finally, a third variant is considered analogously to eGACo, but having combinations in tier B trying to minimize (rather than maximize) the entropy of combinations in tier A. This can be regarded as competitive co-evolution [18], aimed to provide a constant thrust in the search of new feasible solutions. We denote this competitive variant as eGACm.

4 Experimental Results

The evolutionary approaches considered use a population of 50 solutions (divided in two subpopulations of 25 solutions in the case of eGACo and eGACm), binary tournament selection, one-point crossover ($p_X = .9$), random-substitution mutation ($p_m = 1/\ell$), and an elitist generational replacement policy. The algorithms are run for a minimum number of 500 evaluations in each move. If no feasible solution is in the (sub)population(s) at this point, the algorithm keeps on running until one is found.

Table 1. Comparison of the evolutionary approaches with random (R) and seeded (S) initialization. Underlined results are statistically significant.

		eGA		eGACo		eGACm	
		played	evals	played	evals	played	evals
$\kappa = 6$	R	4.489 ± .011	1872 ± 10	4.448 ± .011	2315 ± 24	4.425 ± .011	2252 ± 20
	S	4.438 ± .011	1792 ± 9	4.439 ± .011	1977 ± 14	4.425 ± .011	1982 ± 15
$\kappa = 8$	R	5.279 ± .013	2501 ± 20	5.207 ± .013	3456 ± 51	5.207 ± .013	3465 ± 50
	S	5.240 ± .013	2348 ± 19	5.222 ± .013	2804 ± 38	5.207 ± .013	2810 ± 37

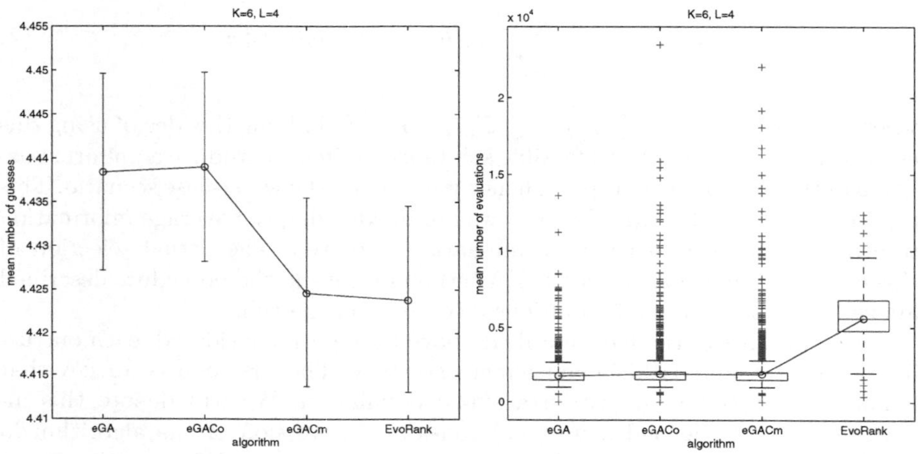

(a) (b)

Fig. 1. Results for $\kappa = 6$ colors and $\ell = 4$ pegs. (a) Mean number of guesses required. The error bars indicate the standard deviation of the mean. (b) Distribution of the number of evaluations required.

The experiments have been performed on two Mastermind problems, namely the classical version defined by $\ell = 4$ pegs and $\kappa = 6$ colors, and a harder version involving the same number of pegs but more colors ($\kappa = 8$). In both cases a problem-generator approach has been considered: 5,000 runs of each algorithm have been carried out, each one of a randomly generated instance (i.e., hidden combination). To maximize the breadth of the benchmark, instances have been generated so that any instance in the test set is used at most once more that any other existing instance.

The first set of experiments is devoted to test the impact of the seeded initialization of the population. For this purpose, the algorithms are run both using random initial populations and seeded initialization. The results are shown in Table 1. As it can be seen, there is no remarkable difference in the mean number of guesses required, but the computational effort is clearly better (with statistical significance at 0.05 level using a Wilcoxon signed rank test).

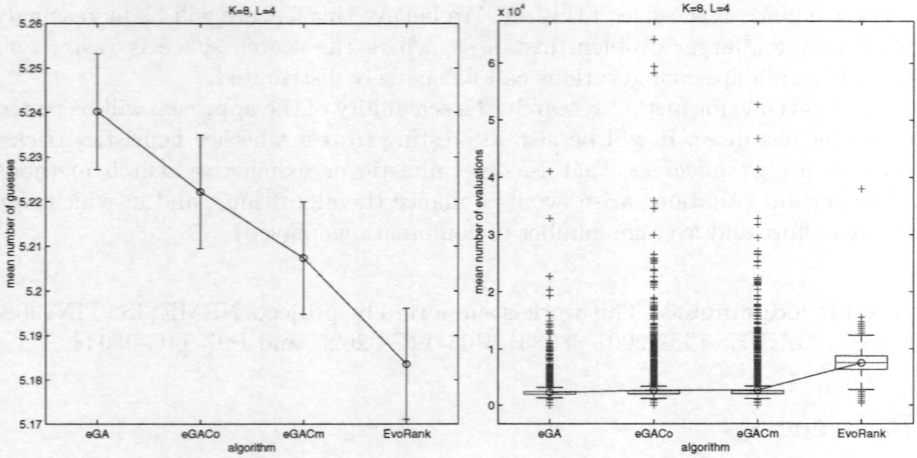

(a) (b)

Fig. 2. Results for $\kappa = 8$ colors and $\ell = 4$ pegs. (a) Mean number of guesses required. The error bars indicate the standard deviation of the mean. (a) Distribution of the number of evaluations required.

Subsequently, the EGA* variants are compared with EvoRank, a state-of-the-art algorithm for this problem, on the same set of instances. The results of the comparison are shown in Figs. 1 and 2. Notice all algorithms perform very similarly in number of guesses for $\kappa = 6$ (statistically not significant results). EvoRank seems to be slightly better for $\kappa = 8$, but note that the difference is only statistically significant for EGA and EGACo, and there is no statistically significant difference between EGACM and EvoRank (p-value=.18). The situation is different in the case of the number of evaluations required: all EGA* variants are much computationally cheaper than EvoRank (with statistical significance). Among EGA* variants, the computational cost of EGA* is significantly lower than that of EGACo and EGACM for $\kappa = 6$ and $\kappa = 8$.

5 Conclusions and Future Work

Mastermind is a prime example of dynamically constrained problem which offer an excellent playground for optimization techniques. We have presented a new evolutionary approach for this problem in which in addition to the hard constraint of consistency with previous guesses, solutions are also evaluated in terms of maximizing the information potentially returned by the CM. Seeded initialization of the population with feasible solutions for the previous move turns out to be an important ingredient for reducing the computational effort of the algorithm. As a result, it can perform at state-of-the-art level, at lower cost than other algorithms. This also indicates that entropy-guided construction of a variable-size feasible set Φ is competitive against other procedures that impose a lower bound on the size of this set. We believe this feature will be increasingly important for larger problem instances, where the search space is vaster and feasible candidate configurations can be sparsely distributed.

As an avenue for further research, the scalability of the approach will be tested on larger instances. It will be also interesting to test whether heuristics tricks, such as using *endgames*, that is, deterministic or exhaustive search methods when certain situations arise, would enhance the algorithms, and in which way (search effort and average number of combinations played).

Acknowledgements. This work is supported by projects NEMESIS (TIN2008-05941), NoHNES (TIN2007-68083), P06-TIC-02025, and P07-TIC-03044.

References

1. Chen, S.T., Lin, S.S., Huang, L.T.: A two-phase optimization algorithm for mastermind. Computer Journal 50(4), 435–443 (2007)
2. Goodrich, M.: On the algorithmic complexity of the Mastermind game with black-peg results. Information Processing Letters 109(13), 675–678 (2009)
3. Merelo-Guervós, J.J., Castillo, P., Rivas, V.: Finding a needle in a haystack using hints and evolutionary computation: the case of evolutionary MasterMind. Applied Soft Computing 6(2), 170–179 (2006)
4. Stuckman, J., Zhang, G.Q.: Mastermind is NP-complete. INFOCOMP J. Comput. Sci. 5, 25–28 (2006)
5. Kendall, G., Parkes, A., Spoerer, K.: A survey of NP-complete puzzles. ICGA Journal 31(1), 13–34 (2008)
6. Merelo, J.J.: Genetic Mastermind, a case of dynamic constraint optimization. GeNeura Technical Report G-96-1, Universidad de Granada (1996)
7. Bernier, J.L., Herráiz, C.I., Merelo-Guervós, J.J., Olmeda, S., Prieto, A.: Solving *mastermind* using GAs and simulated annealing: a case of dynamic constraint optimization. In: Voigt, H.M., et al. (eds.) PPSN 1996. LNCS, vol. 1141, pp. 553–563. Springer, Heidelberg (1996)
8. Berghman, L., Goossens, D., Leus, R.: Efficient solutions for Mastermind using genetic algorithms. Computers and Operations Research 36(6), 1880–1885 (2009)
9. Knuth, D.E.: The computer as Master Mind. J. Recreational Mathematics 9(1), 1–6 (1976-77)
10. Irving, R.W.: Towards an optimum Mastermind strategy. Journal of Recreational Mathematics 11(2), 81–87 (1978-79)
11. Neuwirth, E.: Some strategies for Mastermind. Zeitschrift fur Operations Research 26(8), B257–B278 (1982)
12. Bestavros, A., Belal, A.: Mastermind, a game of diagnosis strategies. Bulletin of the Faculty of Engineering, Alexandria University (December 1986)
13. Kooi, B.: Yet another Mastermind strategy. ICGA Journal 28(1), 13–20 (2005)
14. Bento, L., Pereira, L., Rosa, A.: Mastermind by evolutionary algorithms. In: Procs SAC 1999, pp. 307–311 (1999)
15. Kalisker, T., Camens, D.: Solving Mastermind using genetic algorithms. In: Cantú-Paz, E., et al. (eds.) GECCO 2003. LNCS, vol. 2724, pp. 1590–1591. Springer, Heidelberg (2003)

16. Runarsson, T.P., Merelo, J.J.: Adapting heuristic Mastermind strategies to evolutionary algorithms. In: González, J., et al. (eds.) Nature Inspired Cooperative Strategies for Optimization 2010, Granada, Spain. Studies in Computational Intelligence, vol. 284, pp. 255–267. Springer, Heidelberg (2010)
17. Merelo, J.J., Runarsson, T.P.: Finding better solutions to the mastermind puzzle using evolutionary algorithms. In: Di Chio, C., et al. (eds.) EvoApplicatons 2010. LNCS, vol. 6024, pp. 120–129. Springer, Heidelberg (2010)
18. Angeline, P., Pollack, J.: Competitive environments evolve better solutions for complex tasks. In: Forrest, S. (ed.) Proceedings of the Fifth International Conference on Genetic Algorithms, pp. 264–270. Morgan Kaufmann, San Francisco (1993)

Evolutionary Detection of New Classes of Equilibria: Application in Behavioral Games

Dumitru Dumitrescu[1], Rodica Ioana Lung[1], Réka Nagy[1], Daniela Zaharie[2], Attila Bartha[1], and Doina Logofătu[3]

[1] Babes-Bolyai University, Cluj Napoca, Romania
[2] West University, Timisoara, Romania
[3] University of Applied Sciences, Munich, Germany

Abstract. Standard game theory relies on the assumption that players are rational decision makers that try to maximize their payoffs. Experiments with human players show that real people rarely follow the predictions of normative theory. Our aim is to model the human behavior accurately. Several classes of equilibria (Nash, Pareto, Nash-Pareto and fuzzy Nash-Pareto) are considered by using appropriate generative relations. Three versions of the centipede game are used to illustrate the different types of equilibrium.

Based on a study of how people play the centipede game, an equilibrium configuration that models the human behavior is detected. This configuration is a joint equilibrium obtained as a fuzzy combination of Nash and Pareto equilibria. In this way a connection between normative theory, computational game theory and behavioral games is established.

Keywords: Games, Equilibrium, Evolutionary Detection.

1 Introduction

Standard (non-cooperative) game theory is built on the assumption that players are rational decision makers acting for maximizing their profit. Moreover each player understands the other agents behavior. The underlying solution concept is Nash equilibrium or one of its variants, backward induction, iterated dominance, etc. This theory has some limitations when applied to explain social life and fails to account for some aspects of economic transitions. Even very simple well-defined games are played by real people in manners that significantly differ from those prescribed by the normative theory.

One simple step towards a more realistic and flexible approach is to relax the rationality principle and look for the corresponding equilibrium concepts. According to a meta-rationality principle each agent is characterized by its bias towards a certain type of equilibrium (Nash, Pareto, etc.) This way several new types of equilibria, like Nash-Pareto, can be obtained [4].

Our aim is to investigate if there are some equilibrium configuration explaining how people play games. Experimental and computational aspects of game theory

R. Schaefer et al. (Eds.): PPSN XI, Part II, LNCS 6239, pp. 432–441, 2010.

are linked together. Each equilibrium concept is described (non necessarily in a unique manner) by a generative relation on the strategy space.

Heterogenous equilibria are invoked for generating hypothesis about the manner real people act in playing games and making decisions. A new fuzzy Nash-Pareto equilibrium is defined. Our intuition is the new equilibrium concept is able to capture the manner real people play games. Experimental results seem to support this assumption.

1.1 Game Theory Prerequisites

For better understanding, some basic notions related to game theory are presented.

A finite strategic game is defined as a system $\Gamma = ((N, S_i, u_i), i = 1, n)$ where:
- $N = \{1, ..., n\}$ is a set of n players;
- S_i represents the set of actions (pure strategies) available;
- $S = S_1 \times S_2 \times ... \times S_n$ is the set of all possible situations of the game. An element of S is a strategy profile (or strategy) of the game;
- for $i \in N$, $u_i : S \to R$ represents the payoff function.

Denote by (s_{i_j}, s^*_{-i}) the strategy profile obtained from s^* by replacing the strategy of player i with s_{i_j} i.e. $(s_{i_j}, s^*_{-i}) = (s^*_1, s^*_2, ..., s^*_{i-1}, s_{i_j}, s^*_{i+1}, ..., s^*_n)$.
The concept of game solution is usually described as a game equilibrium, where each player has adopted a strategy that they are unlikely to change.

A strategy is a Nash equilibrium [10][1][8] if each player has no incentive to unilaterally deviate i.e. she can not improve the payoff by modifying her strategy while the others do not modify theirs.
More formal, the strategy s^* is a Nash equilibrium if and only if the inequality $u_i(s_i, s^*_{-i}) - u_i(s^*) \leq 0$, $\forall s_i \in S_i, \forall i \in N$ holds.

A strategy s' Pareto-dominates the strategy s'' if and only if each player has a better payoff for s' than for s''. We write $s' \leq_p s''$ or $(s', s'') \in P_d$.
Formally: $(s', s'') \in P_d$ if and only if $u_i(s') \geq u_i(s''), \forall i \in \{1, ..., n\}$ and $\exists j \in \{1, ..., n\} : u_j(s') > u_j(s'')$.

A strategy s'' is Pareto-non dominated, or Pareto-efficient, if does not exist $s' \in S$ such that $(s', s'') \in P_d$. The Pareto equilibrium of the game is represented by the set of Pareto-efficient strategies.

2 Centipede Game

The centipede game, introduced by Rosenthal in [11], is a two person, extensive-form game. Consider two players playing the game. At the first round player one can either take the larger portion of a fixed pot, or she can pass the pot to the other player. Player two has the same choices of either taking or passing the pot. In each round the pot increases according to a previously defined rule, but the larger portion of the pot in a round will always be smaller than the smaller portion in the next round. This way passing strictly decreases a player's payoff if the opponent takes on the next round. The game continues until one player

decides to end the game by taking her portion of the pot or for a finite number of rounds, which is known in advance for both players.
Although the traditional centipede game had a limit of 100 rounds, any game with this structure but a different number of rounds is called a centipede game.

The centipede game is a very challenging game from a game theoretical point of view. Rational play suggests that the first player should take the pot right at the start (corresponding to the Nash equilibrium strategy). However, experiments show, that regular people almost never choose to stop in the first round.

McKelvey and Palfrey [9] studied actual behavior in different versions of the centipede game: a four move (Figure 1), a six move (Figure 2) and high payoff versions (Figure 3), and found that subjects did not follow the theoretical predictions. Experiments involve human players without a game theoretical background (they are not informed and they do not make inferences about Nasg equilibrium). Very few players stopped in the first round, choosing a secure win, and despite the high payoffs, very few players risked to wait until the last round of the game. Detailed results about the players' choices are presented in Figure 4.

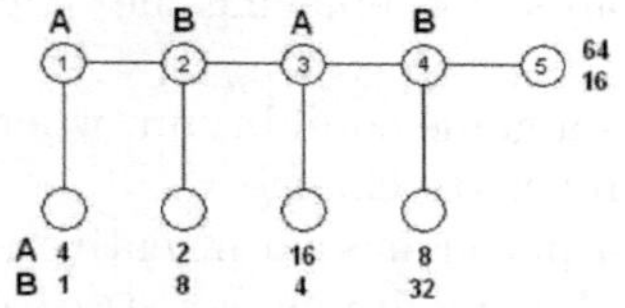

Fig. 1. The four-move version of the centipede game

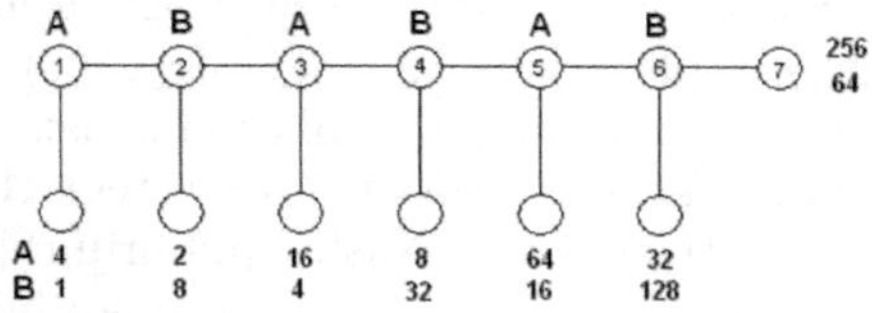

Fig. 2. The six-move version of the centipede game

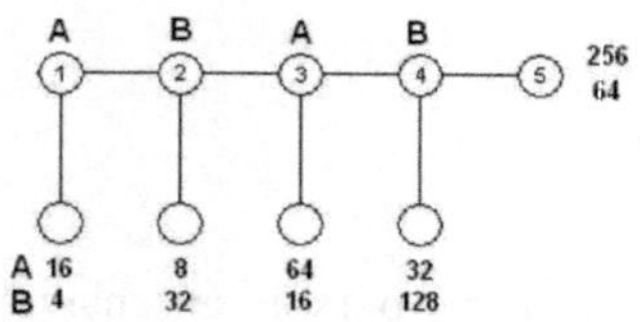

Fig. 3. The high payoff version of the centipede game

Terminal node	Four move version	High payoff version	Six move version
node 1	7%	15%	1%
node 2	36%	37%	6%
node 3	37%	32%	20%
node 4	15%	11%	38%
node 5	5%	5%	25%
node 6			8%
node 7			2%

Fig. 4. The frequencies of each of the terminal nodes

Our goal is to find a proper equilibrium concept, that would lead to an equilibrium corresponding to the experienced results.

3 Evolutionary Equilibria Detection

In this section we describe a general method for detecting different types of equilibria. Each equilibrium concept is described (non necessarily in a unique manner) by a generative relation [4] on the strategy space.

An appealing technique is the use of generative relations and evolutionary algorithms for detecting equilibrium strategies. The payoff of each player is treated as an objective and the generative relation includes an appropriate dominance concept, which is used for fitness assignment purpose. Evolutionary multi-objective algorithms (EMOAs) are thus suitable tools in searching for game equilibria.

A population of strategies is evolved and a chromosome is an n-dimensional vector representing a strategy $s \in S$. The initial population is randomly generated. Strategy population at iteration t may be regarded as the current equilibrium approximation. Subsequent application of the search operators (like the simulated binary crossover (SBX) [2] and real polynomial mutation [3]) is guided by a specific selection operator induced by the generative relation.

The population of strategies is sorted based on nondomination, where the dominance concept is induced by the generative relation. The diversity among nondominated strategies is introduced by using the crowding comparison. [3] In the case when two strategies have the same domination rank, the one located in a lesser crowded region is preferred. In this way successive populations produce new approximations of the equilibrium front, which hopefully are better than the previous ones.

In case the game does not have a certain equilibrium (in the sense of strict mathematical characterization) the proposed evolutionary technique may allow to detect a game situation which is a suitable approximation of this equilibrium.

In the next sections several generative relations for different types of equilibria are presented. The considered equilibria types are evolutionary detected for the different versions of the centipede game introduced in Section 2. Based on a study of how people play the centipede game [9], an equilibrium configuration that models the human behavior is detected. This configuration is a joint equilibrium obtained as a fuzzy combination of Nash and Pareto equilibria. In this way a connection between normative theory, computational game theory and behavioral games is established.

In our numerical experiments a population of 100 strategies has been evolved. In all experiments the process converges in less then 20 generations.

4 Equilibria and Their Generative Relations

Game equilibria may be characterized by generative relations [4]. The idea is the non-dominated strategies with respect to the generative relation equals (or approximate) the equilibrium set. The generative relation is not unique. For Nash equilibrium we consider two generative relations. Both of them are obtained from quality measures.

4.1 Binary Generative Relation for Nash Equilibrium

Let x and y be two pure strategies and $k(y,x)$ denotes the number of players which benefit by deviating from y towards x: $k(y,x) = card\{i|u_i(x,y_{-i}) > u_i(y)\}$. $k(y,x)$ is a relative quality measure of y and x – with respect to the Nash equilibrium.

Strategy y is better than strategy x with respect to Nash equilibrium, and we write $y \prec_N x$, if and only if $k(y,x) < k(x,y)$. We may consider $\prec_N$ as a generative relation of Nash equilibrium, i.e. the set of non-dominated strategies with respect to $\prec_N$ equals the Nash equilibrium [5]. Unfortunately this relation is not transitive. A transitive generative relation may be obtained from an unary quality measure.

4.2 A Generative Relation for Nash Equilibrium Induced by an Unary Quality Measure

Let us denote by $c(x)$ the number of players that can improve the strategy x by unilateral deviation (or the number of positions the strategy x can be improved). We may thus write: $c(x) = card\{i|\exists s_{ij} \in S_i, u_i(s_{ij}, x_{-i}) > u_i(x)\}$.

We may define the relation $\prec_{N1}$ as $x \prec_{N1} y$ if and only if $c(x) < c(y)$. It is easy to prove, that if a Nash equilibrium exists then there is a strategy s^* non-dominated with respect to $\prec_{N1}$.
Therefore, using the unary quality measure c, we have defined a transitive generative relation for the Nash equilibrium.

4.3 Nash Equilibrium and the Centipede Game

In case of the centipede game, defection by the first player in the first round is the unique Nash equilibrium of the game. This can be established by backward induction. Suppose two players would reach the final round of the game (and suppose, that at the last round player one would win the larger pot). Player two would win more, if she chooses to stop the game one round earlier. Since player one supposes player two would defect, she wins more defecting in the second to last round. This reasoning proceeds backwards through the game tree until one concludes that the best action is for the first player to defect in the first round. The same reasoning can apply to any node in the game tree.

As depicted in Figure 4, Nash equilibrium is rarely played by real people. In fact in only 7% of the four-move games, 1% of the six-move games, and 10% of the high payoff game the first player stopped at the first move.
Table 1 describes the Nash equilibrium evolutionary detected in the considered versions of the centipede game.

4.4 Generative Relation for Quasi Pareto Equilibrium

Let us define the quantity $P(y,x)$ as the number of players having a better payoff for x than for y: $P(y,x) = card\{i|u_i(y) < u_i(x)\}$.

Table 1. The evolutionary detected Nash equilibrium

Game version	Terminal nodes	Payoffs
Four-move	1	(4, 1)
Six-move	1	(4, 1)
High payoff	1	(16, 4)

Table 2. The evolutionary detected quasi-Pareto equilibrium

Game version	Terminal nodes	Payoffs
Four-move	4	(8, 32)
	5	(64, 16)
Six-move	6	(32, 128)
	7	(256, 64)
High payoff	4	(32, 128)
	5	(256, 64)

Consider the relation $\prec_P$ defined as $y \prec_P x$ if and only if $P(y,x) < P(x,y)$.

We may admit that the relation $\prec_P$ expresses a certain type of Pareto rationality. Otherwise stated the non-dominated strategies with respect to the relation $\prec_P$ represents a variety of Pareto equilibrium, called *quasi-Pareto equilibrium.*

4.5 Quasi-pareto Equilibrium and the Centipede Game

The Pareto equilibrium of the game is reached when players win as much as possible. In the case of the centipede game the payoffs increase towards the end of the game, so the maximum payoffs can be won at the final two rounds. In the case of the studied versions of the centipede game, player A's maximal payoff is at the last round, while player B's is at the round before the final round. From the results presented in Figure 4 we can conclude, that very few people choose to go until the maximal profit, they more likely choose a smaller, but more sure payoff.

The detected quasi-Pareto equilibrium for the studied versions is presented in Table 2.

4.6 Generative Relation for Joint Nash-Pareto Equilibrium

Let us consider two strategies: $x = (x_1, x_2, ..., x_n)$, $y = (y_1, y_2, ..., y_n)$ and two meta-strategies $M_1 = (x_1|r_1, x_2|r_2, ..., x_n|r_n)$, $M_2 = (y_1|r_1, y_2|r_2, ..., y_n|r_n)$. Let us denote by I_N the set of Nash biased players (N-players) and by I_P the set of Pareto biased players (P-players). Therefore we have $I_N = \{i \in \{1, ..., n\}|r_i = \text{Nash}\}$, and $I_P = \{j \in \{1, ..., n\}|r_j = \text{Pareto}\}$.

Let us introduce an operator E, measuring the relative efficiency of meta-strategies: $E : M \times M \to \mathbf{N}$, defined as
$E(M_1, M_2) = card(\{i \in I_N|u_i(x_i, y_{-i}) \geq u_i(y), x_i \neq y_i\} \cup$
$\{j \in I_P|u_j(y) < u_j(x), x \neq y\})$,
where $(x_i, y_{-i}) = (y_1, ..., y_{i-1}, x_i, y_{i+1}, ..., y_n)$.

$E(M_1, M_2)$ measures the relative efficiency of the meta-strategy M_1 with respect to the meta-strategy M_2.

Consider the relation $\prec_{NP}$ defined as $M_1 \prec_{NP} M_2$ if and only if $E(M_2, M_1) < E(M_1, M_2)$. We may consider the relation $\prec_{NP}$ as a generative relation for *joint Nash-Pareto equilibria.*

4.7 Joint Nash-Pareto Equilibrium and the Centipede Game

Table 3 and Table 4 describes the evolutionary detected Nash-Pareto and Pareto-Nash equilibria. These results seem to indicate that in the established joint equilibria the player with a Pareto rationality has a greater payoff. Also, the detected equilibria is closer to the Pareto equilibrium.

Table 3. The evolutionary detected Nash-Pareto equilibrium

Game version	Terminal nodes	Payoffs
Four-move	2	(2, 8)
	4	(8, 32)
Six-move	4	(8, 32)
	6	(32, 128)
High payoff	2	(8, 32)
	4	(32, 128)

Table 4. The evolutionary detected Pareto-Nash equilibrium

Game version	Terminal nodes	Payoffs
Four-move	3	(16, 4)
	5	(64, 16)
Six-move	5	(64, 16)
	7	(256, 64)
High payoff	3	(64, 16)
	5	(256, 64)

4.8 Generative Relation for Fuzzy Nash-Pareto Equilibrium

Each concept of equilibrium may be associated with a rationality type. We consider games with players having several rationality types [5]. Each player can be more or less biased towards a certain rationality type. This bias may be expressed by a fuzzy membership degree. A player may have for instance the membership degree 0.7 to Nash and the membership 0.3 to Pareto. We may also say that the player has a Nash rationality defect of 0.3.

Let us consider a fuzzy set A_N on the player set N i.e. $A_N : N \to [0, 1]$ $A_N(i)$ expresses the membership degree of the player i to the class of Nash-biased players. Therefore A_N is the class of Nash-biased players. Similar a fuzzy set $A_P : N \to [0, 1]$ may describe the fuzzy class of Pareto-biased players.

A fuzzy Nash-Pareto equilibrium concept [6] using an appropriate generative relation is considered in this section. The concept is a natural generalization of the sharp Nash-Pareto equilibrium characterization by generative relations and it is completely different from the notion considered in [7].

Let us consider a game involving both Nash and Pareto-biased players. It is natural to assume that $\{A_N, A_P\}$ represents a fuzzy partition of the player set. Therefore the condition $A_N(i) + A_P(i) = 1$ holds for each player i.

The relative quality measure of two strategies has to involve the fuzzy membership degrees. Let us consider the threshold function:

$$t(a) = \begin{cases} 1, & \text{if } a > 0, \\ 0, & \text{otherwise} \end{cases}$$

The fuzzy version of the quality measure $k(y, x)$ is denoted by $E_N(y, x)$ and may be defined as $E_N(y, x) = \sum_{i=1}^{n} A_N(i)t(u_i(x, y_{-i}) - u_i(y))$.

$E_N(y, x)$ expresses the relative quality of the strategies y an x with respect to the fuzzy class of Nash-biased players.

The fuzzy version of $P(y, x)$ may be defined as $E_P(y, x) = \sum_{i=1}^{n} A_P(i)t(u_i(x) - u_i(y))$, where A_P is the fuzzy set of the Pareto-biased players.

A fuzzy version of the sharp Nash-Pareto equilibrium introduced in [4] can be considered. The relative quality measure of the strategies y and x with respect to fuzzy Nash-Pareto rationality may be defined as $E(y, x) = E_N(y, x) + E_p(y, x)$.

Using the relative quality measure E we can compare two strategy profiles. Let us introduce the relation $\prec_{fNP}$ defined as $y \prec_{fNP} x$ if and only if the strict inequality $E(y, x) < E(x, y)$ holds.

Fuzzy Nash-Pareto equilibrium is the set of non-dominated strategies with respect to the relation $\prec_{fNP}$.

4.9 Fuzzy Nash-Pareto Equilibrium and the Centipede Game

As depicted in Figure 4, when regular people play, Nash equilibrium is rarely observed. Instead, human subjects regularly choose partial cooperation (choose to pass the pot to the opponent) before eventually defect. It is also rare for human players to cooperate trough the whole game, so people neither follow a pure Pareto-type rationality. We can conclude that, in case of the centipede game, the human rationality is between the Nash and Pareto rationality (and can be modeled as a fuzzy combination of Nash and Pareto rationality).

According to Figure 4, the most preferred terminal nodes in case of the four move and high payoff versions of the game are the nodes two and three. In case of the four move version of the game, 35% of the subjects chose to stop at step two and 37% stopped at step three. In case of the high payoff version of the game, these percentages are 37% and 32% respectively. In the case of the six move version of the game, the most often chosen terminal node is the node four with a percentage of 38%.

In order to model the human behavior, we need a new equilibrium concept. Our hypothesis is that the fuzzy Nash-Pareto equilibrium would supply a model for human behavior. For the studied versions of the game, we search for proper types of rationality for the two players that would lead to an equilibrium consisting of the terminal nodes chosen by people. The rationality types are modeled as fuzzy memberships to the Nash and Pareto classes.

Our intuition is that assigning Pareto-type rationality to one player and Nash-type rationality to the other player would lead to the desired equilibrium.

The desired nodes in case of the four move and high payoff version of the game, the nodes two and three, have been obtained with the following parameter settings: player A has any membership degree to Pareto from the interval (0.5, 1). This also means that the membership degree to Nash is in the interval (0, 0.5), meaning that rationality of player A is closer to Pareto. For player B we have assigned a pure Nash rationality (the membership degree to Nash is one, and to Pareto is zero). The same parameter settings worked for both the four move version and the high payoff version of the centipede game.

In case of the six move version of the game, the proper membership degree to Nash for player A is 1 (the membership degree to Pareto is 0), so the player A has a pure Nash rationality. For player B the proper membership degree to Pareto is from the interval (0.5, 1) (the membership degree to Nash is from the interval (0, 0.5)), meaning that the rationality of player B is closer to Pareto. The detected fuzzy Nash-Pareto equilibrium with these memberships contains the node four, which is the most frequently chosen stopping node.

Table 5 describes the evolutionary detected equilibria using the fuzzy Nash-Pareto relation with the above described parameters.

Table 5. Evolutionary detected fuzzy Nash-Pareto equilibria

Game version	Fuzzy memberships	Terminal nodes	Payoffs
Four-move	$A_N(1) \in (0, 0.5)$, $A_P(1) \in (0.5, 1)$, $A_N(1) + A_P(1) = 1$ $A_N(2) = 1$; $A_P(2) = 0$	2 3	(2, 8) (16, 4)
Six-move	$A_N(1) = 1$; $A_P(1) = 0$ $A_N(2) \in (0, 0.5)$ $A_P(2) \in (0.5, 1)$ $A_N(2) + A_P(2) = 1$	4	(8, 32)
High payoff	$A_N(1) \in (0, 0.5)$ $A_P(1) \in (0.5, 1)$ $A_N(1) + A_P(1) = 1$ $A_N(2) = 1$; $A_P(2) = 0$	2 3	(8, 32) (64, 16)

Based on the results presented above, we may conclude that the fuzzy Nash-Pareto equilibrium is a suitable equilibrium (solution) concept for modeling the human decision making in case of the studied versions of the centipede game.

5 Conclusion

Evolutionary equilibria detection techniques based on generative relations are considered. A new transitive generative relation for the Nash equilibrium is introduced.

A connection between equilibrium detection models and experimental game theory is established. A fuzzy combination of equilibria is proposed and it is used to explain the manner people play trust games. The fuzzy equilibrium modeling human behavior does not necessary mean that real players play a fuzzy mixture of Pareto and Nash. The reason of their behavior might be different.

The discrete centipede games are investigated. Particular combinations of fuzzy Nash-Pareto equilibria (explaining observed human decisions) have been detected. We conjecture the proposed evolutionary method would be of interest in modeling human players behavior in a large class of behavioral games. This represents a new connection between computational and behavioral games.

Acknowledgment

The third author wish to thank for the financial support provided from programs co-financed by the Sectoral Operational Programme Human Resources Development, Contract POSDRU/88/1.5/S/60185 Innovative Doctoral Studies in a Knowledge Based Society.

References

1. Bade, S., Haeringer, G., Renou, L.: More strategies, more Nash equilibria, Working Paper 2004-15, School of Economics University of Adelaide University (2004)
2. Deb, K.: Multi-objective optimization using evolutionary algorithms. Wiley, Chichester (2001)
3. Deb, K., Agrawal, S., Pratab, A., Meyarivan, T.: A Fast and Elitist Multi-Objective Genetic Algorithm: NSGA-II KanGAL Report No. 200001, Indian Institute of Tehnology Kanpur (2000)
4. Dumitrescu, D., Lung, R.I., Mihoc, T.D.: Evolutionary Equilibria Detection in Non-cooperative Games. In: EvoStar 2009. LNCS, vol. 5484, pp. 253–262. Springer, Heidelberg (2009)
5. Lung, R.I., Dumitrescu, D.: Computing Nash Equilibria by Means of Evolutionary Computation. Int. J. of Computers, Communications & Control, 364–368 (2008)
6. Dumitrescu, D., Lung, R.I., Mihoc, T.D., Nagy, R.: Fuzzy Nash-Pareto Equilibrium: Concepts and Evolutionary Detection. In: Di Chio, C., Cagnoni, S., Cotta, C., Ebner, M., Ekárt, A., Esparcia-Alcazar, A.I., Goh, C.-K., Merelo, J.J., Neri, F., Preuß, M., Togelius, J., Yannakakis, G.N. (eds.) EvoApplicatons 2010. LNCS, vol. 6024, pp. 71–79. Springer, Heidelberg (2010)
7. Jean-Jacques Herings, P., Mauleon, A., Vannetelbosch, V.J.: Fuzzy play, matching devices and coordination failures. International Journal of Game Theory 32(4), 519–531 (2004)
8. McKelvey, R.D., McLennan, A.: Computation of equilibria in finite games. In: Amman, H.M., Kendrick, D.A., Rust, J. (eds.) Handbook of Computational Economics. Elsevier, Amsterdam (1996)
9. McKelvey, R.D., Palfrey, T.R.: An Experimental Study of the Centipede Game. Econometrica 60, 803–836 (1992)
10. Nash, J.F.: Non-cooperative games. Annals of Mathematics 54, 286–295 (1951)
11. Rosenthal, R.W.: Games of perfect information, predatory pricing and the chain-store paradox. Journal of Economic Theory 25, 92–100 (1981)

Design and Comparison of two Evolutionary Approaches for Solving the Rubik's Cube

Nail El-Sourani and Markus Borschbach

University of Applied Sciences,
Faculty of Computer Science, Chair of Optimized Systems,
Hauptstr. 2, D-51465 Bergisch Gladbach, Germany
nail@elsourani.com, markus.borschbach@fhdw.de

Abstract. Solutions calculated by Evolutionary Algorithms have come to surpass exact methods for solving various problems. The Rubik's Cube multiobjective optimization problem is one such area. In this paper we design, benchmark and compare two different evolutionary approaches to solve the Rubik's Cube. One is based on the work of Michael Herdy using predefined swapping and flipping algorithms, the other adapting the Thistlethwaite Algorithm. The latter is based on group theory, transforming the problem of solving the Cube into four subproblems. We give detailed information about realizing those Evolutionary Algorithms regarding selection method, fitness function and mutation operators. Finally, both methods are benchmarked and compared to enable an interesting view of solution space size and exploration/exploitation in regard to the Rubik's Cube.

1 Introduction

Since the Rubik's Cube's invention by Erno Rubik in 1974 and its commercialization in 1980 it has been the interest of not only hobbyists but also scientific research. Primarily mathematicians found themselves working on the Rubik's Cube as a discrete optimization problem trying to find efficient ways to solve it. With its simple structure the classic Rubik's Cube can reach a total number of $4.3 \cdot 10^{19}$ different configurations which induces an underlying complex optimization problem. To this day it is impossible to calculate all of those configurations. Also, the shortest length of sequences to reach any of those configurations (*God's Number*) is still unknown and subject to ongoing research [8]. In this paper we use two different evolutionary approaches to solve the Rubik's Cube as a discrete optimization problem. First, we briefly describe some characteristics and notation for the classic Rubik's Cube puzzle. We then introduce our two evolutionary approaches, one extending a method by Micheal Herdy [5], the other building upon Thistlethwaite's group-theoretic approach [2], [3], [10]. In contrast to [3] where we introduced our Thistlethwaite Evolution Strategy in detail, this work concentrates on our improvement of the Herdy Evolution Strategy and a thorough examination of both ES' mechanics. The careful benchmark and comparison of both algorithms provide an interesting analysis of solution space size and relation between exploration and exploitation in regard to the Rubik's Cube.

R. Schaefer et al. (Eds.): PPSN XI, Part II, LNCS 6239, pp. 442–451, 2010.

2 The Rubik's Cube

2.1 Structure

The classic 3^3 Rubik's Cube is widely known and the one subject to this paper. It consists of 26 pieces: 8 corner pieces, 12 edge pieces and 6 center pieces, distributed equally on the six sides of the Cube. Each side of the Cube will be called *face*, each 2-dimensional square on a face will be referred to as *facelet*. Corners, edges and centers are all *cubies* - representing the physical object. A corner shows 3 facelets, an edge 2 and a center 1. Each side of the Rubik's Cube can be rotated clockwise (CW) and counterclockwise (CCW). Every such single move changes the position of 4 edges and 4 corners - note that the center facelet on every of the Cube's faces always stays in the same position. Thus, the color of a solved face is always determined by its center color. For each edge and corner it is of great importance to distinguish between *position* and *orientation*: i.e. an edge can be in its right position (defined by the two adjacent center colors) but in the wrong orientation (flipped).

2.2 Notation

There are several known notations [6] for applying single moves on the Rubik's Cube. We will use F, R, U, B, L, D to denote a clockwise quarter-turn of the front, right, up, back, left, down face and Fi, Ri, Ui, Bi, Li, Di for a counterclockwise quarter-turn. Every such turn is a *single move*. In Cube related research half-turns ($F2, R2, U2, B2, L2, D2$) are also counted as single move, we will do so as well. This notation is dependent on the users viewpoint to the cube rather than the center facelets' colors. However, as a convention used for this research work we assume the classic Rubik's Cube color configuration which is $white : yellow, red : orange, blue : green$ where : denotes $opposite\ of$. The starting orientation for the scrambles will be $F = white, R = red, U = blue, B = yellow, L = orange, D = green$.

2.3 Characteristics

Obviously the Rubik's Cube fulfills the characteristics of a mathematical group [4], [9]. The number of all attainable states $4.3 \cdot 10^{19}$ depicts the order of the Cube group $G_C =< F, R, U, B, L, D >$. All configuration of the Rubik's Cube can be reached by using combinations of single moves in this group, thus the single moves *generate* G_C. Further, there is always a neutral element, i.e. $F \cdot FFFF = FFFFF = F$ and $F^4 = 1$ (also showing the order of each generator in G_C is 4) and an inverse: $Fi \cdot F = 1$ and $Fi = FFF$.

The inverse of any operation is quickly calculated by reversing the order of single moves and their direction. For example the inverse of $FRiDLBUi$ would be $UBiLiDiRFi$. Further we can define subgroups of the group G_C. Let $H =< R, U >$ be a such a subgroup. If we only use moves from H there are just $2^6 \cdot 3^8 \cdot 5^2 \cdot 7 = 73483200$ different configurations we can attain [7]. This significantly reduces the number of possible states a Cube can reach, but induces certain constraints like unchangeable edge orientations.

3 Related Work

Only a few evolutionary approaches dedicated to solve the Rubik's Cube exist. In 1994 Herdy devised a method which successfully solves the Cube [5] using pre-defined sequences as mutation operators that only alter few cubies, resulting in very long solutions. Another approach by Castella could not be verified due to a lack of documentation. Recently Borschbach and Grelle [1] devised a 3-stage Genetic Algorithm based on a common human "SpeedCubing" [6] method, first transforming the Cube into a 2x2x3 solved state, then into a subgroup where it can be completed using only two adjacent faces (two-generator group).

4 Groundwork

When designing an ES to solve the Rubik's Cube a few things come to mind. While ES turn out to be very useful problem-solving algorithms for complex optimization problems, typically those problems are described in a system of continuous variables. Small variations of these variables usually lead to small changes of the value of the fitness function (called *strong causality*). This is a fundamental behavior needed in ES to ensure a steady search for better solutions in the solution space. It is of high importance to transfer this behavior to discrete optimization problems. This can be done by adapting suitable mutation operators that work well with the fitness function used. Obviously, the individuals that will be evolved are actual representations of a Rubik's Cube. Starting from the solved state, a certain configuration of a Cube is defined through a sequence of moves applied. A distinct state most certainly has multiple sequences leading to it, the state itself however is unique and one of the $4.3 \cdot 10^{19}$ possible states. A scrambled Cube can be evolved by applying moves that hopefully near it to the solved state. This will be the ground principle of the two forthcoming ES.

4.1 Individual Representation

To enable us a better opportunity for comparison all ES designed for this work use the same representation of a Rubik's Cube. Each face is implemented as a 3×3 2-dimensional matrix containing values from 1 to 6 where each value depicts one color. Thus, one individual is described by 6 matrices, describing each facelets color configuration. Every quarter- and half-turn can be applied to this representation, yielding a total of 18 different single moves while leaving the Rubik's Cube integrity intact. Additionally the whole Cube can be rotated clockwise and counterclockwise. This guarantees easy human verification with a real physical Cube.

4.2 Mutation

Mutation being the primary search operator of ES is easily realized by not modifying a single facelet's color but applying a sequence of moves to the Cube. This guarantees that the Cube's integrity stays intact at all times and makes a separate integrity test superfluous. Allowing mutations which operate on single facelets and change their color could yield non-existent Cube states, due to its structure. Every Cube saves the mutations it has undergone, i.e. a list of moves that have been applied. To keep this list as small as possible, redundant moves are removed automatically. To clarify: we assume a Cube where only F has been applied. Let the next mutation be $FRRiB$. This will automatically yield in $F \cdot FRRiB = F2B$. We will go into further detail when describing each of the Evolution Strategies.

5 Two Evolutionary Approaches to the Rubik's Cube Puzzle

5.1 Herdy Evolution Strategy

In 1994 Michael Herdy presented an (1,50) Evolution Strategy solving the Rubik's Cube in a mean of 38.7 generations, calculating only a mean of 1935 of all possible configurations [5]. This algorithm was enhanced and implemented as follows.

Fitness Calculation. To calculate the fitness of an individual the standard fitness function proposed by Michael Herdy was used. Three qualities q_1, q_2, q_3 are defined:

- q_1 is increased by 1 for each facelet whose color differs from the center facelet on the same face
- q_2 is increased by 4 for each wrong positioned edge, orientation is not considered
- q_3 is increased by 6 for each wrong positioned corner, orientation is not considered

As we see, each of those qualities can reach a maximum of 48, leading us to $max\{q_1 + q_2 + q_3\} = 144$. Obviously the Cube is in a solved state when the sum's value reaches 0.

Mutation Operators. Herdy proposed the following mutation operators which change the fitness of each individual just slightly: two-edge-flip, two-corner-flip, three-edge-swap, two-edge/two-corner-swap, three-corner-swap, two-corner-swap and two-edge-swap - each in both directions (meaning the inverse sequence). Those were slightly extended using mirrors and adding three-inslice-edge-swap [2]. Note that each such mutation operator is depicted by a sequence of $x, 6 \leq x \leq 14$ single moves. The operators in Table 1 change the state of a Cube just slightly by only affecting 2-4 positions while leaving the rest of the Cube intact. This guarantees a slow and steady exploration of the solution space,

Table 1. Herdy ES Mutation Operators (omitting Mirrors)

mutation	sequence	length
two edge flip cw	FRBLULiUBiRiFiLiUiLUi	14
two edge flip ccw	FiLiBiRiUiRUiBLFRURiU	14
two corner flip cw	LDiLiFiDiFUFiDFLDLiUi	14
two corner flip ccw	RiDRFDFiUiFDiFiRiDiRU	14
three edge swap cw	UF2UiRiDiLiF2LDR	10
three edge swap ccw	UiF2ULDRF2RiDiLi	10
two edge/corner swap cw	RiURUiRiUFRBiRBRFiR2	14
two edge/corner swap ccw	LUiLiULUiFiLiBLiBiLiFL2	14
three corner swap cw	FiUBUiFUBiUi	8
three corner swap ccw	FUiBiUFiUiBU	8
three inslice edge swap cw	RLiU2RiLF2	6
three inslice edge swap ccw	LiRU2LRiF2	6

slightly improving the population per generation. To fully utilize the above operators potential, the face being the physical front of the Cube has to be randomly chosen before each mutation as well as the orientation of the Cube. Thus an actual mutation step looks like this:

1. choose a random face to become the new front, this involves the entire Cube to be rotated, there are 6 faces to choose from
2. choose a random orientation of the front by rotating the whole Cube cw/ccw 1 or 2 times

Looking at the mutation we see how only very specialized operators are used. Tests with random sequences or single moves turned out to always get stuck in local optima which corresponds to the discovery made by Borschbach and Grelle. Particularly with this fitness function just a single quarter turn of one face will drastically change the fitness of an individual.

Selection Method. After having applied random mutations to every individual in the population, they are sorted by their fitness in ascending order (remember 0=solved). Now, the μ best individuals are selected for duplication. There are several methods to select an individual for duplication from the selection pool, ranging from very simple ones (gaussian random) to quite complex e.g. non-linear fitness-scaling. It turns out that simply choosing random individuals (with a random function that generates values that are equally distributed within the specified range) is the most effective.

5.2 Thistlethwaite Evolution Strategy

We can say that the grade of specialization of mutation operators corresponds to the grade of restriction of an Evolutionary Strategy. With the Herdy ES turning out to be a very restricted ES we wanted to design an ES with the slightest possible grade of restriction i.e. truly random mutation operators. To achieve this, a common method is to break very complex optimization problems down into smaller parts which are solved independently, at best returning a solution for each subproblem that can finally be joined to get a solution for the original problem. This is exactly how the Algorithm devised by Thistlethwaite in 1981 works. In the original algorithm each subproblem is solved by finding a solution in precalculated lookup-tables. Our ES does not use those lookup-tables to solve those subproblems, but rather evolves Cubes solving each problem by using a dedicated fitness function.

Fitness Calculation. The Thistlethwaite Algorithm (TWA) starts with the Cube in the G_o group, a group where all moves are allowed to be applied to the Cube. This starts a chain of nested groups, i.e. $G_o > G_1 > G_2 > G_3$, with the order of magnitude of each subsequent group being smaller than the one before. The groups are defined as follows [10]:

- $G_0 =< F, R, U, B, L, D >, |G_0| = 4.33 \cdot 10^{19}$
- $G_1 =< F, R2, U, B, L2, D >, |G_1| = 2.11 \cdot 10^{16}$
- $G_2 =< F2, R2, U, B2, L2, D >, |G_2| = 1.95 \cdot 10^{10}$
- $G_3 =< F2, R2, U2, B2, L2, D2 >, |G_3| = 6.63 \cdot 10^{5}$

As we can see, with every subsequent group the total number of possible states the Cube can achieve (and therefore the number of states the Cube has to be in, to be solved) by only using moves from this distinct group, is greatly reduced (detailed group order calculation in [3]). Naturally the higher the group index, the more constraints we have to fulfill. Once a Cube is maneuvered into a subgroup, we can freely apply all moves of this subgroup and it will stay in this group. Furthermore we must only use moves of this subgroup to put it into the next group and so forth. We can not use any moves from previous groups which are not part of the current subgroup without destroying what we have reached so far.

Assuming we have some random scrambled Cube. Naturally this Cube will be in G_0 which allows us to apply any move we want. Now, by using moves from G_0 we put the Cube into G_1. Applying R would put it back into G_0 again, as this move is not part of G_1 and so forth. Thus, we choose to calculate fitness separately for each group transition. This is done via

$$phase_i = weight \cdot v + c, i = 0, 1, 2, 3 \tag{1}$$

where v is a count for wrong positioned/oriented cubies in regard to the constraints induced by the particular subgroup and c the length of moves already applied to the current individual. Thus, not only the necessary group transition but also a short solution sequence length is promoted as the desired goal.

To clarify the constraints induced by different subgroups let us take a closer look at the group G_1. Changing an edge's orientation on the Rubik's Cube implicates the ability to use quarter-turns of adjacent faces. Evidently, this is not possible once in G_1. Thus, G_1 induces the constraint that each edge cubie must be oriented (not necessarily positioned) right. Further subgroups induce further constraints, detailed in [3].

Mutation Operators. Dividing the problem of solving the Rubik's Cube into 5 phases by splitting it into 4 subproblems gives us the ability to use truly random sequences of single moves as mutation operators. The only restriction is given by the group transition the Cube is undergoing. Conveniently the maximum sequence length needed to transform the Cube from one subgroup to another is given by Thistlethwaite: $G_0 \rightarrow G_1 : 7$, $G_1 \rightarrow G_2 : 13$, $G_2 \rightarrow G_3 : 15$ and $G_3 \rightarrow solved : 17$.

In each phase, a sequence of random length between 0 and *max. sequence length (=i)* is generated by filling each position with a random single move of the according group. By allowing the sequence length to be 0, we are artificially inducing characteristics of an $(\mu+\lambda)$-ES while actually being an (μ, λ)-ES. Unlike the Herdy ES where with each mutation the number of applied moves increases, here a decrease is possible induced by the automatic cleanup of move sequences.

Selection Method. Tests have shown no need for a dedicated Selection method. Standard EA selection methods except for random choice tend to decrease the population's diversity after phase transitions. In early versions this would sometimes lead to a local optimum, thus not solving the Cube. This results from the TW ES being very sensitive about selection pressure. To counter this property a gaussian random function is used to choose from the selection pool for duplication.

6 Benchmark and Comparison

For benchmark and comparison a total of 100 random scrambles between 10 and 50 single moves were generated and solved by both ES in 5 repetitions. These tests were conducted on the same hardware and conditions using the shared Rubik's Cube framework proposed earlier (individual representation and mutation mechanic). The numbers used for μ and λ were chosen after preleminary benchmarks found them to be the most effective. The high parent population is necessary to provide the possibility for a broad spectrum of different sequence combinations in later generations. In short, large parent populations enable a high individual diversity - which is crucial in a solution space of this size. This is even more critical to the Thistlethwaite ES due to the random mutation operators used, contrary to the hard-coded sequences in the Herdy ES.

6.1 Herdy Evolution Strategy

The Herdy Evolution Strategy with $(\mu, \lambda) = (100, 5000)$ is used. Calculations were repeated 5 times to enable a statistically more precise evaluation. Fast performance, a small number of generations needed and long solution sequences seem to be typical properties of the Herdy ES. Results for this particular Evolution Strategy are very predictable and perfectly fit the mechanics described above.

Table 2. Solutions of 100 random scrambles, 5 repetitions, Herdy ES

	run 1	run 2	run 3	run 4	run 5
avg. generations	18.13	18.28	18.06	18.42	18.11
avg. moves	232.27	234.50	233.43	236.62	236.21
avg. time(s)	5.66	5.12	5.00	4.66	4.75

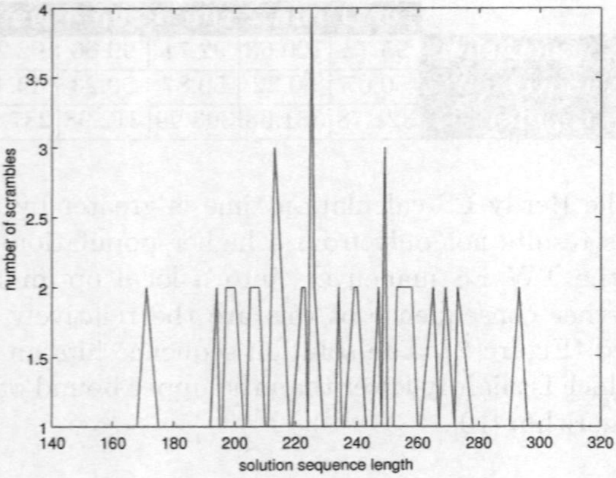

Fig. 1. 100 Random Scramble Test: Distribution of scrambles on solution length, Herdy ES

96% of the scrambles (Figure 1) could be solved in x moves with $180 < x < 280$, roughly resembling a normal distribution around the x value with $y = max(x)$. This is when providing a sufficient number and/or versatility of scrambles.

Comments. The Herdy ES is really a theoretically simple ES for solving the Rubik's Cube. It performs amazingly fast, only calculating a tiny fraction of the total number of possible Cube states, solving the Cube in only few generations (Table 2). On the downside the mean length of the solution sequences calculated by this algorithm can be estimated by

$$11 \cdot g \,|\, g := number\,of\,generations\,needed \tag{2}$$

which is quite huge. Solution sequences with a length of well above 100 are quite normal. In addition the "random element" which makes up the "evolutionary-ness" of an ES is quite small - only affecting which cubies will be swapped or flipped as the mutation sequences are all hard-coded. Tests were conducted that added the length of the applied moves in each generation to the fitness function. No improvements could be observed which is not surprising regarding the above arguments.

6.2 Thistlethwaite Evolution Strategy

The following settings were used: $(\mu, \lambda) = (1000, 50000)$, weighing factors are $(5, 5, 5, 5, 5)$, mutation lengths $(5, 5, 13, 15, 17)$ and maximum generations before reset (250).

Table 3. Solutions of 100 random scrambles, 5 repetitions, Thistlethwaite ES

	run 1	run 2	run 3	run 4	run 5
avg. generations	95.72	100.63	92.71	99.66	92.22
avg. moves	50.67	50.32	50.87	50.23	49.46
avg. time(s)	321.78	381.68	393.99	312.98	287.93

Compared to the Herdy ES calculation time is greater by a factor of about 60 (Table 3). This results not only from a higher population size but also, for some scrambles, the TW ES maneuvers into a local optimum which triggers a self-reset. A further consequence of this are the relatively high numbers of generations needed (Figure 2). The solution sequence hits an average of about 50 single moves which is slightly lower than the upper bound of 52 for the classic Thistlethwaite Algorithm [10].

Comments. The Thistlethwaite ES yields results that resemble the theoretical solution lengths of the classic Thristlethwaite's Algorithm without any use of precalculated lookup-tables. Unlike the Herdy ES which is efficient with a small population size, the TW ES needs much larger populations to guarantee a successful solve. This directly affects calculation time. Even though the Cube solving is broken down into 4 smaller subproblems, each subproblem is still of great order.

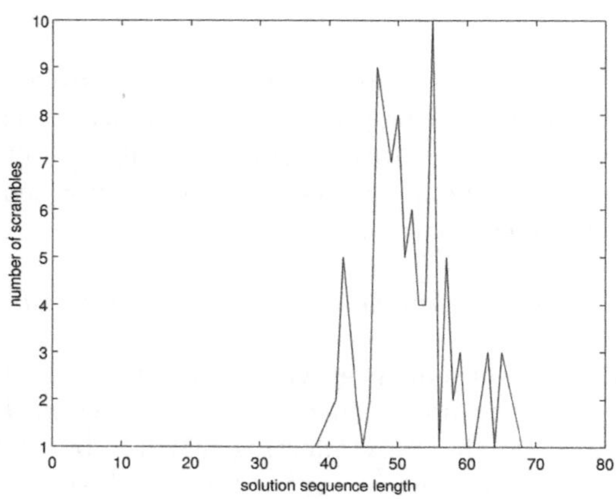

Fig. 2. 100 Random Scramble Test: Distribution of scrambles on solution length, Thistlethwaite ES

7 Conclusion

When comparing the results of both Evolution Strategies we notice a tradeoff between computing time and quality of solution. Although very different in nature both ES solve the same optimization problem. Obviously there are some important differences. The Herdy ES is a $(100, 5000)$ ES which uses predefined mutation operators and incorporates a very simple fitness function.

Each of those are reasons for the faster performance of the Herdy ES. We can look at it the opposite way. The Thistlethwaite ES is a $(1000, 50000)$ ES which uses randomly generated mutation operators and incorporates a complex fitness function dividing the solving process into four subproblems.

The differences seem obvious. While the Herdy ES provides fast but long solutions to any scramble, the Thistlethwaite ES delivers better results in terms of solution length. This is easily explained with the mutation operators used. While the TW ES could solve the very simple scramble F by simply returning Fi as the solution, the Herdy ES would need to perform 2 edge and 2 corner swaps, resulting in a very long solution for such a simple scramble. This disadvantage becomes less important when solving more complex scrambles, however the TW ES performs a shorter solution at all times. Running the TW ES with smaller μ, λ yields faster calculation but can not guarantee a successful solve. More tests will have to be conducted for further examination of the TW ES's behavior with different parameters and sequence optimization which could result in faster computation and/or shorter solution lengths and occurrence reduction of self resets.

References

1. Borschbach, M., Grelle, C.: Empirical Benchmarks of a Genetic Algorithm Incorporating Human Strategies. Technical Report, University of Applied Sciences, Bergisch Gladbach (2009)
2. El-Sourani, N.: Design and Benchmark of different Evolutionary Approaches to Solve the Rubik's Cube as a Discrete Optimization Problem. Diploma Thesis, WWU Muenster, Germany (2009)
3. El-Sourani, N., Hauke, S., Borschbach, M.: An Evolutionary Approach for Solving the Rubik's Cube Incorporating Exact Methods. In: Di Chio, C., Cagnoni, S., Cotta, C., Ebner, M., Ekárt, A., Esparcia-Alcazar, A.I., Goh, C.-K., Merelo, J.J., Neri, F., Preuß, M., Togelius, J., Yannakakis, G.N. (eds.) EvoApplicatons 2010. LNCS, vol. 6024, pp. 80–89. Springer, Heidelberg (2010)
4. Frey, A., Singmaster, D.: Handbook of Cubic Math. Enslow, Hillside (1982)
5. Herdy, M., Patone, G.: Evolution Strategy in Action, 10 ES-Demonstrations. Technical Report, International Conference on Evolutionary Computation (1994)
6. Petrus, L.: Solving Rubik's Cube for speed, `http://lar5.com/Cube/`
7. Reid, M.: Cube Lovers Mailing List, `http://www.math.rwth-aachen.de/ Martin.Schoenert/Cube-Lovers/Index_by_Author.html`
8. Rokicki, T.: Twenty-Three Moves Suffice, `http://Cubezzz.homelinux.org/drupal`
9. Singmaster, D.: Notes on Rubik's Magic Cube. Enslow, Hillside (1981)
10. Thistlethwaite, M.B.: The 45-52 Move Strategy. London CL VIII (1981)

Statistical Analysis of Parameter Setting in Real-Coded Evolutionary Algorithms

Maria I. García Arenas, Pedro Ángel Castillo Valdivieso, Antonio M. Mora García, Juan J. Merelo Guervós, Juan L. Jiménez Laredo, and Pablo García-Sánchez

Department of Architecture and Computer Technology
University of Granada, Spain
mgarenas@atc.ugr.es

Abstract. When evolutionary algorithm (EA) applications are being developed it is very important to know which parameters have the greatest influence on the behavior and performance of the algorithm. This paper proposes using the ANOVA (ANalysis Of the VAriance) method to carry out an exhaustive analysis of an EA method and the different parameters it requires, such as those related to the number of generations, population size, operators application and selection type. When undertaking a detailed statistical analysis of the influence of each parameter, the designer should pay attention mostly to the parameter presenting values that are statistically most significant. Following this idea, the significance and relative importance of the parameters with respect to the obtained results, as well as suitable values for each of these, were obtained using ANOVA on four well known function optimization problems.

1 Introduction

When using search heuristics such as evolutionary algorithms (EAs), simulated annealing and local search algorithms, components such as genetic operators, selection and replacement mechanisms, and the initial population, must first be chosen. The parameters used to apply some of these elements determine the way they operate and influence the results obtained. Obtaining suitable values for them is an expensive, time-consuming and laborious task.

One of the most common ways of setting these parameters is by hand, after intensive experimentation with different values [21]. As Eiben states, the straightforward approach is to generate and test [8,28]. An alternative is to use a meta-algorithm to optimise the parameters [15], that is, to run a higher level algorithm that searches for an optimal and general set of parameters to solve a wide range of optimisation problems.

However, as some authors remark, solving specific problems requires specific parameter value sets [3,18,10] and, as Harik [17] claims, nobody knows the "optimal" parameter settings for an arbitrary real-world problem. Therefore, establishing the optimal set of parameters for a sufficiently general case is a difficult problem.

R. Schaefer et al. (Eds.): PPSN XI, Part II, LNCS 6239, pp. 452–461, 2010.

Current best practices are based on intensive test, ad-hoc choices and conventions [29,8,28,9], that is why new practices, based on solid tuning methods (i.e. robust mathematical methods), are needed. Such a methodology is what we intend to present in this paper.

Genetic algorithm users adjust their main design parameters (crossover probability, mutation probability, population size, number of generations, selection rate) by hand [6,19]. The decision as to which values are best is usually made in terms of the most common values or experimental formulae given in the bibliography, or by trial and error [15,22].

However, other researchers have proposed determining a good set of evolutionary algorithm parameters by analogy, undertaking a theoretical analysis [2,13,14,16,25,31]. Establishing parameters by analogy means using suitable sets of parameters to solve similar problems. However they do not explain how to measure the similarity between problems. Also, a clear protocol has not been proposed for situations when the similarity between problems implies that the most suitable sets of parameters are also similar [18,10]. Weyland has described a theoretical analysis of both an evolutionary algorithm [20] and simulated annealing algorithm [33] to search for the optimal parameter setting to solve the longest common subsequence problem. However, Weyland does not carry out this approach to practice.

Some authors have proposed practical approaches that eliminate the need for a parameter search in genetic algorithms [17]. In these works, a set of parameters is found, but instead of finding them by means of intense experimentation, the parameter settings are backed up with theoretical work - meaning that these settings are robust.

New approaches to the problem of establishing parameter values [23] have been proposed. Several proposals are based on setting parameter values on-line (during the run) instead of testing and comparing different values before the run (parameter tuning). In this sense, some authors propose self-adaptation of parameters (coding those parameters in the individual's genome); others propose non-static parameter settings techniques (controlled by feedback from the search and optimization process) [9,28,30]. However, control strategies also have parameters and there are indicators that good tuning works better than control [8].

Finally, authors proposed in [4] using the ANOVA (ANalysis Of the VAriance) [12] statistical method to analyze the main parameters involved in the design of a neuro-genetic algorithm.

It is very important to know which parameter values involved in the design of an optimization method have the greatest influence on its behaviour and performance and to obtain accurate values for those parameters. In any case, after performing a detailed statistical analysis of the influence of each parameter, the designer should pay attention mostly to the parameter providing the values that are statistically most significant. In this paper, we propose using the ANOVA statistical method as a powerful tool to analyze a real-coded EA to solve function approximation problems.

The ANOVA method allows us to determine whether a change in the results (responses) is due to a change in a parameter (variable or factor) or due to a random effect. Thus it is possible to determine the variables that have the greatest effect on the method that is being evaluated.

The theory and methodology of ANOVA was mainly developed by R.A. Fisher during the 1920s [12]. ANOVA examines the effects of one, two or more quantitative or qualitative variables (called factors) on one quantitative response. ANOVA is useful in a range of disciplines when it is suspected that one or more factors might affect a response. ANOVA is essentially a method used to analyse the variance to which a response is subjected, dividing it into the various components corresponding to the sources of variation, which can be identified.

With ANOVA, we test a null hypothesis that all the population means are equal against the alternative hypothesis that there is at least one mean that is not equal to the others. We find the sample mean and variance for each level (value) of the main factor. Using these values, we obtain a significance value (Sig. Level). If this level is lower than 0.05, then the influence of the factor is statistically significant at the confidence level of 95%.

After applying ANOVA (to determine if means are different), another tests must be used to determine which are different; that will give information about which parameter values are more accurate. In this sense, either TukeyHDS (Tukey's Honestly Significant Difference) [7] or Bonferroni [24] tests can be used.

In this paper, ANOVA will be used to determine the most important parameters of an EA (in terms of their influence on the results), and to establish the most suitable values for such parameters (thus obtaining an optimal operation).

The rest of this paper is structured as follows: Section 2 describes the EA and the parameters we propose to be evaluated. This section contains an exhaustive analysis of the method, showing how the parameters are determined. Section 3 details the experimental setup and the statistical study using ANOVA. Obtained results are analysed in order to establish the most suitable values. Finally, a brief conclusion and future work is presented in section 4.

2 The EA Method and the Experimental Setup

The purpose of this study is to analyze the dynamics of a typical EA [3,11], to determine which parameters influence the obtained fitness and to find an adequate value for the following parameters:

- **Generations** (G)**:** number of generations.
- **Population Size** (P)**:** number of individuals in the population.
- **Selector** (S)**:** Selection operator to generate the offspring. In this paper a roulette wheel selector, a random selector and a selector based on always taking the best individual in the popultation are proposed.
- **Operators Combination** (O)**:** This parameter refers to the percentage of offspring generated using either an uniform mutator or a BLX-α crossover operator.

Table 1 shows the different levels used to evaluate these parameters using ANOVA. The response variable used to perform the statistical analysis is the fitness in the last generation. The changes in the response variable are produced when a new combination of parameters is considered.

Table 1. Parameters (factors) and the abbreviation used as reference later that determine the EA behaviour and values used to apply ANOVA

Generations (G)	Population Size (P)	Operators Combination (O)	Selector (S)
160 320 640 1280 2560 5120	100 200 400 800 1600 3200	- Crossover only (C) - Mutation only (M) - Crossover (80%) and Mut. (20%) (C8M2) - Crossover (90%) and Mut. (10%) (C9M1)	- Roulette Wheel - Random - Always the best individual

Thus, 12960 runs were carried out for each problem (30 times * 6 levels for G * 6 levels for P * 4 levels for O * 3 levels for S, that represent the possible combinations) to obtain the fitness for each combination.

The application of ANOVA consisted in running an EA using these parameter combinations to obtain the best fitness. Then R [1] was used to obtain the ANOVA table. In this paper a simplified table is shown, including for each factor, the sum of squares (Sum Sq), the value of the statistical F (F value) and its significance level (Sig. Level). As previously stated, if the latter is smaller than 0.05, then the factor effect is statistically significant at a 95% confidence level (what means that some means are different for these parameter values).

In order to evaluate the EA and its parameters, four function approximation problems are used:

- The Griewangk function [1] is a continuous multimodal function with a high number of local optima. Its global optimum is located at $(0, ..., 0)$ [5]. This problem has been used with vectors of 100 real numbers in the interval $[-512, 512]$.
- The Rastrigin function [32] is a multimodal real function optimisation problem, whose global optimum is located at point 0 and whose minimum value is 0. This problem has been addressed with vectors of 100 real numbers in the interval $[-512, 512]$.
- The Normalized Schwefel function [26] is a multimodal separable real function optimisation problem, whose global optimum is located at point $(x_0, ..., x_d)$ with $x_i = 420.96$, and whose minimum value is $(y_0, ..., y_d)$ with $y_i = -418.9828$. Vectors of 100 real numbers in the interval $[-512, 512]$ were used.
- The Shekel function [27] is a multimodal real function optimisation problem, whose optimum for dimension 5 is located at point $x_1 = 8.02$, $x_2 = 9.15$,

[1] `http://www.r-project.org`

$x_3 = 5.11$, $x_4 = 7.62$, $x_5 = 4.56$ and whose minimum value is -10.4056. This problem has been addressed with vectors of 5 real numbers in the interval $[-10, 10]$.

In all cases, the fitness of an individual is calculated as the distance to the optimum for that function (the optimum is known).

3 Statistical Study and Results Obtained

In this section, the ANOVA statistical tool is applied to determine whether the influence on parameter values (factors) is significant in the obtained fitness, (to obtain an optimal operation). The set of tests carried out to apply the ANOVA method and thus to determine the most suitable parameter values, is described in detail. In all cases, the goal is to obtain the smallest fitness for the optimised function.

Table 2. ANOVA tables for the fitness (response) with the EA parameters as factors. Those parameters with a significance level over 95% are in bold. Although the full ANOVA table includes combinations of 2, 3 and 4 parameters, only results related to single parameters are shown.

Param.	Sum Sq	F value	Pr(> F)
G	71278	2.4251	0.1202
O	4540135	154.4671	**<2e**$^{-16}$
P	29353	0.9987	0.3182
S	2781525	94.6346	**<2e**$^{-16}$

Griewangk

Param.	Sum Sq	F value	Pr(> F)
G	$1e^{+12}$	2.1689	0.1416
O	$7.4e^{+13}$	154.4393	**<2e**$^{-16}$
P	$5.3e^{+11}$	1.1018	0.2945
S	$4.6e^{+13}$	94.9568	**<2e**$^{-16}$

Rastrigin

Param.	Sum Sq	F value	Pr(> F)
G	6	0.0162	0.898701
O	52393	135.3265	**<2.2e**$^{-16}$
P	3857	9.9611	**0.002**
S	36652	94.6682	**<2.2e**$^{-16}$

Normalized Schwefel

Param.	Sum Sq	F value	Pr(> F)
G	9.79	2.3170	0.12873
O	18.73	4.4313	**0.036**
P	79.09	18.7111	**1.9e**$^{-05}$
S	118.79	28.1017	**1.9e**$^{-07}$

Shekel

We will determine if the influence on a parameter value is significant in the value of the approximation function (fitness).

Table 2 shows the result of applying ANOVA on proposed approximation function problems. Parameters with a signification level over 95% are highlighted in boldface. The ANOVA analysis shows that O and S parameters influence the obtained fitness, which indicates that changes in these parameters influence the results significantly. However, this influence is not as important for the rest of the parameters in all the cases (problems).

Table 3. Griewangk function: obtained error for the different parameter levels. This table shows the effect each level has on the fitness.

Param.	Means					
G	160	320	640	1280	2560	5120
	129.5	95.3	81.6	74.3	74.9	**70.2**
O	C	M	C8M2		C9M1	
	289.0	57.8	**1.7**		2.1	
P	100	200	400	800	1600	3200
	104.5	93.1	91.1	83.1	79.1	**75.0**
S	roulette		random		best	
	21.7		23.0		218.3	

Table 4. Rastrigin function: obtained error for the different parameter levels. This table shows the effect each level has on the fitness.

Param.	Means					
G	160	320	640	1280	2560	5120
	525797	387416	321216	296155	**288133**	295020
O	C	M	C8M2		C9M1	
	1167593	228796	**5560**		7211	
P	100	200	400	800	1600	3200
	419618	382480	359795	337218	316751	**297876**
S	roulette		random		best	
	85272		90843		880754	

In Normalized Schwefel and Shekel, the P parameter is significant too. This fact shows how for each problem different set of parameters can influence results in a different manner.

Once the parameters with greater influence on the results are determined, accurate parameter values should be established in order to obtain an optimal operation. To do so, tables of means are calculated to show the effect each level has on the approximation error.

The obtained error for the Griewangk, Rastrigin, Normalized Schwefel and Shekel functions and the different parameter levels are shown in Tables 3, 4, 5 and 6.

Lets examine each one of the parameters in turn:

- Focusing attention to the operator combinations (O), using either only mutation or only crossover leads to worse fitness results, which was only to be expected. According to the tables, using a low crossover and high mutation probabilities to generate offspring in each generation is the more accurate. However, as the mutation role is to generate diversity, reducing too much the mutation probability leads to premature convergence of the population.

Table 5. Normalized Schwefel function: obtained error for the different parameter levels. This table shows the effect each level has on the fitness.

Param.	Means					
G	160	320	640	1280	2560	5120
	8.146208	7.848331	8.151095	9.062385	8.641955	**7.645236**
O	C	M	C8M2		C9M1	
	$3.3e^{+01}$	$2.2e^{-01}$	$8.1e^{-04}$		$6.6e^{-03}$	
P	100	200	400	800	1600	3200
	14.9	11.0	8.0	6.8	5.1	**3.8**
S	roulette		random		best	
	0.8		**0.7**		23.3	

Table 6. Shekel function: obtained error for the different parameter levels. This table shows the effect each level has on the fitness.

Param.	Means					
G	160	320	640	1280	2560	5120
	7.32	6.90	6.78	6.70	6.64	**6.63**
O	C	M	C8M2		C9M1	
	8.10	**4.14**	7.49		7.60	
P	100	200	400	800	1600	3200
	7.57	7.30	6.94	6.65	6.34	**6.19**
S	roulette		random		best	
	6.37		**6.48**		7.65	

On the other hand, applying too much mutation affects exploration, leading to random search.

- Both P and G have not been reported as significant according to the ANOVA table in all problems. However, as it can be observed, using the higher values yields better fitness values, although those values lead to a higher number of evaluations and time needed to run the algorithm. Logically, the greater the number of generations or population size, the more possibilities there are of achieving a good individual from the current population, as there exists a greater variety of individuals.
- Taking into account the S parameter, using the roulette wheel selector yields much better results than using a random selector or always taking the best individual. The reason for this is that the roulette selector produces the greatest diversification in the EA solutions. The third selector is the most elitist; obtained results using this one selector (taking the best individual) are much worse, which indicates that too much selective presion is not appropriate (the population must be diverse, otherwise a premature convergence of the algorithm might occur).

These conclusions have been confirmed by applying the Bonferroni statistical test [24]. As can be seen, these results are in agreement with those available in bibliography. In any case, in this paper we have verified the adequacy of these parameter values through a rigorous statistical study.

4 Conclusions and Work in Progress

A statistical study of the different parameters involved in the design of an EA has been carried out by using ANOVA, which consists of a set of statistical techniques that analyze and compare experiments by describing the interactions and interrelations between the variables (factors) of the system. The motivation of the present statistical study lies in the great variety of alternatives that a designer has to take into account when designing an EA.

Proposed methodology has been applied to four well known function approximation problems, widely used by practitioners, having determined which parameters have a higher influence on obtained results (a change on those parameters will affect the fitness). Likewise, the more accurate value has been determined for those parameters in order to obtain an optimal operation.

Accurate results are obtained using the higher value tested for population size and number of generations, although this increases the number of evaluations and time needed to run the algorithm. Obviously, as the number of generations or individuals in the population increases, there is a greater probability that the fitness of the best individual will be higher.

Paying attention to the selector operator, a roulette wheel selector yields much better results than using a random selector or always taking the best individual, due to the fact that the roulette selector produces the greatest diversification in the EA solutions.

As far as the genetic operator application rates are concerned, reducing too much the mutation probability leads to premature convergence of the population, while applying too much mutation is like a random search. According to the tables, using a low mutation and high crossover probabilities to generate offspring in each generation is the more accurate.

This methodology based on ANOVA and Bonferroni statistical tests could be helpful for practitioners in analyzing and adjusting parameters of any optimisation method.

Our work in progress includes the analysis of modified EA considering other selection schemes, new genetic operators and other meta-heuristics. As future work, the implementation of a parameter control method would be of interest, as proposed by Eiben et al. [9]. In this case, ANOVA could be used to analyse not only the optimisation method parameters but also the control strategy parameters.

Acknowledgments

This work has been supported in part by the Spanish MICYT project NoHNES (Spanish Ministerio de Educacion y Ciencia - TIN2007-68083), the Junta de

Andalucía P06-TIC-02025, P07-TIC-03044, P08-TIC03928 projects and the Jaén University UJA-08-16-30 project.

References

1. Griewank, A.O.: Generalized descent for global optimization. Journal of Optimization Theory and Applications 34(1), 11–39 (1981)
2. Bäck, T.: Optimal mutation rates in genetic search. In: Forrest, S. (ed.) Proceedings of the 5th International Conference on Genetic Algorithms, pp. 2–8. Morgan Kaufmann, San Francisco (1993)
3. Bäck, T., Fogel, D., Michalewicz, Z.: Handbook of Evolutionary Computation. Institute of Physics Publishing Ltd. Bristol and Oxford University Press, NY (1997)
4. Castillo, P.A., Merelo, J.J., Romero, G., Prieto, A., Rojas, I.: Statistical Analysis of the Parameters of a Neuro-Genetic Algorithm. IEEE Trans. on Neural Networks 13(6), 1374–1394 (2002) ISSN:1045-9227
5. Cho, H., Olivera, F., Guikema, S.D.: A derivation of the number of minima of the griewank function. Applied Mathematics and Computation 204(2), 694–701 (2008)
6. Davis, L.: Handbook of genetic algorithms. Van Nostrpand Reinhold, NY (1991)
7. Dickinson, P., Chow, B.: Some properties of the Tukey test to Duckworth's specification by Peter Dickinson, Bryant Chow. Office of Institutional Research, University of Southwestern Louisiana, Lafayette, Louisiana, USA (1971)
8. Eiben, A.E.: Principled Approaches to tuning EA parameters. In: Tutorials - IEEE Congress on Evolutionary Computation (CEC 2009) (2009), `http://www.few.vu.nl/~gusz/papers/eiben-cec-2009-tutorial-corrected.pdf`
9. Eiben, A.E., Michalewicz, Z., Schoenauer, M., Smith, J.E.: Parameter control in evolutionary algorithms. In: Parameter Setting in Evolutionary Algorithms. Studies in Computational Intelligence, vol. 54, pp. 19–46. Springer, Heidelberg (2007) ISBN 978-3-540-69431-1, ISSN 1860-949X
10. Eiben, A.E., Hinterding, R., Michalewicz, Z.: Parameter Control in Evolutionary Algorithms. IEEE Transactions on Evolutionary Computation 3(2), 124–141 (1999), doi:10.1109/4235.771166, ISSN: 1089-778X
11. Eiben, A.E., Smith, J.E.: Introduction to Evolutionary Computing. Springer, Heidelberg (2003) ISBN 3-540-40184-9
12. Fisher, R.A.: Theory of Statistical Estimation. In: Proceedings of the Cambridge Philosophical Society, vol. 22, pp. 700–725 (1925)
13. Goldberg, D.E.: Genetic Algorithms in Search, Optimization and Machine Learning. Addison-Wesley Longman Publishing Co., Inc., Boston (1989)
14. Goldberg, D.E., Deb, K., Theirens, D.: Toward a better understanding of mixing in genetic algorithms. In: Belew, R.K., Booker, L.B. (eds.) Proc. of the 4th Int. Conf. on Genetic Algorithms, pp. 190–195. Morgan Kaufmann, San Francisco (1991)
15. Grefenstette, J.J.: Optimization of control parameters for genetic algorithms. IEEE Trans. Systems, Man, and Cybernetics, SMC- 16(1), 122–128 (1986)
16. Harik, G., Cantú-Paz, E., Goldberg, D.E., Miller, B.L.: The gambler's ruin problem, genetic algorithms, and the sizing of populations. In: Proceedings of the 4th IEEE Conference on Evolutionary Computation, pp. 7–12. IEEE Press, Los Alamitos (1997)
17. Harik, G.R., Lobo, F.G.: A parameter-less genetic algorithm. In: Banzhaf, W., Daida, J., Eiben, A.E., Garzon, M.H., Honavar, V., Jakiela, M., Smith, R.E. (eds.) Proceedings of the Genetic and Evolutionary Computation Conference, Orlando, Florida, USA, 13-17, vol. 1, pp. 258–265. Morgan Kaufmann, San Francisco (1999)

18. Hinterding, R., Michalewicz, Z., Eiben, A.E.: Adaptation in Evolutionary Computation: A Survey. In: Proceedings of the 4th IEEE Conference on Evolutionary Computation, pp. 65–69. IEEE Press, Los Alamitos (1997)
19. Jagielska, I., Matthews, C., Whitfort, T.: An investigation into the application of neural networks, fuzzy logic, genetic algorithms, and rough sets to automated knowledge acquisition problems. Neurocomputing 24(1-3), 37–54 (1999)
20. Jansen, T., Weyland, D.: Analysis of evolutionary algorithms for the longest common subsequence problem. In: GECCO 2007: Proceedings of the 9th Annual Conference on Genetic and Evolutionary Computation, pp. 939–946. ACM, New York (2007)
21. De Jong, K.A.: An analysis of the behavior of a class of genetic adaptive systems. Ph.D. thesis, University of Michigan, Ann Arbor (1975)
22. Kim, D., Kim, C.: Forecasting time series with genetic fuzzy predictor ensemble. IEEE Transactions on Fuzzy Systems 5(4), 523–535 (1997)
23. Lobo, F.G., Lima, C.F., Michalewicz, Z.: Parameter Setting in Evolutionary Algorithms. Springer Publishing Company, Incorporated, Heidelberg (2007)
24. Savin, N.E.: The bonferroni and the scheff multiple comparison procedures. Review of Economic Studies (XLVII), pp. 255–273 (1980)
25. Schaffer, J.D., Morishima, A.: An adaptive crossover distribution mechanism for genetic algorithms. In: Grefenstette, J.J. (ed.) Proceedings of the 2nd International Conference on Genetic Algorithms and Their Applications, pp. 36–40. Lawrence Erlbaum Associates, Mahwah (1987)
26. Schwefel, H.-P.: Numerical Optimization of Computer Models. John Wiley & Sons, Inc., New York (1981)
27. Shekel, J.: Test functions for multimodal search techniques. In: Fifth Annual Princeton Conference on Information Science and Systems, pp. 354–359 (1971)
28. Smit, S.K., Eiben, A.E.: Comparing parameter tuning methods for evolutionary algorithms. In: Haddow, P., et al. (eds.) CEC 2009: Proc. of the Eleventh conference on Congress on Evolutionary Computation, Piscataway, NJ, USA, pp. 399–406. IEEE Press, Los Alamitos (2009)
29. Smit, S.K., Eiben, A.E.: Using Entropy for Parameter Analysis of Evolutionary Algorithms. In: Beielstein, B., et al. (eds.) Empirical Methods for the Analysis of Optimization Algorithms. Natural Computing Series. Springer, Heidelberg (2009)
30. Smit, S.K., Eiben, A.E.: Parameter tuning of evolutionary algorithms: Generalist vs. specialist. In: Di Chio, C., et al. (eds.) EvoApplicatons 2010. LNCS, vol. 6024, pp. 542–551. Springer, Heidelberg (2010)
31. Thierens, D.: Dimensional analysis of allele-wise mixing revisited. In: Ebeling, W., Rechenberg, I., Voigt, H.-M., Schwefel, H.-P. (eds.) PPSN 1996. LNCS, vol. 1141, pp. 255–265. Springer, Heidelberg (1996)
32. Torn, A., Zilinskas, A.: Global Optimization. LNCS, vol. 350, p. 124. Springer, Heidelberg (1989)
33. Weyland, D.: Simulated annealing, its parameter settings and the longest common subsequence problem. In: GECCO 2008: Proceedings of the 10th Annual Conference on Genetic and Evolutionary Computation, pp. 803–810. ACM, New York (2008)

Performance of Network Crossover on NK Landscapes and Spin Glasses

Mark Hauschild and Martin Pelikan

Missouri Estimation of Distribution Algorithms Laboratory, 320 CCB;
University of Missouri at St. Louis; One University Blvd., St. Louis, MO 63121
mwh308@umsl.edu, pelikan@cs.umsl.edu

Abstract. This paper describes a network crossover operator based on knowledge gathered from either prior problem-specific knowledge or linkage learning methods such as estimation of distribution algorithms (EDAs). This operator can be used in a genetic algorithm (GA) to incorporate linkage in recombination. The performance of GA with network crossover is compared to that of GA with uniform crossover and the hierarchical Bayesian optimization algorithm (hBOA) on 2D Ising spin glasses, NK landscapes, and SK spin glasses. The results are analyzed and discussed.

1 Introduction

It has been argued that to solve many difficult classes of problems in a robust and scalable manner, variation operators of GAs must respect linkages between variables [1]. Unfortunately, conventional variation operators of GAs often break up important linkages [1]. To help remedy this problem, competent linkage learning GAs such as EDAs [2–5] have been developed. EDAs work by building a probabilistic model of promising solutions and sampling new candidate solutions from the build model. While EDAs have many advantages over standard GAs [4, 6], the model building is often computationally intensive, and much of the work in EDAs focuses on efficiency enhancement techniques [7–9].

Often, though, practitioners have prior information about the problem being solved. For example, in graph-based problems we are implicitly given a guide to the strongest dependencies in a problem. In addition, EDAs can also be used to find the structure of the problem by mining their generated models [10, 11]. The key question is, how do we exploit this information in practice?

One way to do this is to modify the crossover operator in a GA to better respect the strongest linkages in the underlying problem structure. This paper discusses a network crossover operator that works with a user-specified network graph to determine which bits are exchanged. We compare the performance of GA with network crossover to that of GA with uniform crossover and the hierarchical Bayesian Optimization Algorithm [9, 12], which is one of the most powerful linkage learning GAs. All algorithms are tested on a broad range of problems of different structure known to be hard for standard evolutionary algorithms.

R. Schaefer et al. (Eds.): PPSN XI, Part II, LNCS 6239, pp. 462–471, 2010.

The paper is organized as follows. Section 2 describes the network crossover operator used in this paper. Section 3 outlines the algorithms tested. Section 4 describes the test problems used in this paper. Section 5 presents the experimental results. Finally, section 6 summarizes and concludes the paper.

2 Network Crossover

While uniform crossover is effective at solving many problems [13], it can often become highly ineffective because it processes each bit independently of others and is unable to consider dependencies between problem variables [14]. In many cases it is desirable that crossover preserves combinations of values of variables that depend on each other while it effectively exploits independencies of other variables to ensure effective mixing [14]. The operator in this paper tries to do this by using a network of assumed dependendies between problem variables.

Any two-parent crossover operator starts by creating a binary mask, which defines what bits to exchange. In this paper we show how we can create a specialized mask that respects problem structure encoded in a network of dependencies.

The network crossover requires an $n \times n$ incidence matrix G that specifies the strongest dependencies between bits. Specifically, denoting the element in the ith row and jth column of G by $G_{i,j}$, bits in locations i and j depend on each other if $G_{i,j} = 1$, whereas they are independent if $G_{i,j} = 0$. Note that a similar representation of dependencies in a problem is used in the dependency structure matrix GA (DSM-GA) [15]. While the network does require information from the practitioner, it does not require in-depth knowledge of the strength of interactions. For graph problems, such as graph-bipartitioning or graph coloring, this graph G is inherent to the problem definition. For problems like MAXSAT and Ising Spin glasses, it is straightforward to specify such a structure from the additive decomposition of these problems. One may also run an EDA on trial instances of the problem to learn promising network structures as suggested in [10]. The key point to remember when constructing G is that it is not necessary to specify the entire problem structure, but only the strongest dependencies.

Given the network G, a crossover mask m is then built as follows. First a random bit i is selected. This bit is then added to the crossover mask by setting $m_i = 1$. Then a randomized breadth-first search is performed on the network G, setting each corresponding bit in m to 1, until m reaches the desired size. If the breadth-first search ends before the desired size is reached, an additional random starting point is selected and the process repeated. The end result is a crossover mask that should most often disrupt bits that are not connected in G.

The idea of using a network of dependencies to modify two-parent recombination operators is not new and has been inspired by past work [16–18]. Our variant of network crossover most closely resembles the work by Stonedahl [18] where the mask was built using a random walk through the network structure. We use the breadth-first search to emphasize the short range dependencies, since these dependencies were found to be strongest in prior work [10]. However, our work is *not* an attempt to improve on this operator, but rather to perform a systematic study of network crossover on a broad range of problems, as well as to compare it to a linkage-learning GA.

3 Tested Algorithms

The genetic algorithm (GA) [13, 19] evolves a population of candidate solutions typically represented by binary strings of fixed length. The initial population is generated at random according to the uniform distribution over all binary strings. Each iteration starts by selecting promising solutions from the current population; in this work we use binary tournament selection without replacement. New solutions are created by applying variation operators to the population of selected solutions. These new candidate solutions are then incorporated into the population using a replacement operator. The run is terminated when either the optimum has been found or after a maximum number of iterations.

Instead of using crossover and mutation to create new candidate solutions, hBOA learns a Bayesian network with local structures [9, 12] as a model of the selected solutions and generates new candidate solutions from the distribution encoded by this model. Using Bayesian networks to generate new candidate solutions ensures effective processing of partial solutions in the class of nearly decomposable and hierarchical problems [9], including many problems that cannot be efficiently solved with standard two-parent crossover operators [9, 12].

In both GA and hBOA runs, a deterministic hill climber (DHC) was incorporated to improve performance. In each iteration, DHC evaluates all possible one-bit flips and chooses the one that leads to the maximum improvement of solution quality. DHC is terminated when no single-bit flip improves solution quality and the solution is thus locally optimal.

4 Test Problems

In concatenated traps of order 5 (trap-5) [20, 21], the input string is partitioned into independent groups of 5 bits. This partitioning is unknown to the algorithm and it does not change during the run. A 5-bit fully deceptive trap function is applied to each group of 5 bits and the contributions of all trap functions are added to form the fitness. The contribution of each group of 5 bits is given by

$$trap_5(u) = \begin{cases} 5 & \text{if } u = 5 \\ 4 - u & \text{otherwise} \end{cases}, \tag{1}$$

where u is the number of 1s in the input string of 5 bits. The task is to maximize the function. An n-bit trap5 function has one global optimum in the string of all 1s and $(2^{n/5} - 1)$ other local optima. Traps of order 5 necessitate that all bits in each group are treated together, because statistics of lower order are misleading.

An NK fitness landscape [22] is fully defined by the following components: (1) The number of bits, n, (2) the number of neighbors per bit, k, (3) a set of k neighbors $\prod(X_i)$ for the i-th bit, X_i for every $i \in \{0, \ldots, n-1\}$, and (4) a subfunction f_i defining a real value for each combination of values of X_i and $\prod(X_i)$ for $i \in \{0, \ldots, n-1\}$, with each subfunction defined as a lookup table.

In this paper we consider two classes of NK landscape instances: (1) Unrestricted NK landscapes and (2) nearest-neighbor NK landscapes. In unrestricted NK landscapes, thet set of k neighbors for each string position X_i is selected at random according to the uniform distribution of all subsets of k bits. Then, a lookup table defining each f_i is generated using the uniform distribution over $[0, 1)$. In this paper we consider unrestricted NK landscapes with $k = 5$; the considered class of NK landscapes is NP-complete [23]. In nearest-neighbor NK landscapes, the bits are arranged on a circle and the neighbors of each bit are restricted to the k bits that follow this bit in the circle. The bit positions are shuffled randomly in order to eliminate tight linkage. Nearest-neighbor NK landscapes are solvable in polynomial time. The algorithm used to solve nearest-neighbor NK instances is based on refs. [24, 25]. The branch and bound algorithm used to solve unrestricted NK landscapes is based on ref. [26].

Ising spin glasses [27] are prototypical models for disordered systems. A simple model to describe a finite-dimensional Ising spin glass is typically arranged on a regular 2D or 3D grid where each node i corresponds to a spin s_i and each edge $\langle i, j \rangle$ corresponds to a coupling between two spins s_i and s_j. Each edge has a real value $J_{i,j}$ associated with it that defines the relationship between the two connected spins.

For the classical Ising model, each spin s_i can be in one of two states: $s_i = +1$ or $s_i = -1$. Given a set of coupling constants $J_{i,j}$, and a configuration of spins C, the energy can be computed as

$$E(C) = -\sum_{\langle i,j \rangle} s_i J_{i,j} s_j \,, \tag{2}$$

where the sum runs over all couplings $\langle i, j \rangle$. The task is to find a spin configuration for a given set of coupling constants that minimizes the energy of the spin glass. The states with minimum energy are called *ground states*. Here we consider the $\pm J$ spin glass, where each spin-spin coupling is set randomly to either $+1$ or -1 with equal probability. The ground states of the instances were obtained from the Spin Glass Ground State Server at the Univ. of Cologne [28].

The Sherrington-Kirkpatrick (SK) spin glass [29] is described by a set of spins s_i and a set of couplings $J_{i,j}$ between all pairs of spins. The SK model does not limit the range of spin-spin interactions to only neighbors in the lattice. The goal is to find ground states for the given coupling constants.

Here we consider two types of random instances of the SK model. The first type uses couplings generated from the Gaussian distribution with zero mean and unit variance $N(0, 1)$. The second set of SK instances are one-dimensional spin glasses with power-law interactions [30] where the spins are arranged equidistantly and numbered counterclockwise on a circle with circumference n. While all spins interact with each other, the interactions between spins located further from each other are weaker. The effects of distance are controlled with a parameter σ; in this paper we examine instances with $\sigma = 2$, which gives the model short ranged behavior. Both types of SK spin glasses are NP-complete [31].

To find guaranteed ground states, a branch-and-bound algorithm adopted from ref. [32] was used for the smaller instances. For larger systems, the population-doubling approach from ref. [33] was used.

5 Experiments

5.1 Experimental Setup

For trap-5, problem sizes of $n = 100$ to 300 with step 10 were examined and G was constructed by setting $G_{i,j} = 1$ if and only if i and j were in the same trap partition. For nearest-neighbor NK landscapes, problem sizes of $n = 30$ to 210 with step 30 were considered and for unrestricted NK landscapes, problem sizes of $n = 20$ to 38 with step 2 were used. For NK landscapes of all types, $k = 5$ and G was constructed by setting $G_{i,j} = 1$ if and only if i and j were neighbors. For 2D Ising spin glasses, problem sizes of $n \in \{256, 324, 400, 484, 576\}$ were examined and G was constructed by setting $G_{i,j} = 1$ if and only if i and j were neighbors in the underlying spin glass lattice.

For one-dimensional spin glasses with power-law interactions, problem sizes of $n \in \{100, 150, 200, 300\}$ were examined with $\sigma = 2$. For the standard SK spin glass, problem sizes of $n = 300$ and $n = 400$ were considered. For all SK instances, G was constructed by connecting each node i to a fixed number m of spins with the strongest magnitude of couplings with i. For example, for $m = 4$, i was connected to four other spins that have the strongest couplings with i. For the experiments with network crossover, $m \in \{1, \ldots, 7\}$ were examined.

hBOA and GA with both uniform crossover and network crossover were applied to all problem instances. For GA runs, bit-flip mutation was used with a probability $p_m = 1/n$. The probability of crossover was set to $p_c = 0.6$. To effectively maintain population diversity, new solutions were incoroporated into the old population using restricted tournament replacement (RTR) [9, 34].

For all problem types except for trap-5, 1000 random instances for each problem size were tested. Bisection [9] was used to determine the minimum population size to ensure convergence to the global optimum in 10 out of 10 independent runs for each instance. For trap-5, bisection was run 10 times to obtain more reliable results. Each run was terminated when the global optimum had been found or when the maximum number of generations $n \times 4$ had been reached. Three results were compared for all solved instances: the number of evaluations, the number of steps of DHC, and the total execution time in seconds.

5.2 Experimental Results

The number of evaluations, the number of DHC flips, and the execution time to solve trap-5 is shown in figure 1. GA with network crossover performed the best. This is to be expected as the network crossover will disrupt much fewer trap partitions than other algorithms and, unlike hBOA, network crossover is given the correct problem decomposition on input. hBOA scales similarly as GA with

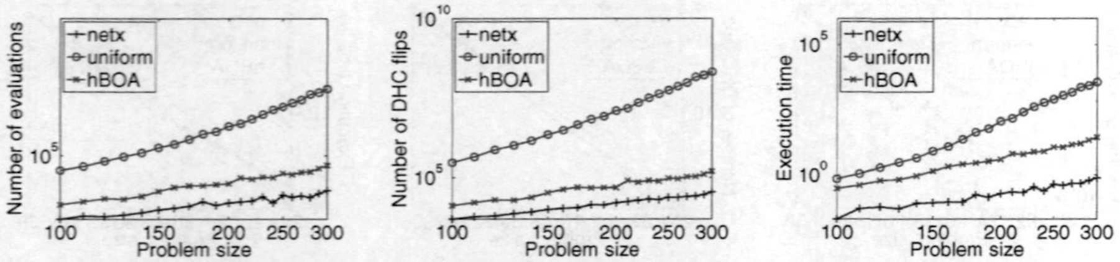

(a) Number of evaluations (b) Number of DHC flips (c) Execution time

Fig. 1. Results on trap-5

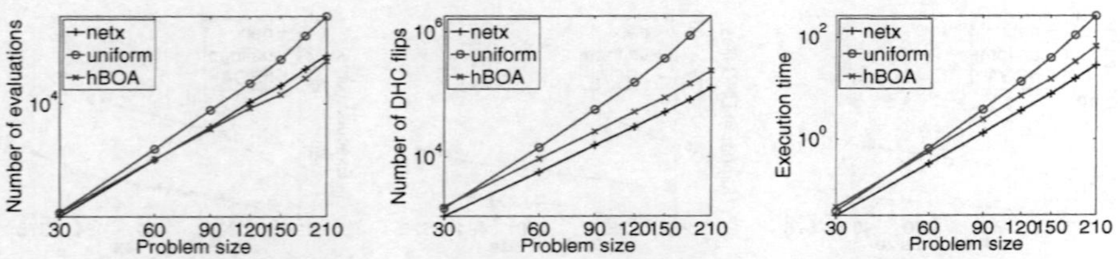

(a) Number of evaluations (b) Number of DHC flips (c) Execution time

Fig. 2. Results on nearest-neighbor NK landscapes with $k = 5$ neighbors

network crossover, indicating that hBOA can learn the problem decomposition scalably. GA with uniform crossover performs very poorly.

Figure 2 shows the performance of the three compared algorithms on nearest-neighbor NK landscapes. hBOA is the best performing algorithm with respect to the number of evaluations. However, GA with network crossover is best with respect to the total execution time and the number of DHC steps. GA with uniform crossover is the worst scaling algorithm.

The results on unrestricted NK landscapes are shown in figure 3. With respect to the number of evaluations, hBOA is clearly the best performing algorithm and GA performs approximately the same regardless of the crossover operator used. With respect to the number of local search steps, while hBOA starts off performing poorly, it scales better than GA with uniform or network crossover. With respect to the total execution time, the results are mixed and all algorithms seem to scale about the same.

The performance of the three compared algorithms on 2D Ising spin glasses is shown in Figure 4. We see that hBOA strongly outperforms the other algorithms with respect to the number of evaluations and local search steps. However, due to less overhead, GA with network crossover has very similar execution times, although it scales slightly worse compared to hBOA as the problem size increases. GA with uniform crossover is shown to have quite poor performance.

Selected experimental results on the one-dimensional SK spin glasses are shown in figure 5. We see that GA with uniform crossover is the worst performing algorithm. GA with network crossover with $m = 3$ (each node in the network is connected to 3 other bits), performs worse than hBOA. For $m = 5$,

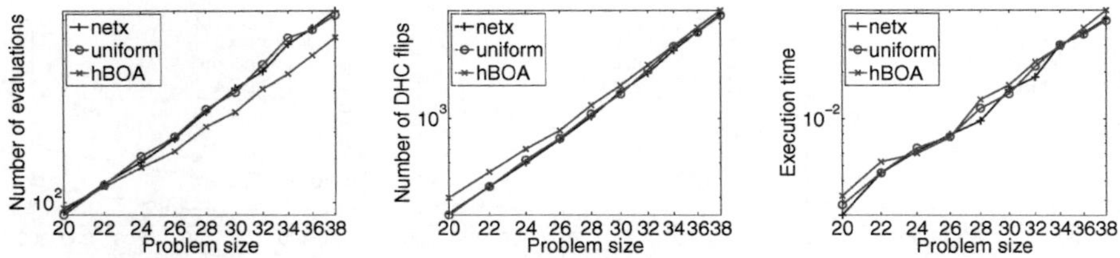

(a) Number of evaluations (b) Number of DHC flips (c) Execution time

Fig. 3. Results on unrestricted NK landscapes with $k = 5$ neighbors

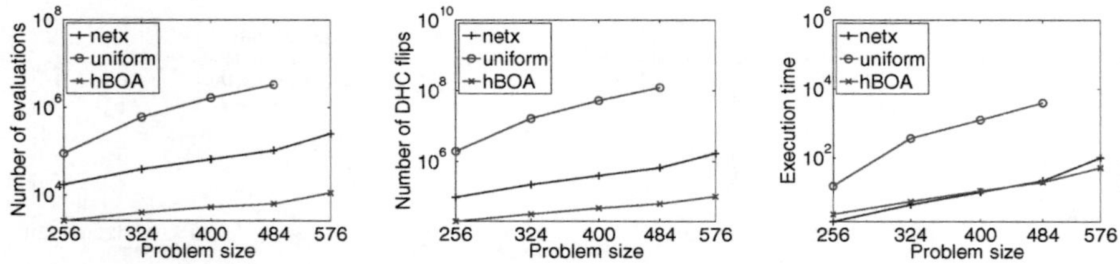

(a) Number of evaluations (b) Number of DHC flips (c) Execution time

Fig. 4. Results on 2D Ising spin glasses

while GA with network crossover is the best performer in terms of the execution time for the smaller problem sizes, hBOA shows superior scaling. The results for other values of m were omitted as $m = 5$ was the best performing value found.

Table 1 shows the results of the three compared algorithms on unrestricted SK spin glasses. For network crossover, the best value of m is used for each value of n. In contrast to all previous results, for SK spin glasses hBOA is the worst performing algorithm with respect to the growth of the execution time with problem size. For $n = 300$ GA with network crossover ($m = 3$) performed best, whereas for $n = 400$ GA with uniform crossover performed best.

6 Summary and Conclusions

This paper described a network crossover operator which can be used to incorporate problem-specific knowledge about dependencies between problem variables into a GA. Performance of GA with the described crossover operator was then compared to GA with uniform crossover and hBOA on a number of challenging problem instances from several problem classes.

On most problems, hBOA and GA with network crossover outperformed GA with uniform crossover. This result is not surprising because all test problems were additively decomposable and it has been argued that to effectively solve additively decomposable problems, recombination must often exploit problem structure to ensure proper mixing and juxtaposition of important partial solutions [1, 14]. GA with uniform crossover was only slightly better on one problem

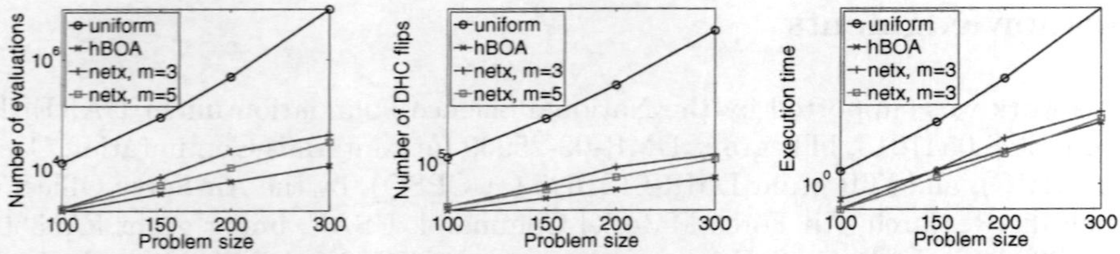

(a) Number of evaluations (b) Number of DHC flips (c) Execution time

Fig. 5. Results on the one-dimensional SK spin glass with power-law interactions, $\sigma = 2$

Table 1. Results on the general SK spin glasses. The table includes the best found value of m for network crossover.

Size	Algorithm	Evaluations	Execution time	Number of DHC flips
300	hBOA	6,615	27.1	284,482
300	GA with uniform crossover	11,579	13.5	53,0150
300	GA with netx, $m = 3$	9,028	10.8	45,0834
400	hBOA	30,021	215.5	1,236,547
400	GA with uniform crossover	26,601	56.8	1,767,551
400	GA with netx, $m = 3$	32,672	66.7	2,077,639

that contained little structure. The more regular the structure of the problem was, the better hBOA and GA with network crossover performed.

Network crossover received information about the structure of the problem on input. That is why it is expected that GA with network crossover would typically outperform hBOA, which must learn the problem structure on its own. Nonetheless, the results were somewhat surprising because in most cases, hBOA showed superior scaling of execution times with problem size. This is in spite of the fact that hBOA is given no information about the structure of the problem on input and it is required to obtain such information itself.

While there is no doubt that using information about the structure of the problem can help in solving many difficult optimization problems more efficiently, learning the problem structure is a computationally intensive task. That is why it should certainly be beneficial if one could incorporate prior problem-specific knowledge into the GA if such knowledge is available, and use specialized crossover operators such as the network crossover discussed here. However, the mixed results presented in this paper show that this is certainly not a straightforward task, and that there are several questions that must be addressed in future research in this area.

Most importantly, future research in this area should examine whether the mixed results are a consequence of using two-parent recombination as opposed to gene-pool recombination, or whether the main reason for these results was in the specific dependencies used for constructing masks in network crossover. Variations of network crossover should also be explored, using different approaches for constructing network masks. Finally, GA with network crossover should be tested on additional classes of problems.

Acknowledgments

This work was supported by the National Science Foundation under CAREER grant ECS-0547013, ITR grant DMR-03-25939 (at Materials Computation Center, UIUC), and ITR grant DMR-0121695 (at CPSD), by the Air Force Office of Scientific Research, Air Force Materiel Command, USAF, under grant FA9550-06-1-0096, by the Dept. of Energy under grant DEFG02-91ER45439 (at Fredrick Seitz MRL), by the European Commission contract No. FP6-511931, and by the Univ. of Missouri in St. Louis through the High Performance Computing Collaboratory sponsored by Information Technology Services, and the Research Award and Research Board programs. Some experiments presented in this work were done using the hBOA software developed by Martin Pelikan and David E. Goldberg at the Univ. of Illinois at Urbana-Champaign.

References

1. Goldberg, D.E.: The design of innovation: Lessons from and for competent genetic algorithms. Kluwer, Dordrecht (2002)
2. Baluja, S.: Population-based incremental learning: A method for integrating genetic search based function optimization and competitive learning. Tech. Rep. No. CMU-CS-94-163, Carnegie Mellon University, Pittsburgh, PA (1994)
3. Mühlenbein, H., Paaß, G.: From recombination of genes to the estimation of distributions I. Binary parameters. Parallel Problem Solving from Nature, 178–187 (1996)
4. Larrañaga, P., Lozano, J.A. (eds.): Estimation of Distribution Algorithms: A New Tool for Evolutionary Computation. Kluwer, Boston (2002)
5. Pelikan, M., Goldberg, D.E., Lobo, F.: A survey of optimization by building and using probabilistic models. Computational Optimization and Applications 21(1), 5–20 (2002)
6. Pelikan, M., Sastry, K., Cantú-Paz, E. (eds.): Scalable optimization via probabilistic modeling: From algorithms to applications. Springer, Heidelberg (2006)
7. Cantú-Paz, E.: Efficient and Accurate Parallel Genetic Algorithms. Kluwer, Boston (2000)
8. Schwarz, J., Ocenasek, J.: A problem-knowledge based evolutionary algorithm KBOA for hypergraph partitioning, Personal communication (2000)
9. Pelikan, M.: Hierarchical Bayesian optimization algorithm: Toward a new generation of evolutionary algorithms. Springer, Heidelberg (2005)
10. Hauschild, M., Pelikan, M., Sastry, K., Goldberg, D.E.: Using previous models to bias structural learning in the hierarchical BOA. In: Genetic and Evolutionary Comp. Conf (GECCO 2008), pp. 415–422 (2008)
11. Hauschild, M.W., Pelikan, M.: Intelligent bias of network structures in the hierarchical boa, pp. 413–420. ACM, New York (2009)
12. Pelikan, M., Goldberg, D.E.: Escaping hierarchical traps with competent genetic algorithms. In: Genetic and Evolutionary Comp. Conf. (GECCO 2001), pp. 511–518 (2001)
13. Goldberg, D.E.: Genetic algorithms in search, optimization, and machine learning. Addison-Wesley, Reading (1989)

14. Thierens, D.: Scalability problems of simple genetic algorithms. Evolutionary Computation 7(4), 331–352 (1999)
15. Yu, T.L., Goldberg, D.E., Sastry, K., Lima, C.F., Pelikan, M.: Dependency structure matrix, genetic algorithms, and effective recombination. Evolutionary Computation 17(4), 595–626 (2009)
16. Drezner, Z., Salhi, S.: Using hybrid metaheuristics for the one-day and two-way network design problem. Naval Research Logistics 49(5), 449–463 (2002)
17. Drezner, Z.: A new genetic algorithm for the quadratic assignment problem. INFORMS Journal on Computing 15(3), 320–330 (2003)
18. Stonedahl, F., Rand, W., Wilensky, U.: CrossNet: a framework for crossover with network-based chromosomal representations. In: Genetic and Evolutionary Comp. Conf. (GECCO 2008), pp. 1057–1064. ACM, New York (2008)
19. Holland, J.H.: Adaptation in natural and artificial systems. University of Michigan Press, Ann Arbor (1975)
20. Ackley, D.H.: An empirical study of bit vector function optimization. Genetic Algorithms and Simulated Annealing, 170–204 (1987)
21. Deb, K., Goldberg, D.E.: Analyzing deception in trap functions. IlliGAL Report No. 91009, University of Illinois at Urbana-Champaign, Illinois Genetic Algorithms Laboratory, Urbana, IL (1991)
22. Kauffman, S.: Adaptation on rugged fitness landscapes. In: Stein, D.L. (ed.) Lecture Notes in the Sciences of Complexity, pp. 527–618. Addison Wesley, Reading (1989)
23. Wright, A.H., Thompson, R.K., Zhang, J.: The computational complexity of n-k fitness functions. IEEE Trans. Evolutionary Computation 4(4), 373–379 (2000)
24. Pelikan, M., Sastry, K., Butz, M.V., Goldberg, D.E.: Performance of evolutionary algorithms on random decomposable problems. In: PPSN, pp. 788–797 (2006)
25. Pelikan, M., Sastry, K., Goldberg, D.E., Butz, M.V., Hauschild, M.: Performance of evolutionary algorithms on NK landscapes with nearest neighbor interactions and tunable overlap. MEDAL Report No. 2009002, Missouri Estimation of Distribution Algorithms Laboratory, University of Missour–St. Louis, St. Louis, MO (2009)
26. Pelikan, M.: Analysis of estimation of distribution algorithms and genetic algorithms on NK landscapes, pp. 1033–1040 (2008)
27. Mezard, M., Parisi, G., Virasoro, M.: Spin glass theory and beyond. World Scientific, Singapore (1987)
28. Spin Glass Ground State Server. University of Köln, Germany (2004), `http://www.informatik.uni-koeln.de/ls_juenger/research/sgs/sgs.html`
29. Kirkpatrick, S., Sherrington, D.: Infinite-ranged models of spin-glasses. Phys. Rev. B 17(11), 4384–4403 (1978)
30. Katzgraber, H.G.: Spin glasses and algorithm benchmarks: A one-dimensional view. Journal of Physics: Conf. Series 95(012004) (2008)
31. Barahona, F.: On the computational complexity of Ising spin glass models. Journal of Physics A: Mathematical, Nuclear and General 15(10), 3241–3253 (1982)
32. Hartwig, A., Daske, F., Kobe, S.: A recursive branch-and-bound algorithm for the exact ground state of Ising spin-glass models. Computer Physics Communications 32(2), 133–138 (1984)
33. Pelikan, M., Katzgraber, H.G., Kobe, S.: Finding ground states of Sherrington-Kirkpatrick spin glasses with hierarchical BOA and genetic algorithms. In: Genetic and Evolutionary Comp. Conf. (GECCO 2008), pp. 447–454. ACM, New York (2008)
34. Harik, G.R.: Finding multimodal solutions using restricted tournament selection. In: International Conf. on Genetic Algorithms (ICGA 1995), pp. 24–31 (1995)

Promoting Phenotypic Diversity in Genetic Programming

David Jackson

Dept. of Computer Science, University of Liverpool
Liverpool L69 3BX, United Kingdom
djackson@liverpool.ac.uk

Abstract. Population diversity is generally seen as playing a crucial role in the ability of evolutionary computation techniques to discover solutions. In genetic programming, diversity metrics are usually based on structural properties of individual program trees, but are also sometimes based on the spread of fitness values in the population. We explore the use of a further interpretation of diversity, in which differences are measured in terms of the behaviour of programs when executed. Although earlier work has shown that improving behavioural diversity in initial GP populations can have a marked beneficial effect on performance, further analysis reveals that lack of behavioural diversity is a problem throughout whole runs, even when other diversity levels are high. To address this, we enhance phenotypic diversity via modifications to the crossover operator, and show that this can lead to additional performance improvements.

1 Introduction

It is generally accepted that a problem associated with genetic programming and other forms of evolutionary computation is that population diversity tends to diminish over time [1, 2], and that this may then lead to convergence on local minima from which the evolutionary process cannot escape to explore the search landscape more widely for solutions. However, more rigorous probing into the whole issue of diversity and its possible impact reveals that there is little consensus as to how it should be defined, measured, analysed or promoted.

Broadly speaking, diversity in genetic programming refers to how different each individual is from other members of the host population, and although this notion of 'difference' may be interpreted in a variety of ways, previous work has tended to categorise diversity as falling into two camps. The first is *genotypic*, or structural, diversity, which is based upon differences between the structure of the trees or other representations used to encode individual programs. The second category is *phenotypic* diversity. Ostensibly, this is defined in terms of the behaviour rather than the structure of programs, but in practice it has usually corresponded to the spread of fitness values amongst members of the population.

We shall say more about these previously established views of diversity in the next section on related work. Following that, in Section 3 we will describe an alternative

R. Schaefer et al. (Eds.): PPSN XI, Part II, LNCS 6239, pp. 472–481, 2010.

view of phenotypic diversity which is based not on fitness values, but on the behaviour of individuals when they are executed. In Section 4 we will compare the extent of the various forms of diversity in several problem domains, with particular emphasis on how it changes during the lifetime of GP runs. In Section 5 we build on earlier work concerned with creating more diverse initial populations [3] by describing methods for maintaining that diversity in subsequent generations, then go on to describe and compare the effects of these techniques in experimentation. Finally, Section 6 draws some conclusions and gives pointers to further work.

2 Related Work

As it relates to genetic programming, the term 'diversity' has a variety of interpretations, and hence a number of different ways have been proposed for measuring it, creating it and maintaining it. Overviews of diversity measures can be found in [4] and [5], while Burke et al [6] give a more extensive analysis of these measures and of how they relate to fitness.

The most common usage of the term is concerned with differences in the *structure* of individual program trees; that is, in their size, their shape, and in the functions and terminals used at individual nodes. Recognizing the importance of including a wide range of structures in the initial population, Koza [7] proposed the use of a 'ramped half-and-half' algorithm, and many implementations have continued to follow his advice. The approach is claimed to give good diversity in the structure of program fragments which can then be combined and integrated to produce more complex and hopefully fitter programs.

Measurements of structural diversity may involve nothing more than simple node-for-node comparison of program trees;. More sophisticated structural diversity metrics may be based on edit distance [8], where the similarity between two individuals is measured in terms of the number of edit operations required to turn one into the other.

A difficulty with comparing individuals based on their apparent structure is that program trees which are seemingly very different in appearance may in fact compute identical functions. Seeing beyond these surface differences requires the use of graph isomorphism techniques, but these are computationally expensive and become even more so as program trees grow larger over time. A simpler, less costly alternative is to check for pseudo-isomorphism [5], in which the possibility of true isomorphism is assessed based on characteristics such as tree depth and the numbers of terminals and functions present. However, the accuracy of this assessment may be subject to the presence of introns in the code; Wyns et al [9] describe an attempt to improve on this situation through the use of program simplification techniques to remove redundant code.

In contrast, *behavioural* or *phenotypic* diversity metrics are based on the functionality of individuals, i.e. the execution of program trees rather than their appearance. Usually, behavioural diversity is viewed in terms of the spread of fitness values obtained on evaluating each member of the population [10]. One way of measuring such diversity is by considering the fitness distribution as an indicator of entropy, or disorder, in the population [11, 9]. Other approaches consider sets or lists of fitness values and use them in combination with genotypic measures [12, 13]. For certain types of problem it may be possible to achieve the effect of behavioural diversity without invoking the fitness function, via the use of semantic sampling schemes [14]. Semantic analysis of programs is also used in the diversity enhancing techniques described by Beadle and Johnson [15, 16].

3 Phenotypic Diversity

Our own approach to diversity differs from others in that it does not involve structural considerations, fitness values or semantic analysis of programs. Instead, it focuses on the observed behaviour of individuals when they are executed. To investigate this fully, we have applied it to a variety of problem domains; these comprise two Boolean problems (6-multiplexer and even-4 parity), two navigation problems (Santa Fe and maze traversal), and one numeric problem (symbolic regression of a polynomial).

The 6-mux, even-4 parity and Santa Fe problems are all standard benchmarks in GP, and further details can be found elsewhere (e.g. Koza [7]). In our symbolic regression problem we attempt to evolve a program equivalent to the polynomial $4x^4 - 3x^3 + 2x^2 - x$. The fitness cases consist of 32 x-values in the range [0,1), starting at 0.0 and increasing in steps of 1/32, plus the corresponding y-values. Other than this, the problem is again fairly standard. Our second navigation problem is that of finding a route through a maze. Although less well-known than the ant problem, it has been used as the subject for research on introns in several studies [17-19], and again details can be found in those papers and in our earlier paper on this topic [3]. Other parameters as they apply to the experiments described in the remainder of this paper are shown in Table 1.

Table 1. GP system parameters common to all experiments

Population size	500
Initialisation method	Ramped half-and-half
Evolutionary process	Steady state
Selection	5-candidate tournament
No. generations	51 generational equivalents (initial+50)
No. runs	100
Prob. crossover	0.9
Mutation	None
Prob. internal node used as crossover pt.	0.9

In the case of the Boolean problems, the behaviour of an individual is measured in terms of the outputs it produces; this is recorded as a string of binary values for each of the test cases used during fitness evaluation. So, for the 6-mux problem, there is a 64-bit string associated with each member of the population, while for the even-4 parity problem only a 16-bit string need be recorded. To save memory, these can be packed into 64-bit and 16-bit integers, respectively. We say that two individuals exhibit phenotypic differences if they differ in any of the corresponding bits in their output strings, and that an individual is phenotypically unique in a population if there are no other members of that population with exactly the same binary output string.

For the symbolic regression problem, we again record the outputs produced for each test case, but this time it is a vector of 32 floating point results. In comparing phenotypes we choose not to look for exact matches, but instead check whether the differences between corresponding outputs lie within some pre-defined value epsilon. Hence, two individuals are said to be behaviourally identical if, for each x-value, the absolute difference between the corresponding y-values is less than epsilon. The value for epsilon was arbitrarily chosen to be the same as that used to check for 'hits' in fitness assessment, i.e. 0.01.

For the two navigation problems, the situation is complicated by the fact that the evolving programs do not produce outputs as such: their behaviour is measured in terms of the movements produced by function and terminal execution. Because of this, the record we make of an individual's behaviour is the sequence of moves it makes in the grid or maze during execution. We are not concerned with recording any left or right turns that are executed while remaining on a square, nor with any decision making via execution of statements such as IF_FOOD_AHEAD or WALL_AHEAD.

To record the path histories, we associate with each individual in the population a vector of {north, east, south, west} moves. Each time the executing program moves to a different square, the heading is added to the end of the vector. Since a program times-out after a fixed number of steps (600 for the ant problem, 1000 for the maze), we know that the vector cannot be longer than that number of elements per individual, and so memory occupancy is not a huge issue. Determining behavioural differences between individuals becomes simply a matter of comparing these direction vectors.

4 Measuring Diversity

The first thing we wish to explore is the extent to which diversity is present in populations, and how it alters over the lifetime of each run. For this, we need to define appropriate metrics.

We start with structural diversity (SD). This is found by counting the number of distinct program structures present in the population and then dividing this by the population size. Thus, SD = 1 if and only if every member of the population has a different structure, whereas a value of SD close to zero indicates poor structural diversity (i.e. much duplication). Behavioural diversity (BD) is defined in a very similar way, by counting up the number of distinct behaviours and again dividing by the size of the population. As before, BD = 1 only when each individual's behaviour is unique.

Fitness diversity (FD) is calculated in a slightly different manner. Unlike either structural or behavioural diversity, in which every member of the population can be unique, the range of possible fitness values tends to be much smaller than the population size, and varies from problem to problem. We therefore define fitness diversity (FD) to be the number of distinct fitness values found in the population divided by the number of possible values. For example, if a given population in the even-4 parity problem contained a total of 5 different fitness values, then its FD would be 5/17 = 0.294, since the problem allows for 17 possible fitness values (0-16). On this scale, a value of 1.0 would indicate full fitness diversity (which would also imply that the population contained a solution!).

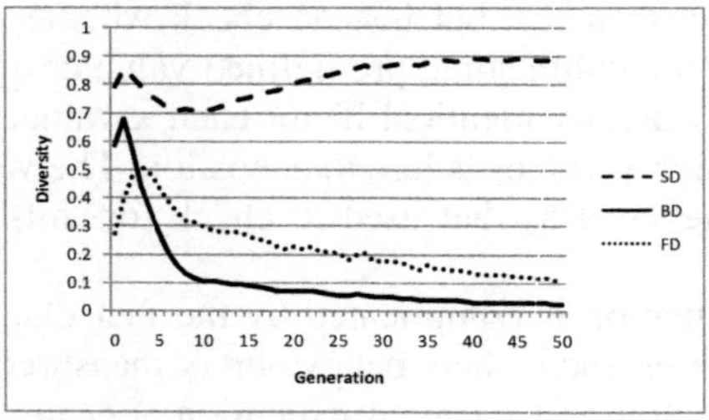

(a) Ant

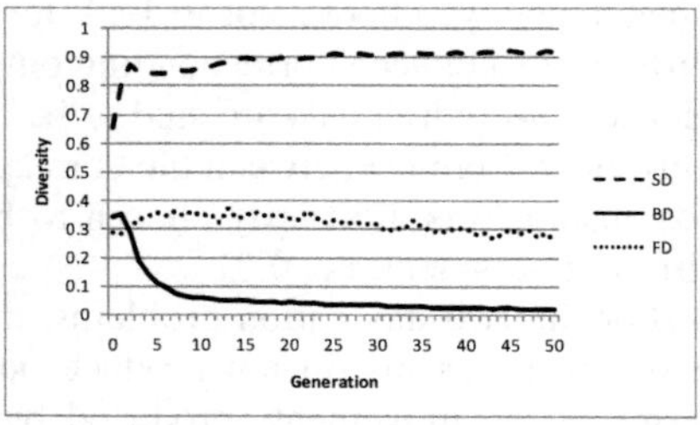

(b) Parity

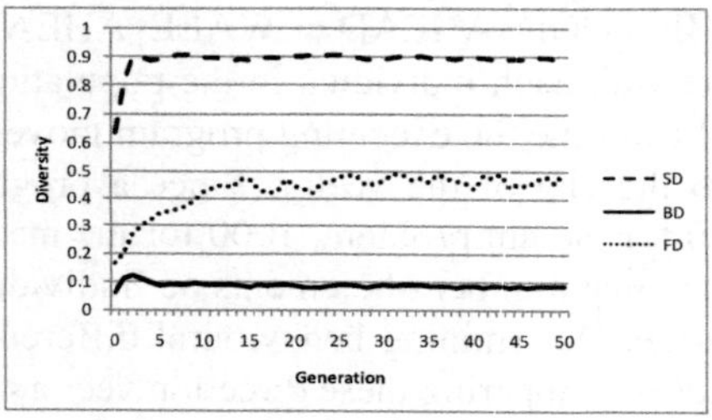

(c) Maze

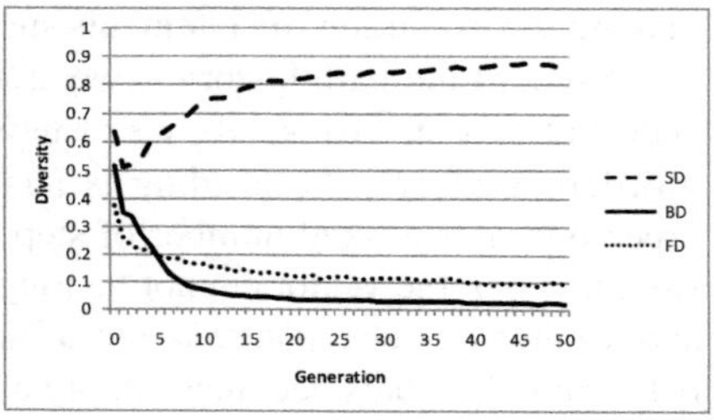

(d) Mux

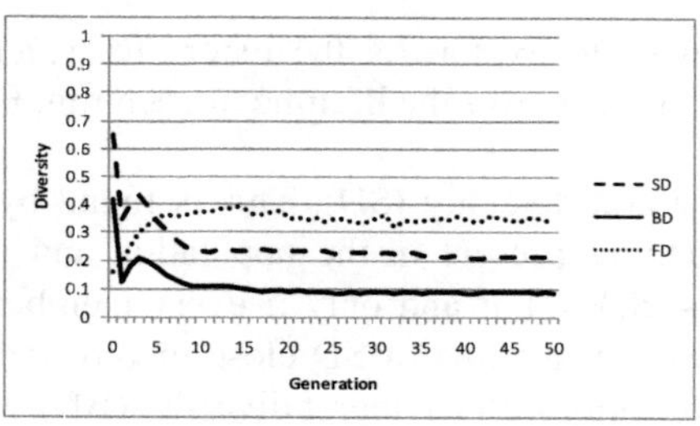

(e) Regression

Fig. 1. Diversity changes for each test problem, averaged over 100 runs

Given these definitions, we can plot the changes that take place in each form of diversity. The graphs of Figure 1 show this for each of our benchmark problems. The figures are averaged over 100 runs. The first observation that can be made about these graphs is that, in four of the five problems, structural diversity increases rather than decreases, and tends to reach quite a high level of about 0.9. In other words, only about 10% of individuals have structural clones elsewhere in the population. This suggests that crossover is successful at generating structurally unique program trees, and that efforts to increase structural diversity in these problems may not have a substantial impact. Only in the case of the symbolic regression problem does structural diversity fall significantly, dropping to just over 0.2 from generation 10 onwards.

The picture is more mixed for fitness diversity: in the ant problem it rises and then falls to below its initial level; in the maze and regression problems it rises steadily and then remains fairly constant; in the parity problem it tends to hover around a value of 0.3; and in the multiplexer problem it falls from 0.4 down to about 0.1.

In all of the problems it is behavioural diversity that comes off worst. Whatever its initial value (which in some cases can be comparatively good), it always drops to below 0.1. In the even parity and mux problems it reaches as low as 0.02 in generation 50; this corresponds to only 10 distinct behaviours in a population of 500. For most of the problems there is a wide gap between the SD and BD graphs, indicating that many programs that are structurally dissimilar do not differ at all in their behaviour. The number of distinct behaviours in the parity and mux problems dip to only 10 despite the number of distinct structures in the population climbing to about 450.

5 Preserving Diversity

The experiments of the previous section show that there is relatively little opportunity to increase the structural diversity of a population, but much greater scope for increasing behavioural diversity. Fitness diversity also remains low, but doing something about this is far more problematic. In initial GP populations, most members will have poor fitness, and so the number of distinct fitness values can be expected to be quite small. Increasing diversity in these early stages of evolution would involve introducing some fitter individuals into the population, but this begs the question of how one goes about finding these fitter programs without resorting to the evolutionary process that subsequent generations are meant to implement!

Our approach to promoting diversity is a two-step process. The first phase is to establish increased diversity in the initial population. This can be done by making changes to the ramped half-and-half algorithm mentioned earlier so that duplicates (either structural or behavioural) are eliminated. This approach was covered in detail in an earlier paper [3].

The next task is to attempt to preserve diversity throughout the remainder of the evolutionary process. There will be pressure to reduce diversity because of the reproduction operator, which simply clones relatively fit individuals. This will be counterbalanced to an extent by the recombination operator, which creates new individuals via subtree crossover. In most problems, the nature of the terminal and function sets, and the way in which crossover points are chosen, is sufficient to ensure that entirely novel structures are created. This explains the continually high SD levels in the earlier graphs (the regression problem being an exception).

Unfortunately, crossover does not always introduce new behaviours. A primary reason for this is that crossover often takes place at the site of introns, particularly during the latter stages of evolution in which bloat is extensive. These introns are often non-executed pieces of code [19], and so the crossover operation creates a child which does not differ behaviourally from the parent receiving the subtree. It is for these reasons that we have chosen the crossover operator as the focal point for changes to maintain at least some of the diversity already introduced during initialisation. The changes are simple, as the following pseudo-code shows:

```
function crossover
   childnum = 0
   select parents by tournament
   select member to be replaced
   do
      establish crossover points
      create child
      childnum = childnum + 1
   while (child same as a parent
          and childnum < MAX_BROOD)
endfunction
```

The parents and the individual to be replaced are selected by tournament. The code then enters a loop, choosing crossover points and generating offspring until a child is found which differs from both its parents. The comparison between a child and its parents can be either structural or behavioural. In both cases the variable *childnum* is used to prevent an excessive number of offspring from being produced. We used a value of 20 for MAX_BROOD, but found in practice that the actual number of children generated in each crossover usually fell far short of that, even when searching for behavioural differences.

Table 2. Effects of diversity promotion on solution discovery rate, given as number of solutions found in 100 runs

Problem	**Standard GP**	**SD-Initial**	**SD-All**	**BD-Initial**	**BD-All**
6-mux	56	66	70	79	99
Even-4	14	11	11	23	54
Regress	10	10	18	24	36
Ant	13	9	12	18	25
Maze	14	18	17	51	95

In Table 2 we show the effects of these algorithms on the rate of success at finding solutions to our benchmark problems. For each problem, we first of all give the success rate of our standard GP system, measured as the number of solutions found in 100 runs. We also wish to distinguish between the effects of using our first algorithm in isolation (i.e. eliminating duplicates in the initial population only), and using both algorithms together to promote diversity throughout the lifetime of each run. Hence, the column labelled SD-Initial shows what happens when structural clones are eliminated in the initial population only, while the column labelled SD-All shows the effects of augmenting this initialisation phase with the subsequent attempts to prevent structural duplicates being created during crossover. The columns headed BD-Initial and BD-All should be interpreted in the same manner, but based on the promotion of behavioural rather than structural diversity.

What we can see from Table 2 is that the removal of structural duplicates does not always have a beneficial effect on solution finding performance. It is worth noting, however, that the biggest positive impact of the combined approach to structural diversity is on the symbolic regression problem, where the success rate is almost doubled. Interestingly, it was the regression problem which the graphs of Fig. 1 revealed to have the most scope for improvement in the structural diversity levels.

The situation with regard to behavioural diversity appears more promising. Even if we restrict our diversity enhancing measures to the initial population, a substantial increase in performance follows for most of our problems, particularly so for maze traversal, where the success rate jumps from 14% to 51%. Once we add in the second algorithm to preserve behavioural diversity throughout the remainder of each run, performance improves even further, with near perfect records being seen for the maze and 6-mux problems. A t-test ($p < 0.05$) performed on the best fitness values found at the end of each run indicates that the improvements are statistically significant.

To make the comparison fair, we have to ask at what cost our improvements are obtained. One commonly used method of comparing cost is Koza's computational effort metric [7]. However, this assumes a fixed number of fitness evaluations per generation, which for our purposes is not applicable because of the additional effort required both to create the initial populations and to assess each member of the broods created in our amended crossover operator.

The cost metric we shall use instead is a count of the number of fitness evaluations performed over all 100 runs, divided by the number of solutions found. This gives us a measure of effort in terms of the number of evaluations per solution. Table 3 gives these figures for each of our problems and diversity enhancing mechanisms.

Table 3. Effects of diversity promotion on computational effort, measured as number of fitness evaluations per solution

Problem	Standard GP	SD-Initial	SD-All	BD-Initial	BD-All
6-mux	23263	16416	18261	12140	7703
Even-4	151518	195781	208667	101570	58662
Regress	217612	217486	142659	88430	71465
Ant	158498	240285	193817	118068	119118
Maze	150959	115998	130956	8329286	4482422

In most cases, it would seem that when it comes to counting the computational cost of finding solutions, our approaches to behavioural diversity enhancement offer significant improvements over standard GP, and over systems that attempt to increase structural diversity. There is, however, a notable exception. In the maze problem, the effort involved in creating the initial behaviourally diverse population is immense; and although a dramatic increase in the success rate is observed, this is still not enough to bring the number of evaluations per solution down to a level that is competitive with standard GP.

6 Conclusions

It is accepted wisdom that lack of diversity in evolutionary computing techniques such as genetic programming can hamper the search for solutions, and that diversity tends to diminish over the lifetime of a run. The inference to be drawn from this is that efforts to promote and maintain diversity should have a beneficial effect. Our research supports this, but with the caveat that it very much depends on how the notion of diversity is interpreted.

We have distinguished in this paper between genotypic and phenotypic diversity in GP. The former is associated with the structure of program trees making up the population; the latter is usually defined as corresponding to the spread of fitness values in a population. We have introduced a further interpretation of phenotypic diversity, defined in terms of the recorded behaviour of programs as they execute. We have also shown that, despite great differences in the nature of problems for which GP may be used, the approach is a general one applicable to a wide range of domains.

When assessing the extent of the various forms of diversity experimentally we find that, far from diminishing, structural diversity tends to increase early on in each run, and then remain comparatively high throughout the remainder of the run. By contrast, behavioural diversity (in the form we have described it) does fall and remain at very low levels. Accordingly, when we introduce algorithms to promote structural diversity, we see little in the way of performance gains. The greatest improvement is seen in the symbolic regression problem; it is perhaps no coincidence that it is this problem which has the biggest scope for increasing its structural diversity levels.

Again in contrast, when the algorithms are used to create and preserve behavioural diversity, we see a substantial improvement in the solution finding performance for all the problems we studied. Although additional fitness evaluations are required to generate an initially diverse population, and then to assess broods of children to preserve that diversity during crossover, this is outweighed by the improved success rate, so that the computational costs per solution are significantly reduced. The one exception to this in our problem set is maze traversal. The reason for this is that the physical constraints of the maze make it difficult to generate a set of unique behaviours when using an unintelligent initialization algorithm.

One way of combating this problem would be to place an upper limit on the number of new programs that are created for each member that enters the population. This would enhance diversity but allow some duplication. Indeed, there is a general research issue here as to how different levels of diversity may affect performance, and we hope to explore this further. More generally, we plan to investigate the potential for exploiting phenotypic diversity in the form we have described it.

References

1. McPhee, N.F., Hopper, N.J.: Analysis of Genetic Diversity through Program History. In: Banzhaf, W., et al. (eds.) Proc. Genetic and Evolutionary Computation Conf., Florida, USA, pp. 1112–1120 (1999)
2. Daida, J.M., Ward, D.J., Hilss, A.M., Long, S.L., Hodges, M.R., Kriesel, J.T.: Visualizing the Loss of Diversity in Genetic Programming. In: Proc. IEEE Congress on Evolutionary Computation, Portland, Oregon, USA, pp. 1225–1232 (2004)

3. Jackson, D.: Phenotypic Diversity in Initial Genetic Programming Populations. In: Esparcia-Alcázar, A.I., Ekárt, A., Silva, S., Dignum, S., Uyar, A.Ş. (eds.) EuroGP 2010. LNCS, vol. 6021, pp. 98–109. Springer, Heidelberg (2010)
4. Hien, N.T., Hoai, N.X.: A Brief Overview of Population Diversity Measures in Genetic Programming. In: Pham, T.L., et al. (eds.) Proc. 3rd Asian-Pacific Workshop on Genetic Programming, Hanoi, Vietnam, pp. 128–139 (2006)
5. Burke, E., Gustafson, S., Kendall, G., Krasnogor, N.: Advanced Population Diversity Measures in Genetic Programming. In: Guervos, J.J.M., et al. (eds.) PPSN 2002. LNCS, vol. 2439, pp. 341–350. Springer, Heidelberg (2002)
6. Burke, E., Gustafson, S., Kendall, G.: Diversity in Genetic Programming: An Analysis of Measures and Correlation with Fitness. IEEE Transactions on Evolutionary Computation 8(1), 47–62 (2004)
7. Koza, J.R.: Genetic Programming: On the Programming of Computers by Means of Natural Selection. MIT Press, Cambridge (1992)
8. de Jong, E.D., Watson, R.A., Pollack, J.B.: Reducing Bloat and Promoting Diversity using Multi-Objective Methods. In: Spector, L., et al. (eds.) Proc. Genetic Evolutionary Computation Conf., San Francisco, CA, USA, pp. 11–18 (2001)
9. Wyns, B., de Bruyne, P., Boullart, L.: Characterizing Diversity in Genetic Programming. In: Collet, P., et al. (eds.) EuroGP 2006. LNCS, vol. 3905, pp. 250–259. Springer, Heidelberg (2006)
10. Rosca, J.P.: Genetic Programming Exploratory Power and the Discovery of Functions. In: McDonnell, J.R., et al. (eds.) Proc. 4th Conf., Evolutionary Programming, San Diego, CA, USA, pp. 719–736 (1995)
11. Rosca, J.P.: Entropy-Driven Adaptive Representation. In: Rosca, J.P. (ed.) Proc. Workshop on Genetic Programming: From Theory to Real-World Applications, Tahoe City, CA, USA, pp. 23–32 (1995)
12. D'haeseleer, P., Bluming, J.: Effects of Locality in Individual and Population Evolution. In: Kinnear, K.E., et al. (eds.) Advances in Genetic Programming, ch. 8. pp. 177–198. MIT Press, Cambridge (1994)
13. Ryan, C.: Pygmies and Civil Servants. In: Kinnear, K.E., et al. (eds.) Advances in Genetic Programming, ch.11, pp. 243–263. MIT Press, Cambridge (1994)
14. Looks, M.: On the Behavioural Diversity of Random Programs. In: Thierens, D., et al. (eds.) Proc. Genetic and Evolutionary Computing Conf. (GECCO 2007), London, England, UK, pp. 1636–1642 (2007)
15. Beadle, L., Johnson, C.G.: Semantic Analysis of Program Initialisation in Genetic Programming. Genetic Programming and Evolvable Machines 10(3), 307–337 (2009)
16. Beadle, L., Johnson, C.G.: Semantically Driven Crossover in Genetic Programming. In: Proc. IEEE Congress on Evolutionary Computation (CEC), Hong Kong, pp. 111–116 (2008)
17. Soule, T.: Code Growth in Genetic Programming. PhD Thesis, University of Idaho (1998)
18. Langdon, W.B., Soule, T., Poli, R., Foster, J.A.: The Evolution of Size and Shape. In: Spector, L., et al. (eds.) Advances in Genetic Programming, vol. 3, pp. 163–190. MIT Press, Cambridge (1999)
19. Jackson, D.: Dormant Program Nodes and the Efficiency of Genetic Programming. In: Beyer, H.-G., et al. (eds.) Proc. Genetic and Evolutionary Computing Conf. (GECCO 2005), Washington DC, USA, pp. 1745–1751 (2005)

A Genetic Programming Approach to the Matrix Bandwidth-Minimization Problem

Behrooz Koohestani and Riccardo Poli

School of Computer Science and Electronic Engineering,
University of Essex, CO4 3SQ, UK
{bkoohe,rpoli}@essex.ac.uk

Abstract. The bandwidth of a sparse matrix is the distance from the main diagonal beyond which all elements of the matrix are zero. The bandwidth minimisation problem for a matrix consists of finding the permutation of rows and columns of the matrix which ensures that the non-zero elements are located in as narrow a band as possible along the main diagonal. This problem, which is known to be NP-complete, can also be formulated as a vertex labelling problem for a graph whose edges represent the non-zero elements of the matrix. In this paper, a Genetic Programming approach is proposed and tested against two of the best-known and widely used bandwidth reduction algorithms. Results have been extremely encouraging.

Keywords: Bandwidth Minimization Problem; Genetic Programming; Graph Labelling; Sparse Matrices; Combinatorial Optimisation.

1 Background

The Bandwidth Minimization Problem (BMP) is a very well-known problem, familiar to applied mathematicians and arising in many applications in science and engineering [18]. BMP consists of finding the permutation of rows and columns of a matrix which ensures that the non-zero elements are located in as narrow a band as possible along the main diagonal. One of the most common applications of bandwidth-minimisation algorithms arises from the need to efficiently solve large systems of equations [19]. In such a scenario, more efficient solutions are obtained if the rows and columns of the matrix representing the set of equations can be permuted in such a way that the bandwidth of the matrix is minimized [19]. BMP has also connections with a wide range of other problems, including: finite element analysis of mechanical systems, large scale power transmission systems, circuit design, VLSI design, data storage, chemical kinetics, network survivability, numerical geophysics, industrial electromagnetics, saving large hypertext media and topology compression of road networks.

The BMP is NP-complete [18] and, hence, it is highly unlikely that there exists an algorithm which finds the minimum bandwidth of a matrix in polynomial time. It has also been proved that the BMP is NP-complete even for trees with

R. Schaefer et al. (Eds.): PPSN XI, Part II, LNCS 6239, pp. 482–491, 2010.

a maximum degree of three and only in very special cases it is possible to find the optimal ordering in polynomial time [5].

The first direct method for the BMP was proposed by Harary [9]. Cuthill and McKee [3] introduced the first heuristic approach to the problem. Their method is still one of the most important and widely used methods to (approximately) solve the problem. In this method, the nodes in the graph representation of a matrix are partitioned into equivalence classes based on their distance from a given root node. The partition is known as *level structure* for the given node. In Cuthill and McKee's algorithm, the root node for the level structure is chosen from the peripheral nodes in the graph (i.e., nodes which have the highest distance to any other node in the graph). The permutation chosen to reduce the bandwidth of the matrix is then simply obtained by visiting the nodes in the level structure in increasing-distance order. A few years later, Gibbs et al. [7] proposed an algorithm, known as GPS (which stands for "Gibbs, Poole and Stockmeyer"), that makes more extensive use of level structures. The algorithm is substantially faster than the Cuthill and McKee algorithm, and it can occasionally outperform it. However, the algorithm is significantly more complex to implement. More recently, Barnard *et al.* [1] have proposed the use of spectral analysis of the Laplacian matrix associated with the graph representing the non-zero elements in a sparse matrix as an effective method for the reduction of the envelope of a graph. In particular, the method permutes a matrix based on the eigenvector associated with the first non-zero eigenvalue of the Laplacian matrix. While the envelope is only indirectly related to the bandwidth of a matrix, this algorithm is very effective at reducing it. Further information on these and other classic methods for the BMP can be found in [2, 8].

Recently meta-heuristic approaches have been tested to see if they can be viable alternatives to solve the BMP. For example, Tabu search was employed by Marti *et al.* [17] while Lim *et al.* [14, 16] used a hybrid between genetic algorithms and hill-climbing to solve this problem. Lim *et al.* also introduced two other hybrid algorithms to solve BMP: one combining ant colony optimization with hill-climbing [12] and one combining particle swarm optimization with hill-climbing [13]. Recently also simulated annealing has been used to attack the problem [21].

In this paper, a Genetic Programming (GP) [11, 20] approach is proposed and tested against a high-performance version of the Cuthill and McKee algorithm [3] and a version of the spectral analysis proposed by Barnard *et al.* [1]. To the best of our knowledge, no prior attempt to use GP to solve BMP has been reported in the literature. Despite this lack of prior art, the results of our first investigations have been very encouraging and suggest this is a fertile area for further research.

The paper is organised as follows. In Sect. 2, we provide two equivalent formulations of the bandwidth-minimisation problem. In Sect. 3, we describe our GP system for the solution of BMP. In Sect. 4, we report the results of our experiments. Finally, in Sect. 5, we provide some conclusions.

2 Graph-Theoretic and Matrix Formulations of the BMP

Let $G = (V, E)$ be a finite undirected graph, such that V is the set of vertices, E is the set of edges and $f : V \to \{1, \ldots, n\}$ is a labeling of its nodes where $n = |V|$, then the *bandwidth* of G under f can be defined as:

$$B_f(G) = \max_{(u,v)\in E} |f(u) - f(v)| \ , \tag{1}$$

i.e., as the maximum absolute difference between the labels of the adjacent nodes (i.e., nodes connected by an edge). The *bandwidth minimisation problem* consists in finding a labeling f which minimises $B_f(G)$ while the easier *bandwidth reduction problem* requires finding any labeling which reduces $B_f(G)$. Since there are $n!$ possible labellings for a graph with n vertices, it stands to reason that the BMP is, in general, a very difficult combinatorial optimisation problem.

The BMP can also be stated in the context of matrices. If $A = [a_{ij}]_{n\times n}$ is a sparse matrix, its bandwidth is defined as

$$B(A) = \max_{(i,j):a_{ij}\neq 0} |i - j| \ . \tag{2}$$

The matrix bandwidth minimisation problem consists of finding a permutation of rows and columns which brings all non-zero elements of A into the smallest possible band around the diagonal. More formally, if σ is a permutation of $(1, 2, ..., n)$, and A_σ is the matrix obtained by permuting the rows and columns of A according to σ (i.e., $A_\sigma = [a_{\sigma_i \sigma_j}]$), then the problem can be formulated as

$$\min_\sigma B(A_\sigma) \ . \tag{3}$$

An important concept related to the bandwidth is the notion of *profile*. Given a matrix A, its profile is:

$$P(A) = \sum_{i=1}^{n} \max_{j:j<i,a_{ij}\neq 0} (i - j) \ . \tag{4}$$

Naturally, also the profile is influenced by permutations of A.

3 GP for BMP

We used a tree-based GP system implemented in C# with some additional decoding steps required by BMP and our particular choice of representation for solutions. A high-level description of the operations of the system is given in Algorithm 1. Below, the elements of the system are described in detail.

3.1 GP Setup

Individuals in our GP system are tree-like expressions (which, for efficiency reasons, are internally stored as linear arrays using a flattened representation for trees).

Algorithm 1. Pseudo-code of our GP system for BMP

```
1: Randomly generate an initial population of programs from the available primitives.
2: repeat
3:     Execute each program in the population.
4:     Create the permutation represented by each program.
5:     Apply each permutation to the adjacency list of the initial graph/matrix and
       generate new adjacency lists.
6:     Compute the bandwidth and profile for the adjacency lists obtained in step 5.
7:     Calculate the fitness value of each program in the population using Equation (5).

8:     Perform selection to choose individual program(s) from the population based on
       fitness to participate in genetic operations.
9:     Create a new generation of individual program(s) by applying genetic operations
       with specified probabilities.
10: until the maximum number of generations is reached.
11: return the permutation represented by the best program in the last generation.
```

We used the function set $\{+, -, \times, \%(\text{protected division}), \text{Sin}, \text{Abs}\}$ and the terminal set $\{X, Y, Z, c_1, ..., c_{100}\}$ where X, Y and Z are floating-point input variables whose role will be described later and $c_1, ..., c_{100}$ are uniformly-distributed random constants with floating-point values in the interval $[-5.0, +5.0]$.

The initial population was generated randomly using a modified version of the ramped half-and-half method [11, 20]. In our system, during the process of tree initialisation, terminals are not chosen with equal probability from the terminal set. Instead, we artificially increase the chance of selecting the variables X, Y and Z ensuring that every tree contains at least one variable.[1]

Tournament selection was used to choose individual program(s) from the population based on fitness to participate in genetic operations. New individual programs were created by applying the genetic operations of reproduction, sub-tree crossover and point mutation with specified probabilities. These and other parameters of the runs are presented in Table 1. We used elitism to ensure the best individual in one generation was transferred unaltered to the next. The termination criterion used was based on the predetermined maximum number of generations to be run. The permutation extracted from the best program tree appearing in the last generation was designated as the final result of a run.

3.2 The Generation of Permutations

The GP system described in the previous section is effectively quite similar to a system one might want to use to solve symbolic regression problems. One may wonder then how we transform program trees into the permutations which are needed to actually solve the BMP. We do this using a simple trick: we interpret

[1] Trees without variables are a problem for our system because they produce a constant output and, thus, they represent the permutation $\sigma = (1, 2, ..., n)$. Such a permutation is useless since it produces no change in bandwidth.

Table 1. Parameters used in our GP experiments

Parameter	Value
Maximum Number of Generations	100
Maximum Length of any GP Program	500
Maximum Depth of Initial Programs	3
Population Size	1000
Tournament Size	5
Elitism Rate	0.1%
Reproduction Rate	0.9%
Crossover Rate	70%
Mutation Rate	29%
Mutation Per Node	0.05%

the outputs produced by a program tree when executed over a set of fitness cases as a permutation. In this section, we explain this decoding process. The process is exemplified for a 5×5 array (or a 5-node graph) in Fig. 1.

For each program tree, the interpreter is called n times where n is the number of nodes of a given graph or the dimension of a given matrix. In other words, the number of fitness cases used depends on the dimension of the permutation to be generated. Each call of the interpreter executes the selected program with respect to the different values of the independent variables X, Y and Z.

The fitness cases in our system are somehow peculiar. Firstly, no target output is specified in the fitness cases. In addition, the values for X, Y and Z for each fitness case are carefully chosen during the initialisation of the system in such a way as to maximise the chance that they represent good permutations. In particular, following the ideas of [1] (see Sect. 1) the X values were set to be the components of the eigenvector associated with the first non-zero eigenvalue of the Laplacian matrix associated with the matrix to be optimised. The values of Y and Z were obtained by using the notion of level structure, which is at the basis of other high-performance BMP solvers, namely: the GPS algorithm and the reverse Cuthill-McKee algorithm which construct the level structure of peripheral nodes. Since the identification of peripheral nodes is expensive, we used the approach followed in [16] which picked random nodes in a graph and built permutations using their associated level structures. In other words, to initialise Y and Z we picked two random nodes and built their level structures. By traversing them we constructed two permutations. Then we converted these permutations into two floating point arrays such that, if sorted, they would produce those permutations back (see below). Finally, we used such arrays to provide the fitness case inputs associated with the Y and Z variables. It should be noted that the time required for the process of the initialization of the independent variables X, Y and Z is negligible compared to the total computational time.

The outputs obtained from each execution of the given tree were stored in a one dimensional array. This array was then sorted in ascending order while also recording the position that each element originally had in the unsorted

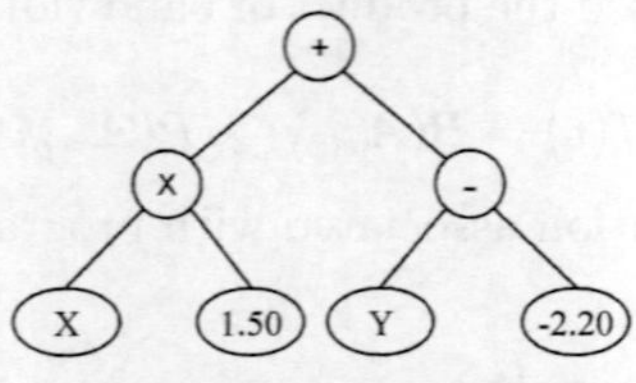

$X \in \{0.11, -0.25, -0.33, 0.96, 0.54\}$
$Y \in \{0.88, 0.76, 0.18, -0.64, -0.22\}$

Order	X	Y	Execution Results
1	0.11	0.88	3.245
2	-0.25	0.76	2.585
3	-0.33	0.18	1.885
4	0.96	-0.64	3.000
5	0.54	-0.22	2.790

1	2	3	4	5
3.245	2.585	1.885	3.000	2.790

Initial Execution Results

1	2	3	4	5
1.885	2.585	2.790	3.000	3.245

Sorted Results

1	2	3	4	5
3	2	5	4	1

Permutation

Fig. 1. The process of generating a permutation from a program tree

array. Reading such positions sequentially from the sorted array produced the permutation associated with the original tree.

3.3 The Calculation of Fitness Values

After the creation of the permutation representing a program, this permutation is applied to the adjacency list of the initial graph (or to a sparse matrix) in order to generate a new adjacency list which is used for calculating the bandwidth via (1) or (2).

The standard objective function for the BMP is simply that given in (2) and its calculation is straightforward. However, although this is precisely the value that needs to be minimized, it is not an ideal fitness function for a metaheuristic search. The main reason for this is that there may be many candidate solutions which have formally the same bandwidth, but which are quite different, with some being much closer to better solutions than others. A more effective approach is to use a fitness function which incorporates information about how close the candidate solution is to better areas of the search space.

As suggested in [10], the incorporation of the profile (see Sect. 2) into the fitness function used for the BMP can help capture this information. For these reasons, in this work we used the product of bandwidth and profile as our fitness function, i.e.,

$$f(p) = B(A_{\sigma(p)}) \times P(A_{\sigma(p)}) \tag{5}$$

where $\sigma(p)$ is the permutation associated with program tree p.

4 Experimental Results

In the experiments conducted in this study, we used a set of 15 instances from the well-known Harwell-Boeing sparse matrix collection (available from `http://math.nist.gov/MatrixMarket/data/Harwell-Boeing`). This collection includes benchmark matrices arising from problems in linear systems, least squares and eigenvalue calculations.

In order to evaluate the performance of the proposed GP method, we compared it against two high-performance methods. Firstly, we used the Reverse Cuthill-McKee (RCM) algorithm contained in the MATLAB library (RCMM). This is based closely on the SPARSPAK implementation described by George and Liu [6]. As indicated in Sect. 1, the algorithm first finds a (pseudo) peripheral vertex of the graph of the matrix. It then generates a level structure by breadth-first search and orders the vertices by decreasing distance from the chosen vertex. This algorithm is still one of the best and most widely used methods for the BMP. Indeed, the results reported in [4] and [15] indicate that RCMM

Table 2. Comparison of our GP approach with the RCMM and Spectral algorithms. Best results are shown in boldface, while ties are shown in italics.

Harwell-Boeing Matrix	Dimension	RCMM $B_f(G)$	Spectral $B_f(G)$	GP Approach Mean $B_f(G)$ (10 runs)
ash85	85 × 85	13	17	**12.0**
bcspwr01	39 × 39	*5*	11	*5.0*
bcspwr02	49 × 49	13	12	**10.4**
bcsstk01	48 × 48	27	**20**	24.5
can_24	24 × 24	7	6	**5.6**
can_61	61 × 61	19	14	**13.4**
can_62	62 × 62	9	10	**8.0**
can_73	73 × 73	27	27	**23.3**
can_96	96 × 96	23	18	**15.6**
dwt_59	59 × 59	8	10	**7.2**
dwt_66	66 × 66	*3*	*3*	*3.0*
dwt_72	72 × 72	7	12	**6.0**
dwt_87	87 × 87	17	19	**13.3**
lap_25	25 × 25	9	7	**6.0**
nos4	100 × 100	12	**11**	11.3
Mean of $B_f(G)$ values		13.27	13.13	10.97
Standard deviation		7.73	5.98	6.16
Standard error of the mean		2.00	1.54	1.59

Table 3. Run-by-run results of our GP system for BMP

	Benchmark Matrix														
Run	ash85	bcspwr01	bcspwr02	bcsstk01	can_24	can_61	can_62	can_73	can_96	dwt_59	dwt_66	dwt_72	dwt_87	lap_25	nos4
1	11	5	10	24	6	13	9	23	16	8	3	8	13	6	12
2	12	5	10	26	5	14	8	23	16	7	3	8	13	6	11
3	12	5	10	25	6	13	8	23	17	7	3	8	14	6	12
4	12	5	12	24	7	13	8	23	15	7	3	8	13	6	11
5	12	5	10	23	7	13	8	23	16	7	3	8	14	6	12
6	13	5	10	25	5	14	7	25	15	7	3	8	13	6	11
7	11	5	10	25	5	14	8	23	16	7	3	8	13	6	11
8	14	5	10	25	5	13	8	24	15	7	3	9	13	6	11
9	12	5	10	25	5	14	8	23	15	8	3	8	14	6	11
10	11	5	12	23	5	13	8	23	15	7	3	8	13	6	11
Best	11	5	10	23	5	13	7	23	15	7	3	8	13	6	11
Worst	14	5	12	26	7	14	9	25	17	8	3	9	14	6	12
Mean	12.0	5.0	10.4	24.5	5.6	13.4	8.0	23.3	15.6	7.2	3.0	8.1	13.3	6.0	11.3
Standard deviation	0.943	0	0.843	0.972	0.843	0.516	0.471	0.675	0.699	0.422	0	0.316	0.483	0	0.483
Standard error	0.298	0	0.267	0.307	0.267	0.163	0.149	0.213	0.221	0.133	0	0.1	0.153	0	0.153

is also superior to the well-known GPS [7] algorithm. In addition, we compared our method with the spectral analysis approach described in [1], summarised in Sect. 1 and used to set up the variable X in Sect. 3.2.

Each method was tested on our set of 15 benchmark matrices, which had sizes of up to 100×100. Considering the non-deterministic nature of GP, 10 independent runs were executed for each of the selected benchmark instances and performance values were averaged across such runs. Table 2 shows a performance comparison of the algorithms under test.

A simple inspection of the results of the experiments reported in Table 2 reveals that our GP approach is superior to both the RCMM algorithm and the spectral algorithm with respect to the mean of the bandwidth values and the number of the best results obtained. We performed a paired t-test in order to determine the statistical significance of the results obtaining a two-tailed P value of 0.0009 for the RCMM algorithm, and 0.0113 for the Spectral algorithm, which indicates that performance differences between our approach and these algorithms are statistically significant.

To give an idea of the reliability of the GP system, in Table 3 we provide its run-by-run results over the 15 benchmark problems. As one can see the algorithm produced very good results in all runs.

5 Conclusions

In this paper, a genetic programming approach has been introduced for solving the problem of the reduction of the bandwidth of a graph or a matrix. The proposed method was compared to the RCMM and Spectral algorithms which are among the best methods for solving the BMP. The results obtained on 15 standard benchmark matrices show that the proposed approach performs very favourably compared to these algorithms. However, it should be noted that the RCMM and Spectral methods are much faster than our GP-based method: they process a matrix in our dataset in seconds while our method requires approximately 3 minutes. So, the presented method is particularly appropriate when execution speed is not critical.

In future work, we will explore the possibility of building a parallel implementation of the system in order to improve its execution speed. We will also see if performance can be further enhanced by mutating constants, by using non-random populations and by hybridising the system by incorporating a local optimiser as was done in [12–14]. In addition, we will need to assess the generality and scalability of the proposed method by testing it with larger datasets and datasets including larger matrices.

References

1. Barnard, S.T., Pothen, A., Simon, H.D.: A spectral algorithm for envelope reduction of sparse matrices. In: Supercomputing 1993: Proceedings of the 1993 ACM/IEEE Conference on Supercomputing, pp. 493–502. ACM, New York (1993)

2. Corso, G.D., Manzini, G.: Finding exact solutions to the bandwidth minimization problem. Computing 62(3), 189–203 (1999)
3. Cuthill, E., McKee, J.: Reducing the bandwidth of sparse symmetric matrices. In: ACM National Conference, pp. 157–172. Association for Computing Machinery, New York (1969)
4. Esposito, A., Malucelli, F., Tarricone, L.: Bandwidth and profile reduction of sparse matrices: An experimental comparison of new heuristics. In: ALEX 1998, Trento, Italy, pp. 19–26 (1998)
5. Garey, M., Graham, R., Johnson, D., Knuth, D.: Complexity results for bandwidth minimization. SIAM Journal on Applied Mathematics 34(3), 477–495 (1978)
6. George, J.A., Liu, J.W.H.: Computer Solution of Large Sparse Positive Definite Systems. Prentice-Hall, Englewood Cliffs (1981)
7. Gibbs, N.E., Poole, W.G., Stockmeyer, P.K.: An algorithm for reducing the bandwidth and profile of a sparse matrix. SIAM Journal on Numerical Analysis 13(2), 236–250 (1976)
8. Gurari, E., Sudborough, I.: Improved dynamic programming algorithms for bandwidth minimization and the min-cut linear arrangement problem. Journal of Algorithms 5, 531–546 (1984)
9. Harary, F.: Graph Theory. Addison-Wesley, Reading (1969)
10. Koohestani, B., Corne, D.: An improved fitness function and mutation operator for metaheuristic approaches to the bandwidth minimization problem. In: Proceedings of the 1st International Conference on Bio-Inspired Computational (BICS), AIP Conference Proceedings, vol. 1117, pp. 21–28 (2009)
11. Koza, J.R.G.P.: On the Programming of Computers by Means of Natural Selection. MIT Press, Cambridge (1992)
12. Lim, A., Lin, J., Rodrigues, B., Xiao, F.: Ant colony optimization with hill climbing for the bandwidth minimization problem. Applied Soft Computing 6(2), 180–188 (2006)
13. Lim, A., Lin, J., Xiao, F.: Particle swarm optimization and hill climbing for the bandwidth minimization problem. Applied Intelligence 26(3), 175–182 (2007)
14. Lim, A., Rodrigues, B., Xiao, F.: Integrated genetic algorithm with hill climbing for bandwidth minimization problem. In: Cantú-Paz, E., et al. (eds.) GECCO 2003. LNCS, vol. 2724, pp. 1594–1595. Springer, Heidelberg (2003)
15. Lim, A., Rodrigues, B., Xiao, F.: A centroid-based approach to solve the bandwidth minimization problem. In: 37th Hawaii International Conference on System Sciences (HICSS), Big Island, Hawaii, p. 30075a (2004)
16. Lim, A., Rodrigues, B., Xiao, F.: A genetic algorithm with hill climbing for the bandwidth minimization problem, available from Citeseer-X (2002)
17. Marti, R., Laguna, M., Glover, F., Campos, V.: Reducing the bandwidth of a sparse matrix with tabu search. European Journal of Operational Research 135(2), 450–459 (2001)
18. Papadimitriou, C.H.: The NP-completeness of the bandwidth minimization problem. Computing 16(3), 263–270 (1976)
19. Pissanetskey, S.: Sparse Matrix Technology. Academic Press, London (1984)
20. Poli, R., Langdon, W.B., McPhee, N.F.: A Field Guide to Genetic Programming with contributions by J. R. Koza (2008), Published via `http://lulu.com`
21. Rodriguez-Tello, E., Jin-Kao, H., Torres-Jimenez, J.: An improved simulated annealing algorithm for bandwidth minimization. European Journal of Operational Research 185(3), 1319–1335 (2008)

Using Co-solvability to Model and Exploit Synergetic Effects in Evolution

Krzysztof Krawiec and Paweł Lichocki

Institute of Computing Science, Poznan University of Technology, Poznań, Poland
krawiec@cs.put.poznan.pl, pawel.lichocki@gmail.com

Abstract. We introduce, analyze, and experimentally examine *co-solvability*, an ability of a solution to solve a pair of fitness cases (tests). Based on this concept, we devise a co-solvability fitness function that makes solutions compete for rewards granted for solving pairs of tests, in a way analogous to implicit fitness sharing. We prove that co-solvability fitness function is by definition synergistic and imposes selection pressure which is qualitatively different from that of standard fitness function or implicit fitness sharing. The results of experimental verification on eight genetic programming tasks demonstrate that evolutionary runs driven by co-solvability fitness function usually converge faster to well-performing solutions and are more likely to reach global optima.

1 Introduction

Fitness function in evolutionary algorithms is a technical means to express experimenter's expectations with respect to the final outcome of the search process. It is typically designed so as to return a maximum value for an optimal design – an ideal solution. Unfortunately, a definition of fitness function that is appropriate from experimenter's viewpoint is not necessarily also the best one for *guiding* the search in solution space.

This issue becomes more evident when one confronts the concept of fitness in evolutionary computation to its counterpart in natural evolution. In biology, fitness is an artificial gauge introduced to model the *a posteriori* probability of individual's reproduction or the changes in relative frequencies of genotypes. The particular value of such an indicator stems from innumerable interactions between the organism and its environment, including other co-evolving individuals. As such, it is inevitably a very crude derivative of individual's characteristic and cannot fully reflect the richness of all its aspects.

By an analogy to the aforementioned multiple interactions, in evolutionary computation one often simulates the behavior of a solution (its phenotype) in multiple 'environmental conditions'. This can boil down to, for instance, testing an evolved machine learning classifier on various examples, simulating an evolved robot controller in different settings, or querying an evolved function on different combinations of inputs. Each such environment, typically referred to as *fitness case* or *test*, verifies the solution on a single instance or aspect of the problem.

R. Schaefer et al. (Eds.): PPSN XI, Part II, LNCS 6239, pp. 492–501, 2010.

The outcomes of interactions with particular tests are usually additively aggregated into scalar fitness value. Analogously to natural evolution, also here such aggregation is usually simplistic and implies inevitable loss of information.

Theoreticians and practitioners of evolutionary computation have been long aware of this problem and observed its aftermaths in various undesirable phenomena, including loss of diversity and premature convergence. Diverse countermeasures has been proposed, some of which avoid aggregation into scalar fitness measure by resorting to multiple objectives, either defined explicitly by a human (evolutionary multi-objective optimization [3]), or automatically derived from problem structure (multi-objectivization [5] and underlying objectives [2]). Switching to multi-objective perspective brings however other problems, like weakened selection pressure resulting from solution incomparability (mutual non-dominance).

The method proposed in this paper relies on scalar fitness and tries to improve search convergence by adjusting the rewards assigned to solutions for coping with particular fitness cases, in a way related to implicit fitness sharing [11]. In particular, our contribution is a method that focuses on individual's ability to properly handle *pairs* of fitness cases, and treats such pairs as elementary competences (skills) for which solutions can be awarded.

2 Preliminaries

We consider here the class of iterative search problems in which solutions are evaluated on a fixed set of fitness cases. This setup is typical for, among others, genetic programming (GP), where individuals are programs (procedures) that cannot be assessed otherwise than by applying them to some external input data. This mode of evaluation can be considered as a special case of a *test-based problem.* This term has been introduced in [1] to delineate a class of coevolutionary algorithms, particularly two-population coevolution, where a population of solutions co-evolves along the population of tests. A *test* in such scenario corresponds to a fitness case in genetic programming, with the major difference being that in GP tests typically do not evolve. Because of this analogy, we will identify these notions in the remaining part of this paper and borrow some terminology from coevolutionary algorithms.

The implementation of a single act of confronting a solution s with a test t, termed *interaction* in coevolution, is problem-dependent, and can boil down to testing an evolved entity s in particular environmental conditions t, testing an evolved logical or arithmetic expression s on a specific input-output pair t, or testing a machine learning classifier s on a specific training example t. Clearly, this class of problems pertains to a great share of real-world applications.

In following, we focus on problems with binary interaction outcomes: a solution either *solves* (passes) a test or not, a fact we denote using logical predicate $s(t)$ that returns *true* if solution s solves test t, and $false$ otherwise. Given set T of tests used to evaluate individuals in population, let $s(T) = \{t \in T : s(t)\}$ denote the subset of tests solved by s.

For this class of problems, the most straightforward way of defining fitness of a solution is to simply count the number of solved tests (a.k.a. *hits* in GP):

$$f(s) = |s(T)| \tag{1}$$

This definition, rational if no extra information on tests is available, suffers from substantial drawback: all tests contribute equally to fitness, so f can assume at most $|T|+1$ distinct values, which can result in numerous plateaus in the fitness landscape, particularly when T is small. This in turn weakens selection pressure: two solutions are likely to be indiscernible in terms of f.

A simple way of improving this state of matter is to *weigh* the rewards granted for solving particular tests. Such weights can be sometimes provided by a human expert and express his/her subjective assessment of test difficulty, test importance, or both. This, however, requires a substantial amount of domain knowledge and an extra effort. *Implicit fitness sharing* introduced by Smith *et al.* [11] and further explored for genetic programming by McKay [10,9] offers a more appealing alternative, by letting the evolution alone assess the difficulty of particular tests. Assuming that individual s is a member of population P, its fitness is here defined as:

$$f_s(s) = \sum_{t \in s(T)} \frac{1}{|P(t)|} \tag{2}$$

where $P(t) \subseteq P$ denotes the set of population members that solve test t. Thus, implicit fitness sharing simulates limits imposed on resources: individuals *share* the rewards for solving particular tests, each of which can vary from $\frac{1}{|P|}$ to 1 inclusive. Higher rewards are provided for solving tests that are rarely solved by population members (small $P(t)$), while importance of tests that are easy (large $P(t)$) is diminished. Additionally, because $P(t)$ typically pertains to the current population only, the assessed difficulties of tests change with time, which can help the search process escape local minima (as opposed to fixed weighting).

As such, fitness sharing can be perceived as a simple form of coevolution, where individuals compete for tests and their fate depends on the performance of other individuals (though there are no direct, face-to-face interactions between individuals). From yet another perspective, fitness sharing is a diversity maintenance technique: an individual that solves a low number of tests can still survive if its competence is rare. In this way, implicit fitness sharing helps reducing crowding and premature convergence; it shares this objective with explicit fitness sharing proposed in [4], where population diversity is enforced by monitoring genotypic or phenotypic distances between individuals.

3 Co-solvability

In broader terms, implicit fitness sharing enables an evolutionary process to assess the relative importance of *skills*, where skill is identified with the ability to solve a particular test. In real world however, it is often the *combination* of skills that matters. For an animal, the skill of digging and the skill of navigation can bring substantial benefits independently. However, when combined, they enable finding the previously buried prey and survive when food is scarce, a benefit which can be greater than the sum of benefits of its constituents. As another example, the overall performance of a mobile robot that is intended to move around a building depends on multiple skills, like the ability to move straight, the ability to make precise turns, and the ability to estimate its position. Again, each of these skills alone is not enough for the completion of the task, but together they make it possible.

Implicit fitness sharing cannot model such nonlinear accumulation of skills: the reward for simultaneous mastering of two or more skills amounts to the sum of rewards obtained for each skill individually. To enable synergy between pairs of skills, we introduce the notion of *co-solvability*. We call a pair of tests (t_i, t_j) *co-solvable by* s if and only if $s(t_i) \wedge s(t_j)$. The *co-solvability matrix* for a population P evaluated on set of tests T is a $|T| \times |T|$ matrix, with elements defined as

$$c_{ij} = \begin{cases} |\{s \in P : s(t_i) \wedge s(t_j)\}|, & i \leq j \\ 0, & otherwise \end{cases}$$

This matrix allows us to define the *co-solvability fitness function* f_c that rewards individuals for solving pairs of *distinct* tests:

$$f_c(s) = \sum_{t_i, t_j \in T : s(t_i) \wedge s(t_j), i<j} \frac{1}{c_{ij}} \tag{3}$$

Similarity of this formula to Formula (2) is not coincidental: co-solvability can be viewed as second-order fitness sharing. Let us notice that sharing of rewards for co-solving particular pairs of tests is an essential component here: simply *counting* the co-solvable tests ($|\{(t_i, t_j) : s(t_i) \wedge s(t_j),\ i < j\}|$) orders solutions in the same way as the standard fitness measure (Formula (1)), yielding precisely the same proceeding of evolution under any rank-based selection (e.g., tournament selection).

4 Properties of Co-solvability

Let us start from noticing that co-solvability fitness function f_c, similarly to f and f_s, fulfills the fundamental property we expect from any test-based fitness function, i.e., it is monotonous with respect to inclusion of sets of tests solved: for any $s_1, s_2, s_1(T) \subset s_2(T)$, it holds that $f_c(s_1) < f_c(s_2)$.

Let us consider four solutions s_1, s_2, s_3, s_4 such that when tested on four tests t_1, t_2, t_3, t_4, they perform as shown in Table 1a. Next, let us assume that the population contains a copies of s_1, b copies of s_2, c copies of s_3, and d copies of s_4[1]. The co-solvability matrix C for this population is shown in Table 1b.

[1] In other words, we work here with equivalence classes of solutions rather than with single solutions.

Table 1. An exemplary problem: the performances of solutions on tests (a) and the corresponding co-solvability matrix (b). Empty cells denote zeroes.

(a)

	t_1	t_2	t_3	t_4
s_1	1	1	0	0
s_2	0	0	1	1
s_3	0	1	1	0
s_4	1	0	0	1

(b)

	t_1	t_2	t_3	t_4
t_1	a+d	a		d
t_2		a+c	c	
t_3			b+c	b
t_4				b+d

Table 2 presents the fitness values for solutions $s_1 \ldots s_4$ as assigned by particular fitness functions: standard fitness f (Eq. (1)), fitness sharing f_s (Eq. (2)), and co-solvability fitness function f_c (Eq. (3)). We note that f does not discern any pair of solutions, no matter how often they occur in population. The ability of f_s and f_c to discern solutions depends on the actual values of a, b, c and d.

Let us use the solutions s_1 and s_3 from the above example to demonstrate that f_c can produce different ordering of individuals than fitness sharing. Technically, we want to check whether it is possible for $f_s(s_1) < f_s(s_3)$ and $f_c(s_1) > f_c(s_3)$ to hold simultaneously. As it follows from Table 2, these two conditions are respectively equivalent to $a+d > c+b$ and $a < c$, which are fulfilled by infinitely many quadruples of $a, b, c, d \geq 0$. Therefore, f_c is able to order solutions differently from f_s. Quite interestingly, it can be proven that four is the minimal number of tests required to produce such difference.

Let us now translate this observation into evolutionary context and give examples of scenarios when f_c produces substantially different results than f_s:

1. Consider two individuals s_1, s_2 such that $s_1(T) \cap s_2(T) = \emptyset$. Assume they undergo crossover and produce offspring s such that $s(T) = s_1(T) \cup s_2(T)$. Under f_c it has to hold $f_c(s) > f_c(s_1) + f_c(s_2)$, whereas for f and f_s equalities would hold. Thus, co-solvability is not additive and *enforces* synergy: for parents that exhibit mutually exlusive skills, their offspring that adopts all their skills is by definition better than both of them taken together. Also, f_c can be considered non-Markovian with respect to the changes observed in $s(T)$ as s undergoes evolutionary modifications: the increase of individual's fitness resulting from acquiring an ability of solving another test depends on the set of tests already solved by that individual.

2. Consider two individuals s_1, s_2 such that $s_1(T) \neq s_2(T)$ and $f_c(s_1) > f_c(s_2)$, and a test t such that $t \notin s_1(T) \cup s_2(T)$. Then, let us assume that, as a result of genetic modification both s_1 and s_2 acquire the skill of solving t, so that for the resulting solutions s'_1 and s'_2 it holds $s'_1(T) = s_1(T) \cup \{t\}$ and $s'_2(T) = s_2(T) \cup \{t\}$. As a conclusion from the above analysis, it is possible that $f_c(s'_1) < f_c(s'_2)$. In other words, s_2 can gain more from the same modification than s_1. Neither standard fitness nor implicit fitness sharing allow such possibility (the offspring of s_1 would be better than the offspring of s_2).

Table 2. Fitness values assigned to individuals from Table 1 by particular fitness functions

Fitness definition	s_1	s_2	s_3	s_4
Standard fitness f	2	2	2	2
Implicit fitness sharing f_s	$\frac{1}{a+d}+\frac{1}{a+c}$	$\frac{1}{b+c}+\frac{1}{b+d}$	$\frac{1}{a+c}+\frac{1}{b+c}$	$\frac{1}{a+d}+\frac{1}{b+d}$
Co-solvability fitness f_c	$\frac{1}{a}$	$\frac{1}{b}$	$\frac{1}{c}$	$\frac{1}{d}$

Co-solvability fitness is usually less discrete than f and f_s. For $n = |T|$ tests, standard fitness function f can return only $n+1$ distinct values. For the fitness sharing method, that number amounts to $n^{|P|}$ if the population is sufficiently large (precisely: if $|P|$ is greater than the nth prime number[2]). For co-solvability, $[\frac{n(n-1)}{2}]^{|P|}$ distinct values are possible.

In [7], Lasarczyk *et al.* proposed a method for selection of fitness cases based on a concept similar to co-solvability. The method maintains a weighted graph that spans fitness cases, where the weight of an edge reflects the historical frequency of a pair of tests being solved simultaneously. Fitness cases are selected based on a sophisticated analysis of that graph. Compared to that, our co-solvability is a simpler, parameter-free approach, which does not *select* the fitness cases but *weighs* pairs of them, individually for each solution (in [7], the same selected subset of fitness cases is used for all solutions).

5 The Experiment

The above analysis proves that co-solvability measure can produce different orderings of solutions and thus potentially steer evolution in other directions than conventional fitness measure f and fitness sharing f_s. It is however far from obvious whether this change is beneficial for effectiveness of search. In this section, we verify whether the fitness pressure imposed by co-solvability improves the performance of an evolutionary run applied to typical problems of logical function synthesis: multiplexer, parity, and two types of comparators. To this aim, we apply the approach of genetic programming (GP), with individuals being programs (procedures) encoded as expression trees.

For each problem, we prepared two instances, small and large, differing in the number of inputs. Table 3 summarizes the four considered problems, listing for each problem instance the number of inputs (independent variables, bits), the number of tests ($|T|$), and the proportion of tests for which the output of the program should be 1 and such for which the output should be 0. In the *Parity-odd* problem, the task is to evolve an expression that returns true if an odd number of ones appears on its inputs. *Multiplexer* should return the same value as the state of the addressed input (6-bit multiplexer uses two inputs to

[2] Sketch of proof: if $|P(t)|$ is a distinct prime number for each t, every combination of rewards received for particular tests yields a unique value of fitness f_s (Formula 2).

Table 3. The summary of problems and problem instances

	Small instance			*Large instance*		
Problem	*Inputs*	*Tests*	*Proportions*	*Inputs*	*Tests*	*Proportions*
Parity-odd	5	32	16:16	6	64	32:32
Multiplexer	6	64	32:32	11	2048	1024:1024
Eq	6	64	8:56	8	256	16:240
Cmp	6	64	28:36	8	256	120:136

Table 4. Success rates of best-of-run individuals produced by the methods, defined as the probability of run producing an ideal solution estimated from the total of 30 runs

	Small instance			*Large instance*		
Problem	f	f_s	f_c	f	f_s	f_c
Parity-odd	0.133	0.800	0.967	0.000	0.000	0.067
Multiplexer	1.000	1.000	1.000	0.433	0.700	0.567
Eq	0.000	1.000	1.000	0.000	0.700	0.933
Cmp	0.633	1.000	1.000	0.133	0.800	0.933

address the remaining four inputs, 11-bit multiplexer uses three inputs to address the remaining eight inputs). m-bit comparator *Eq* returns one if the $\frac{m}{2}$ least significant input bits encode a number that is equal to the number represented by the $\frac{m}{2}$ most significant bits. The *Cmp* comparator does the same but only when the former of these numbers is smaller than the latter.

Concerning GP-specifics, we use the Koza-I-style setup [6] with some modifications. The most important settings include: population of 1024 individuals initialized using the standard ramped half-and-half method, tournament selection with tournament of size 7, tree-swap crossover engaged with probability 0.9, subtree-replacing mutation applied with probability 0.1, no elitism. Evolution lasts for 200 generations. The software testbed has been implemented with help of ECJ [8] and is available at http://www.cs.put.poznan.pl/kkrawiec.

Table 5. Success effort (the expected number of generations to find the ideal)

	Small instance			*Large instance*		
Problem	f	f_s	f_c	f	f_s	f_c
Parity-odd	1448	159	102	∞	∞	2999
Multiplexer	9	8	8	370	164	224
Eq	∞	38	33	∞	192	121
Cmp	186	32	31	1453	181	136

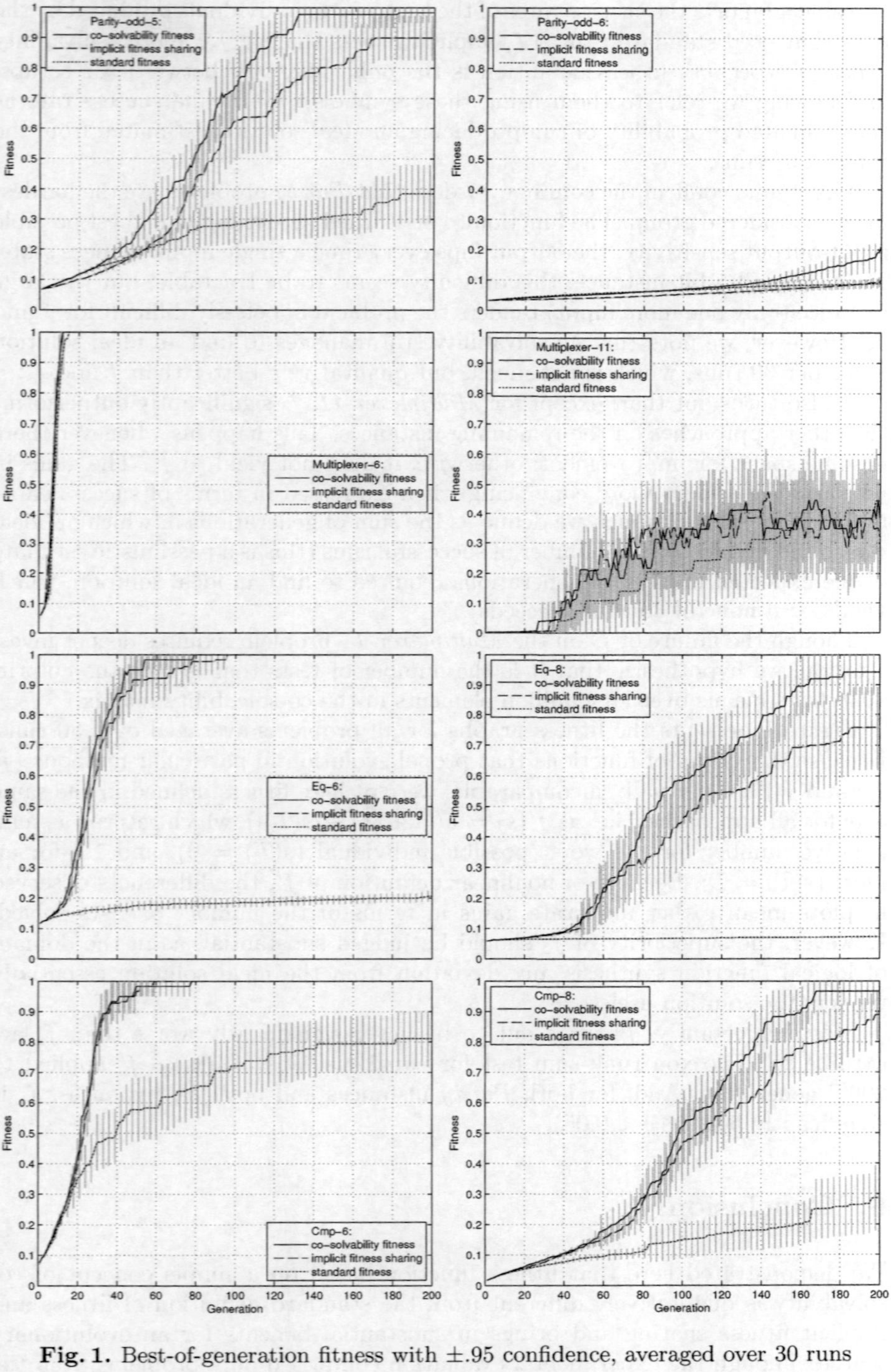

Fig. 1. Best-of-generation fitness with ±.95 confidence, averaged over 30 runs

Table 4 reports the success rates of the best-of-run individuals produced by the runs that used standard fitness f, implicit fitness sharing f_s, and co-solvability fitness f_c (because fitness definition is the only difference between the setups, in following we refer to them using these symbols). We define success rate as the estimated probability of run producing an ideal solution estimated from the total of 30 runs.

The results confirm the common wisdom that *Parity* problems are the hardest in the considered group. The function to be learned here exhibits highest possible input-output sensitivity: the output flips every time a single input changes state. For the small 5-bit instance, this difficulty seems to be tractable, but the large instance (only one more input) renders the problem hopelessly difficult for f and f_s. However, we note that co-solvability still manages to find an ideal solution twice per 30 runs, which is not much, but qualitatively better than f and f_s.

It also turns out that, except for *Multiplexer-11*, f_c significantly outperforms the other approaches on the remaining instances. This happens whenever there is some space for improvement; otherwise, it does not yield to f_s. The gains in performance appear more convincing when expressed in terms of success effort reported in Table 5, which we define as the sum of generations in which an ideal was found divided by the number of successful runs (this is a pessimistic estimate of the expected number of generations required to find an ideal solution, which yields ∞ if none of 30 runs succeeds).

Though the failure of f_c on the *Multiplexer-11* problem requires deeper investigation, we hypothesize that it is the number of tests that is here the culprit: 2048 tests means over four million elements in the co-solvability matrix.

Figure 1 presents the fitness graphs for all problems averaged over 30 runs. Because the values of functions that propel evolution in particular methods (f, f_s, and f_c) are mutually incomparable, we plot here fitness defined in the same way for all approaches, i.e. as $f'(s) = 1/(1 + |T| - |s(T)|)$, which returns a small positive number for the worst possible individual ($s(T) = \emptyset$), and 1.0 for an ideal ($s(T) = T$). Because of nonlinear definition of f', the differences observed in plots mean rather moderate gains in terms of the number of tests solved. However, the superiority of f_c should be judged substantial, as in the domain of logical function synthesis any deviation from the ideal solution essentially renders the solution useless.

Most importantly, f_c turns out to be never significantly worse than f_s according to Wilcoxon runk-sum test for equal medians of fitness f' applied to 200th generation. And, for both *Parity* instances and both *Eq* instances, f_c is significantly better ($p < 0.05$).

6 Conclusion

We demonstrated here that fitness function based on a simple concept of co-solvability is qualitatively different from the standard definition of fitness and implicit fitness sharing and brings in substantial benefits for an evolutionary search. Though the experimental evaluation comprised only problems from the realm of genetic programming, the proposed method abstracts from representation of solutions and is thus applicable to any test-based search problem. It is likely then that similar gains could be observed for other classes of problems.

The method is straightforward to implement and its computational overhead is basically the same as in case of fitness sharing. This extra cost can be considered negligible when compared to the actual cost of testing a solution on a test (i.e., determining the outcome of $s(t)$), which is typically much higher.

Acknowledgment. This work was supported in part by Ministry of Science and Higher Education grant # N N519 3505 33.

References

1. Bucci, A., Pollack, J.B., de Jong, E.: Automated extraction of problem structure. In: Deb, K., et al. (eds.) GECCO 2004. LNCS, vol. 3102, pp. 501–512. Springer, Heidelberg (2004)
2. de Jong, E.D., Pollack, J.B.: Ideal Evaluation from Coevolution. Evolutionary Computation 12(2), 159–192 (Summer 2004)
3. Deb, K.: Multi-Objective Optimization using Evolutionary Algorithms. Wiley-Interscience Series in Systems and Optimization. John Wiley & Sons, Chichester (2001)
4. Goldberg, D.: Genetic algorithms in search, optimization and machine learning. Addison-Wesley, Reading (1989)
5. Knowles, J.D., Watson, R.A., Corne, D.: Reducing local optima in single-objective problems by multi-objectivization. In: Zitzler, E., Deb, K., Thiele, L., Coello Coello, C.A., Corne, D.W. (eds.) EMO 2001. LNCS, vol. 1993, pp. 269–283. Springer, Heidelberg (2001)
6. Koza, J.R.: Genetic Programming: On the Programming of Computers by Means of Natural Selection. MIT Press, Cambridge (1992)
7. Lasarczyk, C.W.G., Dittrich, P., Banzhaf, W.: Dynamic subset selection based on a fitness case topology. Evolutionary Computation 12(2), 223–242 (Summer 2004)
8. Luke, S.: ECJ evolutionary computation system (2002), `http://cs.gmu.edu/eclab/projects/ecj/`
9. McKay, R.I.B.: Committee learning of partial functions in fitness-shared genetic programming. In: 26th Annual Confjerence of the IEEE Third Asia-Pacific Conference on Simulated Evolution and Learning 2000, Industrial Electronics Society, IECON 2000, Nagoya, Japan, October 22-28, vol. 4, pp. 2861–2866. IEEE Press, Los Alamitos (2000)
10. McKay, R.I.B.: Fitness sharing in genetic programming. In: Whitley, D., Goldberg, D., Cantu-Paz, E., Spector, L., Parmee, I., Beyer, H.-G. (eds.) Proceedings of the Genetic and Evolutionary Computation Conference (GECCO 2000), Las Vegas, Nevada, USA, July 10-12, pp. 435–442. Morgan Kaufmann, San Francisco (2000)
11. Smith, R., Forrest, S., Perelson, A.: Searching for diverse, cooperative populations with genetic algorithms. Evolutionary Computation 1(2) (1993)

Fast Grammar-Based Evolution Using Memoization

Martin Luerssen and David Powers

Artificial Intelligence Laboratory,
School of Computer Science, Engineering and Mathematics,
Flinders University, Adelaide, Australia
{martin.luerssen,david.powers}@flinders.edu.au

Abstract. A streamlined, open-source implementation of Shared Grammar Evolution represents candidate solutions as grammars that can share production rules. It offers competitive search performance, while requiring little user-tuning of parameters. Uniquely, the system natively supports the memoization of return values computed during evaluation, which are stored with each rule and also shared between solutions. Significant improvements in evaluation time, up to 3.9-fold in one case, were observed when solving a set of classic GP problems – and even greater improvements can be expected for computation-intensive tasks. Additionally, the rule-based caching of intermediate representations, specifically of the terminal stack, was explored. It was shown to produce significant, although lesser speedups that were partly negated by computational overhead, but may be useful in dynamic and memory-bound tasks otherwise not amenable to memoization.

Keywords: Evolutionary algorithms, genetic programming, grammatical evolution, shared grammar evolution, memoization.

1 Introduction

Genetic Programming (GP) [1] operates on variable-length n-ary syntax trees, which offer greater representational flexibility than the fixed-length strings used in the canonical genetic algorithm. One of the main constraints of GP is the need for closure, i.e., the requirement that any combination of arguments is syntactically legal. With CFG-GP, Whigham [2] not only introduced the use of context-free grammars (CFGs) for this purpose, but also included an automatic mechanism for modifying the grammar based on the fittest solutions, an idea that has since found further refinement in Grammar-Model-based Program Evolution [3]. Adapting the CFG in this way improves the search bias towards not just valid, but better performing solutions.

Grammars can also assist in another GP challenge: that of achieving scalability. For a variable length representation, search spaces are effectively unbounded; solutions can exist in higher-dimensional spaces, even if they are structurally simple. Scalability can be facilitated here by a "divide and conquer" approach, which

R. Schaefer et al. (Eds.): PPSN XI, Part II, LNCS 6239, pp. 502–511, 2010.

decomposes the larger problem into weakly correlated subproblems that can be dealt with independently. Software engineering practice has inspired techniques such as Automatically Defined Functions (ADFs) [4] that extend GP by enabling the creation and reuse of discovered modules. Nature, however, has its own means of promoting modularity. Embryogenesis efficiently captures the difference between a compact, searchable representation (genotype) and the functional, high-dimensional solution (phenotype). This mapping process can be modelled with a generative grammar; L-systems have been particularly popular in this context [5].

Shared Grammar Evolution (SGE) [6] combines the three uses of grammars for closure, bias, and modularity. Solutions obtained through SGE comply with a user-defined CFG – but are also represented directly as simple, deterministic CFGs. One of the notable side-benefits of this is an intrinsic support of memoization, that is, the caching of return values for parts of the evolved solution. As an evolved population converges, we can expect the similarity between members to increase, so memoization can accelerate evolution by reducing or even eliminating evaluation time of shared modules. This paper explores the performance benefits of memoization on solving several classic GP problems with SGE. Toward this purpose, we present a streamlined version of SGE that is now publicly available under an open source license.

1.1 Grammatical Evolution

Grammatical Evolution (GE) [7] is currently the most well-published technique of applying grammars to evolution. GE and SGE can be used towards the same tasks and share in common that solutions are derived from context-free grammars (CFGs) defined in Backus-Naur Form (BNF). However, the underlying mechanics differ substantially. GE employs a genetic algorithm, where the genotype is a variable-length bit string that is read left to right to generate 8-bit integers (so-called codons). The modulus of each codon and the total number of production rules in the CFG specifies the rule to be applied to the currently replaced nonterminal, starting from the axiom (starting symbol) and ending when all nonterminals have been replaced. If all codons are read but not all nonterminals are replaced, the expansion wraps and continues from the start of the genome, unless a pre-determined maximum number of wrappings has already been exceeded.

2 Shared Grammar Evolution

With SGE, the user must likewise provide an initial *template* CFG, which delimits the space of valid solution candidates. If we had knowledge of what made a good solution, we could perhaps define a grammar that only contains good solutions. The user rarely has this information in advance, but it is possible to sample this space. We do so by deriving a solution candidate from the original grammar, which again involves choosing between alternative production rules according to the original CFG. However, in SGE, these choices specify an *i*-grammar (individual grammar; see Figure 1) that produces just that one solution. *i*-grammar rules that originate from the same template rule are legal alternatives to each other, just as *i*-grammar axioms constitute alternative solution candidates. Applying selection to the solution population implies selection of *i*-grammar rules through their contribution to the selected solutions. The rule alternatives that survive are those that contribute to better solutions. As our search converges towards the optimum, so should the suitability of the production rules and hence the building blocks from which we build the solutions.

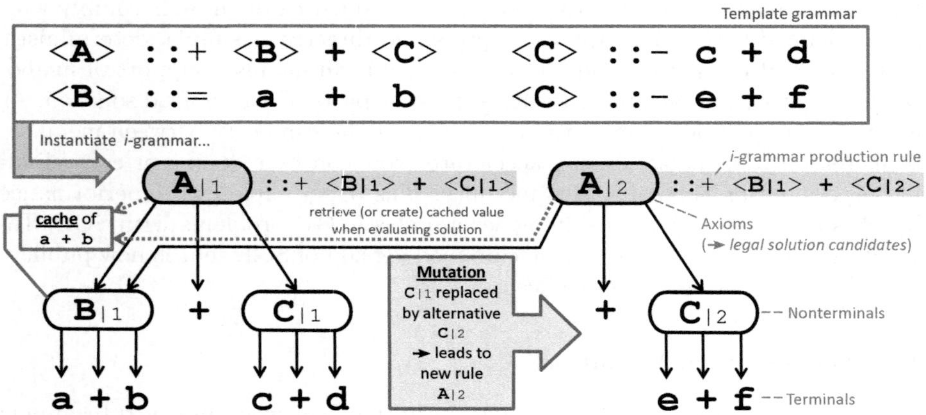

Fig. 1. Solution candidates are described by production rules initially derived from the user-defined template grammar. Mutation involves replacing an existing rule with another rule, either part of another solution (a *reuse*) or generated from the template grammar (a *reinstantiation*). In this example, rule $B_{|1}$ is shared between two solutions; memoized results of evaluating this rule can be reused by both solutions.

2.1 Shared Representation

Due to the overhead of having to describe each rule, representing a solution as a series of production rules is not very efficient. However, if multiple rules have identical successors, they can be represented by a single rule. Instead of keeping duplicates, *i*-grammar rules are shared across multiple solution candidates. If these solutions are similar, then the population of solutions and rules can be represented more compactly. Moreover, since *i*-grammar rules are context-free and deterministic, every rule leads to a specific derivation. We can therefore treat an *i*-grammar rule as an encapsulated module or subroutine that may be shared and reused among solution candidates. Once it is not required any longer, the rule is automatically eliminated. Unlike in [6], our new implementation relies solely on a reference-counting scheme; e.g., a rule that is only referred to by one other rule, even if that rule is referred to many times, will now have a usage count of only one.

2.2 Mutation

Each i-grammar rule has syntactically valid alternatives based on the template rule it relates to. A new solution candidate can be created by choosing an existing candidate, then choosing a rule from it, and replacing it with an alternative, which can either be from any other i-grammar (a reuse of a rule) or by deriving a new sequence from the template grammar (a reinstantiation of a rule). The chance of a particular rule being reused is proportional to its reference count; a rule that is part of many solutions, but only called by few other rules – i.e., rarely added successfully to a new solution on its own terms – will be less likely to be chosen. The reinstantiation probability is decided by the user. A notable exception is that reinstantiation will be automatically chosen instead of reuse if there is no valid alternative in existence, or if the production by which a replacement is to occur is the same as the replacement. During reinstantiation, the choice between reuse and reinstantiation is separately made for every rule considered.

If we want to maintain both parent and offspring solution, the production rules of the parent, between the starting rule and where the replacement occurs, will need to be copied and pointed to the replacement or the likewise modified successor. The new implementation is simplified compared to [6], in that a production rule only affects one point in the derivation tree, not all points where that rule is called. No recursion can arise from such singular changes, so for problem tasks that benefit from a recursive description, recursion must be represented within the solution space by providing recursion operators in the terminal set. Furthermore, we rely on non-elitist selection to account for cases where progress is only possible through multiple changes on separate branches of the derivation tree.

2.3 Memoization

The shared representation does not merely lead to a potentially smaller memory footprint. As each solution candidate is derived in a series of rule expansion steps, we can also interpret these as nested subroutine calls. If the same rule is used by many candidates, then this is equivalent to calling the same subroutine, which will return the same result. Why not store the result of the first call and re-use it afterwards? Memoization is a well-established technique that enables a subroutine to record, in a local table, values that have previously been calculated. The term was coined by Donald Michie [8] and is frequently encountered in the context of functional programming languages. Memoization lowers a function's time cost in exchange for space cost; that is, memoized functions are optimized for speed in exchange for greater use of memory space. This kind of caching is highly applicable to the evolution of programs and other hierarchies that include common modules that are reused many times. Its benefits have been exploited on only a few problems with expensive evaluation functions, such as robot evolution [9] and chess playing [10]. In these instances, however, memoization is applied only at the level of the terminals, whereas in SGE every rule that can have a result associated with it (a *cacheable* rule) can be memoized.

To be able to look up a result in a table, rather than recompute it, a number of constraints must be satisfied. Firstly, there needs to be a deterministic outcome that, preferably, is isolated from its context. If a production A leads to an expression such as $X + X$, then $2A$ would be $2X + X$, so knowing the result of $X + X$ is not helpful here (unlike if A were $(X + X)$). Secondly, memoization becomes impractical for tasks that involve side-effects, e.g., dynamic control tasks, because of the indirect dependencies that arise between program parts. Memoizing the evaluated results is therefore not useful in all cases.

As an alternative, we could instead cache the expanded terminal string. It may be represented as, or compiled into, a form that is easy to compute, i.e., through an intermediate representation, such as bytecode. The table look-up would then involve retrieving not the value of a particular evaluation, but essentially a subprogram that accomplishes the same task as the derivation of the production rule would – but, ideally, faster. Separating the evaluated representation from the grammar in this way can bring about further opportunities, e.g., evaluating solutions on specialized hardware. In this paper, we limit ourselves to testing the idea of using the terminal stack as the intermediate representation. Thus, rather than repeatedly deriving from rules and evaluating operator precedences, the ordered symbol sequence will be shared, with higher-level sequences constructed from encapsulated, lower-level sequences (or so-called fragments).

3 Implementation

SGE as presented in this paper has been implemented as an open source software package written in C++, called COGENT [11]. COGENT reads a template grammar as well as a set of parameters from a text file, evolves solution candidates to the problem, and evaluates these on user-defined objective functions. We can optimize towards any number of objectives at the same time; the objectives can be organized according to their importance. Objectives of identical importance are sorted according to their Pareto-domination using the NSGA-II [13]. If, for selection purposes, two solutions are ranked identically, further user-defined objectives of increasingly lesser importance are taken into account, possibly describing other aspects of the solution, such as crowding-distance. Grammar definitions are BNF-like, extended by allowing you to set multiple production rules as starting rules (rather than just the first) when following the left-hand side with a "::+"". Such starting rules are automatically isolated for memoization; other cacheable rules can be denoted with "::=". Rules that have only terminals as their successors are regarded as constant and are represented by one rule for each successor combination.

Memory for memoization is obtained from a Boost memory pool [12] for fast allocation when creating a cacheable rule. This is not feasible for variable size allocations, such as for memoization of intermediate representations. In those cases, the memory only contains pointers to the allocations, which are allocated conventionally via *malloc*. COGENT supports multiple threads for deriving, memoizing, and evaluating solutions, but not for modifying the grammar, due to the extensive and costly use of the mutexes that would be required.

Table 1. COGENT parameters for the five problem tasks

<table>
<tr><th>Task</th><th>Description</th><th>Template CFG</th><th>Evolution Parameters</th></tr>
<tr><td>Quintic Regression</td><td>Symbolic regression of $x^5 + x^4 + x^3 + x^2 + x$ for 20 equidistant points in $x = [-1, 1)$</td><td><EXPR> ::+ (<EXPR> <OP> <EXPR>)
| <VAR>
<OP> ::- add | sub | mul
<VAR> ::- VAR | CONST<1></td><td rowspan="5">Generations: 100

Population size:
100 (200 for 6th-order and multiplexer)

Selection:
Applied to the combined set of parents and offspring; elite of 10, with remaining 90 solutions chosen by tournament of size 3

Solution size limit:
Maximum size of 800 rules per solution (if exceeded, discard solution and retry)

Objectives:
Minimize error; if equal, minimize solution size, then minimize solution age (generations)</td></tr>
<tr><td>6th-order Regression</td><td>Symbolic regression of $x^6 - 2x^4 - x^2$ for 20 equidistant points in $x = [-1, 1)$</td><td><EXPR> ::+ (<EXPR> <OP> <EXPR>)
| <VAR>
<OP> ::- add | sub | mul | div
<VAR> ::- VAR | CONST<1></td></tr>
<tr><td>Even-5 Parity</td><td>Returns true if an even number of 5 Boolean inputs is true (over all 32 combinations of inputs)</td><td><EXPR> ::+ <EXPR> <OP> <EXPR>
| (<EXPR> <OP> <EXPR>)
| <VAR>
| <PREOP> <VAR>
<OP> ::- or | and | xor
<PREOP> ::- not
<VAR> ::- var<0> | var<1>
| var<2> | var<3>
| var<4></td></tr>
<tr><td>6-bit Multiplexer</td><td>Return the value of a data register (d0, d1, d2, d3) specified by the binary address (a0, a1) (over all 64 combinations of inputs)</td><td><A> ::+ (<A>)
<A> ::- <A> and <A>
| <A> or <A>
| not <A>
| if (<A>) (<A>) (<A>)
| var | var<1>
| var<2> | var<3>
| var<4> | var<5></td></tr>
<tr><td>Santa Fe Ant Trail</td><td>Collect 89 food pellets laid out on a toroidal 32 × 32 grid</td><td><A> ::+ move | left | right
| iffoodahead <A> <A>
| prog <A> <A>
| prog <A> <A> <A></td></tr>
</table>

4 Experiment

The experimental aim is to determine whether evolution can be accelerated using memoization, but the extent to which this is possible may be highly dependent on the problem task and the exact evolutionary parameters. Five classic GP problems were chosen that are already well-understood and also quick to evaluate. The observed results should therefore be closer to the lower bound of what can be achieved with memoization, rather than artificially inflated. As shown in Table 1, the experimental setup contains few surprises, but note that SGE does not have mutation or recombination parameters in the traditional sense; the reinstantiation probability controls the changes that are made to solutions. Since rules that are freshly derived from the template grammar are unshared, we expect this to directly affect the memoization benefit. We hence investigated the effect of having no reinstantiation, to reinstantiating in 25% of cases, 50% of cases, and always. Furthermore, the evolutionary runs are timed for four different memoization scenarios:

1. **no caching:** solution is explicitly derived for each problem case
2. **solution caching:** derived solution (i.e., the terminal stack; see section 2.3) is stored and reused between problem cases
3. **fragment caching:** each cacheable rule stores its own intermediate terminal stack
4. **value caching:** each cacheable rule stores the return value for its evaluation (i.e., classic memoization)

Finally, the quintic regression, parity problem, and ant trail were evolved using GEVA [14], an open-source implementation of Grammatical Evolution, where they are included as example problems (with equivalent parameters to the above).

4.1 Results

Table 2 lists performance statistics for the best evolved solutions as well as the total computation time, which is split into an evaluation time and evolution time. The former specifies how long the program spent on deriving and evaluating all solutions, whereas the latter includes the remaining computing time, mostly of the evolutionary algorithm itself and the collection of experimental statistics. Results are averaged over 100 runs. Random seeds remain constant across different memoization options so that timings are based on the same evolutionary outcomes.

There is surprisingly little objective performance variation between the different reinstantiation probabilities; only on the two regression problems does the 100% setting perform significantly worse than the alternative configurations ($p < 0.001$ on a two-tailed t-test). It suggests that one need not be particularly careful with this parameter. However, reinstantiation is theoretically very limited, as it derives changes directly from the template grammar, i.e., a fixed distribution. Its apparent effectiveness may simply reflect the ineffectiveness of the opposite: if we just draw building blocks from within the population, overall entropy will drop and premature convergence may happen. Conversely, random draws from the template grammar increase entropy, so a trade-off is ultimately necessary, especially for deceptive problems; i.e., some – but not too much – reinstantiation is needed. Since it appears to work well for a broad range of values, however, we cannot deduce much more from this experiment.

The memoization results offer greater clarity. Firstly, evaluating a solution by deriving it for every variation of a terminal value is about an order of magnitude slower than if we retain an intermediate representation (the terminal stack) across all the problem cases. If we break the representation further up by storing substacks with each evaluable rule, we gain a 50+% improvement in evaluation speed on quintic and parity problems, over 100+% on the 6th-order polynomial (all significant to $p < 0.001$), but just a few percent on the multiplexer (not significant). One of the explanations for this lies in the evolved solution size. For the 6th-order polynomial, it is around 300 rules by the 100th generation, but for the multiplexer, it is only 20-30 rules. The extent to which productions are shared is directly affected by this; the share ratio indicates that there is 6× as much sharing happening with the polynomial than with the multiplexer. Additionally, most of the rules of the multiplexer grammar are not defined as cacheable (and are not designed to be), whereas all of the rules for the polynomial are. These factors influence how worthwhile it is to use memoization.

Storing the intermediate representation with each cacheable rule involves a notable overhead, as can be observed from the evolution time. On all the cacheable problems tested here, it negates the entire performance benefit of this strategy. However, on real-world problems with more costly evaluation functions, we

Table 2. Evolution statistics for each problem, averaged over 100 runs. *Best* refers to the fittest solution. The *Share Ratio* is the number of expressed rules across all solutions divided by the rules in the global grammar and averaged over all generations. Indicated run times (in milliseconds) are for each complete run and given separately for the evaluation and evolution components of the system.

Experiment	Success Rate	Best Error	Best Size	Share Ratio	Total Evaluation + *Evolution* time (ms) for memoization scenario: None	Solution	Fragment	Value
Quintic								
0% reinst.	99%	0.003	39.2	5.47	3,575+ *104*	372+ *105*	241+ *266*	254+ *111*
25% reinst.	94%	0.008	58.5	6.02	4,638+ *117*	495+ *118*	289+ *353*	296+ *128*
50% reinst.	90%	0.009	90.5	6.99	5,223+ *127*	591+ *126*	327+ *430*	323+ *135*
100% reinst.	83%	0.015	80.4	4.81	5,338+ *208*	531+ *208*	322+ *296*	559+ *225*
6th-order								
0% reinst.	41%	0.009	361.2	9.95	40,091+ *485*	5,039+ *495*	2,135+ *3233*	1,297+ *541*
25% reinst.	38%	0.010	315.6	9.23	36,000+ *460*	4,384+ *470*	1,962+ *2883*	1,206+ *515*
50% reinst.	44%	0.011	315.6	9.09	36,068+ *473*	4,395+ *483*	2,000+ *2911*	1,225+ *529*
100% reinst.	14%	0.020	296.4	6.91	29,592+ *540*	3,242+ *550*	1,552+ *2331*	1,455+ *598*
Even-5 Parity								
0% reinst.	79%	1.34	35.1	6.11	7,284+ *102*	596+ *103*	385+ *296*	339+ *109*
25% reinst.	80%	0.99	38.7	6.89	8,058+ *107*	656+ *108*	414+ *311*	351+ *113*
50% reinst.	81%	0.97	37.0	5.67	8,056+ *113*	678+ *114*	420+ *341*	373+ *119*
100% reinst.	78%	1.20	36.4	3.73	7,376+ *183*	615+ *186*	414+ *443*	641+ *197*
6-bit M.plex								
0% reinst.	75%	1.36	29.5	2.81	22,696+ *376*	2,468+ *378*	2,202+ *703*	1,546+ *388*
25% reinst.	85%	0.57	24.7	2.60	20,283+ *364*	2,280+ *366*	2,054+ *662*	1,458+ *375*
50% reinst.	82%	0.63	20.6	2.23	17,973+ *351*	1,997+ *354*	1,813+ *613*	1,383+ *361*
100% reinst.	71%	1.10	19.9	1.91	9,992+ *320*	1,224+ *315*	1,185+ *441*	1,195+ *322*
Ant Trail								
0% reinst.	11%	21.2	75.5	4.56	147+ *157*	N/A		
25% reinst.	7%	22.6	80.3	4.48	148+ *168*	N/A		
50% reinst.	15%	20.3	85.1	4.58	147+ *176*	N/A		
100% reinst.	19%	21.6	32.7	2.91	138+ *449*	N/A		

should still expect a substantial net benefit. A more convincing result is produced by memoization in the classic sense of caching evaluation results. Here, the evolution cost is only marginally higher than baseline, but the evaluation improvement is even greater: up to 3.9× on the 6th-order polynomial, and between 1.4× to 1.9× on the other problems (all significant $p < 0.001$). These numbers exclude the 100% reinstantiation setting, as hardly any improvement is observed there. 100% reinstantiation should lead to less speedup, because any rules obtained via reinstantiation are new and unshared, and we can accordingly observe a much lower share ratio for 100% than for any other setting. It is not clear, however, why this has a much greater impact with value caching than with stack caching – we may be hitting an implementation bottleneck of some kind here.

The Santa Fe ant trail turned out to be a task not suitable for memoization, as it is not only recurrent but also a single case problem that can be evaluated directly and more efficiently from its *i*-grammar than through any other means. We include it solely for comparative purposes. Likewise intended for informal comparison are the GEVA outcomes: 20% success rate (with error 0.727) for the quintic regression, 76% success rate (with error 1.61) for the parity problem, and 8% success rate (with error 23.88) for the ant trail. COGENT appears to perform slightly better on these problems, significantly so for the regression ($p < 10^{-10}$), but since a GE expert could likely improve upon this, we can only say that our system appears to be competitive without much fine-tuning involved.

5 Conclusion

In this paper we have presented a streamlined implementation of a novel scheme for evolving solutions from a user-defined CFG, which is performance competitive with GE, requires few parameters to be tuned, and is available as open source [11]. A particularly noteworthy feature is its support for memoization, through which we achieved significant improvements to evaluation speed between $1.4-3.9\times$ over just caching the expanded solution between sample presentations. As the tested problems were quite simple, we expect this to be a lower bound; real-world problems with expensive evaluation functions should benefit much more so. Memoization was also noted to be more effective for larger solutions that share many production rules.

Connected to this, we also explored the reinstantiation probability, which defines the balance between creating new rules from the user-defined CFG and exploiting existing rules in the population. It affects both sharing and the objective error of evolved solutions, although significant changes (for the worse) were only observed when reuse of existing rules was discouraged completely. The observed tolerance to parameter changes is postulated to be due to its complex relationship to diversity in the population and the associated exploration-exploitation balance, which needs to be investigated further.

At present, our scheme is used only at its most basic; there are many opportunities for enhancement. For instance, the rule to be changed is chosen randomly, but one could instead impose and modify probabilities for each rule, e.g., in an ant-system like manner. Experimental evaluation of such ideas will greatly profit from faster evolution. Multicore support is currently limited to shared-memory multithreading for derivation and evaluation of solutions. We intend to expand this to a distributed memory, multi-grammar approach, but this has implications on the applicability of memoization that need to be explored. We also hope to expand on the notion of intermediate representations, especially for more complex problems that share correspondingly complex modules, with more selective caching to emphasize performance benefits (over the drawbacks) of memoization, and compilation to faster representations that can run on specialized hardware, such as GPUs. We will continue to make these changes publicly available; fellow researchers are welcome to check them out and contribute.

Acknowledgments

We acknowledge and appreciate the support of the *Thinking Head* SRI grant TS0669874 of the Australian Research Council (ARC) and the National Health and Medical Research Council (NH&MRC).

References

1. Koza, J.: Genetic programming: on the programming of computers by means of natural selection. MIT Press, Cambridge (1992)
2. Whigham, P.: Grammatically-based genetic programming. In: Rosca, J. (ed.) Workshop on Genetic Programming: From Theory to Real-World Applications, pp. 33–41. Morgan Kaufmann Publishers, San Mateo (1995)
3. Shan, Y., McKay, R., Baxter, R., Abbass, H., Essam, D., Nguyen, H.: Grammar Model-based Program Evolution. In: IEEE Congress on Evolutionary Computation (CEC 2004), pp. 478–485. IEEE Press, Los Alamitos (2004)
4. Koza, J.: Genetic programming II: automatic discovery of reusable programs. MIT Press, Cambridge (1994)
5. Hornby, G.: Generative representations for evolutionary design automation. Ph.D. thesis, Brandeis University Dept. of Computer Science (2003)
6. Luerssen, M., Powers, D.: Evolving encapsulated programs as shared grammars. Genetic Programming and Evolvable Machines 9(3), 203–228 (2008)
7. O'Neill, M., Ryan, C.: Grammatical evolution. IEEE Transactions on Evolutionary Computation 5(4), 349–358 (2001)
8. Michie, D.: Memo functions and machine learning. Nature 218, 19–22 (1968)
9. Suwannik, W., Chongstitvatana, P.: On-Line Evolution of Robot Arm Control Programs for Visual-Reaching Tasks using Memoized Function. ECTI Trans. on Electrical Eng., Electronics, and Communications 4(2), 145–155 (2006)
10. Sipper, M., Azaria, Y., Hauptman, A., Shichel, Y.: Designing an Evolutionary Strategizing Machine for Game Playing and Beyond. IEEE Transactions on Systems, Man, and Cybernetics, Part C: Applications and Reviews 37(4), 583–593 (2007)
11. COGENT: Concurrent Grammar Exploration and Transformation, `http://code.google.com/p/cogent`
12. Boost Pool Library, `http://www.boost.org/doc/libs`
13. Deb, K., Pratab, A., Agrawal, S., Meyarivan, T.: A Fast and Elitist Multiobjective Genetic Algorithm: NSGA-II. IEEE Transactions on Evolutionary Computation 6(2), 182–197 (2002)
14. GEVA: Grammatical Evolution in Java, `http://ncra.ucd.ie/Site/GEVA.html`

Evolution of Conventions and Social Polarization in Dynamical Complex Networks

Enea Pestelacci and Marco Tomassini

Information Systems Department, HEC, University of Lausanne, Switzerland
{enea.pestelacci,marco.tomassini}@unil.ch

Abstract. Effective coordination is very important in society. Starting from the observation that individuals interact through networks of acquaintances, we study the co-evolution of the agents' strategies and of the network itself using pure coordination games. We find that coordination among agents emerges through the self-organization of the social network into strong and stable clusters in which individuals follow the same convention.

1 Introduction

Norms and conventions in society can be seen as the product of a gradual evolution of behaviors such that, in the end, adherence to the convention is a globally advantageous strategy. This kind of evolution can be modeled in an abstract way with evolutionary game theory (EGT) concepts [1]. The population dynamics in EGT is such that those strategies that do better than average increase their share in the population, while those that do worse decline. The rest points of the dynamics are the equilibrium states, some of which are called *evolutionarily stable strategies* (ESS) and are a subset of the set of Nash equilibria (NE) of the game, i.e. those ensembles of strategies, one for each player, such that each strategy is a best response to the strategy of the other players [1]. This framework has allowed to satisfactorily explain a number of aspects and behaviors in human and animal societies. However, there exist games in which either the equilibrium posited by the theory is logically and socially unsatisfying, or there is more than one equilibrium and no way to rationally choose between them, although some equilibrium clearly appears to be socially more efficient. This situation is well illustrated by the class of *coordination games*. In their simplest form, these games are two-person games in which choosing the same strategy leads to socially efficient outcomes, while miscoordination is harmful. Here we study in particular the sub-class of *pure coordination games* which, for a two-strategy game, have the normal form depicted in Table 1. The Nash equilibria in pure strategies correspond to diagonal elements in the matrix where the two players coordinate on the same strategy, while there is a common lower uniform payoff for all other strategy pairs which is set to 0 here. Obviously, the payoff table can be generalized to any finite number k of strategies. We assume $a \geq b > 0$ and thus strategy α is said to be *weakly dominant* since a player obtains at least

R. Schaefer et al. (Eds.): PPSN XI, Part II, LNCS 6239, pp. 512–522, 2010.

Table 1. A general two-person, two-strategies pure coordination game

	α	β
α	a, a	$0, 0$
β	$0, 0$	b, b

the same payoff as the other player playing α rather than β. For mathematical convenience, standard EGT is based on infinite mixing populations where pairs of individuals are drawn uniformly at random at each step and play the given game. Correlations are absent by definition and the population has an homogeneous structure. However, everyday observation tells us that in animal and human societies, individuals usually tend to interact more often with some specified subset of partners; for instance, teenagers tend to adopt the fashions of their close friends group; closely connected groups usually follow the same religion, and so on. Likewise, in the economic world, a group of firms might be directly connected because they share capital, technology, or otherwise interact in some way. In short, social interaction is mediated by networks, in which vertices identify people, firms etc., and edges identify some kind of relation between the concerned vertices such as friendship, collaboration, economic exchange and so on. Thus, locality of interaction plays an important role. This kind of approach was pioneered in EGT by Nowak and May [2] by using simple two-dimensional regular grids. Recently, the dynamical and evolutionary behavior of games on networks that are more likely to represent actual social interactions than regular grids has been investigated (see [3] for a comprehensive review). However, most of these studies have assumed a fixed population size and structure. But real social networks, such as friendship or collaboration networks, are not in an equilibrium state, they are open systems that continually evolve with new agents joining or leaving the network, and relationships being made or dismissed by agents already in the network. In the present work we take into account some of these coupled dynamics and we investigate under which conditions coordinate behavior may emerge and be stable.

Some previous work has been done on evolutionary games on dynamic networks essentially dealing with the Prisoner's Dilemma, e.g. [4–6]. For a recent review, see [7]. The present study follows our own model described in [8, 9] and differs from previous ones in the way in which links between agents are represented and interpreted.

2 Model Description

The model has already been presented in [8, 9] and thus we only give a brief description here in order to make the paper as self-contained as possible. The model is strictly local: a player only uses information about the strength of the links with her first neighbors and the knowledge of her own payoff plus the strategies of her immediate neighbors.

Network and Interaction Structure. The network of agents is represented by a directed weighted graph $G(V, E)$, where the set of vertices V represents the agents, while the set of oriented edges (or links) E represents their weighted interactions. The population size N is the cardinality of V. For network structure description purposes, we shall also use an unoriented version $G^{'}$ of G having exactly the same set of vertices V but only a single unoriented unweighted edge ij between any pair of connected vertices i and j of G. A neighbor of an agent i is any other agent j such that there is an edge ij. The set of neighbors of i is called V_i. For G' we shall define the degree k_i of vertex $i \in V$ as the number of neighbors of i. The average degree of the network $G^{'}$ will be called $\bar{k}$.
Each link in G has a weight or "force" f_{ij} that represents in an indirect way the "trust" player i places in player j and, in general, $f_{ij} \neq f_{ji}$. This weight may take any value in $[0, 1]$ and its variation is dictated by the payoff earned by i in each encounter with j. We define a quantity s_i called *satisfaction* of an agent i as the mean weight of i's links: $s_i = \frac{\sum_{j \in V_i} f_{ij}}{k_i}$, with $0 \leq s_i \leq 1$. The link strengths can be seen as a kind of "memory" of previous encounters. However, they must be distinguished from the memory used in iterated games, in which players "remember" a certain number of previous moves and can thus conform their future strategy on the analysis of those past encounters [1, 10]. Our interactions are strictly one-shot, i.e. players "forget" the results of previous rounds and cannot recognize previous partners and their playing patterns over time.

Strategy and Link Evolution. The dynamics is discrete in time t: $t = \{t_0, t_1, \ldots\}$. At each time step an individual i is chosen uniformly at random with replacement to play the game with its neighbors. It updates its strategy according to a local *myopic best response* rule with probability $1 - q$ or, with probability q, agent i may delete a link with a given neighbor j and creates a new 0.5 force link with another node k. The parameter $0 < q < 1$ has the role of a "temperature": the higher q the more often links are broken and rewired. At the end of the step the forces between i and its neighbors V_i are updated. Myopic best response means that a player chooses, among the available actions, the one that would give her the best payoff assuming that her neighbors do not change their previous strategy. Calling G_t the population graph at time t, where each node is labeled with its present strategy, the resulting stochastic process $\{G_0, G_1, \ldots\}$ is a Markov chain since the probability of state G_t only depends on the previous step: $P(G_t|G_{t-1}, G_{t-2}, \ldots) = P(G_t|G_{t-1})$. To verify the stability of the process we introduced a small amount of noise equal to 0.01. When noise is present, a player that decides to update his strategy has a small probability to chose the wrong strategy. These small random effects are meant to capture various sources of uncertainty such as deliberate and involuntary decision errors.

The active agent i will, with probability q, attempt to dismiss an interaction with one of its neighbors, say j, selected proportionally to $1 - f_{ij}$, i.e. the higher f_{ij}, the lower the probability of the link being selected for rewiring. Likewise, the lower s_i the higher the probability of dismissing the ij link. If the link is finally severed (in both directions), i asks $k \in V_i \setminus \{j\}$ to select one of its neighbors $l \in V_k \setminus \{i\}$ and attempts to create a new link il. Links ik and kl are selected such that edges with high forces are probabilistically favored. Obviously, this bias will cause the clustering coefficient of the network to increase over time due to the transitive closure it causes, i.e. triangles will be more likely in the evolving graph. The solution adopted here is inspired by the observation that, in social networks, links are usually created more easily between people who have a satisfactory mutual acquaintance than those who do not. If the new link already exists, the process is repeated with l's neighbors. If this also fails, a new link between i and a randomly chosen node is created. In all cases the new link is initialized with a strength of 0.5 in both directions.

3 Simulation Results

3.1 Simulation Settings

The constant size of the network during the simulations is $N = 1000$. The initial graph G'_0 is generated randomly with a mean degree $\bar{k} = 6$. The companion oriented graph G_0 is trivially built from G'_0 and forces between any pair of neighboring players are initialized at 0.5.

The non-zero diagonal payoff a has been varied in the range $[0.5, 1]$ in steps of 0.05 with $b = 1 - a$; the range $[0, 0.5]$ is symmetrically equivalent. Each value in the phase space reported in the following figures is the average of 50 independent runs. Each run has been performed on a fresh realization of the corresponding initial random graph.

To detect steady states of the dynamics, i.e. those states with little or no fluctuation over extended periods of time, we first let the system evolve for a transient period of $5000 \times N$ times steps ($= 5 \times 10^6$ time steps when $N = 1000$). After a quasi-equilibrium state is reached past the transient, averages are calculated during $500 \times N$ additional time steps. A steady state has always been reached in all simulations performed within the prescribed amount of time, for most of them well before the limit.
We have experimented with different proportions of uniformly randomly distributed initial strategies α belonging to the set $\{0.05, 0.25, 0.5, 0.75\}$.

3.2 Discussion of Results

Figure 1 reports the amount of α-strategists in the population when a quasi-equilibrium state has been reached as a function of the rewiring frequency q. The upper light part of the plots indicate the region of the parameters space where the α-strategists are able to completely take over the population. This can happen because α strategy offers the best payoff since $a - b$ is positive, therefore β-strategists are prone to adapt in order to improve their wealth.
Figure 1(a) shows the case where both α and β strategies are present in the same ratio at the beginning of the simulation. The darker region indicates the

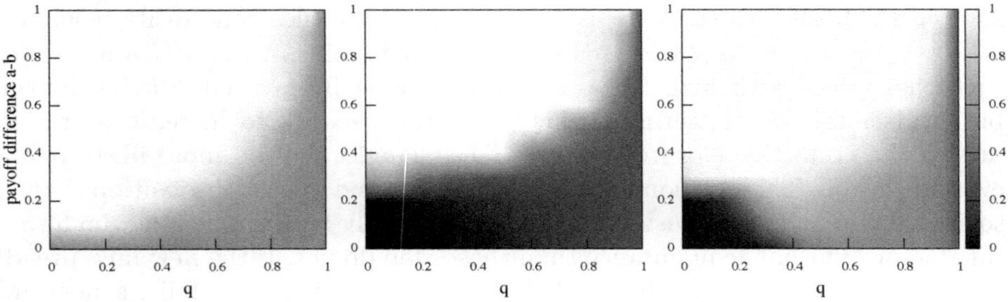

(a) (b) (c)

Fig. 1. Average fraction of α-strategists in the population when the quasi-equilibrium has been reached. (a) shows the case where the initial fraction of α is 0.5 and noise is absent. In (b) and (c) the initial fraction of α is 0.25. (b) shows the noiseless case and (c) the case with a noise of 0.01. The light areas denote a higher proportion of α. Results are averages over 50 independent runs.

situations where diversity is able to resist. This clearly happens when the payoff difference $a - b$ is zero. In this case both α and β are winning strategies and the players tends to organize in two big clusters to minimize the links with the opposing faction. More surprisingly even when one of the two strategies has a payoff advantage, the evolution of the topology of the interaction allows the less favorable strategy to resist. The faster the network evolution is (greater q), the greater the payoff difference that can be tolerated by the agents.

In figures 1(b) the case when α represent only 25% of the initial population is presented. When no noise is present the stronger strategy needs an increased payoff advantage to take over the population. When $a-b < 0.3$ the payoff-inferior strategy β is able to maintain the majority.

To confirm the stability of the evolution process we did some series of simulations using a noisy version of strategy evolution rule. This noise is rather small and does note change the results obtained when the two populations are equally represented in the initial network, the graphic representation is almost the same of the one in fig. 1(a) with respect to stochastic fluctuations. However when the initial share is not the same, the presence of noise allows a considerable increase in the performance of the Pareto-superior strategy when this strategy is less represented in the beginning. Figure 1(c) shows the case when the initial ratio of α-strategists is 25% of the initial population. We can clearly see that the strategy that offers the higher payoff can recover a considerable amount of the parameters space even when it starts from an unfavorable situation. The coexistence of stochastic errors and network plasticity allows the more advantageous strategy to improve its share. In this case, when $q > 0.4$ the situation is almost the same as when the initial shares are the same. The same phenomenons happen when the initial ratio of α is smaller. The case of an initial ratio of 5% has been verified but it's not shown here.

Figures 2, and 3 show the evolution of the network G' and of the strategy distribution from the initial state in which strategies are distributed uniformly

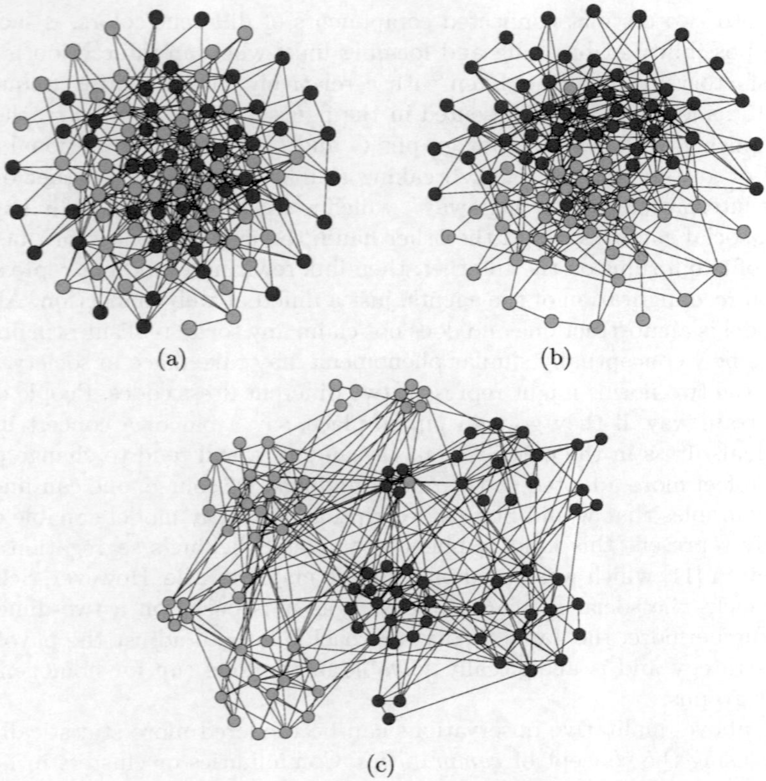

Fig. 2. (a) The simulation starts from a random network with $N = 100$ and 50 players for each type. (b) In the first short part of the simulation ($\sim$ 500 time steps) the strategy fractions reach a steady state, the network however is still unorganized. (c) The community structure starts then to emerge, many small clusters with nearly uniform strategy appears.

at random up to a final quasi-equilibrium steady state. These results have been obtained for a symmetric payoff of the strategies $a = b$ and for an equal initial fraction of α-strategists and β-strategists. It is visually clear that the system goes from a random state of both the network and the strategy distribution to a final one in which the network is no longer completely random[1] and, even more important, the strategies are distributed in a completely polarized way. In other words, the system evolves toward an equilibrium where individuals following the same convention are clustered together. Since both norms are equivalent in the sense that their respective payoffs are the same, agents tend to pair-up with other agents playing the same strategy since playing the opposite one is a dominated strategy. The process of polarization and, in some cases, even the splitting of the

[1] We have checked that the network's degree distribution function $p(k)$ is no longer Poissonian but rather long-tailed to the right, and the clustering coefficient is much larger than that of the corresponding random graph.

graph into two distinct connected components of different colors, is facilitated by the possibility of breaking and forming links when an interaction is judged unsatisfactory by an agent. Even with a relatively small rewiring frequency of $q = 0.15$ as for the case represented in the figures, polarization is reached relatively quickly. In fact, since our graphs G and G' are purely relational entities devoid of any metric structure, breaking a link and forming another one may also be interpreted as "moving away", which is what would physically happen in certain social contexts. If, on the other hand, the environment is say, belonging to one of two forums on the Internet, then link rewiring would not represent any physical reconfiguration of the agents, just a different web connection. Although our model is an abstract one and does not claim any social realism, still one could imagine how conceptually similar phenomena may take place in society. For example, the two norms might represent two different dress codes. People dressing in a certain way, if they go to a public place, say a bar or a concert in which individuals dress in the other way in the majority, will tend to change place in order to feel more adapted to their surroundings. Of course, one can find many other examples that would fit this description. An early model capable of qualitatively represent this kind of phenomena was Schelling's segregation cellular automaton [11] which was based on a simple majority rule. However, Schelling's model lacks the social network dimension as it is based on a two-dimensional grid. Furthermore, the game theory approach allows to adjust the payoffs for a given strategy and is analytically solvable in the long run for homogeneous or regular graphs.

The above qualitative observations can be rendered more statistically rigorous by using the concept of *communities*. Communities or clusters in networks can be loosely defined as being groups of nodes that are strongly connected between them and poorly connected with the rest of the graph. These structures are extremely important in social networks and may determine to a large extent the properties of dynamical processes such as diffusion, search, and rumor spreading among others. Several methods have been proposed to uncover the clusters present in a network (for a recent review see, for instance, [12]). To detect communities, here we have used the divisive method of Girvan and Newman [13] which is based on iteratively removing edges with a high value of edge betweennes. A commonly used statistical indicator of the presence of a recognizable community structure is the *modularity* Q. According to Newman [14] modularity is proportional to the number of edges falling within clusters minus the expected number in an equivalent network with edges placed at random. In general, networks with strong community structure tend to have values of Q in the range $0.4-0.7$. In the case of our simulations $Q = 0.19$ for the initial random networks with $N = 100$ like the one shown in fig. 2(a). Q progressively increases and reaches $Q = 0.29$ for fig. 2(c) and $Q = 0.45$ for the final polarized network of fig. 3. In the case of the larger networks with $N = 1000$ the modularity is slightly higher during the evolution, $Q \sim 0.3$ at the beginning of the simulation and $Q \sim 0.5$ when the network has reached a polarized state. This is due to the more sparse structure of these networks.

To confirm the stability of this topological evolution we performed several simulation using the noisy strategy update rule. Even in this situation the network will attain a polarized state but due to the stochastic strategy fluctuations the two main clusters almost never reach a completely disconnected state and the modularity remains slightly lower (~ 0.4) compared to the noiseless case.

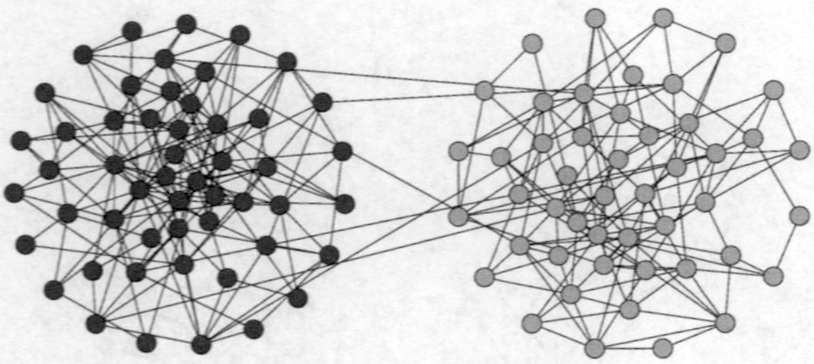

Fig. 3. In the last phase the network is entirely polarized in two homogeneous clusters. If the simulation is long enough all the link between the two poles will disappear.

A few comments on the results shown are in order. The case depicted in the figures is not a particular one; we have performed 50 independent runs for each parameter set and the results are always qualitatively the same; thus, the behavior shown here is typical. Also, while in the figures we show results for systems of size $N = 100$, this is entirely for illustration purposes. Our simulations have been always run for $N = 1000$ but the resulting graphs are a mess and cannot be shown properly in an image. However, in spite of the relatively small size, the phenomena are qualitatively the same for $N = 100$ and $N = 1000$, the major difference is just the time to convergence which is much shorter for $N = 100$.

As a second kind of numerical experiment, we asked how the population will react when in a polarized social situation a few connected players of one of the clusters suddenly switch to the opposite strategy. The results of a particular but typical simulation are shown in Figs. 4. Starting from the clusters obtained as a result of the co-evolution of strategies and network described above, a number of "dark" individuals replace some "pale" ones in the light gray cluster. The evolution is very interesting: after some time the two-cluster structure disappears and is replaced by a different network in which several clusters with a majority of one or the other strategies coexist. However, these intermediate structures are unstable and, at steady state one recovers essentially a situation close to the initial one, in which the two poles form again but with small differences with respect to the original one. Clearly the size of the clusters is different from that of before the invasion. Even in this case, if the evolution time is long enough, the two components can become disconnected at the end. This means that, once formed, polar structures are rather stable, except for noise and stochastic errors. Moreover if we analyze more rigorously the evolution of the structure we can see that at the beginning the invasion process the modularity drops slightly due

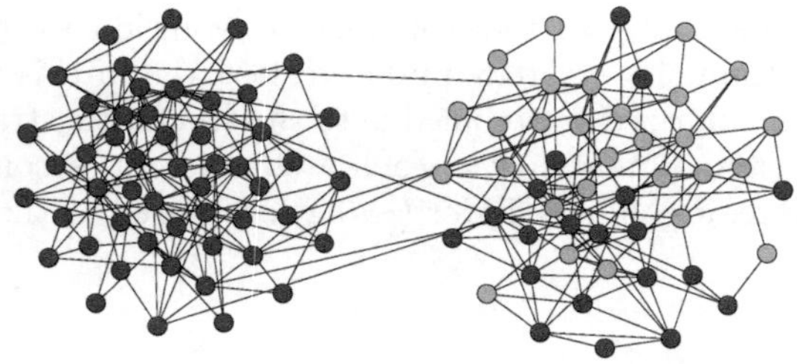

(a)

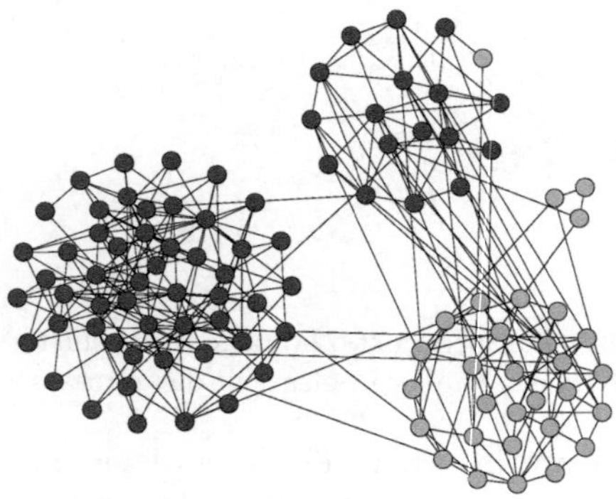

(b)

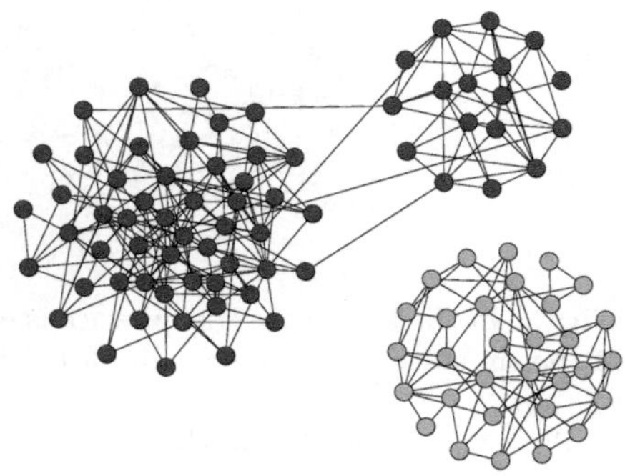

(c)

Fig. 4. (a) A consistent amount of mutant is inserted in one of the two clusters. (b) This invasion perturbs the structure of the population that starts to reorganize. (c) With enough evolution time the topology reaches a new polarized quasi-equilibrium.

to the strong reorganization of the network but then increases again and often reaches a higher value in respect of the previous state. In the case shown here, the final modularity is 0.56. The same happens in the larger networks where, after the invasion process Q reaches values of $Q \sim 0.55$.

4 Conclusions

In this paper we have studied the behavior of network structured populations where individuals play a pure coordination game by using numerical simulations based on the EGT model with best response as a strategy update rule and with co-evolution of the network of contacts itself as a further degree of freedom. In a fixed network the stable equilibrium would be for the payoff-dominant strategy to take over and occupy the whole population. In social terms this means that a single norm or convention is followed by everybody. However, when links in the network are allowed to be cut and rewired, even with relatively slow network evolution one observes that co-existence of stable clusters of both convention-followers becomes the normal outcome of the dynamics. In social terms, this means that the possibility of changing neighbors allows individuals to find a more satisfactory situation in which, being surrounded by agents of the same type, their common payoff is higher. Although the model is too abstract to claim any social realism, still it contains the essential ingredients to explain why individuals following the same convention or cultural norm often find themselves interacting with agents of the same type. To gauge the stability of the results we have added some small noise in the strategy-update decision which is meant to represent the possibility of rare decision errors by the agents. We find that the quasi equilibria are robust with respect to that kind of stochastic fluctuations, which may also favor a payoff-weaker norm to be maintained in its own clusters. Finally, we have simulated a more radical process in which a small but sizable amount of agents playing the opposite strategy replaces the same number of agents in a stable cluster. The typical behavior is that, after extensive rearrangement of the network, the system separates again in clusters following different norms, albeit with possibly different shapes and different numbers of individuals with respect to the initial situation. In any case, this behavior in the face of massive perturbations shows that the separation in clusters, or even separated components, is the typical and stable behavior of the system. In future work, we would like to investigate other strategy update rules and use agents with more heterogeneity which are capable of learning to some extent beyond the simple best response or imitation learning rules.

Acknowledgments. E. Pestelacci and M. Tomassini are grateful to the Swiss National Science Foundation for financial support under contract number 200020-119719.

References

1. Vega-Redondo, F.: Economics and the Theory of Games. Cambridge University Press, Cambridge (2003)
2. Nowak, M.A., May, R.M.: Evolutionary games and spatial chaos. Nature 359, 826–829 (1992)
3. Szabó, G., Fáth, G.: Evolutionary games on graphs. Physics Reports 446, 97–216 (2007)
4. Luthi, L., Giacobini, M., Tomassini, M.: A minimal information Prisoner's Dilemma on evolving networks. In: Rocha, L.M. (ed.) Artificial Life X, pp. 438–444. The MIT Press, Cambridge (2006)
5. Santos, F.C., Pacheco, J.M., Lenaerts, T.: Cooperation prevails when individuals adjust their social ties. PLOS Comp. Biol. 2, 1284–1291 (2006)
6. Zimmermann, M.G., Eguíluz, V.M.: Cooperation, social networks, and the emergence of leadership in a prisoner's dilemma with adaptive local interactions. Phys. Rev. E 72, 056118 (2005)
7. Perc, M., Szolnoki, A.: Coevolutionary games - A mini review. Biosystems 99, 109–125 (2010)
8. Pestelacci, E., Tomassini, M.: Cooperation in coevolving networks: the prisoneros dilemma and stag-hunt games. In: Rudolph, G., et al. (eds.) PPSN 2008. LNCS, vol. 5199, pp. 539–548. Springer, Heidelberg (2008)
9. Pestelacci, E., Tomassini, M., Luthi, L.: Evolution of cooperation and coordination in a dynamically networked society. J. Biol. Theory 3(2), 139–153 (2008)

10. Axelrod, R.: The Evolution of Cooperation. Basic Books, Inc., New-York (1984)
11. Schelling, T.: Micromotives and Macrobehavior. Norton (1978)
12. Fortunato, S.: Community detection in graphs. Physics Reports 486, 75–174 (2010)
13. Newman, M.E.J., Girvan, M.: Finding and evaluating community structure in networks. Phys. Rev. E 69, 026113 (2004)
14. Newman, M.E.J.: Modularity and community structure in networks. Proc. Natl. Acad. Sci. USA 103, 8577–8582 (2006)

Evolving Strategies for Updating Pheromone Trails: A Case Study with the TSP

Jorge Tavares[1] and Francisco B. Pereira[1,2]

[1] CISUC, Department of Informatics Engineering, University of Coimbra
[2] Polo II - Pinhal de Marrocos, 3030 Coimbra, Portugal
ISEC, Quinta da Nora, 3030 Coimbra, Portugal
jorge.tavares@ieee.org, xico@dei.uc.pt

Abstract. Ant Colony Optimization is a bio-inspired technique that can be applied to solve hard optimization problems. A key issue is how to design the communication mechanism between ants that allows them to effectively solve a problem. We propose a novel approach to this issue by evolving the current pheromone trail update methods. Results obtained with the TSP show that the evolved strategies perform well and exhibit a good generalization capability when applied to larger instances.

1 Introduction

Ant Colony Optimization (ACO) draws its inspiration from pheromone-based strategies of ant foraging. Initially, it was conceived to find the shortest path in the well-known Traveling Salesman problem (TSP), but soon it was applied to several different types of combinatorial optimization problems [1]. Examples of situations addressed include both static and dynamic variants of academic and real world problems. Usually, the problem is mapped into a fully connected graph. When seeking for a solution, ants deposit pheromone while traveling across the graph edges, thus creating a virtual trail. A solution to the problem will emerge from the interaction and cooperation made by the ants. Different types of ACO architectures were proposed in the literature [1], with differences, e.g., in what concerns the way ants deposit pheromone in the graph edges. Designing new variants of ACO algorithms is an active area of research, as this may allow the application of these methods to new situations and/or enhance its effectiveness on problems that it usually addresses. Typically, the design and extension of the existing algorithms is carried out manually.

In this paper our approach will be different. We aim to automatically develop and improve the current components by using Evolutionary Algorithms (EA). Our framework relies on a Genetic Programming algorithm (GP) [2] to evolve the way an Ant algorithm updates its pheromone trails. We investigate the evolution of different strategies and study how they perform when compared to standard methods. Different instances of the TSP will be used to access the effectiveness of the proposed approach.

R. Schaefer et al. (Eds.): PPSN XI, Part II, LNCS 6239, pp. 523–532, 2010.

The paper is structured as follows. In section 2 we present the system to evolve the pheromone trails update strategies. Section 3 contains the experimentation and analysis, followed by a discussion. Finally, in section 4 we summarize the main conclusions and highlight directions for future work.

2 Evolving Pheromone Trail Update Methods

There are many research efforts for granting bio-inspired approaches the ability to self adapt their strategies. On-the-fly adaptation may occur just on the parameter settings or be extended to the algorithmic components. One remarkable example of the first type of self-adaptation is the well-known 1/5 success rule used to control the mutation strength for the (1+1)-ES. In what concerns the adaptation of the optimization algorithm, Diosan and Oltean recently proposed an evolutionary framework that aims to evolve a full-featured EA [3].

As for the Swarm Intelligence area, there are also some reports describing the self-adaptation of parameter settings (see, e.g., [4,5]). Still, there are a couple of approaches that resemble the framework proposed in this paper. Poli et. al [6] use a GP algorithm to evolve the update equation that controls particle movement in a Particle Swarm Optimization (PSO) algorithm. Diosan and Oltean also did some work regarding PSO structures [7]. Also, Runka [8] applies GP to evolve the probabilistic rule used by an ACO algorithm to select the solution components in the construction phase.

2.1 Overview of the System

The framework to evolve pheromone trails update strategies is composed of two main components: a Genetic Programming (GP) engine and an Ant System (AS) algorithm. The main task of the GP component is to evolve individuals that encode effective trail update strategies, whereas the AS is required to assign fitness to each generated solution. The GP engine adopts a standard architecture: individuals are encoded as trees and ramped half-and-half initialization is used for creating the initial population. The algorithm follows a steady-state model, tournament selection chooses parents and standard genetic operators for manipulating trees are used to generate descendants. As for the AS framework, it closely follows the variants proposed in [1]. It allows the application of the standard AS, Elite AS (EAS) and Rank-based AS for the TSP, as described in the aforementioned book. The activation of the pheromone update methods was defined in a way to allow an easy integration with the GP engine: when an individual (i.e., an evolved pheromone trail update strategy) needs to be evaluated, the GP executes the AS algorithm for a given TSP instance. The result of the optimization is assigned as the fitness value of the GP individual.

A key decision in the development of the framework is the definition of the function and terminal sets used by the GP, as they will determine which components can be used in the design of pheromone update strategies. In this paper our aim is to show that the proposed approach is able to evolve such strategies.

Therefore, to validate our ideas we keep the definition of the function and terminal sets as simple as possible. Our choice is to provide these sets with ingredients that allow the replication of standard AS and EAS methods. By providing the self-adaptive algorithm with these components we ensure that a valid strategy exists and, at the same time, we explicitly verify if the evolutionary process is able to determine it. The function and terminal sets are composed by:

- *(prog2 p1 p2)* and *(prog3 p1 p2 p3)*: They allow the sequential execution of two or three functions/terminals. The result of the last one is returned;
- *(evaporate rate)*: Runs the standard evaporation formula with a given *rate*.
- *(deposit ants amount)*: *ants* deposit a given *amount* of pheromone. The parameter *ants* can be an array of ants or a single one.
- *(all-ants)*, *(best-ant)*, *(rho)*: They return: 1) the array with all the ants; 2) the best ant found so far in a run; 3) a fixed learning rate.
- *(integer)* and *(real)*: Ephemeral constants.

The evaporation and deposit formulas are the ones defined in the literature [1], with the exception of the given parameters. Since standard GP needs the closure property [2], all the functions and terminals are protected to ensure that no invalid combination can be formed.

3 Experiments and Analysis

Selected instances from the TSPLIB[1] are used in the experiments. For all tests, the GP settings are: Population size: 100; Maximum tree depth: 5; Crossover rate: 0.9; Mutation rate: 0.05; Tourney size: 3. For the AS algorithms we used the standard parameters found in the literature [1]. The number of runs for all experiments is 30. To determine the existence of statistical differences in the results, we apply the Wilcoxon rank sum and the Kruskalwallis tests ($\alpha = 0.05$).

3.1 Evolution of the Update Strategies

In the first set of experiments we aim to detect the evolution of feasible update strategies. A crucial issue that needs to be addressed is the evaluation step of each evolved strategy, as it requires the execution of an AS algorithm. Two parameters, the number of runs and the number of iterations per run, define the optimization effort of the AS and can have a major impact on the behavior of the self-adaptive framework. On the one hand, if we grant the AS a small optimization period to evaluate a GP individual, then it might not be enough to correctly estimate the quality of an update strategy. On the other hand, a longer evaluation period can increase computational costs to insupportable levels. For the eil51 instance (a TSP instance with 51 cities adopted for these experiments), AS variants find a near-optimal solution within approximately 100 iterations. With this value in mind, we defined three different configurations for

[1] `http://comopt.ifi.uni-heidelberg.de/software/TSPLIB95/`

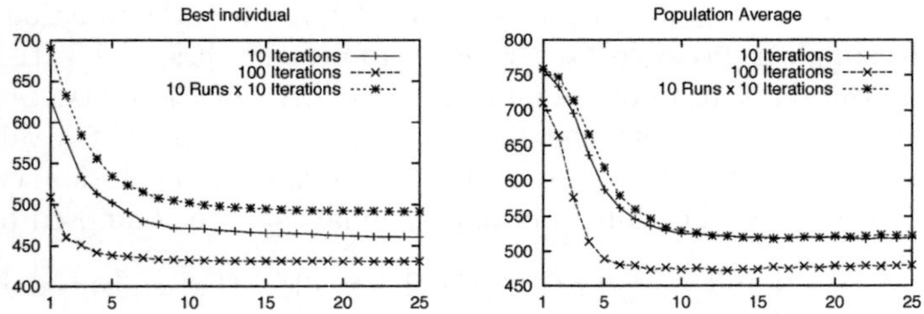

Fig. 1. Evolution plots with generations on the x axis and fitness values on the y axis

the evaluation component of the GP individuals: 1) a single AS run with 100 iterations; 2) a single AS run with 10 iterations; 3) 10 AS runs with 10 iterations. Configuration 1 grants enough time for a strategy to find good solutions. The second one will test the increase of evolutionary pressure, even knowing that finding good solutions in 10 iterations is hard. In the third case we want to see the effect of multiple runs.

The left plot on Figure 1 shows the evolution of the mean best fitness (MBF - average of the best individuals generated by the GP over 30 runs) for the three evaluation configurations. The plot on the right displays the evolution of the average fitness of the population. It is clear that evolution proceeds as in a typical EA. Although individuals from the early stages of the optimization are unable to find good solutions for the TSP, after some time the GP starts to discover effective update strategies (the best solution for eil51 has length 426).

By comparing the three evaluation configurations, we conclude that single runs perform better, particularly when using 100 iterations. A single run with 10 iterations also enables the discovery of sporadic good quality solutions, but, in this case there is a larger deviation from the global optimum. This is consistent with the expected outcome. The time for the AS to build a good solution is short and the additional evolutionary pressure is not enough to eliminate the bad solutions. Using multiple runs attained the worst results. The reason is simple: the fitness assigned to the update strategy is the average of 10 solutions and, initially, most of these are bad solutions. The number of iterations to counterbalance this effect would need to be longer. In the subsequent analysis we will not consider the best tree from the multiple runs configuration since it performs worse than the other strategies.

3.2 Validation and Analysis of the Evolved Update Strategies

To validate the best evolved solutions we will compare their optimization performance with existing AS variants (AS and EAS). We selected the best three trees evolved by the system in the experiments described in the previous section: GP10 is the best update strategy found in 10 iterations, whereas GP100A and GP100B are the best ones found in 100 iterations. The TSP instance used in

Table 1. Comparison of the best solution between the Strategies, with 100 iterations and 1000 iterations for 30 runs

Strategies	Iterations	Best	Worst	Mean Best Fitness	Deviation	Branching
AS	100	442	471	459.10	7.46	5.25
EAS	100	432	467	447.37	9.76	3.54
GP10	100	**426**	469	446.17	8.83	2.77
GP100A	100	**426**	**456**	**443.00**	9.50	2.50
GP100B	100	**426**	472	445.70	10.58	2.96
AS	1000	431	454	444.03	5.06	5.17
EAS	1000	428	451	440.40	5.72	3.45
GP10	1000	**426**	448	435.40	5.89	2.75
GP100A	1000	**426**	446	435.23	5.35	2.21
GP100B	1000	**426**	**443**	**434.20**	4.78	2.82

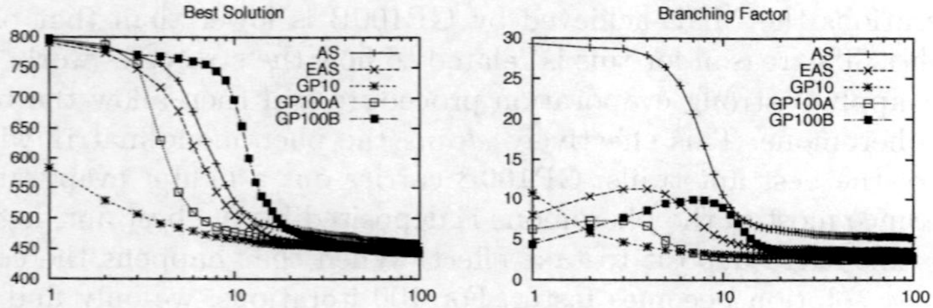

Fig. 2. Plots showing the evolution of the best solution and the branching factor for 100 iterations (x axis). The same pattern is observed with 1000 iterations.

these tests is the same one adopted for the evolution of the update strategies. Table 1 contains the results obtained by the five AS strategies (3 evolved and 2 standard) for two different optimization scenarios: the upper part of the table shows the outcomes obtained when the AS was allowed to run for 100 iterations and the lower part is from experiments that ran for 1000 iterations. In the last column we include the branching factor (with $\lambda = 0.05$), a measure to determine the convergence of the pheromone matrix. Results clearly show that all evolved solutions perform better than the standard methods. The optimal solution was always found by the evolved strategies, whereas AS and EAS were unable to discover it. An inspection of the MBF values confirms that the evolved strategies tend to achieve better results. This trend is more evident in longer runs.

The evolved strategies perform well and are competitive with the standard variants. Even with restrict function and terminal sets it is possible to evolve behaviors that are not exact copies of the ones adopted by standard AS. The evolved strategies are elitist in nature (this is not surprising given the components at the disposal of the GP algorithm), but there is some originality in the way they manipulate the pheromone matrix. Figure 2 confirms the distinctiveness of the evolved strategies. They mostly resemble the EAS, especially with regard to the search space exploration (given by the branching factor). Note one important difference: tree GP100B. This solution starts to act more like AS

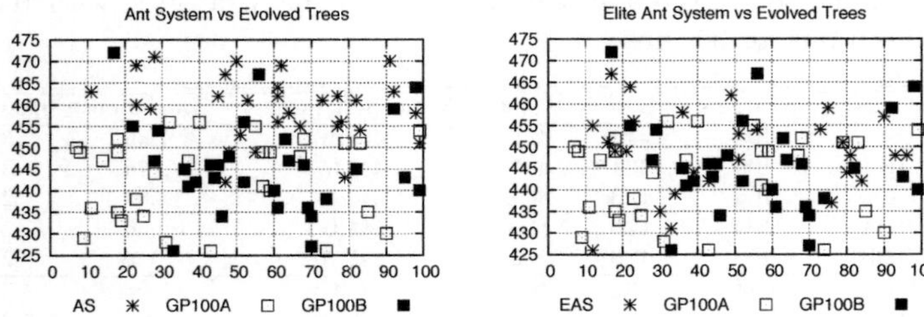

Fig. 3. Scatter plots of the best solution found in 30 runs using 100 iterations

and, after some time, quickly converges to better solutions. For example, with 1000 iterations the MBF achieved by GP100B is lower than that of all other approaches. The reason for this is related to how the strategies work. GP10 and GP100A apply a strong evaporation procedure and then allow the best ant to deposit pheromone. This effectively *cleans* the pheromone matrix with the exception of the best ant trails. GP100B carries out a lighter evaporation of the matrix. Since most of the pheromone is deposited by the best ant, it takes some time for this extra deposit to take effect. When that happens the convergence to the best solution becomes faster. For 100 iterations, we only find statistical significant differences between AS and the evolved strategies, whilst, for 1000 iterations, significant differences also exist with EAS.

To conclude, Figure 3 presents scatter plots of the best solutions found in the 30 runs (y axis) with the iteration when they were built (x axis). We compare both GP100A and GP100B strategies with the AS and EAS. The plots show that most of the best solutions found by the evolved strategies have better quality than the ones produced by the AS. There is a clear separation between both evolved update methods and AS. This is not so evident when comparing with EAS. Although there are more solutions from the evolved strategies in the lower left corner of the plot (which indicates better solutions found in less time) than solutions from EAS, overall the dispersion is similar. This reinforces the elitist nature of the evolved strategies and that in spite of being similar, they still present different performance values.

3.3 Using Different Function and Terminal Sets

Even though the self-adaptive framework evolved moderately different strategies, the function and terminal sets used in the previous sections allow the exact replication of the standard AS and EAS trail update strategies. Now we will slightly change this, as we want to test the ability of the system to evolve effective solutions with component sets that do not allow to write the standard methods. The changes are simple: 1) we remove the function *(all-ants)* thus removing the possibility of all ants to deposit pheromone; 2) we add the function *(rank-ants n)* which ranks the ants by solution quality and returns the best n ants. With these changes, the evolution of any of the standard AS variants (AS, EAS and

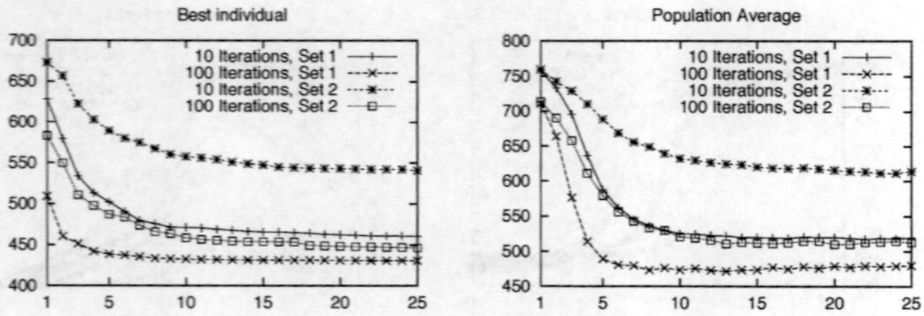

Fig. 4. Evolution with the new sets, plotting generations (x axis) and fitness (y axis)

Table 2. Comparison of the best solution between the Strategies, with 100 iterations and 1000 iterations for 30 runs

Strategies	Iterations	Best	Worst	Mean Best Fitness	Deviation	Branching
Rank-AS	100	429	460	439.87	6.66	2.00
GP10s2	100	437	468	453.70	8.62	2.00
GP100As2	100	**426**	443	432.20	5.13	2.00
GP100Bs2	100	**426**	**437**	**431.23**	2.68	2.01
Rank-AS	1000	428	449	436.63	5.43	2.00
GP10s2	1000	435	490	454.73	13.06	2.00
GP100As2	1000	427	**440**	431.73	3.56	2.00
GP100Bs2	1000	**426**	442	**430.87**	3.52	2.00

Rank-AS) is no longer possible. We also want to see if the evolved strategies can achieve the same level of success as the previous ones, and if the evolved behaviors compare to the standard Rank-AS.

We fed the GP with the new sets and repeated the experiments described in section 3.2 (for configurations 1 and 2). The plots from figure 4 present a comparison between the results achieved by both sets. A brief perusal of the charts reveals that the evolution of update strategies is slower with the new test set, particularly when 10 AS iterations are used to evaluate GP individuals. On the contrary, the number of GP runs that evolved strategies with the ability to find the optimal solution increased from 2 to 6.

Once again we performed some additional tests to measure the optimization performance of three evolved strategies: GP100As2, GP100Bs2, GP10s2. The first two are examples of strategies that discovered the optimal solution when 100 iterations were performed, whereas the last is the best strategy found with 10 iterations. Table 2 presents an overview of the optimization results. The three evolved strategies and also the standard Rank-AS were applied to solve eil51, both for 100 and 1000 iterations. The outcomes confirm that evolved strategies (with the exception of GP10s2) are effective. Results are even slightly better than those achieved by the evolved strategies described in section 3.2. The comparison with the standard methods is also favorable to the evolved strategies. GP100As2 and GP100Bs2 act on a similar way. They pick the best 8 to 10 ants and allow them to deposit a large amount of pheromone. The evaporation step is performed before and/or after. No specific deposit for the best ant is done.

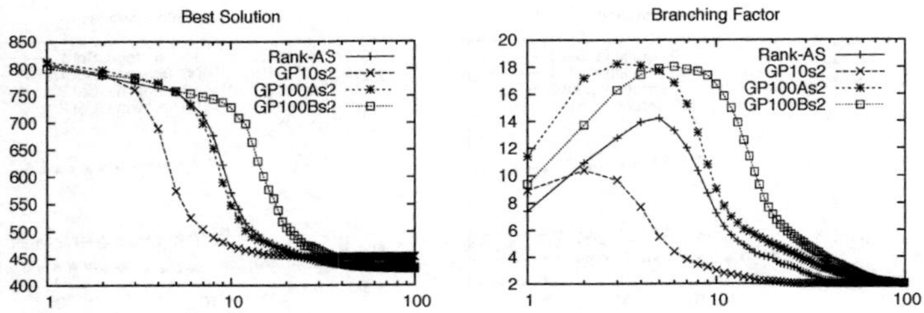

Fig. 5. Plots showing the evolution of the best solution and the branching factor of the second function and terminal sets for 100 iterations (x axis). The same pattern is observed with 1000 iterations.

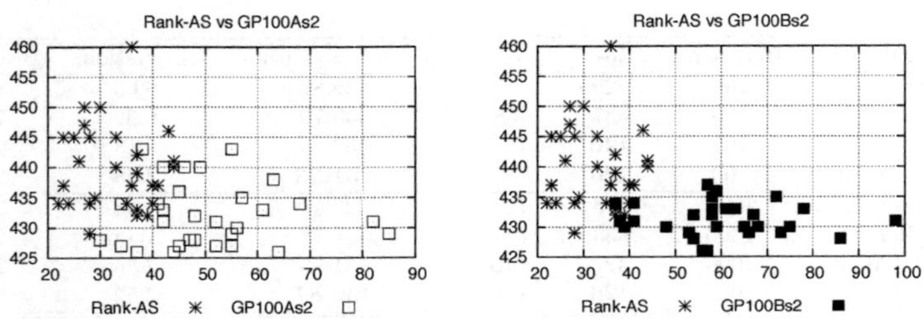

Fig. 6. Scatter plots of the best solution found in 30 runs using 100 iterations

This is a surprising consequence. The standard Rank-AS allows all the ants to deposit, picks the 6 best ants and allows them to deposit a weighted sum of pheromone and, finally, the best ant deposits the largest amount of pheromone. In the best evolved strategies, the deposit of the best ant disappears, therefore removing some greediness. For both 100 and 1000 iterations, we find statistically significant differences between the evolved trees and all the standard methods.

Figure 5 plots the behavior of these new strategies and the Rank-AS. The curves are similar in shape, especially in terms of best solution quality, but different in size. The evolution of the branching factor is interesting since it shows a larger exploration ability by the ants adopting GP100As2 and GP100Bs2, followed by a gradual convergence to the best solution. This demonstrates that the evolved strategies keep more options during the initial stage of the search and converge to a good solution in the later steps. The scatter plots in Figure 6 help us to see this. They tell us that Rank-AS converges faster but is unable to reach the best solutions. Evolved strategies take longer, but are consistently able to find better solutions.

3.4 Generalization to Other TSP Instances

The previous sections dealt with the ability of the self-adaptive framework to evolve an effective strategy for updating pheromone trails. All tests were done

Table 3. Results with larger TSP instances using 100 iterations for 30 runs

Strategies	Instance	Size	Best	Worst	Mean Best Fitness	Deviation
AS	kroA100	100	22759	24258	23649.90	367.90
EAS			22339	24646	23457.07	535.40
Rank-AS			22004	24285	23113.00	544.00
GP100A			22062	25165	23310.80	726.26
GP100As2			**21632**	**23994**	**22778.80**	560.45
AS	d198	198	17487	18533	18098.37	242.21
EAS			17515	18871	18138.77	369.62
Rank-AS			16922	18310	**17486.50**	309.34
GP100A			**16698**	18689	17859.80	408.62
GP100As2			16853	**18179**	17696.13	276.97
AS	lin318	318	49075	52449	50742.73	780.21
EAS			47688	54174	50919.20	1553.24
Rank-AS			45804	**49505**	47665.87	962.95
GP100A			45323	51152	48392.23	1244.16
GP100As2			**44534**	49782	**47037.73**	1034.50
AS	pcb442	442	64535	69335	67244.40	991.15
EAS			60140	68257	64274.20	2435.61
Rank-AS			60637	70377	66843.77	2070.65
GP100A			**58858**	**67038**	**63129.54**	2215.41
GP100As2			65338	71367	68808.16	1477.87

with a single instance and this raises the question whether the evolved strategies are general enough so that they can be useful in other situations. To address this issue, we applied the evolved strategies to solve larger instances. By comparing the results of the standard approaches and the evolved methods we can obtain some evidence regarding the generalization ability of the system.

Table 3 contains the results obtained by GP100A and GP100As2 on four larger TSP instances (other evolved strategies follow a similar pattern). In short, the evolved strategies generalize. For most cases they attain the best results in terms of absolute quality and MBF. Although the differences are not large, it proves that a strategy evolved from a particular instance can be used to solve different and larger ones. There are significant statistical differences between the evolved strategies and the other methods for all instances with just a few exceptions (for kroA100 between both evolved strategies and EAS; and for pcb442 between EAS and GP100A).

3.5 Discussion

The evolved strategies are effective for solving the instance for which they were evolved, while they also exhibit a good generalization capability. However, it must be pointed out that the system uses high-level function and terminal sets, which resemble the actual standard AS methods. This choice rules out the possibility of evolving strategies that strongly deviate from standard ones. Moreover, the evolution is not too difficult as the key components are provided and GP just needs to find the proper arrangement. Nevertheless, the results show that evolution did not converge to the standard methods (in spite of, for example, the first sets permitting it) and found different designs in structure and in parameters. This indicates an open space for improvement of the actual methods used in Ant-based algorithms. The study described in this paper is a first approach to this effort. Still, the study and use of high-level sets is important. Finally, as an example, we show one of the evolved trees GP100A:

```
(prog2 (prog3 (evaporate 0.064867854)
              (deposit (all-ants) 0.064867854)
              (prog3 (evaporate (rho))
                     (evaporate 0.6832942)
                     (deposit (best-ant) 0.699499)))
       (evaporate (rho)))
```

4 Conclusions

In this paper we proposed a system to evolve strategies for updating pheromone trails in the standard Ant System architecture. The TSP was used as a test case and two different function and terminal sets were considered. The evolved strategies have proven to perform good when compared to the standard AS, EAS and Rank-AS update methods. Moreover, they showed a generalization capability when applied to different and larger instances. In the future, we plan to design and use different function and terminal sets to evolve strategies with a non-elitist and ranking behavior. Additionally, the generalization of the evolved strategies to different problems will be addressed.

References

1. Dorigo, M., Stützle, T.: Ant Colony Optimization. MIT Press, Cambridge (2004)
2. Poli, R., Langdon, W.B., McPhee, N.F.: A field guide to genetic programming. Published via and freely (With contributions by J. R. Koza) (2008), http://lulu.com, http://www.gp-field-guide.org.uk
3. Diosan, L., Oltean, M.: Evolutionary design of evolutionary algorithms. Genetic Programming and Evolvable Machines 10, 263–306 (2009)
4. Botee, H., Bonabeau, E.: Evolving ant colony optimization. Advances in Complex Systems 1, 149–159 (1998)
5. White, T., Pagurek, B., Oppacher, F.: ASGA: Improving the ant system by integration with genetic algorithms. In: Proc. of the Third Genetic Programming Conference, pp. 610–617. Morgan Kaufmann, San Francisco (1998)
6. Poli, R., Langdon, W.B., Holland, O.: Extending particle swarm optimisation via genetic programming. In: Keijzer, M., Tettamanzi, A.G.B., Collet, P., van Hemert, J., Tomassini, M. (eds.) EuroGP 2005. LNCS, vol. 3447, pp. 291–300. Springer, Heidelberg (2005)
7. Diosan, L., Oltean, M.: Evolving the structure of the particle swarm optimization algorithms. In: Gottlieb, J., Raidl, G.R. (eds.) EvoCOP 2006. LNCS, vol. 3906, pp. 25–36. Springer, Heidelberg (2006)
8. Runka, A.: Evolving an edge selection formula for ant colony optimization. In: GECCO 2009 Proceedings, pp. 1075–1082 (2009)

The Role of Syntactic and Semantic Locality of Crossover in Genetic Programming

Nguyen Quang Uy[1], Nguyen Xuan Hoai[2], Michael O'Neill[1], and Bob McKay[3]

[1] Natural Computing Research & Applications Group, University College Dublin, Ireland
[2] Department of Computer Science, Le Quy Don University, Vietnam
[3] School of Computer Science and Engineering, Seoul National University, Korea
quanguyhn@gmail.com, nxhoai@gmail.com,
m.oneill@ucd.ie, rimsnucse@gmail.com

Abstract. This paper investigates the role of syntactic locality and semantic locality of crossover in Genetic Programming (GP). First we propose a novel crossover using syntactic locality, *Syntactic Similarity based Crossover* (SySC). We test this crossover on a number of real-valued symbolic regression problems. A comparison is undertaken with Standard Crossover (SC), and a recently proposed crossover for improving semantic locality, *Semantic Similarity based Crossover* (SSC). The metrics analysed include GP performance, GP code bloat and the effect on the ability of GP to generalise. The results show that improving syntactic locality reduces code bloat, and that leads to a slight improvement of the ability to generalise. By comparison, improving semantic locality significantly enhances GP performance, reduces code bloat and substantially improves the ability of GP to generalise. These results comfirm the more important role of semantic locality for crossover in GP.

Keywords: Genetic Programming, Semantics, Syntaxtic, Crossover.

1 Introduction

Locality is important in all search methods. Our only justification for non-random methods is the assumption that there is some correlation between distance and fitness in the semantic space (otherwise random search is provably optimal). If we use a separate syntactic representation for the search algorithm, this requirement carries over: if there is no correlation between syntactic and semantic representation, we might as well use pure random search. Thus locality (continuity – small changes in genotype corresponding to small changes in phenotype) has long been seen as a crucial property of Evolutionary Computation (EC) representations [9,10,20,21].

Assuming a continuous genotype-phenotype mapping, one may then ask, whether it is better to design operators to control locality in genotype or phenotype space. On the side of the genotype space lies the advantage of simplicity: it is easy to measure and control locality directly in the space where the operators are applied. Thus virtually all such work has relied on genotypic distance through syntactic metrics. On the other hand, at the cost of greater complexity, one might argue that phenotypic distances, being (presumably) more closely correlated with fitness, might lead to better metrics.

R. Schaefer et al. (Eds.): PPSN XI, Part II, LNCS 6239, pp. 533–542, 2010.

Is this so? Is it worth the extra complication of designing semantically-based control of operators? This paper examines this question, comparing our own recent work [24] on semantics-based control of crossover in Genetic Programming (GP) with a new, syntactically-based form.

Recent GP research has paid much attention to incorporating semantics in search [26,11,12,3,5,13,15,15,2,23]. In recent work [24], Uy et al. presented *Semantic Similarity based Crossover* (SSC), which improves the semantic locality of crossover by paying attention to the scale of semantic differences between two subtrees. The results reported in [24] show that SSC significantly improves the performance of GP in solving a family of real-valued symbolic regression problems. However it also raises an important question, of the relationship between syntactic and semantic locality. Which (semantic or syntactic locality) is more important? We compare a crossover designed to directly improve syntactic locality with one relying on semantic locality for the effects on three aspects of GP: bloat, performance and generalisation. We show that syntactic locality plays a role in GP code bloat, but semantic locality is even more important in improving GP performance and in GP's ability to generalise.

The remainder of the paper is organised as follows. In the next section, we give a review of related work on semantic based crossovers in GP and a brief review of locality in Evolutionary Computation (EC). Section 3 describes SSC and a novel crossover for improving syntactic locality. The experimental settings are detailed in Section 4. The results of the experiments are presented and discussed in section 5. Section 6 concludes the paper and highlights some potential future work.

2 Related Work

2.1 Semantics in Genetic Programming

Recently, semantics in GP has been addressed by a number of researchers. The work falls into three main strands:

1. using formal methods [11,12,13,15,14]
2. using grammars [26,3,5]
3. using structures such as GP trees [2,23,24]

The first approach was advocated by Johnson in a series of papers [11,12,13]. In these methods, semantic information extracted from formal methods (e.g., Abstract Interpretation and Model Checking) is used to quantify fitness in problems where it is difficult to measure by sample point fitness. Katz and Peled consequently used model checking to measure fitness of individuals in solving the Mutual Exclusion problem [15,14]. These formal methods have a strict mathematical foundation, that potentially may aid GP. Perhaps because of high complexity, however, they have seen only limited research. Their main application to date has been in evolving control strategies.

The second category presents semantics by using Attribute Grammars. Attributes added to a grammar can generate some useful semantic information about individuals, which can be used to eliminate bad individuals [5], or to prevent generating semantically invalid ones [26,3]. The attributes used to represent semantics are, however, problem dependent, and it is not always easy to design such attributes for a new problem.

In the last category, semantics has mainly been used to control the GP operators. In [2], the authors investigated the effect of semantic diversity on Boolean domains, checking the semantic equivalence between offspring and parents by transformation to a canonical form, Reduced Ordered Binary Decision Diagrams (ROBDDs) [6]. This information is used to decide whether the offspring are copied to the next generation. The method improved GP performance, presumably because it increased semantic diversity.

Uy et al. [23] proposed Semantics Aware Crossover (SAC), another crossover operator promoting semantic diversity, based on checking semantic equivalence of subtrees. It showed limited improvement on some real-value problems. This crossover was then extended to Semantic Similarity based Crossover (SSC) [24] by improving its semantic locality. The experimental results showed improved performance of SSC over both SC and SAC [24]. Our aim here is to investigate the effectiveness of semantic locality through a comparison of SSC with a crossover designed to improve syntactic locality.

2.2 Locality in Evolutionary Computation

In the field of GP in particular and Evolutionary Computation in general, locality (small change in genotype corresponding to small change in phenotype) plays a crucial role in the efficiency of an algorithm [9,10,20,21]. Rothlauf [21] investigated the locality of representations in Evolutionary Computation (EC). To determine the locality of a genotype-phenotype mapping, we must define two metrics, in the genotype and phenotype spaces. He argued that a representation with high locality is necessary for efficient evolutionary search. Although a representation with high locality is desirable, it may be very difficult to achieve. Thus many current GP representations are of low-locality, so that small syntactic changes in genotype can cause large change in phenotype. In this paper, we extend the concept of representation locality [21] to locality of an operator. We consider locality in both syntactic and semantic domains.

3 Methods

This section briefly presents SSC before giving details of the new crossover based on syntactic locality.

3.1 Semantic Similarity Based Crossover

SSC as used here is almost identical to that of Uy et al. [24], with the exception of a slight change in the definition of the distance measure. We start with a clear defintion of (sub)tree semantics. Formally, the *Sampling Semantics* (SS) of a (sub)tree is defined as follows:

Let F be a function expressed by a (sub)tree T on a domain D. Let P be a sequence of points sampled from domain D, $P = (p_1, p_2, ..., p_N)$. Then, the *Sampling Semantics* of T on P in domain D is the corresponding sequence $S = (s_1, s_2, ..., s_N)$ where $s_i = F(p_i), i = 1, 2, ..., N$.

The optimal choice of N and P depend on the problems; we follow the approach of [24] in setting the number of points for evaluating the semantics equal to the number of fitness cases (20 points – Section 4) and in choosing the sequence of points P uniformly randomly from the problem domain.

Based on SS, we define a *Sampling Semantic Distance* (SSD) between two subtrees. It differs from that in [24] in using the mean absolute difference in SS values, rather than (as before) the sum of absolute differences. Let $U = (u_1, u_2, ..., u_N)$ and $V = (v_1, v_2, ..., v_N)$ represent the SSs of two subtrees, S_1 and S_2; then the SSD between S_1 and S_2 is defined in equation 1:

$$\mathrm{SSD}(S_1, S_2) = \frac{\sum_{i=1}^{N} |u_i - v_i|}{N} \tag{1}$$

We follow [24] in defining a semantic relationship, *Semantic Similarity* (SSi), on the basis that the exchange of subtrees is most likely to be beneficial if they are not semantically identical, but also not too different. Two subtrees are semantically similar if their SSD lies within a positive interval. The formal definition of SSi between subtrees S_1 and S_2 is given in the following equation:

$$\mathrm{SSi}(S_1, S_2) = \mathbf{TruthValue}(\alpha < \mathrm{SSD}(S_1, S_2) < \beta)$$

where α and β are two predefined constants, the *lower* and *upper* bounds for semantics sensitivity. In general, the best values for these semantic sensitivity bounds are problem dependent. In this work we set $\alpha = 10^{-4}$ and several values of β were tested.

The primary objective of SSC was to improve the locality of crossover. Algorithm 1 (adapted from [24]) shows the detailed operation of SSC. The value of *Max_Trial* was set at 12, a value which was determined through experiment.

3.2 Syntactic Similarity-Based Crossover

For our syntactic crossover, we require a syntactic distance. For this we use the Levenshtein tree distance [17], using the procedure described by Ekart and Nemeth [7] to compute it. From this, a syntactic similarity relationship between two (sub)trees is defined in a similar way to the semantic counterpart. In other words, two subtrees are said to be syntactically similar if the syntactic distance (SyD) between them lies in a specific range. Formally, two subtrees S_1 and S_2 are syntactically similar (SySi) if

$$\mathrm{SySi}(S_1, S_2) = \mathbf{TruthValue}(\alpha < \mathrm{SyD}(S_1, S_2) < \beta)$$

where α and β are two predefined constants, the *lower* and *upper* bounds for syntactic sensitivity. In this paper, α was set to 0, and several values of β were tested.

Based on SyS, a syntactic similarity crossover is proposed. *Syntactic Similarity-based crossover* (SySC), is inspired by SSC, using the same algorithm except for the distance metric. The *Max_Trial* of SySC is set at 4, this value again being experimentally determined.

Algorithm 1: Semantic Similarity based Crossover

```
select Parent 1 P1;
select Parent 2 P2;
Count=0;
while Count<Max_Trial do
    choose a random crossover point Subtree1 in P1;
    choose a random crossover point Subtree2 in P2;
    generate a number of random points (P) on the problem domain;
    calculate the SSD between Subtree1 and Subtree2 on P
    if Subtree1 is similar to Subtree2 then
        execute crossover;
        add the children to the new population;
        return true;
    else
        Count=Count+1;
if Count=Max_Trail then
    choose a random crossover point Subtree1 in P1;
    choose a random crossover point Subtree2 in P2;
    execute crossover;
    return true;
```

4 Experimental Settings

To investigate the impact of these operators on GP performance, code bloat, and ability to generalise, we used six real-valued symbolic regression problems. The problems, training and testing data are shown in Table 1. The training data is used to train and measure GP performance and the testing data is used to measure the ability of GP to generalise. These functions were taken from previous work on GP learning generalisation [19]. We note that the testing sets are larger than the training sets and contain values lying outside the training intervals – i.e. the system is required to extrapolate, not merely interpolate, requiring greater generalisation capability from GP.

Table 1. Symbolic Regression Functions

Functions	Training Data	Testing Data
$F_1 = x^4 + x^3 + x^2 + x$	20 random points $\subseteq$ [-1,1]	30 points $\subseteq$[0:0.05:1.5]
$F_2 = x^3 - x^2 - x - 1$	20 random points $\subseteq$ [-1,1]	30 points $\subseteq$[0:0.05:1.5]
$F_3 = arcsin(x)$	20 random points $\subseteq$ [-1,0]	30 points $\subseteq$[-1:0.67:1]
$F_4 = \sqrt{x}$	20 random points $\subseteq$ [0,2]	30 points $\subseteq$[0:0.1:3]
$F_5 = 0.3sin(2\pi x)$	20 random points $\subseteq$ [-1,1]	30 points $\subseteq$[0:0.05:1.5]
$F_6 = cos(3x)$	20 random points $\subseteq$ [-1,1]	30 points $\subseteq$[0:0.05:1.5]

Table 2. Run and Evolutionary Parameter Values

Parameter	Value
Population size	500
Generations	50
Selection	Tournament
Tournament size	3
Crossover probability	0.9
Mutation probability	0.05
Initial Max depth	6
Max depth	15
Max depth of mutation tree	5
Non-terminals	+, -, *, / (protected version), sin, cos, exp, log (protected version)
Terminals	X, 1
Raw fitness	mean absolute error on all fitness cases
Trials per treatment	100 independent runs for each value

The GP parameters used for our experiments are shown in Table 2. Despite this being an experiment purely concerned with crossover, we have retained a low rate of mutation with an aim to study crossover in the context of a normal GP run.

For SySC, the the upper syntactic sensitivities were set at 6, 8, 10. These three values were calibrated from our experiments as good values for the performance of SySC. In total, three configurations of SySC were tested, denoted as SySCX with X=6, 8, 10. Three upper semantic sensitivities were tested for SSC: 0.4, 0.5, 0.6. These semantic sensitivities were also used in [24], and are denoted as SSCX with X=04, 05, 06.

5 Results and Dicussion

This section presents the comparative results of three crossovers on the GP performance, GP code bloat effect and the ability of GP to generalise.

5.1 Performance

To compare the performance of the three operators, we recorded a classic performance metric, the mean best fitness. The results are shown in Table 3. It can be seen from this table that syntactically-bounded crossover (SySC) does not improve GP performance. The mean best fitness of SySC is often slightly worse than SC, with exceptions on function F_6. Conversely, the mean best that found by SSC is much better than both SC and SySC. This is consistent with the results in [24]..

We also statistically tested the performance of SSC versus SC and SySC using Wilcoxon's signed-rank test with a confidence level of 99% (Table 3), confirming the significant improvement of SSC over SC and SySC. Thus despite some advantages in implementation of SySC, SSC is a better bet in improving perforance.

Table 3. Comparison of the effects of SC, SSC and SySC on GP performance (mean of the best fitness). The values are scaled by 10^2.

Xovers	F_1	F_2	F_3	F_4	F_5	F_6
SC	1.51	3.07	0.37	0.96	4.36	1.48
SySC6	1.63	3.20	0.46	1.06	4.42	1.46
SySC8	1.49	3.50	0.43	0.99	4.36	1.98
SySC10	1.56	3.08	0.39	1.18	4.41	2.04
SSC04	0.78	1.30	0.20	0.58	3.36	0.67
SSC05	0.85	1.40	0.21	0.61	3.28	0.81
SSC06	0.87	1.70	0.22	0.38	3.44	0.92

Table 4. Comparison of the effects of SC, SSC and SySC on code bloat (average tree size over the population)

Xovers	F_1	F_2	F_3	F_4	F_5	F_6
SC	53.9	64.7	51.2	54.1	73.2	56.3
SySC6	45.3	49.9	40.0	43.1	56.7	44.8
SySC8	44.6	51.0	39.2	44.9	58.7	45.8
SySC10	45.3	52.0	42.8	45.9	60.8	48.2
SSC04	49.2	59.2	50.1	52.2	66.3	47.2
SSC05	50.1	58.2	50.0	52.1	68.7	50.4
SSC06	50.5	62.2	49.1	52.2	72.1	53.6

5.2 Code Bloat

It has been known since the early days of GP that the average size of programs inexorably grow [1], and after some generations this may become exponential; the phenomenon is known as code bloat. Given the negative effects of code bloat, many methods for reducing it, and thus simplifying GP inviduals, have been proposed [16,22]. While bloat was not the primary focus of this work, we decided to examine how syntactically and semantically-limited crossovers affected bloat?

We measured the average size of individuals (number of nodes) over 50 generations, averaged over 100 runs. This metric is presented in Table 4. Table 4 reveals that improving syntactic and semantic locality both tend to reduce code bloat, and in this respect, improving syntactic locality has a greater effect than improving semantic locality. We also investigated whether reducing the scale of change could further reduce GP code bloat, and indeed this was the case – but at the cost of poorer performance of SySC; the syntactic sensitivities used in this paper are some of the best values found for the SySC operator.

5.3 Ability to Generalise

Strong generalisation is one of the most desirable properties for learning machines [18]. As GP is a form of (evolutionary) machine learning, it is important to test its generali-

Table 5. Comparison of the effects of SC, SSC and SySC on GP's ability to generalise (the number of good solutions)

Xovers	F_1	F_2	F_3	F_4	F_5	F_6
SC	18	26	9	75	10	52
SySC6	22	21	10	80	17	56
SySC8	26	28	8	78	21	47
SySC10	21	24	9	77	12	53
SSC04	28	46	15	83	22	65
SSC05	22	41	17	81	20	58
SSC06	31	32	12	88	16	59

sation ability [4]. For most real learning problems, we are most interested in the ability of an algorithm to generalise over unseen data.

Previous research on improving GP generalisation largely focused on reducing solution size [25,8]. These high-complexity solutions are often poor in their ability to generalise, resulting from their failure to observe the principle of Ockham's razor [18] (simple solutions are prefered). As the previous sections have shown, improving locality helps to both improve GP performance and reduce GP code bloat, so it is important to test the effect of these crossovers on GP's generalisation ability.

To measure the generalisation ability, we tested the best individual found using the training set, for its ability to generalise over independent test sets (see Table 1). Given ε=0.1, we define a solution 'good' if the fitness on this set is less than ε. We counted the number of good solutions out of 100 runs; the results are shown on Table 5. We can see that improving syntactic locality slightly improves generalisation: the number of good solutions found by SySC is only slightly greater than for SC on some functions, perhaps as the result of reducing code bloat. However improving the semantic locality, as in SSC, substantially improves generalisation. The number of good solutions found by SSC are usually substantially greater than those found by SC and SySC. Generally, these results confirm the importance of improving semantic, rather than merely syntactic, crossover locality in GP.

6 Conclusion and Future Work

In this paper, we proposed a new crossover based on syntactic similarity, as a means of improving syntactic locality. We compared it with similar crossovers based on semantic similarity, *Semantic Similarity based Crossover*, and with standard crossover. We based the comparison on a number of aspects.

In our results, improving semantic locality significantly improves GP performance, reduces code bloat, and substantially enhances generalisation; in comparison, improving syntactic locality gives greater control over code bloat, but leads to only a slight improvement in generalisation over standard crossover. The results confirm the greater importance of semantic rather than syntactic locality. We hope these results will attract GP researchers to pay greater attention to semantic mechanisms in future research.

In the near future, we will study operators incorporating both syntactic and semantic locality, as these might benefit further from the combined effects of both mechanisms. We also plan to examine the relationship of these operators to the search landscape.

Acknowledgment

This paper was funded under a Postgraduate Scholarship from the Irish Research Council for Science Engineering and Technology (IRCSET). The second authors were partly funded by The Vietnam National Foundation for Science and Technology Development (NAFOSTED) under grant number 102.01.14.09 for doing this work. The Seoul National University Institute for Computer Technology provided some facilities supporting this research.

References

1. Banzhaf, W., Langdon, W.B.: Some considerations on the reason for bloat. Genetic Programming and Evolvable Machines 3(1), 81–91 (2002)
2. Beadle, L., Johnson, C.: Semantically driven crossover in genetic programming. In: Proceedings of the IEEE World Congress on Computational Intelligence, pp. 111–116. IEEE Press, Los Alamitos (2008)
3. Cleary, R., O'Neill, M.: An attribute grammar decoder for the 01 multi-constrained knapsack problem. In: Proceedings of the Evolutionary Computation in Combinatorial Optimization, pp. 34–45. Springer, Heidelberg (April 2005)
4. Costelloe, D., Ryan, C.: On improving generalisation in genetic programming. In: Vanneschi, L., Gustafson, S., Moraglio, A., De Falco, I., Ebner, M. (eds.) EuroGP 2009. LNCS, vol. 5481, pp. 61–72. Springer, Heidelberg (2009)
5. de la Cruz Echeanda, M., de la Puente, A.O., Alfonseca, M.: Attribute grammar evolution. In: Mira, J., Álvarez, J.R. (eds.) IWINAC 2005. LNCS, vol. 3562, pp. 182–191. Springer, Heidelberg (2005)
6. Bryant, R.E.: Graph-based algorithms for Boolean function manipulation. IEEE Transactions on Computers C-35, 677–691 (1986)
7. Ekart, A., Nemeth, S.Z.: A metric for genetic programs and fitness sharing. In: Poli, R., Banzhaf, W., Langdon, W.B., Miller, J.F., Nordin, P., Fogarty, T.C. (eds.) EuroGP 2000. LNCS, vol. 1802, pp. 259–270. Springer, Heidelberg (2000)
8. Gagne, C., Schoenauer, M., Parizeau, M., Tomassini, M.: Genetic programming, validation sets, and parsimony pressure. In: Collet, P., Tomassini, M., Ebner, M., Gustafson, S., Ekárt, A. (eds.) EuroGP 2006. LNCS, vol. 3905, pp. 109–120. Springer, Heidelberg (2006)
9. Gottlieb, J., Raidl, G.: The effects of locality on the dynamics of decoder-based evolutionary search. In: Proceedings of the Genetic and Evolutionary Computation Conference, pp. 283–290. ACM, New York (2000)
10. Hoai, N.X., McKay, R.I., Essam, D.: Representation and structural difficulty in genetic programming. IEEE Transection on Evolutionary Computation 10(2), 157–166 (2006)
11. Johnson, C.: Deriving genetic programming fitness properties by static analysis. In: Foster, J.A., Lutton, E., Miller, J., Ryan, C., Tettamanzi, A.G.B. (eds.) EuroGP 2002. LNCS, vol. 2278, pp. 298–308. Springer, Heidelberg (2002)
12. Johnson, C.: What can automatic programming learn from theoretical computer science. In: Proceedings of the UK Workshop on Computational Intelligence, University of Birmingham (2002)

13. Johnson, C.: Genetic programming with fitness based on model checking. In: Proceedings of the 10th European Conference on Genetic Programming (EuroGP 2002), pp. 114–124. Springer, Heidelberg (2007)
14. Katz, G., Peled, D.: Genetic programming and model checking: Synthesizing new mutual exclusion algorithms. In: Cha, S(S.), Choi, J.-Y., Kim, M., Lee, I., Viswanathan, M. (eds.) ATVA 2008. LNCS, vol. 5311, pp. 33–47. Springer, Heidelberg (2008)
15. Katz, G., Peled, D.: Model checking-based genetic programming with an application to mutual exclusion. In: Ramakrishnan, C.R., Rehof, J. (eds.) TACAS 2008. LNCS, vol. 4963, pp. 141–156. Springer, Heidelberg (2008)
16. Langdon, W.B.: Size fair and homologous tree genetic programming crossovers. Genetic Programming and Evolvable Machines 1(1), 91–1119 (2000)
17. Levenshtein, V.I.: Binary Codes Capable of Correcting Deletions, Insertions, and Reversa ls. Soviet Physics Doklady 10, 707 (1966)
18. Mitchell, T.: Machine Learning. McGraw-Hill, New York (1996)
19. Nguyen, Q.U., Nguyen, T.H., Nguyen, X.H., O'Neill, M.: Improving the generalisation ability of genetic programming with semantic similarity based crossover. In: Esparcia-Alcazar, A.I., Ekart, A., Silva, S., Dignum, S. (eds.) EuroGP 2010. LNCS, vol. 6021, pp. 184–195. Springer, Heidelberg (2010)
20. Rothlauf, F., Goldberg, D.: Redundant Representations in Evolutionary Algorithms. Evolutionary Computation 11(4), 381–415 (2003)
21. Rothlauf, F., Oetzel, M.: On the locality of grammatical evolution. In: Collet, P., Tomassini, M., Ebner, M., Gustafson, S., Ekárt, A. (eds.) EuroGP 2006. LNCS, vol. 3905, pp. 320–330. Springer, Heidelberg (2006)
22. Silva, S., Almeida, J.: Dynamic maximum tree depth: A simple technique for avoiding bloat in tree-based gp. In: Proceedings of the 5th Annual Conference on Genetic and Evolutionary Computation (GECCO), pp. 1776–1787. ACM Press, New York (2003)
23. Uy, N.Q., Hoai, N.X., O'Neill, M.: Semantic aware crossover for genetic programming: the case for real-valued function regression. In: Vanneschi, L., Gustafson, S., Moraglio, A., De Falco, I., Ebner, M. (eds.) EuroGP 2009. LNCS, vol. 5481, pp. 292–302. Springer, Heidelberg (2009)
24. Uy, N.Q., O'Neill, M., Hoai, N.X., McKay, B., Lopez, E.G.: Semantic similarity based crossover in GP: The case for real-valued function regression. In: Collet, P. (ed.) 9th International Conference on Evolution Artificielle. LNCS, pp. 13–24. Springer, Heidelberg (October 2009)
25. Vanneschi, L., Gustafson, S.: Using crossover based similarity measure to improve genetic programming generalization ability. In: GECCO 2009: Proceedings of the 11th Annual Conference on Genetic and Evolutionary Computation, Montreal, July 8-12, pp. 1139–1146. ACM, New York (2009)
26. Wong, M.L., Leung, K.S.: An induction system that learns programs in different programming languages using genetic programming and logic grammars. In: Proceedings of the 7th IEEE International Conference on Tools with Artificial Intelligence (1995)

The *Layered Learning* Method and Its Application to Generation of Evaluation Functions for the Game of Checkers

Karol Walędzik and Jacek Mańdziuk

Warsaw University of Technology,
Faculty of Mathematics and Information Science,
Plac Politechniki 1, 00-661 Warsaw, Poland
{K.Waledzik,J.Mandziuk}@mini.pw.edu.pl

Abstract. In this paper we describe and analyze a Computational Intelligence (CI)-based approach to creating evaluation functions for two player mind games (i.e. classical turn-based board games that require mental skills, such as chess, checkers, Go, Othello, etc.). The method allows gradual, step-by-step training, starting with end-game positions and gradually moving towards the root of the game tree. In each phase a new training set is generated basing on results of previous training stages and any supervised learning method can be used for actual development of the evaluation function.

We validate the usefulness of the approach by employing it to develop heuristics for the game of checkers. Since in previous experiments we applied it to training evaluation functions encoded as linear combinations of game state statistics, this time we concentrate on development of artificial neural network (ANN)-based heuristics.

Games provide cheap, reproducible environments suitable for testing new search algorithms, pattern-based evaluation methods or learning concepts. Since the seminal papers devoted to programming chess [1–3] and checkers [4] in the 1950s., games remained through decades an interesting topic for both classical AI and CI-based approaches.

Most examples of application of CI methods to mind game playing make use of either reinforcement learning methods, neural networks-based approaches, evolutionary methods or hybrid neuro-genetic solutions, e.g. in chess [5–7], checkers [8–11], Go [12], Othello [13], or give-away-checkers [14, 15].

The main focus of this paper is on testing the efficacy of what we call *Layered Learning* - a generally-applicable approach to building the evaluation function for two-player games (checkers in here) which can be implemented either in the evolutionary mode or as a gradient backpropagation-type neural network training. The method, originally proposed in [16], was used previously by the authors in the case of linear heuristic composed of checkers-specific components [17, 18]. In this paper a more detailed description of the method is provided along with some modifications to the previously used version. Furthermore, as opposed to [17] and [18], where evolutionary learning methods were employed, this work concentrates on applicability of Layered Learning to the case of artificial neural networks (ANNs) based evaluation functions with much lesser use of pre-defined domain knowledge.

R. Schaefer et al. (Eds.): PPSN XI, Part II, LNCS 6239, pp. 543–552, 2010.

The reminder of the paper is organized as follows: in section 1 the basic idea of the proposed Layered Learning method is described along with its several modifications and enhancements. The next two sections present the experimental setup and the results of experiments, respectively. Conclusions and summary of possible research prospects are placed in section 4.

1 Layered Learning

1.1 Learning Method

Layered Learning (LL), schematically depicted in fig. 1, is an end-game first method. Similarly to TD(λ) [19] it attempts to propagate the knowledge of final game results from endgame positions up the game tree. Still, we believe that it differs enough to be worth separate analysis and evaluation.

This learning scheme starts with division of the game tree into a number of disjoint stages, depending on the game progress. The simplest criterion that can be employed here in case of checkers is the number of moves performed. The whole process starts with positions expected to be very close to the end of the game. They are analyzed by a minimax algorithm (typically employing alpha-beta pruning) with a null evaluation function. It is assumed that in most cases the analysis will be able to reach the leaves of the game tree and the heuristic evaluation function will not be needed. The other cases are treated as draws and have neutral value assigned by the evaluation function.

Once a set of assessed game positions is obtained, it can be used as training data for any supervised learning approach, so as to create an evaluation function able to assess those endgame positions. In our experiments we employed both evolutionary methods (with various representations of heuristic evaluation functions) and backpropagation learning methods (in case of ANNs).

Having trained an evaluation function for one stage, the algorithm moves to the next stage, closer to the beginning of the game. A number of game states from this new stage are generated and, again, they are analyzed with a minimax algorithm. It is expected that its search depth will be enough to always reach positions from the previously trained stage. In that case, the result of previous stage can be used as the evaluation function for this analysis and another training set can easily be generated. This process, repeated for all game stages, should lead to creation of an evaluation function capable of assessing positions from all game stages (or an ensemble of such functions covering the whole game).

1.2 Method Variations

The general approach described in previous section can be implemented in several different ways and its quality may be influenced by a number of fine details of the algorithm. Implementors of LL scheme have, of course, to tackle all typical hurdles of CI methods application, such as choosing learning coefficients, designing ANN architecture, or defining evolutionary operators etc. There are, however, also several decisions typical for LL learning that must be made.

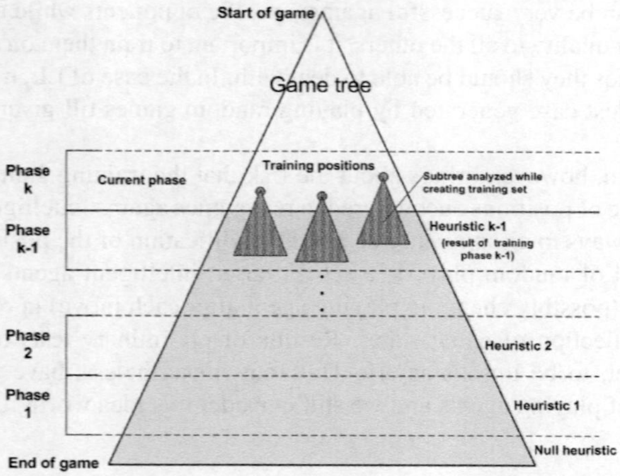

Fig. 1. *Layered Learning* method - an overview

First of all, implementors should settle on supervised learning method. Our first experiments concentrated on using evolutionary methods - initially with evaluation functions represented as linear combinations of simple game state features. More sophisticated checkers position description features were introduced afterwards, and the definition of evaluation function was modified so as to allow dynamic switching of linear combinations' coefficients depending on game progress.

In the experiments described in this document we concentrated on evaluation functions represented by ANNs in the form of fully connected feed-forward multi-layer perceptrons. Input vectors would contain either only board content representation (with no preprocessing applied) or, alternatively, also values of a number of simple game state features. The ANNs would be trained either by backpropagation (RPROP [20]) or evolutionary methods.

Another problem faced by implementors of the LL method may be the risk of trained evaluators (be it ANN or any other representation) 'forgetting' knowledge learned during previous stages. There are several ways this issue can be dealt with. It is possible to train a separate evaluator in each stage and treat the resulting ensemble as the output of the training process. Each evaluator would then be used only in the stage it was trained for. Otherwise, special care must be taken to ensure that no (or next to no) 'forgetting' takes place. One way to achieve that is to make sure, that in each phase, evaluator is trained not only on positions from the current stage but also a number of game states from previous stages. These historical positions can be either regenerated each time they are needed (to improve the diversity of training positions), or reused in all subsequent training phases (to save time required for their regeneration and minimax analysis). Whatever the choice, the current stage positions should be slightly over-represented in the training set, as they introduce new knowledge not yet acquired by trained evaluator.

One of the problems obvious in most training approaches in the domain of CI application to mind games is the selection of training games, positions or opponents. Since some players can be very successful against specific opponents while at the same time being of inferior quality to all the others, it is important to train them on a wide selection of game strategies they should be able to deal with. In the case of LL, training positions are in the simplest case generated by playing random games till given depth in game tree is reached.

This approach, however, brings about the risk that the training game states will not be representative of positions encountered in real games against intelligent players. One of the possible ways to circumvent this risk is modification of the positions generation process. Instead of random players, a set of varied intelligent agents can be used to play the games (possibly changing playing agent after each move) in order to generate the required collection of game states. Results of preliminary tests of this approach proved, however, to be unsatisfactory. This may, nevertheless, have been caused by poor selection of playing agents and we still consider this idea worth further testing.

2 Experiments Setup

First of all, it should be stressed that the aim of the experiment was not to create a master level player capable of competing with commercial checkers applications. Our solution was not fully optimized for speed, employed only basic alpha-beta pruning algorithm with no further modifications and used only simple fully-connected ANNs for evaluation function representation.

During our experiments we tested several different sets of control parameters in combination with varied evaluation function architectures, which makes it impossible to list all of them in such a short document. Still, we will point out the most typical values (or intervals) of the coefficients used during training and indicate whenever an atypical value was employed. We also believe that there is a huge potential for results improvement by further tuning all the learning parameters and coefficients, considering how little attention has yet been devoted to the LL method.

We concentrate in this paper on analysis of the learning method itself and try to prove its applicability to mind games such as checkers, especially in zero-initial-knowledge training scheme. We hope to present LL method to wider audience and point out possible directions for further research.

2.1 Neural Networks Architecture

All our ANN-based experiments (as opposed to earlier experiments described in [17, 18]) involved feed-forward fully-connected multi-layered perceptrons. Ideally, we wanted their input vectors to contain board description only. They would, therefore, consist of 32 neurons representing individual board squares, each with one of five values: -2, -1, 0, 1 and 2 representing, respectively, opponent's king, opponent's checker, empty square and current player's checker or king. In some of the experiments the input layer would further be extended to contain a number of simple game state description features:

- differences between player's and opponent's checkers and kings counts;
- differences between player's and opponent's safe (i.e. adjacent to the edge of the board) checkers and kings counts;

– differences between player's and opponent's moveable (i.e. able to perform move other than capturing, ignoring capturing priority) checkers and kings counts;
– difference between player's and opponent's aggregated distances of checkers to promotion line;
– difference between player's and opponent's unoccupied fields on promotion line count.

For comparison, some experiments with input vectors containing only the game state features (without raw game state representation) were performed as well. In most experiments, the neural networks would contain one hidden layer of up to 10 neurons. Output layer would always contain a single neuron expected to output game state evaluation within the interval $[-1,1]$.

2.2 RPROP Training

During the first phase of our experiments evaluators were trained using RPROP backropopagation method. In order to minimize the chance of random factors hindering the learning process, learning process was augmented by elements of evolutionary training procedures. At every stage of the algorithm 8 networks were trained simultaneously on the same training set. Each training phase consisted of 4 generations. After each generation, the quality of all candidate evaluators was tested by computing their mean square errors on a test set. Test set used for this task was separate from training sets and contained a number of game positions from all game stages trained on so far (including the current one). Candidate solutions were afterwards sorted based on their thus measured quality and the worse half of them was replaced by mutated copies of the best networks.

During the RPROP learning, training patterns were presented to the networks in random order (independent for each network). After each training phase, a more significant modification of the population took place. Only two best networks survived intact to the next phase. Additional 4 were generated by mutating them - once with lower (0.01 to 0.03) and once with higher (0.1) mutation probabilities. Further two candidate solutions were created with fully random weight values.

Mutation was applied independently to each weight in the mutated network, with each connection having equal probability to mutate. Once it was decided that given connection value should change, one of four possible mutations would be applied to it (each with equal probability): multiplication by 2, division by 2, sign change (multiplication by -1) or replacement with random value.

Results of preliminary experiments comparing training for varied number of epochs and with varied training set sizes (not presented here in details due to lack of space), suggest that one of the main problems hindering further improvement of the solutions generated by backpropagation method was the risk of overtraining. With high ANN capacities, small training sets and long training the networks would quickly loose their generalization capabilities. This would result in low training set errors but higher test set errors and poor performance in actual comparison games.

In order to overcome this hurdle, in the subsequent experiments we limited the ANNs' sizes and attempted to use training sets as big as possible, which, of course, resulted in slower training process. This meant, however, that in order to keep the training time within reasonable limits we had to reuse the same training boards across multiple training phases (continuously increasing the training set size with positions from lower depths in game tree).

Finally, we decided to employ early stopping routine as a way to define stop condition for training, so that the probability of overtraining is reduced. It proved, however, less successful than we expected. It turned out that the specificity of the problem caused the validation set error to fluctuate significantly - sometimes rising for several epochs only to drop afterwards. Early stopping, even modified to accept temporary rise of validation set error, was prone to ceasing the training too early, which forced us to train all networks for a preset number of epochs before the technique was employed.

2.3 Evolutionary Training

Second phase of our experiments made use of a simple evolutionary approach. The trained population would, in this case, contain several dozen (up to 100, depending on individual experiment settings; 40 in most runs) candidate networks that would be modified over several hundred generations in each training phase.

In each generation mean square error of training boards assessment was calculated for each candidate ANN. Afterwards, a number (55% of population size in the most successful experiments) of the worst performing solutions were discarded. The remaining individuals were replicated with mutation, with a subset of them (typically the top 5% of the original population size) being replicated twice (once with lower and once with higher mutation probability coefficients). In case the resulting number of individuals was still lower than the requested population size, additional candidate solutions were generated randomly.

After each phase, population was refreshed in similar manner but with different control parameters. In that case, only 30% of the population would survive. Thus, after each generation a significant number of candidate solutions was regenerated randomly.

In most experiments the evolutionary method was additionally augmented by element of RPROP training. Namely, each newly created (be it randomly or via replication and mutation) network was first once trained on all training patterns. The RPROP training was also repeated for all networks after each training set change, i.e. at the beginning of each training phase.

3 Results Analysis

3.1 Evaluators Comparison

In order to analyze the results of our experiments we first decided to perform direct comparison of resulting evaluation functions by means of tournament, in which each agent played 20 games against every other one (with sides swapped after each game). Search depth limit for alpha-beta algorithm used in these comparison was set to relatively low limit of 4 - thus increasing the influence of evaluation function quality on the final score (in the case of greater search depths, score differences might prove less significant). Games played by each selected pair of agents were pairwise different thanks to the fact that, in each position, available moves were considered by the alpha-beta algorithm in random order. In order to make the results of the tournament as representative of the true quality as possible, it included a significant number of various agents trained in these and earlier experiments. Two scoring schemes were used in the tournament: games-based and clashes-based. In the former, agents were assigned points for each individual game: 2 points for a win and 1 point for draw. In the latter, contestants were scored analogically based on 20-game clashes (series of games against a single opponent).

Figure 2 presents results of the tournament for selected most important evaluation functions:

- *HG-Expert3Phase* the most successful evaluation function generated in the first Layered Learning experiment described in [18], consisting of 3 linear combinations of advanced game position features - each applied to one of disjoint phases of the game;
- *BoardsAndFeatures5N* - ANN trained with backpropagation method (RPROP), with 5 neurons in its single hidden layer and input vector containing both plain board description and basic game position numerical features;
- *BoardsAndFeatures10N* - ANN similar to the previous one but with doubled number of neurons in hidden layer;
- *PlainBoard5N* - ANN, differing from *BoardsAndFeatures5N* only in size of its input layer, as it did not include precalculated checkers position features;
- *PlainBoard10N* - ANN similar to the previous one but with doubled number of neurons in hidden layer;
- *EvoPlainBoard10NWithRPROP* - ANN trained using evolutionary approach (augmented by RPROP procedure), with hidden layer of 10 neurons;
- *EvoPlainBoard10NNoRPROP* - ANN with architecture identical to the previous one, but trained with pure evolutionary approach (with no backpropagation learning component);
- *EvoPlainBoard10NWithRPROPNoBoardsReuse* - yet another identical ANN, but this time trained with training boards set fully regenerated after each phase.

Based on the results of the described tournament and several minor comparisons performed independently, several conclusions can be drawn. First of all, it can easily be spotted that none of the evaluation functions generated in the current experiment managed to surpass the best results of the original experiment based on game state description features. The explanation of this fact is twofold. Firstly, the experiments differed in their focus and amount of learning parameters tuning applied. More importantly, however, it should not be forgotten that the most successful linear heuristics operated on manually defined advanced game positions characteristics. At the same time ANN had access to either raw board position or the simplest game position features only.

What is interesting, no significant difference in quality was observed between evaluators being provided with raw board representation only and those having additionally access to simple game state statistics. Since earlier experiments confirmed that those statistics are actually important in evaluation function building, it can be inferred that the trained neural networks were actually able to successfully learn to compute at least some of them basing on the raw position description only.

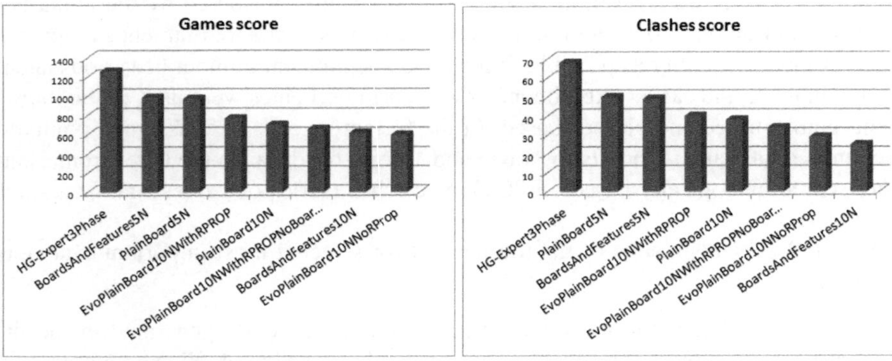

(a) Individual games (b) 20-game clashes

Fig. 2. Comparison of evaluators based on individual games scores or clashes scores

Better results of smaller networks confirmed our preliminary expectations of overfitting being a common and serious problem for our training process. It should also not be omitted here that our decision to mix elements of evolutionary and backpropagation training was a highly successful one. Both training methods yielded far weaker solutions when used independently. In case of evolutionary method this fact is clearly visible in figure 2 with *EvoPlainBoard10NNoRPROP* being scored more than 20% lower than *EvoPlainBoard10NWithRPROP*. At the same time, analysis of the training logs of the backpropagation-based processes clearly indicates that in this case a significant number of mutations led to improvement of ANNs' mean square errors.

3.2 Training Progress Analysis

In order to verify the training process itself and evaluators' quality improvement from phase to phase, for several selected individual experiments we decided to perform further tournaments comparing solutions generated in subsequent phases of the same training process. We were aware that poor choice of training configuration might cause the evaluators to loose during the training knowledge gathered in earlier phases. What is more, any errors in heuristic generated in one of the early phases, would be repeated or even magnified during the training process, because the results of previous phases are used to generate training sets for further training.

It should also be stressed that, even if none of the above dangers actually applied, in case of such a comparison we had no reason to expect a monotonous increase in scores, as all but the last few evaluation functions were in no way prepared to play full games, having been trained only on their final stages. This fact might have lead them to choosing seemingly random moves in the first parts of games which could in consequence cause the objectively better end-game players to arrive at very disadvantageous positions. We expected, however, the heuristics generated in last training phases to play significantly better than the earlier ones.

Since in our experiments we decided to divide the game into 14 stages, we expected the winner to be one of evaluation functions generated in stages 11 to 14. This assumption proved, in general, to be true. The results of further classification proved, however, sometimes surprising. In some cases (for example for *PlainBoard5N* as visible in figure 3) one or more of the early stage evaluators turned out to be unexpectedly strong players as well. This can be attributed to the fact that it can be expected for such early heuristics to rely heavily on material differences and such a simple approach may be enough to beat opponents using evaluation functions being more sophisticated but applicable to mid-game positions only (with no ability to play any reasonable opening moves).

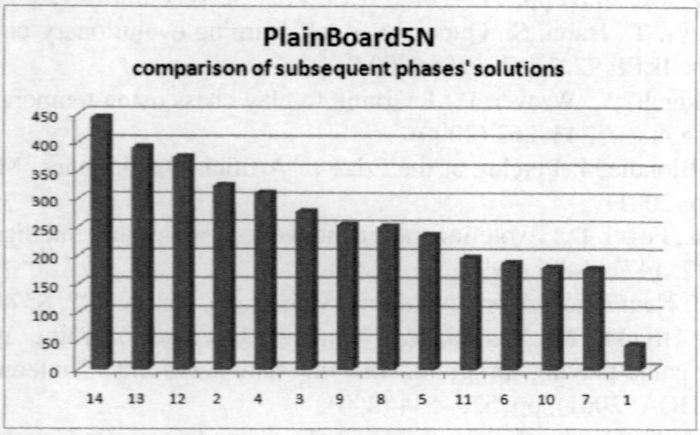

Fig. 3. Comparison of results of subsequent phases

4 Conclusions

In this paper a generally-applicable game learning approach to creating evaluation function for two-player games has been described.

To verify usefulness of this training method, we decided to apply it to the game of checkers. Following our previous experiments, in which we evolved linear-combination-based heuristics, this time we concentrated on training ANNs.

We believe that our experiments prove the method is worth further analysis, testing its various aspects and applicability to other mind games. We identified some of the most troublesome aspects of the approach and proposed several modifications to it, that we think are worth further research.

Although the method was introduced some time ago [16] it hasn't been extensively researched yet. Since only few experiments utilizing LL method have been performed so far, we think that its true potential is still yet to be discovered. We also believe that the name Layered Learning coined in this paper aptly describes the general idea of the method.

Our current research is focused on direct comparison of the LL method with the TD(λ) learning scheme.

References

1. Shannon, C.E.: Programming a computer for playing chess. Philosophical Magazine 41 (7th series), 256–275 (1950)
2. Turing, A.M.: Digital computers applied to games. In: Bowden, B.V. (ed.) Faster than Thought: a Symposium on Digital Computing Machines. Pitman, London (1953)
3. Newell, A., Shaw, J., Simon, H.: Chess-playing programs and the problem of complexity. IBM Journal of Research and Development 2, 320–335 (1958)
4. Samuel, A.L.: Some studies in machine learning using the game of checkers. IBM Journal of Research and Development 3, 210–229 (1959)
5. Kendall, G., Whitwell, G.: An evolutionary approach for the tuning of a chess evaluation function using population dynamics. In: Proceedings of the 2001 Congress on Evolutionary Computation, CEC 2001, pp. 995–1002. IEEE Press, Los Alamitos (2001)
6. Fogel, D., Hays, T., Hahn, S., Quon, J.: A self-learning evolutionary chess program. Proceedings of the IEEE 92, 1947–1954 (2004)
7. Baxter, J., Tridgell, A., Weaver, L.: Learning to play chess using temporal differences. Machine Learning 40(3), 243–263 (2000)
8. Fogel, D.B.: Blondie24: Playing at the Edge of Artificial Intelligence. Morgan Kaufmann, San Francisco (2001)
9. Chellapilla, K., Fogel, D.: Evolution, neural networks, games, and intelligence. Proceedings of the IEEE 87, 1471–1496 (1999)
10. Aleksander, I.: Neural networks - evolutionary checkers. Nature 402, 857 (1999)
11. Schaeffer, J., Hlynka, M., Jussila, V.: Temporal difference learning applied to a high-performance game-playing program. In: International Joint Conference on Artificial Intelligence (IJCAI 2001), pp. 529–534 (2001)
12. Schraudolph, N.N., Dayan, P., Sejnowski, T.J.: Learning to evaluate Go positions via Temporal Difference methods. In: Baba, N., Jain, L.C. (eds.) Computational Intelligence in Games, vol. 62, pp. 77–98. Springer, Berlin (2001)
13. Moriarty, D.E., Miikkulainen, R.: Discovering complex othello strategies through evolutionary neural systems. Connection Science 7, 195–209 (1995)
14. Mańdziuk, J., Osman, D.: Temporal difference approach to playing give-away checkers. In: Rutkowski, L., Siekmann, J.H., Tadeusiewicz, R., Zadeh, L.A. (eds.) ICAISC 2004. LNCS (LNAI), vol. 3070, pp. 909–914. Springer, Heidelberg (2004)
15. Osman, D., Mańdziuk, J.: Comparison of tdleaf(λ) and td(λ) learning in game playing domain. In: Pal, N.R., Kasabov, N., Mudi, R.K., Pal, S., Parui, S.K. (eds.) ICONIP 2004. LNCS, vol. 3316, pp. 549–554. Springer, Heidelberg (2004)
16. Borkowski, M.: Analysis of algorithms for two-player games. M.Sc. Thesis, Warsaw University of Technology (2000) (in Polish)
17. Kusiak, M., Walędzik, K., Mańdziuk, J.: Evolution of heuristics for give-away checkers. In: Duch, W., Kacprzyk, J., Oja, E., Zadrożny, S. (eds.) ICANN 2005. LNCS, vol. 3697, pp. 981–987. Springer, Heidelberg (2005)
18. Mańdziuk, J., Kusiak, M., Walędzik, K.: Evolutionary-based heuristic generators for checkers and give-away checkers. Expert Systems 24(4), 189–211 (2007)
19. Sutton, R.: Learning to predict by the methods of temporal differences. Machine Learning 3, 9–44 (1988)
20. Riedmiller, M., Braun, H.: Rprop- a fast adaptive learning algorithm (1992), `http://citeseer.ist.psu.edu/riedmiller92rprop.html`

Author Index

GPSR Compliance

The European Union's (EU) General Product Safety Regulation (GPSR) is a set of rules that requires consumer products to be safe and our obligations to ensure this.

If you have any concerns about our products, you can contact us on ProductSafety@springernature.com

In case Publisher is established outside the EU, the EU authorized representative is:

Springer Nature Customer Service Center GmbH
Europaplatz 3
69115 Heidelberg, Germany

Batch number: 10704693

Printed by Printforce, the Netherlands